Emerging Photon Technologies for Chemical Dynamics

Sheffield, UK

9–11 July 2014

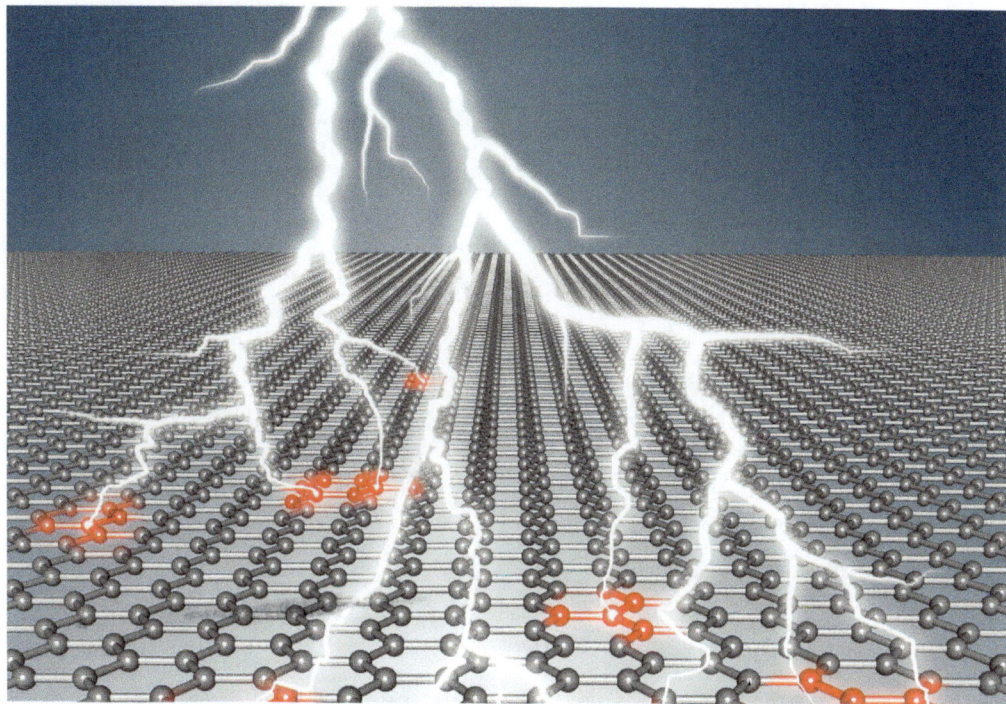

FARADAY DISCUSSIONS

Volume 171, 2014

ROYAL SOCIETY
OF CHEMISTRY

The Faraday Division of the Royal Society of Chemistry, previously the Faraday Society, founded in 1903 to promote the study of sciences lying between Chemistry, Physics and Biology.

Editorial staff

Executive Editor
Robert Eagling

Deputy editor
Heather Montgomery

Development editor
Alessia Millemaggi

Editorial Production Manager
Philippa Ross

Publishing editors
T anya Smekal, Carla Pegoraro

Publishing assistants
Victoria Bache, Bethany Johnson, Kate McCallum, Ruba Miah

Faraday Discussions (Print ISSN 1359-6640, Electronic ISSN 1364-5498) is published 8 times a year by the Royal Society of Chemistry, Thomas Graham House, Science Park, Milton Road, Cambridge, UK CB4 0WF.

Volume 171 ISBN-13: 978-1-78262-172-0

2014 annual subscription price: print+electronic £827, US $1542; electronic only £785, US $1465. Customers in Canada will be subject to a surcharge to cover GST. Customers in the EU subscribing to the electronic version only will be charged VAT. All orders, with cheques made payable to the Royal Society of Chemistry, should be sent to RSC Order Department, Royal Society of Chemistry, Thomas Graham House, Science Park, Milton Road, Cambridge, CB4 0WF, UK.
Tel +44 (0) 1223 432398; E-mail www.orders@rsc.org

If you take an institutional subscription to any RSC journal you are entitled to free, site-wide web access to that journal. You can arrange access via Internet Protocol (IP) address at www.rsc.org/ip. Customers should make payments by cheque in sterling payable on a UK clearing bank or in US dollars payable on a US clearing bank.

Printed in the UK

Faraday Discussions documents a long-established series of Faraday Discussion meetings which provide a unique international forum for the exchange of views and newly acquired results in developing areas of physical chemistry, biophysical chemistry and chemical physics.

Chemical Dynamics

Faraday Discussions

www.rsc.org/faraday_d

A General Discussion on Chemical Dynamics was held in Sheffield, UK on the 9th, 10th and 11th of July 2014.

RSC Publishing is a not-for-profit publisher and a division of the Royal Society of Chemistry. Any surplus made is used to support charitable activities aimed at advancing the chemical sciences. Full details are available from www.rsc.org

CONTENTS

ISSN 1359-6640;   ISBN 978-1-78262-172-0

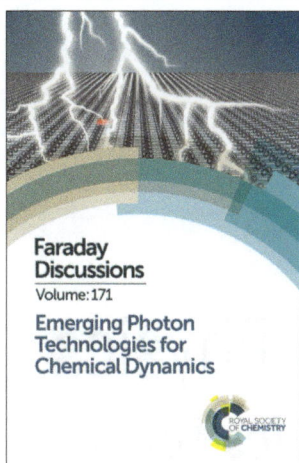

Cover
Image courtesy of Dr Joerg M. Harms, MPI for the Structure and Dynamics of Matter, Hamburg, Germany.

Faraday Discussions
Volume: 171
Emerging Photon Technologies for Chemical Dynamics

INTRODUCTORY LECTURE

PAPERS AND DISCUSSIONS

Lotte Holmegaard, Per Johnsson, Jens S. Kienitz, Thomas Kierspel,
Faton Krasniqi, Kai-Uwe Kühnel, Jochen Maurer, Marc Messerschmidt,
Robert Moshammer, Nele L. M. Müller, Benedikt Rudek, Evgeny Savelyev,
Ilme Schlichting, Carlo Schmidt, Frank Scholz, Sebastian Schorb,
Joachim Schulz, Jörn Seltmann, Mauro Stener, Stephan Stern,
Simone Techert, Jan Thøgersen, Sebastian Trippel, Jens Viefhaus,
Marc Vrakking, Henrik Stapelfeldt, Jochen Küpper, Joachim Ullrich,
Artem Rudenko and Daniel Rolles

CONCLUDING REMARKS

ADDITIONAL INFORMATION

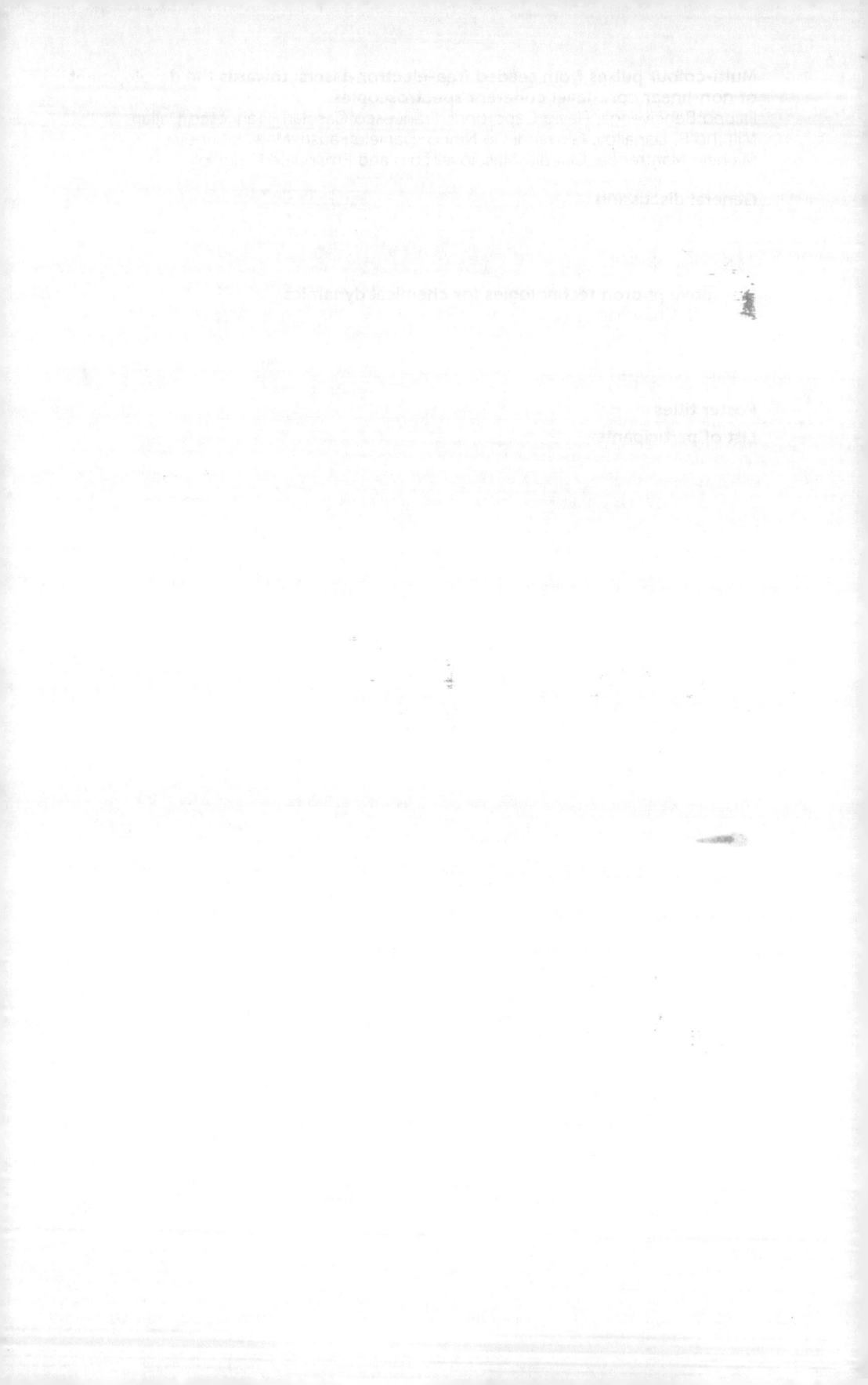

PAPER

Emerging photon technologies for chemical dynamics

Majed Chergui*

Received 12th August 2014, Accepted 13th August 2014

DOI: 10.1039/c4fd00157e

I discuss the recent developments concerning emerging ultrafast sources of short wavelength radiation, as well as the new methods they generate. I then dwell on a few examples of recent results and the way the new sources will bring major advancements in our understanding of fundamental chemical dynamics.

1   Introduction

Ultrafast femtosecond–picosecond linear and non-linear optical spectroscopies emerged some 25 years ago with the development of lasers delivering ultrashort pulses of light.[1-3] These tools have had a huge impact on our understanding of chemical reactions, biological functions and lattice dynamics in materials owing to their ability to probe, in real-time, the nuclear motion within these different types of systems. However, while optical (*i.e.* infrared, visible and ultraviolet) spectroscopy provides detailed information about the evolution of the system, it lacks elemental and structural specificity as the wavelength of radiation is several 100 nm. Thus connecting the observed changes to the structural dynamics is possible only for small systems, such as diatomic or triatomic molecules for which potential curves or surfaces, respectively, are known. The awareness of this limitation in the early days of ultrafast optical-domain spectroscopy steered efforts to combine the atomic resolution of time (femtoseconds) with that of space (tenths of ångströms). Towards this goal, various groups adopted diffraction methods based on the use of ultrashort pulses of X-rays[4-8] or electrons,[9-12] while others opted for time-resolved X-ray absorption spectroscopy (XAS).[13-15]

We are now witnessing the results of the huge efforts made throughout the 2000s to reach these goals, many of which were presented at this meeting. These results have taken several directions, mostly concurrently, in terms of the development of new sources of ultrashort photon pulses in the hard and soft X-ray, extreme ultraviolet (XUV) and vacuum ultraviolet (VUV) ranges, and along with them the development of new experimental methods and strategies for data

Ecole Polytechnique Fédérale de Lausanne, Laboratoire de Spectroscopie Ultrarapide, ISIC, FSB, Station 6, CH-1015 Lausanne, Switzerland. E-mail: Majed.Chergui@epfl.ch; Fax: +41 21 6930365; Tel: +41 21 6930447/0457

acquisition and analysis. In terms of sources, two classes of technologies have emerged:

(i) Large-scale installations, firstly synchrotrons, with the implementation of high repetition rate schemes for picosecond experiments[16-18] and the development of the slicing scheme that allows the generation of femtosecond X-ray pulses.[19-21] Secondly, the appearance of X-ray free electron lasers (XFELs) represents a real revolution in the field of structural dynamics, with their several orders of magnitude increase in photon flux and femtosecond pulse durations in the soft[22] and hard X-ray range.[23,24]

(ii) Laboratory sources for the hard X-ray range that are plasma-based[4-7,25] or for the VUV–XUV range that are based on high-harmonic generation (HHG) and reach the attosecond time domain.[26]

In photoinduced phenomena it is the electronic rearrangements that drive the structural changes, thus the need to probe the photoinduced electronic structure changes that lead to a change of geometry was one of the main reasons why certain groups opted for ultrafast X-ray absorption spectroscopy (XAS),[13,14,15,27] which is now being extended to other spectroscopic methods such as ultrafast X-ray emission spectroscopy (XES).[28-30] The availability of ultrashort sources of VUV/XUV radiation has also opened the possibility to extend core-level spectroscopies to ultrafast photon-in/electron-out spectroscopies such as photoelectron spectroscopy[31,32] or photoemission (for solids).[33-35]

The range of systems, scientific problems, sources and methods that all these emerging technologies are covering is impressive and this issue is a perfect snapshot of the state-of-the-art. In terms of systems, papers are presented which concern isolated gas phase molecules,[31,32,36-41] clusters,[42] metal complexes,[30,43,44] quantum dots,[45] biosystems,[31,46-49] surfaces/interfaces,[43-45] Mott–Peierls insulators,[33] superconductors,[34] ionic crystals,[50] and graphene.[35] These deal with scientific issues such as the structural dynamics of molecules,[30,32,36-39,41] the charge and spin dynamics in molecules,[30,31,37,38] solvation dynamics,[30,46] surface and interface phenomena,[43-45] charge density waves, magnetism and melting in materials.[33-35,50] It is interesting to note that the vast majority of these issues were discussed in the 2009 Basic Energy Sciences Advisory Committee Report on Solving Science and Energy Grand Challenges with Next-Generation Photon

Table 1 Comparison of the main properties of ultrafast sources of short wavelength radiation. Photons/pulse relate to monochromatic photons. Brilliance is in photons per s mrad^2 mm^2 0.1%BW. 1 attosecond = 1 As

	Energy range	Pulse width	Photons/ pulse	Peak brilliance	Average brilliance	Repetition rate
Synchrotrons	10 eV–20 keV	50–100 ps	10^4–10^6	10^{23}–10^{25}	10^{21}	MHz
Slicing scheme at synchrotrons	10 eV–10 keV	<150 fs	10	10^{19}		KHz
X-ray free electron lasers	<10 keV	<50 fs	10^{12}	10^{33}	10^{20}–10^{21}	100 Hz
High harmonic generation	10–<400 eV	10–100 As to fs	10^3			1–10 kHz

Sources.[51] Many of the photon attributes for science drivers listed in Table 1 of this report are met, not only by the XFELs, but also by several lab-based sources.

In this article, I will first briefly review the state-of-the-art of the available methods before dwelling on some scientific cases brought up during this meeting. Therefore, this paper is by no means exhaustive and does not fully reflect the diversity and wealth of discussions that took place during the meeting.

2   Emerging technologies

Fig. 1 shows the timescale of typical dynamical processes in nature and the light sources that are used to probe them. Table 1 is a comparison of the most important properties of the above mentioned light sources of short wavelength/ short pulse radiation relevant for chemical dynamics. This comparison is not easy because not all parameters are well established and some properties are difficult to compare between large-scale installations and lab-based sources.

The use of a given source depends on the scientific question to address. Synchrotrons provide pulses of 50–100 ps duration, which do not probe dynamics (in the sense of nuclear motion) but provide invaluable information about excited state structures with lifetimes from 100 ps to milliseconds. With the advent of the femtosecond (fs) slicing scheme,[19-21] fs X-ray diffraction (XRD) and fs XAS experiments have become possible at synchrotrons to probe the ultrafast dynamics of condensed matter[52-56] and of molecular systems[57-59] in the hard and soft X-ray regimes. It should be mentioned though that coherent phonon oscillations and melting phenomena of crystals had already been reported in fs XRD studies using plasma-based lab sources of fs X-ray pulses.[4-6,60-62]

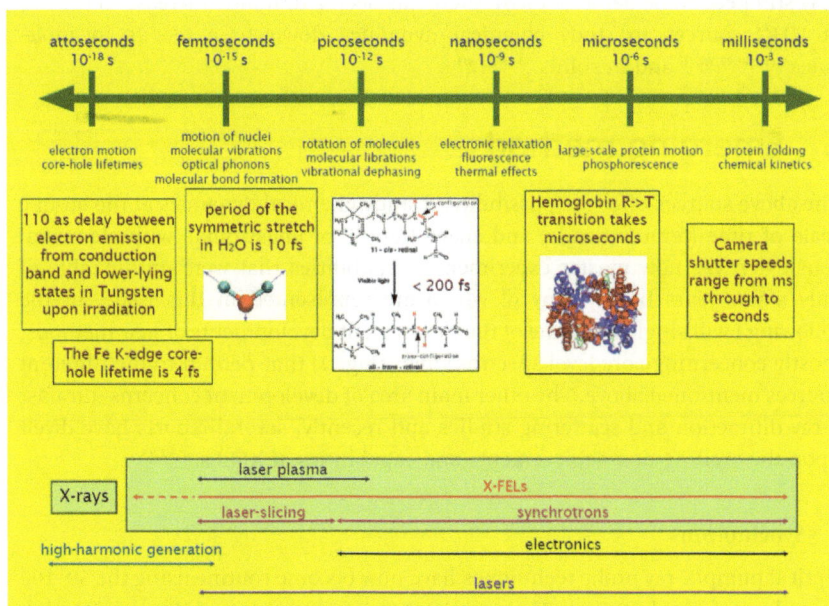

Fig. 1   Time scales of fundamental processes in matter and (below) types of sources to probe them in real-time.

The advent of the first hard X-ray FEL in Stanford (LCLS) in 2009 has represented a game changer due to the huge photon flux they provide (10 orders of magnitude larger than with the slicing scheme) in a pulse of a few 10s of fs duration.[23] This is a major advancement not only for the study of materials and chemical dynamics, but also for protein crystallography and other applications that do not necessarily require ultrashort pulses. In addition, coherent diffractive imaging (CDI)[63–65] exploits the coherence properties of XFEL radiation and may lead to time-resolved CDI in the near future.[66,67] The launching of an XFEL in Japan (SACLA) and the construction of others in Germany (Hamburg, European-FEL), Switzerland (SwissFEL) and Korea (PAL-XFEL) reflects the high demand for such instruments. All these sources operate on the basis of the self-amplified spontaneous emission (SASE), which implies that the lasing process is chaotic, leading to a large temporal laser/X-ray jitter, as well as energy and intensity fluctuations (as large as 15%). One way to overcome these limitations is by seeding with a UV laser pulse having the appropriate wavelength. This is what was implemented at the XUV free electron laser FERMI@Elettra, which is the world's first seeded short wavelength FEL.[68–72] I will come back to the possibilities offered by this type of machine. In the meantime, self-seeding schemes are also being explored and/or implemented at the LCLS.[73] Finally, in terms of lab-sources of ultrashort X-ray and XUV pulses, I already mentioned plasma-based sources, which were first used for ultrafast Bragg diffraction experiments and have recently been extended to powder diffraction, so that the ultrafast chemical dynamics of ionic[50] and molecular[74] crystals can now be investigated.

However, the most important developments of the past 10 years in lab-sources have been in the area of high harmonic generation (HHG), which allows for the generation of VUV to soft X-ray pulses with durations down to a few tens of attoseconds. While the need for attosecond pulses for chemical dynamics *per se* may still need to be demonstrated, there has been a dramatic increase in the use of HHG sources to study chemical dynamics down to a few fs in molecules[31,32,40,42,75,76] and in solids.[33–35,77–81]

3   Emerging methods

The above sources offer the possibility to study chemical dynamics at the atomic scale of time (femtoseconds) and space (tenths of ångströms), with elemental sensitivity. We now master experimental capabilities that were unthinkable of only a few years back, many of which are represented in this issue. In the following I will single out some of the most recent developments of new methods, mostly concerning core-level spectroscopies (Fig. 2) that pertain to the different sources mentioned above. The other main area of development concerns ultrafast X-ray diffraction and scattering studies and recently, several reports have dwelt upon the exciting new achievements and capabilities at XFELs.[48,82,83]

A   Synchrotrons

Optical pump/X-ray probe techniques have now become routine using the 50–100 ps pulses of synchrotrons. There are numerous examples of X-ray diffraction studies with 100 ps resolution of protein crystals,[84–88] and more recently, X-ray scattering studies of proteins in solutions.[89–92] XRD with 100 ps resolution has

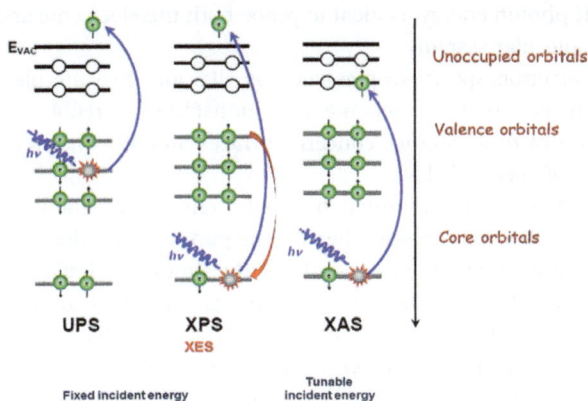

Fig. 2   Different types of core-level spectroscopies: ultraviolet photoelectron spectroscopy (UPS), X-ray photoelectron spectroscopy (XPS), X-ray absorption spectroscopy (XAS) and X-ray emission spectroscopy (XES).

also been used to study solid materials[93–95] and molecular crystals.[96,97] Time-resolved X-ray absorption studies were also carried out on solids,[98,99] molecular systems in solutions[13,14,15] and biological systems.[16,100]

Until recently, amplified femtosecond lasers were used for excitations that run at a few kHz repetition rates, while the storage ring operates at MHz rates. In order to exploit more of the X-ray pulses from the storage ring, we implemented a high repetition rate scheme replacing the 1 kHz fs pump laser by a ps one that runs at half the repetition rate of the Swiss Light Source (SLS) synchrotron (i.e. 0.52 MHz).[16] This ensures that all X-ray pulses are used as the detection scheme is based on the difference absorption spectra (excited sample minus unexcited sample absorption) on a pulse-to-pulse basis, already developed for the kHz scheme.[16] This provides an increase of signal-to-noise ratio corresponding to the square root of the increase in repetition rate. Soon after, similar schemes were announced at the APS,[17] Elettra[18] and the ALS.[44] These developments are crucial for several applications:

(i) Compared to molecular systems in solution, the concentration of biological systems in physiological media is 1 to 2 orders of magnitude lower, i.e. typically in the millimolar level. This has hindered the study of such systems using the kHz repetition rate schemes but now, this has become possible in absorption spectroscopy[16,101] and it is perfectly adequate for diffraction and scattering experiments.

(ii) While XAS probes the density of unoccupied states, XES probes that of the occupied states (Fig. 2). It is highly attractive in time domain studies as it senses the changes in electronic and spin structure that are induced by photoexcitation of the valence electrons. It is particularly attractive in the case of spin state changes,[102] as recently demonstrated first in the ps,[28,30] then the fs domain.[29] XES also provides information about the nature of chemical bonds and their distances around a given atom.[103,104] The quantum yield for XES is typically <0.1%, therefore signals are very weak. The use of the high repetition rate scheme has greatly improved the signal-to-noise ratio, as demonstrated in ref. 30 and 105. These studies also show how the combined use of XES and XRD, which both operate at fixed incident photon energy, is ideal to probe both the electronic and structural changes of molecular systems.

(iii) Photoelectron spectroscopies such as UPS and XPS are also benefiting from the high repetition rate scheme as demonstrated in ref. 44, 106 and 107. Indeed, several of these studies concern surfaces and adsorbates in which the concentration of species is low.

Synchrotrons will remain indispensable for chemical dynamics despite the advent of XFELs because: firstly, they fill the gap between the 100 ps and the nanosecond time domain studies, while XFELs should be dedicated to shorter times; secondly, they are multiuser machines and thus allow a thorough characterization of the system under study prior to implementing an experiment at the XFEL; thirdly, their large and flexible tunability range allows the exploration of various spectral ranges, prior to focussing on the more demanding experiments at XFELs that are usually constrained to a limited spectral range.

B   X-ray free electron lasers

All of the above applications are being implemented at XFELs with success. Ref. 30 and 108 show the results of fs XAS experiments of dilute molecular systems in solution at XFELs following their demonstration using the slicing scheme at synchrotrons.[57-59] The huge flux increase per pulse at XFELs have made fs-XES possible as very recently demonstrated at the LCLS on spin cross-over Iron complexes.[29] Likewise, the implementation of fs-XPS has also been demonstrated very recently on dye-sensitized solar cells.[109]

Ultrashort X-ray pulses from XFELs are not only attractive for studies of chemical dynamics, but are also useful to monitor sample damage during steady state X-ray diffraction (XRD) measurements. In a recent paper, Yachandra et al.[110] combined XES and XRD using 50 fs, 7 keV pulses from the LCLS showing by XES that the chemical integrity of the sample at room temperature was maintained during the measurement, whilst doing the same experiment using the 100 ps long X-ray pulses from a synchrotron leads to sample damage unless the latter is cooled (Fig. 3).

The other major development at XFELs is of course that of serial crystallography with new sample delivery schemes.[48,82,83,111-117] In particular, the liquid injection system allows a liquid stream of a few μm in diameter containing nanocrystals of proteins. This holds the promise of studying chemical dynamics at the XFEL, especially in the μsec to msec range, which is particularly relevant to biological systems.

Last, but not least, seeded-XFEL will enable the X-ray equivalent of ultrafast non-linear optical techniques, such as transient grating experiments, four-wave mixing, photon echo,[118] etc., which are already being explored at FERMI@Elettra XUV FEL as discussed in ref. 71 and 72 and use two (or more) XUV pulses. X-ray and optical wave mixing has already been reported[119] while the first 2-colour FEL emission was demonstrated in view of X-ray/X-ray wave-mixing techniques.[70,120,121] The developments at FERMI are crucial and are preparing the ground for their extension to the shorter wavelength range, when seeding of hard X-ray FEL will be implemented.

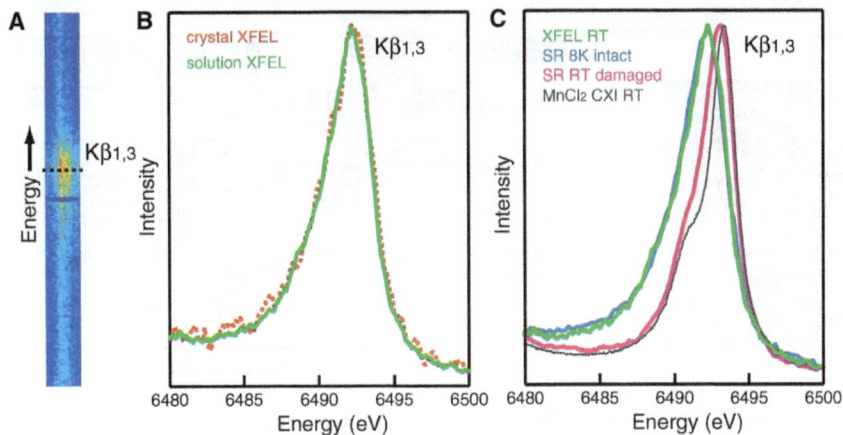

Fig. 3 Femtosecond X-ray emission spectra (XES) of photosystem II (PSII): (A) two-dimensional (2D) $K\beta_{1,3}$ X-ray emission spectrum of PS II microcrystals using sub-50 fs pulses of about 2 to 3 × 10^{11} photons per pulse and μm^2. (B) XES of a solution of PS II (green curve) and single crystals of PS II (red dashed curve) in the dark state obtained from the 2D plot in (A) by integration along the horizontal axis. (C) XES of PS II solutions in the dark state at RT (green) or collected using SR under cryogenic conditions with a low X-ray dose ("8K intact," light blue) or SR at RT under photoreducing conditions ("RT damaged", pink). The spectrum from Mn^{II}Cl$_2$ in aqueous solution collected at RT is shown (gray) for comparison (reproduced from ref. 110).

C Laboratory-based sources

I already mentioned the implementation of fs-powder diffraction using plasma-based sources. In the past 10 years or so, the most important developments in lab-based methods have been around HHG-based sources of VUV–XUV pulses. This meeting highlighted several new schemes for the alignment or orientation of molecules in the gas phase and the characterization of their dynamics to a great level of detail, especially for systems of greater complexity.[39,116] These alignment/orientation experiments are also strongly connected to developments at XFELs. Among the new methods for chemical dynamics that are being developed, photoelectron diffraction[39] is promising because of its possible extension to molecules adsorbed on surfaces.

As far as condensed phases are concerned, the recent implementation of VUV photoelectron spectroscopy of solutions based on the technology of the liquid microjets (Fig. 4, left panels) is a very promising development for chemical dynamics. It was first demonstrated for steady state soft X-ray spectroscopy,[123,124] and recently extended to the ultrashort time domain using VUV pulses from an HHG source.[125–127] This now opens the door to finer studies of chemical reactions in solution with elemental specificity, while delivering precise information about the electronic structure of the species involved in the reaction. Many studies have already been carried out in the UV,[128–131] which probe the valence orbitals and therefore lack the element-specificity, but the extension to the VUV, demonstrated by Abel, Faubel and co-workers has overcome this limitation.[125–127] My group has recently achieved the construction of such a set-up for ultrafast optical pump/ultraviolet photoelectron spectroscopy (UPS) of liquid microjets using

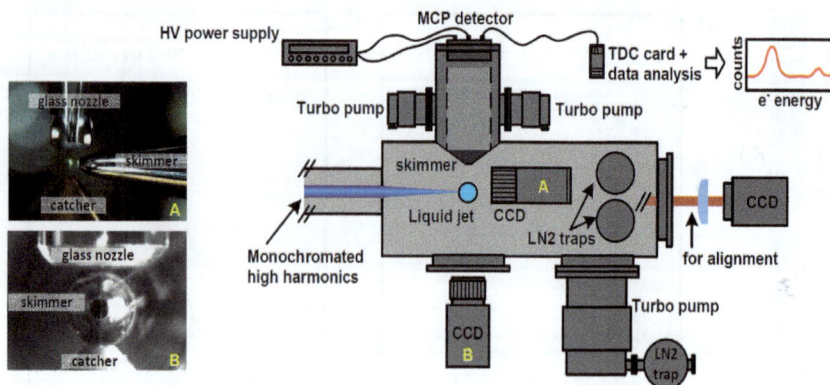

Fig. 4   The Lausanne set-up for liquid phase VUV photoelectron spectroscopy using liquid microjet technology. (A) and (B) are side views of the microjet (glass nozzle), the catcher of the liquid and on the right the skimmer (entrance to the electron detector). See ref. 122 for details.

monochromatized VUV pulses from an HHG source (Fig. 4). A simple, differentially pumped electron time-of-flight spectrometer was built for conducting ultrafast photoelectron spectroscopy from liquid samples probed with monochromatic VUV photons from an HHG source. The monochromatization is done by a time-preserving monochromator.[122] The set-up will have three beam lines, one for the liquid phase experiments, one for the gas phase and a third for the study of condensed matter systems.

In the latter case, a major methods development of the past few years has been the extension of angle-resolved photoemission spectroscopy (ARPES) to the ultrashort time domain of which three very nice examples are presented in this issue.[33-35] While not directly relevant to chemical dynamics and requiring single crystals, they do offer a deep level of insight into the electronic structure of solids and the photoinduced changes therein.

4   Scientific questions

In the following I would like to dwell on some of the scientific and technical issues that were raised during this meeting. This part is by no means exhaustive and does not reflect the rich discussions and exchanges that took place. I refer the reader to the respective articles and their appended discussions.

A   Aligning molecules

As already mentioned several papers discussed the issue of aligning and orienting molecules in the gas phase. These papers aimed at imaging molecular structural dynamics using tools as diverse as photoelectron diffraction,[39] X-ray diffraction[36] and Coulomb explosion,[41] or to describe electron and hole dynamics in molecules.[31]

The first ideas to align molecules in the gas phase came from simulations by Williamson and Zewail[132] for ultrafast electron diffraction studies (Fig. 5). They are actually not an alignment as such but are based on the principle of

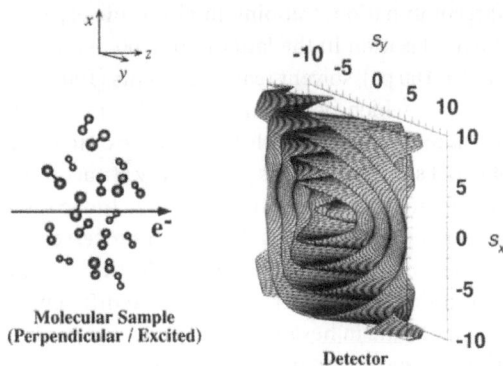

Fig. 5 Calculated electron diffraction pattern for an ensemble of molecules containing photoselected ones (*i.e.* which are more or less parallel along the x-axis) and randomly oriented unexcited ones. The former give rise to a fringe pattern while the latter give rise to a ring pattern. The calculated pattern is a superposition of both with the intensity of the fringe pattern being determined by the excitation yield (reproduced from ref. 132).

photoselection, initially developed for molecules in solids.[133] In this approach, one exploits the relative orientation of the transition dipole moment of the molecule with respect to the incident laser electric field. The degree of photo-selection is not very high as it follows the $\cos^2\theta$ law, where θ is the angle between the dipole moment and the electric field vector of the laser. Theoretical studies on photoselected or photoaligned molecules were further undertaken by Wilson and co-workers, who extended them to the case of ultrafast X-ray diffraction.[134,135] In the meantime, several papers have discussed in a more elaborate way (including anisotropy) the case of ultrafast electron diffraction[136,137] and X-ray diffraction of diatomics and polyatomics.[138–140] More elaborate schemes have been developed, many of which are discussed in this meeting,[39] for aligning and orienting molecules, with very successful outcomes in terms of probing their structure and dynamics.

Most natural and preparative chemistry takes place in the condensed phases (solutions and interfaces) for which none of these schemes would be applicable. Thus one is left with photoselection as the only way to have an excited sample that is aligned, though as already mentioned, the degree of alignment is not very high and will quickly be destroyed by the fluctuations of the solvent. Exploiting pho-toselection to probe the dynamics of molecular systems is perfectly applicable with 100 ps synchrotron pulses if the object is large enough such that its reor-ientational dynamics is slow. This was exploited early on in the study of the magnetic field induced realignment behaviour of smectic-A liquid crystals, which was probed by time-resolved (on the time scale of seconds) small angle X-ray scattering (SAXS) experiments.[141,142] More recently, time-resolved anisotropic X-ray scattering (XRS) using photoselection by a polarized pump laser pulse was realized by Kim *et al.*,[143] who studied photo-excited myoglobin.

When it comes to the chemical dynamics of small molecules, time-resolved X-ray scattering studies of solutions with 100 ps time resolution have been per-formed for several years now.[144] However, photoexcitation leads to several processes such as dissociation (10s to 100s fs), geminate (<1 ps) and non-geminate (>10 ps) recombination, trapping in electronic excited states (1–10 ps), electronic-vibrational relaxation in the latter (1 ps–100 ps), vibrational relaxation in the ground state (10–100 ps), solvent rearrangements (100 fs–ps), heating of the solvent (10s ps–ns), *etc.*, which occur on time scales from fs to ns,[145] and are therefore very difficult to disentangle with 100 ps pulses. However, solution phase X-ray scattering of small solutes becomes particularly attractive in the sub-ps to ps time domain, which is now possible with the XFELs. Indeed, the <100 fs pulses provided by these allows for the various dynamical processes to be separated in time, in particular when combined with a photoselective excitation.

We simulated the anticipated scattering patterns at different time delays in the case of the excited I_2 molecule in hexane.[146] The molecule was impulsively excited to the B-state that is known to undergo wavepacket dynamics over several ps in the gas phase.[147] We performed molecular dynamics simulations without and with the anisotropy of the solvent and the simulated difference (excited minus unexcited) diffraction patterns are shown in Fig. 6: (i) at $t = 0$, the population of excited molecules has an alignment defined by $\cos^2\theta$. At 100 fs, the wave packet dynamics have started to evolve in the excited B-state of the molecule and the resulting difference pattern shows the fringes expected for a population of nearly parallel double-slits, represented by the photoselected molecules; (ii) the fringe pattern persists up to 400 fs, with changes in their spacings due to the oscillatory motion of the wave packet, as the population of molecules stretches and contracts in phase along the I–I distance; (iii) during this time, collisions with the solvent molecules gradually blur the pattern due to dephasing of the wave packet and loss of the initial anisotropy; (iv) as time evolves, the ring pattern fades away allowing one to follow the evolution of the solvent response at wide angles for early times, then at small angles for later ones.

These simulations were carried out using realistic parameters for the excitation yield (3%) and the X-ray probe pulse, delivering a signal-to-noise ratio of 4–5 under single shot operation. This is a very promising prediction, which we hope to verify in future experiments at an XFEL.

B Structural dynamics of large molecules

Because time and length scales go together, X-ray absorption spectroscopy turns out to be ideal to probe ultrafast processes which take place at short times of sub-ps to a few ps, in addition to delivering information about the electronic structure of the excited molecular system. In ref. 30 and 105 it was shown how combining ps XAS with ps XES and XRS, one can get a fairly complete picture of the excited state structure and its associated solvation shell structure. Pushing these techniques to the fs time domain would deliver great insight into the associated dynamics leading to the high spin state.

One class of systems that has received much attention in recent years by the ultrafast X-ray community, is the Fe(II) spin cross over (SCO) complexes. Fs hard,[57] then soft XAS[59] previously demonstrated on the Fe(II) SCO complex using the slicing scheme at synchrotrons has recently been implemented at XFELs.[30,108] These studies, which saw [Fe(bpy)$_3$]$^{2+}$ emerging as the archetypal SCO complex, were steered by the need to understand the nature of the ultrafast spin transition (Fig. 7) from the singlet metal-to-ligand-charge-transfer (MLCT) state to the quintet state (^{5}T). It was found to occur in <150 fs, with unity quantum yield, as a

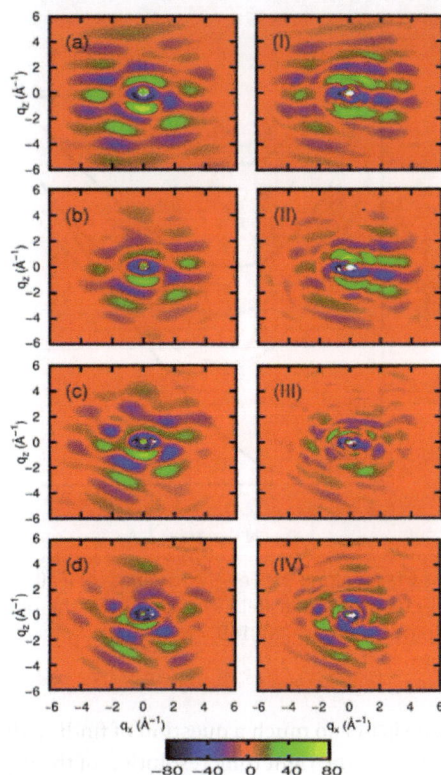

Fig. 6 Difference scattering patterns of photo-excited I_2 in n-hexane for time delays of 100, 200, 300 and 400 fs (descending order). The polarization of the pump laser pulse is along the y-axis in these plots. For panels (a)–(d) we have assumed that no anisotropy is induced in the solvent, while for panels (I)–(IV) we assume that anisotropy is induced in the first solvation shell and therefore the 54 closest carbon atoms are included in the explicit atomic description. The colour key is unitless, but the range shown corresponds to 3.5% of the total scattering signal. The white areas are regions of high signal.

^{1}MLCT-^{3}MLCT-^{5}T cascade.[148] A bypass via intermediate metal-centred ^{1,3}T states was not excluded[57] but the time resolution was deemed insufficient to confirm or exclude it.

A ^{1}MLCT-^{3}MLCT-^{5}T cascade raises however a number of questions: (a) the transition from MLCT states to the ^{5}T state is a two-electron transition, which has low probability and; (b) the spin-orbit coupling (SOC) constant between ^{3}MLCT and ^{5}T is too weak;[149] (c) the calculated SOC cannot explain why the process occurs with unity quantum yield. It was argued that the metal-centred (MC) states, ^{3}T$_2$ and ^{1}T$_1$, may be part of the relaxation cascade from the MLCT states to the ^{5}T state.[149] Recently, fs XES experiment at the LCLS XFEL with 150 fs resolution confirmed this hypothesis.[29] By fitting the rise kinetics of the ^{5}T XES signal, it was concluded that an intermediate state is involved in the SCO, which was proposed to be the ^{3}T$_2$ state (Fig. 7). The authors extracted a 150 ± 50 fs time constant for the decay of the MLCT manifold to the ^{3}T state and a 70 ± 30 fs time constant for ^{3}T decay to the ^{5}T$_2$ state. However, the lack of a specific spectroscopic signature of the intermediate state, along with the large XFEL time jitter raises doubts about this conclusion.

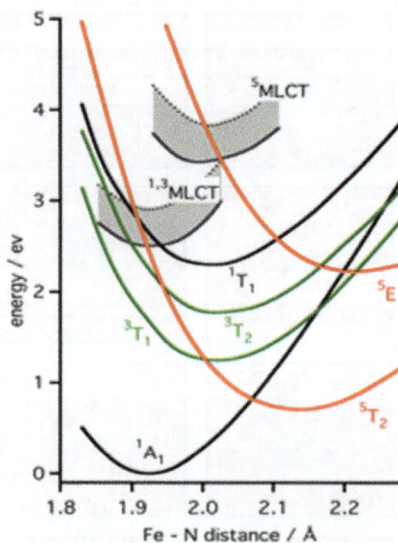

Fig. 7 Potential energy surfaces along the Fe–N coordinate of [Fe(bpy)$_3$]$^{2+}$. The gray areas are the metal-to-ligand-charge-transfer (MLCT) states, while the other states are metal-centred (MC) states. Reproduced from ref. 149.

Addressing this issue is not so much a question of finding the right observable of the spin state, but more so the right time resolution of the experiment, since all previous studies were carried out with resolutions >130 fs,[57,150,151] whilst our conclusions in ref. 57 and those of the fs-XES study[29] imply that a resolution of <60 fs would be needed to observe the intermediate states, if at all.

We recently carried out a visible pump/visible probe and UV probe study of the spin dynamics of [Fe(bpy)$_3$]$^{2+}$ at a high time resolution (<40 fs in the visible, <60 fs in the UV). These spectral ranges probe both the departure of the population from the MLCT states and its arrival in the $^5T_2$ state. It was found that the SCO process occurs in <50 fs![152] This would rule out a bypass *via* one of the MC states, unless the latter occurs in <20 fs. In addition, the 50 fs time scale corresponds to barely ¼ of the Fe–N stretch oscillation of the system,[149] such that the process is impulsive and in a strongly non-adiabatic regime.

These results show that a "traditional" pump-probe experiment, with the adequate time resolution, can yield more insight than using new schemes with an inadequate time resolution. Beyond that, they also raise a fundamental issue about the spin dynamics in metal complexes, which seems to violate any hierarchy based on SOC constants,[153] as traditionally observed in organic photophysics.

The observation of an extremely fast (at sub-vibrational time scales) internal conversion (IC), intramolecular vibrational redistribution (IVR) and intersystem crossing (ISC) processes in a growing class of complex molecules[154] calls for element-specific studies at an even higher time resolution. Among the approaches we are planning is ultrafast liquid jet UPS with our HHG source (Fig. 4) at <20 fs time resolution.[122]

C Charge carrier dynamics in metal oxides

The above metal complexes are also ideal candidates for solar energy conversion in so-called dye sensitized solar cells (DSSCs),[155,156] which were discussed at this meeting.[44,45] In DSSCs, the most popular dyes are ruthenium–polypyridine complexes, adsorbed onto a transition metal oxide substrate (TiO$_2$ or ZnO). The ruthenium complexes serve as visible light harvesters and upon excitation of their ^{1}MLCT states, injection of an electron occurs to the conduction band (CB) of the substrate. The fate of the electrons in the latter has been the centre of investigations for over two decades using ultrafast laser techniques from the THz to the UV.[157–162] It is known that trapping of the electrons occurs in the CB but the nature, the structure of these traps, the time scales for trapping and the lifetime of the traps could not be determined. These questions call for time-resolved tools with element- and structure-sensitivity, such as those presented at this meeting.[43–45]

Optical pump/XAS probe techniques with 100 ps resolution were recently implemented in the case of ruthenium[163] and copper[164] dyes adsorbed on TiO$_2$ allowing to probe the oxidized dyes at their metal K-edges, after injection. However, no probing of the electrons in the metal oxide substrate was done. Previously, Katz et al.[165,166] reported an Fe K-edge absorption study of iron oxide NPs upon electron injection from adsorbed organic dyes. By comparing their transients with simulated ones, they concluded that the reduced metal sites formed small polarons on a 100 ps time scale.

In order to investigate the fate of the electron prior to and after injection into the substrate, we recently carried out a laser pump/XAS probe study of bare and ruthenium dye-sensitized anatase and amorphous TiO$_2$ nanoparticles (NPs, typically 10–20 nm diameter) in colloidal solutions, with 100 ps resolution, at both the Ru L-edges and the Ti K-edge.[167] The anatase form of TiO$_2$ is the most frequently used for solar energy applications. For the bare NPs, excitation at 355 nm was carried out above the band gap of the material, while for the dye-sensitized NPs, excitation was at 532 nm into the ^{1}MLCT state.

Fig. 8A shows the steady-state Ti K-edge XAS spectra of colloidal solutions of the 20 nm bare anatase NPs and 10 nm amorphous NPs. There are striking differences between the two (the same was found for the powder X-ray diffraction patterns), pointing to a clear structural difference at short range. The pre-edge XAS region (Fig. 8A) exhibits four peaks, labelled A1–3 and B. The A2 and A3 peaks are, relative to the A1 and B peaks, stronger in the amorphous than in the anatase case. These peaks are dipole-forbidden transitions from the 1s core orbital to unoccupied Ti 3d valence orbitals, and they become partially allowed because of symmetry reduction in the lattice, by virtue of 3d–4p mixing, which introduces a dipole component. Stronger deviations from the octahedral symmetry of anatase would further enhance such mixings, which is the reason why these peaks are stronger in the amorphous case.

Fig. 8B and C show the transients (excited minus unexcited sample absorption) of the bare and dye-sensitized NPs recorded 100 ps after laser excitation. These transients point to notable photoinduced spectral changes. In particular, the strong feature at 4.982 keV is due to a red shift of the edge, resulting from the reduction of Ti atoms to form Ti^{3+} centres. The pre-edge peaks are significantly enhanced, which points to a change of local symmetry around the reduced

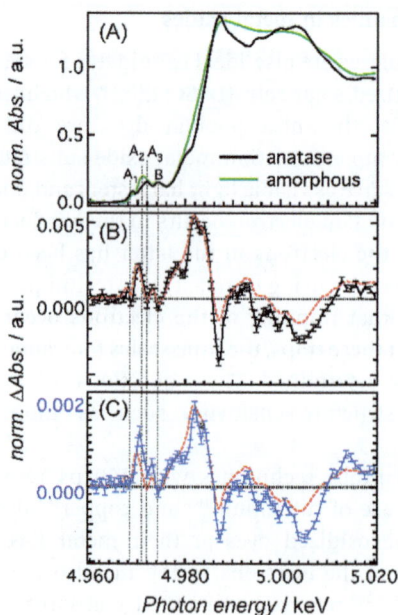

Fig. 8 (A) Ti K-edge X-ray absorption near edge spectra (XANES) of bare anatase and amorphous TiO_2 in colloidal aqueous solutions. (B) Transient X-ray absorption spectra of bare anatase TiO_2 NPs excited at 355 nm with a 100 ps time delay (black) together with the calculated difference spectrum (red dashed line) taking the 1 eV shifted amorphous steady state spectrum minus the anatase spectrum (scaled by 0.016, by normalization at the edge at 4.982 keV, see text). (C) Transient XAS of Ruthenium N719 dye-sensitized anatase TiO_2 NPs with a 100 ps time delay (blue), excited at 532 nm. The dashed red trace is the same as in (B), but is scaled by 0.006. Reproduced from ref. 167.

centres, as are also the changes above the edge. There are some subtle differences between the bare and dye-sensitized cases, in that the pre-edge, as well as the above edge changes, are more pronounced relative to the signal at 4.982 keV compared to the bare case. In order to confirm that the changes in the dye-sensitized NPs are really due to injection from the dye, we also recorded the changes at the Ru L_3-edge. The transient (not shown here, see ref. 167) reflects the blue shift of the edge due to oxidation of the Ru atom, from 2^+ to 3^+, as well as the creation of a hole in the lower Ru $d(t_{2g})$ orbitals.

The transients in Fig. 8B and C could be well reproduced by taking the difference of the amorphous steady-state spectrum shifted by −1 eV (to account for the reduction) minus the anatase steady-state spectrum (red dashed traces). The agreement is somewhat poorer in the case of sensitized NPs. This overall good agreement between the transients and the calculated difference spectra does not mean that the NPs undergo a change from anatase (ordered) to amorphous, since the signal decays to zero in ns time scales.[167] Rather, it reflects a shift of spectral weight from anatase sites to defects. In anatase NPs, defects are mostly found in the surface shell,[167] although they also occur in the core. The fact that the edge shifts by −1 eV suggests that a full electron charge has localised (a ±1 eV oxidation shift typically corresponds to a localization of a ∓1e charge) at the Ti centres. The conclusion from this is that small polarons are formed, as predicted by Selloni *et al.*[168] Large polarons would distribute the charge over several centres and therefore yield a smaller oxidation shift. The differences between the bare and sensitized cases has been analysed in ref. 167 and is due to the fact that in the former, the electrons are trapped deep in the surface shell, while upon injection, they remain on the outer surface. This suggests the occurrence of a charge transfer exciton between the cationic dye and the electron in DSSCs.

These results along with those of ref. 43–45 show that time-resolved core-level spectroscopies are providing deep insight into the behaviour of charge carriers in metal oxides, and prepare the ground for studies with fs resolution at the XFELs. Pending such studies, we have recently demonstrated fs Ti K-edge XAS to probe the localization dynamics of electrons in bare TiO_2 and found a time scale of <300 fs, which points to the electrons being trapped very close to where they are generated.[56] Complementary studies on single crystals of TiO_2 are being planned using fs ARPES with our set-up (Fig. 4).

5   Outlook

The above three examples were meant to illustrate some recent examples of novel problems whose study is enabled by emerging photon technologies. Many other examples have been discussed at this meeting, among which those of the structural dynamics of biological systems on which I have not much dwelt.

The past ten years or so have been marked by *disruptive* (in a positive sense, of course!) progress in photon technologies. Indeed, we have seen the development of the fs-slicing scheme and the ps high repetition rate scheme at synchrotrons, the advent of X-ray free electron lasers, then of their seeded counterparts, and the amazing improvements in lab-based XUV/VUV sources based on HHG sources,[26] *etc.* These new sources have in turn triggered a major development of methods, such as ultrafast X-ray absorption and X-ray emission, picosecond WAXS, ultrafast ARPES of solids, ultrafast XPS/UPS, diffractive imaging, liquid jet photoelectron spectroscopies, ultrafast powder diffraction, femtosecond serial crystallography of biological systems, non-linear XUV/X-ray optics, *etc.*, all of which have been discussed at this meeting.

The present trend in terms of photon technologies will go on with the several improvements that are planned. For example, the seeding of XFEL is solving the issues of time, energy and intensity,[68,169] as well as offering the possibility to tune photon energy. In the case of lab-based sources, the introduction of new disk laser technology (high repetition rate, high peak energy) along with optical parametric chirped-pulse amplification (OPCPA) is a game changer for HHG and plasma generation.[170-172] In spite, or rather, because of these developments, synchrotrons will remain indispensable. Not only do they provide a testing ground for complex experiments to be carried out at XFELs, but also they allow for more flexible experiments due to their large spectral tunability. They also fill the gap between the tens of ps and the msec range where important phenomena occur in biology, catalysis, *etc.*

Emerging methods that promise new insights once pushed into the time domain are:

(i) Femtosecond serial crystallography: its recent successes thanks to the Spence microjet[111,117] holds great promise. While the method is still being benchmarked by proof-of-principle experiments of known static structures, it will ultimately be judged against the extent to which new structural insights emerge that would not be accessible using synchrotron radiation or single particle cryo-electron microscopy.[173] One advantage of XFEL sources is that diffraction data recorded from radiation-sensitive proteins do not show signs of radiation damage[174,175] and this can be very important for some proteins such as photo-system II.[110] The other major advantage of XFELs is their high time resolution. However, the classical Laue diffraction approach, of collecting both dark refer-ence and light-activated images for each and every oscillation from the same crystal, may be difficult to apply at XFELs because SASE produces an X-ray spec-trum with large pulse-to-pulse variation. Spectral fluctuations are difficult when processing Laue diffraction data due to the need to normalize the measured scattering intensities against the X-ray fluence through the crystal. Spence dis-cussed some solutions at this meeting, such as using attosecond X-ray pulses.[48] Seeded FELs avoid the problems of SASE fluctuations[73,169] but at the cost of a very narrow XFEL bandwidth, which is advantageous in most applications but not appropriate for time-resolved Laue diffraction. Indeed, a narrow bandwidth results in many observations being partials and demands that many angles be sampled, making the experiment sensitive to laser pump and X-ray probe induced damage. The pioneering experiments at the XFEL with μsec resolution on complexes of photosystem I and ferodoxin[176] and those on photosystem II showing no photoreduction by X-rays[110] lay the ground for future work on the chemical dynamics of biosystems.

(ii) Ultrafast wide-angle X-ray scattering (WAXS): solution phase WAXS is a very promising technique for the chemical dynamics of molecular and biological solutes. This method was first developed for the study of small molecules,[8,144] and was then extended to probe the dynamics of proteins with 100 ps resolution,[89] recording WAXS data following the photo-dissociation of carbon monoxide from tetrameric hemoglobin and cytochrome c. Later studies probed the reaction dynamics of light-triggered reactions such as the photo-dissociation of CO from the heme protein groups of myoglobin,[89,92,143] as well as chromophore isomeri-zation-driven reactions within photoactive yellow protein[177,178] and retinal proteins.[179,180] Depending on the time scale of the process under study, the 100 ps resolution of synchrotrons may well be sufficient for biological systems, but the higher fluxes of XFELs provide higher signal-to-noise ratios of the difference patterns. On the other hand, this resolution is wholly inadequate for the study of the chemical dynamics of small molecular systems *via* WAXS. XFELs with their high time resolution and high fluxes are the only possible routes, especially in combination with photoselection, as discussed above and in ref. 146. It will be particularly exciting to establish it for small molecules and move to larger systems. In this respect, our ultrafast optical studies on diplatinum complexes were aimed at solution fs-WAXS[181] and this plan is still on the agenda.

(iii) Femtosecond high energy resolution off-resonant spectrum (HEROS): the idea of using off-resonant excitations to probe the density of unoccupied states of a system was developed in the early 1980s.[182] It was found that for incident beam energies tuned below an X-ray absorption edge (called the off-resonance region), the shape of the XES spectrum is proportional to the unoccupied density of states of an atom. Until recently, the potential of extracting the electronic structure from a single XES spectrum recorded at off-resonant excitations was not explored, likely due to the extremely weak scattering cross section as compared to the resonant

XES spectrum. By combining off-resonant excitation and an X-ray spectrometer operating in a dispersive geometry at a synchrotron, HEROS was experimentally demonstrated. It represents an alternative to XAS, with significant advantages when used with pulsed X-ray sources. HEROS requires monochromatic photon energies for the incoming beam and in a recent work, Szlachetko et al.[183] used the "self-seeding" method at the LCLS to produce a narrow energy bandwidth beam with more stable beam characteristics than during normal SASE operation with a monochromator. They could thus record single short HEROS spectra of copper and copper oxide using fs X-ray pulses. Their results are very promising in view of fs-HEROS experiments that would offer an alternative to fs-XAS, without the problems of normalization.

(iv) Ultrafast non-linear XUV/X-ray optics: the advent of seeded FELs is now rendering non-linear 4-wave mixing (4WM) experiments possible in the XUV range, and hopefully soon, in the harder X-ray domain. The pioneering optical/X-ray non-linear experiment at the LCLS[120] was carried out with the assistance of an optical laser pulse. The upcoming multi-pulse, multicolour FEL emission can be profitably exploited in nonlinear wave mixing applications.[184,185] The FERMI@Elettra seeded FEL[68] offers many of the required properties for such experiments, in terms of pulse duration, flux, wavelength and coherence properties. In particular, the time-coherence is mandatory for 4WM experiments. Several promising experiments[71,72] are planned with this source, which hopefully set the stage for an extension into the hard X-ray range.

(v) Ultrafast photoelectron spectroscopy: the improvements in HHG sources of VUV/XUV radiation are also offering new opportunities for chemical dynamics, while the tools of photoelectron spectroscopy of gas phase molecules and photoemission of solids with energy and angular detection are well established now in the ultrashort time domain. Their recent implementation for the study of liquids[125–127,186,187] is very promising and is offering a complementary tool to other core-level spectroscopies in the description of the electronic structure of solutes undergoing a chemical process.[122]

In conclusion, the "disruptive" improvements of the past 10 years or so in terms of new sources and the associated methods are making the future look very bright!

Acknowledgements

I am grateful to all my co-workers for the studies reported in this paper. Many of these were supported by the Swiss NSF via the NCCR MUST.

References

1 A. H. Zewail, *J. Phys. Chem. A*, 2000, **104**, 5660–5694.
2 G. R. Fleming, G. S. Schlau-Cohen, K. Amarnath and J. Zaks, *Faraday Discuss.*, 2012, **155**, 27–41.
3 G. A. Voth and R. M. Hochstrasser, *J. Phys. Chem.*, 1996, **100**, 13034–13049.
4 C. Rose-Petruck, R. Jimenez, T. Guo, A. Cavalleri, C. W. Siders, F. Raksi, J. A. Squier, B. C. Walker, K. R. Wilson and C. P. J. Barty, *Nature*, 1999, **398**, 310–312.

5 K. Sokolowski-Tinten, C. Blome, J. Blums, A. Cavalleri, C. Dietrich, A. Tarasevitch, I. Uschmann, E. Forster, M. Kammler, M. Horn-von-Hoegen and D. von der Linde, *Nature*, 2003, **422**, 287–289.

6 A. Cavalleri, C. W. Siders, C. Rose-Petruck, R. Jimenez, C. Toth, J. A. Squier, C. P. J. Barty, K. R. Wilson, K. Sokolowski-Tinten, M. H. von Hoegen and D. von der Linde, *Phys. Rev. B: Condens. Matter Mater. Phys.*, 2001, **6319**, 193306.

7 T. Elsaesser and M. Woerner, *Acta Crystallogr., Sect. A: Found. Crystallogr.*, 2010, **66**, 168–178.

8 Q. Y. Kong, J. H. Lee, M. Lo Russo, T. K. Kim, M. Lorenc, M. Cammarata, S. Bratos, T. Buslaps, V. Honkimaki, H. Ihee and M. Wulff, *Acta Crystallogr., Sect. A: Found. Crystallogr.*, 2010, **66**, 252–260.

9 A. H. Zewail, *Annu. Rev. Phys. Chem.*, 2006, **57**, 65–103.

10 D. Shorokhov and A. H. Zewail, *Phys. Chem. Chem. Phys.*, 2008, **10**, 2879–2893.

11 G. Sciaini, M. Harb, S. G. Kruglik, T. Payer, C. T. Hebeisen, F. J. M. Z. Heringdorf, M. Yamaguchi, M. H. V. Hoegen, R. Ernstorfer and R. J. D. Miller, *Nature*, 2009, **458**, 56–U52.

12 M. Chergui and A. H. Zewail, *ChemPhysChem*, 2009, **10**, 28–43.

13 L. X. Chen, X. Y. Zhang, J. V. Lockard, A. B. Stickrath, K. Attenkofer, G. Jennings and D. J. Liu, *Acta Crystallogr., Sect. A: Found. Crystallogr.*, 2010, **66**, 240–251.

14 M. Chergui, *Acta Crystallogr., Sect. A: Found. Crystallogr.*, 2010, **66**, 229–239.

15 C. J. Milne, T. J. Penfold and M. Chergui, *Coord. Chem. Rev.*, 2014, DOI: 10.1016/j.ccr.2014.02.013.

16 F. A. Lima, C. J. Milne, D. C. V. Amarasinghe, M. H. Rittmann-Frank, R. M. van der Veen, M. Reinhard, V. T. Pham, S. Karlsson, S. L. Johnson, D. Grolimund, C. Borca, T. Huthwelker, M. Janousch, F. van Mourik, R. Abela and M. Chergui, *Rev. Sci. Instrum.*, 2011, **82**, 063111.

17 A. M. March, A. Stickrath, G. Doumy, E. P. Kanter, B. Krassig, S. H. Southworth, K. Attenkofer, C. A. Kurtz, L. X. Chen and L. Young, *Rev. Sci. Instrum.*, 2011, **82**, 073110.

18 L. Stebel, M. Malvestuto, V. Capogrosso, P. Sigalotti, B. Ressel, F. Bondino, E. Magnano, G. Cautero and F. Parmigiani, *Rev. Sci. Instrum.*, 2011, **82**, 123109.

19 R. W. Schoenlein, S. Chattopadhyay, H. H. W. Chong, T. E. Glover, P. A. Heimann, C. V. Shank, A. A. Zholents and M. S. Zolotorev, *Science*, 2000, **287**, 2237–2240.

20 S. Khan, K. Holldack, T. Kachel, R. Mitzner and T. Quast, *Phys. Rev. Lett.*, 2006, **97**, 074801.

21 P. Beaud, S. L. Johnson, A. Streun, R. Abela, D. Abramsohn, D. Grolimund, F. Krasniqi, T. Schmidt, V. Schlott and G. Ingold, *Phys. Rev. Lett.*, 2007, **99**, 174801.

22 W. Ackermann, G. Asova, V. Ayvazyan, A. Azima, N. Baboi, J. Bahr, V. Balandin, B. Beutner, A. Brandt, A. Bolzmann, R. Brinkmann, O. I. Brovko, M. Castellano, P. Castro, L. Catani, E. Chiadroni, S. Choroba, A. Cianchi, J. T. Costello, D. Cubaynes, J. Dardis, W. Decking, H. Delsim-Hashemi, A. Delserieys, G. Di Pirro, M. Dohlus, S. Dusterer, A. Eckhardt, H. T. Edwards, B. Faatz, J. Feldhaus, K. Flottmann, J. Frisch, L. Frohlich, T. Garvey, U. Gensch, C. Gerth, M. Gorler, N. Golubeva, H. J. Grabosch, M. Grecki, O. Grimm,

K. Hacker, U. Hahn, J. H. Han, K. Honkavaara, T. Hott, M. Huning, Y. Ivanisenko, E. Jaeschke, W. Jalmuzna, T. Jezynski, R. Kammering, V. Katalev, K. Kavanagh, E. T. Kennedy, S. Khodyachykh, K. Klose, V. Kocharyan, M. Korfer, M. Kollewe, W. Koprek, S. Korepanov, D. Kostin, M. Krassilnikov, G. Kube, M. Kuhlmann, C. L. S. Lewis, L. Lilje, T. Limberg, D. Lipka, F. Lohl, H. Luna, M. Luong, M. Martins, M. Meyer, P. Michelato, V. Miltchev, W. D. Moller, L. Monaco, W. F. O. Muller, A. Napieralski, O. Napoly, P. Nicolosi, D. Nolle, T. Nunez, A. Oppelt, C. Pagani, R. Paparella, N. Pchalek, J. Pedregosa-Gutierrez, B. Petersen, B. Petrosyan, G. Petrosyan, L. Petrosyan, J. Pfluger, E. Plonjes, L. Poletto, K. Pozniak, E. Prat, D. Proch, P. Pucyk, P. Radcliffe, H. Redlin, K. Rehlich, M. Richter, M. Roehrs, J. Roensch, R. Romaniuk, M. Ross, J. Rossbach, V. Rybnikov, M. Sachwitz, E. L. Saldin, W. Sandner, H. Schlarb, B. Schmidt, M. Schmitz, P. Schmuser, J. R. Schneider, E. A. Schneidmiller, S. Schnepp, S. Schreiber, M. Seidel, D. Sertore, A. V. Shabunov, C. Simon, S. Simrock, E. Sombrowski, A. A. Sorokin, P. Spanknebel, R. Spesyvtsev, L. Staykov, B. Steffen, F. Stephan, F. Stulle, H. Thom, K. Tiedtke, M. Tischer, S. Toleikis, R. Treusch, D. Trines, I. Tsakov, E. Vogel, T. Weiland, H. Weise, M. Wellhoffer, M. Wendt, I. Will, A. Winter, K. Wittenburg, W. Wurth, P. Yeates, M. V. Yurkov, I. Zagorodnov and K. Zapfe, *Nat. Photonics*, 2007, **1**, 336–342.

23 P. Emma, R. Akre, J. Arthur, R. Bionta, C. Bostedt, J. Bozek, A. Brachmann, P. Bucksbaum, R. Coffee, F. J. Decker, Y. Ding, D. Dowell, S. Edstrom, A. Fisher, J. Frisch, S. Gilevich, J. Hastings, G. Hays, P. Hering, Z. Huang, R. Iverson, H. Loos, M. Messerschmidt, A. Miahnahri, S. Moeller, H. D. Nuhn, G. Pile, D. Ratner, J. Rzepiela, D. Schultz, T. Smith, P. Stefan, H. Tompkins, J. Turner, J. Welch, W. White, J. Wu, G. Yocky and J. Galayda, *Nat. Photonics*, 2010, **4**, 641–647.

24 T. Ishikawa, H. Aoyagi, T. Asaka, Y. Asano, N. Azumi, T. Bizen, H. Ego, K. Fukami, T. Fukui, Y. Furukawa, S. Goto, H. Hanaki, T. Hara, T. Hasegawa, T. Hatsui, A. Higashiya, T. Hirono, N. Hosoda, M. Ishii, T. Inagaki, Y. Inubushi, T. Itoga, Y. Joti, M. Kago, T. Kameshima, H. Kimura, Y. Kirihara, A. Kiyomichi, T. Kobayashi, C. Kondo, T. Kudo, H. Maesaka, X. M. Marechal, T. Masuda, S. Matsubara, T. Matsumoto, T. Matsushita, S. Matsui, M. Nagasono, N. Nariyama, H. Ohashi, T. Ohata, T. Ohshima, S. Ono, Y. Otake, C. Saji, T. Sakurai, T. Sato, K. Sawada, T. Seike, K. Shirasawa, T. Sugimoto, S. Suzuki, S. Takahashi, H. Takebe, K. Takeshita, K. Tamasaku, H. Tanaka, R. Tanaka, T. Tanaka, T. Togashi, K. Togawa, A. Tokuhisa, H. Tomizawa, K. Tono, S. K. Wu, M. Yabashi, M. Yamaga, A. Yamashita, K. Yanagida, C. Zhang, T. Shintake, H. Kitamura and N. Kumagai, *Nat. Photonics*, 2012, **6**, 540–544.

25 F. Zamponi, Z. Ansari, M. Woerner and T. Elsaesser, *Opt. Express*, 2010, **18**, 947–961.

26 G. Sansone, L. Poletto and M. Nisoli, *Nat. Photonics*, 2011, **5**, 656–664.

27 C. Bressler, R. Abela and M. Chergui, *Z. Kristallogr.*, 2008, **223**, 307–321.

28 G. Vanko, P. Glatzel, V. T. Pham, R. Abela, D. Grolimund, C. N. Borca, S. L. Johnson, C. J. Milne and C. Bressler, *Angew. Chem., Int. Ed.*, 2010, **49**, 5910–5912.

29 W. K. Zhang, R. Alonso-Mori, U. Bergmann, C. Bressler, M. Chollet, A. Galler, W. Gawelda, R. G. Hadt, R. W. Hartsock, T. Kroll, K. S. Kjaer, K. Kubicek,

H. T. Lemke, H. Y. W. Liang, D. A. Meyer, M. M. Nielsen, C. Purser, J. S. Robinson, E. I. Solomon, Z. Sun, D. Sokaras, T. B. van Driel, G. Vanko, T. C. Weng, D. L. Zhu and K. J. Gaffney, *Nature*, 2014, **509**, 345–348.

30 C. Bressler, W. Gawelda, A. Galler, M. M. Nielsen, V. Sundstrom, A.-M. March, S. H. Southworth, L. Young, G. Doumy and G. Vanko, *Faraday Discuss.*, 2014, DOI: 10.1039/C4FD00097H.

31 J. P. Marangos, B. Cooper, P. Kolorenc, L. Frasinski and V. Averbukh, *Faraday Discuss.*, 2014, DOI: 10.1039/C4FD00051J.

32 D. Baykusheva, P. Kraus, S. B. Zhang, N. Rohringer and H. J. Worner, *Faraday Discuss.*, 2014, DOI: 10.1039/C4FD00018H.

33 C. Sohrt, A. Stange, M. Bauer and K. Rossnagel, *Faraday Discuss.*, 2014, DOI: 10.1039/C4FD00042K.

34 L. Rettig, J. H. Chu, I. R. Fisher, U. Bovensiepen and M. Wolf, *Faraday Discuss.*, 2014, DOI: 10.1039/C4FD00045E.

35 I. Gierz, S. Link, U. Starke and A. Cavalleri, *Faraday Discuss.*, 2014, DOI: 10.1039/C4FD00020J.

36 M. P. Minitti, *et al.*, *Faraday Discuss.*, 2014, DOI: 10.1039/C4FD00030G.

37 Z. Li, *et al.*, *Faraday Discuss.*, 2014, DOI: 10.1039/C4FD00078A.

38 K. Schnorr, A. Senftleben, G. Schmid, A. Rudenko, M. Kurka, K. Meyer, L. Foucar, M. Kubel, M. F. Kling, Y. H. Jiang, S. Dusterer, R. Treusch, C. D. Schroter, J. Ullrich, T. Pfeifer and R. Moshammer, *Faraday Discuss.*, 2014, DOI: 10.1039/C4FD00031E.

39 R. Boll, A. Rouzee, M. Adolph, D. Anielski, A. Aquila, S. Bari, C. Bomme, C. Bostedt, J. D. Bozek, H. N. Chapman, L. Christensen, R. N. Coffee, N. Coppola, P. Decleva, S. Epp, B. Erk, F. Filsinger, L. Foucar, L. Gumprecht, L. Holmegaard, P. Johnsson, R. Moshammer, B. Rudek, I. Schlichting, F. Scholz, J. Schulz, J. Seltmann, M. Stener, S. Stern, S. Trechert, J. ThOgersen, S. Trippel, J. Viefhaus, M. Vrakking, H. Stapelfeldt, J. Kupper, J. Ullrich, A. Rudenko and D. Rolles, *Faraday Discuss.*, 2014, DOI: 10.1039/C4FD00037D.

40 M. Negro, M. Devetta, D. Facciala, S. De Silvestri, C. Vozzi and S. Stagira, *Faraday Discuss.*, 2014, DOI: 10.1039/C4FD00033A.

41 N. Berrah, L. Fang, T. Osipov, Z. Jurek, B. F. Murphy and R. Santra, *Faraday Discuss.*, 2014, DOI: 10.1039/C4FD00015C.

42 G. Galinis, *et al.*, *Faraday Discuss.*, 2014, DOI: 10.1039/C4FD00099D.

43 G. Smolentsev, A. A. Guda, M. Janousch, C. Frieh, G. Jud, F. Zamponi, M. Chavarot-Kerlidou, V. Artero, J. van Bokhoven and M. Nachtegaal, *Faraday Discuss.*, 2014, DOI: 10.1039/C4FD00035H.

44 S. Neppl, *et al.*, *Faraday Discuss.*, 2014, DOI: 10.1039/C4FD00036F.

45 B. F. Spencer, M. J. Cliffe, D. M. Graham, S. J. O. Hardman, E. A. Seddon, K. L. Syres, A. G. Thomas, F. Sirotti, M. G. Silly, J. Akhtar, P. O'Brien, S. M. Fairclough, J. M. Smith, S. Chattopadhyay and W. R. Flavell, *Faraday Discuss.*, 2014, DOI: 10.1039/C4FD00019F.

46 T. Globus, I. Sizov and B. Gelmont, *Faraday Discuss.*, 2014, DOI: 10.1039/C4FD00029C.

47 R. Mak, M. Lerotic, H. Fleckenstein, S. Vogt, S. M. Wild, S. Leyffer, Y. Sheynkin and C. Jacobsen, *Faraday Discuss.*, 2014, DOI: 10.1039/C4FD00023D.

48 J. C. H. Spence, *Faraday Discuss.*, 2014, DOI: 10.1039/C4FD00025K.

49 J. J. van Thor, M. M. Warren, C. N. Lincoln, M. Chollet, H. T. Lemke, D. M. Fritz, M. Schmidt, J. Tenboer, Z. Ren, V. Srajer, K. Moffat and T. Graber, *Faraday Discuss.*, 2014, DOI: 10.1039/C4FD00011K.

50 M. Woerner, M. Holtz, *et al.*, *Faraday Discuss.*, 2014, DOI: 10.1039/C4FD00026A.

51 *Solving Science and Energy Grand Challenges with Next-Generation Photon Sources*, ed. W. Eberhardt and F. Himpsel, U.S. Department of Energy, Rockville, Maryland, 2009.

52 A. Cavalleri, M. Rini, H. H. W. Chong, S. Fourmaux, T. E. Glover, P. A. Heimann, J. C. Kieffer and R. W. Schoenlein, *Phys. Rev. Lett.*, 2005, **95**.

53 C. Stamm, T. Kachel, N. Pontius, R. Mitzner, T. Quast, K. Holldack, S. Khan, C. Lupulescu, E. F. Aziz, M. Wietstruk, H. A. Durr and W. Eberhardt, *Nat. Mater.*, 2007, **6**, 740–743.

54 P. Beaud, S. L. Johnson, C. J. Milne, F. S. Krasniqi, E. Vorobeva and G. Ingold, *Springer Ser. Chem. Phys.*, 2009, **92**, 104–106.

55 B. Mansart, M. J. G. Cottet, G. F. Mancini, T. Jarlborg, S. B. Dugdale, S. L. Johnson, S. O. Mariager, C. J. Milne, P. Beaud, S. Grubel, J. A. Johnson, T. Kubacka, G. Ingold, K. Prsa, H. M. Ronnow, K. Conder, E. Pomjakushina, M. Chergui and F. Carbone, *Phys. Rev. B: Condens. Matter Mater. Phys.*, 2013, **88**, 054507.

56 F. G. Santomauro, A. Lübcke, J. Rittmann, E. Baldini, A. Ferrer, M. Silatani, P. Zimmermann, S. Grübel, J. A. Johnson, S. O. Mariager, P. P. Beaud, D. Grolimund, C. Borca, G. Ingold, S. L. Johnson and M. Chergui, *Nat. Chem.*, 2014, submitted.

57 C. Bressler, C. Milne, V. T. Pham, A. ElNahhas, R. M. van der Veen, W. Gawelda, S. Johnson, P. Beaud, D. Grolimund, M. Kaiser, C. N. Borca, G. Ingold, R. Abela and M. Chergui, *Science*, 2009, **323**, 489–492.

58 V. T. Pham, T. J. Penfold, R. M. van der Veen, F. Lima, A. El Nahhas, S. L. Johnson, P. Beaud, R. Abela, C. Bressler, I. Tavernelli, C. J. Milne and M. Chergui, *J. Am. Chem. Soc.*, 2011, **133**, 12740–12748.

59 N. Huse, H. Cho, K. Hong, L. Jamula, F. M. F. de Groot, T. K. Kim, J. K. McCusker and R. W. Schoenlein, *J. Phys. Chem. Lett.*, 2011, **2**, 880–884.

60 C. W. Siders, A. Cavalleri, K. Sokolowski-Tinten, C. Toth, T. Guo, M. Kammler, M. H. von Hoegen, K. R. Wilson, D. von der Linde and C. P. J. Barty, *Science*, 1999, **286**, 1340–1342.

61 K. Sokolowski-Tinten, C. Blome, C. Dietrich, A. Tarasevitch, M. H. von Hoegen, D. von der Linde, A. Cavalleri, J. Squier and M. Kammler, *Phys. Rev. Lett.*, 2001, **87**(22), 225701.

62 M. Bargheer, N. Zhavoronkov, Y. Gritsai, J. C. Woo, D. S. Kim, M. Woerner and T. Elsaesser, *Science*, 2004, **306**, 1771–1773.

63 A. Barty, S. Marchesini, H. N. Chapman, C. Cui, M. R. Howells, D. A. Shapiro, A. M. Minor, J. C. H. Spence, U. Weierstall, J. Ilavsky, A. Noy, S. P. Hau-Riege, A. B. Artyukhin, T. Baumann, T. Willey, J. Stolken, T. van Buuren and J. H. Kinney, *Phys. Rev. Lett.*, 2008, **101**, 055501.

64 M. J. Bogan, W. H. Benner, S. Boutet, U. Rohner, M. Frank, A. Barty, M. M. Seibert, F. Maia, S. Marchesini, S. Bajt, B. Woods, V. Riot, S. P. Hau-Riege, M. Svenda, E. Marklund, E. Spiller, J. Hajdu and H. N. Chapman, *Nano Lett.*, 2008, **8**, 310–316.

65 H. N. Chapman, S. Bajt, A. Barty, W. H. Benner, M. J. Bogan, S. Boutet, A. Cavalleri, S. Dusterer, M. Frank, J. Hajdu, S. F. Hau-Riege, B. Iwan, S. Marchesini, K. Sokolowski-Tinten, M. M. Siebert, R. Treusch and B. W. Woods, *Springer Series in Chemical Physics*, 2009, **92**, 143–145.

66 H. N. Chapman, A. Barty, M. J. Bogan, S. Boutet, M. Frank, S. P. Hau-Riege, S. Marchesini, B. W. Woods, S. Bajt, H. Benner, R. A. London, E. Plonjes, M. Kuhlmann, R. Treusch, S. Dusterer, T. Tschentscher, J. R. Schneider, E. Spiller, T. Moller, C. Bostedt, M. Hoener, D. A. Shapiro, K. O. Hodgson, D. Van der Spoel, F. Burmeister, M. Bergh, C. Caleman, G. Huldt, M. M. Seibert, F. R. N. C. Maia, R. W. Lee, A. Szoke, N. Timneanu and J. Hajdu, *Nat. Phys.*, 2006, **2**, 839–843.

67 A. V. Martin, N. D. Loh, C. Y. Hampton, R. G. Sierra, F. Wang, A. Aquila, S. Bajt, M. Barthelmess, C. Bostedt, J. D. Bozek, N. Coppola, S. W. Epp, B. Erk, H. Fleckenstein, L. Foucar, M. Frank, H. Graafsma, L. Gumprecht, A. Hartmann, R. Hartmann, G. Hauser, H. Hirsemann, P. Holl, S. Kassemeyer, N. Kimmel, M. Liang, L. Lomb, F. R. N. C. Maia, S. Marchesini, K. Nass, E. Pedersoli, C. Reich, D. Rolles, B. Rudek, A. Rudenko, J. Schulz, R. L. Shoeman, H. Soltau, D. Starodub, J. Steinbrener, F. Stellato, L. Struder, J. Ullrich, G. Weidenspointner, T. A. White, C. B. Wunderer, A. Barty, I. Schlichting, M. J. Bogan and H. N. Chapman, *Opt. Express*, 2012, **20**, 13501–13512.

68 E. Allaria, C. Callegari, D. Cocco, W. M. Fawley, M. Kiskinova, C. Masciovecchio and F. Parmigiani, *New J. Phys.*, 2010, **12**, 075002.

69 E. Allaria, D. Castronovo, P. Cinquegrana, P. Craievich, M. Dal Forno, M. B. Danailov, G. D'Auria, A. Demidovich, G. De Ninno, S. Di Mitri, B. Diviacco, W. M. Fawley, M. Ferianis, E. Ferrari, L. Froehlich, G. Gaio, D. Gauthier, L. Giannessi, R. Ivanov, B. Mahieu, N. Mahne, I. Nikolov, F. Parmigiani, G. Penco, L. Raimondi, C. Scafuri, C. Serpico, P. Sigalotti, S. Spampinati, C. Spezzani, M. Svandrlik, C. Svetina, M. Trovo, M. Veronese, D. Zangrando and M. Zangrando, *Nat. Photonics*, 2013, **7**, 913–918.

70 T. Hara, *Nat. Photonics*, 2013, **7**, 851–854.

71 F. Bencivenga, F. Capotondi, F. Casolari, F. Dallari, M. B. Danailov, D. Fausti, M. Kiskinova, M. Manfredda, C. Masciovecchio and E. Pedersoli, *Faraday Discuss.*, 2014, DOI: 10.1039/C4FD00100A.

72 F. Bencivenga, S. Baroni, C. Carbone, M. Chergui, M. B. Danailov, G. De Ninno, M. Kiskinova, L. Raimondi, C. Svetina and C. Masciovecchio, *New J. Phys.*, 2013, **15**, 123023.

73 J. Amann, W. Berg, V. Blank, F. J. Decker, Y. Ding, P. Emma, Y. Feng, J. Frisch, D. Fritz, J. Hastings, Z. Huang, J. Krzywinski, R. Lindberg, H. Loos, A. Lutman, H. D. Nuhn, D. Ratner, J. Rzepiela, D. Shu, Y. Shvyd'ko, S. Spampinati, S. Stoupin, S. Terentyev, E. Trakhtenberg, D. Walz, J. Welch, J. Wu, A. Zholents and D. Zhu, *Nat. Photonics*, 2012, **6**, 693–698.

74 B. Freyer, F. Zamponi, V. Juve, J. Stingl, M. Woerner, T. Elsaesser and M. Chergui, *J. Chem. Phys.*, 2013, **138**, 144504.

75 M. F. Lin, D. M. Neumark, O. Gessner and S. R. Leone, *J. Chem. Phys.*, 2014, **140**, 064311.

76 Z.-H. Loh and S. R. Leone, *J. Phys. Chem. Lett.*, 2012, **4**, 292–302.

77 R. Cortes, L. Rettig, Y. Yoshida, H. Eisaki, M. Wolf and U. Bovensiepen, *Phys. Rev. Lett.*, 2011, **107**, 097002.

78 J. Vura-Weis, C.-M. Jiang, C. Liu, H. Gao, J. M. Lucas, F. M. F. de Groot, P. Yang, A. P. Alivisatos and S. R. Leone, *J. Phys. Chem. Lett.*, 2013, **4**, 3667–3671.

79 S. Mathias, H. Kapteyn and M. Murnane, in *Ultrafast Nonlinear Optics*, ed. R. Thomson, C. Leburn and D. Reid, Springer International Publishing, 2013, pp. 149–175.

80 H. C. Kapteyn and M. M. Murnane, *Frontiers in Optics 2013*, Orlando, Florida, 2013.

81 L. X. Yang, G. Rohde, T. Rohwer, A. Stange, K. Hanff, C. Sohrt, L. Rettig, R. Cortés, F. Chen, D. L. Feng, T. Wolf, B. Kamble, I. Eremin, T. Popmintchev, M. M. Murnane, H. C. Kapteyn, L. Kipp, J. Fink, M. Bauer, U. Bovensiepen and K. Rossnagel, *Phys. Rev. Lett.*, 2014, **112**, 207001.

82 R. Neutze, *Philos. Trans. R. Soc., B*, 2014, **369**, 20130318.

83 J. C. H. Spence and H. N. Chapman, *Philos. Trans. R. Soc., B*, 2014, **369**, 20130309.

84 B. Perman, V. Srajer, Z. Ren, T. Y. Teng, C. Pradervand, T. Ursby, D. Bourgeois, F. Schotte, M. Wulff, R. Kort, K. Hellingwerf and K. Moffat, *Science*, 1998, **279**, 1946–1950.

85 F. Schotte, M. H. Lim, T. A. Jackson, A. V. Smirnov, J. Soman, J. S. Olson, G. N. Phillips, M. Wulff and P. A. Anfinrud, *Science*, 2003, **300**, 1944–1947.

86 S. Anderson, V. Srajer, R. Pahl, S. Rajagopal, F. Schotte, P. Anfinrud, M. Wulff and K. Moffat, *Structure*, 2004, **12**, 1039–1045.

87 H. Ihee, S. Rajagopal, V. Srajer, R. Pahl, S. Anderson, M. Schmidt, F. Schotte, P. A. Anfinrud, M. Wulff and K. Moffat, *Proc. Natl. Acad. Sci. U. S. A.*, 2005, **102**, 7145–7150.

88 G. Hummer, F. Schotte and P. A. Anfinrud, *Proc. Natl. Acad. Sci. U. S. A.*, 2004, **101**, 15330–15334.

89 M. Cammarata, M. Levantino, F. Schotte, P. A. Anfinrud, F. Ewald, J. Choi, A. Cupane, M. Wulff and H. Ihee, *Nat. Methods*, 2008, **5**, 881–886.

90 H. Ihee, K. H. Kim, K. Y. Oang, J. Kim, J. H. Lee and Y. Kim, *Chem. Commun.*, 2011, **47**, 289–291.

91 T. K. Kim, J. H. Lee, M. Wulff, Q. Y. Kong and H. Ihee, *ChemPhysChem*, 2009, **10**, 1958–1980.

92 S. Ahn, K. H. Kim, Y. Kim, J. Kim and H. Ihee, *J. Phys. Chem. B*, 2009, **113**, 13131–13133.

93 A. M. Lindenberg, I. Kang, S. L. Johnson, R. W. Falcone, P. A. Heimann, Z. Chang, R. W. Lee and J. S. Wark, *Opt. Lett.*, 2002, **27**, 869–871.

94 D. A. Reis, M. F. DeCamp, P. H. Bucksbaum, R. Clarke, E. Dufresne, M. Hertlein, R. Merlin, R. Falcone, H. Kapteyn, M. M. Murnane, J. Larsson, T. Missalla and J. S. Wark, *Phys. Rev. Lett.*, 2001, **86**, 3072–3075.

95 M. Kozina, T. Hu, J. S. Wittenberg, E. Szilagyi, M. Trigo, T. A. Miller, C. Uher, A. Damodaran, L. Martin, A. Mehta, J. Corbett, J. Safranek, D. A. Reis and A. M. Lindenberg, *Structural Dynamics*, 2014, **1**, 034301.

96 P. Coppens, J. Benedict, M. Messerschmidt, I. Novozhilova, T. Graber, Y. S. Chen, I. Vorontsov, S. Scheins and S. L. Zheng, *Acta Crystallogr., Sect. A: Found. Crystallogr.*, 2010, **66**, 179–188.

97 H. Cailleau, M. Lorenc, L. Guerin, M. Servol, E. Collet and M. Buron-Le Cointe, *Acta Crystallogr., Sect. A: Found. Crystallogr.*, 2010, **66**, 189–197.

98 A. M. Lindenberg, I. Kang, S. L. Johnson, T. Missalla, P. A. Heimann, Z. Chang, J. Larsson, P. H. Bucksbaum, H. C. Kapteyn, H. A. Padmore, R. W. Lee, J. S. Wark and R. W. Falcone, *Phys. Rev. Lett.*, 2000, **84**, 111–114.

99 S. L. Johnson, P. A. Heimann, A. G. MacPhee, A. M. Lindenberg, O. R. Monteiro, Z. Chang, R. W. Lee and R. W. Falcone, *Phys. Rev. Lett.*, 2005, **94**, 057407.

100 A. B. Stickrath, M. W. Mara, J. V. Lockard, M. R. Harpham, J. Huang, X. Y. Zhang, K. Attenkofer and L. X. Chen, *J. Phys. Chem. B*, 2013, **117**, 4705–4712.

101 M. Silatani, F. A. Lima, R. Monni, T. J. Penfold, J. Rittman, I. Tavernelli, U. Röthlisberger, G. Auböck, C. J. Milne and M. Chergui, *J. Am. Chem. Soc.*, 2014, submitted.

102 G. Vanko, A. Bordage, P. Glatzel, E. Gallo, M. Rovezzi, W. Gawelda, A. Galler, C. Bressler, G. Doumy, A. M. March, E. P. Kanter, L. Young, S. H. Southworth, S. E. Canton, J. Uhlig, G. Smolentsev, V. Sundstrom, K. Haldrup, T. B. van Driel, M. M. Nielsen, K. S. Kjaer and H. T. Lemke, *J. Electron Spectrosc. Relat. Phenom.*, 2013, **188**, 166–171.

103 U. Bergmann and P. Glatzel, *Photosynth. Res.*, 2009, **102**, 255–266.

104 G. Smolentsev, A. V. Soldatov, J. Messinger, K. Merz, T. Weyhermuller, U. Bergmann, Y. Pushkar, J. Yano, V. K. Yachandra and P. Glatzel, *J. Am. Chem. Soc.*, 2009, **131**, 13161–13167.

105 K. Haldrup, G. Vanko, W. Gawelda, A. Galler, G. Doumy, A. M. March, E. P. Kanter, A. Bordage, A. Dohn, T. B. van Driel, K. S. Kjaer, H. T. Lemke, S. E. Canton, J. Uhlig, V. Sundstrom, L. Young, S. H. Southworth, M. M. Nielsen and C. Bressler, *J. Phys. Chem. A*, 2012, **116**, 9878–9887.

106 T. Giessel, D. Brocker, P. Schmidt and W. Widdra, *Rev. Sci. Instrum.*, 2003, **74**, 4620–4624.

107 W. Widdra, D. Brocker, T. Giessel, I. V. Hertel, W. Kruger, A. Liero, F. Noack, V. Petrov, D. Pop, P. M. Schmidt, R. Weber, I. Will and B. Winter, *Surf. Sci.*, 2003, **543**, 87–94.

108 H. T. Lemke, C. Bressler, L. X. Chen, D. M. Fritz, K. J. Gaffney, A. Galler, W. Gawelda, K. Haldrup, R. W. Hartsock, H. Ihee, J. Kim, K. H. Kim, J. H. Lee, M. M. Nielsen, A. B. Stickrath, W. K. Zhang, D. L. Zhu and M. Cammarata, *J. Phys. Chem. C*, 2013, **117**, 735–740.

109 K. R. Siefermann, C. D. Pemmaraju, S. Neppl, A. Shavorskiy, A. A. Cordones, J. Vura-Weis, D. S. Slaughter, F. P. Sturm, F. Weise, H. Bluhm, M. L. Strader, H. Cho, M.-F. Lin, C. Bacellar, C. Khurmi, J. Guo, G. Coslovich, J. S. Robinson, R. A. Kaindl, R. W. Schoenlein, A. Belkacem, D. M. Neumark, S. R. Leone, D. Nordlund, H. Ogasawara, O. Krupin, J. J. Turner, W. F. Schlotter, M. R. Holmes, M. Messerschmidt, M. P. Minitti, S. Gul, J. Z. Zhang, N. Huse, D. Prendergast and O. Gessner, *J. Phys. Chem. Lett.*, 2014, **5**, 2753–2759.

110 J. Kern, R. Alonso-Mori, R. Tran, J. Hattne, R. J. Gildea, N. Echols, C. Glockner, J. Hellmich, H. Laksmono, R. G. Sierra, B. Lassalle-Kaiser, S. Koroidov, A. Lampe, G. Y. Han, S. Gul, D. DiFiore, D. Milathianaki, A. R. Fry, A. Miahnahri, D. W. Schafer, M. Messerschmidt, M. M. Seibert, J. E. Koglin, D. Sokaras, T. C. Weng, J. Sellberg, M. J. Latimer, R. W. Grosse-Kunstleve, P. H. Zwart, W. E. White, P. Glatzel, P. D. Adams,

M. J. Bogan, G. J. Williams, S. Boutet, J. Messinger, A. Zouni, N. K. Sauter, V. K. Yachandra, U. Bergmann and J. Yano, *Science*, 2013, **340**, 491–495.

111 D. P. DePonte, U. Weierstall, K. Schmidt, J. Warner, D. Starodub, J. C. H. Spence and R. B. Doak, *J. Phys. D: Appl. Phys.*, 2008, **41**, 195505.

112 K. E. Schmidt, J. C. H. Spence, U. Weierstall, R. Kirian, X. Wang, D. Starodub, H. N. Chapman, M. R. Howells and R. B. Doak, *Phys. Rev. Lett.*, 2008, **101**, 115507.

113 H. N. Chapman, P. Fromme, A. Barty, T. A. White, R. A. Kirian, A. Aquila, M. S. Hunter, J. Schulz, D. P. DePonte, U. Weierstall, R. B. Doak, F. R. N. C. Maia, A. V. Martin, I. Schlichting, L. Lomb, N. Coppola, R. L. Shoeman, S. W. Epp, R. Hartmann, D. Rolles, A. Rudenko, L. Foucar, N. Kimmel, G. Weidenspointner, P. Holl, M. N. Liang, M. Barthelmess, C. Caleman, S. Boutet, M. J. Bogan, J. Krzywinski, C. Bostedt, S. Bajt, L. Gumprecht, B. Rudek, B. Erk, C. Schmidt, A. Homke, C. Reich, D. Pietschner, L. Struder, G. Hauser, H. Gorke, J. Ullrich, S. Herrmann, G. Schaller, F. Schopper, H. Soltau, K. U. Kuhnel, M. Messerschmidt, J. D. Bozek, S. P. Hau-Riege, M. Frank, C. Y. Hampton, R. G. Sierra, D. Starodub, G. J. Williams, J. Hajdu, N. Timneanu, M. M. Seibert, J. Andreasson, A. Rocker, O. Jonsson, M. Svenda, S. Stern, K. Nass, R. Andritschke, C. D. Schroter, F. Krasniqi, M. Bott, K. E. Schmidt, X. Y. Wang, I. Grotjohann, J. M. Holton, T. R. M. Barends, R. Neutze, S. Marchesini, R. Fromme, S. Schorb, D. Rupp, M. Adolph, T. Gorkhover, I. Andersson, H. Hirsemann, G. Potdevin, H. Graafsma, B. Nilsson and J. C. H. Spence, *Nature*, 2011, **470**, 73–U81.

114 M. M. Seibert, T. Ekeberg, F. R. N. C. Maia, M. Svenda, J. Andreasson, O. Jonsson, D. Odic, B. Iwan, A. Rocker, D. Westphal, M. Hantke, D. P. DePonte, A. Barty, J. Schulz, L. Gumprecht, N. Coppola, A. Aquila, M. N. Liang, T. A. White, A. Martin, C. Caleman, S. Stern, C. Abergel, V. Seltzer, J. M. Claverie, C. Bostedt, J. D. Bozek, S. Boutet, A. A. Miahnahri, M. Messerschmidt, J. Krzywinski, G. Williams, K. O. Hodgson, M. J. Bogan, C. Y. Hampton, R. G. Sierra, D. Starodub, I. Andersson, S. Bajt, M. Barthelmess, J. C. H. Spence, P. Fromme, U. Weierstall, R. Kirian, M. Hunter, R. B. Doak, S. Marchesini, S. P. Hau-Riege, M. Frank, R. L. Shoeman, L. Lomb, S. W. Epp, R. Hartmann, D. Rolles, A. Rudenko, C. Schmidt, L. Foucar, N. Kimmel, P. Holl, B. Rudek, B. Erk, A. Homke, C. Reich, D. Pietschner, G. Weidenspointner, L. Struder, G. Hauser, H. Gorke, J. Ullrich, I. Schlichting, S. Herrmann, G. Schaller, F. Schopper, H. Soltau, K. U. Kuhnel, R. Andritschke, C. D. Schroter, F. Krasniqi, M. Bott, S. Schorb, D. Rupp, M. Adolph, T. Gorkhover, H. Hirsemann, G. Potdevin, H. Graafsma, B. Nilsson, H. N. Chapman and J. Hajdu, *Nature*, 2011, **470**, 78–U86.

115 D. Starodub, A. Aquila, S. Bajt, M. Barthelmess, A. Barty, C. Bostedt, J. D. Bozek, N. Coppola, R. B. Doak, S. W. Epp, B. Erk, L. Foucar, L. Gumprecht, C. Y. Hampton, A. Hartmann, R. Hartmann, P. Holl, S. Kassemeyer, N. Kimmel, H. Laksmono, M. Liang, N. D. Loh, L. Lomb, A. V. Martin, K. Nass, C. Reich, D. Rolles, B. Rudek, A. Rudenko, J. Schulz, R. L. Shoeman, R. G. Sierra, H. Soltau, J. Steinbrener, F. Stellato, S. Stern, G. Weidenspointner, M. Frank, J. Ullrich, L. Struder, I. Schlichting, H. N. Chapman, J. C. H. Spence and M. J. Bogan, *Nat. Commun.*, 2012, **3**, 1276.

116 J. Kupper, S. Stern, L. Holmegaard, F. Filsinger, A. Rouzee, A. Rudenko, P. Johnsson, A. V. Martin, M. Adolph, A. Aquila, S. Bajt, A. Barty, C. Bostedt, J. Bozek, C. Caleman, R. Coffee, N. Coppola, T. Delmas, S. Epp, B. Erk, L. Foucar, T. Gorkhover, L. Gumprecht, A. Hartmann, R. Hartmann, G. Hauser, P. Holl, A. Homke, N. Kimmel, F. Krasniqi, K. U. Kuhnel, J. Maurer, M. Messerschmidt, R. Moshammer, C. Reich, B. Rudek, R. Santra, I. Schlichting, C. Schmidt, S. Schorb, J. Schulz, H. Soltau, J. C. H. Spence, D. Starodub, L. Struder, J. Thogersen, M. J. J. Vrakking, G. Weidenspointner, T. A. White, C. Wunderer, G. Meijer, J. Ullrich, H. Stapelfeldt, D. Rolles and H. N. Chapman, *Phys. Rev. Lett.*, 2014, **112**, 083002.

117 U. Weierstall, J. C. H. Spence and R. B. Doak, *Rev. Sci. Instrum.*, 2012, **83**, 035108.

118 D. Healion, Y. Zhang, J. D. Biggs, W. Hua and S. Mukamel, *Structural Dynamics*, 2014, **1**.

119 T. E. Glover, D. M. Fritz, M. Cammarata, T. K. Allison, S. Coh, J. M. Feldkamp, H. Lemke, D. Zhu, Y. Feng, R. N. Coffee, M. Fuchs, S. Ghimire, J. Chen, S. Shwartz, D. A. Reis, S. E. Harris and J. B. Hastings, *Nature*, 2012, **488**, 603–608.

120 A. A. Lutman, R. Coffee, Y. Ding, Z. Huang, J. Krzywinski, T. Maxwell, M. Messerschmidt and H. D. Nuhn, *Phys. Rev. Lett.*, 2013, **110**, 134801.

121 T. Hara, Y. Inubushi, T. Katayama, T. Sato, H. Tanaka, T. Tanaka, T. Togashi, K. Togawa, K. Tono, M. Yabashi and T. Ishikawa, *Nat. Commun.*, 2013, **4**, 2919.

122 C. A. Arrell, J. Ojeda, M. Sabbar, W. Okell, T. Witting, T. Siegel, Z. Diveki, S. Hutchinson, L. Gallmann, U. Keller, F. van Mourik, R. Chapman, C. Cacho, N. Rodrigues, I. C. E. Turcu, J. W. G. Tisch, E. Springate, J. P. Marangos and M. Chergui, *Rev. Sci. Instrum.*, 2014, submitted.

123 M. Faubel, B. Steiner and J. P. Toennies, *J. Chem. Phys.*, 1997, **106**, 9013–9031.

124 B. Winter and M. Faubel, *Chem. Rev.*, 2006, **106**, 1176–1211.

125 O. Link, E. Lugovoy, K. Siefermann, Y. Liu, M. Faubel and B. Abel, *Appl. Phys. A: Mater. Sci. Process.*, 2009, **96**, 117–135.

126 O. Link, E. Vohringer-Martinez, E. Lugovoj, Y. X. Liu, K. Siefermann, M. Faubel, H. Grubmuller, R. B. Gerber, Y. Miller and B. Abel, *Faraday Discuss.*, 2009, **141**, 67–79.

127 K. R. Siefermann, Y. X. Liu, E. Lugovoy, O. Link, M. Faubel, U. Buck, B. Winter and B. Abel, *Nat. Chem.*, 2010, **2**, 274–279.

128 F. Buchner, A. Lubcke, N. Heine and T. Schultz, *Rev. Sci. Instrum.*, 2010, **81**, 113107.

129 A. Lubcke, F. Buchner, N. Heine, I. V. Hertel and T. Schultz, *Phys. Chem. Chem. Phys.*, 2010, **12**, 14629–14634.

130 Y. Tang, H. Shen, K. Sekiguchi, N. Kurahashi, T. Mizuno, Y. I. Suzuki and T. Suzuki, *Phys. Chem. Chem. Phys.*, 2010, **12**, 3653–3655.

131 Y. Tang, Y. I. Suzuki, H. Shen, K. Sekiguchi, N. Kurahashi, K. Nishizawa, P. Zuo and T. Suzuki, *Chem. Phys. Lett.*, 2010, **494**, 111–116.

132 J. C. Williamson and A. H. Zewail, *J. Phys. Chem.*, 1994, **98**, 2766–2781.

133 A. C. Albrecht, *J. Mol. Spectrosc.*, 1961, **6**, 84–108.

134 M. BenNun, J. S. Cao and K. R. Wilson, *J. Phys. Chem. A*, 1997, **101**, 8743–8761.

135 J. S. Cao and K. R. Wilson, *J. Phys. Chem. A*, 1998, **102**, 9523–9530.

136 P. Reckenthaeler, M. Centurion, W. Fuß, S. A. Trushin, F. Krausz and E. E. Fill, *Phys. Rev. Lett.*, 2009, **102**, 213001.

137 J. Yang, V. Makhija, V. Kumarappan and M. Centurion, *Structural Dynamics*, 2014, **1**, 044101.

138 U. Lorenz, K. B. Moller and N. E. Henriksen, *New J. Phys.*, 2010, **12**, 113022.

139 U. Lorenz, K. B. Moller and N. E. Henriksen, *Phys. Rev. A: At., Mol., Opt. Phys.*, 2010, **81**, 023422; U. Lorenz, K. B. Moller and N. E. Henriksen, *Erratum Phys. Rev. A*, 2010, **82**, 069901.

140 A. Debnarova, S. Techert and S. Schmatz, *J. Chem. Phys.*, 2011, **134**, 054302.

141 W. Bras, *Nucl. Instrum. Methods Phys. Res., Sect. B*, 2003, **199**, 90–97.

142 B. J. Lemaire, P. Davidson, P. Panine and J. P. Jolivet, *Phys. Rev. Lett.*, 2004, **93**, 267801.

143 J. Kim, K. H. Kim, J. G. Kim, T. W. Kim, Y. Kim and H. Ihee, *J. Phys. Chem. Lett.*, 2011, **2**, 350–356.

144 K. Hwan Kim, J. Kim, J. Hyuk Lee and H. Ihee, *Structural Dynamics*, 2014, **1**, 011301.

145 A. L. Harris, J. K. Brown and C. B. Harris, *Annu. Rev. Phys. Chem.*, 1988, **39**, 341–366.

146 T. J. Penfold, I. Tavernelli, R. Abela, M. Chergui and U. Rothlisberger, *New J. Phys.*, 2012, **14**, 113002.

147 R. M. Bowman, M. Dantus and A. H. Zewail, *Chem. Phys. Lett.*, 1989, **161**, 297–302.

148 A. Cannizzo, C. J. Milne, C. Consani, W. Gawelda, C. Bressler, F. van Mourik and M. Chergui, *Coord. Chem. Rev.*, 2010, **254**, 2677–2686.

149 C. Sousa, C. de Graaf, A. Rudavskyi, R. Broer, J. Tatchen, M. Etinski and C. M. Marian, *Chem. - Eur. J.*, 2013, **19**, 17541–17551.

150 C. Consani, M. Premont-Schwarz, A. ElNahhas, C. Bressler, F. van Mourik, A. Cannizzo and M. Chergui, *Angew. Chem., Int. Ed.*, 2009, **48**, 7184–7187.

151 A. L. Smeigh, M. Creelman, R. A. Mathies and J. K. McCusker, *J. Am. Chem. Soc.*, 2008, **130**, 14105–14107.

152 G. Aubock and M. Chergui, *Nat. Chem.*, 2014, submitted.

153 M. Chergui, *Dalton Trans.*, 2012, **41**, 13022–13029.

154 O. Bram, F. Messina, A. M. El-Zohry, A. Cannizzo and M. Chergui, *Chem. Phys.*, 2012, **393**, 51–57.

155 M. Gratzel, *Nature*, 2001, **414**, 338–344.

156 M. Gratzel, *Acc. Chem. Res.*, 2009, **42**, 1788–1798.

157 N. Serpone, D. Lawless and R. Khairutdinov, *J. Phys. Chem.*, 1995, **99**, 16646–16654.

158 N. Serpone, D. Lawless, R. Khairutdinov and E. Pelizzetti, *J. Phys. Chem.*, 1995, **99**, 16655–16661.

159 G. Benko, J. Kallioinen, J. E. I. Korppi-Tommola, A. P. Yartsev and V. Sundstrom, *J. Am. Chem. Soc.*, 2002, **124**, 489–493.

160 G. Benko, P. Myllyperkio, J. Pan, A. P. Yartsev and V. Sundstrom, *J. Am. Chem. Soc.*, 2003, **125**, 1118–1119.

161 E. Hendry, F. Wang, J. Shan, T. F. Heinz and M. Bonn, *Phys. Rev. B: Condens. Matter Mater. Phys.*, 2004, **69**, 081101.

162 O. Bram, A. Cannizzo and M. Chergui, *Phys. Chem. Chem. Phys.*, 2012, **14**, 7934–7937.

163 X. Y. Zhang, G. Smolentsev, J. C. Guo, K. Attenkofer, C. Kurtz, G. Jennings, J. V. Lockard, A. B. Stickrath and L. X. Chen, *J. Phys. Chem. Lett.*, 2011, **2**, 628–632.

164 J. Huang, O. Buyukcakir, M. W. Mara, A. Coskun, N. M. Dimitrijevic, G. Barin, O. Kokhan, A. B. Stickrath, R. Ruppert, D. M. Tiede, J. F. Stoddart, J. P. Sauvage and L. X. Chen, *Angew. Chem., Int. Ed.*, 2012, **51**, 12711–12715.

165 J. E. Katz, B. Gilbert, X. Y. Zhang, K. Attenkofer, R. W. Falcone and G. A. Waychunas, *J. Phys. Chem. Lett.*, 2010, **1**, 1372–1376.

166 J. E. Katz, X. Y. Zhang, K. Attenkofer, K. W. Chapman, C. Frandsen, P. Zarzycki, K. M. Rosso, R. W. Falcone, G. A. Waychunas and B. Gilbert, *Science*, 2012, **337**, 1200–1203.

167 M. H. Rittmann-Frank, C. J. Milne, J. Rittmann, M. Reinhard, T. J. Penfold and M. Chergui, *Angew. Chem., Int. Ed.*, 2014, **53**, 5858–5862.

168 C. Di Valentin and A. Selloni, *J. Phys. Chem. Lett.*, 2011, **2**, 2223–2228.

169 E. Allaria, R. Appio, L. Badano, W. A. Barletta, S. Bassanese, S. G. Biedron, A. Borga, E. Busetto, D. Castronovo, P. Cinquegrana, S. Cleva, D. Cocco, M. Cornacchia, P. Craievich, I. Cudin, G. D'Auria, M. Dal Forno, M. B. Danailov, R. De Monte, G. De Ninno, P. Delgiusto, A. Demidovich, S. Di Mitri, B. Diviacco, A. Fabris, R. Fabris, W. Fawley, M. Ferianis, E. Ferrari, S. Ferry, L. Froehlich, P. Furlan, G. Gaio, F. Gelmetti, L. Giannessi, M. Giannini, R. Gobessi, R. Ivanov, E. Karantzoulis, M. Lonza, A. Lutman, B. Mahieu, M. Milloch, S. V. Milton, M. Musardo, I. Nikolov, S. Noe, F. Parmigiani, G. Penco, M. Petronio, L. Pivetta, M. Predonzani, F. Rossi, L. Rumiz, A. Salom, C. Scafuri, C. Serpico, P. Sigalotti, S. Spampinati, C. Spezzani, M. Svandrlik, C. Svetina, S. Tazzari, M. Trovo, R. Umer, A. Vascotto, M. Veronese, R. Visintini, M. Zaccaria, D. Zangrando and M. Zangrando, *Nat. Photonics*, 2012, **6**, 699–704.

170 M. Gerrity, S. Brown, T. Popmintchev, M. Murnane, H. Kapteyn and S. Backus, *CLEO: 2014*, San Jose, California, 2014.

171 K.-H. Hong, S.-W. Huang, J. Moses, X. Fu, C.-J. Lai, G. Cirmi, A. Sell, E. Granados, P. Keathley and F. X. Kärtner, *Opt. Express*, 2011, **19**, 15538–15548.

172 A. Harth, P. Rudawski, C. Guo, M. Miranda, E. Lorek, E. Marsell, E. Witting Larsen, C. Heyl, J. Matyschok, T. Binhammer, U. Morgner, A. Mikkelsen, A. L'Huillier and C. Arnold, *Research in Optical Sciences*, Messe Berlin, Berlin, 2014.

173 M. T. J. Smith and J. L. Rubinstein, *Science*, 2014, **345**, 617–619.

174 S. Boutet, L. Lomb, G. J. Williams, T. R. M. Barends, A. Aquila, R. B. Doak, U. Weierstall, D. P. DePonte, J. Steinbrener, R. L. Shoeman, M. Messerschmidt, A. Barty, T. A. White, S. Kassemeyer, R. A. Kirian, M. M. Seibert, P. A. Montanez, C. Kenney, R. Herbst, P. Hart, J. Pines, G. Haller, S. M. Gruner, H. T. Philipp, M. W. Tate, M. Hromalik, L. J. Koerner, N. van Bakel, J. Morse, W. Ghonsalves, D. Arnlund, M. J. Bogan, C. Caleman, R. Fromme, C. Y. Hampton, M. S. Hunter, L. C. Johansson, G. Katona, C. Kupitz, M. N. Liang, A. V. Martin, K. Nass, L. Redecke, F. Stellato, N. Timneanu, D. J. Wang, N. A. Zatsepin, D. Schafer, J. Defever, R. Neutze, P. Fromme, J. C. H. Spence, H. N. Chapman and I. Schlichting, *Science*, 2012, **337**, 362–364.

175 L. C. Johansson, D. Arnlund, G. Katona, T. A. White, A. Barty, D. P. DePonte, R. L. Shoeman, C. Wickstrand, A. Sharma, G. J. Williams, A. Aquila, M. J. Bogan, C. Caleman, J. Davidsson, R. B. Doak, M. Frank, R. Fromme, L. Galli, I. Grotjohann, M. S. Hunter, S. Kassemeyer, R. A. Kirian, C. Kupitz, M. N. Liang, L. Lomb, E. Malmerberg, A. V. Martin, M. Messerschmidt, K. Nass, L. Redecke, M. M. Seibert, J. Sjohamn, J. Steinbrener, F. Stellato, D. J. Wang, W. Y. Wahlgren, U. Weierstall, S. Westenhoff, N. A. Zatsepin, S. Boutet, J. C. H. Spence, I. Schlichting, H. N. Chapman, P. Fromme and R. Neutze, *Nat. Commun.*, 2013, **4**.

176 A. Aquila, M. S. Hunter, R. B. Doak, R. A. Kirian, P. Fromme, T. A. White, J. Andreasson, D. Arnlund, S. Bajt, T. R. M. Barends, M. Barthelmess, M. J. Bogan, C. Bostedt, H. Bottin, J. D. Bozek, C. Caleman, N. Coppola, J. Davidsson, D. P. DePonte, V. Elser, S. W. Epp, B. Erk, H. Fleckenstein, L. Foucar, M. Frank, R. Fromme, H. Graafsma, I. Grotjohann, L. Gumprecht, J. Hajdu, C. Y. Hampton, A. Hartmann, R. Hartmann, S. Hauriege, G. Hauser, H. Hirsemann, P. Holl, J. M. Holton, A. Homke, L. Johansson, N. Kimmel, S. Kassemeyer, F. Krasniqi, K. Kuhnel, M. N. Liang, L. Lomb, E. Malmerberg, S. Marchesini, A. V. Martin, F. R. N. C. Maia, M. Messerschmidt, K. Nass, C. Reich, R. Neutze, D. Rolles, B. Rudek, A. Rudenko, I. Schlichting, C. Schmidt, K. E. Schmidt, J. Schulz, M. M. Seibert, R. L. Shoeman, R. Sierra, H. Soltau, D. Starodub, F. Stellato, S. Stern, L. Struder, N. Timneanu, J. Ullrich, X. Y. Wang, G. J. Williams, G. Weidenspointner, U. Weierstall, C. Wunderer, A. Barty, J. C. H. Spence and H. N. Chapman, *Opt. Express*, 2012, **20**, 2706–2716.

177 T. W. Kim, J. H. Lee, J. Choi, K. H. Kim, L. J. van Wilderen, L. Guerin, Y. Kim, Y. O. Jung, C. Yang, J. Kim, M. Wulff, J. J. van Thor and H. Ihee, *J. Am. Chem. Soc.*, 2012, **134**, 3145–3153.

178 P. L. Ramachandran, J. E. Lovett, P. J. Carl, M. Cammarata, J. H. Lee, Y. O. Jung, H. Ihee, C. R. Timmel and J. J. van Thor, *J. Am. Chem. Soc.*, 2011, **133**, 9395–9404.

179 M. Andersson, E. Malmerberg, S. Westenhoff, G. Katona, M. Cammarata, A. B. Wohri, L. C. Johansson, F. Ewald, M. Eklund, M. Wulff, J. Davidsson and R. Neutze, *Structure*, 2009, **17**, 1265–1275.

180 E. Malmerberg, Z. Omran, J. S. Hub, X. W. Li, G. Katona, S. Westenhoff, L. C. Johansson, M. Andersson, M. Cammarata, M. Wulff, D. van der Spoel, J. Davidsson, A. Specht and R. Neutze, *Biophys. J.*, 2011, **101**, 1345–1353.

181 R. M. van der Veen, A. Cannizzo, F. van Mourik, A. Vlcek and M. Chergui, *J. Am. Chem. Soc.*, 2011, **133**, 305–315.

182 J. Tulkki and T. Aberg, *J. Phys. B: At., Mol. Opt. Phys.*, 1982, **15**, L435–L440.

183 J. Szlachetko, C. J. Milne, J. Hoszowska, J.-C. Dousse, W. Błachucki, J. Sà, Y. Kayser, M. Messerschmidt, R. Abela, S. Boutet, C. David, G. Williams, M. Pajek, B. D. Patterson, G. Smolentsev, J. A. van Bokhoven and M. Nachtegaal, *Structural Dynamics*, 2014, **1**, 021101.

184 E. Allaria, F. Bencivenga, R. Borghes, F. Capotondi, D. Castronovo, P. Charalambous, P. Cinquegrana, M. B. Danailov, G. De Ninno, A. Demidovich, S. Di Mitri, B. Diviacco, D. Fausti, W. M. Fawley, E. Ferrari, L. Froehlich, D. Gauthier, A. Gessini, L. Giannessi, R. Ivanov, M. Kiskinova, G. Kurdi, B. Mahieu, N. Mahne, I. Nikolov, C. Masciovecchio, E. Pedersoli,

G. Penco, L. Raimondi, C. Serpico, P. Sigalotti, S. Spampinati, C. Spezzani, C. Svetina, M. Trovo and M. Zangrando, *Nat. Commun.*, 2013, **4**.

185 G. De Ninno, B. Mahieu, E. Allaria, L. Giannessi and S. Spampinati, *Phys. Rev. Lett.*, 2013, **110**, 064801.

186 B. Abel, U. Buck, A. L. Sobolewski and W. Domcke, *Phys. Chem. Chem. Phys.*, 2012, **14**, 22–34.

187 K. R. Siefermann and B. Abel, *Angew. Chem., Int. Ed.*, 2011, **50**, 5264–5272.

Faraday Discussions

ROYAL SOCIETY OF CHEMISTRY

PAPER

Multiple ionization and fragmentation dynamics of molecular iodine studied in IR–XUV pump–probe experiments

K. Schnorr,[a] A. Senftleben,[ab] G. Schmid,[a] A. Rudenko,[c] M. Kurka,[a] K. Meyer,[a] L. Foucar,[d] M. Kübel,[ef] M. F. Kling,[ef] Y. H. Jiang,[g] S. Düsterer,[h] R. Treusch,[h] C. D. Schröter,[a] J. Ullrich,[ai] T. Pfeifer[a] and R. Moshammer[a]

Received 4th March 2014, Accepted 14th April 2014

DOI: 10.1039/c4fd00031e

The ionization and fragmentation dynamics of iodine molecules (I_2) are traced using very intense ($\sim 10^{14}$ W cm^{-2}) ultra-short (~ 60 fs) light pulses with 87 eV photons of the Free-electron LASer at Hamburg (FLASH) in combination with a synchronized femtosecond optical laser. Within a pump–probe scheme the IR pulse initiates a molecular fragmentation and then, after an adjustable time delay, the system is exposed to an intense FEL pulse. This way we follow the creation of highly-charged molecular fragments as a function of time, and probe the dynamics of multi-photon absorption during the transition from a molecule to individual atoms.

1. Introduction

How does intense XUV (extreme ultra-violet) or X-ray radiation couple with atoms, molecules, and matter in general? This question is in the focus of research with free-electron lasers (FELs) that provide femtosecond light pulses with unprecedented intensities of 10^{15} W cm^{-2} and more. Under these conditions atoms are ionized by multi-photon transitions, and charge states are reached that are inaccessible by single-photon absorption.[1,2] Moreover, atomic resonances were found to further enhance the efficiency of high charge-state production in heavy atoms.[3,4] In contrast to the atomic case, experiments with rare-gas clusters yielded a more subtle picture. On the one hand, a large fraction of only singly charged

[a]Max-Planck-Institut für Kernphysik, 69117, Heidelberg, Germany

[b]Universität Kassel, 34132, Kassel, Germany

[c]J.R. MacDonald Laboratory, Department of Physics, Kansas State University, Manhattan, KS 66506, USA

[d]Max-Planck-Institut für medizinische Forschung, Heidelberg, Germany

[e]Max-Planck-Institut für Quantenoptik, 85748, Garching, Germany

[f]Physics Department, Ludwig-Maximilians-Universität, 85748, Garching, Germany

[g]Shanghai Advanced Research Institute, Chinese Academy of Sciences, 201210, Shanghai, China

[h]DESY, Notkestrasse 85, 22607, Hamburg, Germany

[i]Physikalisch-Technische Bundesanstalt, 38116, Braunschweig, Germany

monomers were observed.[5] The occurrence of these low charge states was assumed to be due to charge redistribution involving neighbouring neutral atoms as well as nano-plasma formation and recombination prior to fragmentation. On the other hand, intensity selective experiments in the soft X-ray regime revealed a complete absence of low charge states at highest intensities, a feature which is still under debate.[6] The processes involved in radiation-induced fragmentation are, thus, strongly dependent not only on the atomic structure, the actual wavelength and the intensity, but also on the direct environment of a specific atom of interest embedded in a molecule, a cluster, or any complex system in general.

The understanding of the fragmentation dynamics of molecules irradiated by X-ray photon pulses is crucial for the realization of single-molecule imaging with FELs which relies on the "diffract before destroy" concept.[7] It must be ensured that the reconstructed structure is still that of the intact molecule. Therefore, the dynamics of absorption and charge redistribution as well as the accompanied nuclear motion is of interest on the femtosecond time scale, *i.e.* the typical duration of the FEL pulses.[8] Particularly, the localized photon absorption at heavy constituents with large cross sections efficiently triggers electron rearrangement throughout the molecule within a few femtoseconds. For small molecules, this has been investigated in two recent studies on methylselenol[9] and ethylselenol,[10] which contain selenium atoms as the heavy absorption centre. Even for 5 fs FEL pulses charge rearrangement and considerable displacement of the atomic constituents were observed.

Here, we trace the XUV absorption dynamics of iodine molecules (I_2) during the transition from a molecule to individual atoms. Besides the advantageous slow nuclear motion of the heavy iodine atoms the system exhibits strong similarities with Xe concerning photon-induced excitation and absorption cross sections in the XUV. Both, xenon and iodine are very efficiently excited with $\sim$90 eV photons *via* shape or giant resonances which results in a highly effective coupling of intense FEL radiation.[2] In the present experiment we use ultra-short XUV pulses of the Free-electron LASer at Hamburg (FLASH) in combination with a synchronized femtosecond optical laser. Within a pump–probe scheme the nuclear dynamics as well as the charge state of I_2 were followed as a function of time. The IR pulse initiates a molecular fragmentation and then, after an adjustable time delay, the system is exposed to an intense FEL pulse. By measuring the kinetic energies and the charge states of coincident ions that emerge from dissociating I_2 molecules we monitor the inter-nuclear distance dependent absorption of intense FEL radiation. Our results confirm, and largely extend recent IR–XUV pump–probe studies with I_2 at FLASH.[11,12] In addition, changes in the ionization dynamics along the transition from molecular to atomic iodine are presented and discussed.

2. Experimental setup

The experiment was performed at FLASH using the focused beam ($\sim$30 µm diameter) of beamline BL2. Pulses with 87 eV photons were delivered at a repetition rate of 10 Hz, and pulse durations of approximately 60 fs (FWHM) and intensities up to 3×10^{14} W cm^{-2} were reached. The FEL beam was focused into a supersonic gas jet containing iodine molecules. Solid iodine was heated up to 400 K inside a reservoir containing helium gas at a stagnation pressure of about

1 bar resulting in a He : I_2 ratio of approximately 50 : 1. The gas mixture was expanded through a heated 30 μm diameter nozzle to create a supersonic jet. By means of a reaction microscope[13] we detected in coincidence the ionic fragments that emerged from the ionization of target molecules by the FEL. While the XUV beam was directly focused into the gas jet, the IR laser was focused with a lens outside of the vacuum system, deflected by 90 degree with a holey mirror and overlapped collinearly with the FEL beam. Both foci were merged at the same point in the target (see Fig. 1). The IR laser provided 800 nm pulses synchronized with the FEL (10 Hz) with pulse durations of ~80 fs (FWHM).[14] After focusing the IR beam onto a diameter of 50 μm (FWHM) we reached a maximum intensity of approximately 3.4×10^{14} W cm^{-2}.

During the IR–XUV pump–probe experiments the arrival time of the IR laser was varied and continuously scanned over a range of ~10 ps in total. Unfortunately, due to technical reasons, some of the usually available IR–XUV synchronization tools for offline pulse arrival-time correction at FLASH were not running properly. Therefore, the pulse-to-pulse timing-jitter as well as long time drifts resulted in an overall temporal resolution for the arrival times between XUV and IR pulses on the order of one picosecond or slightly better. The spatial overlap was adjusted by simultaneously monitoring both beams on a screen that was moved into the interaction point. In order to visualize the XUV beam the screen is coated with a fluorescing powder (ZnS). The temporal overlap of the IR and XUV pulses was found by scanning the delay stage and monitoring the emission characteristic of the ions, which is different for a preceding or a delayed IR pulse.

Both beams, XUV and IR, were polarized along the gas-jet direction (*i.e.* in the drawing plane of Fig. 1). The ionic fragments created by ionization in the centre of the reaction microscope were accelerated perpendicular to the gas-jet beam and the FEL beam (*i.e.* out of the plane in Fig. 1) by a homogeneous electric field of

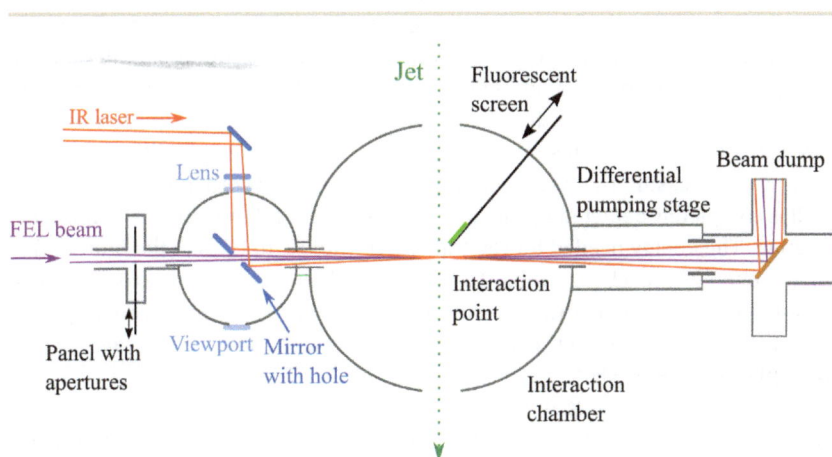

Fig. 1 Sketch of the experimental setup. The FEL beam passes the IR deflection mirror from behind through a 5 mm hole. The XUV and IR beam are focused into the gas target and dumped onto a tilted copper blade to suppress stray light. Differential pumping stages before and behind the interaction chamber ensure XHV conditions (<10^{-11} mbar) at the interaction point. The main chamber comprises a reaction microscope (not shown, see text) for coincident ion detection.

35 V cm^{-1} onto a large area channel-plate detector (120 mm diameter) with position readout (0.1 mm resolution). The time-of-flight (TOF) and position information allows us to reconstruct the three-dimensional momentum of each ion. As most of them emerge from the Coulomb explosion of multiply charged I_2 molecules they carry a significant amount of kinetic energy (up to 100 eV). This complicates the unique assignment of ion pairs I^{Q1+} and I^{Q2+} that emerge from one molecule with a total charge of $Q1 + Q2$ because the flight-time distributions for different species start to overlap. The charge state separation by the ions' time-of-flight alone is hardly possible for such high kinetic energies. Therefore, we use the method of coincident ion detection and the application of momentum conservation to unambiguously identify the two ions that belong to one molecule. In the analysis we request that the vector sum of the momenta of two iodine ions is equal to zero (within the experimental resolution). In practice this means that for a detected particle A we search the list of additionally registered ions for a particle B such that their momentum vectors cancel each other. This is done for every FEL pulse. In order to calculate the momentum vectors we need to assume a charge $Q1$ for ion A and a charge $Q2$ for ion B. If we guess wrong the momentum conservation is likely to fail in at least one spatial coordinate. In this way we find the charge combinations of coincident detected pairs of ions, even if more than one molecule has been ionized by the laser. The confidence that the correct ion pairs are assigned is estimated to be 98% for $I^{1+} + I^{1+}$ and 80% for $I^{7+} + I^{8+}$.

3.   Ionization of I_2 with intense XUV radiation

Ionization of iodine at a photon energy of 87 eV proceeds predominantly by electron emission from the 4d shell,[15] which exhibits a large cross section (more than 10 Mb), in close analogy to the 4d giant resonance in Xe. To understand the pathways leading to the observed charge states in I_2 we first will briefly discuss the ionization of atomic iodine. The absorption of one photon creates a 4d inner-shell vacancy that relaxes by one or two Auger processes[16] resulting in I^{2+} or I^{3+}, respectively. The cross section for ionization of I^{2+} is also dominated by the 4d absorption, which initiates another Auger process leading to I^{4+}.[17] The typical Auger decay times on the order of 20 fs are shorter than the FEL pulse duration used here. For charge states I^{3+} and higher, 5p valence-shell ionization becomes dominant increasing the charge state by only one per absorbed photon. The 5p shell is completely depleted by the time I^{5+} is reached and thus the 5s shell will be emptied. The energy needed for the transition from I^{6+} to I^{7+} is at the limit of the photon energy of 87 eV (see Table 1), but may still be accessible due to the broad bandwidth of the FEL. For the transition from I^{7+} to I^{8+} two photons are required to overcome the ionization threshold, and finally, for the step from I^{9+} to I^{10+} even three photons are needed.

For the discussion of XUV ionization of molecular I_2 we again need the ionization potentials up to the highest observed charge states I_2^{16+}. As the experimentally known valence ionization potential only ranges up to I_2^{2+},[19] we use a simple model to estimate the missing values. This is done by approximating the ionization potential of just the least bound electron as this is the most relevant one for the production of the high charge states. Molecular properties are neglected in the following. For an iodine ion I^{Q+}, the influence of a neighbouring ion I^{P+} (with $P > Q$) is taken into account by assuming that the Coulomb potential

Table 1 The lowest ionization potentials E_Q for atomic iodine I^{Q+} and estimated ionization energies E'_Q for the molecule I_2^{Q+}. The latter are calculated for a fixed inter-nuclear distance of $R_{eq} = 5.0$ a.u., i.e. that of the neutral molecule (see text). The atomic ionization potentials are taken from ref. 18. Additionally listed are the minimum numbers of photons (87 eV) needed for the indicated ionization step

$Q+ \rightarrow (Q+1)+$	E_Q (eV) $I^{Q+} \rightarrow I^{(Q+1)+}$	Photon Number	E'_Q (eV) $I_2^{Q+} \rightarrow I_2^{(Q+1)+}$	Photon Number
$0 \rightarrow 1$	10.4	1	9.35	1
$1 \rightarrow 2$	19.1	1	15.7	1
$2 \rightarrow 3$	29.6	1	24.4	1
$3 \rightarrow 4$	40.4	1	29.7	1
$4 \rightarrow 5$	51.2	1	40.2	1
$5 \rightarrow 6$	74.4	1	45.5	1
$6 \rightarrow 7$	87.6	1	56.3	1
$7 \rightarrow 8$	150.8	2	61.6	1
$8 \rightarrow 9$	171.0	2	72.4	1
$9 \rightarrow 10$	197.0	3	77.7	1
$10 \rightarrow 11$	—	—	100.9	2
$11 \rightarrow 12$	—	—	106.2	2
$12 \rightarrow 13$	—	—	119.4	2
$13 \rightarrow 14$	—	—	124.7	2
$14 \rightarrow 15$	—	—	187.9	3
$15 \rightarrow 16$	—	—	193.2	3
$16 \rightarrow 17$	—	—	213.4	3
$17 \rightarrow 18$	—	—	213.4	3

of I^{P+} leads to tighter bound electrons of I^{Q+}. Thus, the binding energy of the least bound electron of I^{Q+} ($E_Q < 0$) is increased to $E'_Q = E_Q - P/R$, where R is the inter-nuclear distance between the two ions (atomic units, a.u., are used throughout

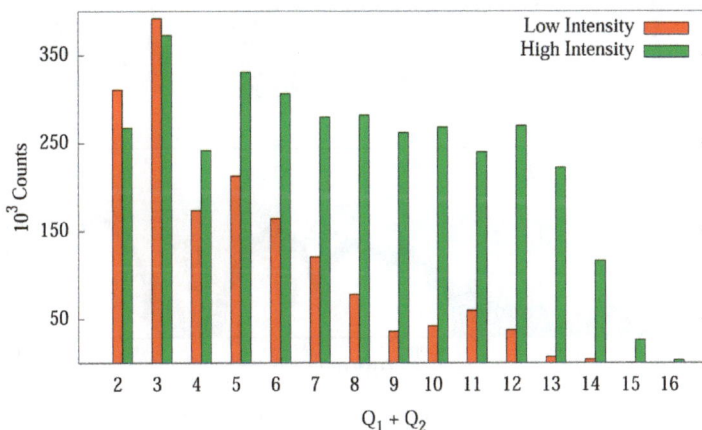

Fig. 2 Measured charge state distributions for multiple ionization of molecular iodine $I_2^{(Q1+Q2)+}$ at a low FEL intensity ($\sim 1 \times 10^{14}$ W cm^{-2}, red bars) and high intensity ($\sim 3 \times 10^{14}$ W cm^{-2}, green bars). The data are taken with XUV pulses only. The total charge state of the molecule is inferred from the charges of coincident ion pairs $I^{Q1+} + I^{Q2+}$. Counts of ion pairs are plotted as a function of their sum charges. The small amount of low charge states in the high-intensity measurement may be attributed to saturation effects in the ion detection system.

with the electron charge, its mass and the Planck constant $e = m_e = \hbar = 1$). The molecule is assumed to fragment into the most symmetric configuration, $e.g.$ I_2^{5+} dissociates into $I^{2+} + I^{3+}$. The estimated ionization potentials for the transition $I_2^{Q+} \rightarrow I_2^{(Q+1)+}$ at the equilibrium inter-nuclear distance $R_{eq} = 5.0$ a.u. are listed in Table 1. To test the quality of our estimate we compare the experimentally determined double-ionization potential $E(I_2 \rightarrow I_2^{2+}) = (24.95 \pm 0.02)$ eV[19] with our calculated value of 25.05 eV from Table 1. Both are in very good agreement, supporting our conclusions.

We will now turn to the ionization pathways in molecular I_2. In analogy to atomic iodine a single photon creates I_2^{2+} or I_2^{3+} as a result of 4d ionization and subsequent Auger emission. We assume that a second photon interacting with I_2^{2+} and I_2^{3+} mainly creates further 4d vacancies which again relax via Auger decay[19] leading to I_2^{4+} and I_2^{5+}, respectively.

From I_2^{4+} onwards 4d inner-shell ionization is competing with valence ionization. This we conclude from the measured charge state ratios of I_2^{4+} and I_2^{5+} ions in Fig. 2, where the distributions of total molecular charge states for two different FEL intensities are shown. The yield of I_2^{5+} ions is enhanced because it features two precursors. However, the exact ratio between inner-shell and valence ionization is not known. Finally, for charge states I_2^{6+} and higher only single valence ionization is energetically allowed, and the charge state increases by one per photon. The highest possible charge state that can be reached by sequential single-photon absorption events is I_2^{10+} according to our estimated ionization potentials in Table 1. If one considers ionization of two individual atoms a

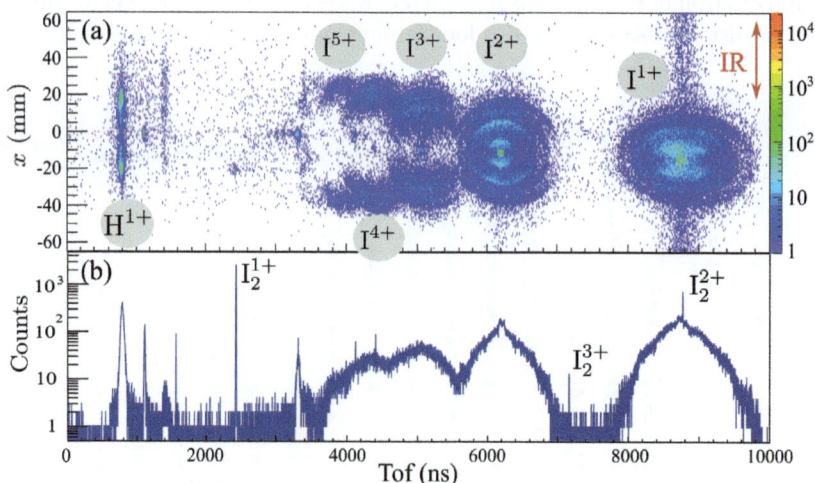

Fig. 3 Ionization of I_2 by IR laser pulses only. (a): Ion time-of-flight (TOF) $versus$ x-coordinate of the hit position on the detector (the direction along the laser polarization). Dominant fragments originating from Coulomb explosion of multiply charged I_2 are indicated as well as protons emerging from residual-gas (H_2) ionization. As the gas jet is propagating into the negative x-direction all fragments emerging from the moving target show a constant x-offset. (b): Projection of (a) onto the TOF axis. Note that due to technical reasons flight times are sorted $modulo$ a $maximum$ time of 10 μs, meaning that the I_2^{1+} peak does not appear at the expected TOF = 12400 ns but at the beginning of the spectrum (at TOF-10000 ns).

maximum charge state of I_2^{14+} is expected. From then on direct two-photon absorption is required for further ionization. In the experiment we see a clear fall-off in the yield above about I_2^{12+} (Fig. 2). To reach this charge state ($Q1 + Q2 = 12$) the stepwise absorption of at least 8 photons is required. At first, this may seem in contradiction to our estimated molecular ionization potentials listed in Table 1. There, direct two-photon absorption is required from I_2^{10+} onwards. However, it should be recalled that these binding energies are estimated for a fixed inter-nuclear distance of $R_{eq} = 5.0$ a.u. This assumption is not valid because ionization often triggers a dissociation of the molecule while more photons get absorbed within the pulse duration. With increasing inter-nuclear distance the ionization potential is lowered. Hence, during the course of sequential photon absorption the electronic binding energies are expected to approach the values of individual atoms or ions. Reaching higher charge states than expected may also be possible due to resonantly excited intermediate states. While a direct ionization with one photon is energetically not allowed from a certain charge state on, it might well be possible for a previously excited ion. If the excited ion is created by a resonant transition with large cross sections, ionization might still be very efficient even though the required energy is above the single-photon ionization threshold. This effect of resonance enhanced ionization was observed for Xe and Kr at higher photon energies.[3,4]

All considerations above are made under the assumption that Auger decay has already occurred before the next photoionization takes place. The situation is different if the first 4d photoionization is followed by a second or third one before Auger decay has happened. This way so-called hollow atoms with multiple core holes are created, where the inner-shells are depleted.[20,21] In general this requires higher intensities, shorter pulses, or larger photon energies where longer, more complicated Auger cascades are initiated. Therefore, we assume that these effects are negligible for our FEL pulse parameters.

4. Ionization of I_2 in strong IR laser fields

While the XUV photons mainly interact with inner-shell electrons, the IR laser field couples most effectively to the least bound electrons in the outer shells. The field imposes a strong perturbation on the molecule that may lead to multi-photon or tunnel ionization. The ionization dynamics of diatomic homo-nuclear molecules in intense femtosecond IR laser pulses has been investigated inten-sively[22-24] including I_2 due to its comparably slow dissociation dynamics.[22,25,26]

Once a molecule is multiply ionized it will most likely dissociate while it may absorb more IR photons that further ionize it. This accelerates the fragments because the Coulomb repulsion increases with the charge. In the literature this process is referred to as multi-electron dissociative ionization (MEDI).[27] Analyzing the fragments' kinetic energies gives insight into the ionization process on a time scale that is comparable to the dissociation time. The MEDI scheme occurs only for some molecules at certain wavelengths but fails for others.[28,29] An alternative scheme, where the ionization probability was found to be drastically enhanced at certain inter-nuclear distances,[30] is the so-called charge-resonance-enhanced ionization (CREI).[24,31] The distance for enhanced ionization of I_2 is approximately two times the equilibrium inter-nuclear distance.[22]

An overview of ion fragments measured only with the IR laser is shown in Fig. 3. The data are recorded at a repetition rate of 10 Hz, the maximum repetition rate of the IR laser system, and an intensity of about 3.4×10^{14} W cm^{-2} (the estimated pulse length was 80 fs). In Fig. 3a the ion times-of-flight (TOF) are plotted against their impact position at the detector. Particles with zero initial momentum appear at the origin of position (x = 0) and at a sharp TOF defined by the charge-to-mass ratio. Ions with a non-zero momentum along the target-point–detector axis exhibit a change in the TOF, and those with a momentum into the gas-jet or polarization direction are displaced along the x-position of detection. Thus, Fig. 3a represents an image of the combined ion charge and momentum distribution.

The molecules in the gas target are randomly oriented, but those with their inter-nuclear axis parallel to the laser polarization are ionized most efficiently by the IR laser, resulting in a predominant ion emission along this axis. This is clearly seen in the data as dipole-like ion distributions. Assuming axial recoiling, the original orientation can be extracted from the measured fragment emission angles after Coulomb explosion of the molecule. As the IR laser couples very efficiently to the least bound valence electrons those from the 5p-like shells are ionized first. A strong indicator for this hypothesis is the charge-state distribution showing a sharp cut-off beyond I^{5+}, which corresponds to a completely depleted 5p shell in atomic iodine. The present intensity is sufficient to create fragments up to I^{5+} in this transition regime between multi-photon absorption and tunnel ionization.

The distribution of each charge state in Fig. 3(a) shows several discrete ring-like structures, with larger radii for higher charge states. The larger the radius, the higher the energy gained from Coulomb explosion. For instance, the distribution of I^{2+} ions features four rings, each from a Coulomb explosion with one fragment being I^{2+}. As the IR laser mostly creates dissociating

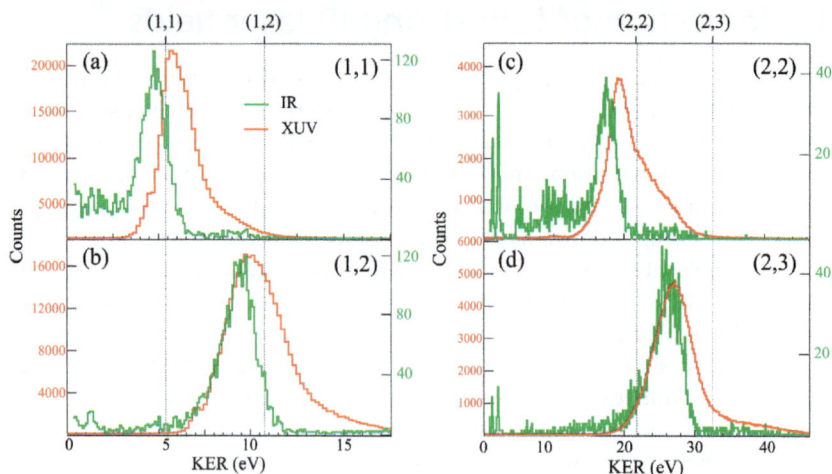

Fig. 4 Comparisons of kinetic energy release (KER) spectra obtained with single XUV (red) and single IR pulses (green). The final charges of coincident ions $I^{Q1+} + I^{Q2+}$ are indicated in brackets (Q1, Q2). The XUV pulses were roughly 60 fs long and the intensity was ~1 × 10^{14} W cm^{-2}. The estimated IR intensity is about 3 × 10^{14} W cm^{-2}.

molecules with at least one neutral fragment,[12] the lowest energy and thus the innermost ring corresponds to the fragmentation channel $I^{2+} + I^0$, the next outer one to $I^{2+} + I^{1+}$, and so forth.

For fragments with charge states up to I^{4+} a maximum in the centre of each Coulomb-explosion ring (see Fig. 3) is visible, which is due to fragments with very low kinetic energies. This is only possible for fragmentation into a multiply charged ion and a neutral atom. At first glance this might be surprising, because the symmetric channels are energetically more favourable. However, for short pulses and at high intensities the IR field bends the molecular potential such that tunnelling ionization leads to asymmetrically charged fragments.[22,26] In addition, we cannot exclude that these events are partly due to a weak pre-pulse or pedestal of the IR pulse that leads to dissociation before the main IR pulse impinges. Energetically, only two 1.55 eV photons are needed to dissociate the neutral I_2 molecule.

Distributions of total ion kinetic energies (KER) obtained with single IR and single XUV pulses are shown in Fig. 4 for various final charge-state combinations. In the following, coincident ion pairs will be denoted as $(Q1, Q2)$ instead of $I^{Q1+} + I^{Q2+}$. For the IR data, the KER spectra of the fragmentation channels $(1, 1)$ and $(2, 2)$ are clearly shifted towards lower values compared to the XUV case. This indicates that also in the IR pulse the charging-up occurs stepwise over a time period that is comparable to the pulse length (80 fs). The molecule is already in the progress of dissociation at the moment when the final charge state is reached resulting in a less energetic Coulomb explosion and, thus, a smaller KER. In fact, several IR photons are needed to reach $(1, 1)$, while a single XUV photon is sufficient. Because the $(1, 1)$ channel is a precursor of $(2, 2)$ a similar trend for the KER peak is expected for the $(2, 2)$ channel. In principle, the same argumentation holds for the channels $(1, 2)$ and $(2, 3)$. However, here the IR and XUV spectra share their low-energetic cut-off, indicating a slightly delayed population of the final charge states for the XUV pulses as well. This can be explained again with a sequential, stepwise absorption of two XUV photons during the FEL pulse. Some of the XUV created KER spectra are significantly broader than those for the IR (Fig. 4b and 4c). At the present stage we can only speculate about the possible reason. It might be that the time-delayed emission of Auger electrons after XUV inner valence ionization causes this broadening. While the $(1, 1)$ state is produced after the first Auger decay, the second Auger emission leads to the creation of $(1, 2)$. The exponential nature of the delayed Auger emission may cause a broadening in the case that dissociation has already started before the second Auger decay happens. If the decay occurs very fast we expect a high KER, if it lives longer the KER is small.

5.   IR–XUV pump–probe results

Having introduced the ionization mechanisms in single XUV and single IR pulses we now turn to the discussion of the IR–XUV pump–probe results. The intensity of the IR pulse is identical to that used for the measurement discussed in the previous chapter, and the intensity of the XUV pulses is roughly 3×10^{14} W cm^{-2}. The already introduced spectra of ion TOF *versus* hit position on the detector are presented in Fig. 5 for the two cases of IR-early (top) and IR-late (bottom).

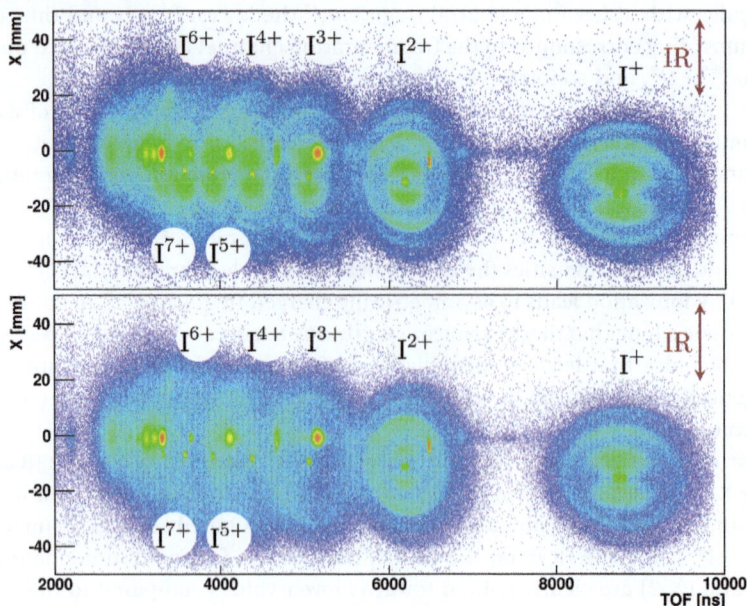

Fig. 5   TOF *versus* the x-position on the detector for (a) when the IR pulse arrives before the XUV pulse and (b) the IR is later than the XUV pulse.

If the IR pulse arrives first it efficiently initiates the dissociation of the molecule. Thus, the laser polarisation determines the emission direction of the coincident fragments. Therefore, the dipole-shaped emission pattern along the polarization axis is preserved even after interaction with the XUV pulse. This is true for all charge states including those that are out of reach with the IR laser only as shown in Fig. 5a. If the XUV pulse arrives first, the emission characteristic is almost spherically symmetric. This is expected because the XUV ionization cross-section is roughly independent of the molecular alignment. For small charge states the emission characteristic remains partially dipole-like because some of those events are created with the IR pulse alone.

The delay-dependent KER spectra for symmetric or almost symmetric coincidence channels are shown in Fig. 6. All time-independent, horizontal features are created within a single pulse, either XUV or IR, and the time-dependent traces stem from the joint interaction of both pulses, XUV and IR. They show a decreasing KER as the delay gets larger, finally converging to an asymptotic value. This is due to the creation of intermediate charge states ($Q1_i$, $Q2_i$) along with dissociation of the molecule after the pump pulse. Both ions move apart until the probe pulse further ionizes the system promoting it onto a steeper Coulomb curve belonging to the final charge states ($Q1_f$, $Q2_f$). The longer the delay the smaller the energy gain on the steeper potential energy curve gets. For very large delays the measured KER is equal to that of the intermediate or precursor state ($Q1_i$, $Q2_i$). Thus, the time-dependent traces allow us to reconstruct the full reaction pathway from the neutral molecule to the intermediate state after the pump pulse ($Q1_i$, $Q2_i$), as well as from this precursor state to the final Coulomb explosion channel ($Q1_f$, $Q2_f$) after the probe pulse. Also shown in Fig. 6 are expected KER values for

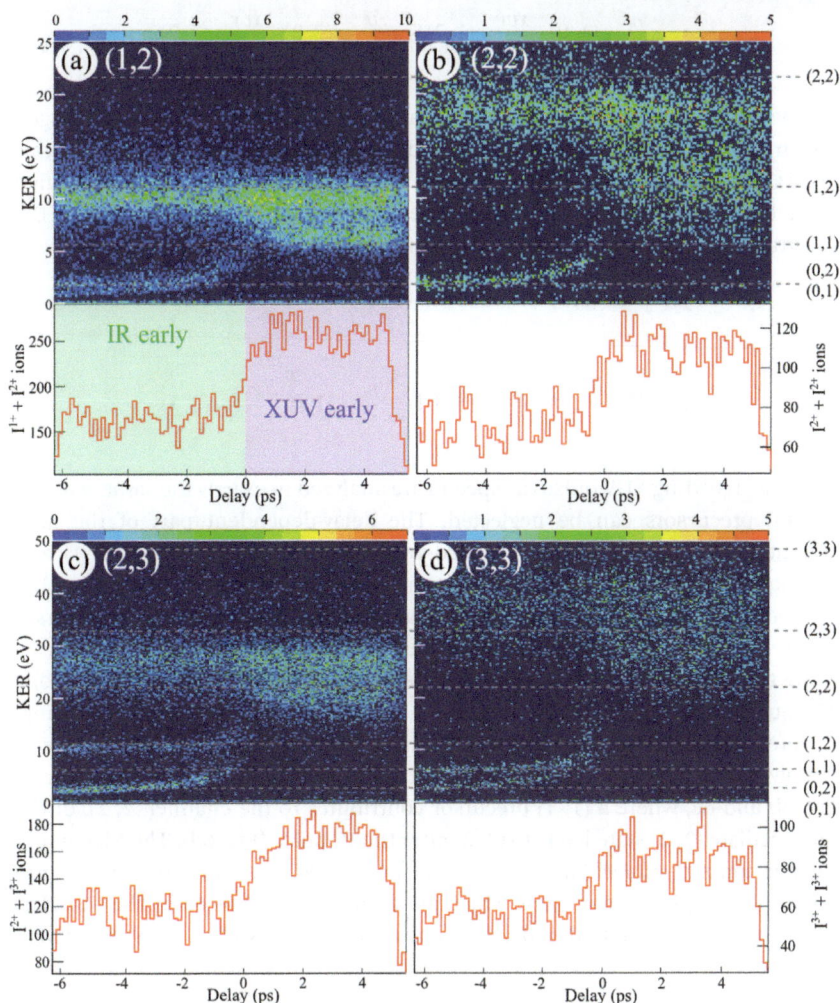

Fig. 6 Delay-dependent KER spectra for the (more) symmetric coincidence channels as indicated along with their projection onto the delay axis. Negative delays denote the case of a preceding IR pulse and positive values stand for a delayed IR pulse. Expected kinetic energy releases calculated with the Coulomb-explosion model are superimposed (grey dashed lines) for channels shown on the right side in order to identify the corresponding precursor states of each coincidence channel.

various fragmentation channels. These values are calculated assuming purely Coulombic potential energy curves $KER = E_{Coul} = (Q1 \cdot Q2)/R$ (in atomic units). For example, for the instantaneous ionization $I_2 \rightarrow (1, 2)$ at the equilibrium internuclear distance $R_{eq} = 5.0$ a.u. the expected KER is 11 eV.

We now discuss the case of negative delays where the IR pulse is preceding. From Fig. 6 we conclude that for the coincidence channels (1, 2) and (2, 2) only low-energy precursors with one neutral fragment are involved. As a delayed XUV pulse will always increase the charge state of one of the fragments by two, the most likely pathway for the production of channel (1, 2) is *via* (1, 0) and for (2, 2) *via* (2, 0):

$$I_2 \xrightarrow{IR} (1,0) \xrightarrow{XUV} (1,2) I_2 \xrightarrow{IR} (2,0) \xrightarrow{XUV} (2,2)$$

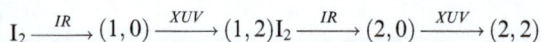

The higher final charge states, $(2, 3)$ and $(3, 3)$, feature neutral and charged precursor states as indicated in Fig. 6c and 6d, respectively. The precursors with a neutral fragment are assigned as above. The charged precursors are identified using the Coulomb explosion model to be $(1, 2)$ and $(1, 1)$, respectively. Thus the spectra with preceding IR laser can be summarized in the following way:

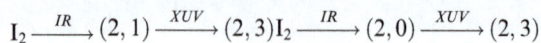

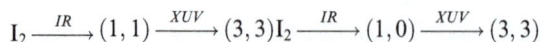

$$I_2 \xrightarrow{IR} (2,1) \xrightarrow{XUV} (2,3) I_2 \xrightarrow{IR} (2,0) \xrightarrow{XUV} (2,3)$$

$$I_2 \xrightarrow{IR} (1,1) \xrightarrow{XUV} (3,3) I_2 \xrightarrow{IR} (1,0) \xrightarrow{XUV} (3,3)$$

For a preceding XUV pulse the spectra are analyzed in exactly the same way, but neutral precursors can be neglected. The delay-dependent part of the final channel $(1, 2)$ for instance is reached by producing the precursor $(1, 1)$ and subsequent removal of a further electron by the IR probe. Similarly, $(2, 2)$ is created *via* the main precursor $(1, 2)$ with a weak contribution from the precursor $(1, 1)$. Interestingly the channel $(2, 3)$ does only show a single precursor, $(2, 2)$, while the next higher coincidence channel $(3, 3)$ again features two precursors. The question arises why $(1, 2)$, the second possible precursor of $(2, 3)$, is missing.

Interestingly, in several cases the IR laser effectively ionizes already singly or doubly charged fragments created by XUV absorption, as can be seen *e.g.* in Fig. 6b and 6c, where a $(1, 1)$ precursor contributes to the channel $(2, 2)$, or the intermediate $(2, 2)$ state leads to $(3, 3)$ after the IR pulse (Fig. 6d). The XUV inner-valence ionized molecular ion or, correspondingly, the ionic fragments may remain in a highly excited state which are easily further ionized by the IR pulse. The results demonstrate that delayed ionization by the IR laser is a very efficient

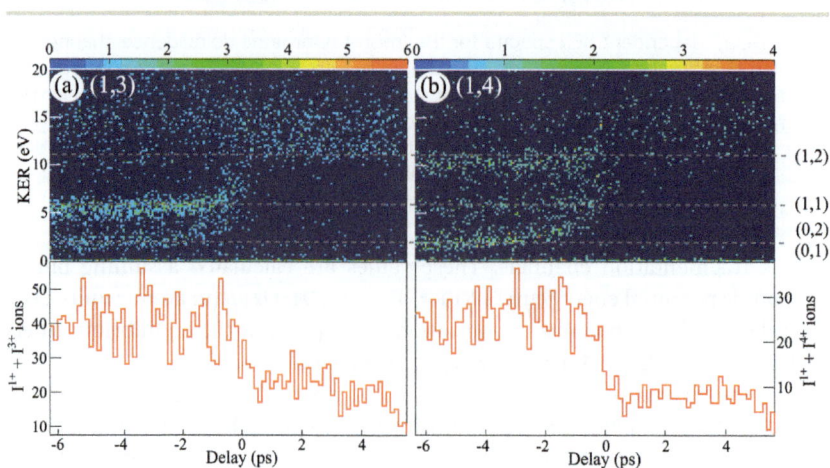

Fig. 7 Delay-dependent KER spectra for the asymmetric coincidence channels $(1, 3)$ and $(1, 4)$ and their projections onto the delay axis. Possible precursor charge states are indicated as dashed lines.

process, because it clearly dominates the ion yield of the delay-dependent KER spectra in Fig. 6. However, this trend of an increased ion yield at positive delays is only present for the more symmetric break-up channels and only up to a certain degree of ionization. For already highly charged molecular ions the IR laser intensity is not sufficient anymore to strip off further electrons, in particular if all electrons are removed from the 5p shell (beyond I^{5+}).

As mentioned, the asymmetric channels exhibit an opposite behaviour with a decrease of the ion yield for positive delays (XUV early). This is shown in Fig. 7, where the delay-dependent KER spectra are plotted for the coincidence channels $(1, 3)$ and $(1, 4)$. From this we draw the conclusion that delayed ionization by the IR laser favours the production of symmetric charge-state distributions. An IR pulse, which couples best to outer-shell electrons, will thus not create asymmetric charge states, except for very strong fields, when the molecular potential is bent so strongly that the asymmetric break-up gets energetically more favourable than the symmetric.[22] Furthermore, even for large positive delays, where the iodine ions are considered as separated, the IR laser does not create asymmetric charge states. If for instance the XUV pulse ionizes I_2 into $(1, 2)$ and the separated ionic fragments are placed into the IR field the number of photons needed to remove another electron from I^+ is much smaller than that for I^{2+}. Therefore the creation of $(1, 3)$ compared to $(2, 2)$ is suppressed. This explanation is valid for the creation of all asymmetric charge states. By contrast, if the XUV pulse arrives late, a single photon is sufficient to even doubly ionize either of the ions and thus asymmetric charge states are created.

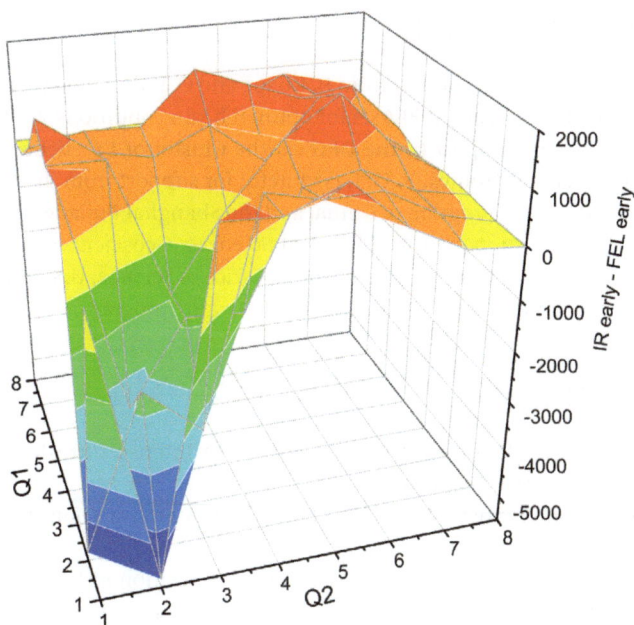

Fig. 8 Yield difference (in arbitrary units) between the two situations of a preceding IR and a successive IR pulse (with respect to the XUV pulse) as a function of $Q1$ and $Q2$. A value larger than zero means that the corresponding channel ($Q1$, $Q2$) exhibits a larger yield if the IR comes first, and *vice versa*. For better visibility the distribution is symmetrized with respect to the diagonal $Q1 = Q2$.

6.   Summary and conclusions

We conclude with an overview of our results and by providing an answer to the question whether higher charge states are reached for ionization of the bound molecule or two separated atoms. The role of the preceding IR pulse is to prepare two isolated neutrals or low charged ions by dissociating the molecule while the subsequent intense XUV pulse produces the high charge states. Thus, depending on whether the IR comes before of after the XUV pulse, we create a situation which allows us to compare the response of isolated atoms or ions with that of a molecule irradiated by exactly the same intense XUV pulse.

Our findings are summarized in Fig. 8, where the yield differences between the two situations IR-early and XUV-early are shown for all coincidence channels ($Q1$, $Q2$). The region of low charge states is clearly dominated by a preceding XUV pulse. Asymmetric and very high charge states are produced more effectively if the IR laser arrives first and dissociates the molecule prior to the removal of additional electrons by the XUV.

In conclusion, we reached the highest charge states in our experiment after dissociation of iodine into separated atoms or ions. This is in contrast to observations with large Xe clusters, where a significant enhancement of XUV absorption was found in comparison to monomers due to collective effects that contribute to cluster ionization.[32] For a system with only two constituents, like the I_2 molecule, such collective phenomena are not active, and the more intuitive picture of individual atoms is still valid.

Acknowledgements

We acknowledge technical support from B. Knape and C. Kaiser, and we thank the scientific and technical team at FLASH for optimal beam conditions. M. Küb. and M. F. K. acknowledge support by the DFG *via* the Cluster of Excellence: Munich Center for Advanced Photonics. Y. H. J. is grateful for support from the NSFC, the National Basic Research Program of China, and the Shanghai Pujiang Program. A. R. acknowledges the support from Chemical Sciences, Geosciences, and Biosciences Division, Office of Basic Energy Sciences, Office of Science, U.S. Department of Energy, and from Kansas NSF EPSCoR "First Award" program.

References

1 G. Doumy, *et al.*, Nonlinear Atomic Response to Intense Ultrashort X Rays, *Phys. Rev. Lett.*, 2011, **106**, 083002.

2 M. Richter, *et al.*, Extreme Ultraviolet Laser Excites Atomic Giant Resonance, *Phys. Rev. Lett.*, 2009, **102**, 163002.

3 B. Rudek, *et al.*, Resonance-enhanced multiple ionization of krypton at an X-ray free-electron laser, *Phys. Rev. A: At., Mol., Opt. Phys.*, 2013, **87**, 023413.

4 B. Rudek, *et al.*, Ultra-efficient ionization of heavy atoms by intense X-ray free-electron laser pulses, *Nat. Photonics*, 2012, **6**, 858.

5 H. Thomas, *et al.*, Shell explosion and core expansion of xenon clusters irradiated with intense femtosecond soft X-ray pulses, *J. Phys. B: At., Mol. Opt. Phys.*, 2009, **42**, 134018.

6 T. Gorkhover, *et al.*, Nanoplasma dynamics of single large Xenon clusters irradiated with superintense X-ray pulses from the linac coherent light source free-electron laser, *Phys. Rev. Lett.*, 2012, **108**, 245005.

7 R. Neutze, *et al.*, Potential for biomolecular imaging with femtosecond X-ray pulses, *Nature*, 2000, **406**, 752.

8 L. Fang, *et al.*, Multiphoton Ionization as a clock to Reveal Molecular Dynamics with Intense Short X-ray Free Electron Laser Pulses, *Phys. Rev. Lett.*, 2012, **109**, 263001.

9 B. Erk, *et al.*, Inner-shell multiple ionization of polyatomic molecules with an intense X-ray free-electron laser studied by coincident ion momentum imaging, *J. Phys. B: At., Mol. Opt. Phys.*, 2013, **46**, 164031.

10 B. Erk, *et al.*, Ultrafast Charge Rearrangement and Nuclear Dynamics upon Inner-Shell Multiple Ionization of Small Polyatomic Molecules, *Phys. Rev. Lett.*, 2013, **110**, 053003.

11 M. Krikunova, *et al.*, Strong-field ionization of molecular iodine traced with XUV pulses from a free-electron laser, *Phys. Rev. A: At., Mol., Opt. Phys.*, 2012, **86**, 043430.

12 M. Krikunova, *et al.*, Ultrafast photofragmentation dynamics of molecular iodine driven with timed XUV and near-infrared light pulses, *J. Chem. Phys.*, 2011, **134**, 024313.

13 J. Ullrich, *et al.*, Recoil-ion and electron momentum spectroscopy: reaction microscopes, *Rep. Prog. Phys.*, 2003, **66**, 1463.

14 H. Redlin, *et al.*, The FLASH pump–probe laser system: Setup, characterization and optical beamlines, *Nucl. Instrum. Methods Phys. Res., Sect. A*, 2011, **635**, 88.

15 B. H. Boo and N. Saito, Dissociative multiple photoionization of Br2, IBr, and I2 in the VUV and X-ray regions: a comparative study of the inner-shell processes involving Br(3d,3p,3s) and I(4d,4p,4s,3d,3p), *J. Electron Spectrosc. Relat. Phenom.*, 2002, **127**, 139.

16 J. Tremblay, *et al.*, Photoelectron spectroscopy of atomic iodine produced by laser photodissociation, *Phys. Rev. A: At., Mol., Opt. Phys.*, 1988, **38**, 3804.

17 H. Kjeldsen, *et al.*, Absolute photoionization cross sections of I+ and I2+ in the 4d ionization region, *Phys. Rev. A: At., Mol., Opt. Phys.*, 2000, **62**, 020702.

18 A. Kramida, Y. Ralchenko, J. Reader, and NIST ASD Team(2013), NIST Atomic Spectra Database (ver. 5.1), Available: http://physics.nist.gov/asd.

19 D. Edvardsson, *et al.*, An experimental and theoretical investigation of the valence double photoionisation of the iodine molecule, *Chem. Phys.*, 2006, **324**, 674.

20 L. Fang, *et al.*, Double Core-Hole Production in N2: Beating the Auger Clock, *Phys. Rev. Lett.*, 2010, **105**, 083005.

21 L. J. Frasinski, *et al.*, Dynamics of Hollow Atom Formation in Intense X-Ray Pulses Probed by Partial Covariance Mapping, *Phys. Rev. Lett.*, 2013, **111**, 073002.

22 J. H. Posthumus, *et al.*, Field-ionization, Coulomb explosion of diatomic molecules in intense laser fields, *J. Phys. B: At., Mol. Opt. Phys.*, 1996, **29**, 5811.

23 D. M. Villeneuve, M. Y. Ivanov and P. B. Corkum, Enhanced ionization of diatomic molecules in strong laser fields: A classical model, *Phys. Rev. A: At., Mol., Opt. Phys.*, 1996, **54**, 736.

24 T. Zuo and A. D. Bandrauk, Charge-resonance-enhanced ionization of diatomic molecular ions by intense lasers, *Phys. Rev. A: At., Mol., Opt. Phys.*, 1995, **52**, R2511.

25 E. Constant, H. Stapelfeldt, and P. Corkum, Enhanced Ionization of Molecular Ions in Intense Laser Fields: Experiments on the Iodine Melecule, in *Ultrafast Phenomena X*, ed. P. Barbara *et al.*, Springer Series in Chemical Physics. Springer, Berlin Heidelberg, 1996, vol. 62, pp. 105–106.

26 G. N. Gibson, *et al.*, Direct evidence of the generality of charge-asymmetric dissociation of molecular iodine ionized by stronglaser fields, *Phys. Rev. A: At., Mol., Opt. Phys.*, 1998, **58**, 4723.

27 L. J. Frasinski, *et al.*, Femtosecond dynamics of multielectron dissociative ionization by use of a picosecond laser, *Phys. Rev. Lett.*, 1987, **58**, 2424.

28 K. Codling, L. J. Frasinski and P. A. Hatherly, On the field ionisation of diatomic molecules by intense laser fields, *J. Phys. B: At., Mol. Opt. Phys.*, 1989, **22**, L321.

29 D. Normand and M. Schmidt, Multiple ionization of atomic and molecular iodine in strong laser fields, *Phys. Rev. A: At., Mol., Opt. Phys.*, 1996, **53**, R1958.

30 T. Seideman, M. Y. Ivanov and P. B. Corkum, Role of Electron Localization in Intense-Field Molecular Ionization, *Phys. Rev. Lett.*, 1995, **75**, 2819.

31 S. Chelkowski and A. D. Bandrauk, Two-step Coulomb explosions of diatoms in intense laser fields, *J. Phys. B: At., Mol. Opt. Phys.*, 1995, **28**, L723.

32 H. Wabnitz, *et al.*, Multiple ionization of atom clusters by intense soft X-rays from a free-electron laser, *Nature*, 2002, **420**, 6915.

Faraday Discussions

PAPER

Imaging molecular structure through femtosecond photoelectron diffraction on aligned and oriented gas-phase molecules

Rebecca Boll,[abc] Arnaud Rouzée,[de] Marcus Adolph,[f] Denis Anielski,[abc] Andrew Aquila,[gh] Sadia Bari,[h] Cédric Bomme,[a] Christoph Bostedt,[i] John D. Bozek,[i] Henry N. Chapman,[gjk] Lauge Christensen,[l] Ryan Coffee,[i] Niccola Coppola,[gh] Sankar De,[mn] Piero Decleva,[o] Sascha W. Epp,[abcp] Benjamin Erk,[abc] Frank Filsinger,[q] Lutz Foucar,[br] Tais Gorkhover,[f] Lars Gumprecht,[g] André Hömke,[ab] Lotte Holmegaard,[gm] Per Johnsson,[s] Jens S. Kienitz,[gk] Thomas Kierspel,[gk] Faton Krasniqi,[bpr] Kai-Uwe Kühnel,[c] Jochen Maurer,[m] Marc Messerschmidt,[i] Robert Moshammer,[bc] Nele L. M. Müller,[g] Benedikt Rudek,[bct] Evgeny Savelyev,[aw] Ilme Schlichting,[r] Carlo Schmidt,[bc] Frank Scholz,[a] Sebastian Schorb,[i] Joachim Schulz,[gh] Jörn Seltmann,[a] Mauro Stener,[o] Stephan Stern,[gj] Simone Techert,[abvw] Jan Thøgersen,[m] Sebastian Trippel,[g] Jens Viefhaus,[a] Marc Vrakking,[de] Henrik Stapelfeldt,[m] Jochen Küpper,[bgjkq] Joachim Ullrich,[bct] Artem Rudenko[bcu] and Daniel Rolles*[abr]

Received 8th March 2014, Accepted 3rd April 2014

DOI: 10.1039/c4fd00037d

[a]Deutsches Elektronen-Synchrotron (DESY), 22607 Hamburg, Germany. E-mail: daniel.rolles@desy.de

[b]Max Planck Advanced Study Group at CFEL, 22607 Hamburg, Germany

[c]Max Planck Institute for Nuclear Physics, 69117 Heidelberg, Germany

[d]Max-Born-Institut, 12489 Berlin, Germany

[e]FOM-Institute AMOLF, 1098 XG Amsterdam, The Netherlands

[f]Technische Universität Berlin, 10643 Berlin, Germany

[g]Center for Free-Electron Laser Science, DESY, 22607 Hamburg, Germany

[h]European XFEL GmbH, 22761 Hamburg, Germany

[i]SLAC National Accelerator Laboratory, Menlo Park, California 94025, USA

[j]Department of Physics, University of Hamburg, 22761 Hamburg, Germany

[k]Center for Ultrafast Imaging, University of Hamburg, 22761 Hamburg, Germany

[l]Department of Physics and Astronomy, Aarhus University, 8000 Aarhus C, Denmark

[m]Department of Chemistry, Aarhus University, 8000 Aarhus C, Denmark

[n]Saha Institute of Nuclear Physics, 700064 Kolkata, India

[o]Dipartimento di Scienze Chimiche e Farmaceutiche, Università di Trieste, 34127 Trieste, Italy

[p]Max Planck Institute for Structural Dynamics, 22607 Hamburg, Germany

[q]Fritz-Haber-Institut der Max-Planck-Gesellschaft, 14195 Berlin, Germany

[r]Max Planck Institute for Medical Research, 69120 Heidelberg, Germany

[s]Department of Physics, Lund University, 22100 Lund, Sweden

[t]Physikalisch-Technische Bundesanstalt (PTB), 38116 Braunschweig, Germany

[u]J.R. MacDonald Laboratory, Kansas State University, Manhattan, Kansas 66506, USA

[v]Max Planck Institute for Biophysical Chemistry, 37077 Göttingen, Germany

[w]Institute of X-ray Physics, 37077 Göttingen University, Germany

This paper gives an account of our progress towards performing femtosecond time-resolved photoelectron diffraction on gas-phase molecules in a pump–probe setup combining optical lasers and an X-ray free-electron laser. We present results of two experiments aimed at measuring photoelectron angular distributions of laser-aligned 1-ethynyl-4-fluorobenzene (C_8H_5F) and dissociating, laser-aligned 1,4-dibromobenzene ($C_6H_4Br_2$) molecules and discuss them in the larger context of photoelectron diffraction on gas-phase molecules. We also show how the strong nanosecond laser pulse used for adiabatically laser-aligning the molecules influences the measured electron and ion spectra and angular distributions, and discuss how this may affect the outcome of future time-resolved photoelectron diffraction experiments.

1   Introduction

1.1   Motivation

The prospect of studying chemical reactions with femtosecond resolution has been an inspiration for many experimental and theoretical investigations ever since the possibility of producing femtosecond light or electron pulses was first discussed.[1,2] Methods such as time-dependent mass spectrometry and absorption spectroscopy[3,4] can provide information on the changes of the molecular structure that occur during chemical reactions by comparing the observed time-dependent signatures to theoretical predictions. More recently, methods aiming at imaging the structural changes more directly, for example by ultrafast X-ray or electron diffraction [2,5] were developed. In most cases however, their interpretation still heavily relies on comparison to theoretical models, and their temporal resolution, in particular for the case of electron diffraction, has, to date, barely broken the one-picosecond mark.[5-7]

Free-electron lasers (FELs) that produce intense, few-femtosecond light pulses in the vacuum ultraviolet (VUV) and X-ray regime,[8-11] along with advances in the generation of (sub-)femtosecond pulses with laser-based high-harmonic genera-tion (HHG) sources[12-14] and with relativistic electron guns,[15] have added new fuel to the long-standing vision of recording *molecular movies* with Ångström spatial and femtosecond temporal resolution. Ideally, these movies would contain real-space images of the changing molecular structure that can be obtained without the necessity of comparison to theoretical modelling.[16]

In this article, we discuss how time-resolved photoelectron diffraction may be used to directly visualize ultrafast structural changes of gas-phase molecules, such as the formation of short-lived intermediate states during photodissociation or isomerization reactions. As an introduction, we discuss the relationship between molecular-frame photoelectron angular distributions and photoelectron diffraction in section 1.2. Section 2 briefly describes the experimental setup used to measure time-resolved photoelectron angular distributions of laser-aligned molecules at an FEL, and section 3 presents the results of these experiments. Here, we focus on data that has not been included in our previous publications[17,18] such as a comparison of ion time-of-flight spectra recorded at an FEL and at a synchrotron (section 3.1), effects of molecular orientation on photoelectron and fragment ion angular distributions (section 3.2), and the influence of both the alignment laser pulse and the femtosecond "pump" laser pulse on the photo-electrons and on the molecular photofragmentation process (sections 3.3 and 3.4, respectively). Our findings are summarized and conclusions for future time-resolved photoelectron diffraction experiments are drawn in section 4.

1.2   Photoelectron diffraction and molecular-frame photoelectron angular distributions

The possibility to measure molecular-frame photoelectron angular distributions (MFPADs) of gas-phase molecules with electron-ion coincidence techniques developed in the 1990s[19-24] led to a breakthrough in the study of molecular photoionization. Measurements of MFPADs allow, for example, the determination of photoionization matrix elements and phases[25,26] as well as investigations of core hole localization[27,28] and of the role of coherence and double-slit interferences in molecular photoemission.[27,29,30] Extending the concept of photoelectron diffraction, which is a well-established method in solid state and surface physics,[31,32] to gas-phase molecules, it was realized early on that MFPADs of inner-shell electrons could also be interpreted in terms of diffraction.[27,33-35] This opens up the possibility to obtain direct information on the geometric structure of the molecule from the photoelectron angular distribution, as illustrated schematically in Fig. 1 for the case of $F(1s)$ inner-shell photoionization of a C_8H_5F molecule. Within the photoelectron diffraction model, the fluorine atom is considered as the source of photoelectrons that may scatter on the neighboring atoms in the molecule. The MFPAD is interpreted as the superposition of direct and scattered waves, creating an interference pattern on a detector in the far field, which contains structural information. This information is usually lost in gas-phase experiments on randomly oriented molecules because the diffraction pattern averages out when integrated over all molecular orientations. It can only be observed when the orientation of the molecule in the laboratory frame at the time of the electron emission is known.

In the surface physics community, scattering and diffraction of inner-shell photoelectrons is used, for example, to determine the geometry of molecules adsorbed on surfaces,[31,32,36] thus providing insights into processes like catalytic reactions. In contrast, the concept of photoelectron diffraction did not gain much

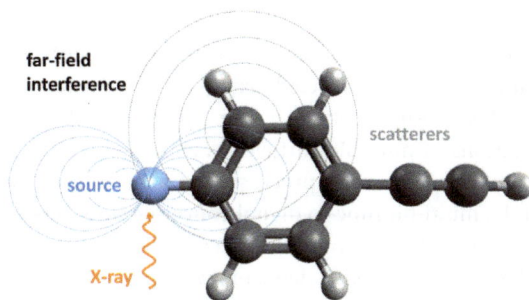

Fig. 1 Schematic illustration of the photoelectron diffraction concept for a gas-phase C_8H_5F molecule: The emitted inner-shell photoelectron wave (blue), here created from the $F(1s)$ level by linearly polarized X-rays, scatters on neighboring atoms inside the molecule. The superposition of direct and scattered photoelectron waves, drawn here only for one of the neighboring carbon atoms, creates an interference pattern in the far field which contains information on the molecular structure.

interest in the gas-phase community, probably because far more precise methods, such as microwave spectroscopy, exist to determine the equilibrium structure of gas-phase molecules. Moreover, angle-resolved photoelectron-photoion coincidence measurements that have, so far, been used to determine the molecular orientation of gas-phase molecules are challenging and often time-consuming.

With the availability of femtosecond VUV and X-ray sources that allow pump–probe studies involving inner-shell ionization, this situation is now changing. Time-resolved measurements of MFPADs and photoelectron diffraction of gas-phase molecules may offer information on ultrafast changes of molecular structure during chemical reactions which is difficult to obtain by other techniques.[16,17,37–39]

In this paper, we give an account of our experimental progress towards performing such femtosecond time-resolved experiments by combining optical lasers with VUV and soft X-ray FELs. The underlying idea is to first initiate a photochemical reaction with a "pump" laser pulse, and then to create an inner-shell photoelectron with an FEL pulse in order to image the molecules *from within*. As a first step, we focus on measuring delay-dependent changes in the photoelectron angular distributions and on linking them to changes in the molecular geometry *via* comparison to density functional theory calculations. The long-term goal is to employ the photoelectron diffraction concept in order to directly image molecular structure, for example by holographic reconstruction.[16]

2 Experimental setup

The experiments were performed at the Atomic, Molecular, and Optical Physics (AMO) beamline[40] of the Linac Coherent Light Source (LCLS)[9] at SLAC National Accelerator Laboratory and at the Variable Polarization† XUV Beamline P04[41] of the synchrotron radiation source PETRA III at DESY using the CFEL-ASG Multi-Purpose (CAMP) endstation.[42] The setup has been described in[17,43,44] and, in detail, in,[18] and is only briefly summarized here. A beam of rotationally cold 1-ethynyl-4-fluorobenzene (C_8H_5F, pFAB) or 1,4-dibromobenzene ($C_6H_4Br_2$, DBB) molecules seeded in helium was created by supersonic expansion into vacuum and crossed with the X-ray beam inside a double-sided velocity map imaging (VMI) spectrometer.

For the PETRA experiments, the molecular beam was operated continuously, and electrons and ions were detected using two microchannel plate (MCP) detectors equipped with RoentDek delay-line anodes, which record the time of flight and hit positions of multiple particles in coincidence. The amplified MCP and delay-line anode signals were processed by a hardware constant fraction discriminator and a multi-hit time-to-digital converter, and were then stored as a listmode event file. At the LCLS, a pulsed molecular beam was used, and electrons and ions were detected using MCP detectors with phosphor screens, that were read out for each FEL shot by 1-Megapixel CCD cameras. For time-of-flight measurements, the MCP signal traces were recorded for each FEL shot with an Acqiris DC282 digitizer. Processing of the single-shot CCD images, including a peak-finding algorithm, data sorting, and filtering on FEL machine parameters

---

† At the time of the experiment, only circular polarization was available.

(photon energy and FEL pulse energy), was performed with the CFEL-ASG Software Suite (CASS).[45] The data shown here were taken during two LCLS experiments in 2010 (DBB) and in 2011 (pFAB) and during two PETRA experiments in 2013.

2.1 Adiabatic laser alignment and orientation

The determination of molecular orientation in an angle-resolved electron-ion coincidence experiment requires an ionization rate of less than one molecule per detection cycle in order to unambiguously correlate electrons and fragment ions. As the currently operating X-ray FELs have a maximum repetition rate of 120 Hz, this technique yields very low count rates in FEL applications. An alternative approach to fix the molecular frame with respect to the laboratory frame is to actively align the molecules in space by using strong laser pulses.[46–49] This allows probing a whole ensemble of molecules with each FEL pulse,[17,18,50–52] thus dramatically increasing the achievable count rate.

At the LCLS, one- or three-dimensional adiabatic alignment was achieved by intersecting the molecular beam with pulses from a 1064 nm, seeded neodymium-doped yttrium aluminum garnet (YAG) laser with a pulse duration of 10–12 ns and a pulse energy of 200–500 mJ. A drilled mirror was used to collinearly propagate the YAG laser beam with the FEL beam, and the timing was set such that the FEL pulse arrived at the maximum of the YAG laser pulse, which corresponds to the maximum of the molecular alignment.[53] When using a linearly polarized YAG pulse, the molecules align such that their most-polarizable axis lies parallel to the laser polarization direction, which is the Br-Br axis in DBB and the F-C axis in pFAB. When using an elliptically polarized laser pulse, the second-most polarizable axis can be fixed in space as well.[54,55] For the molecules used here, the plane of the benzene ring, which freely rotates for the case of one-dimensional alignment, is then also spatially confined.

Moreover, one- or three-dimensional orientation can be achieved for polar molecules when an additional static electric field is present that has a vector component parallel to the polarization direction of the alignment laser field.[46,56,55] In the presented data, the extraction field of the VMI spectrometer was used to define the direction of the fluorine atom in pFAB with respect to the electron and ion detectors.

For the experiments discussed here, the YAG laser operated at a repetition rate of 30 Hz, and the LCLS at 60 Hz in 2010 and at 120 Hz in 2011, respectively. This allowed recording data for aligned and randomly oriented molecules concurrently. The molecular beam was operated at 60 Hz, such that background from residual gas could also be recorded concurrently, in 2011. As shown in the following, this facilitates background subtraction substantially since long-term drifts were equally contained in each data subset.

2.2 Three-color pump–probe experiments

In order to initiate a structral change in the molecules *via* molecular fragmentation by strong-field ionization, an 800 nm (1.55 eV) titanium-sapphire (TiSa) laser synchronized with the FEL was used in the 2010 LCLS experiments to pump the molecules before probing them with the FEL pulse. The TiSa laser beam was co-propagating with the YAG laser beam and the FEL beam, and the relative delay between FEL and TiSa pulses was varied using a delay stage. However, in 2010, the arrival time jitter between the TiSa pulse and the FEL pulse could not yet be corrected by X-ray optical cross-correlation,[57–59] and the temporal resolution of the pump–probe experiment was thus limited to 200–300 fs. While this, among other technical difficulties, prevented the observation of delay-dependent changes in the photoelectron angular distribution, the experiment still demonstrated the feasibility of three-color pump–probe studies at an FEL.[18,60] A subsequent pump–probe experiment at the LCLS in 2012 showed that with cross-correlation, the achievable temporal resolution is, at present, limited by the pulse durations of the TiSa laser and the FEL.[61]

3   Results and discussion

3.1   Fragmentation of pFAB molecules after inner-shell photoionization

In a polyatomic molecule, the core-hole created by inner-shell ionization typically decays within a few femtoseconds *via* single or multiple Auger decay. The resulting multiply charged molecular ion is usually not stable and subsequently dissociates into a variety of fragments. The charged fragments can be characterized by recording an ion time-of-flight (TOF) spectrum as shown in Fig. 2 for the case of pFAB molecules ionized by X-rays from PETRA and the LCLS. The photon energies of 742 and 765 eV lie approximately 50 and 73 eV above the $F(1s)$ ionization threshold, respectively (the $F(1s)$ binding energy in pFAB is assumed to be almost identical to the one in fluorobenzene, which is 692 eV[62]). A large number of fragment ions from pFAB and from residual gas can be identified. However, only a relatively small amount of F^+ ions is produced despite the fact

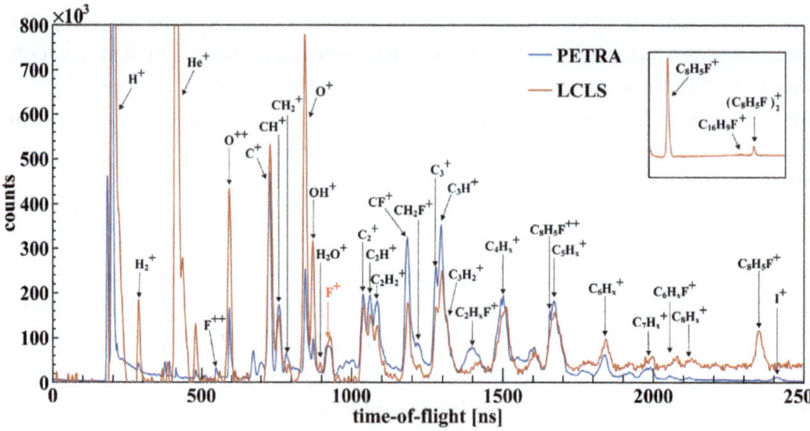

Fig. 2   Ion time-of-flight spectra of pFAB molecules obtained after photoionization with circularly polarized X-rays at a photon energy of 765 eV from the PETRA synchrotron radiation source (blue) compared to the spectrum obtained with linearly polarized X-ray pulses from the LCLS free-electron laser at a photon energy of 742 eV and 80 fs pulse duration. The y-axis shows the total ions counts recorded in the PETRA spectrum, while the LCLS spectrum has been scaled and shifted such as to provide direct comparability with the synchrotron spectrum. The inset shows a zoom on the time-of-flight region with the parent ion and the dimer ion peak in the LCLS spectrum.

that, according to the photoabsorption cross-sections, one third of the absorbed photons are absorbed by the fluorine atom.

Several ten eV above the F($1s$) ionization threshold, far beyond any potential shape resonances or other near-threshold phenomena, the fragmentation of pFAB can be considered to be rather insensitive to the exact photon energy. Therefore, the comparison of the ion TOF spectrum recorded using synchrotron radiation, shown in blue in Fig. 2, with the ion TOF spectrum obtained at the LCLS, shown in red, allows to identify the influence of possible multiphoton ionization that can occur due to the high intensity of the FEL pulse, as well as other influences stemming, for example, from the use of two different molecular beams in the PETRA and the LCLS experiments.‡

Overall, the two spectra are rather similar, showing that multiphoton processes are minor channels contributing to the overall fragmentation of the molecules.§ Besides a stronger contribution of water fragments in the LCLS spectrum, two main differences can be observed: A significantly larger He$^+$ peak in the spectrum recorded at the LCLS, and a relatively large amount of molecular parent ions in the LCLS spectrum, which are almost absent in the PETRA experiment. Whereas a continuous molecular beam with helium as a carrier gas at a relatively low backing pressure (few hundred millibars) was used at PETRA, the pulsed valve at the LCLS was operated with 50 bar helium backing pressure resulting in a large number of helium atoms in the interaction zone. It also appears that for the expansion conditions in the LCLS experiment, a large amount of pFAB clusters was produced in the molecular beam, as indicated by the strong $C_8H_5F^+$ parent ion signal. This is further confirmed by the width of molecular parent ion peak, which indicates that the parent ions are produced with substantial kinetic energy, as well as by the singly charged pFAB dimer peak shown in the inset of Fig. 2. The presence of molecular clusters in the beam is particularly significant since these clusters are, most likely, not well aligned by the YAG pulse. Consequently, they produce a background of unaligned molecules in the ion and electron data recorded for aligned molecules at the LCLS. Unfortunately, the exact ratio of clusters to single molecules cannot be determined from the ion TOF spectra alone.

Additional information on the fragmentation of pFAB molecules can be obtained when two or more charged fragments are recorded in coincidence, which we have done at PETRA and can be represented in a photoion-photoion coincidence (PIPICO) map as shown in Fig. 3. A large number of fragmentation channels can be identified, some of which correspond to the break-up of the pFAB molecules into two fragments, while at least a third fragment (either charged or neutral) must have been present in many of the break-up channels. Channels corresponding to the break-up into two charged fragments generally produce sharp diagonal lines in the PIPICO map as a result of momentum conservation. In contrast, when three or more charged fragments are created that each carry a significant amount of momentum, the corresponding line in the PIPICO map is

---

‡ Note that the high number of ions detected per shot at the LCLS did not allow using a software constant fraction discriminator on the MCP trace to identify individual ion hits. Thus, the averaged MCP signal is shown which exhibits a slightly rising baseline towards higher times of flight.

§ The LCLS experiment was performed outside of the optimum focal position of the beamline, *i.e.* at an FEL spot size of approximately 30 × 30 μm, in order to reduce multi-photon ionization.

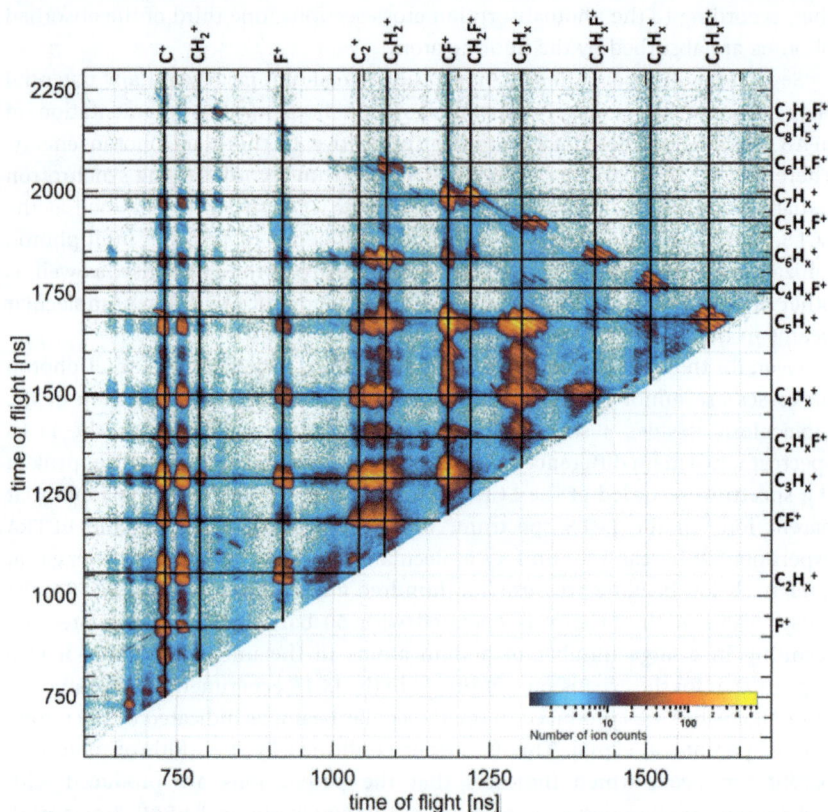

Fig. 3 Photoion-photoion coincidence (PIPICO) spectrum of pFAB molecules obtained after photoionization with circularly polarized X-rays from PETRA at a photon energy of 765 eV.

more washed out.[63] In Fig. 3, sharp PIPICO lines are observed for most of the break-up channels involving C_5H_x, C_6H_x, and C_7H_x fragments, with the exception of C_5H_x-C_2H_x (x denotes varying numbers of H atoms). Most channels involving an F^+ exhibit rather washed out lines, suggesting that these mostly stem from a break-up into at least three charged fragments, each carrying a significant amount of momentum. We note that this does not bode well for using F^+ ions to determine the orientation of the F-C axis in an angle-resolved photoelectron-ion coincidence experiment. In the following, however, the emphasis shall not be put on further interpretations of the wealth of information that can be extracted from the momentum-resolved coincidence data but rather on the effects of the alignment laser on the electron and ion images and spectra recorded at the LCLS.

3.2 Molecular alignment and orientation

In order to characterize the degree of alignment and orientation induced in the beam of pFAB molecules by the combination of the YAG laser pulse and the static electric field of the VMI spectrometer, the emission direction of the F^+ ions can be used as a marker, assuming that they are emitted along the direction of the F-C axis. Fig. 4 shows the F^+ ion images recorded at the LCLS for ionization of pFAB molecules with linearly polarized X-rays at a photon energy of 723 eV, with and without the YAG laser pulses and for different directions of the YAG pulse polarization axis. The ion detector was gated by fast switching of the high voltage such that only hits in the time-of-flight interval corresponding to the arrival time of the F^+ ions were detected. However, when operating the spectrometer in velocity map imaging mode, the signal of F^+ ions (mass of 19 amu) could not be fully separated from the signal of H_2O^+ ions (mass of 18 amu), and a contribution from the ionization of residual water in the vacuum chamber was contained in the detector image. Since this background was continuously recorded, it could be subtracted accurately, resulting in the images in Fig. 4. The sharp dot in the center of Fig. 4(a) corresponds to water that is present in the molecular beam, which was, hence, not removed by the background subtraction.

Without the YAG pulse, the F^+ hits are distributed isotropically, see Fig. 4(a), reflecting the random orientation of the F-C axis and the fact that the photo-ionization probability at this photon energy is almost independent of the molecular orientation with respect to the polarization direction of the X-rays, making the FEL an almost ideal probe for the molecular alignment. If linearly

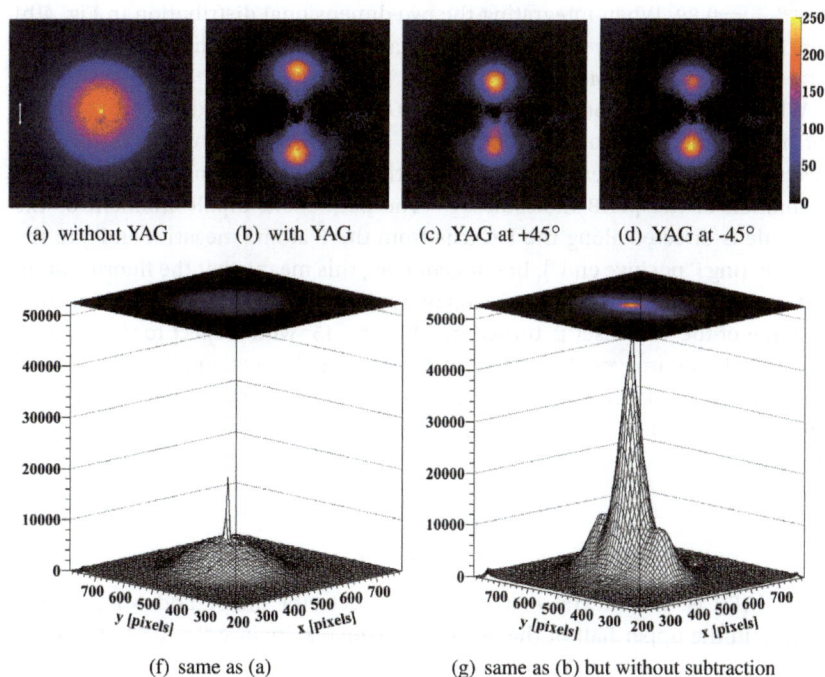

(a) without YAG   (b) with YAG   (c) YAG at +45°   (d) YAG at -45°

(f) same as (a)   (g) same as (b) but without subtraction

Fig. 4 F^+ ion images recorded at the LCLS for ionization of pFAB molecules with linearly polarized X-rays at 723 eV photon energy; (a) without the YAG laser pulse, (b) with the YAG laser pulse linearly polarized parallel to the FEL polarization, and (c) and (d) with the YAG polarization rotated out of the detector plane by +45° and −45°, respectively. The polarization direction of the X-rays is indicated by the arrow in panel (a). The images were obtained by summing up the single-shot CCD camera images after using a peak-finding algorithm. (f) and (g) show the same ion images as (a) and (b) but without subtraction of the low-energy F^+ ions (see text).

polarized YAG laser pulses are present, the F^+ ions are emitted preferentially along the polarization of the YAG pulse, as seen in Fig. 4(b), indicating a strong angular confinement of the F-C axis of the pFAB molecules at the time of the ionization by the FEL pulse.

An additional contribution of isotropically distributed F^+ ions with lower kinetic energies can be seen in Fig. 4(g), which we assume to originate from unaligned molecular clusters that were present in the molecular beam. This contribution was fitted by a two-dimensional Lorentz distribution and subtracted from the ion images recorded with the YAG pulses in order to accurately determine the degree of molecular alignment. Only the resulting distribution of F^+ ions from aligned pFAB molecules is shown in Fig. 4(b) to 4(d). The achieved degree of molecular alignment can be quantified by the ensemble-averaged expectation value of $\cos^2\theta_{2D}$, where θ_{2D} is the angle between the projection of the F^+ ion momentum vector on the detector plane and the polarization axis of the YAG laser pulse. It can be calculated from the integrated ion detector image as

$$\langle \cos^2\theta_{2D} \rangle = \frac{\sum_{i,j} I(R_i, \theta_{2D,j}) \cos^2\theta_{2D,j}}{\sum_{i,j} I(R_i, \theta_{2D,j})} \tag{1}$$

where I is the number of counts at a certain radius R_i, measured from the center of the distribution, and at a certain angle $\theta_{2D,j}$. For Fig. 4(b), the resulting value is $\langle \cos^2\theta_{2D} \rangle = 0.89$. When integrating the two-dimensional distribution in Fig. 4(b) over R and fitting the resulting ion angular distribution with a Gaussian, this corresponds to a FWHM of $47°$.

When the polarization direction of the YAG pulses is rotated such that it does not lie perpendicular to the spectrometer axis, the extraction field of the VMI spectrometer is no longer perpendicular to the YAG polarization and thus induces orientation of the pFAB molecules.[56,55] The permanent dipole moment of the molecule is directed along the F-C axis from the F atom ("negative end") to the benzene ring ("positive end"). In our geometry, this means that the fluorine atom preferentially points away from the ion detector. Therefore, when the polarization direction of the YAG laser is turned by $+45°$ or $-45°$ with respect to the detector plane, the F^+ ion images show an asymmetry, as can be seen in Fig. 4(c) and 4(d). To the best of our knowledge, this is the first realization of mixed-field molecular orientation at an FEL. The degree of molecular orientation can be quantified by the ratio

$$\Delta N = \frac{N\left(F_{up}^+\right)}{N(F^+)} \tag{2}$$

where $N(F^+)$ is the integral of the complete detector image and $N(F_{up}^+)$ is the integral in the upper half of the detector.[56] This results in $\Delta N = 0.61$ and $\Delta N = 0.39$ for Fig. 4(c) and 4(d) respectively.

3.3 Photoelectron angular distributions of aligned and oriented molecules

Simultaneously to the ion imaging, electrons are detected on the other side of the velocity map imaging spectrometer, such that the photoelectron angular distributions can be determined. Fig. 5(a) shows the integrated electron detector image obtained by using a peak-finding algorithm on the single-shot CCD camera images for randomly oriented pFAB molecules ionized by LCLS pulses at a photon energy of 742 eV, resulting in F(1s) photoelectrons of 51 eV kinetic energy. The outer rim of the F(1s) photoline is marked by the white circles. It shows the pronounced angular anisotropy expected for single-photon ionization of an s-orbital. In addition to the F(1s) photoelectrons, a strong electron signal is observed in the center of the image, corresponding to electrons with lower kinetic energy. These electrons are most likely created by multi-electron processes such as Auger cascades, shake-up or shake-off, and inelastic scattering of photoelectrons or Auger electrons inside the molecule. High-energy electrons created from C(1s) and valence ionization as well as fluorine and carbon KLL-Auger electrons have kinetic energies of >240 eV, and are thus collected only in a small solid angle for the chosen spectrometer voltages and appear as a small, almost flat background.

The plots in the bottom row of Fig. 5 show the inverted electron images after applying the pBasex algorithm.[64] The algorithm fits the electron angular distribution by an expansion in Legendre polynomials, which is a valid description of the angular distribution for the case of a cylindrically symmetric system such as one-dimensionally aligned molecules with the axis of alignment parallel to the detector plane. It is then possible to retrieve the full three-dimensional distribution from the experimentally recorded two-dimensional projections. The resulting images in the bottom row show a cut through the three-dimensional electron distribution in the detector plane.

When comparing the electron images recorded with and without the YAG laser pulses in Fig. 5(a) and 5(b) or the inverted images in Fig. 5(e) and 5(f), only small

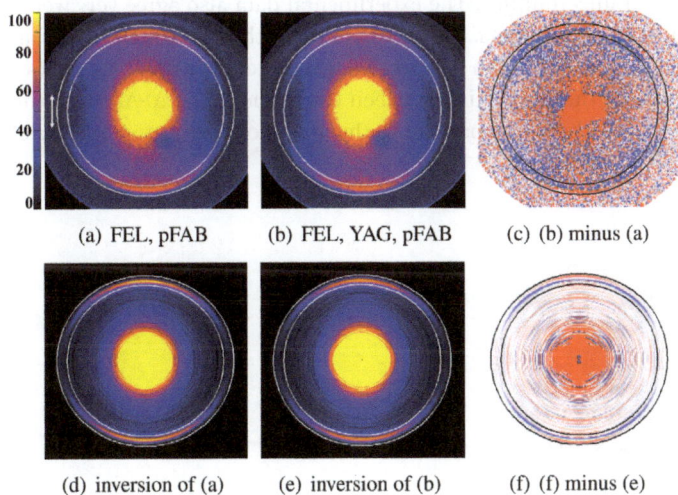

Fig. 5 Electron images from the ionization of randomly oriented and one-dimensionally aligned pFAB molecules by linearly polarized X-rays with 742 eV photon energy. The polarization directions of the FEL and YAG pulses are parallel and are indicated by the arrow. The top row shows the 2-D momentum images, the bottom row the inverted images obtained by applying the pBasex code.[64] The top and bottom right panels show the difference between the images recorded with and without YAG pulses. In the difference plots, red corresponds to positive values, blue to negative values.

differences can be seen in the angular distribution of the F(1s) photoelectrons. This can be explained by contributions from the unaligned molecular clusters to the electron signal, as well as by the averaging over different alignments of the molecular axis, which is confined to the YAG laser polarization axis only within a Gaussian of 47° FWHM. Despite the rather high degree of alignment of $\langle \cos^2\theta_{2D} \rangle$ = 0.89, this averaging smears out possible interference structures, and the photoelectron angular distribution therefore looks very similar to the one for randomly oriented molecules.

Plotting the difference between the images recorded with and without the alignment laser visually enhances the effect of the molecular alignment. An increase of the photoelectron intensity along the polarization direction of the YAG pulses and a decrease at 45° to it is clearly visible in Fig. 5(d) and 5(g). This corresponds to a narrowing of the photoelectron angular distribution for aligned molecules as compared to randomly oriented molecules. We note that there is a radial dependence of this effect even *within* the region of the F(1s) photoline. We tentatively attribute this to the creation of sidebands of the main photoline due to "above-threshold" absorbtion of YAG photons by the photoelectrons,[65] as described in more details in section 3.4. Moreover, an increase of intensity in the center of the image is found when the YAG pulses are present, which we interpret as additional low-energy electrons created by the interaction of the YAG laser pulse with excited molecular fragments, as also explained in section 3.4.

A more quantitative analysis is possible when radially integrating the difference images over the region of interest containing the F(1s) photoline, as defined by the circles in Fig. 5(d) and 5(g). The resulting photoelectron angular distribution differences (ΔPADs)[17] are shown in Fig. 6(a) as polar plots. The ΔPADs obtained from both, the raw projection and the inverted image, agree well within the statistical uncertainties. The experimental data also agree very well with the results of DFT calculations. Further details on the DFT calculations and additional data for other photon energies are presented in a previous publication.[17]

Establishing the connection between the shape of the ΔPADs and the molecular structure without comparison to theory is not straightforward for electrons

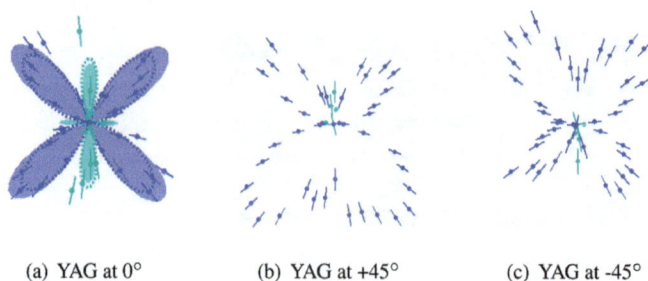

(a) YAG at 0°   (b) YAG at +45°   (c) YAG at -45°

Fig. 6   Fluorine (1s) photoelectron angular distribution differences (ΔPADs), shown as polar plots, for ionization of aligned (a) and oriented (b,c) pFAB molecules recorded for a photoelectron kinetic energy of 51 eV. Positive differences are plotted in cyan, negative differences in blue. The data points are obtained by radial integration in the region of interest in Fig. 5. The shaded areas in (a) are obtained from the inverted data in Fig. 5. Also shown in (a) as a dotted line is the calculated difference obtained from density functional theory.[17]

with kinetic energies of only a few tens of eV, since a direct reconstruction of the molecular geometry in a holographic sense[16] is, in general, not possible. However, the link of the ΔPAD to the molecular geometry becomes clearer when the molecules are oriented in space instead of only being aligned. The resulting ΔPADs for opposite molecular orientations are shown in Fig. 6(b) and 6(c). The distribution is clearly mirrored when the fluorine atom points in opposite directions, as seen in the corresponding ion images in Fig. 4(c) and 4(d). This clearly demonstrates the sensitivity of the photoelectron angular distribution to the molecular frame. An inversion of the VMI image for the case of oriented molecules can not be performed as the cylindrical symmetry is broken when the molecular axis is no longer parallel to the detector surface, thus only raw data are shown in this case.

3.4 Effects of the alignment laser

For the above discussion of the photoelectron angular distributions, it has been implicitly assumed that the alignment laser has no other effect besides fixing the molecular axes in space. Although it has been verified experimentally that the YAG pulse alone does not ionize the molecules, one has to keep in mind that in adiabatic alignment, the laser pulse is present *during* and *after* the X-ray pulse, which means that the ionization as well as all secondary processes happen in the presence of a strong laser field with a field strength on the order of 10^{11} W cm^{-2}. In this section, we will discuss some experimental evidences for resulting two-color effects.

The influence of the alignment laser on the fragmentation of pFAB molecules after inner-shell ionization was investigated by recording ion time-of-flight spectra at a photon energy of 727 eV for different YAG pulse intensities. When comparing these spectra shown in Fig. 7, it is obvious that the YAG pulses indeed influence the molecular fragmentation. Most notably, the largest ionic fragments, including the broad parent ion peak, are strongly suppressed or disappear completely when the YAG pulses are present, while the yield of smaller fragments increases.

A possible explanation for this observation could be that the heavy fragments are produced in excited electronic states. Such excited fragments may occur due to shake-up processes during the photoionization or as intermediates during the following Auger decay, as suggested previously when interpreting HHG-pump infrared-probe experiments on small molecules[66,67] and FEL-pump optical-probe experiments on xenon atoms.[68] Either the photon energy or the intensity of the YAG pulse may thus be sufficient to dissociate or ionize these excited states with a single or a few photons, thereby producing smaller fragments. This is supported by the fact that some smaller fragments, namely C_3H^+, CF^+, C_2^+, and especially C^+, increase in yield when the YAG pulse is present. We note that the ions with the largest masses, most notably the $C_8H_5F^+$ parent ion peak, are produced mainly by X-ray ionization of pFAB clusters, and we cannot conclude from the present data if the post-dissociation or post-ionization by the YAG pulses affects these cluster fragments more strongly than the fragments stemming from individual molecules.

Since the post-ionization of excited fragments should also result in the creation of additional electrons, we now investigate the difference between electron

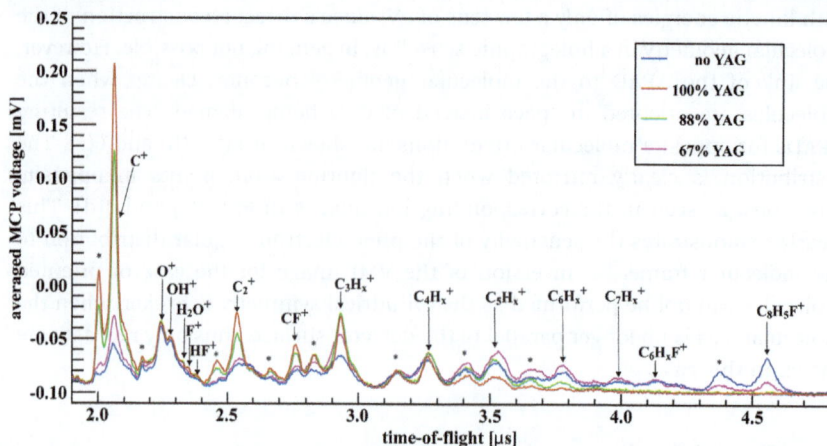

Fig. 7   Ion time-of-flight spectra of pFAB after ionization by 727 eV X-rays from the LCLS in the presence of YAG laser pulses with different intensities. The polarization of the YAG pulses was parallel to the X-ray polarization direction and the full YAG intensity was about 5×10^{11} W cm^{-2}. Contrary to the ion TOF spectra shown in Fig. 2, these spectra were measured in VMI focussing conditions, which results in a decreased time-of-flight resolution. Moreover, for the extraction voltages chosen here, secondary electrons created on the mesh that terminated the ion drift region resulted in additional peaks in the spectrum which are marked by asterisks.

detector images recorded with and without the alignment laser, shown in Fig. 8(a), zoomed in to the central part of the detector. Clearly, two additional contributions of electrons with low energies emerge when the YAG pulse is present. These can also be clearly identified in the electron spectrum shown in Fig. 8(b). The two features are found to have maxima at electron energies of approximately 0.15 and 1.3 eV, as calibrated with a measurement of the above-threshold-ionization in argon performed with the same spectrometer voltages. We note that the difference in kinetic energy between those two lines corresponds, within the uncertainties of our energy calibration, to the YAG photon

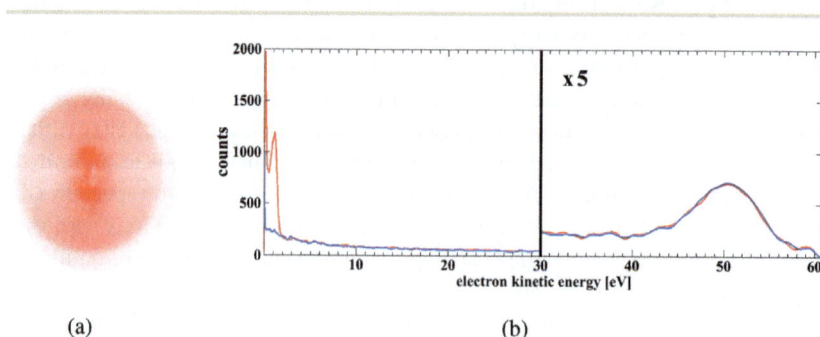

(a)   (b)

Fig. 8   (a) Zoom-in on the central part of the electron difference image shown in Fig. 5. (b) Electron energy spectrum recorded with (red) and without (blue) the YAG alignment laser, obtained from inversion of the detector images with pBasex. The spectrum on the right of the vertical bar is multiplied by a factor of 5.

energy of 1.17 eV, which suggests that the two channels may result from n- and (n + 1)-photon ionization of electronically excited molecules, molecular clusters, or fragments by the YAG pulse, although the exact origin is unclear to us at this point. In particular, it is surprising that two clear lines appear in the electron spectrum rather than a broad feature which one might expected if a series of close-lying Rydberg states was ionized.

Turning to the F(1s)-photoelectron line at 51 eV kinetic energy in Fig. 8(b), we notice that it is rather broad. This can be understood keeping in mind that the FEL pulses at the LCLS are created from self-amplified spontaneous emission (SASE) and therefore have an intrinsic bandwidth of 0.2 - 1.0%.[9] This corresponds to a bandwidth of up to 7.4 eV at an X-ray energy of 742 eV, which cannot be reduced even when sorting on the shot-to-shot photon energy information.¶

Focusing on the photoline in more detail, we can investigate two-color effects on the inner-shell photoelectrons. While the ponderomotive broadening of the photoline due to the field of the YAG pulses is negligible for the given YAG pulse intensity, another possible direct influence of the YAG laser pulses on the photoelectrons is the formation of sidebands.[65,69] When the X-ray and alignment laser pulses are present at the same time, the photoelectron can absorb one or more YAG photons in addition to the X-ray photon in a process referred to as two-color above threshold ionization. Each YAG photon can increase or decrease the nominal electron kinetic energy by 1.17 eV, resulting in a splitting of the photo-line in multiple sub-lines, which is strongest for electron emission parallel to the YAG polarization direction. Given the bandwidth of the FEL pulses, the individual sidebands cannot be resolved in this photoelectron spectrum. Nevertheless, a slight broadening of the photoline recorded in the presence of the YAG pulse is observed when the energy spectrum is analyzed within 10° around the laser polarization direction, see Fig. 9, which may be caused by the formation of sidebands. This broadening and especially its angular dependence can also be seen more clearly in the detector difference image in Fig. 5(d). However, for the analysis of the effects of molecular alignment on the photoelectron angular distributions described in section 3.3, we have assumed that the creation of sidebands does not significantly affect the photoelectron angular distribution as long as the photoelectron intensity is integrated over all sidebands.

3.5 Effects of the pump laser

Although a femtosecond TiSa laser was part of the experimental setup of the pFAB experiment and was used to optimize the molecular alignment, we did not perform a pump–probe experiment for lack of time, thus only static photoelectron angular distributions were investigated.

In the earlier experiment on 1,4-dibromobenzene (DBB) molecules, a TiSa pulse was used to dissociate the molecules before they were ionized by the FEL pulse. The photoelectron angular distributions recorded in that experiment are described elsewhere.[18] Here, we concentrate on the influence of the three different light pulses on the molecular fragmentation as seen in the ion time-of-flight

---

¶ Depending on the operation mode of the linear accelerator, there may also be systematic shifts in photon energy between different 30 Hz sub-sets of the full 120 Hz repetition rate, as we noticed in some of our data recorded in 2011.

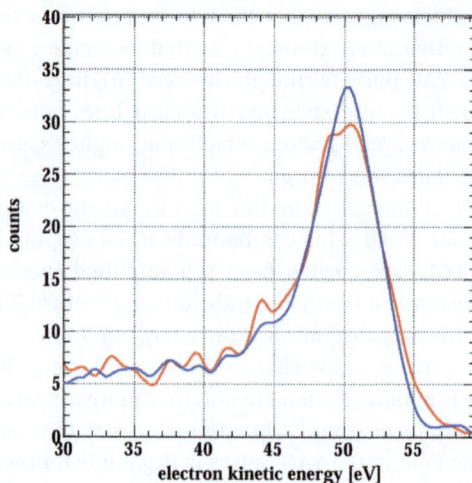

Fig. 9 Photoelectron energy spectra recorded with (red) and without (blue) YAG alignment laser pulse in a cone with an opening angle of 10° around the laser polarization direction.

spectra shown in Fig. 10, which were recorded simultaneously to the electron images reported in.[18] Note that during the DBB experiment, a plate with a 0.5-mm wide slit perpendicular to the FEL beam propagation direction was placed inside of the spectrometer in order to only accept ions that were created in the center of the spectrometer. This limits significantly the angular acceptance for energetic fragment ions, and these spectra therefore only allow a qualitative investigation of

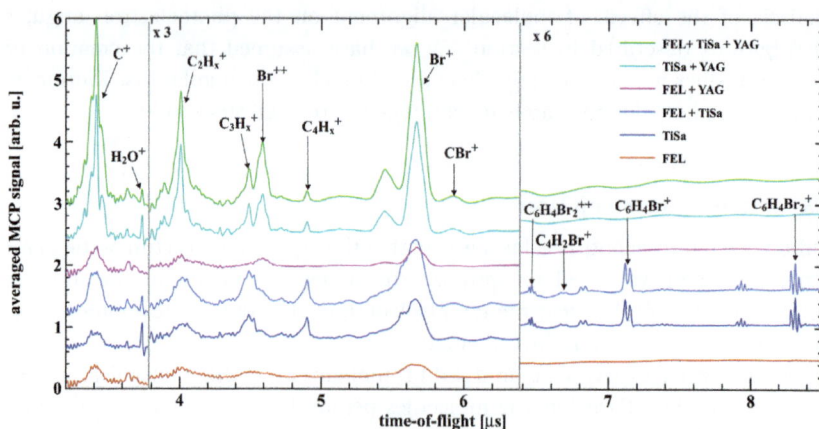

Fig. 10 Ion time-of-flight spectra of DBB measured at the LCLS for a photon energy of 1570 eV and different combinations of FEL, YAG, and TiSa pulses. For the cases with FEL and TiSa pulses present, the TiSa pulse arrives 0.5 ps after the FEL pulse. The YAG pulse alone is non-ionizing. The polarization directions of FEL, YAG, and TiSa pulses are parallel to the detector plane. The traces beyond 3.8 μs are scaled up by a factor of 3, the traces beyond 6.4 μs are scaled up by a factor of 6. As in Fig. 7, some small, additional peaks to the left of each main peak are due to secondary electrons created on the drift tube mesh.

the fragmentation. Furthermore, the first 3 μs of the spectrum are heavily disturbed by high-frequency pickup from the high-voltage switching on the electron detector on the opposite side of the spectrometer, thus the spectra are only shown for mass-to-charge ratios beyond C^+.

As for the case of pFAB discussed above, inner-shell ionization with an X-ray photon alone (red trace), here at a photon energy of 1570 eV, *i.e.* roughly 20 eV above the $Br(2p_{3/2})$ threshold but still below the $Br(2p_{1/2})$ threshold, creates various charged fragments, mostly Br^+ as well as $C_3H_x^+$, $C_2H_x^+$, and C^+ ions. A very small amount of parent ions is also created, either due to valence ionization or due to fluorescent decay of the core-hole. The parent ion peak, which is almost invisible in the FEL spectrum, is very sharp though, so no indications for the formation of clusters in the supersonic expansion are observed in this data.

The TiSa pulse alone (dark blue trace) creates singly charged parent ions (with a triple structure due to the bromine isotopes) along with a variety of other singly charged fragments. A small amount of doubly charged parent ions occurs as well, but most of the doubly charged molecules decay further in smaller fragments, most prominently Br^+. When a TiSa pulse interacts with the molecules after the FEL pulse ionized them (lighter blue trace), only small changes can be seen in the ion TOF spectrum as compared to the spectrum recorded with only the TiSa pulse present. This is understandable since the focus of the TiSa beam was chosen larger than the focus of the FEL beam to ensure that all molecules probed by the X-rays were also in the focus of the pump laser. Furthermore, the cross section for ionization with the TiSa at this intensity is higher than the cross section for ionization with the X-rays. Therefore, significantly more molecules are ionized by the TiSa laser pulse alone and the spectrum is thus dominated by these ions.

When both X-ray and YAG pulses are present (purple trace), the spectrum does not change significantly from the spectrum observed for X-ray pulses alone, although a small increase in the yield of certain ions can be observed. This is very different from what was found in the pFAB data in the previous section. We note, however, that a significant amount of clusters was present in the pFAB experiment, which did not seem to be the case for DBB.

When the strong-field ionization by the TiSa is combined with the pulses from the YAG laser (cyan trace), the changes in the ion TOF spectrum are more dramatic. All fragments heavier then Br^+ disappear, while almost all other peaks are strongly enhanced, indicating that the combination of YAG and TiSa pulses ionizes more strongly than the TiSa pulse alone. We tentatively explain this as the effect of dissociation, single- or multi-photon ionization of excited molecular fragments, which are created by the TiSa pulse, by the YAG pulse. When the X-ray pulse is added to the TiSa and YAG pulses (green trace), the spectrum is again dominated by the fragmentation induced by TiSa and YAG pulses because of the larger focus of TiSa and YAG beams as compared to the X-ray beam and higher cross sections for ionization by the TiSa pulses.

Summarizing our findings for the DBB molecules and the discussion of the effects of the YAG pulse in the pFAB data in the previous section, we can conclude that the field of the YAG laser pulse apparently has a strong influence on the ionization and fragmentation dynamics. At this point, we have no direct evidence that this changes the photoelectron angular distributions, but it certainly gives reason to suspect that the molecular dynamics initiated by a femtosecond pump pulse may be influenced by the presence of the strong field of the YAG pulse. A possibility to circumvent this effect could be to use either impulsive, "field-free" alignment or electron-ion coincidence techniques to align or orient the molecules in space, but as we briefly discuss in the following section, these techniques also have practical limitations.

Concerning the "pump" process, we note that Coulomb explosion by a strong 800-nm TiSa pulse was used here mostly as a proof-of-principle. In order to selectively trigger photochemical reactions, a single-photon transition to a resonant excitation, ideally by a non-ionizing laser pulse, would, in many cases, be more appropriate.

4   Conclusions and outlook

In this paper, along with our previous publications on this subject,[16–18] we have reported the current status of our efforts to perform femtosecond time-resolved photoelectron diffraction experiments on gas-phase molecules in a pump–probe setup combining optical lasers and an X-ray Free-Electron Laser. We have presented results of two photoelectron and ion imaging experiments on laser-aligned 1-ethynyl-4-fluorobenzene (C_8H_5F) and 1,4-dibromobenzene ($C_6H_4Br_2$) molecules conducted at the LCLS and compared some of the results with photoelectron-photoion coincidence data recorded at the PETRA synchrotron radiation facility. We have also discussed the contribution of molecular clusters to our experimental data on 1-ethynyl-4-fluorobenzene as well as the influence of the nanosecond alignment laser pulse and the femtosecond pump laser pulse on the photoelectrons and on the molecular fragmentation.

Our results demonstrate that by combining a strong nanosecond YAG laser pulse with the FEL pulse, it is possible to perform photoionization experiments on adiabatically laser-aligned and mixed-field oriented polyatomic molecules. The corresponding photoelectron angular distributions show a clear dependence on the photoelectron kinetic energy,[17] on the alignment direction of the molecular axis,[18] and on the molecular orientation. While our interpretation was, so far, mostly based on comparison to density function theory calculations[17,18] our long-term goal is to link the observed patterns directly to the molecular structure by applying the concepts of photoelectron diffraction and holography.[16]

Time-resolved photoelectron diffraction and holography has the potential to image the geometric structure of gas-phase molecules with few-femtosecond temporal and sub-Ångström spatial resolution, and offers a complementary approach to time-resolved X-ray and electron diffraction. Using electrons as opposed to X-rays for diffraction has the advantage of much higher elastic scattering cross sections, which is particularly important for targets containing lighter atoms such as carbon, nitrogen, or oxygen, which do not scatter X-rays efficiently. Using photoelectrons instead of an electron beam has the additional benefit of avoiding the problem of velocity mismatch in laser-pump electron-probe experiments on gas-phase targets.

A disadvantage of photoelectron diffraction, however, is the more complicated description of the initial photoelectron wave, which, contrary to the case of X-ray and electron diffraction, does not fulfill the plane-wave approximation. In the case of photoionization of an inner-shell s-orbital, the initial, unscattered photoelectron wave can be described, to a good approximation, by a pure p-wave, while for photoionization of orbitals with an angular momentum quantum number $l \neq 0$, the interference of the $l - 1$ and $l + 1$ partial waves already complicates the description of the unscattered photoelectron wave. Furthermore, the interpretation of the final photoelectron angular distributions in terms of scattering is particularly challenging for low-energy electrons, where the molecular potential can no longer be approximated by the sum of atomic potentials and where multiple scattering can be a significant contribution. Nevertheless, we are convinced that detailed insight into changes of the molecular structure during photochemical reactions can be gained from studying photoelectron angular distributions even in these more difficult cases. Our goal is therefore to establish the photoelectron diffraction concept for gas-phase molecules while developing the experimental tools to perform these experiments in a femtosecond pump–probe setup.

In the experiments we have reported so far, the degree of molecular alignment that we achieved was sufficiently high to observe alignment dependent effects when considering the difference between the photoelectron angular distributions of aligned and unaligned molecules. In order to obtain more direct information on the molecular structure, *e.g.* by holographic reconstruction,[16] a considerably higher degree of alignment is necessary. Such high degrees of alignment up to $\langle \cos^2 \theta_{2D} \rangle = 0.97$ have been achieved for iodobenzene molecules using adiabatic laser alignment in a laboratory setup.[56] Since the presence of the strong laser field used for adiabatic alignment may cause unwanted effects in pump–probe experiments, as discussed in section 3.4 and 3.5, an option to circumvent these effects could be to use field-free alignment techniques. However, these have, so far, not been able to obtain as high degrees of alignment as adiabatic techniques.

For suitable classes of molecules, an alternative way to determine the molecular alignment and orientation very precisely is by means of electron-ion coincidence techniques. This may become a competetive option once higher repetition-rate FEL sources such as the European XFEL are available. However, as our discussion of the fragmentation of pFAB molecules after inner-shell ionization has shown, this may also be challenging for polyatomic molecules, where complicated fragmentation channels and the occurrence of a large number of possible fragments can make it difficult or impossible to find a fragmentation channel that is suitable to define one or several molecular axes.

Finally, the experiments reported here still lack the necessary temporal resolution to resolve dynamics on the order of 100 fs or below, but the use of X-ray optical cross-correlation techniques[57–59] was shown to improve this dramatically. With the lessons learned from our previous experiments, we therefore believe that there is a clear avenue towards a time-resolved photoelectron diffraction experiment that would be able to image the molecular structure during an isomerization reaction or close to transition states in a photochemical reaction by measuring photoelectron angular distributions as suggested, *e.g.*, in.[17]

Acknowledgements

Part of this research was carried out at the Linac Coherent Light Source (LCLS) at the SLAC National Accelerator Laboratory. LCLS is an Office of Science User Facility operated for the U.S. Department of Energy Office of Science by Stanford University. Additional measurements were performed at the PETRA P04 beamline at DESY. We acknowledge the Max Planck Society for funding the development and operation of the CAMP instrument within the ASG at CFEL. D.R. acknowledges support from the Helmholtz Gemeinschaft through the Young Investigator Program. L.C., S.D., and H.S. acknowledge support from the Carlsberg Foundation. J.K. acknowledges support from the excellence cluster The Hamburg Centre for Ultrafast Imaging - Structure, Dynamics and Control of Matter at the Atomic Scale of the Deutsche Forschungsgemeinschaft. N.L.M.M. acknowledges financial support by the Joachim Herz Stiftung. J.K. and T.K. acknowledge financial support from the Helmholtz Virtual Institute "Dynamic Pathways in Multidimensional Landscapes". A.Ro. and M.V acknowledge the research program of the "Stichting voor Fundamenteel Onderzoek der Materie", which is financially supported by the "Nederlandse organisatie voor Wetenschappelijk Onderzoek". P.J. acknowledges support from the Swedish Research Council and the Swedish Foundation for Strategic Research. A.Ru. acknowledges support from the Office of Basic Energy Sciences, US Department of Energy. S.Te. and A.S. acknowledge support through SFB 755 Nanoscale photonic imaging. We are grateful to the entire LCLS and PETRA staff for their support and hospitality during the beamtimes.

References

1 A. H. Zewail, *Angew. Chem., Int. Ed.*, 2000, **39**, 2586–2631.

2 M. Chergui and A. H. Zewail, *ChemPhysChem*, 2009, **10**, 28–43.

3 M. Chergui, *Acta Crystallogr., Sect. A: Found. Crystallogr.*, 2010, **66**, 229–239.

4 W. T. Pollard and R. A. Mathies, *Annu. Rev. Phys. Chem.*, 1992, **43**, 497–523.

5 G. Sciaini and R. J. D. Miller, *Rep. Prog. Phys.*, 2011, **74**, 096101.

6 H. Ihee, V. A. Lobastov, U. M. Gomez, B. M. Goodson, R. Srinivasan, C.-Y. Ruan and A. H. Zewail, *Science*, 2001, **291**, 458–462.

7 C. Hensley, J. Yang and M. Centurion, *Phys. Rev. Lett.*, 2012, **109**, 133202.

8 W. Ackermann, G. Asova, V. Ayvazyan, A. Azima, N. Baboi, J. Bähr, V. Balandin, B. Beutner, A. Brandt, A. Bolzmann, R. Brinkmann, O. I. Brovko, M. Castellano, P. Castro, L. Catani, E. Chiadroni, S. Choroba, A. Cianchi, J. T. Costello, D. Cubaynes, J. Dardis, W. Decking, H. Delsim-Hashemi, A. Delserieys, G. Di Pirro, M. Dohlus, S. Düsterer, A. Eckhardt, H. T. Edwards, B. Faatz, J. Feldhaus, K. Flöttmann, J. Frisch, L. Fröhlich, T. Garvey, U. Gensch, C. Gerth, M. Görler, N. Golubeva, H.-J. Grabosch, M. Grecki, O. Grimm, K. Hacker, U. Hahn, J. H. Han, K. Honkavaara, T. Hott, M. Hüning, Y. Ivanisenko, E. Jaeschke, W. Jalmuzna, T. Jezynski, R. Kammering, V. Katalev, K. Kavanagh, E. T. Kennedy, S. Khodyachykh, K. Klose, V. Kocharyan, M. Körfer, M. Kollewe, W. Koprek, S. Korepanov, D. Kostin, M. Krassilnikov, G. Kube, M. Kuhlmann, C. L. S. Lewis, L. Lilje, T. Limberg, D. Lipka, F. Löhl, H. Luna, M. Luong, M. Martins, M. Meyer, P. Michelato, V. Miltchev and W. D. M, *Nat. Photonics*, 2007, **1**, 336–342.

9 P. Emma, R. Akre, J. Arthur, R. Bionta, C. Bostedt, J. Bozek, A. Brachmann, P. Bucksbaum, R. Coffee, F.-J. Decker, Y. Ding, D. Dowell, S. Edstrom, A. Fisher, J. Frisch, S. Gilevich, J. Hastings, G. Hays, P. Hering, Z. Huang, R. Iverson, H. Loos, M. Messerschmidt, A. Miahnahri, S. Moeller, H.-D. Nuhn, G. Pile, D. Ratner, J. Rzepiela, D. Schultz, T. Smith, P. Stefan, H. Tompkins, J. Turner, J. Welch, W. White, J. Wu, G. Yocky and J. Galayda, *Nat. Photonics*, 2010, **4**, 641–647.

10 T. Ishikawa, H. Aoyagi, T. Asaka, Y. Asano, N. Azumi, T. Bizen, H. Ego, K. Fukami, T. Fukui, Y. Furukawa, S. Goto, H. Hanaki, T. Hara, T. Hasegawa, T. Hatsui, A. Higashiya, T. Hirono, N. Hosoda, M. Ishii, T. Inagaki, Y. Inubushi, T. Itoga, Y. Joti, M. Kago, T. Kameshima, H. Kimura, Y. Kirihara, A. Kiyomichi, T. Kobayashi, C. Kondo, T. Kudo, H. Maesaka, X. M. Maréchal, T. Masuda, S. Matsubara, T. Matsumoto, T. Matsushita, S. Matsui, M. Nagasono, N. Nariyama, H. Ohashi, T. Ohata, T. Ohshima, S. Ono, Y. Otake, C. Saji, T. Sakurai, T. Sato, K. Sawada, T. Seike, K. Shirasawa, T. Sugimoto, S. Suzuki, S. Takahashi, H. Takebe, K. Takeshita, K. Tamasaku, H. Tanaka, R. Tanaka, T. Tanaka, T. Togashi, K. Togawa, A. Tokuhisa, H. Tomizawa, K. Tono, S. Wu, M. Yabashi, M. Yamaga, A. Yamashita, K. Yanagida, C. Zhang, T. Shintake, H. Kitamura and N. Kumagai, *Nat. Photonics*, 2012, **6**, 540–544.

11 E. Allaria, D. Castronovo, P. Cinquegrana, P. Craievich, M. Dal Forno, M. B. Danailov, G. D'Auria, A. Demidovich, G. De Ninno, S. Di Mitri, B. Diviacco, W. M. Fawley, M. Ferianis, E. Ferrari, L. Froehlich, G. Gaio, D. Gauthier, L. Giannessi, R. Ivanov, B. Mahieu, N. Mahne, I. Nikolov, F. Parmigiani, G. Penco, L. Raimondi, C. Scafuri, C. Serpico, P. Sigalotti, S. Spampinati, C. Spezzani, M. Svandrlik, C. Svetina, M. Trovo, M. Veronese, D. Zangrando and M. Zangrando, *Nat. Photonics*, 2013, **7**, 913–918.

12 T. Popmintchev, M.-C. Chen, P. Arpin, M. M. Murnane and H. C. Kapteyn, *Nat. Photonics*, 2010, **4**, 822–832.

13 P. B. Corkum and F. Krausz, *Nat. Phys.*, 2007, **3**, 381–387.

14 P. Agostini and L. F. DiMauro, *Rep. Prog. Phys.*, 2004, **67**, 813–855.

15 H. Delsim-Hashemi, K. Floettmann, M. Seebach and S. Bayesteh, *Proceedings of IBIC2013*, 2013, p. 868.

16 F. Krasniqi, B. Najjari, L. Strüder, D. Rolles, A. Voitkiv and J. Ullrich, *Phys. Rev. A: At., Mol., Opt. Phys.*, 2010, **81**, 033411.

17 R. Boll, D. Anielski, C. Bostedt, J. Bozek, L. Christensen, R. Coffee, S. De, P. Decleva, S. Epp, B. Erk, L. Foucar, F. Krasniqi, J. Küpper, A. Rouzée, B. Rudek, A. Rudenko, S. Schorb, H. Stapelfeldt, M. Stener, S. Stern, S. Techert, S. Trippel, M. Vrakking, J. Ullrich and D. Rolles, *Phys. Rev. A: At., Mol., Opt. Phys.*, 2013, **88**, 061402(R).

18 D. Rolles, R. Boll, M. Adolph, A. Aquila, C. Bostedt, J. D. Bozek, H. Chapman, R. Coffee, N. Coppola, P. Decleva, T. Delmas, S. W. Epp, B. Erk, F. Filsinger, L. Foucar, L. Gumprecht, A. Hömke, T. Gorkhover, L. Holmegaard, P. Johnsson, C. Kaiser, F. Krasniqi, K.-U. Kühnel, J. Maurer, M. Messerschmidt, R. Moshammer, W. Quevedo, I. Rajkovic, A. Rouzée, B. Rudek, I. Schlichting, C. Schmidt, S. Schorb, C.-D. Schroeter, J. Schulz, H. Stapelfeldt, M. Stener, S. Stern, S. Techert, J. Thøgersen, M. Vrakking, A. Rudenko, J. Küpper and J. Ullrich, *J. Phys. B*, 2014, **47**, 124035.

19 A. V. Golovin, N. A. Cherepkov and V. V. Kuznetsov, *Z. Phys. D: At., Mol. Clusters*, 1992, **24**, 371–375.

20 E. Shigemasa, J. Adachi, M. Oura and A. Yagishita, *Phys. Rev. Lett.*, 1995, **74**, 359–362.

21 F. Heiser, O. Gessner, J. Viefhaus, K. Wieliczek, R. Hentges and U. Becker, *Phys. Rev. Lett.*, 1997, **79**, 2435.

22 P. Downie and I. Powis, *Phys. Rev. Lett.*, 1999, **82**, 2864.

23 A. Lafosse, M. Lebech, J. C. Brenot, P. M. Guyon, O. Jagutzki, L. Spielberger, M. Vervloet, J. C. Houver and D. Dowek, *Phys. Rev. Lett.*, 2000, **84**, 5987.

24 R. Dörner, V. Mergel, O. Jagutzki, L. Spielberger, J. Ullrich, R. Moshammer and H. Schmidt-Böcking, *Phys. Rep.*, 2000, **330**, 95–192.

25 S. Motoki, J. Adachi, K. Ito, K. Ishii, K. Soejima, A. Yagishita, S. K. Semenov and N. A. Cherepkov, *J. Phys. B: At., Mol. Opt. Phys.*, 2002, **35**, 3801.

26 O. Gessner, Y. Hikosaka, B. Zimmermann, A. Hempelmann, R. Lucchese, J. Eland, P.-M. Guyon and U. Becker, *Phys. Rev. Lett.*, 2002, **88**, 193002.

27 D. Rolles, *PhD thesis*, TU Berlin, Berlin, 2005.

28 M. S. Schöffler, J. Titze, N. Petridis, T. Jahnke, K. Cole, L. P. H. Schmidt, A. Czasch, D. Akoury, O. Jagutzki, J. B. Williams, N. A. Cherepkov, S. K. Semenov, C. W. McCurdy, T. N. Rescigno, C. L. Cocke, T. Osipov, S. Lee, M. H. Prior, A. Belkacem, A. L. Landers, H. Schmidt-Böcking, T. Weber and R. Dörner, *Science*, 2008, **320**, 920–923.

29 D. Rolles, M. Braune, S. Cvejanovic, O. Geßner, R. Hentges, S. Korica, B. Langer, T. Lischke, G. Prümper, A. Reinköster, J. Viefhaus, B. Zimmermann, V. McKoy and U. Becker, *Nature*, 2005, **437**, 711–715.

30 D. Akoury, K. Kreidi, T. Jahnke, T. Weber, A. Staudte, M. Schöffler, N. Neumann, J. Titze, L. P. H. Schmidt, A. Czasch, O. Jagutzki, R. A. C. Fraga, R. E. Grisenti, R. D. Muino, N. A. Cherepkov, S. K. Semenov, P. Ranitovic, C. L. Cocke, T. Osipov, H. Adaniya, J. C. Thompson, M. H. Prior, A. Belkacem, A. L. Landers, H. Schmidt-Böcking and R. Dörner, *Science*, 2007, **318**, 949–952.

31 D. P. Woodruff and A. M. Bradshaw, *Rep. Prog. Phys.*, 1994, **57**, 1029.

32 C. S. Fadley, *Surf. Interface Anal.*, 2008, **40**, 1579–1605.

33 U. Becker, O. Gessner and A. Rüdel, *J. Electron Spectrosc. Relat. Phenom.*, 2000, **108**, 189–201.

34 A. Landers, T. Weber, I. Ali, A. Cassimi, M. Hattass, O. Jagutzki, A. Nauert, T. Osipov, A. Staudte, M. Prior, H. Schmidt-Böcking, C. Cocke and R. Dörner, *Phys. Rev. Lett.*, 2001, **87**, 013002.

35 B. Zimmermann, D. Rolles, B. Langer, R. Hentges, M. Braune, S. Cvejanovic, O. Geßner, F. Heiser, S. Korica, T. Lischke, A. Reinköster, J. Viefhaus, R. Dörner, V. McKoy and U. Becker, *Nat. Phys.*, 2008, **4**, 649–655.

36 C. S. Fadley, *Synchrotron Radiation Research*, Plenum Press, New York, 1992, pp. 421–518.

37 N. Berrah, J. Bozek, J. Costello, S. Düsterer, L. Fang, J. Feldhaus, H. Fukuzawa, M. Hoener, Y. Jiang, P. Johnsson, E. Kennedy, M. Meyer, R. Moshammer, P. Radcliffe, M. Richter, A. Rouzée, A. Rudenko, A. Sorokin, K. Tiedtke, K. Ueda, J. Ullrich and M. Vrakking, *J. Mod. Opt.*, 2010, **57**, 1015–1040.

38 J. Ullrich, A. Rudenko and R. Moshammer, *Annu. Rev. Phys. Chem.*, 2012, **63**, 635–660.

39 M. Kazama, T. Fujikawa, N. Kishimoto, T. Mizuno, J.-i. Adachi and A. Yagishita, *Phys. Rev. A: At., Mol., Opt. Phys.*, 2013, **87**, 063417.

40 C. Bostedt, J. D. Bozek, P. H. Bucksbaum, R. N. Coffee, J. B. Hastings, Z. Huang, R. W. Lee, S. Schorb, J. N. Corlett, P. Denes, P. Emma, R. W. Falcone, R. W. Schoenlein, G. Doumy, E. P. Kanter, B. Kraessig, S. Southworth, L. Young, L. Fang, M. Hoener, N. Berrah, C. Roedig and L. F. DiMauro, *J. Phys. B: At., Mol. Opt. Phys.*, 2013, **46**, 164003.

41 J. Viefhaus, F. Scholz, S. Deinert, L. Glaser, M. Ilchen, J. Seltmann, P. Walter and F. Siewert, *Nucl. Instrum. Methods Phys. Res., Sect. A*, 2013, **710**, 151–154.

42 L. Strüder, S. Epp, D. Rolles, R. Hartmann, P. Holl, G. Lutz, H. Soltau, R. Eckart, C. Reich, K. Heinzinger, C. Thamm, A. Rudenko, F. Krasniqi, K.-U. Kühnel, C. Bauer, C.-D. Schröter, R. Moshammer, S. Techert, D. Miessner, M. Porro, O. Hölker, N. Meidinger, N. Kimmel, R. Andritschke, F. Schopper, G. Weidenspointner, A. Ziegler, D. Pietschner, S. Herrmann, U. Pietsch, A. Walenta, W. Leitenberger, C. Bostedt, T. Möller, D. Rupp, M. Adolph, H. Graafsma, H. Hirsemann, K. Gärtner, R. Richter, L. Foucar, R. L. Shoeman, I. Schlichting and J. Ullrich, *Nucl. Instrum. Methods Phys. Res., Sect. A*, 2010, **614**, 483–496.

43 J. Küpper, S. Stern, L. Holmegaard, F. Filsinger, A. Rouzée, D. Rolles, A. Rudenko, P. Johnsson, A. V. Martin, M. Adolph, A. Aquila, S. Bajt, A. Barty, C. Bostedt, J. D. Bozek, C. Caleman, R. Coffee, N. Coppola, T. Delmas, S. Epp, B. Erk, L. Foucar, Tais Gorkhover, L. Gumprecht, A. Hartmann, R. Hartmann, G. Hauser, P. Holl, A. Homke, N. Kimmel, F. Krasniqi, K.-U. Kuhnel, J. Maurer, M. Messerschmidt, R. Moshammer, C. Reich, B. Rudek, R. Santra, I. Schlichting, C. Schmidt, S. Schorb, J. Schulz, H. Soltau, L. Strueder, T. Jan, M. J. J. Vrakking, G. Weidenspointner, T. A. White, C. Wunderer, G. Meijer, J. Ullrich, H. Stapelfeldt and H. N. Chapman, *Phys. Rev. Lett.*, 2013, **112**, 083002.

44 S. Stern, *et al.*, *Faraday Discuss.*, 2014, **171**, DOI: 10.1039/C4FD00028E.

45 L. Foucar, A. Barty, N. Coppola, R. Hartmann, P. Holl, U. Hoppe, S. Kassemeyer, N. Kimmel, J. Küpper, M. Scholz, S. Techert, T. A. White, L. Strüder and J. Ullrich, *Comput. Phys. Commun.*, 2012, **183**, 2207–2213.

46 B. Friedrich and D. Herschbach, *J. Phys. Chem.*, 1995, **99**, 15686–15693.

47 H. Stapelfeldt and T. Seideman, *Rev. Mod. Phys.*, 2003, **75**, 543.

48 H. Stapelfeldt, *Eur. Phys. J. D - At. Mol. Opt. Phys.*, 2003, **26**, 15–19.

49 T. Seideman and E. Hamilton, *Adv. At., Mol., Opt. Phys.*, 2006, **52**, 289–329.

50 P. Johnsson, A. Rouzée, W. Siu, Y. Huismans, F. Lépine, T. Marchenko, S. Düsterer, F. Tavella, N. Stojanovic, A. Azima, *et al.*, *J. Phys. B: At., Mol. Opt. Phys.*, 2009, **42**, 134017.

51 J. Cryan, J. Glownia, J. Andreasson, A. Belkacem, N. Berrah, C. Blaga, C. Bostedt, J. Bozek, C. Buth, L. DiMauro, L. Fang, O. Gessner, M. Guehr, J. Hajdu, M. Hertlein, M. Hoener, O. Kornilov, J. Marangos, A. March, B. McFarland, H. Merdji, V. Petrović, C. Raman, D. Ray, D. Reis, F. Tarantelli, M. Trigo, J. White, W. White, L. Young, P. Bucksbaum and R. Coffee, *Phys. Rev. Lett.*, 2010, **105**, 083004.

52 J. P. Cryan, J. M. Glownia, J. Andreasson, A. Belkacem, N. Berrah, C. I. Blaga, C. Bostedt, J. Bozek, N. A. Cherepkov, L. F. DiMauro, L. Fang, O. Gessner, M. Gühr, J. Hajdu, M. P. Hertlein, M. Hoener, O. Kornilov, J. P. Marangos, A. M. March, B. K. McFarland, H. Merdji, M. Messerschmidt, V. S. Petrović, C. Raman, D. Ray, D. A. Reis, S. K. Semenov, M. Trigo, J. L. White, W. White, L. Young, P. H. Bucksbaum and R. N. Coffee, *J. Phys. B: At., Mol. Opt. Phys.*, 2012, **45**, 055601.

53 H. Sakai, C. P. Safvan, J. J. Larsen, K. M. Hilligsøe, K. Hald and H. Stapelfeldt, *J. Chem. Phys.*, 1999, **110**, 10235.

54 J. J. Larsen, K. Hald, N. Bjerre, H. Stapelfeldt and T. Seideman, *Phys. Rev. Lett.*, 2000, **85**, 2470–2473.

55 I. Nevo, L. Holmegaard, J. H. Nielsen, J. L. Hansen, H. Stapelfeldt, F. Filsinger, G. Meijer and J. Küpper, *Phys. Chem. Chem. Phys.*, 2009, **11**, 9912.

56 L. Holmegaard, J. H. Nielsen, I. Nevo, H. Stapelfeldt, F. Filsinger, J. Küpper and G. Meijer, *Phys. Rev. Lett.*, 2009, **102**, 023001.

57 M. R. Bionta, H. T. Lemke, J. P. Cryan, J. M. Glownia, C. Bostedt, M. Cammarata, J.-C. Castagna, Y. Ding, D. M. Fritz and A. R. Fry, *Opt. Express*, 2011, **19**, 21855–21865.

58 S. Schorb, T. Gorkhover, J. Cryan, J. Glownia, M. Bionta, R. Coffee, B. Erk, R. Boll, C. Schmidt, D. Rolles, A. Rudenko, A. Rouzee, M. Swiggers, S. Carron, J.-C. Castagna, J. D. Bozek, M. Messerschmidt, W. F. Schlotter and C. Bostedt, *Appl. Phys. Lett.*, 2012, **100**, 121107.

59 M. Harmand, R. Coffee, M. R. Bionta, M. Chollet, D. French, D. Zhu, D. M. Fritz, H. T. Lemke, N. Medvedev, B. Ziaja, S. Toleikis and M. Cammarata, *Nat. Photonics*, 2013, **7**, 215–218.

60 A. Rouzée, P. Johnsson, L. Rading, A. Hundertmark, W. Siu, Y. Huismans, S. Düsterer, H. Redlin, F. Tavella, N. Stojanovic, A. Al-Shemmary, F. Lépine, D. M. P. Holland, T. Schlatholter, R. Hoekstra, H. Fukuzawa, K. Ueda and M. J. J. Vrakking, *J. Phys. B: At., Mol. Opt. Phys.*, 2013, **46**, 164029.

61 B. Erk, R. Boll, S. Trippel, D. Anielski, L. Foucar, B. Rudek, S. W. Epp, R. Coffee, S. Carron, S. Schorb, K. R. Ferguson, M. Swiggers, J. D. Bozek, M. Simon, T. Marchenko, J. Küpper, I. Schlichting, J. Ullrich, C. Bostedt, D. Rolles and A. Rudenko, *Science*, 2014, **345**, 288–291.

62 D. Davis, D. Shirley and T. Thomas, *J. Am. Chem. Soc.*, 1972, **94**, 6565–6575.

63 J. Eland, *Mol. Phys.*, 1987, **61**, 725–745.

64 G. A. Garcia, L. Nahon and I. Powis, *Rev. Sci. Instrum.*, 2004, **75**, 4989.

65 M. Meyer, J. T. Costello, S. Düsterer, W. B. Li and P. Radcliffe, *J. Phys. B: At., Mol. Opt. Phys.*, 2010, **43**, 194006.

66 A. S. Sandhu, E. Gagnon, R. Santra, V. Sharma, W. Li, P. Ho, P. Ranitovic, C. L. Cocke, M. M. Murnane and H. C. Kapteyn, *Science*, 2008, **322**, 1081–1085.

67 X. Zhou, P. Ranitovic, C. W. Hogle, J. H. D. Eland, H. C. Kapteyn and M. M. Murnane, *Nat. Phys.*, 2012, **8**, 232–237.

68 M. Krikunova, T. Maltezopoulos, A. Azima, M. Schlie, U. Frühling, H. Redlin, R. Kalms, S. Cunovic, N. M. Kabachnik, M. Wieland and M. Drescher, *New J. Phys.*, 2009, **11**, 123019.

69 A. K. Kazansky, I. P. Sazhina and N. M. Kabachnik, *Phys. Rev. A: At., Mol., Opt. Phys.*, 2012, **86**, 033404.

Faraday Discussions

ROYAL SOCIETY
OF CHEMISTRY

PAPER

Toward structural femtosecond chemical dynamics: imaging chemistry in space and time

Michael P. Minitti,*[a] James M. Budarz,[ab] Adam Kirrander,[c]
Joseph Robinson,[a] Thomas J. Lane,[d] Daniel Ratner,[a] Kenichiro Saita,[e]
Thomas Northey,[c] Brian Stankus,[b] Vale Cofer-Shabica,[b]
Jerome Hastings[a] and Peter M. Weber*[b]

Received 4th March 2014, Accepted 28th April 2014

DOI: 10.1039/c4fd00030g

We aim to observe a chemical reaction in real time using gas-phase X-ray diffraction. In our initial experiment at the Linac Coherent Light Source (LCLS), we investigated the model system 1,3-cyclohexadiene (CHD) at very low vapor pressures. This reaction serves as a benchmark for numerous transformations in organic synthesis and natural product biology. Excitation of CHD by an ultraviolet optical pulse initiates an electrocyclic reaction that transforms the closed ring system into the open-chain structure of 1,3,5-hexatriene. We describe technical points of the experimental method and present first results. We also outline an approach to analyze the data involving nonlinear least-square optimization routines that match the experimental observations with predicted diffraction patterns calculated from trajectories for nonadiabatic vibronic wave packets.

Introduction

The exploration of molecular reactions in the ultrafast time regime is an evolving focal point of scientists with diverse and unique interests including physics, material science, biology and chemistry. The common goal of those investigations is to observe the motions of molecules during chemical reactions. The phrase '*molecular movie*' is frequently invoked, suggesting that one day it may be possible to prepare a movie that depicts molecular motions with ultrafast (femtosecond) time resolution and structural resolution on atomic distance scales. New X-ray free-electron laser (FEL) facilities, such as the LCLS,[1] are bringing this dream closer to reality.

[a]SLAC National Accelerator Laboratory, Stanford, CA 94025, USA. E-mail: minitti@slac.stanford.edu
[b]Dept. of Chemistry, Brown University, Providence, RI 02912, USA. E-mail: peter_weber@brown.edu
[c]School of Chemistry, University of Edinburgh, Edinburgh EH9 3JJ, UK
[d]Stanford University, Department of Chemistry, Stanford, CA 94305, USA
[e]School of Chemistry, University of Leeds, Leeds LS2 9JT, UK

Conventional ultrafast methods apply spectroscopic techniques to study chemical dynamics: spectra are taken of the molecule at well-defined times after a pump pulse initiates a reaction. The problem is that, at its core, spectroscopy measures the energies and populations of states, and an inversion of a vibrational or electronic spectrum to yield a molecular structure is difficult for molecules larger than just a few atoms. All reacting molecules are inherently very energetic, so that a multitude of vibrational states are excited on complicated potential energy surfaces, making vibrational spectra difficult, or indeed impossible, to assign. Finally, many chemically interesting molecules have large densities of states which, coupled with broadening on account of the short lifetimes of reaction intermediates, make the spectra fundamentally unresolvable (the 'statistical limit' scenario). In what might be the best current spectroscopic approach, we have recently developed photoionization *via* Rydberg states as a tool to observe structural dynamics.[2-5] In this approach, the vibrational energy content of the molecule does not adversely affect the spectrum, clean spectra can be observed even for large systems, and for most reactions, lifetime broadening is insignificant so that well-resolved spectra result.[6] Yet, while capable and informative, that technique falls short in one important way: it is not yet possible to invert the spectrum to determine the molecular structure. Consequently, the method remains a fingerprint tool rather than a structure determination tool.

Diffraction experiments are fundamentally different from spectroscopic measurements. The molecular structure can be derived from a diffraction pattern by using a back transformation,[7,8] or through a statistical analysis using a cumulant expansion.[9] Diffraction patterns are advantageous also because they measure only the atomic positions, a much smaller parameter space than, for example, the vibrational spectrum of a hot molecule. In principle, the structural dynamics observed by diffraction gives access to the molecular geometry as a function of time, including the spatial distributions of functional groups, steric hindrances, or spatial electrostatic charge distributions.

To observe the dynamics of chemical reactions, gaseous systems are ideal: the background of solvent molecules in a solution can obscure the observation of the chemical reaction and the effects of the solvent complicate the molecular dynamics. Gas phase diffraction methods have traditionally been limited to electron scattering because the scattering cross sections for electrons are many orders of magnitude larger than those for X-rays. However, with the new FELs, the tables have turned: while the scattering cross sections are smaller, the number of photons in a single X-ray pulse is many orders of magnitude larger than the number of electrons in an ultrashort electron pulse.

We have focused our investigations on 1,3-cyclohexadiene (CHD), an important prototypical system that carries great relevance to synthetic organic chemistry and natural product synthesis.[12] For example, the synthesis of vitamin D involves a ring closing reaction that is the reverse of CHD's ring opening. Another advantage to using CHD to develop the ultrafast pump-probe X-ray diffraction method is that the photochemistry of CHD is quite well known, allowing us to calibrate our measurements against a large body of existing work.

We have recently reviewed our fairly extensive knowledge of CHD's photochemistry.[11] The picture shown in Fig. 1 is based on some 30 years of time-resolved studies: the optical pulse excites the molecule to the Franck–Condon region of the 1B state, from where the wave packet rapidly slides past a conical intersection and

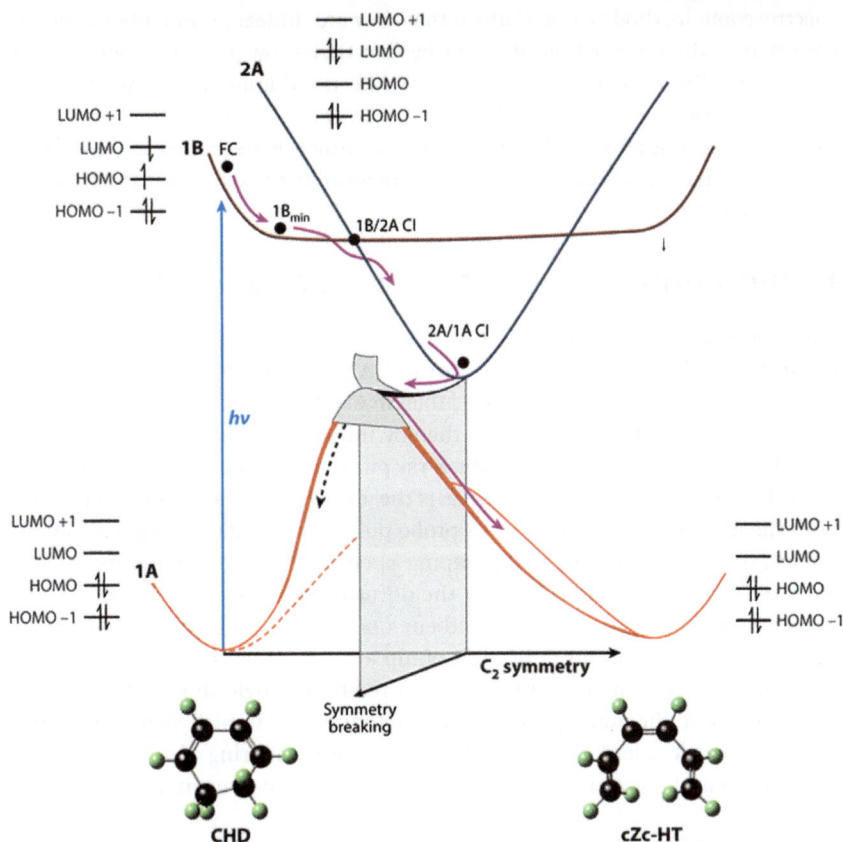

Fig. 1 Ring-opening reaction of 1,3-cyclohexadiene (CHD). Upon excitation with a near UV pump photon (blue), the CHD molecule slides down the 1B surface, crosses to the 2A surface by avoiding a conical intersection that comes from above. Crossing to the ground state along a symmetry-breaking coordinate leads the molecule to the fully opened structure of hexatriene. (After Nenov *et al.*[10] reproduced with permission from ref. 11.)

down the well of the 2A state. At that point (labeled 2A/1A CI in Fig. 1), the conrotatory stereochemistry of the reaction is already decided. A symmetry-breaking displacement through the second conical intersection brings the molecule back to the 1A ground state. In CHD, the geometry of the latter intersection causes the molecule to continue its downhill slide such that it ends up in the open structure, *i.e.* as 1,3,5-hexatriene. Cyclohexadiene is an attractive experimental system because it moves down this path in a ballistic fashion[13] and so 'structure' remains a well-defined and observable concept throughout the reaction.

The time scales of this reaction are such that the molecule crosses into the 2A surface within 55 fs and to the 1A surface after 84 fs.[14] Detailed calculations of the photochemical dynamics are challenging in a molecule the size of CHD, but from multiple computational studies, the nature of the surfaces is well established, and the motions of the wave packet have, at least in coordinate systems of reduced dimensionality, been carefully calculated (please see references in ref. 11).

Yet there still remain unanswered questions and mysteries that the proposed studies aim to address. Specifically, as was pointed out by Deb and Weber,[11] the spectroscopic methods cannot affirm that they are, indeed, observing the dominant part of the wave packet: it might be that they show just a component that remains on the initially excited B state, while the dominant part of the wave packet undergoes the reaction through the 2A state. Clearly, a complementary view, such as that afforded by diffraction, is eminently desirable. In our experiments we aim to settle this question and to determine the molecular structures as the reaction proceeds.

Experimental setup, challenges and results

The experimental concept is illustrated in Fig. 2. The CHD molecules are isotropically distributed in the low-pressure gas. A linearly polarized ultraviolet laser pulse ($\lambda = 266$–275 nm, 65 fs, 4–18 µJ, 100 µm FWHM focal spot) excites the CHD target molecule to the 1B state and thereby initiates the reaction. At a variable time delay from the optical pump, the X-ray pulse from LCLS (8.3 or 20.1 keV, 30–50 fs, 30 µm FWHM focal spot) intersects the gas sample in a collinear geometry. Both the ultraviolet pump and LCLS probe pulses are aligned through a 250 µm upstream alignment/differential-pumping aperture. X-rays scattered from CHD are projected onto a CSPAD.[15] Given the diffusion of the photo-generated 1,3,5-hexatriene through the 1,3-cyclohexadiene one is reasonably assured that fresh CHD molecules are observed by each pump and probe pulse pair. Even though molecules can be excited regardless of their rotational angle about the axis of the dipole moment, the pump pulse is expected to induce a limited alignment of the excited molecules along the laser polarization, with scattering of polarized X-rays into vertical and horizontal directions mapping different structural aspects of the molecules.

For any diffraction experiment the structural resolution increases with the scattering vector, q, range measured. Given the wavelength of the LCLS fundamental at 8.3 keV (1.494 Å) and the third harmonic at 20.1 keV (0.617 Å), we positioned the CSPAD with an active area of 20 cm diameter at a 4 cm distance from the interaction region. This provides ranges of $q = 1$ to 4.2 and 2 to 9 Å^{-1}, respectively, for the two different X-ray energies.

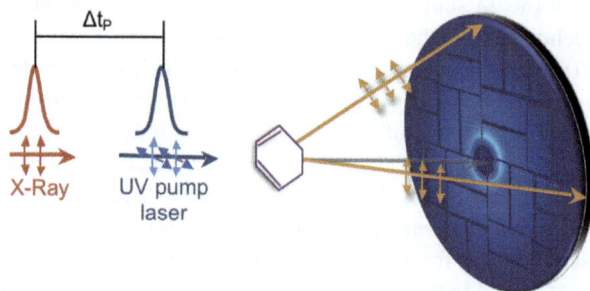

Fig. 2 Schematic of the experiment. The target molecules are intercepted by a linearly polarized UV laser pulse (blue) and, after a fixed time delay (Δt_P), by the polarized X-ray probe pulse (red). Absorption of the UV pump photon initiates the chemical reaction. The X-ray diffraction signal images the evolving reaction. Scattering of polarized X-rays into vertical and horizontal directions maps different structural aspects of the molecules that are partially aligned by the laser polarization. The diffraction pattern superimposed on the detector is from a dilute target of xenon gas recorded on the CSPAD.

The low density of the gas at pressures of only a few torr necessitates using a rather long interaction length of 13 mm. This requirement presents two challenges for the experiment. The first one is that the long interaction length reduces the achievable resolution in q-space because for any point on the detector, the different scattering points on the path of the X-rays through the gas span a range of scattering angles. To address this problem we positioned a beam block and inserted an aperture into the cell such that the low q regions of the detector only see the upstream part of the interaction path; the signal from downstream part of the interaction path is blocked by the beam block. Conversely, the high q regions of the detector only see the downstream part of the interaction path; signal from the upstream part is blocked by the aperture. With this approach, trajectory simulations show that the scattering signal can be measured with a resolution of about 0.2 A^{-1}, depending on the photon energy and the detector position. This approach enables use of a long interaction length while retaining an acceptable resolution, but requires consideration of the detector distance as a function of q during the analysis.

The second challenge arising from the long interaction length is in the absorption of the laser. CHD is a strong absorber, so that the UV laser is attenuated as it passes through the sample. This can be addressed by reducing the pressure of CHD and/or choosing a longer excitation laser wavelength. Additionally, we carefully designed the focal geometry such that the laser beam converges throughout the interaction region, reaching its focal point behind the observed region of the gas. At the centerline of the laser beam, where the X-ray beam propagates, the increases as a result of the tighter focus is compensated by the decreases due to absorption. We carefully balanced these two effects to maintain a reasonably uniform probability of excitation in the 10% range throughout the interaction length.

Given the complexity of the CSPAD and the intricate sample cell geometry, the scattering signal must be calibrated. To achieve this, we used xenon gas that provided a very large and uniform scattering signal. The comparison of the observed scattering intensity with the theoretically calculated signal provides an intensity correction factor that is applied to other gases to obtain a corrected scattering signal that can be compared to theory. Fig. 3 shows the raw CHD signal on the CSPAD as well as the corrected radial diffraction signal for a 20.1 keV X-ray. The comparison to theory shows a satisfying agreement of the experimentally measured intensity with the theoretical pattern.

Observation of pump-probe diffraction patterns places high demands on the signal to noise ratio of a measurement. First, it is important not to excite the target molecules too strongly because invariably, multi-photon absorption would lead to the preparation of the molecules on excited states higher than the target 1B, or cause ionization. This would make impossible an interpretation of the data in terms of chemical dynamics on specific potential energy surfaces. As a rule, therefore, we aim to excite no more than 10% of the target molecules to an excited state.

The diffraction pattern of a reacting molecule is not very different from that of a molecule in the ground state. In the example of the reaction of CHD to hexatriene, most carbon–carbon distances retain a distance typical of a conjugated double bond, while the carbon–hydrogen distances change little and contribute less to the overall diffraction signal. The atom–atom distances across the ring

1,3–CHD CSPAD Image   CHD Radial Average

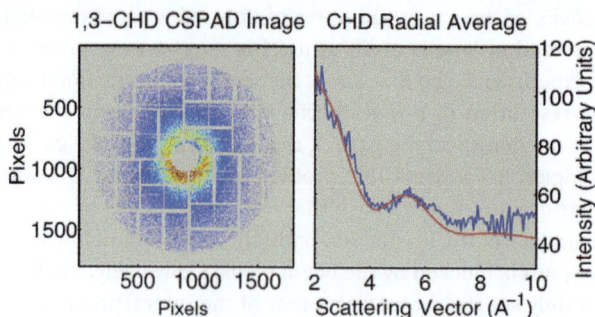

Fig. 3 Experimental results for scattering from 1,3-cyclohexadiene. (Left) A composite of 1000 frames reveal the signal as observed on the CSPAD detector. (Right) After scaling using a separately measured pattern of xenon gas, the radial diffraction pattern of CHD is recovered. The red line shows a theoretical pattern of CHD. In this experiment, the X-ray photon energy was 20.1 keV.

change substantially, but their overall contribution to the signal is comparatively small because their separation is large. Between the small excitation probability and the small change in the molecular diffraction pattern, it is expected that the observed diffraction signal changes will be on the order of 1% of the overall signal as a function of q. Such a small change was beyond the reach of the experimental setup and signal shown in Fig. 3. However, only 1000 shots were summed to provide that pattern and that it was taken with the third harmonic X-ray pulse, which has ~100 times fewer photons per pulse compared with the X-ray fundamental.

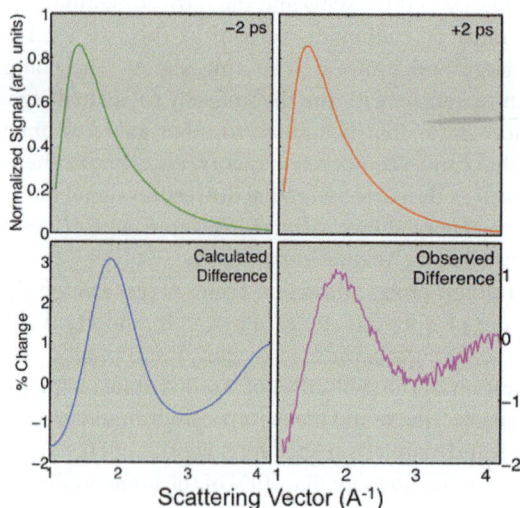

Fig. 4 Preliminary CHD data showing the experimental diffraction signals with the laser pump pulse after the X-ray probe pulse (−2 ps), top left, the laser pump pulse ahead of the X-ray probe pulse (+2 ps), top right. The calculated difference signal at +100 fs pump-probe delay is displayed in the lower left while the experimentally observed difference signal at +2 ps is in the lower right. The difference signals are plotted as a percentage of the signals at each value of the scattering vector.

In a repeat experiment, using the X-ray fundamental, it was possible to attain signal-to-noise ratios that captured extremely clear pump-probe signals. Fig. 4 shows preliminary scattering signals at delay times of −2 ps and +2 ps, respectively. The difference signal, also shown in Fig. 4, is plotted as a percentage of the total signal as a function of the scattering vector. It is evident that the level of noise is on the level of 0.1%, which is expected to be adequate to perform time-resolved pump-probe experiments.

It is well known the ring opening reaction of CHD occurs on timescales far less than 100 fs and could result in several, non-equilibrium structures at this calculated small pump-probe delay time.[11,14] However, even with this, the non-equilibrium pattern calculated at short pump-probe delay time is in remarkable agreement to the observed equilibrium structure at long pump-probe timescales, showing the ultrafast progression of the electrocyclic ring-opening reaction in CHD.

Theory and data analysis

The experimental, time-independent, X-ray scattering data shown in Fig. 3 is reasonably reproduced by a simple elastic X-ray scattering model with the observed signal proportional to the molecular form-factor $f^0(q)$,

$$f^0(q) = \left\langle \Psi_0 \left| \sum_{j=1}^{N} e^{iqr_j} \right| \Psi_0 \right\rangle = \int \rho_0^{(N)}(\mathbf{r}) e^{iqr} d\mathbf{r}, \tag{1}$$

where the momentum transfer vector, $q = k_0 - k$, is defined as the difference between the incident and the scattered wave vectors, with $|k| = |k_0|$ for elastic scattering, and where Ψ_0 is the electronic wave function and $\rho_0^{(N)}(\mathbf{r})$ the corresponding N-electron density. We therefore base our initial, zeroth-order, analysis of anticipated time-dependent 'laser on'-'laser off' data, $I_{exp}(t) = I_{UV}(t) - I_{noUV}(t)$, on time-dependent form-factors, $f^0(q, t)$, where the time-dependence mainly relates to changes in molecular geometry. Predictions of these geometry changes can be made on the basis of ab initio trajectories. Since the structural dynamics is underdetermined given the data available, the trajectories provide sufficient prior information to infer a parsimonious model.

We begin the discussion with the time-dependent trajectories. The photo-induced dynamics of CHD is simulated using ab initio Ehrenfest trajectories.[16] Each trajectory corresponds to a nonadiabatic wave packet,

$$|\Psi(\mathbf{R}, \mathbf{r}, t)\rangle = \left[\sum_i a^{(i)}(t) |\phi^{(i)}(\mathbf{r}; \mathbf{R})\rangle \right] |\chi(\mathbf{R}, t)\rangle, \tag{2}$$

where $a^{(i)}(t)$ are the complex Ehrenfest amplitudes that give the amplitude of the vibrational wave packet on each electronic state i, with $|\varphi^{(i)}(\mathbf{r}; \mathbf{R})\rangle$ the corresponding electronic wave function, and $|\chi(\mathbf{R}, t)\rangle$ the vibrational coherent-state wave packet that propagates on the electronic potential energy surfaces,

$$|\chi(\mathbf{R}, t)\rangle = \left(\frac{\gamma}{\pi}\right)^{M/4} \exp\left(-\frac{\gamma}{2}(\mathbf{R} - \overline{\mathbf{Q}}(t))^2 + \frac{i}{\hbar}\overline{\mathbf{P}}(t)(\mathbf{R} - \overline{\mathbf{Q}}(t)) + \frac{i}{2\hbar}\overline{\mathbf{P}}(t)\overline{\mathbf{Q}}(t)\right), \tag{3}$$

where $M = 3N_{at}$ in Cartesian coordinates with N_{at} the number of atoms. The Ehrenfest trajectory, given by phase-space coordinates $(\overline{\mathbf{P}}(t), \overline{\mathbf{Q}}(t))$, is generated by

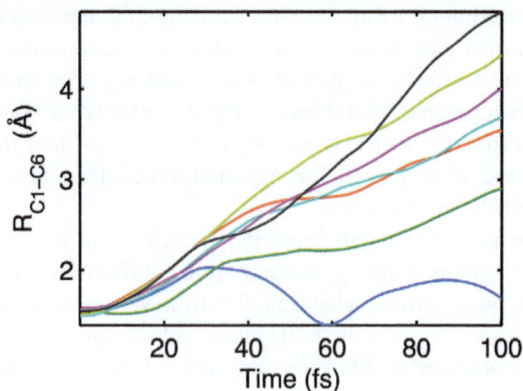

Fig. 5 End-to-end (C1–C6) distances as a function of time for the CHD molecules in the Ehrenfest trajectories. Most trajectories lead to ring opening, but one of the trajectories shown returns to the ring-closed CHD structure.

Hamilton's equations *via* semiclassical propagation on the quantum averaged molecular Hamiltonian. The amplitudes $a^{(i)}$ are calculated from coupled differential equations with the nonadiabatic coupling vectors $\delta^{(ij)} = \langle \varphi^{(i)} | \nabla | \varphi^{(j)} \rangle$. The electronic structure calculations are performed using the *ab initio* package MOLPRO,[17] which supplies the adiabatic electronic states and the nonadiabatic couplings using 3SA-CAS(6,4)-SCF/cc-pVDZ level of theory. An important point is that these calculations are made *on-the-fly* for each trajectory, eschewing the cumbersome task of calculating global potential energy surfaces. A set of calculated Ehrenfest trajectories with initial conditions sampled in the Franck–Condon region are shown in Fig. 5, in terms of the end-to-end (C1–C6) distance in CHD as a function of time for each trajectory. The time-dependent electron density for each trajectory that is required by eqn (1) can be calculated as

$$\rho^{(N)}(\mathbf{r}, t) = \int \left[\sum_i |a^{(i)}|^2 \rho_i^{(N)}(\mathbf{r}, \mathbf{R}) \right] |\chi(\mathbf{R})|^2 d\mathbf{R}, \tag{4}$$

where, to first approximation, the density cross-terms between different configurations have been ignored. An even simpler approximation uses atomic form-factors at the nuclear coordinates $\bar{\mathbf{Q}}(t)$ for each trajectory to calculate the scattering signal.

Fig. 6(a) shows a simulated '*experimental*' difference signal, $I_{exp}(q, t)$, as a function of time t and q, assuming rotational averaging. The calculated Ehrenfest trajectories $\{I_{th}^{(i)}(q, t)\}$ are used to fit the experiment with the target function,

$$F(\mathbf{w}) = \left| \sum_{i=1}^{N_{traj}} w_i I_{th}^{(i)} - I_{exp} \right|^2, \tag{5}$$

where w_i are the weights to be optimized. In practice, it is often better to fit percentage changes in the signal rather than absolute intensities. The weights are calculated using a multi-start nonlinear least-square optimization routine using a trust-region-reflection algorithm with a finite-difference gradient. A best-fit theoretical signal, $I_{opt}(q, t) = \sum w_i I_{th}^{(i)}$ is shown in Fig. 6(b). The trajectories in Fig. 5

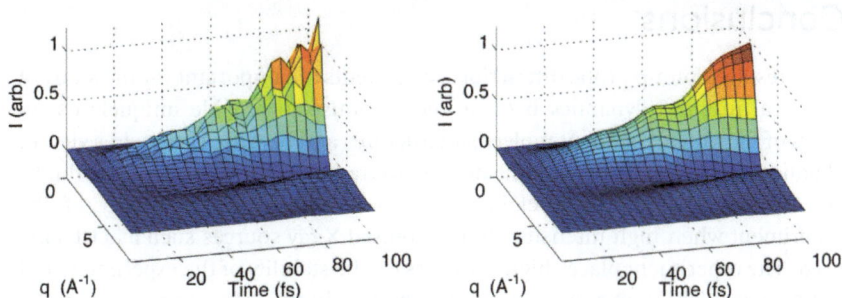

Fig. 6 The intensity of the "laser on"–"laser off" difference signal, $I_{UV} - I_{noUV}$, shown as a function of time (fs) and q (Å^{-1}). (a) The simulated experimental signal, $I_{exp}(q, t)$, (b) the optimized signal from the Ehrenfest trajectories, $I_{opt}(q, t)$.

have been assigned different weights, and, in this instance, the synthetic data is best reproduced by the trajectory that ends with the largest C1–C6 distance at $t = 100$ fs, which receives more than 86% of the weight.

A simple check on the capability of the experimental data to order structures in a sensible time-sequence can be devised by matching the experimental signal to a small number of scattering responses, $\{I_i(\boldsymbol{q})\}$, each representing a molecular structure commensurate with the ring-opening reaction of CHD. The measured signal can be expressed as the weighted sum, $I(\boldsymbol{q}, t) = \sum_i I_i(\boldsymbol{q})p_i(t)$, where $p_i(t)$ is the population of species i at pump-probe delay time t. Our goal is to infer $\mathbf{p}(t)$ from experiment for some small but representative set of $\{I_i(\boldsymbol{q})\}$ to confirm that the species follow the same sequence as observed in the Ehrenfest trajectories. In our present example, shown in Fig. 7, the weights for three structures, corresponding to an early ring-closed structure, an intermediate structure, and a final ring-open structure, appear in correct order following optimization against the simulated, $I_{exp}(q, t)$, signal.

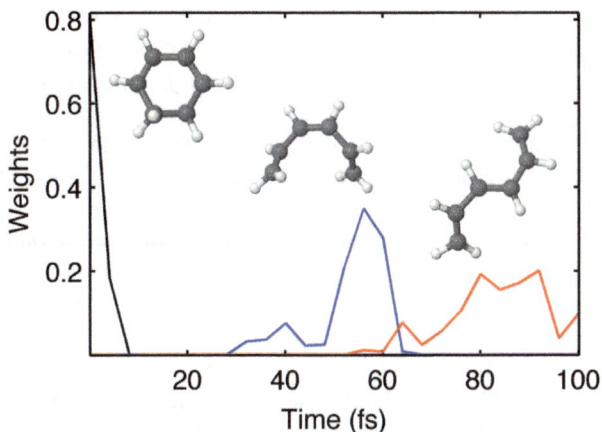

Fig. 7 The weight as a function of time for three dominant structures when fitted against the simulated data shown in Fig. 6a.

Conclusions

Time-resolved pump-probe diffraction experiments are important for the study of chemical reaction dynamics because they promise to provide uniquely useful views of the time-dependent molecular structures and because they yield a view of chemical dynamics that is complementary to traditional spectroscopic methods. Our investigations on 1,3-cyclohexadiene have proven the feasibility of the experiment when high intensity ultrafast pulsed X-ray sources such as LCLS are used. The experiment places high demands on the stability of the experiment and the data acquisition to recover the small changes in the diffraction patterns. Yet with a sufficiently careful setup, those demands can be met. Moreover, it is possible to compute the time dependent diffraction pattern of a wave packet that evolves on an excited electronic surface. Consideration of the electron density of the electronically excited molecule and the time-evolving vibrational wave packet should provide benchmarks against which to compare the experiment.

Acknowledgements

Portions of this research were carried out at the Linac Coherent Light Source (LCLS) at the SLAC National Accelerator Laboratory. LCLS is an Office of Science User Facility operated for the U.S. Department of Energy Office of Science by Stanford University. A.K. and T.N. acknowledge funding from the European Community (FP7-PEOPLE-2013-CIG-NEWLIGHT) and helpful discussions with D. Shalashilin (University of Leeds). K. S. acknowledges the support of EPSRC grant EP/I014500/1. Part of the early phase of this project was supported by the Division of Chemical Sciences, Geosciences, and Biosciences, the Office of Basic Energy Sciences, the U.S. Department of Energy, Grant no. DE-FG02-03ER15452 (PMW).

References

1 C. Bostedt, *et al.*, Ultra-fast and ultra-intense X-ray sciences: first results from the Linac Coherent Light Source free-electron laser, *J. Phys. B: At., Mol. Opt. Phys.*, 2013, **46**, 164003.

2 J. L. Gosselin and P. M. Weber, *J. Phys. Chem. A*, 2005, **109**, 4899–4904.

3 M. P. Minitti and P. M. Weber, *Phys. Rev. Lett.*, 2007, **98**, 253004.

4 S. Deb, B. A. Bayes, M. P. Minitti and P. M. Weber, *J. Phys. Chem. A*, 2011, **115**, 1804.

5 S. Deb, M. P. Minitti and P. M. Weber, *J. Chem. Phys.*, 2011, **135**, 044319.

6 M. P. Minitti, J. D. Cardoza and P. M. Weber, *J. Phys. Chem. A*, 2006, **110**, 10212–10218.

7 L. S. Bartell, in *Stereochemical Applications of Gas-Phase Electron Diffraction*, ed. I. Hargittai and M. Hargittai, Verlag Chemie, New York, 1988, pp. 55–84.

8 R. A. Bonham and M. Fink, *High Energy Electron Scattering*, Van Nostrand Reinhold, New York, 1974.

9 A. A. Ischenko, L. Schäfer and J. D. Ewbank, *J. Mol. Struct.*, 1996, **376**, 157–171.

10 A. Nenov, P. Kölle, M. A. Robb and R. deVivie-Riedle, Beyond the van der Lugt/ Oosterhoff model: when the conical intersection seam and the S1 minimum energy path do not cross, *J. Org. Chem.*, 2010, **75**, 123–129.

11 S. Deb and P. M. Weber, *Annu. Rev. Phys. Chem.*, 2011, **62**, 19–39.

12 R. B. Woodward and R. Hoffmann, *The Conservation of Orbital Symmetry*, Weinheim: Verlag Chemie, 1970, p. 177.

13 M. Garavelli, C. S. Page, P. Celani, M. Olivucci, W. E. Schmid, S. A. Trushin and W. Fuss, *J. Phys. Chem. A*, 2001, **105**, 4458–4469.

14 N. Kuthirummal, F. M. Rudakov, C. Evans and P. M. Weber, Spectroscopy and femtosecond dynamics of the ring-opening reaction of 1,3-cyclohexadiene, *J. Chem. Phys.*, 2006, **125**, 133307.

15 H. T. Philipp, M. Hromalik, M. Tate, L. Koerner and S. M. Gruner, Pixel array detector for X-ray free electron laser experiments, *Nucl. Instrum. Methods Phys. Res., Sect. A*, 2011, **649**, 67–69.

16 K. Saita and D. Shalashilin, *J. Chem. Phys.*, 2012, **137**, 22A506.

17 H.-J. Werner, P. J. Knowles, G. Knizia, F. R. Manby and M. Schütz, *et al.*, *MOLPRO V.2012.1*, a package of ab initio programs.

Faraday Discussions

ROYAL SOCIETY
OF CHEMISTRY

PAPER

Analysis of a measurement scheme for ultrafast hole dynamics by few femtosecond resolution X-ray pump–probe Auger spectroscopy

Bridgette Cooper,[a] Přemysl Kolorenč,[b] Leszek J. Frasinski,[a] Vitali Averbukh[a] and Jon P. Marangos*[a]

Received 25th March 2014, Accepted 30th April 2014

DOI: 10.1039/c4fd00051j

Ultrafast hole dynamics created in molecular systems as a result of sudden ionisation is the focus of much attention in the field of attosecond science. Using the molecule glycine we show through *ab initio* simulations that the dynamics of a hole, arising from ionisation in the inner valence region, evolves with a timescale appropriate to be measured using X-ray pulses from the current generation of SASE free electron lasers. The examined pump–probe scheme uses X-rays with photon energy below the K edge of carbon (275–280 eV) that will ionise from the inner valence region. A second probe X-ray at the same energy can excite an electron from the core to fill the vacancy in the inner-valence region. The dynamics of the inner valence hole can be tracked by measuring the Auger electrons produced by the subsequent refilling of the core hole as a function of pump–probe delay. We consider the feasibility of the experiment and include numerical simulation to support this analysis. We discuss the potential for all X-ray pump-X-ray probe Auger spectroscopy measurements for tracking hole migration.

1 Introduction

Ultrafast hole migration, following sudden ionisation or excitation, is believed to be a universal response of extended molecules and is also likely to occur in some form in many condensed phase systems.[1,2] This process occurs due to the electron correlations within many-electron systems, and is predicted to take place typically on the few to sub-femtosecond timescale (*i.e.* into the attosecond time domain). It is currently a prominent goal in attosecond science to observe and fully characterise processes such as hole migration in order to: (a) deepen our understanding of the process and ascertain the role of hole migration in determining

[a]Imperial College, Blackett Laboratory, Prince Consort Road, London, SW7 2BB, UK. E-mail: b.cooper@imperial.ac.uk
[b]Institute of Theoretical Physics, Faculty of Mathematics and Physics, Charles University in Prague, V Holešovičkách 2, 18000 Prague, Czech Republic

photochemical and photophysical outcomes, (b) to prove attosecond measurement methods can address correlation driven dynamics in extended quantum systems such as biomolecules and polymers.

If we suddenly create a superposition of inner valence hole states in the cation of a moderately sized molecule (*e.g.* by photoionisation) we expect to see rapid evolution of this hole within the molecular frame. The most interesting class of dynamics occurs following the sudden formation of a hole state that is intrinsically a superposition of several eigenstates of the cation and so is expected to display non-exponential decay and time-dependent localisation. This may be due to the coupling of several one-hole (1h) electronic configurations or due to a coupling of a 1h configuration to a series of two-hole one-particle (2h1p) configurations of the cation by electron correlation. The superposition leads to a hole amplitude whose projection onto a single eigenstate may be seen to undergo rapid evolution. This may take the form of hole migration (*i.e.* motion of the hole around the molecular frame) or hole decay typically with some non-exponential (oscillatory) behaviour.[3] This will take place on a timescale that is short (<10 fs) with respect to the timescale of nuclear motion.[4] We expect that the initial stages of the cation evolution to be dominated by electronic only motion (hole migration), with coupling to nuclear modes will become important at later times (charge transfer).

Both hole migration and non-exponential decay therefore stem from electron correlation and are currently of great interest to understanding ultrafast molecular science. Quantifying the role of electron correlation in multi-electron systems in a time-dependent way is a frontier challenge in ultrafast measurement. Moreover the problem of ultrafast hole evolution and motion is of critical importance to many aspects of photochemistry and photophysics where there is hitherto no experimental data at these short timescales.[5,6] An essential task is to find ways to measure time dependent information on the hole evolution and localisation in such situations. This task has up to now been prevented by the extremely fast timescales (few- to sub-femtosecond) of the hole motion. In what follows we will consider a scheme to use an X-ray pump–X-ray probe pulse sequence. The first pulse creates an inner valence hole (IVH) superposition that may undergo non-exponential decay. The second pulse then probes transitions from an inner shell to the hole state formed by the first pulse to give temporal information on the hole survival.

The current methods of attosecond science rely heavily on the process of high harmonic generation (HHG) as a means to generate attosecond pulses in the soft-X-ray photon energy range,[7,8] but despite progress in the generation efficiency and shortness of the pulse duration[9,10] the power levels are currently found to be very low. This has prohibited the widespread application of attosecond pump–probe methods using the soft-X-ray pulses generated by HHG and instead techniques that replace one of the pulses with an infra-red high intensity field, either as the pump[11] or probe,[12] have been commonly adopted. For example, the tracking of the dynamics of suddenly excited holes has recently been achieved in Kr atoms where the dynamics initiated by the strong field ionisation of a pair of spin–orbit split valence states was probed by transient absorption from an inner shell states.[11] In a similar scheme the dynamics of strong field ionisation in a SiO_2 solid state sample was tracked by measuring the holes created in the valence shell through transient absorption probing again from an inner shell state.[13] Triggering ionisation through a strong field can lead, however, to significant complications since; (i) tunnel ionisation primarily accesses only the outer valence hole states, (ii) the tunnel ionisation process in a strong field is not fully understood and our theory for this in larger molecules remains to be adequately developed, (iii) the presence of a strong field may profoundly disturb the electron motion[14] in a molecule and so mask the underlying correlation driven effects.

In the light of the complications caused by strong-field ionisation it therefore seems necessary to resort to using the better understood process of single photon ionisation in a high frequency field if we are to trigger hole dynamics that are to be subject only to the intrinsic processes of a molecule. Pump–probe techniques based on measuring emission of IR or UV radiation emitted after sudden ionisation,[15] using an XUV probe pulse and detection of HHG radiation[16] or measuring a double ionisation count *via* a single-photon laser enabled Auger decay (sp-LEAD)[17] scheme, have been proposed to measure correlation driven electron dynamics that occur in the inner-valence region in molecules. Although single photon ionisation is simple in principle, in practice we know from the discussion above that there is limited intensity from current attosecond pulses based upon HHG. This low intensity means we can in general hope to excite only a minuscule fraction of the sample during the short (few hundred attosecond pulse) thus making pump–probe experiments very challenging. Much effort is currently being put into generating HHG based attosecond sources with higher power.[18] This includes efforts to work with attosecond pulses[19,20] carried at lower frequency *i.e.* around 20 eV (rather than the conventional 90 eV), where the HHG processes is more efficient, and where the photoionisation cross-section in most molecules also tends to be relatively high. Hopefully these developments will permit the use of HHG based attosecond sources in the near future, but it is prudent to examine the potential for using alternative light sources to ensure that we can make progress on this problem.

X-ray free electron lasers (FELs) operating in SASE (self amplified spontaneous emission) mode are making rapid progress towards shorter pulse duration, and although not yet operating in the attosecond regime there are proposals that suggest this will be possible in the future.[21] Moreover X-ray FELs have already been demonstrated to generate pulses as short as a few femtoseconds when operating in the low bunch charge regime[22] or with a slotted emittance spoiler.[23] Given the very much higher pulse power of an X-ray FEL compared to a HHG source (we can expect single pulse photon numbers of $>10^{11}$ for the former and typically $<10^6$ for the latter) it is possible to contemplate efficient pump–probe excitation schemes in a FEL. This coupled to schemes that generate two-pulses, such as X-ray split-and-delay or a double slotted foil emittance spoiler, make possible high temporal resolution pump–probe measurements.

A problem in addition to the power requirements of pump pulse sources is that transient absorption, an attractive probe for attosecond measurements, requires not only a significant excitation fraction induced by the pump, but also a sufficiently dense sample to ensure measurable absorption. This may be readily achieved for condensed phase samples and for many molecular gases if suitable sample delivery is used. The prospects for biomolecules are typically more challenging since these have very low vapour pressures at temperatures that avoid thermally induced dissociation and this prohibits absorption studies. This means we must seek alternative detection methodologies and we identify that Auger spectroscopy is a suitable state specific choice.

We discuss here a generic approach that would use an X-ray free electron laser to overcome the problems highlighted above. To demonstrate the feasibility of measurements of correlation driven dynamics in an amino acid we choose to analyse a concrete example accessible to currently operating X-ray FELs. The example we analyse is of non-exponential hole decay in glycine (rather than charge migration) which serves to demonstrate the feasibility of the scheme, but it would be applicable to a wider class of problem. The general concept, shown in Fig. 1 is that an X-ray FEL pulse is split into two with a small variable delay. Due to the high intensity of the FEL the first pulse can create a sufficient number of inner valence holes to be measured. The X-ray photon energy is, however, kept below the threshold for inner shell ionisation. The probe pulse, which may for convenience be at the same frequency, is tuned to resonance between an inner shell state in a carbon atom of the molecule and an inner valence state. If the inner valence state is filled no transition can occur, but if a hole has been formed by the first pulse then excitation may occur leaving a hole in the inner shell of the carbon atom. The strength of the transition is a probe of the probability of the hole and so can track the hole survival probability. Assuming the sample density is too low to apply absorption techniques we need to find an alternative way of measuring the probability of the inner valence hole. This can be done by registering the appearance of the core hole which will be refilled dominantly (for light elements) by Auger decay that results in a characteristic energy spectrum of emitted electrons. This is the detection scheme adopted in the example discussed below.

In the remainder of this paper we outline the details of the specific scheme and assess the experimental feasibility (Section 2), and then present a theoretical analysis of the hole dynamics in the inner valence states of glycine and our calculations of the effect of the coupling of the strong resonant X-ray fields on the population dynamics in the cation (Section 3). We then discuss more general

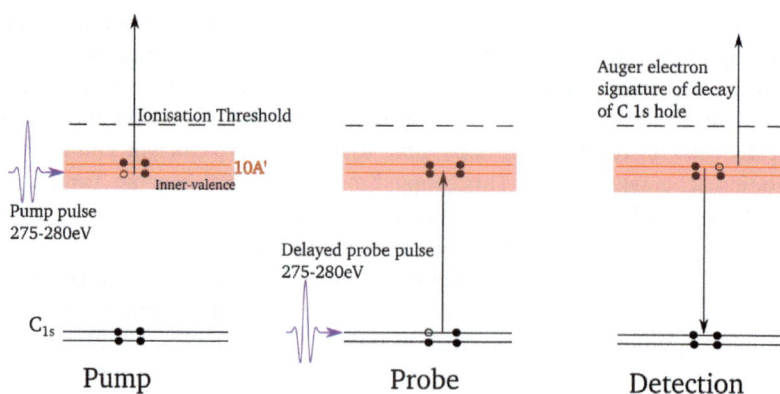

Fig. 1 Pump step opens a number of valence ionisation channels including creation of IVHs. The delayed probe pulse can strongly interact *via* a C 1s IVH transition (a channel only open if inner valence ionisation has occurred). Following this there can be Auger decay back to the C 1s hole with the emission of Auger electrons of characteristic energy that can be detected.

aspects of using X-ray laser pump–probe Auger spectroscopy for measuring hole migration before concluding (Section 5).

2   X-ray laser pump–probe auger spectroscopy scheme in glycine

2.1   Scheme overview

Here we choose a simple molecule, the amino acid glycine (NH_2CH_2COOH) as an example for study. The three most abundant conformers of glycine are commonly referred to as Gly I, Gly II and Gly III[24,25] of which Gly I has the lowest energy. In a temperature range to have a sufficient amount of glycine in the gas phase $\sim$150–170 °C, we expect the Gly I and Gly III conformers to dominate the sample at a ratio of approximately 2 : 1.[24] We begin by calculating the ionisation spectra of the glycine conformers of interest, the results for the Gly I conformer are shown in Fig. 2. For this calculation and all subsequent calculations the inclusion of electron correlation into the wavefunctions to describe both the singly ionised and doubly ionised states is extremely important as electron correlation is driving the dynamics. Consequently where dipole moments, dynamics and Auger spectra are calculated, we use the Greens function technique called algebraic diagrammatic construction (ADC) to second order extended ADC(2)x,[26] further details of which can be found in Section 3.

We concentrate on the inner valence hole associated with the 10A$'$ state at around 20 eV binding energy (*i.e.* accessed by photon energies $\sim$275 eV), although there are other inner valence hole states that we expect to show interesting dynamics in this molecule. There are two eigenstates shown in the right panel of Fig. 2 (corresponding to the 10A$'$ molecular orbital), let us call them $|a\rangle$ and $|b\rangle$, which are each roughly 50% of a single 1h and 50% of a series of 2h1p configurations. In this case the appearance of this doublet ($|a\rangle$ and $|b\rangle$) of mixed 1h-2h1p states and the resulting dynamics is caused purely by electron correlation. Only

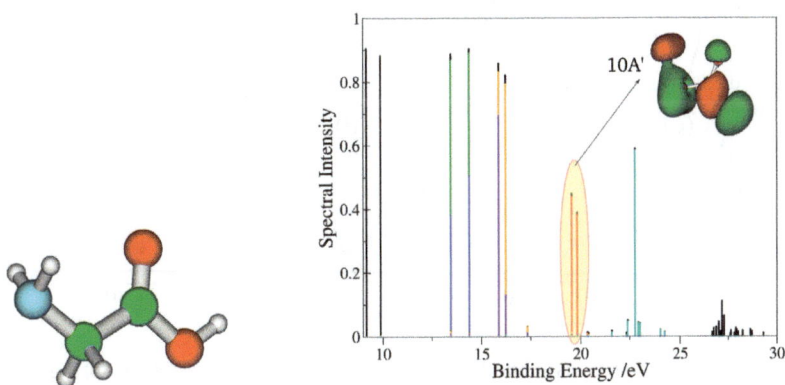

Fig. 2   Left: Glycine molecule NH_2CH_2COOH (blue N, red O, green C) geometry in the Gly I conformer. Right: Calculated energy spectrum of glycine with the states corresponding to ionisation from the 10A$'$ orbital highlighted. Insert shows the 10A$'$ orbital of glycine. Calculation performed using ADC(2)x with cc-pVDZ basis set augmented with 3s3p diffuse functions.[27]

the 1h state can be created directly by sudden photoionisation so that at $t = 0$ we populate the hole which means we have a superposition of $|a\rangle$ and $|b\rangle$, say with "+" relative phase. As time goes by, the difference in the energies of the two states turns "+" into "−" and constructive interference that created a pure hole at $t = 0$ turns into destructive that brings the hole contribution to the wavepacket almost to zero. At this destructive interference time the wavepacket is almost purely composed of many different 2h1p's. Later on we go back to "+" and the hole experiences a revival. And so on until nuclear dynamics finally leads to a significant distortion of the molecular geometry (say, after 20 fs) changing the energies of the involved electronic states and so modifying or even damping the oscillations. This situation of evolution distinguishable over many femtoseconds is rather special to glycine because we only very rarely see breaking of a hole into two states with roughly equal hole contributions to them. Although in principle the same class of dynamics may occur in many other molecules, e.g. trans-butadiene for which we also have detailed calculations, there is not usually breaking of 1h into two states, but rather breaking into tens of states each of which is 90% or more 2h1p (so-called breakdown of molecular orbital picture of ionisation[28]), so to expect a full revival out of such complicated wavepacket is difficult. Nevertheless in these more typical systems it would be possible to follow the dynamics if higher (few hundred attosecond) temporal resolution were available.

Our calculations of the hole survival in the 10A′ state, again using ADC(2)x are shown in Fig. 2 (see Section 3 for a more detailed discussion) for both the Gly I and Gly III conformers and for the the sum of Gly I and Gly III in a ratio of 2 : 1 expected from the operational temperature range. We see that the dynamics are

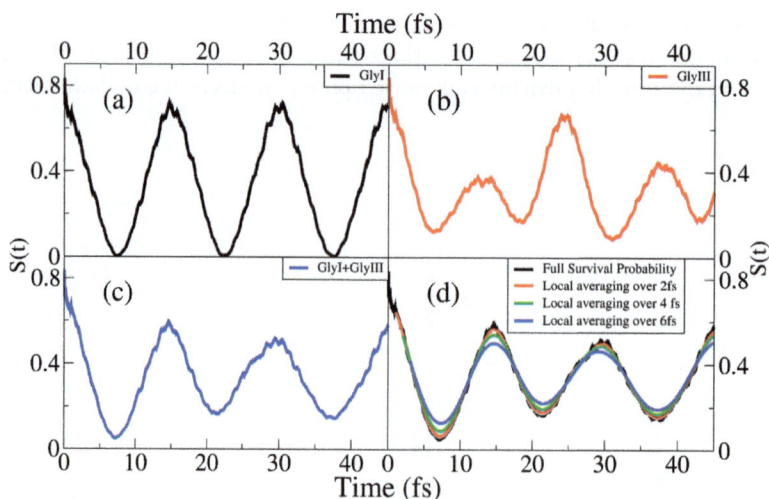

Fig. 3   Glycine 10A′ hole survival probability for (a) the Gly I conformer (black), (b) the Gly III (red), (c) the sum of Gly I and Gly III in a ratio of 2 : 1 expected from the operational temperature range (blue) and (d) the sum of Gly I and Gly III in a ratio of 2 : 1 (black) and normalised signal expected from a probe of the state population with a 2, 4 and 6 fs temporal resolution (red, green, blue). At times longer than 20 fs the signal is expected to be modified significantly by nuclear motion. Calculation of the survival probability was performed using ADC(2)x with cc-pVDZ basis set augmented with 3s3p diffuse functions.[27]

sufficiently slow so that they are accessible to measurement using the temporal resolution available from *e.g.* the LCLS X-ray FEL located at SLAC. Whilst Fig. 3(a) and (b) shows that the dynamics for the Gly I and Gly III conformers show qualitatively similar oscillations there are differences in the periodicity and strength of the revivals, we expect that the signal measured will be dominated by the Gly I conformal evolution as is seen in Fig. 3(c). We will show that in glycine a hole created in the 10A′ inner valence molecular orbital shows an oscillatory evolution on the timescale of ~15 fs in our theoretical calculations (see Fig. 2 and Fig. 3). This dynamics would be easy to resolve with the temporal resolution available from LCLS. We illustrate the robustness of the measurement by simulating the effect of using an X-ray pulse of different durations, 2fs, 4 fs and 6 fs duration whereas we expect the pulse duration to be nearer 3 fs.

In the putative measurements an X-ray pulse from an X-ray FEL is split into a pair of identical X-ray pulses with a precisely controlled variable interpulse delay. Each of the pulses has a duration of only a few femtosecond. The photon energy of the pulses needs to be tuned to near 20 eV below the K edge of carbon in the molecule (requiring photons of around 275 eV from the FEL). The tuning is such that the X-rays cannot cause core shell ionisation or excitation in the molecule and so only valence shell ionisation can occur in the neutral molecule. If the photon energy is tuned to an appropriate value (say ~275 eV) then following valence ionisation of the 10A′ state a K shell (1s) to this inner valence hole (IVH) transition is opened up for the absorption of a second photon. This 1s IVH transition is allowed in the molecular system. We have calculated the dipole moment for the C K shell to IVH transition within the theoretical framework of this work and find a value of 2.8×10^{-31} Cm which is large enough to permit significant excitation probabilities for the X-ray laser intensities likely to be used (see next section).

The extra absorption registers the existence of the inner valence hole and if that hole is evolving in time the excitation probability as a function of pump–probe delay will reflect this. The Auger electron signal associated with the subsequent refilling of the 1s core hole is characteristic of the channel and so we can to use this signature to track the hole dynamics (see Fig. 1). We calculate the spectrum of Auger electrons that will be emitted from the glycine cation when the inner shell hole is refilled using Fano-ADC(2)x method for the partial decay widths,[29] and ADC(2)x method for singly and doubly[30] ionised energies in aug-pCVQZ basis without g functions, uncontracted on the C centres. The calculated spectrum is shown in Fig. 4.

2.2 Experimental implementation: feasibility with current X-ray FEL sources

It is our conjecture that already SASE X-ray FEL light sources offer parameters required to implement this measurement scheme. As a concrete example we choose to consider the LCLS facility at Stanford. There are several alternative operating modes of an X-ray FEL, such as LCLS, that can provide the required X-ray pump–X-ray probe pulses. The simplest method is to use a single short pulse (created in the low bunch charge mode or with a single slotted foil) and then to employ an X-ray split-and-delay apparatus. The requirement on photon energy (275 eV) and the short pulses (~3 fs) will mean the pulse peak power will be relatively low compared to that potentially available with an X-ray FEL. We

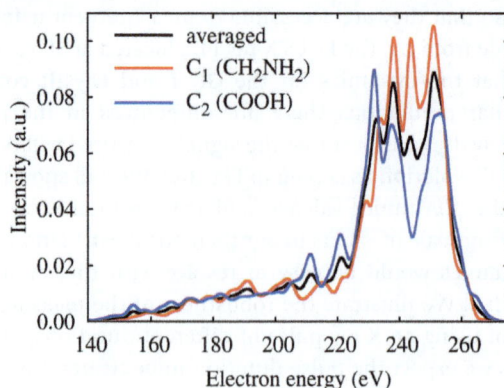

Fig. 4   The calculated Auger spectra for the C1s vacancies in glycine, calculated using Fano-ADC(2)x method for the partial decay widths, and ADC(2)x method for singly and doubly ionised energies in aug-pCVQZ basis without g functions, uncontracted on the C centres.

estimate around 100 μJ can be generated at this pulse energy and accounting for transfer optics $\sim$10 μJ will reach the target in a spot size of $\sim$10 μm to achieve an intensity $\sim 10^{15}$ Wcm^{-2}. In fact the relatively low intensity, compared to the much higher intensities often used in X-ray FEL experiments, is advantageous in this case as it is important to operate in the linear interaction regime for each pulse and to avoid opening additional excitation and ionisation channels in the molecule that might lead to increased background.

The sample can be introduced as a gas to the X-ray interaction region *via* an effusive beam from a nozzle. This requires the sample to be heated in an oven (operating in the temperature range $\sim$150–210 °C) in proximity to the interaction region[31] as has been used successfully in a number of synchrotron based experiments where sufficient densities for NEXAFS and photoelectron experiments in glycine were demonstrated. It is from this earlier work that we obtain the carbon K edge values in this molecule. It is required to work in the lower part of the available temperature range of the oven (150–170 °C) to ensure the sample population is dominated by the Gly I and III conformers which our calculations show sufficiently similar dynamics for the IVH states of interest (see Fig. 3). With an oven operating at these temperatures we would expect to achieve densities[32,33] 10^{13}–10^{15} cm^{-3} which will translate into densities up to $\sim 10^{12}$ cm^{-3} in the interaction region. The use of methods for conformer separation[34] could be used to enable the study of the dynamics of a single conformer.

There will be no core electron ionisation in the glycine molecule from a single photon at an energy of 275 eV. The valence photoionisation (VP) will open up a lot of channels that form the background. However the photoelectron energies from these valence electrons will sit a few tens of eV away from the Auger transitions of interest. In the IPI-CE-AU process (inner-valence photoionisation, core-excitation of 1s to IVH, Auger decay) we will have two correlated electrons: eIPI and eAU. The kinetic energies will be; eIPI $\sim$270–250 eV (for the photon energies planned) and eAU $\sim$260–230 eV. We therefore find much of the Auger spectrum (see calculated spectrum Fig. 4) at photon energies lying well outside of the valence photoelectron range. It should therefore be easy to identify the IPI-CE-AU process which is the signature of the IVH. The Auger photoelectron spectrum must be recorded with sufficient energy discrimination; an electron magnetic bottle TOF spectrometer is a suitable instrument for this in that it provides high energy energy resolution ($\sim$1 eV) and a high collection efficiency that ameliorates the low sample density. The electron energy spectrum will therefore isolate the nearby photoelectrons from the characteristic Auger electrons that arise from the decay of the C K-shell hole.

Additionally one can use covariance mapping to further increase the visibility of the signal. An "island" in the covariance map corresponding to these kinetic energies should be present. It has previously been demonstrated that electron-electron covariance is feasible at FELs (at both FLASH and LCLS).[35,36] Other processes that will give two correlated electrons in the same energy range are two sequential photoionisations (IPI-VP, IPI-IPI, VP-IPI, VP-VP) from valence or inner valence. We can measure these separately by examining shots where the pulse is detuned from resonance, so that it does not induce the core excitation, and so account for their contribution. Moreover the cross section will be lower (at least compared to a IPI-CE-AU process).

The sample density available is sufficiently high (we estimate that there will be up to 10^5 molecules within the 10^{-7}cm^3 interaction region) that the acquisition limit will be set by space charge effects rather than available sample density. The focal spot is 10 μm diameter and the X-rays per 10 μJ pulse will give a total valence ionisation probability of 0.16 so giving $\sim$1.6 $\times$ 10^4 electrons per shot at the higher sample density. Taking a conservative number of 1000 electrons/shot (based upon previous experience with the magnetic bottle instrument) to avoid space charge effects we can proceed with the estimate. Using empirical data the probability of L shell holes (IVH) being created in each ionised molecule is 0.125. Next we estimate the probability of excitation 1s IVH in the second pulse using our calculated dipole moment. Assuming the laser intensity to be $10^{15} Wcm^{-2}$ we can expect the excitation from the inner shell to be at least 0.1 and a probability of the subsequent Auger electrons falling into the suitable energy band can be estimated to be >0.25 (see calculated spectrum Fig. 4).

Next we must consider the intrinsic fluctuations of a SASE FEL generated X-ray pulse. Because we employ an X-ray pump- X-ray probe scheme the delay is immune to the absolute timing jitter of the X-ray pulse. Moreover in the low bunch charge operation the emission is expected to be overwhelmingly in a single X-ray pulse, rather than the more complex X-ray pulse temporal structures commonly encountered when using SASE with a high bunch charge. Nevertheless we can expect significant shot-to-shot spectral fluctuations of the SASE spectrum. Whilst the bandwidth of an individual pulse will be in the region of 1 eV the SASE fluctuation can lead to a spectrum (averaged over fluctuations) around 10 times broader. Fortunately the photon energy spectrum for each shot can be measured via a number of channels, e.g. using an X-ray spectrometer or the electron bunch energy diagnostic. Thus the resonant shots can be selected from the data set and the non-resonant shots used to characterise the VP background.

Collecting the numbers above the total number of counts per shot will be 1000 $\times$ 0.125 $\times$ 0.1 $\times$ 0.25 $\approx$ 3 or 360 per second. Assuming in photon energy binning the effective shot rate is reduced by an order of magnitude, at the operational repetition rate of LCLS of 120 Hz we have in 20 min a count of 4.3 $\times$ 10^4 electrons in the energy region of interest (i.e. a count rate shot noise error of below 0.5%), which indicates the feasibility of measuring the required signals using existing light sources. A detailed analysis of the experimental s/n of course must await the actual execution of an experiment.

3   Theoretical modelling of X-ray the pump–probe measurements of hole survival probability in glycine

Here we model the effects of the strong resonant X-ray probe pulse on the Auger electron yield. The purpose of this is to model the anticipated experimental signals and to confirm that under the likely conditions of a real experiment where the pulse intensities are high enough to give measurable signals the X-ray pump pulse will not significantly perturb the inner valence state dynamics it has excited. We do this by following the methodology set in the work of Rohringer and Santra[37] generalised for the case of two initial states that is relevant to the hole migration dynamics that occurs in glycine. Throughout we will assume the Gly I conformer of glycine, as Fig. 3 shows that in the anticipated experimental conditions the dynamics of this conformer dominates.

At $t = 0$ we assume that the first 275 eV pulse (the pump) has created a superposition of two cationic states $|a\rangle$ and $|b\rangle$ corresponding to ionisation from the 10A$'$ molecular orbital in glycine:

$$\Psi^{(N-1)}(t = 0)\rangle = \tilde{c}_a(t = 0)|a\rangle + \tilde{c}_b(t = 0)|b\rangle. \tag{1}$$

Ionising from the 10A$'$ orbital produces states that are below the double ionisation threshold, and therefore are Auger inactive. However, each state has a large contribution of 2h1p configurations as the calculated spectral intensity spectrum for glycine shows in Fig. 2. The temporal evolution of the superposition of the states created when ionising from the 10A$'$ orbital can be fully characterised by the survival probability of the initial state, $S(t)$:

$$S(t) = \left|\left|\tilde{c}_a\right|^2\exp(-iE_at) + \left|\tilde{c}_b\right|^2\exp(-iE_bt)\right|^2, \tag{2}$$

where E_a and E_b are the energies of the cationic eigenstates $|a\rangle$ and $|b\rangle$ respectively, (atomic units are used throughout). In the case of ionisation from the 10A$'$ the resulting dynamics (see Fig. 3) show an oscillatory evolution of the survival probability involving coupling of a 1h state to a manifold of bound ionic states of the 2h1p type.

At a delay time t after the initial ionisation event, a probe pulse, also of 275 eV, induces a transition from a carbon 1s orbital to fill the inner valence 10A$'$ orbital producing state $|1s\rangle$. In the case of glycine, there are two carbon 1s orbitals: 4A$'$ located on the carboxyl group and 5A$'$ the C centre attached to the amine group. In the following simulation of pump–probe experiment we assume that the width of the probe pulse is insufficient to span both core orbitals, and sum incoherently. In the following, we assume that a core ionised state couples the initial superposition of states only *via* the dipole interaction $\varepsilon(t)\hat{z}$. This assumption excludes any single-photon laser-enabled Auger decay (sp-LEAD) transitions[17] from the initial superposition of states directly to the final states. Thus, if the field free Hamiltonian is denoted as $\hat{H}_0$ the Hamiltonian in the presence of the electric field is given by:

$$\hat{H}(t) = \hat{H}_0 + \varepsilon(t)\hat{z}. \tag{3}$$

The core ionised states will decay *via* Auger transitions into the final states $\sum_i \int d\varepsilon_i \tilde{g}_i(\varepsilon_i, t)|i, \varepsilon_i\rangle$ where i is the ith eigenstate of the glycine cation and ε_i the resonant Auger electron associated with that channel. By using Wigner-Weisskopf theory,[38] the partial decay width Γ_i associated with the irreversible transition from $|1s\rangle$ to $|i, \varepsilon_i\rangle$ can be written as:

$$\Gamma_i = 2\pi\left|\langle i, \varepsilon_i|\hat{V}_C|1s\rangle\right|,$$
$$\Gamma_{1s} = \sum_i \Gamma_i , \tag{4}$$

where $\hat{V}_C$ is the Coulomb operator. In our numerical simulations, the partial widths will be taken from an Fano-ADC theory, where the widths are given by the matrix elements of the total Hamiltonian less the energy of the resonance.[29] The wavefunction to describe the cationic system as a function of time is thus:

$$|\Psi^{(N-1)}(t)\rangle = \tilde{c}_a(t)|a\rangle + \tilde{c}_b(t)|b\rangle + \tilde{c}_{1s}(t)|1s\rangle + \sum_i \int d\varepsilon_i \tilde{g}_i(\varepsilon_i, t)|i, \varepsilon_i\rangle. \tag{5}$$

We consider a symmetric probe pulse, with the detuning of the with respect to the resonant transitions is given by:

$$\delta = E_{1s} - \frac{(E_a + E_b)}{2} - \omega, \tag{6}$$

where E_a and E_b are the eigenstate energies of the two states making up the superposition of the initial state, E_{1s} is the energy of the core ionised state, and the central frequency is given by $E_{1s} - (E_a + E_b)/2$. Thus the amplitude of the field for the transition from the core 1s ionised state to the inner valence state $|a\rangle$ will be the same as for the transition to state $|b\rangle$. The field is assumed to be linearly polarised in the x direction with the field strength given by:

$$\varepsilon(t) = \varepsilon_c(t)\cos(\omega_x t) + \varepsilon_s(t)\sin(\omega_x t), \tag{7}$$

where $\varepsilon_c(t)$ and $\varepsilon_s(t)$ are assumed to be slowly varying compared to $2\pi/\omega_x$.

For the expansion coefficients appearing in the total wavefunction (eqn (5) we make the following ansatz:

$$\tilde{c}_{\{a,b\}}(t) = c_{\{a,b\}}(t)\exp\left[-i\left(E_{1s} - \frac{E_{\{a,b\}} - E_{\{b,a\}}}{2} + \delta/2\right)t\right],$$
$$\tilde{c}_{1s}(t) = c_{1s}(t)\exp\left[-i\left(E_{1s} - \frac{E_a - E_b}{2} - \delta/2\right)t\right], \tag{8}$$
$$\tilde{g}_i(\varepsilon_i, t) = g_i(\varepsilon_i, t)\exp\left[-i\left(E_{1s} - \frac{E_a - E_b}{2} - \delta/2\right)t\right].$$

Equations of motion for the initial states in the superposition can be found by substituting eqn (5) into eqn (3) using the expansion coefficients given in eqn (8) and applying the rotating wave approximation to give:

$$ic_a(t) = -\frac{\delta}{2}c_a(t) + \frac{\mathcal{R}_a^*(t)}{2}c_{1s}(t)\exp\left(-i\frac{E_b - E_a}{2}t\right), \qquad (9)$$

$$ic_b(t) = -\frac{\delta}{2}c_b(t) + \frac{\mathcal{R}_b^*(t)}{2}c_{1s}(t)\exp\left(-i\frac{E_a - E_b}{2}t\right), \qquad (10)$$

where the Rabi frequency $\mathcal{R}_a(t)$ is given by:

$$\mathcal{R}_a(t) = \left\langle 1s\left|\hat{z}\right|a\right\rangle\{\varepsilon_c(t) + i\varepsilon_s(t)\}. \qquad (11)$$

To obtain the equation of motion for the resonant core-ionised state, we treat the decay of this state into the manifold of Auger final states in the standard manner[37,38]:

$$i\dot{c}_{1s}(t) = \frac{\delta}{2}c_{1s}(t) + \frac{\mathcal{R}_a(t)}{2}c_a\exp\left(-i\frac{E_a - E_b}{2}t\right)$$

$$+ \frac{\mathcal{R}_b(t)}{2}c_b\exp\left(-i\frac{E_b - E_a}{2}t\right) - ic_{1s}(t)\frac{\Gamma_{1s}}{2}, \qquad (12)$$

with the total decay width Γ_{1s} given by eqn (4). The equation of motion for the final Auger states are expressed as

$$i\dot{g}_i(\varepsilon_i, t) = \left(E_i^{2+} + \varepsilon_i - E_2 + \frac{\delta}{2}\right)g_i(\varepsilon_i, t) + c_{1s}(t)\sqrt{\frac{\Gamma_i}{2\pi}}, \qquad (13)$$

with E_i^{2+} the energy of the doubly ionised state and ε_i the energy of the Auger electron in the continuum.

Finally the resonant Auger electron line profile associated with the ith decay channel is given as[37]

$$P_i(\varepsilon_i) = \lim_{t \to \infty} |g_i(\varepsilon_i, t)|^2. \qquad (14)$$

Within this model, the area under the resonant Auger electron line profile will evolve with a frequency related to the energy difference between the two states, $|a\rangle$ and $|b\rangle$ that constitute the initial superposition, as can be seen from eqn (12), (13).

3.1 Numerical results

The dynamics of the inner valence hole are driven by electron correlation and therefore the multi-electron states must include configuration interaction. We describe the inner valence ionised initial state, the core ionised decaying (Auger-active) state and the doubly ionised final states using the ADC(2)x ab initio method, which represents the $(N-1)$-electron wavefunction in a basis of inter-mediate states of 1h (ϕ_i) and 2h1p (ϕ_{ij}^a) classes derived from the perturbation theoretically corrected HF ground state of the neutral.[26]

$$\Psi^{(N-1)} = \sum_i c_i\phi_i + \sum_{ija} c_{ij}^a\phi_{ij}^a, \qquad (15)$$

where i, j are hole indices (occupied HF orbitals of the neutral) and a is a particle index (unoccupied HF orbital of the neutral). ADC(2)x describes the 1h-like initial and core-ionised states up to second order in the many-body perturbation theory, while the 2h1p-like final Auger states are described up to first order. The energies of the final states and the partial widths are calculated using Fano-ADC theory, where the widths are given by the matrix elements of the total Hamiltonian minus the energy of the resonance[29] and Stieltjes imaging to energy renormalise the final state wavefunction. The basis sets used are Gaussian $\mathscr{L}^2$ bases supplemented with extra diffuse functions[27] to better describe the continuum states. This methodology is used in all the calculations presented in this discussion.

For our numerical simulations we have assumed a Gaussian temporal envelope for ε_c and set $\varepsilon_s = 0$. The amplitude of the field is 5.14×10^{11} Vm^{-1} with a the root-mean-squared (RMS) width of 2 fs and zero detuning, $\delta = 0$. We assume a uniform intensity distribution; thus spatial averaging over the pulse is not performed. The equations of motion (eqn (9)–(13)) discussed in the previous section are solved using a standard fourth order Runge–Kutta integrator.

In contrast to the survival probability shown in Fig. 3, we first calculate the survival probability of the initial superposition of states when the electric field is applied. This is calculated assuming an equal population of the the state $|a\rangle$ as the state $|b\rangle$, that is $\tilde{c}_a(t=0) = \tilde{c}_b(t=0) = 1/\sqrt{2}$, and inserting the wavefunction at $t' = 0$ (eqn (1)) and the wavefunction at a time t' (eqn (5)), where t' is the propagation time, into the survival probability:

$$S(t') = \left| \langle \Psi^*(t'=0) | \Psi(t') \rangle \right|^2$$
$$= \left| \tilde{c}_a^*(t'=0)\tilde{c}_a(t') + \tilde{c}_b^*(t'=0)\tilde{c}_b(t') \right|^2. \tag{16}$$

Using the above parameters for the pulse, and choosing a probe delay time of 1.2 fs the survival probability of the initial superposition of states is shown in Fig. 5, where the probe pulse transfers the hole to the amino group in (a) and to the carboxyl group in (b). After the probe pulse has passed, population has been transferred to the core-ionised state shown in the reduction of the amplitude, but the remaining population in the initial states shows the same temporal evolution as

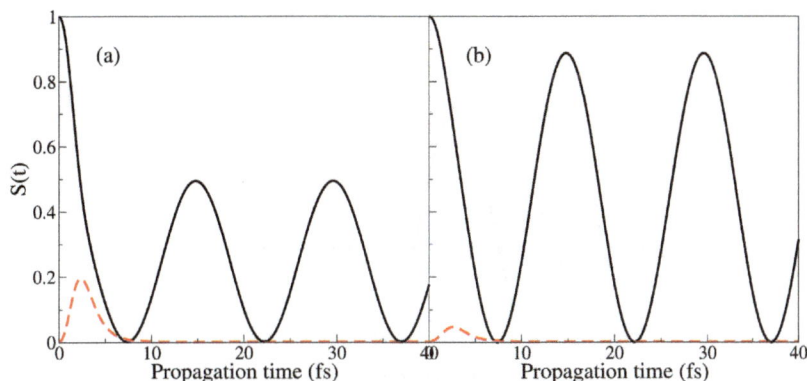

Fig. 5 Survival probability of the initial superposition of states (black), and the probability of populating the core-ionised state $|c_{1s}|^2$ (red dashed line), (a) core ionised from the CH$_2$NH$_2$ group and (b) core ionised from COOH. as a function of propagation time. The probe pulse is centred at 1.2 fs. with an root-mean-squared width of 2 fs.

that shown in Fig. 3 consisting of oscillations with a period of about 15 fs. The underlying dynamics of the initial state have not been destroyed upon application of the the probe pulse, therefore it can be considered as perturbative, even in a field somewhat higher than that required in an experiment. In the equations of motion for the initial states in eqn (9) the dipole coupling between the states, $|a\rangle$ and $|b\rangle$ has been neglected. However, our simulations show that inclusion of this direct high-frequency dipole coupling between these two closely lying states leads to sub-percent corrections with the pulse parameters chosen of 5.14×10^{11} Vm^{-1}. $|c_{2s}|^2$ gives a measure of the transfer of population to the core ionised state resulting from the application of the pulse. Fig. 5 also shows that there is more transfer of population to the core-ionised state where the hole is located on the carbon nearest the amino group. This is because the transition matrix elements are approximately 2.5 times larger for transfer from the amino group to the 10A′ superposition of states compared to transitions from the carboxyl group. The core-ionised state decays to away to zero within 10 fs as a result of coupling to the final Auger states.

The detection part of the pump–probe scheme detailed in Section 2 involves the measurement of the Auger yield as a function of pump–probe delay. We calculate this using eqn (14) where we propagate the equations of motion (eqn (9)–(13)) for 96 fs, considerably further than after the probe pulse has passed. Fig. 6 shows the variation of the Auger yield as a function of pump probe delay. As expected from the analysis of the equations of motion, the modulation in the Auger yield has a frequency related to the energy difference between the two initial states ($|a\rangle$ and $|b\rangle$) and demonstrates that this measurement of the Auger yield successfully captures the dynamics that are characterised by the survival probability shown in Fig. 3. In our present model, we assume that the probe pulse does not have the spectral breadth to couple the two possible core ionised channels. We therefore calculate the Auger yield from each channel separately and sum incoherently. From the survival probabilities Fig. 5 we expect the yield from the amino channel to dominate the expected Auger yield as a function of time because of the differences in the magnitudes of the transition matrix elements.

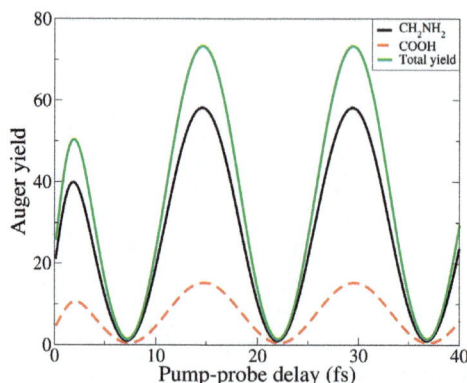

Fig. 6 Auger profile as a function of pump–probe delay. The Auger profile shows the same modulation of signal as is seen in the survival probability of the initial states, which is associated with the energy difference between the two states $|a\rangle$ and $|b\rangle$ that make up the superposition.

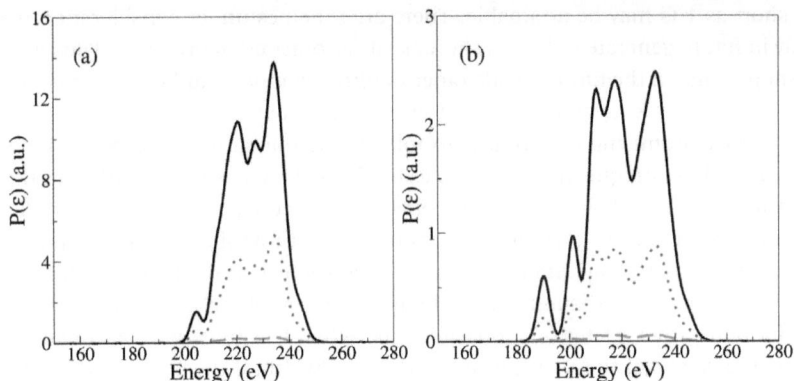

Fig. 7   Calculated Auger profile as a function of energy with vibrational broadening. (a) Spectrum associated with a core vacancy on the CH_2NH_2 and (b) with the vacancy on the COOH carbon centre. The Auger profile at three different pump–probe delays is shown: in black at 14.5 fs corresponding to the first maxima in Fig. 6, in red at 10.3 fs, and in green at 7.3 fs corresponding to just after the first minima in Fig. 6. The entire Auger spectrum is modulated as a function of pump–probe delay.

Finally we show that it is just the magnitude of the Auger profile that changes as a function of the pump–probe delay, as can be seen in Fig. 7. This shows the Auger profile as a function of the energy of the emitted electron for different pump–probe delays. The probe delays correspond to the maxima in Fig. 6 at approximately 14.5 fs, an intermediate time of 10.3 fs, and finally immediately after the minima occurring at 7.3 fs. In each case, the shape of the Auger profile remains the same, and only the amplitude oscillates as a function of pump–probe delay.

4   Discussion

The preceding analysis examines the feasibility of the proposed scheme. The dynamics of conformers Gly I and Gly III are similar and the measured signal will be the result of the weighted sum from these two conformers (plus a small contribution from Gly II). With an expected ratio of 2 : 1 for Gly I and Gly III respectively, we anticipate the dynamics of Gly I to dominate the observed signal. As we noted earlier the concentration of the Gly II conformer is likely to be low for the temperatures likely to be used. Conformer selection is thought to be possible for glycine (J.Kupper, Private Communication) and this would permit signals from a pure sample of a single conformer to be measured.

Of course the simple behaviour that arises purely from the electron dynamics in the glycine will not persist indefinitely. The nuclear dynamics will be revealed by modifications of the oscillation period and amplitude of the hole survival probability that is measured. So far we have not included this into our calculations, but in principle this can be done and this will be required to extract both electronic and nuclear dynamics from the measured signal.

We now turn to the further prospect for X-ray pump - X-ray probe Auger spectroscopy. An extension to measuring faster time-scale dynamics is immediately possible if shorter X-ray pulses can be generated from X-ray FELS. Whilst it is not yet certain, it is likely, that by optimising the operation of a SASE FEL a pulse as short as 1 fs may be attainable. There are schemes under consideration that may in future generate pulses as short as 100 attosecond or even less. Thus it can be hoped that in the future a wide range of ultrafast molecular hole dynamics can be measured by X-ray pump–probe Auger spectroscopy.

Another promising direction is to follow hole migration in space as well as time *via* this technique. In the glycine example we do not expect a significant time dependence in the hole localisation, only in the overall survival probability. The chemical shift of the C K shell at the amine and carboxyl sites could, in principle, be used to register any differences in the hole survival at the two sites. It would allow the separate measurement of the hole survival probability localised at the amine and carboxyl sites of a molecule. The actual feasibility of this will depend upon the details of the X-ray pulse spectral bandwidth. If the pulse bandwidth is near 1 eV or less (*i.e.* much smaller than the C K edge shift between these two sites which is several electron volts) we can anticipate that the pulse will either interact with the amine or the carboxyl site depending upon the tuning of that individual pulse. The data for each channels evolution could thus be separately measured by sorting the data on the photon energy. Such an idea may be extended to measuring hole migration in larger biomolecules.

It is of course possible to use an atom different from the C, N or O common in a biomolecule as a spatial marker to localise the hole survival. Atoms such as S, P and Cl, which have K edges at 2470 eV, 2140 eV and 2820 eV respectively, might occur at only one or a few sites within a larger molecule. Resonant excitation into IVH states from the K shells of these atoms could be used to localise the hole evolution in space and time. These high energy photons will now also cause inner shell ionisation and excitation of C, N and O atoms in the same molecule. However, as the photon energy is significantly higher than the edges of the latter species the cross-sections for these processes will be relatively low. This fact, combined with the anticipated unique Auger spectrum of the inner shell decay of the spatial marker atom, may allow the method to be applied to a wide range of molecules.

5   Conclusion

We have shown that it is possible to measure correlation driven electron dynamics in an amino acid with the X-ray pump–probe technique combined with Auger electron detection. Whilst not all of the correlation driven hole migration processes that one would wish to study have dynamics slow enough to be accessible to the current X-ray FEL pulse duration (3 fs) there are some important classes that are accessible in certain molecules. These are the non-exponential evolutions of superpositions of charge states associated with frustrated Auger decay that occur when inner-valence states are suddenly ionised.[39] Glycine is a molecule that has proved in our theoretical calculations to be highly suitable to observe these somewhat slowed hole migration effects. What is proposed is that a first pulse creates (suddenly) a superposition of hole states in the glycine molecule. Given that the K edge for the two carbons in glycine are at 292.5 eV and 295 eV the X-ray energy needs to be near 270 eV since the binding energy of the states of interest are around 20 eV. The required pulse duration is $\sim$3 fs and the energy bandwidth for a single pulse >1 eV, with the second pulse nominally identical.

The second pulse can then cause photoabsorption from the 1s state into the hole superposition created.

This work indicates that it is feasible to use an X-ray FEL to measure the ultrafast electronic dynamics driven by electron correlation within a typical molecular system (amino acid), a capability that would pave the way to applications to larger molecules (*e.g.* peptides) and condensed phase systems. In principle, the method can be extended to measuring hole migration in larger molecules as the specific atomic core state to inner valence hole transition probability provides time-dependent localisation information for the inner valence hole.

Acknowledgements

The authors acknowledge the financial support of the Engineering and Physical Sciences Research Council (EPSRC, United Kingdom) through the Programme Grant on Attosecond Dynamics (award EP/I032517/1) and the ERC ASTEX project 290467. V.A. acknowledges EPSRC through the Career Acceleration Fellowship (award EP/H003657/1). P.K. acknowledges the support from the Czech Science Foundation (Project GACR P208/12/0521).

References

1 (*a*) F. Remacle and R. D. Levine, *Proceedings of the National Academy of Science*, 2006, **103**, 6793; (*b*) A. I. Kuleff and A. Dreuw, *The Journal of Chemical Physics*, 2009, **130**, 034102; (*c*) H. Eshuis and T. van Voorhis, *Physical Chemistry Chemical Physics*, 2009, **11**, 10293.

2 J. Breidbach and L. S. Cederbaum, *The Journal of Chemical Physics*, 2003, **118**, 3983.

3 (*a*) V. Averbukh, U. Saalmann and J. M. Rost, *Physical Review Letters*, 2010, **104**, 233002; (*b*) J. Craigie, A. Hammad, B. Cooper and V. Averbukh, *J. Chem. Phys.*, 2014, **141**, 014105.

4 D. Mendive-Tapia, M. Vacher, M. J. Bearpark and M. A. Robb, *The Journal of Chemical Physics*, 2013, **139**, 044110.

5 L. S. Cederbaum and J. Zobeley, *Chemical Physics Letters*, 1999, **307**, 205.

6 R. Weinkauf, E. W. Schlag, T. J. Martinez and R. D. Levine, *The Journal of Physical Chemistry A*, 1997, **101**, 7702.

7 M. Drescher, M. Hentschel, R. Kienberger, G. Tempea, C. Spielmann, G. A. Reider, P. B. Corkum and F. Krausz, *Science*, 2001, **291**, 1923–1927.

8 F. Frank, C. Arrell, T. Witting, W. A. Okell, J. McKenna, J. S. Robinson, C. A. Haworth, D. Austin, H. Teng, I. A. Walmsley, J. P. Marangos and J. W. G. Tisch, *Review of Scientific Instruments*, 2012, **83**, 071101.

9 F. Ferrari, F. Calegari, M. Lucchini, C. Vozzi, S. Stagira, G. Sansone and M. Nisoli, *Nature Photonics*, 2010, **4**, 875–879.

10 T. Popmintchev, M.-C. Chen, P. Arpin, M. M. Murnane and H. C. Kapteyn, *Nature Photonics*, 2010, **4**, 822–832.

11 E. Goulielmakis, Z.-H. Loh, A. Wirth, R. Santra, N. Rohringer, V. S. Yakovlev, S. Zherebtsov, T. Pfeifer, A. M. Azzeer, M. F. Kling, S. F. Leone and F. Krausz, *Nature*, 2010, **466**, 739.

12 A. L. Cavalieri, N. Müller, T. Uphues, V. S. Yakovlev, B. Horvath, B. Schmidt, L. Blümel, R. Holzwarth, S. Hendel, M. Drescher, U. Kleineberg, P. M. Echenique, R. Kienberger, R. Krausz and U. Heinzmann, *Nature*, 2007, **449**, 1029.

13 M. Schultze, E. M. Bothschafter, A. Sommer, S. Holzner, W. Schweinberger, M. Fiess, M. Hofstetter, R. Kienberger, V. Apalkov, V. S. Yakovlev, M. I. Stockman and F. Krausz, *Nature*, 2013, **493**, 75.

14 M. Lezius, V. Blanchet, M. Y. Ivanov and A. Stolow, *The Journal of Chemical Physics*, 2002, **117**, 1575.

15 A. I. Kuleff and L. S. Cederbaum, *Physical Review Letters*, 2011, **106**, 053001.

16 J. Leeuwenburgh, B. Cooper, V. Averbukh, J. P. Marangos and M. Ivanov, *Physical Review Letters*, 2013, **111**, 123002.

17 B. Cooper and V. Averbukh, *Physical Review Letters*, 2013, **111**, 083004.

18 F. Lépine, G. Sansone and M. J. J. Vrakking, *Chemical Physics Letters*, 2013, **578**, 1–14.

19 J. Henkel, T. Witting, D. Fabris, M. Lein, P. L. Knight, J. W. G. Tisch and J. P. Marangos, *Physical Review A*, 2013, **87**, 043818.

20 D. Fabris, W. A. Okell, T. Witting, J. P. Marangos and J. W. G. Tisch, pre-print, arXiv:1311.4738.

21 A. A. Zholents and W. M. Fawley, *Physical Review Letters*, 2004, **92**, 224801.

22 Y. Ding, A. Brachmann, F.-J. Decker, D. Dowell, P. Emma, J. Frisch, S. Gilevich, G. Hays, P. Hering, Z. Huang, R. Iverson, H. Loos, A. Miahnahri, H.-D. Nuhn, D. Ratner, J. Turner, J. Welch, W. White and J. Wu, *Physical Review Letters*, 2009, **102**, 254801.

23 Y. Ding, F.-J. Decker, P. Emma, C. Feng, C. Field, J. Frisch, Z. Huang, J. Krzywinski, H. Loos, J. Welch, J. Wu and F. Zhou, *Physical Review Letters*, 2012, **109**, 254802.

24 J. J. Neville, Y. Zheng and C. E. Brion, *Journal of the American Chemical Society*, 1996, **118**, 10533.

25 A. G. Császár, *Journal of the American Chemical Society*, 1992, **114**, 9568.

26 A. B. Trofimov and J. Schirmer, *The Journal of Chemical Physics*, 2005, **123**, 144115.

27 K. Kaufmann, W. Baumeister and M. Jungen, *Journal of Physics B*, 1989, **22**, 2223.

28 J. S. L. S. Cederbaum, W. Domcke and W. von Niessen, *Adv. Chem. Phys.*, 1986, **65**, 115.

29 V. Averbukh and L. S. Cederbaum, *The Journal of Chemical Physics*, 2005, **123**, 204107.

30 J. Schirmer and A. Barth, *Zeitschrift für Physik A*, 1984, **317**, 267.

31 R. R. Blyth, R. Delaunay, M. Zitnik, J. Krempasky, R. Krempaska, J. Slezak, K. C. Prince, R. Richter, M. Vondracek, R. Camilloni, L. Avaldi, M. Coreno, G. Stefani, C. Furlani, M. de Simone, S. Stranges and M.-Y. Adam, *Journal of Electron Spectroscopy and Related Phenomena*, 1999, **101**, 959.

32 H. J. Svec and D. D. Clyde, *Journal of Chemical and Engineering Data*, 1965, **10**, 151.

33 C. G. de Kruif, J. Voogd and J. C. A. Offringa, *The Journal of Chemical Thermodynamics*, 1979, **11**, 651.

34 F. Filsinger, G. Meijer, H. Stapelfeldt, H. N. Chapman and J. Küpper, *Physical Chemistry Chemical Physics*, 2011, **13**, 2076.

35 O. Kornilov, M. Eckstein, M. Rosenblatt, C. P. Schulz, K. Motomura, A. Rouzée, J. Klei, L. Foucar, M. Siano, A. Lübcke, F. Schapper, P. Johnsson, D. M. P. Holland, T. Schlatholter, T. Marchenko, S. Düsterer, K. Ueda, M. J. J. Vrakking and L. J. Frasinski, *Journal of Physics B*, 2013, **46**, 164028.

36 L. J. Frasinski, V. Zhaunerchyk, M. Mucke, R. J. Squibb, M. Siano, J. H. D. Eland, P. Linusson, P. van der Meulen, P. Salén, R. D. Thomas, M. Larsson, L. Foucar, J. Ullrich, K. Motomura, S. Mondal, K. Ueda, T. Osipov, L. Fang, B. F. Murphy, N. Berrah, C. Bostedt, J. D. Bozek, S. Schorb, M. Messerschmidt, J. M. Glownia, J. P. Cryan, R. Coffee, O. Takahashi, S. Wada, M. N. Piancastelli, R. Richter, K. C. Prince and R. Feifel, *Physical Review Letters*, 2013, **111**, 073002.

37 N. Rohringer and R. Santra, *Physical Review A*, 2008, **77**, 053404.

38 P. Meystre and M. Sargent III, *Elements of Quantum Optics*, Springer, 1991.

39 A. I. Kuleff and L. S. Cederbaum, *Chemical Physics*, 2007, **338**, 320.

Faraday Discussions

PAPER

The sensitivities of high-harmonic generation and strong-field ionization to coupled electronic and nuclear dynamics

Denitsa Baykusheva,[a] Peter M. Kraus,[a] Song Bin Zhang,[bc] Nina Rohringer[bc] and Hans Jakob Wörner[a]

Received 21st February 2014, Accepted 14th April 2014

DOI: 10.1039/c4fd00018h

The sensitivities of high-harmonic generation (HHG) and strong-field ionization (SFI) to coupled electronic and nuclear dynamics are studied, using the nitric oxide (NO) molecule as an example. A coherent superposition of electronic and rotational states of NO is prepared by impulsive stimulated Raman scattering and probed by simultaneous detection of HHG and SFI yields. We observe a fourfold higher sensitivity of high-harmonic generation to electronic dynamics and attribute it to the presence of inelastic quantum paths connecting coherently related electronic states [Kraus et al., Phys. Rev. Lett. 111, 243005 (2013)]. Whereas different harmonic orders display very different sensitivities to rotational or electronic dynamics, strong-field ionization is found to be most sensitive to electronic motion. We introduce a general theoretical formalism for high-harmonic generation from coupled nuclear-electronic wave packets. We show that the unequal sensitivities of different harmonic orders to electronic or rotational dynamics result from the angle dependence of the photorecombination matrix elements which encode several autoionizing and shape resonances in the photoionization continuum of NO. We further study the dependence of rotational and electronic coherences on the intensity of the excitation pulse and support the observations with calculations.

1 Introduction

The motion of electrons determines the basic properties of atoms and molecules. Since electronic motion occurs on femto- down to attosecond time scales, its characterization and control depends on the availability of ultrafast radiation sources combined with efficient detection techniques. Recent advances in this direction include the development of attosecond streaking, which enabled the measurement of Auger decay times[1] and photoemission delays.[2] Resolving

[a]ETH Zürich, Laboratory of Physical Chemistry, 8093 Zurich, Switzerland. E-mail: woerner@phys.chem.ethz.ch
[b]Max Planck Institute for the Physics of Complex Systems, 01187 Dresden, Germany
[c]Center for Free-Electron Laser Science, 22607 Hamburg, Germany

electron dynamics has also been approached employing interferometric techniques,[3] transient absorption[4-6] or strong-field ionization (SFI).[7,8] The methods enumerated so far, however, operate on highly excited states or ionic species. Recently, we have reported the first experiment measuring an electronic wave packet involving the ground electronic state of a neutral molecule.[9]

In the past, the possibility to study coherent electronic wave packets by high-harmonic generation (HHG) has been subject to an intensive theoretical investigation.[10-13] Our new pump–probe technique,[9] relying on impulsive stimulated Raman scattering (ISRS) and HHG, provides unprecedented sensitivity and is thus ideal to study weakly allowed electronic transitions and their coupling to the other motional degrees of freedom in the time domain. The technique is directly sensitive to the electronic coherence and its evolution due to previously unobserved HHG cross channels connecting distinct but coherently related states.[9] The method differs from previous applications of high-harmonic spectroscopy,[14] where a photoexcited molecular wave packet was followed by HHG. The latter technique enabled the resolution of the conical intersection dynamics in NO_2[15,16,17] and the photodissociation of CH_3I and CF_3I[18] but was blind to the cross-channels that reveal the electronic coherence. The present technique also differs from previous measurements of electronic dynamics in molecular ions by HHG[19-22] in the sense that the latter are also blind to electronic coherence between the levels of the cation.

The study of electronic dynamics further requires the development of theoretical frameworks predicting the interaction of a system with an intense laser field and the HHG probe process. Existing methods are based on the density-matrix formalism[23-25] or S-matrix-based approaches.[26-28] Quantitative rescattering theory,[29] which expresses the HHG intensity as a product of a returning electron wave packet and photoionization molecular-frame matrix elements,[30,31] is another popular approach that has been applied to a wide range of diatomic molecules and even polyatomic species.[32]

In the present article, we extend our previous study of electronic wave packets in aligned molecules[9] to longer pump–probe delays, and demonstrate how the complete quantum-level structure of two electronic states can be determined by Fourier transforming the HHG or SFI signals. This approach also enables us to resolve different types of coherences in the frequency domain: rotational, electronic and mixed coherences. We directly compare the sensitivities of high-harmonic generation and strong-field ionization to the electronic dynamics and find that the sensitivity of the former exceeds that of the latter by a factor of $\sim$4. We develop a closed-form theoretical treatment to describe both the excitation and the probing steps. The comparison of theory and experiment shows that the angle dependence of the photorecombination matrix elements is the origin of the different sensitivities of the various harmonic orders to the rotational motion. Several autoionizing and shape resonances in the photoionization continuum of NO around 14 eV are mapped into a pronounced signal modulation at the rotational revivals. Finally, we present a systematic study of the pump-pulse intensity which shows that the rotational excitation saturates at lower intensities than the electronic excitation.

2 Theory

2.1 Field-free rotational structure

The open-shell nature of the ground-state electronic configuration of the NO radical gives rise to two fine-structure components, with $^2\Pi_{1/2}$ being the ground-state and $^2\Pi_{3/2}$ lying $\sim$123 cm^{-1} higher in energy.[33] The field-free Hamiltonian is

$$\hat{H}_0 = \hat{H}_{\mathrm{rot}} + \hat{H}_{\mathrm{SO}} = hcB(\mathbf{J}-\mathbf{L}-\mathbf{S})^2 - hcA\mathbf{L}\cdot\mathbf{S}, \tag{1}$$

whereby $B \approx 1.6961$ cm^{-1} and $A \approx 123.1314$ cm^{-1} designate the ground-state rotational and the spin–orbit coupling constant,[33] respectively. $\mathbf{L}$ and $\mathbf{S}$ stand for the total orbital and spin angular momentum operators, whereas the total angular momentum operator exclusive of nuclear degrees of freedom is denoted by $\mathbf{J}$. Provided that the rotational excitation is low, Hund's coupling scheme (a) is applicable, in which case one uses the quantum numbers Λ, Σ and Ω to quantify the projections of $\mathbf{L}$, $\mathbf{S}$ and $\mathbf{J}$ on the molecule-fixed axis. In this limit, it is convenient to adopt the parity-adapted basis set[34] defined by

$$|J|\Omega|M\varepsilon\rangle = \frac{1}{\sqrt{2}}[|J,|\Omega|,M\rangle + \varepsilon|J,-|\Omega|,M\rangle], \tag{2}$$

wherein $\varepsilon = \pm 1$ represents a symmetry index related to the total parity of the wave function p as $p = \varepsilon(-1)^{J-1/2}$. M quantifies the projection of the total angular momentum vector $\mathbf{J}$ on the reference axis in the laboratory frame and is conserved in the present experiment. The eigenfunctions $|J,\Omega,M\rangle$ relate to the elements of the Wigner rotational matrix[35] as

$$\langle \phi, \theta, \chi | J, \Omega, M \rangle \equiv \sqrt{\frac{2J+1}{4\pi}} D^{J*}_{M\Omega}(\phi, \theta, \chi = 0), \tag{3}$$

where (ϕ, θ, χ) are the Euler angles defining the orientation of the body-fixed frame with respect to the lab frame. The Euler angle χ is redundant for a linear molecule and is therefore set to zero according to the convention used in ref. 34. Under the assumption that interactions with higher-lying Σ-electronic states[36] (Λ-doubling) can be neglected, the wave functions corresponding to the two values of ε can be treated as degenerate. In the case of the NO molecule, $|\Omega|$ assumes the values $\frac{1}{2}$ and $\frac{3}{2}$, and the matrix representation of the field-free Hamiltonian in eqn (1) becomes

$$\hat{\mathbf{H}}_0 = \begin{pmatrix} B\left(J-\frac{1}{2}\right)\left(J+\frac{3}{2}\right) & -B\sqrt{\left(J-\frac{1}{2}\right)\left(J+\frac{3}{2}\right)} \\ -B\sqrt{\left(J-\frac{1}{2}\right)\left(J+\frac{3}{2}\right)} & A-2B+B\left(J-\frac{1}{2}\right)\left(J+\frac{3}{2}\right). \end{pmatrix} \tag{4}$$

Diagonalizing the above expression, one obtains for the eigenbasis:

$$\begin{pmatrix} |JM\varepsilon;1\rangle \\ |JM\varepsilon;2\rangle \end{pmatrix} = \begin{pmatrix} a_J & b_J \\ -b_J & a_J \end{pmatrix} \begin{pmatrix} |J\frac{1}{2}M\varepsilon\rangle \\ |J\frac{3}{2}M\varepsilon\rangle \end{pmatrix}, \tag{5}$$

wherein the coefficients a_J and b_J are functions of the rotational and spin–orbit constants and obey the relationship $a_J^2 + b_J^2 = 1$. The eigenstates in this new basis are labelled F_1 and F_2 and the corresponding eigenenergies are given by

$$\begin{cases} E_{F_1} = B\left(J - \dfrac{1}{2}\right)\left(J + \dfrac{3}{2}\right) - \sqrt{B^2\left(J + \dfrac{1}{2}\right)^2 + \dfrac{A(A - 4B)}{4}} + \dfrac{A - 2B}{2} \\[4mm] E_{F_2} = B\left(J - \dfrac{1}{2}\right)\left(J + \dfrac{3}{2}\right) + \sqrt{B^2\left(J + \dfrac{1}{2}\right)^2 + \dfrac{A(A - 4B)}{4}} + \dfrac{A - 2B}{2} \end{cases} . \tag{6}$$

The limit of a pure Hund's coupling case (a) is characterized by the values $a_J = 1$ and $b_J = 0$ (or *vice versa*). Here, $a_J \sim 1$ and $b_J \sim 0$ for low values of J, thus the F_1 state in NO is dominated by the $^2\Pi_{1/2}$ fine-structure component, whereas the F_2 state is $^2\Pi_{3/2}$-dominated.

2.2 Pump pulse interaction: electronic and rotational Raman transitions

When subject to a short†, intense non-resonant laser pulse, the time evolution of the system obeys the Hamiltonian

$$\hat{H}_{\text{M--L}}(t) = \hat{H}_0 + \hat{H}_{\text{pump}}(t), \tag{7}$$

where the term $\hat{H}_{\text{pump}}(t)$ conveys the interaction between the molecule and the incident electromagnetic field. The linearly polarized pump pulse is modelled as a Gaussian function in the temporal domain

$$\vec{\varepsilon}_{pu}(t) = \hat{\varepsilon}_{pu}\varepsilon_{pu}(t)\cos(\omega_0 t) = \hat{\varepsilon}_{pu}\varepsilon_{pu,0}e^{-2\ln 2(t/\tau_{pu})^2}\cos(\omega_0 t), \tag{8}$$

wherein $\hat{\varepsilon}_{pu}$ is a unit vector parallel to the polarization axis of the field, $\varepsilon_{pu,0}$ is the electric field amplitude, τ_{pu} denotes the duration of the pulse‡ and ω_0 is the fundamental frequency of the carrier field. In the current work, ω_0 corresponds to a wavelength of 800 nm, the pulse duration is estimated to be 60 fs and the peak intensity lies in the range $3{-}6 \times 10^{13}$ W cm^{-2} in the present experiments. The cycle-averaged interaction Hamiltonian reads

$$\begin{aligned} \hat{H}_{\text{pump}}(t) &= -\frac{\varepsilon_{pu}^2(t)}{4}\hat{\varepsilon}_{pu}^{\text{T}}\underline{\underline{\alpha}}\hat{\varepsilon}_{pu} \\[2mm] &= -\frac{\varepsilon_{pu}^2(t)}{4}\left[\frac{2}{3}\Delta\alpha\left(D_{00}^2(\phi,\theta,\chi) + \gamma\left(D_{02}^2(\phi,\theta,\chi) + D_{0-2}^2(\phi,\theta,\chi)\right)\right) \right. \\[2mm] &\quad \left. + \frac{1}{3}\Delta\alpha + \alpha_\perp\right]. \end{aligned} \tag{9}$$

In the static-field limit, the only non-trivial elements of the polarizability tensor $\underline{\underline{\alpha}}$ (evaluated in the principle axis system of the molecule) are $\alpha_\perp \equiv \alpha_{xx} = \alpha_{yy}$ and α_{zz}, with $\Delta\alpha = \alpha_{zz} - \alpha_\perp$. For NO, $\alpha_\perp = 9.715$ a.u. and $\alpha_{zz} = 15.34$ a.u.[37] The

---

† In this context, the term "short" signifies that the pulse duration is significantly smaller than the rotational period of the molecule.

‡ Here and in the remainder of this article, the pulse duration designates the full width at half maximum of the electric-field envelope.

orientation dependence of the interaction is encoded in the Wigner rotation matrices occurring in eqn (9). Using angular momentum algebra arguments, it can be readily proven that matrix elements involving $D^2_{00}(\phi,\theta,\chi)$ capture most of the rotational Raman transitions. The dependence of $\hat{H}_{\text{pump}}$ on $D^2_{0\pm2}(\phi,\theta,\chi)$ accounts for the largest portion of the electronic-rotational Raman excitations. The parameter γ in eqn (9) quantifies the ratio between electronic and purely rotational Raman scattering and has been assigned the empirical value of 0.2.[38] The interaction with the electromagnetic field prepares the system in a superposition of coupled rotational and spin–orbit electronic states and the ensuing dynamics is dictated by

$$i\partial_t|\Phi_{J_0 M_0 \varepsilon_0 i_0}(t)\rangle = \hat{H}_{\text{M–L}}(t)|\Phi_{J_0 M_0 \varepsilon_0 i_0}(t)\rangle. \tag{10}$$

In the above, the fundamental solution $|\Phi_{J_0 M_0 \varepsilon_0 i_0}(t)\rangle$ describes an electronic-rotational wave packet that uniquely evolves from an initially occupied eigenstate $|J_0 M_0 \varepsilon_0; i_0\rangle$ with $i_0 \in \{1, 2\}$ (cp. eqn (5)).[23] Exploiting the orthonormality of the functions $\{|JM_0\varepsilon;i\rangle\}$, the solution of eqn (10) can be obtained by expanding $\Phi_{J_0 M_0 \varepsilon_0 i_0}(t)$ in terms of the basis functions $\{|JM_0\varepsilon;i\rangle\}$

$$|\Phi_{J_0 M_0 \varepsilon_0 i_0}(t)\rangle = \sum_{J\varepsilon i} C^{J_0 M_0 \varepsilon_0}_{F_i}(J\varepsilon; t)|JM_0\varepsilon; i\rangle \tag{11}$$

and solving the resulting coupled differential equations for the expansion coefficients $\left\{C^{J_0 M_0 \varepsilon_0}_{F_i}(J\varepsilon; t)\right\}$ by imposing the initial condition

$$|\Phi_{J_0 M_0 \varepsilon_0 i_0}(t_0)\rangle = |J_0 M_0 \varepsilon_0; 1\rangle, \tag{12}$$

i.e., the entire population resides initially in the F_1 spin–orbit component. The density matrix of the system $\rho(t)$ is formed by summing over the contributions of all initially occupied rotational states

$$\rho(t) = \sum_{J_0 M_0 \varepsilon_0 i_0} w_{J_0}|\Phi_{J_0 M_0 \varepsilon_0 i_0}(t)\rangle\langle\Phi_{J_0 M_0 \varepsilon_0 i_0}(t)|, \tag{13}$$

wherein $\{w_{J_0}\}$ are Boltzmann distribution coefficients corresponding to a rotational temperature of 15 K.

2.3 Calculation of the high-harmonic intensity

In this section, we describe the basic formalism for calculating the high-harmonic emission from a pure state $\Psi(\hat{R},\tau)$ exposed to the field of the probe pulse. $\hat{R}$ is a short-hand notation for the Euler angles (ϕ,θ,χ). The harmonic intensity is proportional to the square of the Fourier transform $|\vec{D}(\hat{R},\omega)|^2$ of the dipole-moment expectation value:

$$\vec{D}(\hat{R},\tau) = \langle\Psi(\hat{R},\tau)|\hat{\vec{\mu}}_{\text{el}}|\Psi(\hat{R},\tau)\rangle. \tag{14}$$

In the framework of the QRS theory, the induced dipole moment representing high-harmonic emission is decomposed into a product of a returning electron wave packet $W(\omega)$ and the photorecombination cross section of the laser-field-free continuum recombining back to the initial ground state.[29,31] For a single molecule whose orientation is defined by $\hat{R}$ with respect to the polarization axis of the driving field, the resulting expression for the dipole moment in the frequency domain reads

$$\vec{D}(\hat{R}, \omega) = \sqrt{\Gamma(\hat{R})}\, W(\omega)\, \vec{d}_{rec}(\hat{R}, \omega), \tag{15}$$

where $\Gamma(\hat{R})$ is the calculated[39] angle-dependent strong-field ionization rate. $W(\omega)$ denotes the complex spectral representation of the recombining photoelectron wave packet. The photorecombination matrix elements $\vec{d}_{rec}(\hat{R},\omega)$ are independent of the laser parameters and encode the dependence of the calculated harmonic spectra on the structure of the target.

2.4 Calculation of photorecombination matrix elements

The photoionization of NO is treated according to the method described by Lucchese et al.[40] and Stratmann et al.[41] including 10 ion-state channels up to an ionization potential of 23.5 eV and all associated interchannel couplings. Spin-orbit interaction is neglected in these calculations. The initial state is the $^2\Pi$ electronic ground state of NO, denoted as $\Psi^{(\Lambda_i)}$ and characterized by the electronic angular momentum projection quantum number $\Lambda_i = \pm1$. Photoionization matrix elements are calculated for ionization to the $X^1\Sigma^+$-ground state of NO^+, $\Phi^{(\Lambda_f)}$, with $\Lambda_f = 0$, and the continuum photoelectron is represented in a single-center expansion in terms of the basis functions Ψ_{klm}, where k denotes the momentum and l and m are the orbital and projection quantum numbers, respectively. The photoionization dipole matrix elements assume the following form in the spherical basis

$$I_{\hat{k},\hat{n}}^{(\Lambda_i,\Lambda_f)} = \sum_{l,m,\mu} \left\langle \Psi^{(\Lambda_i)} |r_\mu| \Phi^{(\Lambda_f)} \Psi_{klm} \right\rangle Y_{l,m}^*(\hat{k})\, Y_{1,\mu}^*(\hat{n}), \tag{16}$$

where $\hat{n}$ denotes the polarization of light and r_μ ($\mu = 0, \pm1$) are the spherical components of the dipole moment operator in the length gauge. The dependence of the photoionization cross section (PICS) on the orientation of the target is captured by the spherical harmonic functions $Y_{l,m}$.

2.5 High-harmonic generation from a coupled electronic-rotational wave packet

Before developing the formalism describing high-harmonic emission from a coupled electronic-rotational wave packet, we introduce two different time scales. The time scale denoted by t labels the time evolution with respect to the pump pulse, whereas τ labels the time scale of high-harmonic generation within the duration of the probe pulse. Since the wave-packet evolution described by the variable t is slow compared to the sub-femtosecond time scale of HHG described by τ, we neglect the time-evolution of the wave packet during HHG and use a parametric dependence on t in the following equations.

Building on the approach outlined by Ramakrishna and Seideman,[23] the quantum state created from the initially-occupied state $|F_{i_0}; J_0 M_0 \varepsilon_0\rangle$ after the strong-field ionization step can be written as

$$|\Phi_{J_0 M_0 \varepsilon_0 i_0}(\tau;t)\rangle = \sum_{J\varepsilon i} C_{F_i}^{J_0 M_0 \varepsilon_0}(J\varepsilon;t)|F_i;JM_0\varepsilon\rangle e^{i\left(I_p - E_J^{F_i}\right)\tau}$$

$$+ \sum_{J_c M_c} \int d^3 k C_{J_c}^{M_c}(\vec{k};\tau)|\vec{k};J_c M_c\rangle e^{iI_p\tau}, \tag{17}$$

where I_p is the vertical ionization potential of the spin-rovibronic ground state and $E_J^{F_i}$ denote the internal energies of the excited spin-rovibronic eigenstates of NO. In the above, $|\vec{k};J_c M_c\rangle = |\vec{k}\rangle \otimes |J_c M_c\rangle$ where $|\vec{k}\rangle$ denotes the electronic continuum associated with the asymptotic momentum $\vec{k}$ and $|J_c M_c\rangle$ specifies the rotational states of the ionic core. The notation $|F_i;JM_0\varepsilon\rangle$ is here to designate a rotational-electronic eigenstate of the molecular Hamiltonian that explicitly contains the complete wave function of the unpaired electron (in a single-active electron approximation). The continuum coefficients $C_{J_c}^{M_c}(\vec{k};\tau)$ are calculated using the strong-field approximation (SFA). Exploiting the fact that $E_J^{F_i}$ is negligible with respect to the magnitude of the ionization potential I_p, eqn (14), (13) and (17) can be combined to yield an expression for the induced dipole moment $d(\tau;t)$:

$$d(\tau;t) = i \sum_{J_0 M_0 \varepsilon_0 i_0} w_{J_0} \int d\hat{R} \sum_{J\varepsilon i} C_{F_i}^{J_0 M_0 \varepsilon_0 *}(J\varepsilon;t) \sum_{J'\varepsilon' i'} C_{F_{i'}}^{J_0 M_0 \varepsilon_0}(J'\varepsilon';t)$$

$$\times \int d^3 k \langle F_i;JM_0\varepsilon|\hat{\vec{\mu}}_{el} \cdot \hat{\varepsilon}_{pr}|\vec{k};\hat{R}\rangle \int_0^\tau d\tau' \langle \vec{k}';\hat{R}|\hat{\vec{\mu}}_{el} \cdot \vec{\varepsilon}_{pr}(\tau')|F_{i'};J'M_0\varepsilon'\rangle e^{-iS(\tau,\tau')} \tag{18}$$

with $S(\tau, \tau')$ being the time-dependent phase in the SFA. Eqn (18) has a transparent physical interpretation. The last two matrix elements encode the tunnel ionization initiating from state $|F_{i'};J'M_0\varepsilon'\rangle$ followed by a recombination to the state labeled $|F_i;JM_0\varepsilon\rangle$, whereas the coefficients $\{C_{F_i}^{J_0 M_0 \varepsilon_0}(J\varepsilon;t)\}$ contain the coupled rotational-electronic dynamics induced by the pump pulse. The completeness relation pertaining to the basis set $|JM\varepsilon;i\rangle$ enables one to project out the rotational degrees of freedom in $|F_i;JM\varepsilon\rangle$, thus arriving at a formal definition of a purely electronic factor that is a function of the electronic real-space coordinate $\{\vec{r}\}$ only

$$\langle \vec{r}| \otimes \langle J'M_0\varepsilon'|F_i;JM_0\varepsilon\rangle = \delta_{JJ'}\delta_{\varepsilon\varepsilon'}\langle \vec{r}|F_i^{J\varepsilon}\rangle. \tag{19}$$

This result allows one to decompose the matrix elements in eqn (18) into a rotational and an electronic factor

$$\langle F_i;JM_0\varepsilon|\hat{\vec{\mu}}_{el} \cdot \hat{\varepsilon}_{pr}|\vec{k};\hat{R}\rangle = \langle F_i^{J\varepsilon}|\hat{\vec{\mu}}_{el} \cdot \hat{\varepsilon}_{pr}|\vec{k}\rangle\langle JM_0\varepsilon;i|\hat{R}\rangle; \tag{20}$$

$$\langle \hat{R};\vec{k}'|\hat{\vec{\mu}}_{el} \cdot \vec{\varepsilon}_{pr}(t)|F_i;JM_0\varepsilon\rangle = \langle \vec{k}'|\hat{\vec{\mu}}_{el} \cdot \vec{\varepsilon}_{pr}(t)|F_i^{J\varepsilon}\rangle\langle \hat{R}|JM_0\varepsilon;i\rangle. \tag{21}$$

For a diatomic molecule with a single unpaired electron, the electronic part of the wavefunction $|F_i^{J\varepsilon}\rangle$ can be formulated in terms of the product $|\Phi_{\text{HOMO}}\rangle e^{i\Lambda\chi_e}$, where $|\Phi_{\text{HOMO}}\rangle$ is a function of all coordinates of the unpaired electron except the cylindrical azimuthal angle χ_e. The latter gives rise to an anisotropy in the electronic charge distribution as discussed in ref. 42. In this particular case, however, the above anisotropy plays a negligible role since the low degrees of rotational excitation ensure the validity of the Hund's case a) limit. Thus the following approximation holds:

$$\left\langle F_1^{J\varepsilon} | \hat{\vec{\mu}}_{el} \cdot \hat{\varepsilon}_{pr} | \vec{k} \right\rangle \simeq \left\langle F_2^{J\varepsilon} | \hat{\vec{\mu}}_{el} \cdot \hat{\varepsilon}_{pr} | \vec{k} \right\rangle, \tag{22}$$

and the resulting expression for the dipole moment expectation value reads:

$$d(\tau; t) = i \sum_{J_0 M_0 \varepsilon_0 i_0} w_{J_0} \int d^3k \left\langle \Phi_{HOMO} | \hat{\vec{\mu}}_{el} \cdot \hat{\varepsilon}_{pr} | \vec{k} \right\rangle \times \int_0^\tau d\tau' \left\langle \vec{k} | \hat{\vec{\mu}}_{el} \cdot \vec{\varepsilon}_{pr}(\tau') | \Phi_{HOMO} \right\rangle e^{-iS(\tau,\tau')}$$

$$\times \int d\hat{R} \sum_{J\varepsilon i} C_{F_i}^{J_0 M_0 \varepsilon_0 *}(J, \varepsilon; t) \left\langle JM_0\varepsilon; i | \hat{R} \right\rangle \sum_{J'\varepsilon' i'} C_{F_i}^{J_0 M_0 \varepsilon_0}(J', \varepsilon'; t) \left\langle \hat{R} | J' M_0 \varepsilon'; i' \right\rangle. \tag{23}$$

By defining a density matrix in Euler-angle space

$$\rho_p(\hat{R}, t) = \sum_{J\varepsilon i} C_{F_i}^{J_0 M_0 \varepsilon_0 *}(J, \varepsilon; t) \left\langle JM_0\varepsilon; i | \hat{R} \right\rangle \sum_{J'\varepsilon' i'} C_{F_i}^{J_0 M_0 \varepsilon_0}(J', \varepsilon'; t) \left\langle \hat{R} | J' M_0 \varepsilon'; i' \right\rangle \tag{24}$$

one obtains for eqn (23)

$$d(\tau; t) = \int d\hat{R} \rho_p(\hat{R}, t) \vec{D}(\hat{R}, \tau), \tag{25}$$

wherein

$$\vec{D}(\hat{R}, t) = i \int d^3k \left\langle \Phi_{HOMO} | \hat{\vec{\mu}}_{el} \cdot \hat{\varepsilon}_{pr} | \vec{k} \right\rangle \int_0^\tau d\tau' \left\langle \vec{k} | \hat{\vec{\mu}}_{el} \cdot \vec{\varepsilon}_{pr}(\tau') | \Phi_{HOMO} \right\rangle e^{-iS(\tau,\tau')} \tag{26}$$

is the dipole-moment expectation value evaluated over a single pure state as defined in eqn (14) and (15). Consequently, $\rho_p(\hat{R},t)$ can be interpreted as a factor that weights the contribution of each of the electronic integrals determining $\vec{D}(\hat{R},\tau)$ for a given molecular orientation $\hat{R}$. The calculations described in what follows were done by replacing the SFA expressions for strong-field ionization and photorecombination in eqn (26) with the strong-field ionization rate from ref. 39 and the photorecombination matrix elements described in Section 2.4. In order to arrive at a final expression for the harmonic intensity as a function of the pump–probe delay, the expressions in eqn (15) and (25) have to be integrated over all possible molecular orientations. This operation is straightforward in case the polarizations of the pump and the probe pulses coincide, and one obtains for the harmonic intensity $I(\omega,t)$

$$I(\omega, t) = \left| \left| \int_0^\pi \sin\theta d\theta \rho_p(\hat{R}, t) \vec{D}_\parallel(\hat{R}, t) \right| \right|^2. \tag{27}$$

The treatment of the case where pump and probe polarizations differ requires the introduction of an additional system of variables (θ', ϕ') denoting the polar and the azimuthal angles of the molecular axis in a frame attached to the probe field.[43] The latter pair of variables is related to the polar and azimuthal angles attached to the pump-pulse frame and the angle between the two polarizations α by

$$\cos\theta = \cos\theta'\cos\alpha + \sin\theta'\sin\alpha\cos\phi'. \tag{28}$$

The intensity of the emitted high-harmonic radiation can be obtained by evaluating

$$I(\omega, t, \alpha) = \left| \int_0^{2\pi} d\phi' \int_0^{\pi} \sin\theta' d\theta' \vec{D}_\parallel(\hat{R}, t)\rho_p(\theta\{\phi', \theta'; \alpha\}, t) \right|^2 +$$

$$+ \left| \int_0^{2\pi} d\phi' \cos\phi' \int_0^{\pi} \sin\theta' d\theta' \vec{D}_\perp(\hat{R}, t)\rho_p(\theta\{\phi', \theta'; \alpha\}, t) \right|^2. \tag{29}$$

3   Experimental

The experimental setup consists of an amplified femtosecond titanium:sapphire laser system (10 mJ, 25 fs, 1 kHz, 800 nm center wavelength), an optical setup and a vacuum chamber for generation and spectral characterization of high-harmonic radiation. The output of the laser system is split into multiple beams. One of the pulses (pump, 60 fs) prepares the coupled electronic-rotational wave packet in NO through impulsive Raman scattering, while the other part (probe, 30 fs) is used to generate high-harmonic radiation with a cut-off at harmonic 27 ($\sim$ 42 eV). A translation stage is employed to control the temporal delay between the pump and the probe pulses. The two pulses impinge on a spherical mirror with a vertical offset of 7 mm and are focused non-collinearly into the molecular beam inside a vacuum chamber. The molecular beam is generated by an expansion of a 5% mixture of NO in He through a pulsed valve with a backing pressure of 9 bars. The total ion yield is measured by recording the electrical current flowing through a wire mesh placed 15 cm below the orifice of the valve and held at a relative potential of -1 kV. The peak intensity of the pump beam was varied in the range $(3.2 \pm 0.3) \times 10^{13}$ W cm^{-2} $- (6.0 \pm 0.5) \times 10^{13}$ W cm^{-2}, whereas typical values for the probe intensity span the range $(1.0 \pm 0.2) \times 10^{14}$ W cm^{-2} $-(1.5 \pm 0.2) \times 10^{14}$ W cm^{-2}. The polarization of the probe beam is kept unchanged, whereas that of the pump beam is varied. The high-harmonic radiation generated by the probe beam propagates into an extreme-ultraviolet (XUV) spectrometer consisting of a 250 μm wide entrance slit, a concave aberration-corrected grating (Shimadzu, 30–002), and a microchannel-plate detector backed with a phosphor screen. The spectral images are recorded by a charge-coupled device camera and subsequently sent to a computer for analysis.

4   Results and discussion

4.1   Electronic wave packets probed by strong-field ionization and high-harmonic generation

We now discuss strong-field ionization and high-harmonic generation from a coherent superposition of electronic states. In Fig. 1a, we show the total strong-field ionization yield as a function of the pump–probe delay for a situation where the polarizations of the two beams coincide (green curve) or are orthogonal to each other (orange curve). The harmonic intensity, integrated over harmonic orders 9 to 23 ($H9 - H23$), is shown in Fig. 1b for the two different polarization configurations. In both excitation schemes, the electronic wave packet prepared by the excitation pulse translates into a modulation with a period of

$$T = \frac{h}{\Delta E_{SO}} \approx 275 \text{ fs},$$ commensurable with the energy difference ΔE_{SO} between the F_1 and F_2 components. Comparing the ionization yield with the harmonic intensity, we find that the modulation depth increases by a factor of ≈ 4 in the latter case. In order to explain this behaviour, we next focus on the characteristic features of each probing mechanism.

Fig. 1 illustrates strong-field ionization (panel (c)) or high-harmonic generation (panel (d)) from the superposition state $\Psi = |c_1|e^{i\phi_1}\psi_1 + |c_2|e^{i\phi_2}\psi_2$, where ψ_1 and ψ_2 denote total normalized wave functions. Since the two initial states have nearly identical electronic structures and are separated by much less than the energy of one photon, the total strong-field ionization yield Y is given by the coherent sum of two contributions S_1 and S_2 as $Y \propto \left| |c_1|e^{i\phi_1}S_1 + |c_2|e^{i\phi_2}S_2 \right|^2$. The high-harmonic yield comprises contributions from four different channels, represented by

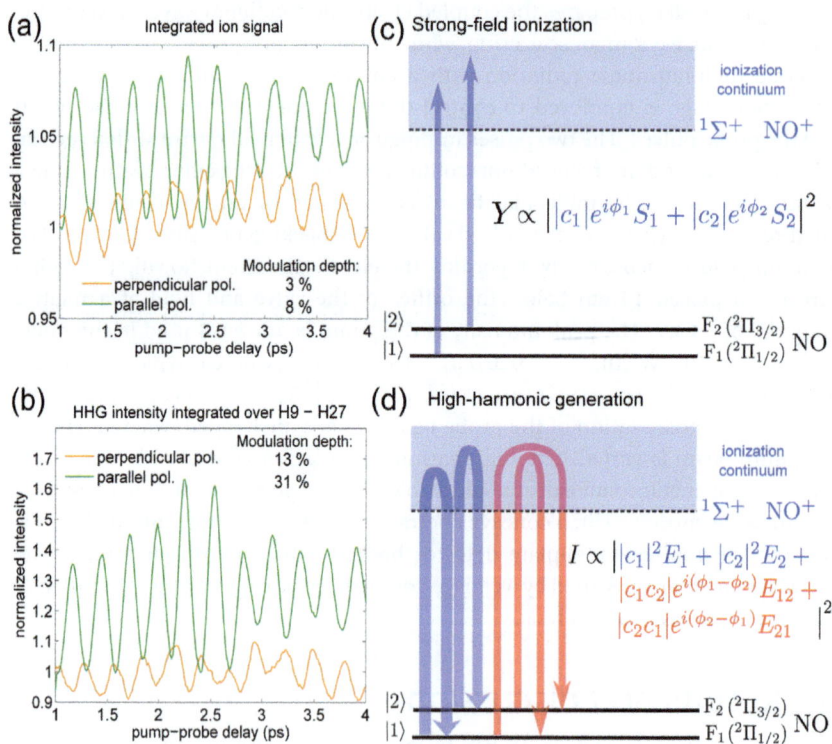

Fig. 1 Strong-field ionization (a) and high-harmonic generation (b) signals from a coherent superposition of F_1 and F_2 electronic states in NO. The plots on the left-hand side show normalized intensities from experiments featuring parallel (green) or crossed (orange) polarizations. The schemes on the right-hand side illustrate the quantum pathways contributing to the observed signals. Panel (d) has been adapted from ref. 9.

arrows in Fig. 1d.[9] The two channels (blue arrows) corresponding to ionization followed by recombination to the same state are sensitive only to the populations of the two states ($|c_1|^2$ and $|c_2|^2$) but not to their quantum phases. Conversely, the two red channels are sensitive to the initial phases and encode their difference in the phase of the emitted radiation. The total intensity of the emitted electric field can be decomposed as a sum of four different contributions as[9]

$$I \propto \left| |c_1|^2 E_1 + |c_2|^2 E_2 + |c_1 c_2| e^{i(\phi_1 - \phi_2)} E_{12} + |c_1 c_2| e^{i(\phi_2 - \phi_1)} E_{21} \right|^2.$$ For the same reasons as mentioned above, one can assume $E_1 \sim E_2 \sim E_{12} \sim E_{21}$. Comparing the expressions for Y and I, we find that the latter is formally similar to the former but is squared once more. This explains the observed higher sensitivity of HHG to the electronic dynamics.

Further, a comparison of the two polarization geometries reveals that the ionization yield as well as the harmonic intensity in the experiment with crossed polarizations display a smaller modulation depth and are shifted in phase by $\sim\pi$ with respect to the parallel configuration. In Fig. 2b we plot the temporal evolution of the harmonic intensity for several harmonic orders (H11-H15), calculated using the theoretical model discussed in section 2. Comparing the calculated results with the experimental data shown Fig. 2a reveals that our theory correctly captures both the different modulation depths and the phase shift between the two configurations. These two aspects encode the temporal evolution of the electronic wave packet, which corresponds to a valence-shell electron current flowing around the internuclear axis.[9] The calculation slightly overestimates the

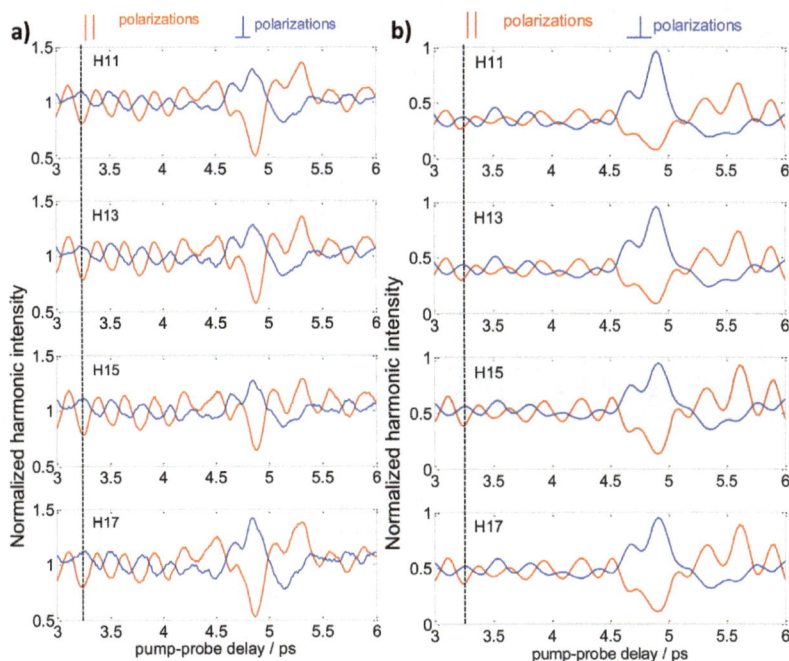

Fig. 2 High-harmonic intensities observed (a) and calculated (b) for different harmonic orders for parallel (red) or perpendicular (blue) polarizations. Note the different vertical scale used in the left and the right panels. The calculation was performed assuming a peak intensity of 4.5×10^{13} W cm^{-2} and a pulse duration of 60 fs for the pump pulse.

modulation depths but is in excellent overall agreement with the experiment. The modulation depth of both the rotational and the electronic modulations is highly sensitive to the parameter γ in eqn (9). The remaining discrepancy between experiment and theory suggests that the empirical value[38] for γ may not be very accurate.

4.2 HHG and SFI as probes of coupled electronic-rotational motion

Fig. 3a and c show the signal intensity as a function of the pump–probe delay for harmonic orders 9 and 15, respectively, whereas panels c) and d) show the Fourier transforms of the signals in the frequency domain. In harmoinc 9, the rotational dynamics manifests itself as a pattern of regularly spaced revival structures, recurring with a period of ~5 ps that corresponds to the revival time of the molecular alignment. An additional signature of the rotational motion is the fractional revival feature that is discernible at each quarter revival time, as is typical of molecules with a π-symmetry HOMO. The maximum (minimum) of the revival structure correspond to the time delays when the wave packet is strongly localized in angular space, either parallel to the axis defined by the pump polarization direction (or delocalized in the plane orthogonal to it). Irregularities

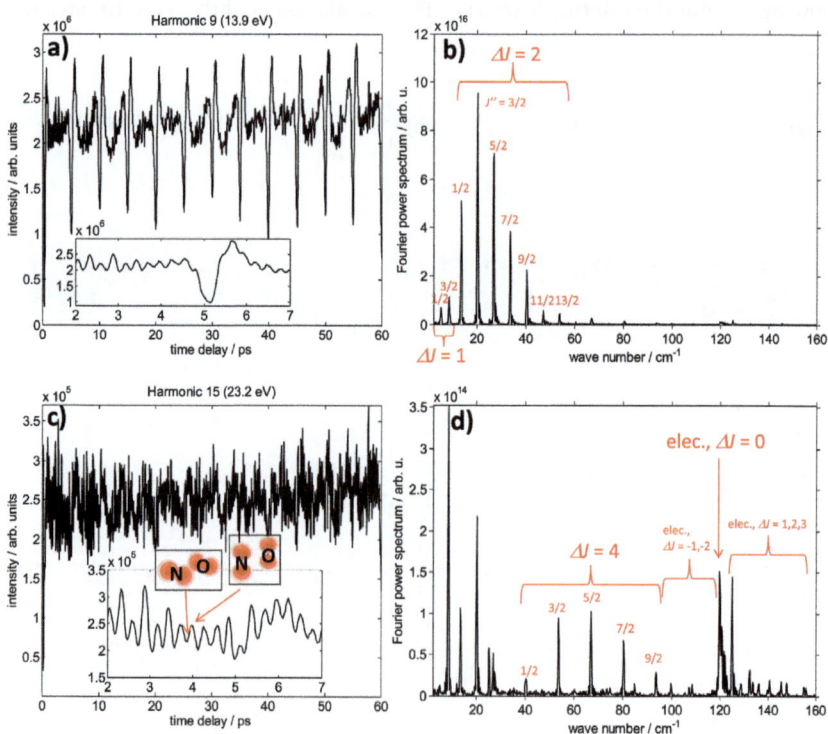

Fig. 3 High-harmonic intensities measured using parallel pump–probe polarizations in H9 (a) and H15 (c) and Fourier-transform power spectra (b and d). The observed coherences are assigned in terms of the angular momentum quantum number J of the lower-lying state, the change in rotational quantum number (ΔJ) and a possible change of electronic state (elec.).

in the signal at long delays hint at progressive dephasing of the electronic quantum beat due to the presence of incommensurate frequencies in the wave packet and coupling to molecular rotation.

The electronic coherence is revealed in the modulations that dominate the signal in between the rotational revivals. These oscillations, shown in the inset of Fig. 3a, represent the electronic beating with a period of $\sim$275 $\pm$ 2 fs, discussed in the preceding section. Both the pure rotational as well as the electronic coherences give rise to characteristic features in the frequency domain. As evident from the Fourier-transformed signals displayed in Fig. 3b, the rotational part of the spectrum consists of a cluster of peaks at low wavenumbers that correspond to pure rotational transitions within the F_1 state and is primarily dominated by contributions arising from exchange of two units of angular momentum ($\Delta J = \pm 2$). The electronic transition is present as a barely distinguishable structure around 120 cm^{-1}. Although similar qualitative arguments can be applied to the spectrum of harmonic 15, the suppression of the pure rotational coherences constitutes a striking difference. As emphasized in the inset of panel c), oscillations arising from the electronic coherence become comparable in amplitude to the rotational revival features, in contrast to the tendency observed in panels a) and b). Moreover, the pure rotational coherences are dominated by higher frequencies ($\Delta J = \pm 4$). The high wave-number part of the spectrum is dominated

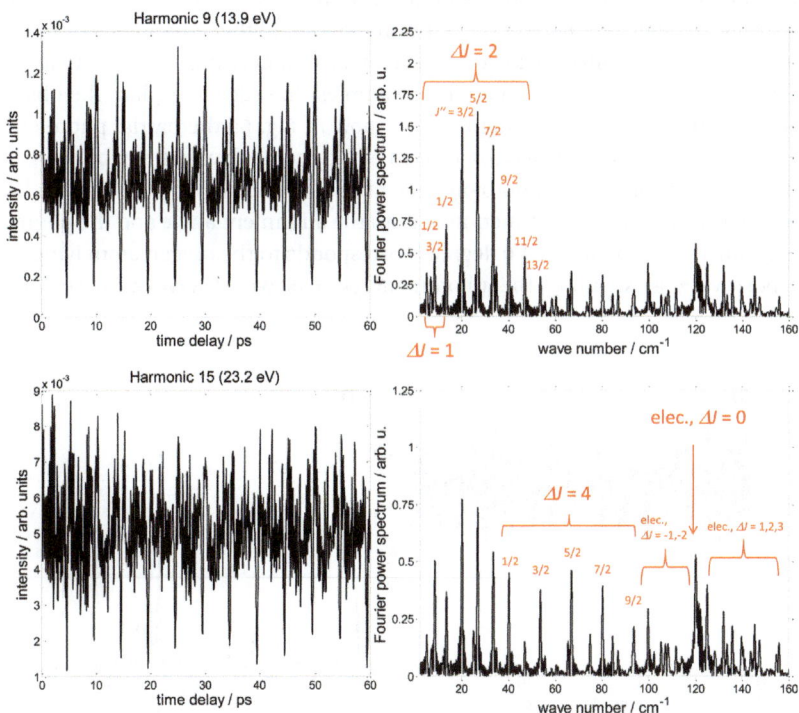

Fig. 4   Calculated high-harmonic intensities for H9 and H15 using a 60 fs pump pulse with a peak intensity of 4.8×10^{13} W cm^{-2}, matching the experimental data shown in Fig. 3. Panels (a) and (c) correspond to the time-domain signals of H9 and H15, respectively. The Fourier-transform power spectra are shown in panels (c) and (d).

by the electronic coherence as well as mixed electronic-rotational transitions ($\Delta J = \pm 1, \pm 2$).

The calculated intensities in both time and frequency domains for harmonic orders 9 and 15 are depicted in Fig. 4. Our model reproduces the main observations such as the increasing importance of the electronic coherences and high-order pure rotational Raman transitions when proceeding from $H9$ to $H15$. However, the calculations tend to overestimate the relative strength of the electronic and mixed coherences in $H9$ and predict a different intensity distribution in $H15$.

For the purpose of comparison, an analogous frequency-domain analysis has been performed on the experimental total ion yield as a function of time (Fig. 5a). The Fourier spectrum of the total ion yield depicted in Fig. 5b reveals the presence of both pure rotational as well as electronic or mixed electronic-rotational excitations. Interestingly, the electronic and mixed coherences strongly dominate over the purely rotational coherences, which is in contrast to the high-harmonic yields. Strong-field ionization is thus found to be relatively more sensitive to electronic rather than rotational motion in NO, while the opposite is true for HHG.

We thus conclude that while harmonic 9 is most sensitive to rotational motion, harmonics of order 15 (and higher) are more sensitive to the electronic motion. This observation might be a general effect because the near-treshold region in photoionization is usually rich in continuum resonances which can cause a strong angular dependence of the photoionization cross section and thus give rise to a strong angle-dependence of the harmonic signal. In the remaining part of the current section, we investigate the mechanisms underlying the sensitivity of different harmonic orders to rotational dynamics. In order to explain the observations reported in the preceding text, we show in Fig. 6a the partial photoionization cross section for an electron ejected along the positive z-direction defined by the polarization of the photoionizing radiation and leaving NO^+ in its $^1\Sigma^+$ ground electronic state, as a function of both the alignment angle and the photon energy. An alignment angle of 0 degrees corresponds to the oxygen atom lying on the positive z axis. Along the photon-energy axis we observe a sharp local

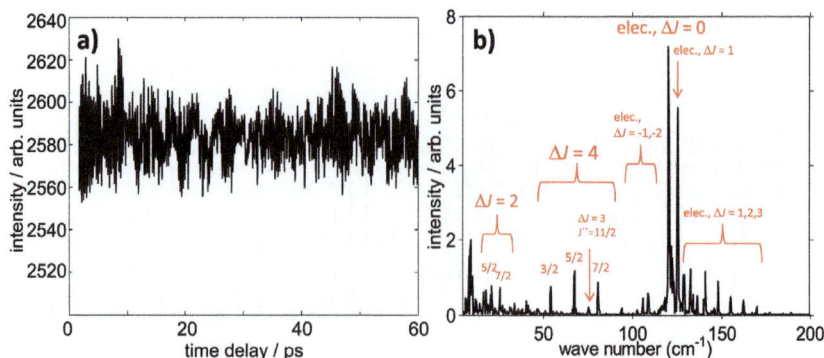

Fig. 5 Measured variation of the total ion yield as a function of the pump–probe delay (a) and Fourier-transform power spectrum (b) with coherences labeled by the corresponding change in rotational angular momentum quantum number (ΔJ) and a possible change in electronic state (F_1 to F_2).

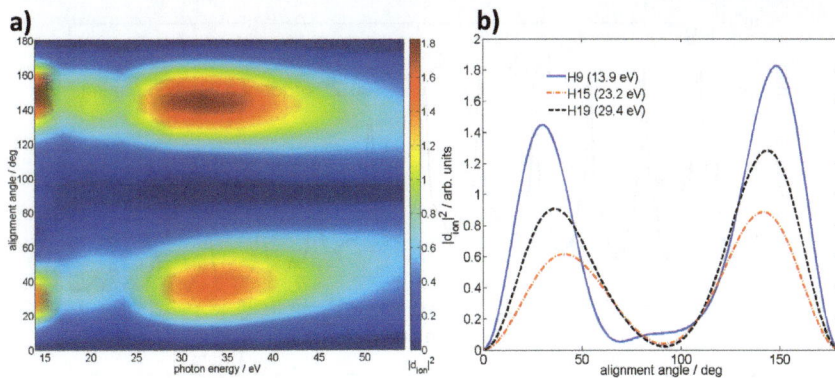

Fig. 6 (a): Calculated molecular-frame photoionization cross section for photoemission along the polarization of the ionizing radiation as a function of the photon energy and alignment angle. (b) Photoionization cross section corresponding to selected harmonic orders as a function of the alignment angle.

maximum close to 14 eV, a minimum close to 23 eV and a subsequent broad maximum in the range of 30–35 eV. The first maximum is caused by the presence of several resonances in the photoionization continuum of the $X^1\Sigma^+$ state. The region of 13–17 eV contains at least two valence autoionizing resonances ($4\sigma \rightarrow 2\pi$) and ($1\pi \rightarrow 2\pi$), in addition to a 3–4 eV broad ($\sigma \rightarrow \sigma^*$) shape resonance.[41] The second broad maximum in the range of 30–35 eV is caused by interchannel coupling to shape resonances in the $(4\sigma)^{-1}$ and $(5\sigma)^{-1}$ channels.

We now turn to the angular dependence of the photoionization cross section. Fig. 6b highlights the angular dependence of three harmonic orders ($H9$, $H15$ and $H19$). The photoionization cross section corresponding to the photon energy of $H9$ exhibits a pronounced variation with alignment angle, which can again be attributed to the presence of the resonances mentioned in the last paragraph. In contrast, the amplitude of $H15$ varies more weakly with the alignment angle. This theoretical result rationalizes the experimental observation from Fig. 3, *i.e.* the fact that H9 is much more sensitive to the rotational dynamics than H15.

The maxima and minima in the photoionization cross section of Fig. 6a are also reflected in the spectral amplitude of the individual harmonic orders. Fig. 7a shows an experimental high-harmonic spectrum emitted from aligned NO molecules and Fig. 7b shows the intensities of all harmonic orders on a linear scale. The spectral amplitude exhibits a maximum at $H9$ and a local minimum at $H15$. These features are also reproduced in the high-harmonic spectrum calculated according to eqn (15) (using $W(\omega) = 1$ and averaging over the calculated axis distribution) which is shown in panel *c*). This observation corroborates the fact that structures of photoionization continua such as Cooper minima[31] or shape resonances[44–46] become observable in high-harmonic spectra even when they arise through inter-channel coupling as in the case of the giant resonance in xenon.[47,48]

4.3 Intensity scaling

Next, we exploit the sensitivity of HHG to rotational or mixed electronic-rotational motion to study the influence of the pump intensity. As is evident from Fig. 8a–c, which shows the Fourier transform of $H11$ at three different pump intensities,

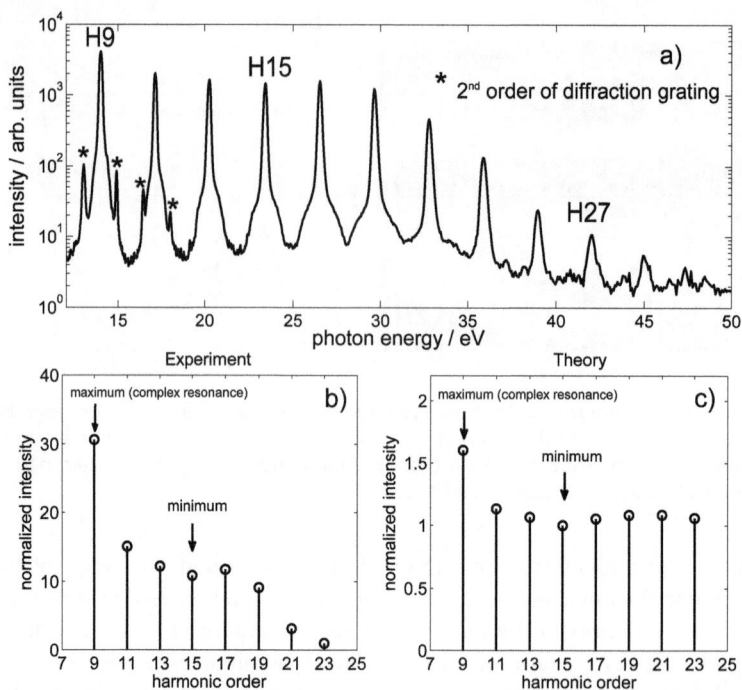

Fig. 7 Observed high-harmonic spectrum at a pump–probe delay corresponding to maximal alignment (5.25 ps, panel (a)), extracted intensity stick spectrum (b) and calculated high-harmonic stick spectrum (c). The decreasing intensity of harmonic orders above 21 (cutoff region) in panel (c) is not reproduced in the calculations because the spectral amplitude of the electron wave packet has been set to unity.

increasing the pump intensity leads to an enhancement of the electronic coherence, whose spectral signature is present as the pronounced peak at $\sim$120 cm^{-1}. Simultaneously, higher pump intensities permit the observation of weak spectral signatures associated with mixed electronic-rotational coherences. Remarkably, the pure rotational Raman transitions follow a different trend. While increasing the pump beam intensity from 3.2×10^{13} W cm^{-2} to 4.8×10^{13} W cm^{-2} stimulates the rotational Raman process, as evident in Fig. 8b and c, a further increase results in an observable decay. Concomitantly, the rotational distribution becomes broader due to consecutive Raman excitations. Thus, we may conclude that an increase of the pump intensity favors the electronic Raman scattering process and leads to a more effective population transfer from the F_1 to the F_2 state while rotational Raman transitions within the F_1 manifold are saturated at lower excitation energies. In Fig. 8d–f, the pump intensity scaling is studied theoretically by solving the time-dependent Schrödinger equation of the system with the methods outlined in Section 2. Although the model captures correctly the experimentally observed intensity dependence, it visibly overestimates the role of the higher-order Raman transitions while at the same time underestimating the growth of the electronic coherence. In particular, the predicted amplitudes of the mixed electronic-rotational coherences and rotational Raman transitions involving the exchange of four units of angular momentum ($\Delta J = 4$) are

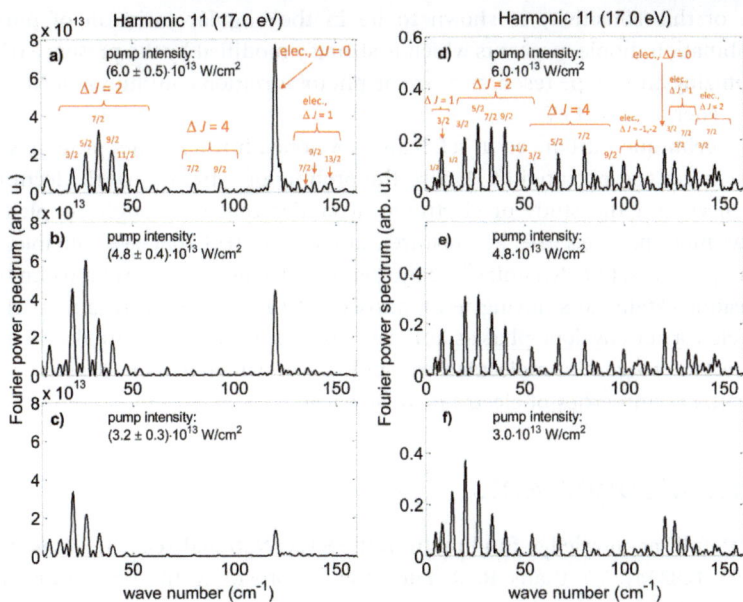

Fig. 8 a–c) Fourier-transform power spectra of experimental pump–probe signals of H11 measured under parallel pump–probe polarizations using different peak intensities of the pump pulse. d–f) Fourier-transform spectra of the calculated high-harmonic intensity of H11 with laser-pulse parameters matching the experimental data shown in a–c.

overestimated with respect to the experimental observations. In addition, the predicted intensity profile of the mixed electronic-rotational coherences is much broader than actually observed. These discrepancies indicate avenues for improving our theory.

5 Conclusions

We have studied a coupled electronic and rotational wave packet by high-harmonic spectroscopy and strong-field ionization. High-harmonic generation from a coherent superposition of electronic states was found to be ~4 times more sensitive to the electronic coherence than SFI. This result can be rationalized as being the consequence of two interfering pathways in the case of SFI and four pathways in the case of HHS. Although high-harmonic generation is usually considered to be a parametric process that leaves the target molecule in the initial quantum state, our experiment clearly demonstrates the existence of inelastic pathways for high-harmonic generation which enable the detection of electronic coherence. We further showed that different high-harmonic orders present a very different relative sensitivity to rotational or electronic motion in NO, while strong-field ionization is most sensitive to the electronic motion.

A theoretical description of HHS for coupled electronic and nuclear dynamics has been developed that quantitatively accounts for most experimental observations. It correctly predicts the detection of purely rotational, purely electronic and mixed coherences. Our theory further explains the surprisingly different sensitivities of different harmonic orders to rotational or electronic dynamics. The origin of this sensitivity is shown to lie in the angular variation of photo-recombination dipole moments which is strongly modified by the presence of an autoionizing and shape resonances in the photoionization continuum of NO at a photon energy of ~14 eV.

The technique introduced in ref. 9 and developed further in the present work demonstrates the potential of HHS to the probing of extremely weak electronic coherences and the study of electronic dynamics that is strongly coupled to nuclear motion. These are key features unique to HHS that are valuable for studying excited-state dynamics in polyatomic molecules. The present technique will readily extend to studying few-femtosecond to attosecond dynamics when few-cycle carrier-envelope-phase-stable laser pulses are used. It will also benefit from recent progress in molecular orientation[44,49,50] which will enable studies of the spatial asymmetries of electronic wave packets.

Acknowledgements

We gratefully acknowledge funding from the Swiss National Science Foundation (PP00P2_128274). We thank R. R. Lucchese for providing the photoionization matrix elements displayed in Fig. 6.

References

1 M. Drescher, M. Hentschel, M. Uiberacker, V. Yakovlev, S. A., T. Westerwalbesloh, U. Heinzmann and F. Krausz, *Nature*, 2002, **419**, 803–807.

2 M. Schultze, M. Fie, N. Karpowicz, J. Gagnon, M. Korbman, M. Hofstetter, S. Neppl, A. L. Cavalieri, Y. Komninos, T. Mercouris, C. A. Nicolaides, R. Pazourek, S. Nagele, J. Feist, J. Burgdörfer, A. M. Azzeer, R. Ernstorfer, R. Kienberger, U. Kleineberg, E. Goulielmakis, F. Krausz and V. S. Yakovlev, *Science*, 2010, **328**, 1658–1662.

3 J. Mauritsson, P. Johnsson, E. Mansten, M. Swoboda, T. Ruchon, A. L'Huillier and K. J. Schafer, *Phys. Rev. Lett.*, 2008, **100**, 073003.

4 E. Goulielmakis, Z.-H. Loh, A. Wirth, R. Santra, N. Rohringer, V. S. Yakovlev, S. Zherebtsov, T. Pfeifer, A. M. Azzeer, M. F. Kling, S. R. Leone and F. Krausz, *Nature*, 2010, **466**, 739–743.

5 C. Ott, A. Kaldun, P. Raith, K. Meyer, M. Laux, J. Evers, C. H. Keitel, C. H. Greene and T. Pfeifer, *Science*, 2013, **340**, 716–720.

6 J. Herrmann, M. Weger, R. Locher, M. Sabbar, P. Rivière, U. Saalmann, J.-M. Rost, L. Gallmann and U. Keller, *Phys. Rev. A: At., Mol., Opt. Phys.*, 2013, **88**, 043843.

7 H. J. Wörner and P. B. Corkum, *J. Phys. B: At., Mol. Opt. Phys.*, 2011, **44**, 041001.

8 A. Fleischer, H. J. Wörner, L. Arissian, L. R. Liu, M. Meckel, A. Rippert, R. Dörner, D. M. Villeneuve, P. B. Corkum and A. Staudte, *Phys. Rev. Lett.*, 2011, **107**, 113003.

9 P. M. Kraus, S. B. Zhang, A. Gijsbertsen, R. R. Lucchese, N. Rohringer and H. J. Wörner, *Phys. Rev. Lett.*, 2013, **111**, 243005.

10 T. Millack and A. Maquet, *J. Mod. Opt.*, 1993, **40**, 2161–2171.

11 F. I. Gauthey, C. H. Keitel, P. L. Knight and A. Maquet, *Phys. Rev. A: At., Mol., Opt. Phys.*, 1995, **52**, 525–540.

12 J. B. Watson, A. Sanpera, X. Chen and K. Burnett, *Phys. Rev. A: At., Mol., Opt. Phys.*, 1996, **53**, R1962–R1965.

13 H. Niikura, D. M. Villeneuve and P. B. Corkum, *Phys. Rev. Lett.*, 2005, **94**, 083003.

14 H. J. Wörner, J. B. Bertrand, D. V. Kartashov, P. B. Corkum and D. M. Villeneuve, *Nature*, 2010, **466**, 604–607.

15 H. J. Wörner, J. B. Bertrand, B. Fabre, J. Higuet, H. Ruf, A. Dubrouil, S. Patchkovskii, M. Spanner, Y. Mairesse, V. Blanchet, E. Mével, E. Constant, E. Constant, P. B. Corkum and D. M. Villeneuve, *Science*, 2011, **334**, 208–212.

16 P. M. Kraus, Y. Arasaki, J. B. Bertrand, S. Patchkovskii, P. B. Corkum, D. M. Villeneuve, K. Takatsuka and H. J. Wörner, *Phys. Rev. A: At., Mol., Opt. Phys.*, 2012, **85**, 043409.

17 P. M. Kraus and H. J. Wörner, *Chem. Phys.*, 2013, **414**, 32–44.

18 A. Tehlar and H. J. Wörner, *Mol. Phys.*, 2013, **111**, 2057–2067.

19 O. Smirnova, Y. Mairesse, S. Patchkovskii, N. Dudovich, D. Villeneuve, P. Corkum and M. Y. Ivanov, *Nature*, 2009, **460**, 972–977.

20 S. Haessler, J. Caillat, W. Boutu, C. Giovanetti-Teixeira, T. Ruchon, T. Auguste, Z. Diveki, P. Breger, A. Maquet, B. Carre, R. Taieb and P. Salières, *Nat. Phys.*, 2010, **6**, 200–206.

21 A. Rupenyan, P. M. Kraus, J. Schneider and H. J. Wörner, *Phys. Rev. A: At., Mol., Opt. Phys.*, 2013, **87**, 033409.

22 A. Rupenyan, P. M. Kraus, J. Schneider and H. J. Wörner, *Phys. Rev. A: At., Mol., Opt. Phys.*, 2013, **87**, 031401.

23 S. Ramakrishna and T. Seideman, *Phys. Rev. Lett.*, 2007, **99**, 113901.

24 S. Ramakrishna and T. Seideman, *Phys. Rev. A: At., Mol., Opt. Phys.*, 2008, **77**, 053411.

25 S. Ramakrishna, P. A. J. Sherratt, A. D. Dutoi and T. Seideman, *Phys. Rev. A: At., Mol., Opt. Phys.*, 2010, **81**, 021802.

26 A. Abdurrouf and F. H. M. Faisal, *Phys. Rev. A: At., Mol., Opt. Phys.*, 2009, **79**, 023405.

27 F. H. M. Faisal, A. Abdurrouf, K. Miyazaki and G. Miyaji, *Phys. Rev. Lett.*, 2007, **98**, 143001.

28 F. H. M. Faisal and A. Abdurrouf, *Phys. Rev. Lett.*, 2008, **100**, 123005.

29 A.-T. Le, R. R. Lucchese, S. Tonzani, T. Morishita and C. D. Lin, *Phys. Rev. A: At., Mol., Opt. Phys.*, 2009, **80**, 013401.

30 T. Morishita, A.-T. Le, Z. Chen and C. D. Lin, *Phys. Rev. Lett.*, 2008, **100**, 013903.

31 H. J. Wörner, H. Niikura, J. B. Bertrand, P. B. Corkum and D. M. Villeneuve, *Phys. Rev. Lett.*, 2009, **102**, 103901.

32 A.-T. Le, R. R. Lucchese and C. D. Lin, *Phys. Rev. A: At., Mol., Opt. Phys.*, 2013, **87**, 063406.

33 J. M. Brown and A. Carrington, *Rotational Spectroscopy of Diatomic Molecules*, Cambridge University Press, 1st edn, 2003.

34 R. N. Zare, *Angular Momentum: Understanding Spatial Aspects in Chemistry and Physics*, John Wiley & and Sons, 1st edn, 1988.

35 A. R. Edmonds, *Angular Momentum in Quantum Mechanics*, Princeton University Press, 1st edn, 1957.

36 J. H. Van Vleck, *Rev. Mod. Phys.*, 1951, **23**, 213–227.

37 P. U. Manohar and S. Pal, *Chem. Phys. Lett.*, 2007, **438**, 321–325.

38 D. W. Lepard, *Can. J. Phys.*, 1970, **48**, 1664–1674.

39 H. Li, D. Ray, S. De, I. Znakovskaya, W. Cao, G. Laurent, Z. Wang, M. F. Kling, A. T. Le and C. L. Cocke, *Phys. Rev. A: At., Mol., Opt. Phys.*, 2011, **84**, 043429.

40 R. R. Lucchese, G. Raseev and V. McKoy, *Phys. Rev. A: At., Mol., Opt. Phys.*, 1982, **25**, 2572–2587.

41 R. E. Stratmann, R. W. Zurales and R. R. Lucchese, *J. Chem. Phys.*, 1996, **104**, 8989–9000.

42 M. H. Alexander and J. Dagdigian, *J. Chem. Phys.*, 1984, **80**, 4325.

43 M. Lein, R. De Nalda, E. Heesel, N. Hay, E. Springate, R. Velotta, M. Castillejo, P. L. Knight and J. P. Marangos, *J. Mod. Opt.*, 2005, **52**, 465–478.

44 A. Rupenyan, J. B. Bertrand, D. M. Villeneuve and H. J. Wörner, *Phys. Rev. Lett.*, 2012, **108**, 033903.

45 X. Ren, V. Makhija, A.-T. Le, J. Troß, S. Mondal, C. Jin, V. Kumarappan and C. Trallero-Herrero, *Phys. Rev. A: At., Mol., Opt. Phys.*, 2013, **88**, 043421.

46 P. Kraus, D. Baykusheva and H. J. Worner, *Phys. Rev. Lett.*, 2014, **113**, 023001.

47 A. D. Shiner, B. Schmidt, C. Trallero-Herrero, H. J. Wörner, S. Patchkovskii, P. B. Corkum, J.-C. Kieffer, F. Légaré and D. M. Villeneuve, *Nat. Phys.*, 2011, **7**, 464–467.

48 S. Pabst and R. Santra, *Phys. Rev. Lett.*, 2013, **111**, 233005.

49 P. M. Kraus, A. Rupenyan and H. J. Wörner, *Phys. Rev. Lett.*, 2012, **109**, 233903.

50 P. M. Kraus, D. Baykusheva and H. J. Wörner, *J. Phys. B: At. Mol. Opt. Phys.*, 2014, **47**, 124030.

Faraday Discussions

ROYAL SOCIETY
OF CHEMISTRY

PAPER

High-order harmonic spectroscopy for molecular imaging of polyatomic molecules

M. Negro,[a] M. Devetta,[a] D. Faccialà,[b] S. De Silvestri,[b] C. Vozzi*[a] and S. Stagira[b]

Received 7th March 2014, Accepted 12th May 2014

DOI: 10.1039/c4fd00033a

High-order harmonic generation is a powerful and sensitive tool for probing atomic and molecular structures, combining in the same measurement an unprecedented attosecond temporal resolution with a high spatial resolution of the order of an angstrom. Imaging of the outermost molecular orbital by high-order harmonic generation has been limited for a long time to very simple molecules, like nitrogen. Recently we demonstrated a technique that overcame several of the issues that have prevented the extension of molecular orbital tomography to more complex species, showing that molecular imaging can be applied to a triatomic molecule like carbon dioxide. Here we report on the application of such a technique to nitrous oxide (N_2O) and acetylene (C_2H_2). This result represents a first step towards the imaging of fragile compounds, a category which includes most of the fundamental biological molecules.

1   Introduction

High order harmonic generation (HHG) occurs when atoms or molecules exposed to an intense femtosecond laser pulse are ionized by tunneling. The freed electron is then accelerated in the external electric field. Because of the periodic oscillation of the laser field the electron is brought back to the parent ion where it may recombine emitting an XUV photon.[1] This XUV radiation has been shown to contain information on the electronic structure of the emitting molecule and on its internal dynamics. Attosecond nuclear[2] and electronic dynamics[3,4] have been extracted from HHG in simple molecules and spectral features in the harmonic emission have been related to the molecular electronic structure and have been used for imaging the highest occupied molecular orbital (HOMO).

The idea of exploiting HHG for the tomographic reconstruction of molecular orbitals was first introduced by Itatani *et al.* in 2004 for the nitrogen molecule.[5] Since then, numerous experiments have been realized addressing the role of the

[a]Istituto di Fotonica e Nanotecnologie, CNR, 20133 Milan, Italy. E-mail: caterina.vozzi@ifn.cnr.it
[b]Dipartimento di Fisica, Politecnico di Milano, 20133 Milan, Italy

HOMO in the harmonic spectral intensity,[6,7] in the molecular-frame photo-ionization[8] and in the subsequent attosecond XUV emission,[9] as well as in the polarization state of the emitted radiation.[10] The dependence of the HHG process on the HOMO structure has also been exploited for the characterization in the time domain of the rotational[11] and vibrational[12] molecular excitations.

All these studies rely on two major assumptions: (i) the molecular HHG is dominated by the HOMO structure; (ii) the relationship between molecular structure and emitted XUV spectrum is simple and completely captured by the Strong Field Approximation (SFA), *i.e.* the electron quiver motion is not perturbed by the Coulomb potential of the ion. Both these assumptions have been recently put into question. Recent experiments have enlightened the role of multiple orbital contributions to HHG emission.[3,4] Furthermore, the influence of the Coulomb field of the parent ion in the generation of high order harmonics from molecules has been considered as a serious hindrance to a clear HOMO reconstruction.[13] To perform molecular tomography of more complex species one has to go beyond these assumptions.

As well as these two more fundamental obstacles, there are also additional, more technical difficulties. In order to retrieve the HOMO structure, one has to record the XUV harmonic spectra for different molecular orientations with respect to the laser field. Hence, it is necessary to fix the molecular orientation in space and change the polarization direction of the HHG-driving field.[5] Laser-assisted molecular alignment is a widespread technique able to accomplish this task,[14] but the molecular alignment achieved in this way is not ideal. Hence the experimental results and the corresponding HOMO tomography are affected by angular averaging effects. Moreover, in the case of non-linear molecules, the tomographic procedure requires to fix two or three angular coordinates of the molecule under investigation. For instance, the study of linear polar molecules requires the head–tail direction in space to be fixed. The feasibility of the laser assisted molecular orientation has been recently demonstrated[15] and exploited in HHG spectroscopy,[16–18] but no direct application to molecular imaging has been yet realized.

The amount of information that can be extracted from the harmonic emission depends on the spectral extension of the XUV radiation, that is known to scale with the so-called cut-off law: $E_{max} = I_p + 3.17U_p$, where I_p is the ionization potential of the molecule and U_p is the ponderomotive energy of the electron in the laser field. This poses another important problem when HHG molecular imaging is extended to species with a low ionization potential (*i.e.* all organic molecules, and in particular those having important biological functions) as the extension emission spectrum is reduced. Since $U_p \propto \lambda^2 I$, where I is the peak intensity and λ the wavelength of the driving laser pulse, the emission cut-off may be extended by both increasing the field intensity or the laser wavelength. In this respect, standard Ti:sapphire laser sources generally used in HHG are not ideal candidates for tomography in fragile molecules, since the intense optical fields needed completely ionize the molecule before a well-developed XUV spectrum is generated.

To overcome the limitations posed by ionization saturation, the exploitation of mid-infrared driving sources has been demonstrated to be a powerful tool to extend harmonic emission far in the XUV.[19–23] With a mid-IR source[24] we recently demonstrated that it is possible to extend the spectral investigation in carbon dioxide beyond 100 eV in the absence of multielectron effects, thus avoiding any ambiguity in the reconstructed wavefunction. In addition, by exploiting an all-optical non-interferometric technique, it was possible to trace both the spectral intensity and phase of the high order harmonics generated by single molecules as a function of the emitted photon energy and molecular angular orientation, without averaging effects. Furthermore, the tomographic procedure was generalized in order to take into account the Coulomb potential seen by the re-colliding electron wavepacket.[25]

In this work, we extend that approach to more complex molecules, such as N_2O and C_2H_2 pointing out some strengths and weaknesses of this investigation technique.

2 Experimental setup

We exploited an optical parametric amplifier (OPA) pumped by an amplified Ti:sapphire laser system (60 fs, 20 mJ, 800 nm). The OPA is based on difference frequency generation and provides driving pulses with a 1450 nm central wavelength, a pulse duration of 20 fs and a pulse energy of 1.2 mJ.[24] High harmonics were generated by focusing the mid-IR pulse in a supersonic gas jet under vacuum, due to the strong absorption exhibited by air in the XUV spectral region. The molecules in the jet were impulsively aligned with a portion of the fundamental 800 nm beam which was spectrally broadened by optical filamentation in an argon-filled gas cell and temporally stretched up to 100 fs by propagation through a glass plate. Such duration is required for achieving a good alignment of the molecular sample. In our experimental setup, the driving and aligning pulses were collinear and their polarizations were parallel. The delay between the two pulses was adjusted by means of a fine-resolution translation stage. The XUV radiation was acquired by means of a flat-field spectrometer and a multi-channel plate detector coupled to a CCD camera.[26]

3 Results

Harmonic spectra were acquired in N_2O and C_2H_2 as a function of the delay τ between the aligning and driving pulse around the first rotational half revival ($\tau_{N2O} = 19.95$ ps and $\tau_{C2H2} = 7.08$ ps). The results are shown in Fig. 1(a) and 2(a) for N_2O and C_2H_2 respectively. Fig. 1(b) and 2(b) show the corresponding calculated alignment factor for the experimental conditions.

In both molecules, the sequence of harmonic spectra shows a strong modulation with the delay τ that can be ascribed to the dependence of the harmonic yield on the molecular orbital structure. In particular, a reduction of the harmonic emission can be observed for the delay corresponding to the maximum of the alignment factor and an enhancement of the harmonic yield appears for the minimum of the alignment factor. A major difference between the two cases is the presence of a region of harmonic enhancement at high photon energy, that appears in N_2O at maximum alignment.

These effects can be naively interpreted in terms of a two-center interference occurring in the re-collision step.[7,27] If one considers a diatomic homo-nuclear molecule with a symmetric electronic state with respect to the nuclei exchange and assumes the re-colliding electron as a plane wave, the condition for

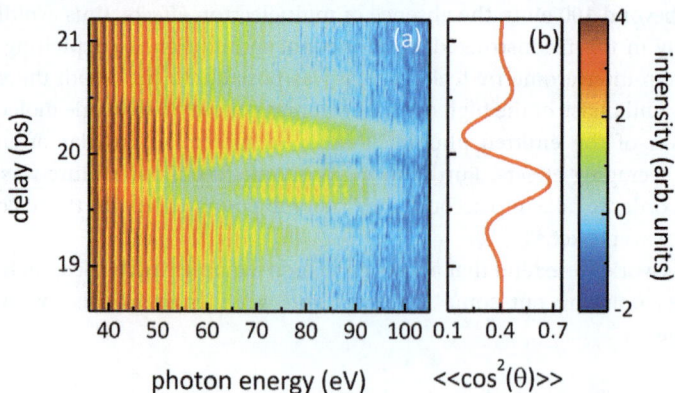

Fig. 1   (a) Sequence of harmonic spectra measured in N_2O as a function of emitted photon energy and delay between the aligning and the driving pulse (log scale). (b) Calculated alignment factor for N_2O in the experimental conditions (rotational temperature 75 K, aligning pulse duration 100 fs, aligning pulse intensity 3.32×10^{13} W cm^{-2}).

constructive interference reads $R\cos(\theta) = n\lambda_B$, where R is the internuclear separation, θ is the angle between the molecular axis and the electron wave-vector, n is an integer number and λ_B is the de Broglie wavelength associated to the recolliding electron wave-packet. Similarly the condition for destructive interference is $R\cos(\theta) = (n + 1/2)\lambda_B$ and the first destructive interference occurs for $n = 0$. The conditions become reversed for molecules with an antisymmetric electronic structure.

This concept can be extended to the molecules subject of our investigation. The acetylene molecule has a symmetric π HOMO in which the separation between the carbon atoms is $R_{C\equiv C} = 1.2$ Å. This is the distance that should be considered for the evaluation of the interference condition. The N_2O HOMO does

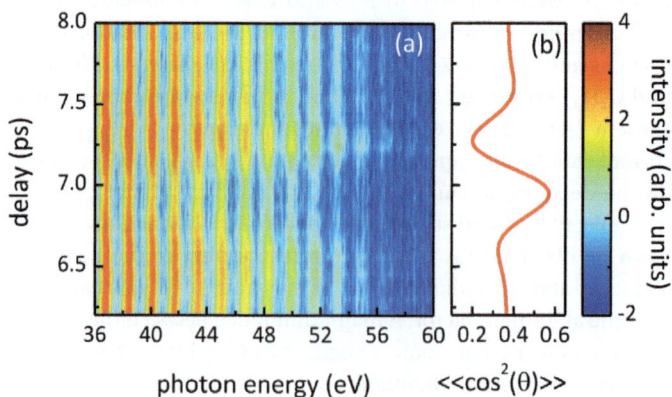

Fig. 2   (a) Sequence of harmonic spectra measured in C_2H_2 as a function of emitted photon energy and delay between the aligning and the driving pulse (log scale). (b) Calculated alignment factor for C_2H_2 in the experimental conditions (rotational temperature 75 K, aligning pulse duration 100 fs, aligning pulse intensity 2.16×10^{13} W cm^{-2}).

not have a clear symmetry, however in our experimental condition the harmonic spectra are acquired in aligned molecules and correspond to the average between the two possible orientation. The resulting signal can be interpreted in terms of emission from an effective molecular orbital similar to the anti-symmetric π orbital of CO_2. In this view the overall length of this "effective" orbital is $R_{N_2O} =$ 2.3 Å. Since $R_{N_2O} \approx 2R_{C\equiv C}$, a destructive interference occurs in the same spectral region for both molecules, corresponding to $n = 1$ for N_2O and $n = 0$ for C_2H_2.

Fig. 1(a) and 2(a) show two peculiar advantages related to the exploitation of mid-IR driving pulses for HHG. Indeed the harmonic cutoff extension related to the increase in the ponderomotive energy with respect to standard Ti:sapphire sources allows the observation of spectral features as the harmonic enhancement for high photon energy visible in N_2O in correspondence of the revival peak. In the framework of the above mentioned two-center model, this feature can be attributed to the appearance of a constructive interference in that spectral region. Moreover, for the same emitted photon energy, mid-IR driving wavelengths require a lower pulse peak intensity thus reducing the ionization saturation in species with a relatively low ionization potential, such as C_2H_2 ($I_p = 11.4$ eV).

4   Reconstruction of single molecule XUV emission

From the experimental data reported in Fig. 1(a) and 2(a) it is possible to retrieve structural information on the target molecule following the approach introduced by Vozzi et al.[25] Fig. 3(a) and 4(a) show the same experimental results presented in Fig. 1(a) and 2(a), in which the harmonic structure due to the periodic re-collision of the electron wave-packet has been filtered out. These results have been exploited for the reconstruction of the XUV field emitted from a single molecule and projected on the polarization direction of the aligning field as a function of the angle between the molecular axis and the driving polarization direction. The reconstruction is based on a combination of a phase-retrieval algorithm and a Kaczmarz algorithm.[28] The main idea behind this approach is that the macroscopic XUV emission is the coherent superposition of the XUV field emitted by all molecules weighted with their angular distribution. This distribution changes along the revival in a predictable way, hence the sequence of harmonic emission contains enough information for the reconstruction of the harmonic electric field in amplitude and phase.

The result of this reconstruction is shown in Fig. 5 for N_2O and in Fig. 6 for C_2H_2. In both figures, panel (a) reports the amplitude of the XUV field and panel (b) shows the corresponding phase. In N_2O there is a clear phase jump of about 2 rad, that changes its position with the photon energy and molecular alignment. This phase jump corresponds to a minimum in the XUV amplitude and its position is in good agreement with the prediction of the naive two-center model introduced above, which is shown as a dashed line in the figure. It is worth noting that the reconstruction technique is based on the interference of XUV emission from different molecular orientations, thus the phase can be retrieved as a function of θ at a fixed XUV photon energy. In order to retrieve the phase relationship between contributions at neighboring energies it is necessary to

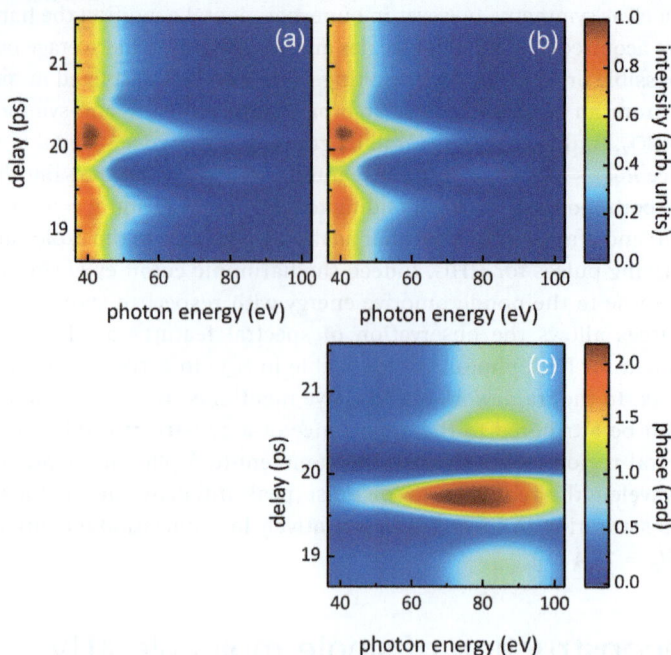

Fig. 3 (a) Sequence of XUV spectra measured in N_2O as a function of the emitted photon energy and delay between the aligning and the driving pulse; the harmonic structure has been filtered out. Retrieved macroscopic harmonic emission amplitude (b) and phase (c) corresponding to the data reported in (a).

introduce an *a priori* condition that can be derived from theoretical consider-ations or experimental measurements. In the case of N_2O we imposed a flat spectral phase of the macroscopic harmonic emission for the delay correspond-ing to the molecular anti-alignment. This condition was chosen in analogy with the CO_2 case[25] due to the similarity between the two HOMOs as discussed in the previous section. The results of this assumption can be observed in Fig. 3, where the reconstructed amplitude (b) and phase (c) of the macroscopic XUV emission from N_2O are reported. The retrieved amplitude is in good agreement with the experimental data (Fig. 3(a)). The phase of the macroscopic emission shows a steep change of about 2 rad around 50 eV at the delay τ corresponding to the maximum alignment.

In the case of C_2H_2 we followed the same approach in the retrieval procedure. We imposed in this case a flat spectral phase for the macroscopic harmonic emission at the delay τ corresponding to the molecular alignment. This assumption was necessary in order to complete the retrieval procedure, but it is arbitrary and not supported by theoretical models; it could be however improved by changing the retrieving condition according to an experimental spectral phase measurement. This kind of experiment can be performed for example by RABBIT technique at a given alignment delay.[9] The retrieved single molecule XUV emis-sion in C_2H_2, shown in Fig. 6, is very different from the one reported for N_2O. In particular a strong contribution comes from molecules with perpendicular orientation with respect to the driving field polarization direction. In the retrieved

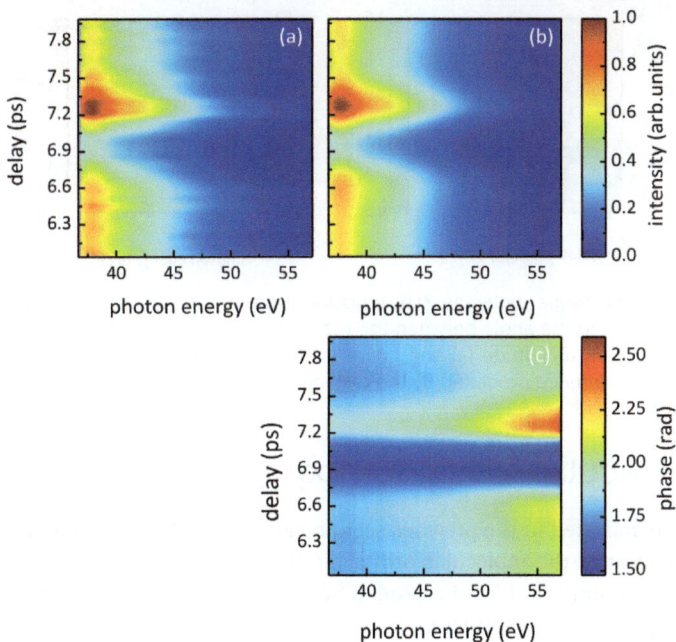

Fig. 4 (a) Sequence of XUV spectra measured in C_2H_2 as a function of the emitted photon energy and delay between the aligning and the driving pulse; the harmonic structure has been filtered out. Retrieved macroscopic harmonic emission amplitude (b) and phase (c) corresponding to the data reported in (a).

phase (Fig. 6(c)) two phase jumps are clearly observed. The first one appears for small alignment angles and roughly follows the prediction of the two-center model. The second jump appears at large alignment angles and may be attributed to the shape of the HOMO seen by the re-colliding electron. However, since the reconstruction is based on the arbitrary assumption of a flat macroscopic spectral phase at the alignment delay, the retrieved outcomes should be considered preliminary. In spite of this, the retrieved macroscopic XUV amplitude (Fig. 4(b)) is in fair agreement with the experimental results.

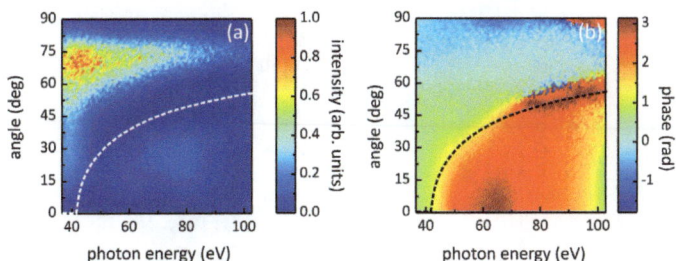

Fig. 5 Retrieved single molecule XUV emission map in N_2O as a function of emitted photon energy and the angle between the molecular axis and the aligning beam polarization direction in amplitude (a) and phase (b). Dashed lines show the position of the destructive interference predicted by the two-center model.

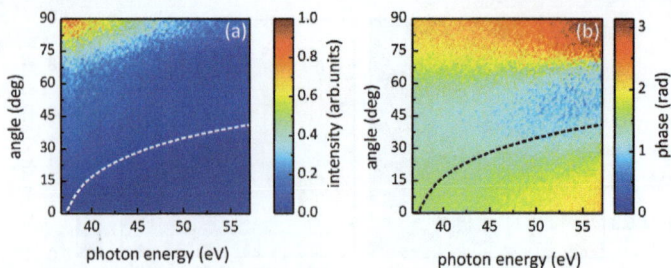

Fig. 6   Retrieved single molecule XUV emission map in C_2H_2 as a function of emitted photon energy and the angle between the molecular axis and the aligning beam polarization direction in amplitude (a) and phase (b). Dashed lines show the position of the destructive interference predicted by the two-center model.

5   Molecular orbital tomography

The results reported in the previous section can be used for the two-dimensional reconstruction of the molecular orbitals, following the tomographic procedure proposed by Itatani et al.[5] and extended by Vozzi et al.[25] However to proceed with this tomographic reconstruction, it is necessary to rule out the occurrence of multi-electron effects in HHG. A simple experimental procedure to check whether spectral modulations in harmonic emission are due to multi-electron effects is to change the driving field intensity. As shown by Smirnova et al.,[3] one expects all the features due to multi-electron effects to shift with the driving field intensity. Fig. 7 shows the harmonic spectra acquired in aligned N_2O for a delay τ corresponding to the maximum of the alignment for different values of the driving intensity. The spectral minimum associated to the phase change retrieved in Fig. 5(b) appears always around 55 eV and does not shift with the driving intensity, as already observed by Rupenyan et al.[29] in similar experimental conditions. This behavior guarantees that the main spectral features in the harmonic emission are mainly dictated by the HOMO structure. This consistency check allowed us to exploit the retrieved single molecule harmonic emission for the reconstruction of the N_2O

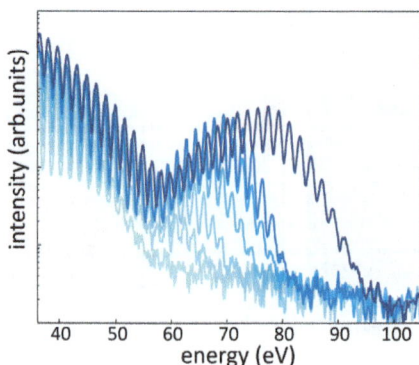

Fig. 7   Harmonic spectra generated in N_2O at the delay τ corresponding to the maximum molecular alignment for several driving peak intensities I between 1 and 1.7 × 10^{14} W cm^{-2}.

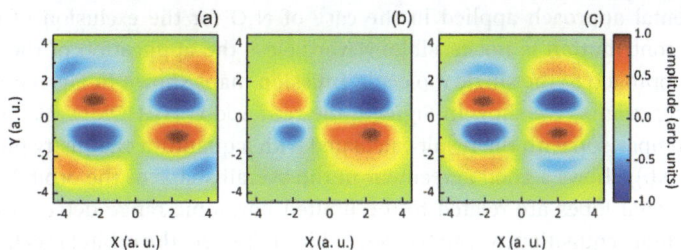

Fig. 8 (a) Highest occupied molecular orbital of N_2O as retrieved from the single molecule XUV emission map. (b) Highest occupied molecular orbital of N_2O calculated with a quantum chemistry program.[30] (c) N_2O HOMO calculated averaging over the two possible orientations of the molecular axis and considering the filtering in spectral domain corresponding to the experimental conditions.

orbital. The result is shown in Fig. 8(a). Fig. 8(b) shows the N_2O orbital calculated with a quantum chemistry program.[30] Even if the overall dimension of the molecular orbital is well reproduced, the asymmetry of this orbital is very clear and cannot be addressed by the tomographic reconstruction, since in the experiment the molecules were aligned but not oriented. Another departure of the retrieved orbital with respect to the calculated one is the presence of side lobes, that can be attributed to the limited working range of the XUV spectrometer used in these experiments. Since there is a correspondence between the energy range of harmonic emission and the spatial frequency domain, the limited spectral range collectible in the experiment corresponds to a spatial filtering in the Fourier domain, which gives raise to such lobes. These observations are further confirmed by Fig. 8(c), which shows the calculated HOMO corresponding to the average between the two possible orientations of N_2O molecular axis and takes into account the limited spectral bandwidth available in the experiment. The features of this fictitious orbital are in very good agreement with the reconstruction of Fig. 8(a). It is worth noting that such limitations can be overcome by extending the acquired spectral range over all the XUV emission and by exploiting all-optical impulsive techniques for the orientation of polar molecules, such as the one demonstrated by Frumker et al.[16,17]

Differently from the case of N_2O, in C_2H_2 it is not possible to easily rule out the multi-electron contributions. Because of the smaller cutoff energy, the

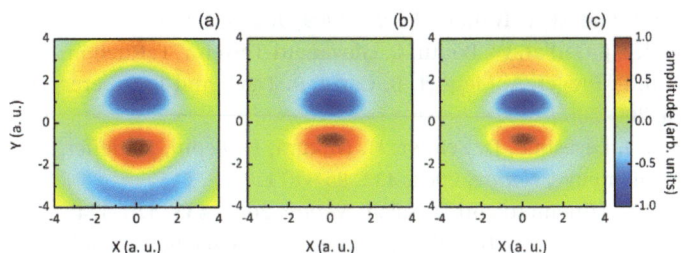

Fig. 9 (a) Highest occupied molecular orbital of C_2H_2 as retrieved from the single molecule XUV emission map. (b) Highest occupied molecular orbital of C_2H_2 calculated with a quantum chemistry program.[30] (c) C_2H_2 HOMO calculated considering filtering in the spectral domain corresponding to the experimental conditions.

experimental approach applied in the case of N_2O for the exclusion of multi-electron contribution is not feasible. Nevertheless the application of the tomographic approach to the single molecule emission maps shown in Fig. 6 provides interesting results. We show in Fig. 9(a) the retrieved C_2H_2 HOMO. Also in this case, a comparison with the result calculated with a quantum chemistry program (see Fig. 9(b)) shows a good agreement in the overall shape of the orbital. Again the additional lobes are related to the limited harmonic range detected in the experimental acquisition, as can be seen in Fig. 9(c) where the orbital is calculated taking into account spectral filtering.

6 Conclusions

Since the pioneering work of Itatani *et al.* on molecular orbital imaging, the impressive advances in laser technologies have given access to new mid-IR sources for driving HHG and pushing the harmonic emission far towards the soft-X-ray range. These sources allowed the application of HHG spectroscopy to fragile molecules, such as hydrocarbons, which can be considered prototypes for the study of ubiquitous phenomena in chemistry and materials science. In this work we showed the application of molecular orbital reconstruction based on HHG to non-trivial samples, such as N_2O and C_2H_2. These results, though requiring further improvements, demonstrate the capability of molecular orbital tomography and represent the first step towards the imaging of dynamical processes in complex molecules.

Acknowledgements

The research leading to these results has received funding from the LASERLAB-EUROPE (grant agreement no. 284464, EC Seventh Framework Programme), from the ERC Starting Research Grant UDYNI (grant agreement no. 307964, EC Seventh Framework Programme) and from the Italian Ministry of Research and Education (ELI project – ESFRI Roadmap).

References

1 P. B. Corkum, *Phys. Rev. Lett.*, 1993, **71**, 1994–1997.
2 S. Baker, J. Robinson, C. Haworth, H. Teng, R. Smith, C. Chirila, M. Lein, J. Tisch and J. Marangos, *Science*, 2006, **312**, 424–427.
3 O. Smirnova, Y. Mairesse, S. Patchkovskii, N. Dudovich, D. Villeneuve, P. Corkum and M. Y. Ivanov, *Nature*, 2009, **460**, 972–977.
4 S. Haessler, J. Caillat, W. Boutu, C. Giovanetti-Teixeira, T. Ruchon, T. Auguste, Z. Diveki, P. Breger, A. Maquet, B. Carre, R. Taieb and P. Salieres, *Nat. Phys.*, 2010, **6**, 200–206.
5 J. Itatani, J. Levesque, D. Zeidler, H. Niikura, H. Pepin, J. Kieffer, P. Corkum and D. Villeneuve, *Nature*, 2004, **432**, 867–871.
6 T. Kanai, S. Minemoto and H. Sakai, *Nature*, 2005, **435**, 470–474.
7 C. Vozzi, F. Calegari, E. Benedetti, J. Caumes, G. Sansone, S. Stagira, M. Nisoli, R. Torres, E. Heesel, N. Kajumba, J. Marangos, C. Altucci and R. Velotta, *Phys. Rev. Lett.*, 2005, **95**, 153902.
8 A.-T. Le, R. R. Lucchese, M. T. Lee and C. D. Lin, *Phys. Rev. Lett.*, 2009, **102**, 203001.

9 W. Boutu, S. Haessler, H. Merdji, P. Breger, G. Waters, M. Stankiewicz, L. J. Frasinski, R. Taieb, J. Caillat, A. Maquet, P. Monchicourt, B. Carre and P. Salieres, *Nat. Phys.*, 2008, **4**, 545–549.

10 J. Levesque, Y. Mairesse, N. Dudovich, H. Pépin, J.-C. Kieffer, P. B. Corkum and D. M. Villeneuve, *Phys. Rev. Lett.*, 2007, **99**, 243001.

11 K. Miyazaki, M. Kaku, G. Miyaji, A. Abdurrouf and F. H. M. Faisal, *Phys. Rev. Lett.*, 2005, **95**, 243903.

12 W. Li, X. Zhou, R. Lock, S. Patchkovskii, A. Stolow, H. C. Kapteyn and M. M. Murnane, *Science*, 2008, **322**, 1207–1211.

13 Z. B. Walters, S. Tonzani and C. H. Greene, *J. Phys. Chem. A*, 2008, **112**, 9439–9447.

14 H. Stapelfeldt and T. Seideman, *Rev. Mod. Phys.*, 2003, **75**, 543–557.

15 S. De, I. Znakovskaya, D. Ray, F. Anis, N. G. Johnson, I. A. Bocharova, M. Magrakvelidze, B. D. Esry, C. L. Cocke, I. V. Litvinyuk and M. F. Kling, *Phys. Rev. Lett.*, 2009, **103**, 153002.

16 E. Frumker, C. T. Hebeisen, N. Kajumba, J. B. Bertrand, H. J. Woerner, M. Spanner, D. M. Villeneuve, A. Naumov and P. B. Corkum, *Phys. Rev. Lett.*, 2012, **109**, 113901.

17 E. Frumker, N. Kajumba, J. B. Bertrand, H. J. Wörner, C. T. Hebeisen, P. Hockett, M. Spanner, S. Patchkovskii, G. G. Paulus, D. M. Villeneuve, A. Naumov and P. B. Corkum, *Phys. Rev. Lett.*, 2012, **109**, 233904.

18 M. Spanner, S. Patchkovskii, E. Frumker and P. Corkum, *Phys. Rev. Lett.*, 2012, **109**, 113001.

19 E. J. Takahashi, T. Kanai, K. L. Ishikawa, Y. Nabekawa and K. Midorikawa, *Phys. Rev. Lett.*, 2008, **101**, 253901.

20 C. Vozzi, R. Torres, M. Negro, L. Brugnera, T. Siegel, C. Altucci, R. Velotta, F. Frassetto, L. Poletto, P. Villoresi, S. De Silvestri, S. Stagira and J. P. Marangos, *Appl. Phys. Lett.*, 2010, **97**, 241103.

21 C. Vozzi, F. Calegari, M. Negro, F. Frassetto, L. Poletto, G. Sansone, P. Villoresi, M. Nisoli, S. De Silvestri and S. Stagira, *J. Mod. Opt.*, 2010, **57**, 1008–1013.

22 C. Vozzi, F. Calegari, F. Frassetto, M. Negro, L. Poletto, G. Sansone, P. Villoresi, M. Nisoli, S. Silvestri and S. Stagira, *Laser Phys.*, 2010, **20**, 1019–1027.

23 T. Popmintchev, M.-C. Chen, D. Popmintchev, P. Arpin, S. Brown, S. Ališauskas, G. Andriukaitis, T. Balčiunas, O. D. Mücke, A. Pugzlys, A. Baltuška, B. Shim, S. E. Schrauth, A. Gaeta, C. Hernández-García, L. Plaja, A. Becker, A. Jaron-Becker, M. M. Murnane and H. C. Kapteyn, *Science*, 2012, **336**, 1287–1291.

24 C. Vozzi, F. Calegari, E. Benedetti, S. Gasilov, G. Sansone, G. Cerullo, M. Nisoli, S. De Silvestri and S. Stagira, *Opt. Lett.*, 2007, **32**, 2957–2959.

25 C. Vozzi, M. Negro, F. Calegari, G. Sansone, M. Nisoli, S. De Silvestri and S. Stagira, *Nat. Phys.*, 2011, **7**, 822–826.

26 L. Poletto, G. Tondello and P. Villoresi, *Rev. Sci. Instrum.*, 2001, **72**, 2868–2874.

27 M. Lein, N. Hay, R. Velotta, J. Marangos and P. Knight, *Phys. Rev. A: At., Mol., Opt. Phys.*, 2002, **66**, 023805.

28 C. Popa and R. Zdunek, *Math. Comput. Simul.*, 2004, **65**, 579–598.

29 A. Rupenyan, P. M. Kraus, J. Schneider and H. J. Wörner, *Phys. Rev. A: At., Mol., Opt. Phys.*, 2013, **87**, 031401(R).

30 DALTON, a molecular electronic structure program Release 2.0 (2005) see http://www.kjemi.uio.no/software/dalton/dalton.html.

ROYAL SOCIETY OF CHEMISTRY

DISCUSSIONS

Chemical reaction dynamics I and electron dynamics in molecules: general discussion†

Oriol Vendrell, Jochen Küpper, Martin Wolf, Henry Chapman, Majed Chergui, Katharine Reid, Klaus von Haeften, Robert Moshammer, Gwyn Williams, Andres Tehlar, Gopal Dixit, Hans Jakob Wörner, Jonathan Underwood, Jonathan Marangos, Michael Woerner, Christian Bressler, Michael Minitti, Adam Kirrander, Caterina Vozzi and Daniel Rolles

DOI: 10.1039/c4fd90014f

Christian Bressler opened the discussion of the paper by **Majed Chergui**: You mentioned the different results on Fe(bpy)$_3$ dynamics obtained *via* our LCLS experiment and your in-house laser-only studies. Actually, both results contradict each other, according to your assessments. Where exactly did the LCLS experiment go wrong in interpretation against your laser-only studies? And why? In other words: how do the X-ray observables fail against your optical-laser observables? Follow-up: can we as scientists decide on the correct tool to choose, or is this choice—to your opinion—merely a case-to-case decision, which always has to depend on circumstances of the scientific question?

Majed Chergui answered: I fully agree that our fs-laser experiment[1] on the Fe(bpy)$_3$ spin cross-over complex contradicts the results of the fs-XES one at the LCLS.[2] I cannot judge the latter, but as you know XFELs do have a problem of timing jitter, which becomes particularly serious in the case of the scientific question raised here.

The issue is not so much that of the observables. It is true that XES is a 'universal' probe of the spin state of metal-containing systems, but the ones previously used, although specific to Fe(II) complexes, also represented unambiguous signatures of the quintet state: XANES,[3] UV transient absorption,[4] resonance Raman.[5]

The real issue is that of time resolution, in all these studies including the fs-XES the time resolution was 130 fs or more. This was already clear in the conclusions of ref. 3, stressing that if an intermediate (^{1,3}T) state were present in the SCO cascade, then its lifetime would need to be <60 fs. The fs-XES experiment concludes the same. Therefore, the ideal way to be clear on the issue is to carry out an experiment that has a resolution of <60 fs, regardless of which of the above

---

† Electronic supplementary information (ESI) available. See DOI: 10.1039/c4fd90014f

observables one chooses. This is what we did in the lab, and we had a resolution of 60 fs in the UV, which probes the absorption of the quintet state ^{5}T, and <40 fs in the visible. The spectral width of the white light continuum in the latter case was such that we could detect the departure of the dynamics from the MLCT manifold, but also its arrival in the quintet state (^{5}T). The result is that the two times coincide and are < 50 fs. This still does not rule out an intermediate state, but it would need to live < 20 fs.

I think we can decide on the tool to choose, but it will always depend on the scientific question one raises. In the case of SCO complexes, the issue of time resolution was clear already when we first published our fs-XANES work.[3] Even if the fs-XES experiment[2] did not have a sufficient time resolution, it is still a major breakthrough because it demonstrated the feasibility of such experiments for the first time. With the improvements in stability and time resolution of the XFELs, repeating the same experiment with sub-50 fs X-ray pulses would be really a major achievement, as it opens the door to the study of a very wide class of systems undergoing remarkably fast spin transitions.[6]

1 G. Auböck and M. Chergui, *Nature Chem.*, 2014, under review.
2 W. K. Zhang, R. Alonso-Mori, U. Bergmann, C. Bressler, M. Chollet, A. Galler, W. Gawelda, R. G. Hadt, R. W. Hartsock, T. Kroll, K. S. Kjaer, K. Kubicek, H. T. Lemke, H. Y. W. Liang, D. A. Meyer, M. M. Nielsen, C. Purser, J. S. Robinson, E. I. Solomon, Z. Sun, D. Sokaras, T. B. van Driel, G. Vanko, T. C. Weng, D. L. Zhu and K. J. Gaffney, *Nature*, 2014, **509**, 345.
3 W. K. Zhang, R. Alonso-Mori, U. Bergmann, C. Bressler, M. Chollet, A. Galler, W. Gawelda, R. G. Hadt, R. W. Hartsock, T. Kroll, K. S. Kjaer, K. Kubicek, H. T. Lemke, H. Y. W. Liang, D. A. Meyer, M. M. Nielsen, C. Purser, J. S. Robinson, E. I. Solomon, Z. Sun, D. Sokaras, T. B. van Driel, G. Vanko, T. C. Weng, D. L. Zhu and K. J. Gaffney, *Nature*, 2014, **509**, 345.
4 W. K. Zhang, R. Alonso-Mori, U. Bergmann, C. Bressler, M. Chollet, A. Galler, W. Gawelda, R. G. Hadt, R. W. Hartsock, T. Kroll, K. S. Kjaer, K. Kubicek, H. T. Lemke, H. Y. W. Liang, D. A. Meyer, M. M. Nielsen, C. Purser, J. S. Robinson, E. I. Solomon, Z. Sun, D. Sokaras, T. B. van Driel, G. Vanko, T. C. Weng, D. L. Zhu and K. J. Gaffney, *Nature*, 2014, **509**, 345.
5 W. K. Zhang, R. Alonso-Mori, U. Bergmann, C. Bressler, M. Chollet, A. Galler, W. Gawelda, R. G. Hadt, R. W. Hartsock, T. Kroll, K. S. Kjaer, K. Kubicek, H. T. Lemke, H. Y. W. Liang, D. A. Meyer, M. M. Nielsen, C. Purser, J. S. Robinson, E. I. Solomon, Z. Sun, D. Sokaras, T. B. van Driel, G. Vanko, T. C. Weng, D. L. Zhu and K. J. Gaffney, *Nature*, 2014, **509**, 345.
6 M. Chergui, *Dalton Trans.*, 2012, **41**, 13022–13029.

Christian Bressler responded: Thank you for your extensive answer which underlines the importance and relevance to really understand detailed reaction mechanisms, even in seemingly simple systems. This is an interesting field where new observables show up on the block almost monthly, and one needs to adapt quickly to those findings and implications. My original question remains, however: where did the LCLS fs-XES experiment go awry? You did mention the precise timing issue, and thus also left doubts about the reliability of the LCLS timing tool, since you do not know its detailed implementation. However, the fs-XES results showed— as compared to well-known static reference samples—signatures of (i) the MLCT state, and (ii) an "intermediate" MC state, during the first few hundred femtoseconds, thus not very precisely timed, but still there! Even without a functioning arrival time monitor these findings already nail down, as a "cold fact", the short existence of each aforementioned state, with or without precise timing information. Overall, they apparently showed up in the course of this LS-HS conversion.

My question therefore is/remains: do you still consider the fs-XES results as possibly erroneous against your own optical results? And if so, why exactly?

(I think the paper described the findings clear enough.) Maybe the comparison to reference samples is plainly false? (The entire interpretation is based on the validity of this comparison, so I consider this the weakest point in the entire interpretation.) This could explain your objection to the validity of the interpretation of these fs-XES results.

Due to the emerging (and improving) possibilities at XFEL sources it is paramount that the scientific community uses valid tools and theories to study and interpret their measurements. I have not yet engaged in questioning any details of any optical studies, but I think the same check and balance rules apply there as well. In the end, we have to converge to the same picture, and currently I see pictures from you, which violently contradict the fs-XES results. Therefore your thoughts about the validity of different experimental approaches would be helpful to eventually converge to a common view of such fascinating avenues to study the elementary steps in chemical reaction dynamics.

Jochen Küpper asked: Dear Majed, you mentioned that spin–orbit coupling, or the understanding thereof, does not explain the intersystem crossing (ISC) rates of the investigated systems, often not at all.

Now, spin orbit coupling is, at least for small molecules, a very well tested concept. Therefore, I am wondering whether you are implying that there are dynamical, temporal effects in the spin–orbit interaction that do not manifest in the eigenstate-resolved spectroscopies we base our previous knowledge on?

Moreover, ISC depends on the spin–orbit coupling as well as the related/resulting vibronic coupling between the involved electronic states. The latter does, of course, depend on the actual potential energy surface and the modes on that surface, which clearly are a changing quantity in the ultrafast dynamics experiments. Thus, you would expect a static spin–orbit coupling picture to fail, wouldn't you? Does this explain your discrepancies between experiment and expectations?

Majed Chergui replied: Thanks for this thoughtful comment. You are putting the finger on the core issue: indeed it is the dynamical aspect of ISC I was refering to. ISC is determined by the spin–orbit coupling constant (SOC), but this is only one among many ingredients. We have found variations of the ISC time of over three orders of magnitude going from 30 fs for Ru or Fe complexes[1] to 30 ps for a diplatinum complex.[2] You are quite right when you say that in an ultrafast dynamics experiment the evolving landscape of the dynamics is going to break the SOC picture. We are indeed far from the steady-state description. And of course, I agree that this explains the discrepancy between expectations and observations. Most theoretical models though still reason in terms of the steady-state picture when trying to model ultrafast ISC. Maybe that is due to the fact that there aren't many alternatives.

1 O. Bram, F. Messina, A. M. El-Zohry, A. Cannizzo and M. Chergui, *Chem. Phys.*, 2012, **393**, 51–57.
2 R. M. van der Veen, A. Cannizzo, F. van Mourik, A. Vlcek and M. Chergui, *J. Am. Chem. Soc.*, 2011, **133**, 305–315.

Jochen Küpper opened the discussion of the paper by **Robert Moshammer:** You explain that the molecular ionization is suppressed relative to the ionization cross sections of two individual molecules. This reminds me of dipole blockades for instance, observed as Rydberg blockades in ultracold atom experiments.

Would you agree that your suppression, at least in a simple model, is related to such a dipole blockade, or better a "Coulomb blockade", where ionization of the second atom in the molecule is simply less likely because it sees the neighboring charge? How does this work for the various ionization channels in detail?

Katharine Reid addressed **Robert Moshammer** and **Daniel Rolles:** In paper 2 you consider the formation of high charge states and low charge states in atoms and molecules. In your paper you mention that xenon clusters have been studied previously and that they are able to support high charge states, but your observations show that I_2 molecules support only low charge states. Can you comment on what the likely situation would be for a polyatomic molecule? Also, in paper 3 there is no evidence of the formation of C_8H_5F ions with charges other than +1; were these observed?

Daniel Rolles responded: Our experiment on C_8H_5F was actually performed with a slightly defocused FEL beam in order to minimize the influence of multi-photon ionization. We therefore only observe a small contribution of higher charge states in our ion spectra. Similarly, the paper by Stern *et al.* [*Faraday Discuss.*, 2014, **171**, DOI: 10.1039/c4fd00028e] reports an experiment on $C_7H_3I_2N$, which was performed with the same experimental setup and under similar conditions as C_8H_5F. Here, higher iodine charge states are observed, but still stem mostly from single-photon or few-photon ionization. In contrast, we have also conducted experiments in the tightest possible FEL focus, where the goal was to absorb as many X-ray photons per molecule as possible in order to study multi-photon effects. In these experiments, we observe extremely high charge states, *e.g.* in CH_3SeH, C_2H_5SeH, and CH_3I molecules (see B. Erk *et al.*, *Phys. Rev. Lett.*, 2013, **110**, 053003; *J. Phys. B*, 2013, **46**, 164031; *Science*, 2014, **345**, 288–291).

Jochen Küpper communicated: I strongly believe that we have to develop a new "old" understanding of the interplay of theory and experiment. That is, we should find experiments that provide results independent of the highest level theory. Instead, we should use high-level theory to make predictions of some observables, such as the structural dynamics of a complex chemical reaction, and then benchmark and support or falsify these predictions independently and quantitatively. Therefore, an important question is: how do we get to experiments that can perform and provide these independent benchmarks? What do they look like?

Majed Chergui opened the discussion of the paper by **Daniel Rolles:** By X-ray absorption near-edge spectroscopy (XANES), you have the information on the 3D structure, but at present the theory is the bottleneck and it does not manage to reproduce the experimental data at a quantitative level. We mainly use it on a qualitative level. Photoelectron diffraction experiments of molecules deposited on surfaces is very well established (see *e.g.* the works of A. Bradshaw *et al.*). In these experiments, the molecules are 2D aligned and the photoelectron comes from the substrate, have you considered such experiments?

Daniel Rolles answered: Indeed, the photoelectron diffraction experiments of molecules on surfaces, *e.g.* by Bradshaw *et al.* and by Fadley *et al.*, have been a great inspiration for our gas-phase experiments, which, by the way, were started when I was a PhD student in the group of Uwe Becker in Alex Bradshaw's department at the Fritz-Haber-Institut in Berlin. While it would certainly be very worthwhile to also conduct these surface physics experiments in a femtosecond time-resolved manner, *e.g.* at a Free-Electron Laser, our efforts concentrate on studying ultrafast reactions of molecules in the gas-phase, which would, in most cases, occur differently if the molecules were attached to a surface.

Henry Chapman asked: Are molecules in the gas phase the right model system for developing structure determination methods for chemical dynamics? Does this environment introduce any problems for preparation of samples or interpretation of results?

Daniel Rolles responded: Since molecules in the gas phase are a relatively "clean" model system in the sense that there are no solvation or bulk effects that need to be considered, I think they are an almost ideal target for developing structure determination methods for chemical dynamics. Of course, in many "real life applications", these solvation effects may be the key to understanding a given phenomenon, in which case gas-phase studies can only give part of the answer and have to, most definitely, be complemented by liquid-phase studies. But to develop the tools in the first place and to, ideally, test them against an *ab initio* theory, the gas phase may be the best bet, especially with respect to the theory.

Jonathan Underwood asked: The work presented by Rolles and colleagues elegantly demonstrates the importance of molecular axis alignment for avoiding loss of information arising from orientational averaging. The adiabatic alignment technique used in the work reported has the advantage of achieving high degrees of alignment, and the drawback of the presence of the strong laser field during the measurement which is expected to affect the molecular structure and dynamics. Rolles and co-workers point out in the final section of their paper that non-adiabatic alignment techniques would allow this drawback to be avoided by creating field free alignment, but that to date non-adiabtic techniques "have not been able to be obtained as high degrees of alignment as adiabatic techniques". While this is true of currently reported experiments, there is in the general case no reason why non-adiabatic alignment techniques could not produce equivalent degrees of (field free) alignment as seen with adiabatically applied fields.

With an adiabatically applied field, a maximal alignment is achieved (in the single aligned axis case) through arranging for the phases of the field-free basis functions comprising the field induced superposition to be the same, and thus ensuring the maximal location in the conjugate angular variable. We showed a number of years ago (*Phys. Rev. Lett.*, 2003, **90**, 223001) that it is possible to adiabatically apply a strong laser field, and then rapidly truncate the laser field on a timescale much shorter than molecular rotation. Under such circumstances the field induced superposition is preserved and present immediately after the laser field is switched off. If an experiment were conducted at that instant, it would take advantage of the high degree of field-free alignment produced. Alternatively, it would be possible to carry out an experiment at a later instant when the resulting wavepacket evolution has produced a revival in the alignment. On the other hand, while the field-free alignment produced by a single impulsive laser pulse usually doesn't produce such high degrees of alignment as an adiabatically applied field of the same peak intensity, enhanced alignment may be achieved as a judicious choice of pulse duration (in the "quasi-adiabatic" regime), shaped pulses, and the application of sequences of pulses with varying degrees of duration and polari- zation. Indeed good progress is being made towards achieving high degrees of field free three dimensional alignment using sequences of laser pulses by Kumorappan and colleagues (*Phys. Rev. Lett.*, 2014, **112**, 173602) building on our earlier work (*Phys. Rev. Lett.*, 2006, **97**, 173001). In summary, the future looks bright for carrying out experiments such as those of Rolles and colleagues while exploiting field-free alignment, albeit with extra experimental complexity.

Daniel Rolles responded: We are certainly aware of these exciting develop- ments and have been considering various field-free alignment techniques for our experiments. Since the experimental complexity is a very important factor when planing Free-Electron Laser experiments, which have to be performed within only a few days of beamtime, the prospect of achieving a high degree of alignment with a relatively straightforward adiabatic alignment method has, so far, outweighed the disadvantages that arise from the presence of the laser field. This may, of course, change in the future, and we would be happy to implement other tech- niques or to collaborate with other groups who are specialists for field-free alignment and who feel confident that they can achieve a high degree of align- ment under the environmental constraints of an FEL beamtime.

Michael Woerner said: You mentioned in your talk that your analysis of the photoelectron diffraction data is based on single scattering events, *i.e.*, you neglect multiple scattering events. In contrast to X-ray diffraction the cross section for electron diffraction is extremely large, so that dynamical X-ray diffraction theory (accounting correctly for multiple sacttering) has to be typically applied.

Why is this different in your case?

Daniel Rolles replied: There is definitely a strong contribution of multiple scattering, especially for the relatively low photoelectron kinetic energies that we have used here. The DFT calculations, to which we compare our experimental results in the paper, do include these effects implicitly. However, in order to extract structural information from photoelectron diffraction data in a simple and direct manner, *e.g.* by holographic reconstruction, as we have suggested in *Phys. Rev. A*, 2010, **81**, 033411, it would be favorable to work in a regime where multiple scattering is less prominent or, even better, negligible. We are currently trying to establish, both theoretically and experimentally, under which conditions this would be the case, but our feeling, at the moment, is that if would be enough to go to a few hundred eV kinetic energy rather than using 2 keV photoelectron as suggested in the above PRA from 2010.

Henry Chapman said: One method for X-ray fluorescence holography on crystalline materials is an inverse geometry where the X-ray absorption is measured as a function of molecule orientation to the incident beam and incident photon energy. (X-ray absorption could be measured by the total fluorescence yield in this case.) This could be thought of as angularly-resolved EXAFS. Given that Majed Chergui showed that it is possible to measure EXAFS spectra of dilute systems, could one use this approach on aligned molecules?

Daniel Rolles responded: We have not considered this possibility so far, but it sounds like a very interesting idea that we should discuss further.

Michael Woerner opened the discussion of the paper by **Michael P. Minitti**: A combination of time-resolved X-ray diffraction experiments (information about geometrical structure) and time-resolved X-ray absorption experiments (information about electron energies) would give you more complete information about the chemical reaction.

Do you see any potential in your experimental concept to perform such combined studies?

Michael Minitti answered: Absolutely, we do. Efforts are underway in our group as well as at LCLS to combine both approaches in the hard X-ray regime. Capabilities like this exist in the soft X-ray regime with the new LAMP endstation in AMO, however, the geometrical structure determination (diffractive) studies will suffer do the limit in the obtainable q range.

Gwyn Williams commented: When you are working at the 3rd harmonic, how do you eliminate the fundamental, which for an FEL is usually 1000 times stronger, going as the power of the harmonic number?

Michael Minitti responded: The use of solid foil attenuators in the upstream front end enclosure can knock the transmission of the fundamental X-ray by 10 to 11 orders of magnitude while only mildly diminishing the 3rd harmonic transmission. Moreover, JJ slits can clean up any transmitted beam closer to the sample. The Be lenses needed to focus the 3rd harmonic on the sample do not allow any residual fundamental to be focused near the interaction region.

Gopal Dixit addressed **Michael Minitti** and **Adam Kirrander**: In your experiment, you have chosen 20 KeV photon energy for the imaging of the ring opening reaction. How can you be sure that you are performing coherent (elastic) X-ray scattering? Recently, we have shown the ratio between coherent and incoherent (inelastic/Compton) X-ray scattering as a function of photon energy, fluence and pulse parameter. You could have a look in our recent work about coherent *versus* incoherent X-ray scattering in a single molecule imaging experiment for further discussion: J. M. Slowik *et al.*, *New J. Phys.*, 2014, **16**, 073042, doi: 10.1088/1367-2630/16/7/073042 for your further consideration.

Michael Minitti answered: The structural information is contained in the coherent X-ray scattering and only the coherent scattering. Therefore, it does not need to be accounted for in difference patterns, since the incoherent scattering will necessarily subtract out, leaving only a difference in the molecular portion of the coherent scattering.

Andres Tehlar remarked: Could you determine a quantum yield for the formation of hexatriene after the UV excitation in your experiment? How does it compare to the calculations?

Michael Minitti responded: As for now, no we cannot. The longstanding disagreement between condensed phase *versus* gas phase branching ratios is still unanswered, at least in this study. Since we are dealing with extremely hot molecules, the amount of excess energy contained within all the vibrational and rotational DOFs is high and could cause erroneous observed branching ratios.

Majed Chergui commented: In principle you should have photoselected your sample, but I imagine you are probing the sample too late?

Michael Minitti responded: I'm not sure I fully understand this question. If you are speaking to photoselection by the absorption of the UV pump photon to prepare the excited state wavepacket, then yes we do select the molecule. We can probe the process of the ring-opening reaction to very small time delays (sub-50 fs) with LCLS.

Hans Jakob Wörner asked: The ultraviolet pump pulse creates a transient anisotropy in the molecular-axis distribution of the photoexcited molecules (photoselection). Can this partial alignment be used in your data analysis to obtain additional information? Would your experiment benefit from impulsive, three-dimensional axis alignment?

Michael Minitti answered: Excellent question. An addendum we submitted along with our LCLS beamtime proposal back in early 2013 is available in the Electronic Supplementary Information explaining the importance of varying the polarization of the optical pump laser to obtain the best, most complete view of the reaction given the fact the X-ray probe is a polarized source as well (ESI†). For your second question, our experiment would undoubtedly improve and benefit from field-free alignment techniques. This improvement will be implemented in our upcoming LCLS beamtime later this year.

Jochen Küpper commented: This is a follow-up question to a previous question by Hans Jakob Wörner.

For your conditions—warm sample and weak lasers—you likely do not get any alignment from the optical/NIR lasers, and if at all, it would likely be planar alignment for the molecule investigated. At the same time, you have geometric alignment from the photoselection with your excitation laser. How strong is that under the experimental conditions and how fast does it dephase?

Would you be able to make use of strongly aligned molecules, planar alignment of your molecule or even strong linear or 3D alignment or orientation (Stapelfeldt, *Rev. Mod. Phys.*, 2009; Larsen, *Rev. Mod. Phys.*, 2000; Holemgaard, *Phys. Rev. Lett.*, 2009; Nevo, *Phys. Chem. Chem. Phys.*, 2009). How would you exploit that and how much would it improve your analysis?

Michael Minitti replied: The polarized laser beam preferentially excites molecules with one particular orientation. In the case of this work, the polarization of the optical pump was switched from horizontal to vertical. This is a weak alignment, with a $\cos^2$ distribution. Clearly, a tighter alignment would help to bring out subtle details in the diffraction patterns (S. Ryu *et al.*, *J. Phys. Chem. A*, 2004, **108**). This is a goal of ours in our upcoming beam time at LCLS in October 2014.

Majed Chergui remarked: The Zewail group did ultrafast electron diffraction studies on CHD (see, *e.g.*, *Science*, DOI: 10.1126/science.291.5503.458 and PNAS, DOI: 10.1073/pnas.131192898). How do your results compare with these studies? On a more general level, what are the advantages at using ultrashort X-rays rather than ultrashort electron pulses, which have reached the sub-100 fs limit (see the works of J. Luiten *et al.*) for such studies?

Michael Minitti answered: The electron diffraction results (R. C. Dudek, P. M. Weber, *J. Phys. Chem. A*, 2001, **105**) from that area don't nearly have the time resolution we have now. I am not sure what the 'work of J. Luiten' is; with electrons it is possible to get 100 fs on short samples such as thin foils; it is not yet possible to do that with gas phase samples that have mm length scales, because of the difference in velocity between light and the electrons. Ideally, one would use relativistic electrons as demonstrated by Hastings *et al.*, *A. Phys. Lett.*, 2006, **89**.

Jochen Küpper enquired: Can you directly invert your data to get an instantaneous structure, or at least a pair-correlation function?

Michael Minitti answered: The range of scattering vectors is small in the X-ray diffraction experiment so that the uncertainty in the radial distribution function would be too large to have a meaningful invertible image. Getting the molecule aligned would greatly help the final, structurally resolved image.

Gopal Dixit addressed **Adam Kirrander** and **Michael Minitti**: I am more concerned about the applicability of the formalism which is based on the notion that the time-resolved X-ray scattering encodes time-dependent density of the system (instantaneous electron density). As we have shown, this notion completely breaks down when two or more electronic states are closed, especially in the vicinity of conical intersection and avoided crossing (please see G. Dixit *el al.*, *Proc. Natl Acad. Sci.*, 2012, **109**, 11636–11640; G. Dixit *el al.*, *J. Chem. Phys.*, 2013, **138**, 134311). It is essential to mention the pros and cons of the employed theory, based on the Fourier transform of instantaneous electron density of the system under investigation, in your present work.

Adam Kirrander communicated in reply: The currently proposed analysis uses state-of-the-art calculations for the photochemical ring-opening reaction of the CHD molecule based on the *ab initio* multiconfigurational Ehrenfest method.[1] In terms of the time-resolved X-ray scattering, the various levels of formalism applicable have been discussed in detail by several authors, see for instance ref. 2. At the moment, we make three significant assumptions. The first, which pertains to the question asked above, is that the duration of the coherent X-ray pulse is sufficiently long that scattering cross-terms between different electronic states average out, something which requires that the electronic states are well separated.[2] This approximation therefore breaks down in the direct vicinity of conical intersections and avoided crossings, and has been examined *via* calculations in one- and two-electron atoms by Dixit *et al.*, as referenced in the question.[3] The second assumption is that the nuclear motion is slow as compared to the duration of the X-ray pulse, and this also deserves highlighting given how fast this reaction proceeds. The third assumption, made within the context of the two initial assumptions, is the use of independent atom model (IAM) form factors, which does not account for the valence electron density. All these assumptions warrant further investigation, and should be incorporated in future, more detailed, analysis of the experiment. At the moment, however, the combination of trajectories that realistically map the evolution of the chemical reaction combined with a simplified treatment of the scattering provides a first interpretation of the data.

1 K. Saita and D. V. Shalashilin, *J. Chem. Phys.*, 2012, **137**, 22A506.
2 N. E. Henriksen and K. B. Moller, *J. Phys. Chem. B*, 2008, **112**, 558.
3 G. Dixit, O. Vendrell, R. Santra, *Proc. Natl Acad. Sci.*, 2012, **109**, 11636.

Henry Chapman opened the discussion of the paper by Jon P. Marangos: I have a suggestion for a way of carrying out your single-wavelength X-ray pump X-ray probe experiment with a very high time accuracy, which is to employ the scheme of femtosecond time-delay holography (Chapman *et al.*, *Nature*, 2007, **448**, 676). In a particular place the glycine molecules on a transparent substrate (such as a few nm of silicon nitride or a single graphene layer) are located some distance from a normal-incidence mirror that reflects your 280 eV photons. The sample is pumped by the pulse which then returns to the sample after reflection to probe. The time delay between pump and probe is set by the distance between the sample and mirror. For very short delays, one could deposit a wedged transparent spacer layer directly on the mirror. Film thicknesses can be determined to nanometer precision with correspondingly high time resolution. Normal-incidence multi-layer mirrors have been made with about 10% reflectivity (Leontowich *et al.*, *SPIE*, vol 8777, DOI: 10.1117/12.2022403). Would that be useful for measuring ultrafast hole dynamics?

Jonathan Marangos replied: The proposed scenario is a compact way to achieve the two delayed pulses needed. In fact the proposal already contemplates a split-and-delay set up or alternatively a double-electron pulse to generate the two X-ray pulses. Both these modes have been implemented at LCLS and are in principle capable of delivering the ~ 3 fs pulses with jitter free delays in the range from 0–25 fs.

Oriol Vendrell commented: In the proposed scheme for measuring electron hole dynamics the beating of the electronic wavepacket is mapped onto the Auger yield. I wonder whether the Auger spectrum itself (its shape) still carries some information of the hole dynamics. In view of the fact that the final state (see Fig. 1 in DOI: 10.1039/C4FD00051J) is the same irrespective of the particular state in the valence shell, I would first assume that this is not the case, at least for the particular example presented here. Would there be then ways to modify or generalize the scheme such that the Auger spectrum carries more information on the dynamics of the hole such as, *e.g.*, its position in the molecule?

Jonathan Marangos replied: In principle, there will be signatures in the Auger spectrum that are sensitive to the valence and inner valence hole (IVH) and any excitation. I think it was this type of sensitivity that was used in the recent work led by Markus Guehr on thymine (refer to Nora Berrah who was a co-author of that work). In the current scheme, however, the key to the detection of an IVH is the emission of Auger electrons in an electron energy range where there should be no background without any need to be sensitive to the exact details of the electron spectrum.

Jonathan Underwood enquired: In your theoretical treatment, when considering the probe pulse on page 11 of the manuscript, you assume that the probe pulse is polarized along the *x*-axis of the molecule. Given that your proposed experiment doesn't include any provision for molecular alignment, I would expect that all molecular frame polarizations of the probe laser would be present in the experiment which would access all molecular frame transition dipole moment components. What effect do you expect this to have on a real experiment, compared to the model calculations presented in which only a single transition dipole moment is considered in the probe step?

Jonathan Marangos responded: The calculation was merely to verify that the value of the dipole moment was compatible with the experiment (*i.e.* a sufficient excitation probability whilst not needing a very high intensity that might lead to saturation effects (additional multi-photon channels, strong driving of the transition). In fact the dipole for the other polarisations was checked and found to be very similar so this is not expected to change anything significantly in the scheme.

Jochen Küpper addressed **Jonathan Marangos, Oriol Vendrell** and **Adam Kirrander**: Considering that you now have a detailed theoretical description, at least of a specific ultrafast "charge migration" in glycine, I am wondering how this relates to the underlying findings of Weinkauf and coworkers in the 1990s (Weinkauf, 1996).

As far as I understand your calculations, you propose to measure the calculated dynamics of an electronic wavepacket that, in classical terms, corresponds to the breathing of the electron/hole density.

How is this related and coupled to the actually necessary charge migration and, eventually, charge (and energy) transfer along a molecular chain—as observed by Weinkauf and coworkers (Weinkauf, 1996) and theoretically described ten years later (Remacle, 2006). After all, these experiments of charge migration or transfer along a small peptide chain seem to be the resemblance of energy transport in chemical and biological systems, they triggered the field, and they provide the aim as well as the incentive of what processes we need to understand in the end.

Do you think that current attosecond experiments are the right approach toward the ultimate goal of understanding the charge (and energy) flow along a molecular system?

Jonathan Marangos answered: A few comments first. In the glycine example considered the hole evolution takes the form of oscillatory non-exponential decay rather than charge migration. Nevertheless the scheme is in principle suitable for probing charge migration as the atomic specificity allows the time dependent localisation of the hole to be studied. Of course, any excitation of a hole wavepacket can lead to non-stationary hole evolution (be it valence or inner valence states that are excited). In this case, the superposition involves two inner valence states—one of 1h and the other of 2h1p character. Only the former registers in the detection scheme so the result is an oscillatory hole survival probability.

The early work by Weunkauff stimulated the theoretical analysis (*e.g.* by Cederbaum, Remacle, Levine and others) that predicted charge migration. I would, however, argue that those experiments are rather different in character from those proposed here. They did not involve "sudden" excitation of the hole state and the probe of charge motion was a rather indirect one of registering subsequent laser induced fragmentation. In the experiments proposed here the hole wavepacket would be created by a few fs X-ray pule with a resonant probe (transient absorption) that is directly sensitive to the hole.

Oriol Vendrell responded: Charge migration as triggered by long picosecond or even nanosecond laser pulses as in (Weinkauf, 1996) involves most probably nuclear wavepackets evolving in coupled potential energy surfaces as in usual photochemical processes. In contrast, attosecond pulses create superpositions of several electronic states by virtue of their large bandwidth of several electronvolts. This pure electronic dynamics will disappear though, as soon as the nuclear wavepackets on each of the involved potential energy surfaces lose their overlap, which can be as fast as a few to some tens of femtoseconds. In my opinion, the kind of charge and energy transfer processes occurring naturally in chemistry and biology are related to the former. Nonetheless, the underlying physics involved is of course the same: coherent light creates superpositions of vibrational and/or electronic eigenstates of the system resulting in time evolution of some observable quantities.

Hans Jakob Wörner replied: Let me add a few comments to this discussion. The early work of Weinkauf and colleagues[1,2] established that the observed photofragmentation behavior of photoionized oligopeptides strongly deviated from the expectations of statistical theories. These results lead to the concept of "charge-directed reactivity",[3] *i.e.* that purely electronic charge migration occurs first and largely determines the fate of the molecule, followed by a nuclear motion that localizes the charge and then induces bond breaking. Later, time-resolved work by Weinkauf and colleagues[4] established a 80 ± 28 fs timescale for a similar charge transfer process in the peptide-model compound 2-phenylethyl-*N*,*N*-dimethylamine (PENNA). The timescale showed that this process was clearly not purely electronic, but required nuclear motion. Subsequent theoretical analysis suggested that PENNA does display a few-femtosecond charge migration, but of a very small amplitude only, hardly explaining the observations alone.[5]

What will attosecond spectroscopy contribute to this topic? Clearly, the few- to sub-femtosecond resolution is needed to access the purely electronic dynamics, which is thought to initiate "charge-directed reactivity". However, the large bandwidth of typical attosecond pulses is also a disadvantage because multiple electronic states will be prepared in any medium-sized molecule, leading to electronic wave-packet motion. The dominant part of this dynamics is simply defined by the electronic energy-level spacing of the cation and therefore adds little knowledge as compared to static photoelectron spectra. The truly interesting part of the problem is the dynamics driven by electron correlation.[6] This information can be most readily accessed in systems that display hole-mixing in the cationic manifold and will appear most clearly when we use more selective methods that enable the creation of a well-defined spatially-localized hole such as resonant photoionization or, possibly, strong-field ionization. We have recently completed such an experiment using high-harmonic spectroscopy.

1 R. Weinkauf, P. Schanen, D. Yang, S. Soukara, and E. W. Schlag, *J. Phys. Chem.*, 1995, **99**, 11255- 11265.
2 R. Weinkauf, P. Schanen, A. Metsala, E. W. Schlag, M. Bürgle and H. Kessler, *J. Phys. Chem.*, 1996, **100**, 18567–18585.
3 R. Weinkauf, E. W. Schlag, T. J. Martinez and R. D. Levine, *J. Phys. Chem. A*, 1997, **101**, 7702–7710.
4 L. Lehr, T. Horneff, R. Weinkauf and E. W. Schlag, *J. Phys. Chem. A*, 2005, **109**, 8074–8080.
5 Siegfried Lünnemann, Alexander I. Kuleff, and Lorenz S. Cederbaum, *J. Chem. Phys.*, 2008, **129**, 104305.
6 L. S. Cederbaum and J. Zobeley, *Chem. Phys. Lett.*, 1999, **307**, 205–210.

Adam Kirrander responded: Earlier the discussion touched on the complementarity between time-resolved and frequency-resolved spectroscopies. As a short follow-up to Oriol Vendrell's comment, with which I agree, it might be worthwhile re-iterating that 'dynamics' occurs whether a single state or a superposition of states is excited. The difference is rather in the manner in which this dynamics is observed. A straightforward example is how the linewidth, an apparently static spectral feature, relates to the decay of a system coupled to a continuum. One of the strengths of studies in the time-domain, in particular for complex systems with a high density of states, is that they efficiently capture the dominant aspects of the dynamics,[1] while dense spectra rapidly become too involved to interpret. Even in comparatively simple systems, albeit with complex spectra, does the time-domain provide complementary physical insight.[2]

1 A. Stolow, *Faraday Discuss.*, 2013, **163**, 9.
2 A. Kirrander, Ch. Jungen and H. H. Fielding, *Phys. Chem. Chem. Phys.*, 2010, **12**, 8948–8952.

Katharine Reid asked: In your paper you mention that you did not calculate the nuclear dynamics that would be expected to ensue following the induced hole migration. It strikes me that those nuclear dynamics would be very interesting and I wondered if there was any progress towards calculating them?

Jonathan Marangos responded: Yes, we are currently adding the nuclear dynamics. Preliminary calculations using the Ehrenfest approach have already indicated the influence of nuclear dynamics will be signficant within the first 10 fs. Attempts to use a full quantum calculation are underway.

Martin Wolf opened the discussion of the paper by **C. Vozzi**: I want to comment on the question of orbital imaging and the problem of aligning molecules in the gas phase for such studies. In surface science there is an active field to study "orbital imaging" of adsorbed molecules at surfaces using angle-resolved phozoemission spectroscopy (see, *e.g.*, *Science*, 2009, **326**, 702–706; *Phys. Rev. Lett.*, 2011, **107**, 193002; *Phys. Rev. B*, 2012, **86**, 045417; *Nature Commun.*, 2013, **4**, 1514, *Proc. Natl Acad. Sci.*, 2014, **111**, 605). In certain cases (physisorption and weak electronic coupling to the substrate) the surface acts just to align the molecules in a well defined way. Information about molecular orbitals is obtained from the full angular distribution of photoelectrons, whereby frequently the assumption of a free eelctron final state is used. This is feasible for rather extended molecular systems (*e.g.*, pentacene or PTCDA on silver surfaces). For such Pi-systems the approximation of a free electron final state works fairly well. There is also considerable theoretical work (*e.g.*, *Proc. Natl Acad. Sci.*, 2014, **111**, 605).

A very promising direction would be to study excited states of adsorbates and investigate the (multi)electron dyamics, *e.g.* in organic systems with single triplet conversion. There is a technologocal challenge to perform such time resolved studies at high repetition rates with sufficient excitation densities in the molecular layer. The new generation of XUV pulse generation techniques based on MHz rep rate high harmonic generation (HHG) using OPCPA systems may pave the way to such experiments.

Hans Jakob Wörner answered: The research direction that you propose is very exciting, indeed. We have recently made the first steps towards "orbital imaging" of photochemical reactions of isolated molecules by developing methods to measure (i) the phase and amplitude of high-harmonic emission from photoexcited molecules, relative to ground-state molecules[1–5]; and (ii) the phase and amplitude of aligned ground-state molecules, relative to an atomic reference.[6] The combination of these methods provides the required input for "dynamical orbital imaging". However, electronically excited states are often not well described by a single configuration. When the excited state is multi-determinantal in character, the meaning of a retrieved one-electron orbital becomes less transparent. The weaknesses of the plane-wave approximation for the electronic continuum are another obstacle.[7]

1 H. J. Wörner *et al.*, *Nature*, 2010, **466**, 604.
2 H. J. Wörner *et al.*, *Science*, 2011, **334**, 208.
3 P. M. Kraus and H. J. Wörner, *Chem. Phys.*, 2013, **414**, 32.
4 A. Tehlar and H. J. Wörner, *Mol. Phys.*, 2013, **111**, 2057.
5 A. Tehlar, P. M. Kraus, and H. J. Wörner, *Chimia*, 2013, **67**, 207.
6 J. B. Bertrand *et al.*, *Nat. Phys.*, 2013, **9**, 174.
7 H. J. Wörner *et al.*, *Phys. Rev. Lett.*, 2009, **102**, 103901.

Caterina Vozzi responded: This is a very interesting suggestion. There are anyway some issues that should be solved before photoionization of physisorbed molecules on substrates might be used for dynamical orbital imaging:

1) the XUV pulse may ionize both the molecules and the substrate, thus the contributions from these two sources of photoelectrons should be disentangled. Of course, ionization from the substrate can be characterized in detail; however the photoionization rate as well as the photoelectron energy emitted by the substrate may depend on the angle, hence the subtraction of this background may not be trivial;

2) the target molecules should be replaced with fresh ones before each laser shot; this can be done for instance with moving substrates, but a very precise control of the target position should be applied, in particular with pump–probe configurations. This issue however has already been considered in laser-plasma experiments, like in plasma mirror setups, and can be in principle solved;

3) care should be taken to avoid high space charge effects that might disturb the measurements;

4) experiments involving excited molecules may face an additional problem: molecular excitation may lead to nuclear rearrangement, which could be hindered by the substrate. Hence a detailed study of substrate influence on molecular excited states is required in order to apply this technique in time-resolved experiments.

Klaus von Haeften opened the discussion of the paper by Hans Jakob Wörner: High Harmonics (HH) are frequently generated in the gas phase, and more recently, using laser-aligned molecular targets. Martin Wolf suggested to align molecules on surfaces to establish a maximum degree of alignment. Indeed, it is possible to adsorb molecules on surfaces selectively on well-defined sites. Is it feasible to generate HH using such a target given that the high laser fields may cause desorption of the molecules from the surface or even dissociation of the molecules?

Jonathan Marangos replied: Very challenging to do HHG of molecules at a surface—hard to see how to avoid catastrophic surface damage.

Caterina Vozzi responded: Even this suggestion is intriguing; however, in such a case the strong driving field may affect both the adsorbed molecules and the substrate. Since the density of the substrate is much larger, the XUV signal coming from it may overwhelm the one from molecules. Moreover, two additional issues should be considered:

1) HHG is a coherent process strongly affected by phase matching issues; in the gas phase a detectable macroscopic signal can be obtained only in the direction of the driving laser, since in this direction the emission from all the nonlinear dipoles adds in phase. Hence particular care should be taken in order to obtain a similar coherent sum in reflection geometry, which however may result in being more critical.

2) The excursion of the ionized electron in the continuum is a crucial part in the process of high harmonic generation; any perturbation to electron trajectories due to external factors can alter the emission process. In the worst case, the influence of the substrate may be so important that HHG may be suppressed; in general one could predict a more complex relationship between molecular structure and harmonic spectrum, which may hinder the extraction of useful information from the measurements.

Katharine Reid said: There is an underlying theme in the conference of wishing to record a molecular movie, but there are times when frequency-resolved measurements can provide all the information available in a time-resolved experiment. The determination of electron–nuclear couplings in nitric oxide is a case in point as these have been established using high resolution absorption (or equivalent) spectroscopy. Can you comment on the complementarity of time-resolved and frequency-resolved measurements, particularly when dynamics are induced by a photoabsorption process?

Hans Jakob Wörner responded: Any time-dependent signal can be Fourier-transformed to obtain a frequency-domain spectrum. Similarly, any absorption spectrum can be inverse-Fourier-transformed to obtain an autocorrelation function, which describes the time-dependent survival probability of the impulsively prepared initial state. However, most time-resolved methods go beyond the measurement of auto-correlation functions and each of them has its specificities that make it sensitive to particular aspects of the dynamics. The present experiment was not designed to obtain new information on nitric oxide but rather to explore the sensitivities of high-harmonic generation (HHG) and strong-field ionization (SFI) to coupled electronic and rotational dynamics. We find an extreme sensitivity to electronic dynamics (0.1% excitation leading to ~20% signal modulation) in HHG, which is significantly reduced in SFI. We find a harmonic-order-dependent sensitivity of HHG to rotational dynamics that is traced back to resonances in the photoionization continuum of NO. Comparing time- with frequency-domain information, the present experiment allows us to follow the rotation of the electronic density around the molecular axis (see inset in Fig. 3c in ref. 1 and Fig. 3 in ref. 2). The corresponding frequency-domain information does not allow us to reconstruct the dynamics because it does not measure the relative phases of the populated eigenstates. On longer timescales, the rotational dynamics becomes important and leads to a dephasing of the electronic subsystem (see ref. 2).

More generally, this experiment shows how time-resolved methods can be used to reduce the complexity of molecular dynamics by exploiting the hierarchy of timescales. On short timescales a wavepacket behaves as the solution of a simplified Hamiltonian, containing only the dominant terms. For example, on an attosecond timescale, nuclear motion is frozen, allowing one to measure purely electronic quantum beats, *i.e.* electronic energy intervals at a fixed molecular geometry and orientation. On longer timescales, nuclear dynamics sets in and leads to dephasing of the electronic subsystem. For sufficiently long timescales, a high-resolution spectrum can be recovered from the time-resolved data. Working back from a high-resolution spectrum to electronic energy intervals can be very difficult in congested spectra, especially when vibronic and rovibronic couplings are involved. This situation is typically already encountered in triatomic molecules, such as NO_2, the high-resolution spectrum of which remains largely unassigned around 400 nm. This complexity does not prevent time-resolved methods from providing a clear picture of how energy flows from the electronic excitation to the bending motion, then to asymmetric stretching, which finally leads to bond breaking. These dynamics were recently observed using time-resolved high-harmonic spectroscopy,[3] which is reviewed and compared with time-resolved photoelectron spectroscopy in ref. 4.

1 D. Baykusheva, P. M. Kraus, S. B. Zhang, N. Rohringer, and H. J. Wörner, *Faraday Discuss.*, 2014, DOI: 10.1039/C4FD00018H.
2 P. M. Kraus, S. B. Zhang, A. Gijsbertsen, R. R. Lucchese, N. Rohringer, and H. J. Wörner, *Phys. Rev. Lett.*, 2013, **111**, 243005.
3 H. J. Wörner *et al.*, *Science*, 2011, **334**(6053), 208–212.
4 P. M. Kraus and H. J. Wörner, *Chem. Phys.*, 2013, **414**, 32–44.

Jochen Küpper stated: Following a remark from Katharine Reid I commented that to understand how to look at structure and dynamics of molecules, in the end we will have to do eigenstate resolved as well as time-resolved experiments and combine the data.

Katharine Reid responded: If true eigenstate resolution can be achieved there seems nothing to gain from a time-resolved measurement because any wave-packet evolution can be constructed using the relative amplitudes and phases of the eigenstates. However, for most systems true eigenstate resolution will never be possible, and I agree that a combination of frequency-resolved and time-resolved approaches is likely to be necessary in order to characterize structure and dynamics.

Majed Chergui enquired: Have there been any systematic studies of the coupling between rotational wavepackets and Rydberg wavepackets? Helen Fielding did some on NO, but I haven't seen follow-up studies.

Hans Jakob Wörner replied: Helen Fielding and her group indeed performed a series of pioneering experiments on molecular Rydberg wavepackets, which are summarized in ref. 1. The Rydberg series studied in NO were non-interacting, such that no information on electronic-rotational coupling could be obtained. However, time-dependent multi-channel quantum-defect theory (TD-MQDT)[2,3] has been used to analyze the manifestations of vibronic and rovibronic couplings in Rydberg-wave-packet measurements. Recently-developed methods of atto-second spectroscopy have already led to renewed interest in the dynamics of low atomic Rydberg states,[4] which will almost certainly be extended to molecules in the near future.

1 H. Fielding, *Annu. Rev. Phys. Chem.*, 2005, **56**, 91.
2 F. Texier and Ch. Jungen, *Phys. Rev. A*, 1999, **59**, 412.
3 A. Kirrander, H. H. Fielding, and Ch. Jungen, *J. Chem. Phys.*, 2007, **127**, 164301.
4 H. Wang *et al.*, *Phys. Rev. Lett.*, 2010, **105**, 143002.

Jonathan Underwood asked: The data in Fig. 8 of the manuscript by Wörner and colleagues shows the effect of pump laser intensity on the electronic and rotational coherences produced, and in particular shows an increase in the transfer of population to the upper spin–orbit electronic state (F_2). The splitting between the F_1 and F_2 electronic states of ~123 cm^{-1} lies within the pump pulse bandwidth suggesting that the process for population transfer to the excited state is a stimulated Raman process. Have the authors considered looking at the effect of the pump pulse duration and shape (in other words, relative phases of the optical frequencies of the pump pulse) on the electronic coherence, and would there be an advantage to adding in a second pump laser shifted by ~123 cm^{-1} to further enhance the electronic coherence?

Hans Jakob Wörner responded: This is an excellent suggestion. The excitation process is indeed an electronic stimulated Raman process, mediated by the quadrupolar part of the scattering tensor (see eqn (9) of our manuscript). The electronic population transfer remains however weak, on the order of 1% at the highest applied intensities (6×10^{13} W cm^{-2}). We have so far only studied the effect of the pump-pulse intensity (Fig. 8) but no other parameters. The optimization of the duration, pulse shape or the addition of an additional frequency-shifted pulse would all be very interesting extensions.

Jonathan Marangos addressed **Michael Woerner** and **Caterina Vozzi**: A comment on HHG based methods—there are technical difficulties in creating sufficient gas phase sample densities for many molecules for HHG and of aligning and/or orientating the molecules as required by many HHG spectroscopy (HHGS) techniques. This has so far limited application of the powerful methods of HHG spectroscopy for structural and dynamical measurement to a very limited set of small molecules. We have recently had some success in carrying out HHG measurements in a series of substituted benzenes (*e.g.* toluene, xylene and halobenzenes) in thin samples (suitable for quantitative HHGS) so we do not argue the situation is useless but the technique is not likely to be of application to all but small molecules.

More critically, we must understand the role of the strong laser field intrinsic to HHG on the dynamical processes we are hoping to study. HHG is a non-linear parametric method that results in a macroscopic forward scattered beam of up-converted photons. To coherently contribute to this macroscopic signal the molecules must start and end in the same quantum state (either a single state or a coherent superposition). Any additional changes in the molecule during the process, like nuclear dynamics, hole decay (*e.g.*, by Auger) or multi-electronic excitations (the probability of more than one excited electron recombining back to the initial state is vanishingly small) will result in loss of coherence and thus a loss of signal. Indeed, a key advantage of HHG spectroscopy is that we can use the process to measure the nuclear evolution (Baker, *Science*, 2006) or electronic decay (H. Leeuwenburg, *Phys, Rev. Lett.*, 2013) that occur at the few fs timescale. HHG is a technique highly suited to measuring the evolution of a molecular cation in a strong laser field.

However, in most molecules the cation has many states that can be coupled by the strong near resonant laser field. This has now been thoroughly analysed for CO_2 and the analysis shows that 4 cation channels with numerous cross-channel couplings must be included in any calculation that quantitatively explains the HHG. Most molecules will have many more cation channels that couple, and unravelling their role is a major challenge for theory and experiment that may prove prohibitive to the wider application of HHG spectroscopy techniques.

Hans Jakob Wörner communicated in reply: Following up on Jon Marangos' comment on HHG-based methods, I would like to point out that field-free alignment and orientation at high particle densities will be crucial in extending the methods to more complex molecules. We have recently demonstrated a two-pulse method that achieves macroscopic degrees of orientation, rivalling those accessible through state selection.[1,2]

One important strength of high-harmonic spectroscopy, previously mentioned by Robert Moshammer, is that it offers attosecond resolution. Such resolution is otherwise only achievable in attosecond pump–attosecond probe experiments.

We have recently completed a series of experiments on attosecond charge migration using high-harmonic spectroscopy. Control over the orientation of the molecule enables us to turn on and off the laser-induced dynamics in the molecule and thereby observe both the field-modified and the field-free dynamics, dominated by charge migration.

One further strength of high-harmonic spectroscopy is its sensitivity to extremely weak excitations within coherent superposition states.[3,4] Although we have so far only demonstrated femtosecond resolution, sub-cycle resolution can be achieved by using a pair of phase-locked CEP-stable few-cycle pulses. Such experiments are in progress in our group.

1 P. M. Kraus, D. Baykusheva, and H. J. Wörner, *Phys. Rev. Lett.*, 2014, **113**, 023001.
2 P. M. Kraus, D. Baykusheva, and H. J. Wörner, *J. Phys. B: At. Mol. Opt. Phys.*, 2014, **47**, 124030.
3 P. M. Kraus, S. B. Zhang, A. Gijsbertsen, R. R. Lucchese, N. Rohringer, and H. J. Wörner, *Phys. Rev. Lett.*, 2013, **111**, 243005.
4 D. Baykusheva, P. M. Kraus, S. B. Zhang, N. Rohringer, and H. J. Wörner, *Faraday Discuss.*, 2014, doi: 10.1039/C4FD00018H

Caterina Vozzi communicated in reply: I completely agree with Jon's comment: HHG spectroscopy is not foreseen to be useful for a very large sample due to several experimental and theoretical limitations. Nevertheless, it is a powerful technique for the investigation of a relatively small sample, which can be manipulated (aligned/oriented) in the gas phase with sufficient density. In this perspective I would frame HHG spectroscopy as a complementary technique that can give insight in molecular dynamics when combined with different experimental investigation techniques, such as transient absorption spectroscopy, for example, and appropriate theoretical models.

Concerning the contribution of different channels, this needs for sure to be included in HHG spectroscopy. In simple cases, HHG spectroscopy allows the study of nuclear and electronic dynamics on the attosecond timescale, and this is indeed an exciting perspective. On the other hand, the exploitation of a long wavelength driving source for HHG can in some cases reduce the multi-electron contributions up to the extent of making them negligible, allowing a simpler quantitative interpretation of the results (*Nat. Phys.*, 2011, **7**, 822; *Phys. Rev. A*, 2013, **87**, 031401).

Andres Tehlar opened the discussion of the paper by **Caterina Vozzi**: A simple model of high harmonic emission splits the emission dipole in three parts according to the three-step-model: tunnel ionization, electron propagation and finally recombination. As I understand it, your method to reconstruct the orbital relies only on the last step, ignoring the effect of the angle-dependent tunnel ionization on the signal. Would the inclusion of this effect not significantly change the shape of the reconstructed orbital?

Caterina Vozzi responded: In the method we applied for the orbital reconstruction of a few molecules, namely CO_2, N_2O and C_2H_2, we assumed that the tunnel ionization is independent of the molecular alignment. In the case of these relatively simple molecules this assumption worked reasonably well, but it is indeed a rough approximation and we will need to include the angular dependence of tunnel ionization to refine the method and apply it to more complex species. The angular dependent ionization rate can be measured (see, for example, *Phys. Rrev. Lett.*, 2014, **112**, 253001 and references therein) or calculated (see, for example, the work of Michael Spanner and Serguei Patchkovskii) and the inclusion of a more detailed and credible description of the ionization step is needed for the generalization of molecular orbital tomography.

Jochen Küpper addressed **Caterina Vozzi, Hans Jakob Wörner** and **Jonathan Marangos**: Dear Caterina, you said that in your experiment you can measure the orbital. Now, as far as I have followed the discussion after the original orbital tomography by Itatani *et al.* (Itatani, *Nature*, 2004), in HHG you actually measure a "Dyson orbital" or, in simple words, the difference of the ionic and neutral electron densities. Is that correct, and in how far is that directly related to the orbital of the neutral parent molecule you are trying to investigate?

Moreover, for the analysis of your data you have to make many assumptions, simple ones as the nuclear structure or the chemical composition, but also more detailed once that goes into the description of the strong-field ionization process.

Especially for the tomographic approach, there are effects in SFI due to the alignment angle (Meckel, *Nat. Phys.*, 2014; Küpper, *Nat. Phys.*, 2014]. Moreover, there are the well-known problems of necessary Coulomb corrections (Goreslavski, *Phys. Rev. Lett.*, 2004), contributions from lower energy orbitals [add], electron emission out of the laser plane [add], observation of low-energy electrons [add], and so forth (Küpper, *Nat. Phys.*, 2014).

In that light, do you think you will be able to observe ultrafast dynamics, using HHG, in previously not-so-well understood (complex) molecules?

Hans Jakob Wörner answered: The Dyson orbital is a one-electron wave function obtained by projecting the N-1 electron wave function of the relevant cationic state onto the N-electron wave function of the neutral molecule. When configuration interaction can be neglected in both states, the Dyson orbital is identical to the corresponding Hartree–Fock orbital. In the experiments presented by Caterina Vozzi, the high-harmonic signal is dominated by the electronic ground state of the cation (see also ref. 1 and 2 for a detailed analysis). Because configuration interaction is very small in both the neutral and ionic ground states, the Dyson and Hartree–Fock orbitals are very similar.

Concerning the second part of your question, I would just like to add one comment. Research on photoionization[3] has clearly established which phenomena are well described by single-electron theories (*e.g.*, Cooper minima, shape resonances) and which are not (*e.g.* channel interactions, correlation effects). When a single-electron description of photoionization (*i.e.* also photorecombination) is valid and accurate, one may try to solve the inverse scattering problem and obtain an effective potential and the bound-state orbital. This approach overcomes the limitations of the plane-wave approximation and is currently being pursued in our group.

1 A. Rupenyan, P. M. Kraus, J. Schneider and H. J. Wörner, *Phys. Rev. A*, 2013, **87**(3), 031401.
2 A. Rupenyan, P. M. Kraus, J. Schneider, and H. J. Wörner, *Phys. Rev. A*, 2013, **87**(3), 033409.
3 *VUV and soft X-ray photoionization*, ed. U. Becker and D. A. Shirley, Springer Science and Business Media, 1996.

Caterina Vozzi answered: The exploitation of HHG as a general spectroscopic tool requires of course to take into account all the mentioned aspects; in principle, those aspects may require considerable theoretical and experimental efforts in order to deeply understand their role in HHG. Nevertheless, one can interpret them not as simple hindrances, but as additional opportunities in extracting more information on the target molecule. For instance:

1) it has already been demonstrated by the works of Smirnova *et al.* and by Guehr *et al.* that the role of different orbitals in HHG can be disentangled and HHG may be used to trace the dynamics of the hole left in the molecular ion;

2) departures of the harmonic spectra from the calculated ones in the framework of the SFA can be used to understand how the Coulomb field affects the harmonic emission, although one may predict tiny effects on high-energy colliding electrons. In the same view, additional effects as the ones recently published in *Nat. Phys.* by Meckel and coworkers (DOI: 10.1038/nphys3010) concerning the phase of the continuum electron wavepacket, can be studied with HHG or by other approaches and add further knowledge on the molecular physics in strong fields that may then be exploited in HHG spectroscopy;

3) the role of the angle-dependent SFI in HHG can be disentangled by careful studies on the photoionization of aligned molecules.

Concerning the relationship between the molecular orbital of the neutral molecule and the harmonic spectra: although the Dyson orbital is the real physical quantity involved in the HHG process, it is often well approximated by the molecular orbital of the leaving electron (see, for instance, this link http://iopenshell.usc.edu/research/projects/dyson-orbitals/) hence in several cases the issue of what one is measuring might not be so crucial.

Robert Moshammer addressed **Caterina Vozzi, Jonathan Marangos** and **Hans Jakob Wörner**: The method of High Harmonic spectroscopy of molecules provides, beside several conceptual difficulties and problems, unique information about the dynamics on very short attosecond timescales. The time between excitation (ionization) and probe (recombination), however, is kind of fixed at a given IR wavelength, are there ideas or concepts of how to freely adjust the "pump–probe" delay time?

Jonathan Marangos responded: In fact, the intrinsic chirp of HHG emission means there is a well defined mapping between time and emission frequency. This is the concept of PACER (probing attosecond dynamics by chirp encoded recollision). The time- frequency mapping can be manipulated by changing the laser wavelength and intensity (and adding an extra field).

Hans Jakob Wörner responded: At a given infrared (IR) wavelength driving the high-harmonic-generation process, the transit time of the electron wavepacket in the continuum can be varied by changing the intensity of the field. A different driving intensity leads to a different mapping of transit time to the emitted photon energy. Using an 800-nm driving field limits the transit time to <1.7 fs. The range of accessible transit times can be extended by increasing the wavelength of the driving field. Let me give a few examples from our work. In ref. 1, we have shown that attosecond nuclear dynamics in the ammonia cation can be followed from 0.8 to 3.8 fs using wavelengths from 0.8 to 1.8 microns. In ref. 2–4, we have shown that control over the transit time results in control over the interference of high-harmonic emission from channels associated with different orbitals. These different channels acquire a phase that is proportional to their ionization potential multiplied with the transit time $I_p\tau$.[5] It is however important to mention that such experiments, including ref. 6, do not measure electron-hole dynamics in a quantum-mechanical sense because they are insensitive to the electronic coherence between the electronic states of the cation, see also ref. 7.

1 P. M. Kraus and H. J. Wörner, *ChemPhysChem*, 2013, **14**, 1445–1450.
2 H. J. Wörner, J. B. Bertrand, P. Hockett, P. B. Corkum and D. M. Villeneuve, *Phys. Rev. Lett.*, 2010, **104**, 233904.
3 A. Rupenyan, P. M. Kraus, J. Schneider and H. J. Wörner, *Phys. Rev. A*, 2013, **87**, 031401.
4 A. Rupenyan, P. M. Kraus, J. Schneider and H. J. Wörner, *Phys. Rev. A*, 2013, **87**, 033409.
5 T. Kanai, E. J. Takahashi, Y. Nabekawa and K. Midorikawa, *Phys. Rev. Lett.*, 2007, **98**, 153904.
6 O. Smirnova *et al.*, *Nature*, 2009, **460**, 972.
7 P. M. Kraus, S. B. Zhang, A. Gijsbertsen, R. R. Lucchese, N. Rohringer and H. J. Wörner, *Phys. Rev. Lett.*, 2013, **111**, 243005.

Caterina Vozzi replied: Indeed, it is possible to adjust the time between ionization and recombination in the HHG process in different ways. For instance, it is possible to use a driving pulse with tunable wavelength—such as one based on OPAs—as reported for example in (*J. Mod. Opt.*, 2010, **57**, 1008). Another possibility is to tweak the pump–probe delay by exploiting a two color driving field in which the total driving electric field is composed by a strong fundamental component and a weak second harmonic component. This approach has been successfully applied for the study of the tunneling process and multi-electron rearrangement in carbon dioxide (*Nature*, 2012, **485**, 343).

Hans Jakob Wörner communicated: We have previously shown that the minimum observed in the mid-infrared-driven ($\lambda > 1.4$ μm) high-harmonic spectra of aligned N_2O molecules is a signature of its electronic structure.[1] However, we have also found that the minimum in N_2O appears at higher photon energies than in the isoelectronic CO_2 molecule under the same experimental conditions (7 eV) and even more so when the different axis distributions are taken into account (10 eV). In simplified terms, this observation can be attributed to fact that the HOMO of N_2O has a nonzero amplitude at the central atom and therefore appears "contracted" as compared to CO_2. Can you comment on the degree of axis alignment in your experiment and the comparison between CO_2 and N_2O in your results?

1 A. Rupenyan, P. M. Kraus, J. Schneider, and H. J. Wörner, *Phys. Rev. A*, 2013, **87**, 031401; *Phys. Rev. A*, 2013, **87**, 033409.

Caterina Vozzi communicated in reply: In our experiments we observed the minima in the macroscopic harmonic emission at around 57 eV in the case of N_2O and at around 60 eV for CO_2. We guessed a degree of alignment of about 0.62 in the case of the CO_2 molecules and about 0.68 for N_2O. The difference in the spectral position of the minima is rather small and indeed one expects this position to be influenced by the degree of alignment in the macroscopic sample, since the single molecule emission is convoluted to the alignment distribution.

Even if the experimental conditions were more or less the same, the two experiments were performed in different days and are not suitable for a direct quantitative and systematic comparison as the one performed in ref. 1.

Faraday Discussions

ROYAL SOCIETY
OF CHEMISTRY

PAPER

Solvation dynamics monitored by combined X-ray spectroscopies and scattering: photoinduced spin transition in aqueous [Fe(bpy)$_3$]$^{2+}$

C. Bressler,[*ab] W. Gawelda,[a] A. Galler,[a] M. M. Nielsen,[c] V. Sundström,[d] G. Doumy,[e] A. M. March,[e] S. H. Southworth,[e] L. Young[f] and G. Vankó[f]

Received 6th May 2014, Accepted 9th May 2014

DOI: 10.1039/c4fd00097h

We have studied the photoinduced low spin (LS) to high spin (HS) conversion of aqueous Fe(bpy)$_3$ with pulse-limited time resolution. In a combined setup permitting simultaneous X-ray diffuse scattering (XDS) and spectroscopic measurements at a MHz repetition rate we have unraveled the interplay between intramolecular dynamics and the intermolecular caging solvent response with 100 ps time resolution. On this time scale the ultrafast spin transition including intramolecular geometric structure changes as well as the concomitant bulk solvent heating process due to energy dissipation from the excited HS molecule are long completed. The heating is nevertheless observed to further increase due to the excess energy between HS and LS states released on a subnanosecond time scale. The analysis of the spectroscopic data allows precise determination of the excited population which efficiently reduces the number of free parameters in the XDS analysis, and both combined permit extraction of information about the structural dynamics of the first solvation shell.

A   Introduction

Solvation dynamics of a photoexcited molecule concerns the interactions between the molecule and its bath, *i.e.*, its nearest neighbor solvent molecules. This interplay manifests itself already in steady state studies by the observed Stokes shift between the absorption and emission energies of the solute,[1,2] which reflect the rearrangement of the caging solvent around the excited solute. Quantum

European XFEL, Albert-Einstein-Ring 19, 22761 Hamburg, Germany. E-mail: christian.bressler@xfel.eu
[b]The Hamburg Centre for Ultrafast Imaging, Luruper Chaussee 149, 22761 Hamburg, Germany
[c]Centre for Molecular Movies, Dept. of Physics, Technical University of Denmark, Fysikvej 307, DK-2800 Kongens Lyngby, Denmark
[d]Dept. of Chemical Physics, Lund University, Box 118, 22100 Lund, Sweden
[e]Argonne National Laboratory, 9700 South Cass Avenue, Illinois 60439, USA
[f]Wigner Research Centre for Physics, Hungarian Academy Sciences, H-1525 Budapest, Hungary

chemical calculations have meanwhile advanced and now permit simulating the dynamic response inside a box containing the excited molecule itself and a certain number of solvent molecules.[3-5] In this report we extend the experimental tools conventionally used into the X-ray domain, and apply these to a potentially functional spin transition system, aqueous iron(II)tris(bipyridine), $[Fe(bpy)_3]^{2+}$. It dynamically switches between a low spin (LS) ground to a high spin (HS) excited state upon visible light illumination, starting with a metal to ligand charge transfer (MLCT) process, followed by an ultrafast electron back transfer within 130 fs, eventually generating the spin flip to the HS state.[6-8] The bpy ligands are expected to effectively shield the central metal atom from the solvation shell, so that it appears impossible to extract any information on the solvation dynamics following the ultrafast LS–HS conversion in agreement with the optical-only results in ref. 8. In order to address this property with structural tools we have designed an ultrafast laser-pump/X-ray probe experiment, where the probe exploits complementary structural tools to simultaneously reveal different details of the ongoing process.

A few years ago, we have implemented time-resolved X-ray emission spectroscopy at synchrotron sources using a 1 kHz amplified laser system,[9] and subsequent MHz pump–probe XES studies permitted an accurate analysis of the transient XES and permitted us to extend the tools towards 1s-3p resonant XES studies.[10] Recently, we have extended time-resolved X-ray absorption spectroscopy (XAS) tools to include both X-ray emission spectroscopy (XES) and X-ray diffuse scattering (XDS) at the MHz repetition rates generated at X-ray storage rings.[11] In this contribution we extend XAS to the femtosecond time domain explored with X-ray free electron lasers (XFELs) and link this to the ps studies at storage rings, including the extraction of information content available from these simultaneously recorded X-ray emission spectra and the X-ray diffuse scattering patterns. The combination of these tools allow to acquire a complementary glimpse into the guest–host interactions.

B   Experimental approach

The picosecond experiments were assembled and conducted at beamline 7ID of the Advanced Photon Source. The setup for time-resolved X-ray absorption experiments in fluorescence detection mode at MHz repetition rates were already described in great detail.[11,12] Briefly, in the present experiments, we upgraded the standard X-ray absorption spectroscopy (XAS) equipment in total fluorescence mode by including new X-ray probing tools, which permit simultaneous X-ray emission spectroscopy (XES) and X-ray diffuse scattering (XDS) measurements at energies above the absorption edge of the selected atom of interest (here: for Fe well above 7.1 keV).

For this purpose a secondary X-ray spectrometer in Johann configuration utilizing a spherically bent (radius: 1 m) Si(531) crystal was mounted orthogonally sideways from the incident X-ray beam to suppress the background from scattered light impinging on the crystal and its mount.[10] Just a few centimeters downstream behind the sample a 100k Pilatus detector collected elastically scattered light in forward direction in single-photon counting mode. The exciting laser beam was tuned to the MLCT absorption band of $[Fe(bpy)_3]^{2+}$ ($\lambda = 532$ nm) with adjustable frequencies tailored to the bunch filling patterns of the APS storage ring. The results below utilized the hybrid single-bunch repetition rate of 0.136 MHz, but also higher frequencies matching the 24-bunch filling pattern (6.52 MHz), with the laser tuned one half of this repetition rate (3.26 MHz) to permit laser-on and laser-off measurements in a successive fashion. With such a strategy we successfully eliminated all electronic noise contributions to the recorded signals occurring on frequencies below 1 MHz (and the majority of noise contributions occur at frequencies below one 1 kHz).[13] This combined approach allows us to unravel different contributions in the dynamic processes of the ongoing physicochemical transformations.

In a first attempt to enter the fs time domain we exploited the LCLS for XAFS studies on the same molecule,[14] which we later upgraded to allow the same combined XES and XDS measurements. Below we show the femtosecond XAFS time delay scans and our attempts to descend below the 300 fs time scale. These measurements confirm and extend our previous measurements exploiting time-sliced femtosecond X-ray pulses from the Swiss Light Source.[15]

In the following we demonstrate the capabilities of X-ray absorption spectroscopy, which will serve as dearly needed input for the simultaneously applied XES and XDS tools. This approach nicely demonstrates the added value, when combining different X-ray techniques into one single experiment, as well as the additional insight from combining ps and fs results.

C X-ray absorption spectroscopy: from picosecond to femtosecond time scales

Time-resolved X-ray absorption near edge structure (XANES) spectroscopy relies on detecting photoinduced modulations of the absorption cross section as a function of energy near a core shell absorption edge.[16] The absorption is dependent on both the local atomic configuration as well as on the electronic configuration, thus providing information on unoccupied electronic states and the oxidation state of the absorbing atom, and (to a lesser content) the local molecular structure (e.g., the molecular symmetry, bond angles). The XANES can be measured either in transmission mode or via total fluorescence yield, but fluorescence yield is more suitable for rather dilute samples with concentrations below around 50 mM.

In a time-resolved X-ray absorption experiment, one usually measures the transient absorption changes ΔA due to the exciting laser pulse, which generates—here for a two-level system (see ref. 17 for the general case)—a signal of the type

$$\Delta A(t) = f(t)(A_{\mathrm{exc}} - A_{\mathrm{gr}}) \tag{1}$$

with A_{gr} being the static (ground state) absorption, A_{exc} the excited state absorption spectrum, and $f(t)$ being the fractional excited state population, which changes as a function of time delay between the laser excitation and X-ray probe pulses. Usually there are two unknowns, namely A_{exc} and f, of which the former is the desired new result, and the latter needs to be determined separately (in a separate measurement). In some cases one can determine both unknowns using prior chemical knowledge about the excited state properties.[18]

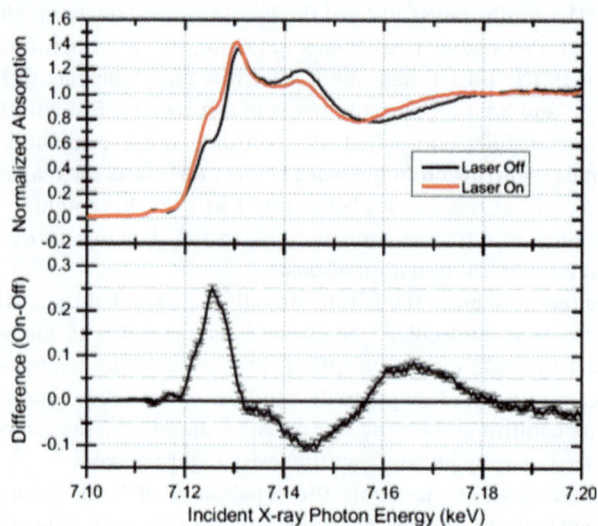

Fig. 1 Time-resolved XANES of photoexcited aqueous Fe(bpy)$_3$ together with the static spectrum; below the recorded transient XANES after 100 ps ($f = 0.4$).[11] Sample concentration was 20 mM, and laser excitation repetition rate was one half of the X-ray acquisition rate of 0.136 MHz.

For the present study we recorded first the TR-XANES, as shown in Fig. 1. This serves to check the overlap conditions on the sample, before switching to the combined XES/XDS experiment. To record its ultrafast temporal behavior in a fs XANES experiment exploiting XFEL radiation[14] the maximum of the transient feature was chosen.[15,19,20] XFEL experiments in general are subject to the jitter between the X-ray pulses from a linac-based SASE source and the (independent) femtosecond laser system located a kilometer downstream at the experiment. Although both light sources have been synchronized there remains an uncertainty between the relative arrival time of the X-ray pulses. When averaging successive shots to acquire sufficient signal-to-noise (S/N) for the measurement, the temporal features of the averaged signal become broadened around the nominal time delay, and this broadening leads to an instrument response function on the order of a few hundred femtoseconds. Since the studies presented here were performed prior to the implementation of single shot timing tools,[21,22] the overall time resolution remained around 200–300 fs.

Fig. 2 shows the analysis of the time delay data, and the black line displays the average value derived, while the false color image displays the individual data points contributing to the averaged signal. Fig. 2a shows the raw data without any time sorting treatment, while Fig. 2b displays the data after applying a rough time sorting using time arrival information given by the so-called phase cavity, which characterizes the relative arrival time of each electron bunch.[23] This information allows us to re-sort the recorded data according to this arrival time information and the false color image in Fig. 2b shows the result of this procedure on the data in Fig. 2a. The (vertically) averaged solid black line displays hereby the same shape and rise time as for the unsorted data in Fig. 2a (around 300 fs), but one can observe a more sharper rise on the order of 150 fs in the 2D data in the same

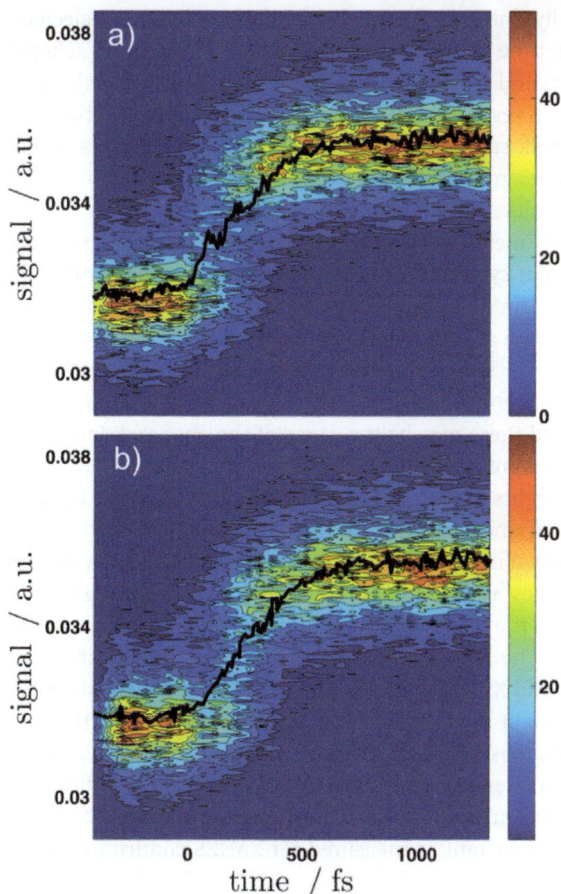

Fig. 2   Femtosecond time trace of the maximum XANES transient (near 7.125 keV, see Fig. 1) recorded at an X-ray free electron (XFEL) source.[14] The dark traces in the upper and lower figures correspond to the averaged signal after 240 pump–probe shots, and the underlying false-color image shows the actual distribution of individual pump–probe shots. The upper image shows the raw recorded data, while the lower figure was generated from the data in the upper after time-sorting each shot following changes in the phase cavity timing, which delivers the steeper rise in the 2D data display than the vertically averaged values (black trace).

Fig. 2b. From this analysis we can conclude that more information can be extracted from such measurements, and exploiting a time-sorting tool in the hard X-ray domain promises to reveal higher resolved dynamic processes in the future.

D   Combined X-ray emission spectroscopy and X-ray diffuse scattering

In a setup with a MHz excitation laser we have used synchrotron radiation to study both the internal molecular processes and the involved guest–host interactions.[11] We focus hereby on the information extracted from the X-ray diffuse scattering patterns, using both XANES and XES recordings as prior input. Fig. 3a shows selected radially integrated transient XDS patterns at selected time delays covering the lifetime of the photoexcited solute. The analysis scheme is described elsewhere,[11,24-26] and it relies on the identification of the major contributions to the transient XDS resulting from[11] (i) the HS population, (ii) the overall deposited heat into the bulk sample, but we had to also include (iii) a density *increase* to achieve adequate agreement with the data at all times within the subpicosecond lifetime of the solute (Fig. 3).

Such a density increase has never before been observed in subnanosecond XDS data for molecular systems in solution, and this hints towards a special case for aqueous Fe(bpy)$_3$: in a theoretical DFT MD simulation of the LS and HS molecules embedded in a cage of water molecules Daku and Hauser determined a dramatic change of the cage structure surrounding the HS [Fe(bpy)$_3$]$^{2+}$ molecule.[3] They determined—on average—an expulsion of two caging water molecules into the bulk solvent, and these may be the cause of the observed density increase in the bulk solvent. In order to underline this interpretation, we have thoroughly analyzed the weights of each contribution, and linked these to our simultaneous findings *via* XES, and our prior findings to the XANES data. The result of this treatment is shown in Fig. 4. Fig. 4a shows the initially recorded XANES transient feature at 7.125 keV as a function of time delay nicely showing the cross correlated rise (100 ps) and the 0.6 ns exponential decay of the excited state molecules.

After moving the X-ray energy to a higher energy above the K edge (here: 7.5 keV) both XES and XDS signals were recorded simultaneously. Fig. 4a shows the time dependence of the transient XES signal together with the previously recorded XANES. One observes a slight time zero shift between both measurements, mostly owing to altered experimental synchronization conditions between both independent measurements. But more importantly, the XES decay displays the exact same time constant as measured for XANES confirming the observation of

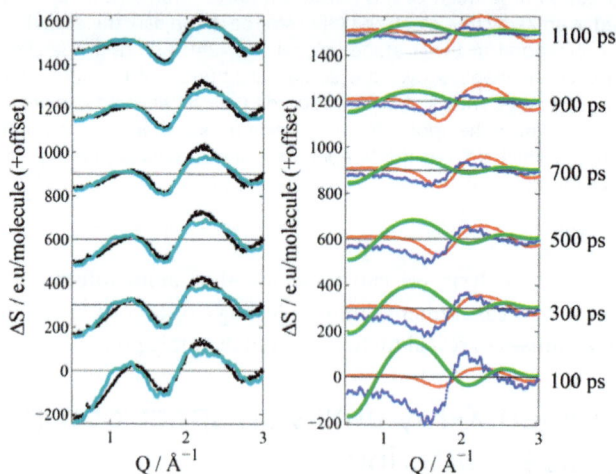

Fig. 3 Transient X-ray diffuse scattering (XDS) patterns after selected time delays (indicated on the far right) of photoexcited aqueous [Fe(bpy)$_3$]$^{2+}$ extracted from the 2D recorded images after radial integration. The best fit curves (cyan) to the data (black) on the left were a linear superposition of the three principal transient components shown on the right: HS population (green), bulk sample heat (red) and density increase (blue).

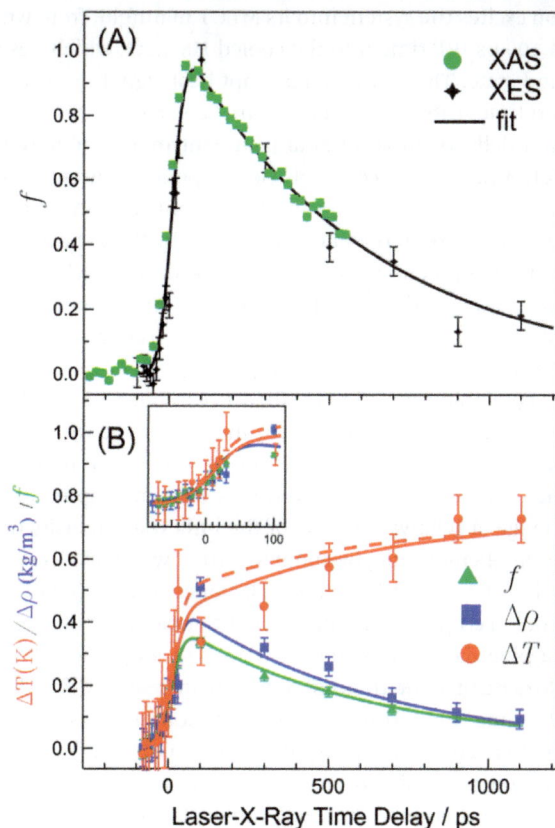

Fig. 4 Analysis of the X-ray diffuse scattering patterns. (A) Transient XANES signal recorded at 7.125 keV together with the XES transient recorded at 7.5 keV incident energy (and the secondary spectrometer tuned to the maximum of the $K_{\alpha 1}$ line). The XES fit curve delivers the IRF and the decay constant used in the XDS analysis in (B) as fixed input (and the maximum amplitude corresponds to $f = 0.4$). (B) Extracted weights from the XDS contributions (data points with error bars) from the superposition shown in Fig. 3 together with kinetic fits for the time traces (solid lines), for HS population (green), deposited heat (red) and density change (blue). For details see text.

the same reaction, but also confirming the identical experimental conditions including laser fluence. Next we use the XES transient curve to fit both time zero and the instrument response function (IRF), which will flow into the XDS analysis as constant input.

The XDS contributions due to population, heat, and bulk density are displayed in Fig. 4b and fitted with a Gaussian-convoluted exponential decay function. Hereby all parameters were locked apart from their amplitudes. Thus the fit curves in Fig. 4b were extracted with constant IRF, time zero, and exponential decay constant (= locked to the previously derived values from the simultaneously recorded XES signal).

The agreement with XDS extracted population is excellent, and the heat curve shows next to an IRF broadened rise a further rise with the 0.6 ns lifetime. This can be rationalized by the energy deposited in the sample: excitation with a 2.3 eV

(532 nm) photon excites the system into its MLCT manifold, from which it relaxes well within the 100 ps IRF time into the cooled HS state, which lies about 0.6 eV above the ground state. This delivers a prompt heat signal rise within the 100 ps cross correlation time. Later, with a 0.6 ns time constant, it relaxes back into the ground state and delivers an additional heat contribution (0.6 eV per molecule into the bulk solvent), which accordingly rises exponentially with the HS lifetime (Fig. 4b, red curves). Locking even the amplitude contributions in the IRF and the exponential associated rise to the energetics given by the exciting photon energy and the intramolecular potential energies yield the dashed red curve in Fig. 4b, in close agreement to the solid curve fitted with free amplitudes. Also the quantitative values extracted from the excited state population correspond nicely to the energy deposited into the bulk slab of sample.

The density signal (blue curve in Fig. 4b) also exhibits a similar behavior as the HS population curve (green curve in Fig. 4b) and—given the DFT results in ref. 3—can be readily explained by the expulsion of two water molecules into the otherwise – at least on a subnanosecond time scale – incompressible liquid. This leads to, on average, a relative increase of the bulk water density (at least in the vicinity around the HS molecule) by about 5×10^{-4}, which is nice agreement with an estimated value of 2 water molecules × HS concentration (8 mM, or $f = 40\%$ in a 20 mM aqueous $[Fe(bpy)_3]^{2+}$ sample). The fit curve itself does not deliver full agreement (the blue data points appear higher in amplitude than the blue fit curve), but it does display the trend of a density increase, which decays with the HS molecular lifetime, supporting the interpretation that this effect is controlled by intramolecular effects (here: the expulsion of two cage molecules into the bulk solvent). In consequence, this observation underlines the importance of guest–host interactions towards understanding molecular reactivity in the condensed phase, and further studies should aim to sharpen these tools towards a more complete understanding. It will also be interesting to see, how the density (but also the other) signals behave on the femtosecond time scale. In a future publication we treat the case of transient femtosecond XDS showing an ultrafast response of both the heat and density increase signals, which also underline the validity of the current interpretation.

E   Conclusions and outlook

The present study demonstrates the power of different, but simultaneously recorded, X-ray measurements. Time-resolved XANES and XES can extract vital electronic structural information (orbitals, spin) from the reacting species, but also deliver quantitative values for the actual fractional excited state population. Since these experiments can be reliably interpreted, they also reliably allow determining time zero and the cross correlation time, which serves as fixed input in the simultaneously recorded transient XDS patterns at different time delays. The XDS analysis can then deliver quantitative information about the global density in the bulk sample as well as about the deposited heat therein. The XDS analysis tools can now be sharpened to permit interpreting femtosecond XDS results, and therefore this study marks a stepping stone towards a more complete understanding of chemical reactivity including dynamic guest–host interactions in the future.

Acknowledgements

We thank our collaborators and coworkers for contributing to the present work: K. Haldrup, A. Dohn, K. S. Kjaer and T. B. van Driel are acknowledged for fruitful discussions and analysis of the XDS data, E. P. Kanter; A. Bordage, H. T. Lemke, S. Canton and J. Uhlig for their help during the experiments, T. Assefa for the analysis shown in Fig. 4. This research was supported by the European XFEL, the European Research Council *via* contract ERC-StG-259709, by the Danish National Research Foundation's Centre for Molecular Movies, DANSCATT, and the "Lendület" (Momentum) Programme of the Hungarian Academy of Sciences. S.H.S., and L.Y. acknowledge support from the U.S. Department of Energy (DOE) Office of Science, Division of Chemical, Geological and Biological Sciences under Contract no. DE-AC02-06CH11357. A.G., W.G., and C.B. acknowledge the support of the German Research Foundation (DFG) *via* contract SFB925 (TP4). Portions of this work were performed at the linac coherent light source (LCLS). Use of the Advanced Photon Source, an Office of Science User Facility operated for DOE Office of Science by Argonne National Laboratory, was supported by the U.S. DOE under Contract no. DE-AC02-06CH11357.

References

1 G. R. Fleming and M. Cho, *Annu. Rev. Phys. Chem.*, 1996, **47**, 109.

2 P. Ball, *Chem. Rev.*, 2008, **108**, 74.

3 L. M. L. Daku and A. Hauser, *J. Phys. Chem. Lett.*, 2010, **1**, 1830.

4 T. J. Penfold, B. F. E. Curchod, I. Tavernelli, R. Abela, U. Rothlisberger and M. Chergui, *Phys. Chem. Chem. Phys.*, 2012, **14**, 9444.

5 T. J. Penfold, C. J. Milne, I. Tavernelli and M. Chergui, *Pure Appl. Chem.*, 2012, **85**, 53.

6 W. Gawelda, A. Cannizzo, V.-T. Pham, F. Van Mourik, C. Bressler and M. Chergui, *J. Am. Chem. Soc.*, 2007, **129**, 8199–8206.

7 C. Consani, M. Prémont-Schwarz, A. El Nahhas, C. Bressler, F. Van Mourik and M. Chergui, *Angew. Chem., Int. Ed.*, 2009, **48**, 7184–7187.

8 O. Bräm, F. Messina, A. M. El-Zohry, A. Cannizzo and M. Chergui, *Chem. Phys.*, 2012, **393**, 51.

9 G. Vankó, P. Glatzel, V. T. Pham, R. Abela, D. Grolimund, C. N. Borca, S. L. Johnson, C. J. Milne and C. Bressler, *Angew. Chem., Int. Ed.*, 2010, **49**(34), 5910.

10 G. Vankó, A. Bordage, P. Glatzel, E. Gallo, M. Rovezzi, W. Gawelda, A. Galler, C. Bressler, G. Doumy, A.-M. March, E. P. Kanter, L. Young, S. H. Southworth, S. E. Canton, J. Uhlig, V. Sundström, K. Haldrup, T. B. van Driehl, M. M. Nielsen, K. S. Kjaer and H. T. Lemke, *J. Electron Spectrosc. Relat. Phenom.*, 2013, **188**, 166–171.

11 K. Haldrup, G. Vankó, W. Gawelda, A. Galler, G. Doumy, A. M. March, E. P. Kanter, A. Bordage, A. Dohn, T. B. van Driel, K. S. Kjær, H. T. Lemke, S. E. Canton, J. Uhlig, V. Sundström, L. Young, S. H. Southworth, M. M. Nielsen and C. Bressler, *J. Phys. Chem. A*, 2012, **116**, 9878.

12 A. March, A. Stickrath, G. Doumy, E. P. Kanter, B. Krässig, S. H. Southworth, K. Attenkofer, C. A. Kurtz, L. X. Chen and L. Young, *Rev. Sci. Instrum.*, 2011, **82**, 073110.

13 W. Gawelda, C. Bressler, M. Saes, M. Kaiser, A. N. Tarnovsky, D. Grolimund, S. L. Johnson, R. Abela and M. Chergui, *Phys. Scr.*, 2005, **115**, 102.

14 H. T. Lemke, C. Bressler, L. X. Chen, D. M. Fritz, K. J. Gaffney, A. Galler, W. Gawelda, K. Haldrup, R. W. Hartsock, H. Ihee, J. Kim, K. H. Kim, J. H. Lee, M. M. Nielsen, A. B. Stickrath, W. Zhang, D. Zhu and M. Cammarata, *J. Phys. Chem. A*, 2013, **117**, 735.

15 C. Bressler, C. Milne, V. T. Pham, A. ElNahhas, R. M. van der Veen, W. Gawelda, S. Johnson, P. Beaud, D. Grolimund, C. N. Borca, G. Ingold, R. Abela and M. Chergui, *Science*, 2009, **323**, 489.

16 C. Bressler and M. Chergui, *Chem. Rev.*, 2004, **104**, 1781.

17 C. Bressler, R. Abela and M. Chergui, *Z. Kristallogr.*, 2008, **223**, 307.

18 W. Gawelda, V.-T. Pham, R. M. van der Veen, D. Grolimund, R. Abela, M. Chergui and C. Bressler, *J. Chem. Phys.*, 2009, **130**, 124520.

19 W. Gawelda, V.-T. Pham, A. El Nahhas, M. Kaiser, Y. Zaushitsyn, S. L. Johnson, D. Grolimund, R. Abela, A. Hauser, C. Bressler and M. Chergui, *AIP Conf. Proc.*, 2007, **882**, 31.

20 W. Gawelda, V.-T. Pham, A. El Nahhas, Y. Zaushitsyn, M. Kaiser, D. Grolimund, S. L. Johnson, R. Abela, A. Hauser, C. Bressler and M. Chergui, *Phys. Rev. Lett.*, 2007, **98**, 057401.

21 M. R. Bionta, H. T. Lemke, J. P. Cryan, J. M. Glownia, C. Bostedt, M. Cammarata, J.-C. Castagna, Y. Ding, D. M. Fritz, A. R. Fry, J. Krzywinski, M. Messerschmidt, S. Schorb, M. L. Swiggers and R. N. Coffee, *Opt. Express*, 2011, **19**, 21855.

22 M. Harmand, R. Coffee, M. R. Bionta, M. Chollet, D. French, D. Zhu, D. M. Fritz, H. T. Lemke, N. Medvedev, B. Ziaja, S. Toleikis and M. Cammarata, *Nat. Photonics*, 2013, **7**, 215.

23 J. M. Glownia, *et al.*, *Opt. Express*, 2010, **18**, 17620.

24 M. Christensen, K. Haldrup, K. Bechgaard, R. Feidenhans'l, Q. Kong, M. Cammarata, M. L. Russo, M. Wulff, N. Harrit and M. M. Nielsen, *J. Am. Chem. Soc.*, 2009, **131**, 502.

25 M. Cammarata, M. Lorenc, T. Kim, J. H. Lee, Q. Y. Kong, E. Pontecorvo, M. L. Russo, G. Schiro, A. Cupane, M. Wulff and H. Ihee, *J. Chem. Phys.*, 2006, **124**, 1245041–1245049.

26 K. Haldrup, M. Christensen and M. M. Nielsen, *Acta Crystallogr., Sect. A: Found. Crystallogr.*, 2010, **66**, 261.

Faraday Discussions

PAPER

Sub-THz specific relaxation times of hydrogen bond oscillations in *E.coli* thioredoxin. Molecular dynamics and statistical analysis

Tatiana Globus,*[ab] Igor Sizov[ab] and Boris Gelmont[ab]

Received 3rd March 2014, Accepted 25th March 2014

DOI: 10.1039/c4fd00029c

Hydrogen bonds (H-bonds) in biological macromolecules are important for the molecular structure and functions. Since interactions *via* hydrogen bonds are weaker than covalent bonds, it can be expected that atomic movements involving H-bonds have low frequency vibrational modes. Sub-Terahertz (sub-THz) vibrational spectroscopy that combines measurements with molecular dynamics (MD) computational prediction has been demonstrated as a promising approach for biological molecule characterization. Multiple resonance absorption lines have been reported. The knowledge of relaxation times of atomic oscillations is critical for the successful application of THz spectroscopy for hydrogen bond characterization. The purpose of this work is to use atomic oscillations in the 0.35–0.7 THz range, found from molecular dynamic (MD) simulations of *E.coli* thioredoxin (2TRX), to study relaxation dynamics of two intra-molecular H-bonds, O⋯H–N and O⋯H–C. Two different complimentary techniques are used in this study, one is the analysis of the statistical distribution of relaxation time and dissipation factor values relevant to low frequency oscillations, and the second is the analysis of the autocorrelation function of low frequency quasi-periodic movements. By studying hydrogen bond atomic displacements, it was found that the atoms are involved in a number of collective oscillations, which are characterized by different relaxation time scales ranging from 2–3 ps to more than 150 ps. The existence of long lasting relaxation processes opens the possibility to directly observe and study H-bond vibrational modes in sub-THz absorption spectra of bio-molecules if measured with an appropriate spectral resolution. The results of measurements using a recently developed frequency domain spectroscopic sensor with a spectral resolution of 1 GHz confirm the MD analysis.

Department of Electrical and Computer Engineering, University of Virginia, 351 McCormick Road, P.O. Box 400743, Charlottesville, VA 22904-4743, USA. E-mail: tg9a@virginia.edu

Vibratess LLC, 104 Chaucer Rd, Charlottesville, VA 22901, USA

1.   Introduction

Hydrogen bonds (H-bonds) in biological macromolecules are important for their structure and functions. Yet there are no simple direct methods to observe and characterize H-bonds. Since interactions *via* hydrogen bonds are weaker than covalent bonds, it can be expected that atomic movements involving H-bonds have low frequency vibrational modes. Sub-Terahertz (sub-THz) vibrational spectroscopy of biological macromolecules, which combines measurements with molecular dynamics (MD) computational prediction, has been demonstrated as a promising approach for studying interactions between low energy radiation and intra-molecular dynamics. It reveals resonance spectroscopic features, vibrational modes or group of modes at close frequencies in absorption (transmission) spectra of biomaterials caused by a fundamental mechanism of interaction of low frequency internal intra-molecular motions *via* hydrogen and other weak bonds with THz radiation. Although multiple resonance absorption lines in sub-THz region have been reported in measurements with appropriate spectral resolution, for example,[1-6] successful application of THz spectroscopy for DNA, RNA and protein characterization requires deep understanding of relaxation processes of atomic dynamics (displacements) within a macromolecule.

The dissipation time is one of the fundamental problems related to THz vibrational modes in biological molecules. The width of individual spectral lines and the intensity of resonance features observed in sub-THz spectroscopy are sensitive to the relaxation processes of atomic movements within a macromolecule. It is clear that the decay (relaxation) time, τ, is the factor that determines the spectral width and the intensity of vibrational transmission/absorption modes, the required spectral resolution, and eventually the discriminative capability of sub-THz spectroscopy. The suggested range of molecular dynamics relaxation times for processes without bio-molecular conformational change varies from approximately 1.5 ps to 1 ns in different studies.[7,8] The corresponding values for the dissipation factor, γ, and the width of spectral lines, W, which are reciprocal to τ, are between 6 and 0.01 cm^{-1}. Values of W above 1 cm^{-1} would result in structureless sub-THz spectra, since vibrational resonances could not be resolved in the case of the high density of low intensity vibrational modes.[3] The existence of long-lasting dynamic processes have been confirmed by relaxation dynamics of side chains in macromolecule thioredoxin observed by time-resolved fluorescence experiments.[9] At the same time the entire mechanism that determines relaxation times in dynamics processes is still not completely understood. There are a number of studies on the problem using MD simulation and Langevin equation along with the analysis of inelastic neutron scattering[10,11] and other experimental techniques.[12] The estimates from inelastic neutron scattering give very large broadening of low frequency motions. Possible reasons for the differences between experiment and simulation results have been discussed in ref. 13, in particular much higher vibrational density of states in simulations compared to neutron scattering experiments. It is known from experiments that "proteins exist in an ensemble of structures, described by an energy landscape",[14] and neutron scattering spectra result from an average over different proteins conformations or substates. These motions, however, are quite different from quasi-harmonic vibrational modes in THz, and especially in sub-THz spectral range, for both time and displacement scales.[15] Weak THz vibrations associate with displacements at distances on the order of only ~0.1–1.0 Angstrom. These oscillations might live for a relatively long time since collision probability is less when displacements are small.

Another problem existing in the analysis of relaxation is that in most cases the data on relaxation cannot be attributed to atomic fluctuations in a certain frequency range. For example, the most common method for relaxation analysis, the autocorrelation function of atomic displacements, often exhibits featureless, non-exponential decays.[16] One explanation for this result is that atoms go through different frictional regions in space and time, and the average produces non-exponential decay.[16] However, another explanation may lie in the fact that the autocorrelation function usually averages oscillations over all frequencies observable in MD simulation. Regarding proteins, it has also been shown that relaxation processes can be a complex and heterogeneous phenomenon.[17]

In this work, statistical analysis of MD data is applied to study relaxation in vibrational dynamics of two intra-molecular H-bonds, $O \cdots H-N$ and $O \cdots H-C$, of *E.coli* thioredoxin. These bonds connecting tryptophan with charged neighbors have been studied in ref. 9 using fluorescence spectroscopy. By studying atomic displacements obtained from MD in time and frequency domains, we found that the atoms in these bonds are involved in a number of collective oscillations, which are characterized by different relaxation time scales ranging from 2–3 ps to more than 150 ps for processes without conformational change. Two different complimentary techniques are used in this study, one is the analysis of the statistical distribution of relaxation time (or dissipation factor) values relevant to low frequency oscillations, and the second is the analysis of the autocorrelation function of low frequency quasi-periodic movements.

The existence of long lasting relaxation processes makes it possible to directly observe and study H-bond vibrational modes in sub-THz absorption spectra of bio-molecules. Until recently we used a Fourier transform spectrometer Bruker IFS 66v with the spectral resolution of 0.25 cm^{-1}.[3] Although significant progress in experimental THz spectroscopy was demonstrated and reliable information was received for transmission/absorption spectra from different biological macromolecules and species, experimental characterization still required milli-gram quantities of material, a detector cooled with liquid helium for reliable characterization, and a system under vacuum or purged with dry gas because of the very low intensity of radiation available from the mercury lamp source and to eliminate disturbances caused by liquid and vapor water.[3] The implementation of THz vibrational spectroscopy was impeded because of the absence of spectro-scopic systems, which simultaneously satisfy the requirements of good spectral and spatial resolution, along with high sensitivity. For our new studies, we utilize a spectroscopic sensor prototype developed by Vibratess.[4] This novel constant wave, frequency-domain spectroscopic instrument with imaging capabilities operates without the need for cryogenic cooling of the detector. The high sensi-tivity, good spectral resolution of 0.03 cm^{-1}, and a spatial resolution below the diffraction limit permitted us to observe intense and narrow spectral resonances in transmission/absorption spectra of nanogram samples of biological materials with spectral line widths as narrow as $W = {\sim}0.1$ cm^{-1}. Transmission spectra were obtained in the sub-THz region between 315 and 480 GHz for both macromolecules and biological species. The results of measurements using this new spectroscopic sensor confirm our MD analysis.

2 Methods

Molecular dynamics

A detailed description of our protocol for MD simulations of protein thioredoxin using Amber 10 can be found in our recent study.[18] We simulated a complex of thioredoxin (PDB code 2TRX) and 8 Å shell of TIP3P water. In preparation steps, a constant volume and temperature (NVT) ensemble is used to raise the temperature to 293 K. The system is heated for ~16 ps, and the protein's atoms are restrained using a 10 kcal mol^{-1} Å^{-2} force constant. During this heating process, bonds involving hydrogen are fixed. Constant pressure periodic boundary conditions are used to scale the system volume during 100 ps to reach a density of ~1 g cm^{-3}. In these procedures, a 10 Å real space cutoff is used with a 2 fs integration time step. Once the system has attained selected values of temperature and density, random velocities from the Maxwellian distribution are assigned to all atoms, followed by another equilibration step (NPT ensemble) for further energy minimization.

After equilibration, a 5.0 ns production run is performed in a constant volume and energy ensemble (NVE) to avoid undesirable effects on atomic motions due to coupling to an external thermal bath. During the simulation, the coordinates of all atoms of the system are recorded every 20 fs for the entire production run.

Atomic trajectories collected in MD simulations are converted to the covariance matrix of atomic displacements $\langle R_i R_k \rangle$. The force-field matrix is found in a quasi-harmonic approximation utilizing the relation between the covariance matrix and the inverse of the force-constant matrix ($\langle R_i R_k \rangle = k_B T [F^{-1}]_{ik}$), where R are the displacements and F are the force constants.[19,20] Diagonalization of the F matrix gives eigenfrequencies (normal mode frequencies) and eigenvectors (displacement vectors- normal modes).

In our study,[18] MD simulations of sub-terahertz vibrational modes of the protein thioredoxin were conducted with the goals of finding the conditions needed for simulation convergence, improving the correlation between experimental and simulated absorption spectra, and ultimately for enhancing the predictive capabilities of computational modeling. We studied the consistency, accuracy and convergence of MD simulations of the sub-THz vibrational modes by comparing simulations performed using different initial conditions, protocols and parameters to the experimental results.

It was demonstrated that the constant energy simulation protocol, NVE, during the production run is more preferable than the constant temperature regime, NPT, for several reasons. Using the NVE ensemble in a production run gives more stable results compared to NPT regime. Constant energy simulations without frequent exchange with the external bath for temperature regulation induce fewer disturbances into trajectories of atoms and, in addition, prevent transition of a protein molecule into a different conformation. At the same time, the NVE protocol requires more attention to the choice of starting energy in the production run. The starting total energy of the system (ETOT) at the beginning of the production run significantly affects results and has to be close to the equilibration minimum.

Absorption spectra calculation

Using atomic trajectories from the constant energy and volume MD simulations, thioredoxin's sub-THz vibrational spectra and absorption coefficients were calculated in a quasi-harmonic approximation. The absorption coefficient spectra $\alpha(\nu)$ as functions of the frequency ν can be calculated through the relationship between α and the imaginary part of the dielectric permittivity:[21]

$$\alpha(\nu) = W\nu^2 \sum_k \frac{S_k}{\left(\nu^2 - \nu_k^2\right)^2 + W^2\nu^2}, \qquad (1)$$

where ν_k are normal mode frequencies calculated by the diagonalization of the force-constant matrix, and S_k are oscillator strengths computed for all vibrational modes k. Two values of line width for all vibrational modes in sub-THz range were earlier suggested from our experimental work: $W = 0.5$ cm^{-1} (moderate spectral resolution in Bruker spectrometer), and $W = 0.1$ cm^{-1} from high resolution spectroscopy using Vibratess spectrometer.

Better simulation convergence and improved consistency between simulated vibrational frequencies and experimental data were obtained using a new procedure for averaging mass-weighted covariance matrices of atomic trajectories in MD simulations.[18] In particular, the open source package ptraj was edited to improve a matrix analyzing function. Averaging of only six matrices gives much more consistent results, with absorption peak intensities exceeding those from the individual spectra, and with a relatively good correlation between simulated vibrational frequencies and experimental data. Reasonably good correlation between absorption spectrum of thioredoxin simulated with $W = 0.5$ cm^{-1} and experimental results as measured with a moderate spectral resolution of 0.25 cm^{-1} have been demonstrated.[18] Experimental spectra taken with much better spectral resolution of 0.03 cm^{-1} compared with computational modeling using $W = 0.1$ cm^{-1} also demonstrated reasonable correlation.[4]

H-bonds studied

Fig. 1(A and B) shows the two H-bonds in thioredoxin that we studied in this work. In one of these bonds, O···H–N, the oxygen atom (OD1, ASP61) from aspartate ASP61 (on the left of Fig.1B) has a weak interaction through the hydrogen atom (HE1, TRP31) to the nitrogen (NE1, TRP31) in the indole ring of tryptophan (TRP31). The location of this bond allows for oscillations of a relatively large amplitude (4–7 Å).

The second hydrogen bond C–H···O connects the 4th carbon atom (CE3, TRP31) from the same indole ring of tryptophan residue TRP31 through its hydrogen atom (HE3, TRP31) to the oxygen atom (O, TRP31) in the peptide bond between TRP31 and CYS32 (shown on the right of Fig. 1B). These two particular H-bonds have been earlier used in experiments on fluorescence spectroscopy[9] to study relaxation times for quenching excited tryptophan TRP31. For the first considered hydrogen bond, O···H–N, we analyzed the dynamics of distances between O and H and between O and N atoms during MD simulation. A similar approach was used for the second hydrogen bond O···H–C.

Two independent complimentary methods for data analysis were used in this study. In the first method of studying atomic displacements, we analyzed statistical distributions of spectral line width, relaxation time values, and vibrational

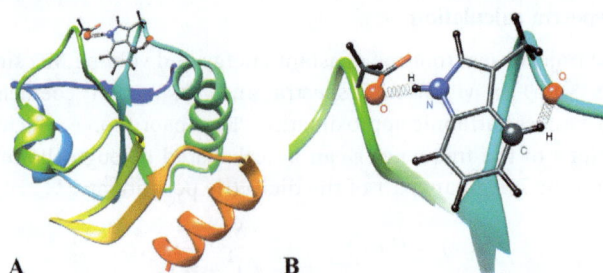

Fig. 1 Two H-bonds in thioredoxin: one is an O⋯H–N interaction formed between ASP61 and TRP31 involving O (OD1, ASP61), H (HE1, TRP31) and N (NE1, TRP31) atoms. The second interaction O⋯H–C is formed between the 4th C atom (CE3, TRP31) of the TPR31 indole ring through H (HE3, TRP31) to an oxygen atom (O, TRP31) involved in the peptide bond between TRP31 and CYS32 residues. A: the entire molecule is shown. B: detailed view of the two bonds. Images were made using Chimera.[22]

frequencies in the sub-THz spectral range. In the second method, an analysis of autocorrelation function of low frequency quasi-periodic movements was used.

MD trajectories and fast Fourier filtering

The coordinates of chosen atoms are extracted from the MD trajectory to calculate the distance between a pair of atoms (1 and 2) for each time point in the 5 ns simulation as

$$d = \sqrt{(x_1 - x_2)^2 + (y_1 - y_2)^2 + (z_1 - z_2)^2}. \tag{2}$$

For better accuracy, a distance trajectory was interpolated to have a 1 fs time step using the Matlab *spline* function.

Since we are interested in low energy vibrational modes in the sub-THz range, to facilitate data analysis and understanding results for fast vibrations with a very small amplitude, the distance fluctuations in time domain are processed using a Fast Fourier Transform (FFT) filter available at the Matlab website.[23] Application of a low frequency rejection filter with the boundary below 0.3 THz provides more uniform motion dynamics. In the FFT filtering procedure, the time domain trajectory is first transferred into the frequency domain, then frequency

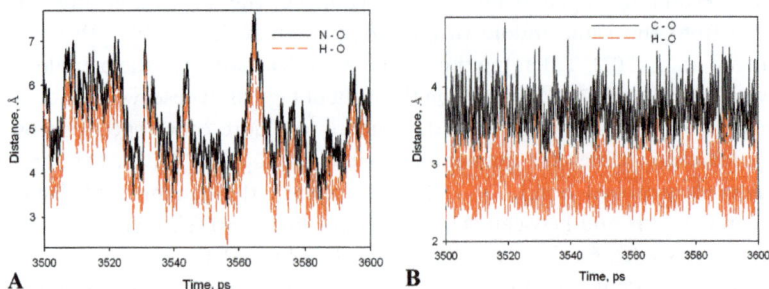

Fig. 2 Dynamics of the distances. A: O⋯H–N hydrogen bond interaction: solid - N and O; dash – H and O; B: O⋯H–C hydrogen bond interaction: solid - C and O; dash – H and O.

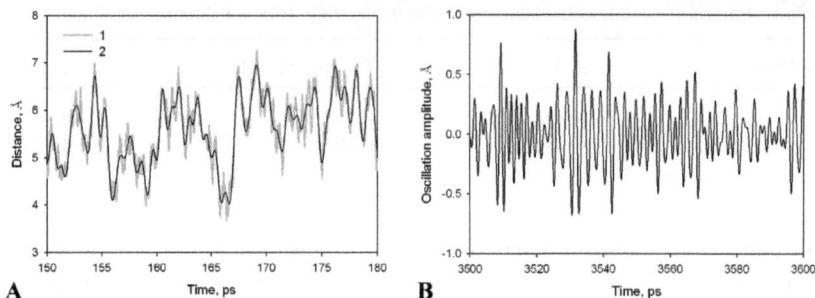

Fig. 3 Application of FFT filter. A: Gray line (1): original N–H–O trajectory from MD simulation, black line (2): fluctuations with frequencies above 1.2 THz were removed using 0–1.2 THz band-pass filter. B: Application of a 0.35–0.7 THz band-pass filter removes both higher frequency and lower frequency movements and reveals periodic oscillations in a sub-THz frequency range of our interest.

components inside filter limits are cut, and after that the inverse FFT procedure is applied to return the data back into the time domain.

3. Results

Statistical distributions of relaxation parameters using fitting to a model of two non-interacting damped oscillators

Fig. 2A and B show fragments of MD trajectories for two atom pairs involved in two H-bond interactions demonstrated in Fig. 1.

Considering the bond O···H–N in Fig. 2A, several important results become obvious immediately: (1) the atoms are involved in heterogeneous movements with different time durations and distance variations up to 3–3.5 Å, and (2) the O–N and O–H distances change with time in a similar way. This second result could be expected since the chemical bond between H and N is very strong and the distance fluctuations between them are significantly smaller compared to the distance O–H. Variations of distances between atoms in the second H-bond, O···H–C, demonstrated in Fig. 2B, are significantly smaller, only ~0.5 Å most of time, although very short variations with an amplitude of ~1 Å are also observed.

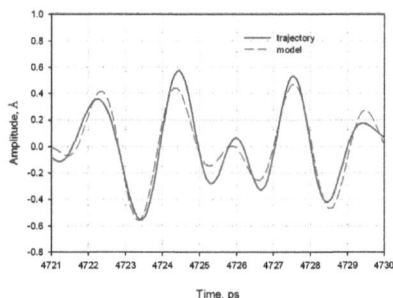

Fig. 4 Example of fitting the filtered trajectory of O–N distance. Solid line: MD data, dash line: model. The portion of unexplained variation in this example is less than 10%.

Table 1 Parameters of two oscillators for the given trajectory window in Fig. 4

Oscillator	A, Å	$f = \omega/(2\pi)$, THz	γ, ps^{-1}
Y_1	−0.26	0.588	0.003
Y_2	−0.34	0.372	0.045

The similarities between C–O and H–O distance variations are, however, preserved. A peak in the time interval 3560–3570 ps observed for H–O distance in the O⋯H–N bond in this particular simulation run is not revealed in the dynamics of O⋯H–C interaction. This peak, involving movement at a distance ~4 A, is probably an indicator of conformational change.

Fig. 2 represents all the movements involving these hydrogen bonds that might include possible vibrations at all frequencies. Since we are interested in a relatively narrow sub-THz frequency range, as a next step, we used a standard Fast Fourier Transform (FFT) filtering technique to separate interfering motions shown in Fig. 2. For example, application of a low frequency reject FFT filter removes movements with amplitudes of several Angstroms and periods of tens and hundreds of picoseconds, similar to the peak at time 3565 ps observed in Fig. 2A. Application of a reject filter above frequency of 1.2 THz removes high frequency oscillations from a raw trajectory (gray line 1, Fig. 3A) and simplifies the trajectory to black line 2. Application of both low frequency and high frequency reject filters, leaves only movements that occur in the frequency range that we are interested in, with much more uniform motion dynamics as demonstrated in Fig. 3B, with obvious quasi-harmonic movements.

The existence of long-lived vibrations with different periods above 1.3 ps, which corresponds to vibration frequencies below ~26 cm^{-1} is clearly demonstrated. The presence of harmonic components indicates that there are local regions in the molecule, with atoms oscillating in this frequency range for at least of several periods. These small amplitude vibrations have been described as "rattling motions in a cage consisting of the neighbouring atoms within a molecule or surrounding solvent".[24]

In the case of quasi-harmonic movement, the distance between two atoms $Y(t)$ as a function of time can be modelled as a damped oscillator:

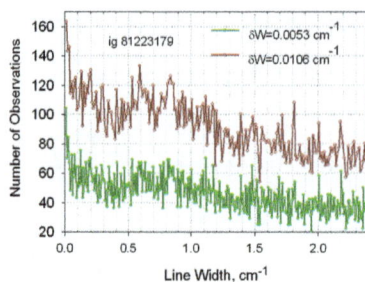

Fig. 5 Statistical distribution of spectral line widths for O⋯H–N bond vibrations (O–N distance) calculated with two values of δW, 0.0106 cm^{-1} and 0.0053 cm^{-1}. Moving window 5 ps, filter 0.35–0.7 THz, threshold 2%.

This journal is © The Royal Society of Chemistry 2014

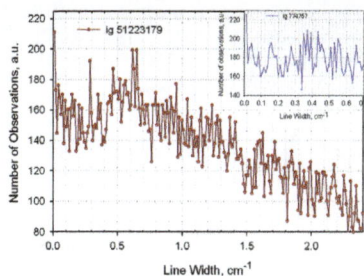

Fig. 6 Statistical distribution of spectral line widths for O⋯H–C bond vibrations (O–C distance). $\delta W = 0.0106$ cm^{-1}, moving window 5 ps, filter 0.35–0.7 THz, threshold 2%. The inset shows a more detailed fragment below 0.7cm^{-1} from another MD simulation.

$$Y(t) = A \cos (\omega t + \varphi) \, e^{-t\gamma}, \tag{3}$$

where A is an amplitude (Å), γ is a dissipation factor or relaxation rate (ps^{-1}), t is time (ps), ω is the angular frequency (2π /ps), and φ is the phase. At the same time the trajectory shown in Fig. 3B is more complicated and can not be described by a model of only one oscillator even for a relatively small time interval. The simplest possible model to describe the filtered trajectory for a quantitative analysis during a relatively short time interval is a model of two non-interacting damped oscillators:

$$Y(t) = Y_1(t) + Y_2(t), \tag{4}$$

where each component is presented by eqn (3).

The fact that relatively long lived vibrations are able to survive without fast dissipation indicates that the damping factors, γ, for at least some vibrational modes, have to be relatively small. A more detail analysis, described below, is required for the identification of other motions that are definitely present and for studying the accurate values of the relaxation parameters.

Using a moving time window (frame) ΔT with a δt step along a filtered trajectory we apply the model of two oscillators to find parameters A, ω, φ and γ for each frame and each step using a fitting procedure. Time frames ΔT of 5,

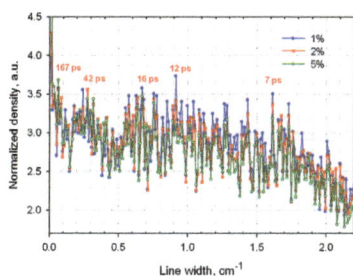

Fig. 7 Comparison of line width distributions calculated with a different percent of unexplained variations. Parameters used: one 5 ns production run, moving time window 5 ps with 50 fs step, $\delta W = 0.0106$ cm^{-1}, filter 0.35–0.7 THz. Numbers above curves are the relaxation time values at local peaks of line width distributions.

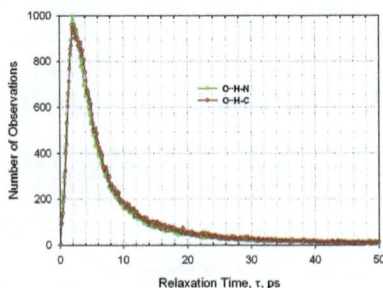

Fig. 8 Distribution of relaxation time values for O···H−N and O···H−C bonds vibrations with small τ. Filter 0.35−0.7 THz.

10 or 20 ps and time steps δt of 50 fs or 100 fs have been used to find optimal values of the parameters that give the clearest presentation of results. The Matlab *lsqcurvefit* function is used to make the fitting to the filtered data. Lower and higher boundaries for relaxation rate were set to 0.001 ps^{-1} and 10 ps^{-1}, which correspond to 0.1–1000 ps relaxation time limits. The quality of fitting is verified by calculation of the Pearson correlation (r) between the fitting curve and the raw data, and the percent of unexplained variation is found as $100(1 - r^2)\%$. This parameter is used as a cutting threshold to remove results with low quality fitting. Fig. 4 gives an example of fitting results in relative MD coordinates after filtering using the model inside a 10 ps window. In this particular example, the portion of unexplained variation in the data is less than 10%. Parameters of two oscillators from fitting in this example are presented in Table 1.

To find the statistical distribution function for a dissipation factor, γ, all fitting results were further grouped into classes using equal intervals, $\delta\gamma$, and the number of observations was calculated for each class. For the purpose of comparing simulation results with experimental data, we also found the distribution function for the absorption Lorentzian line width at half peak (W) using its relationship with the dissipation factor of a damped oscillator, γ :

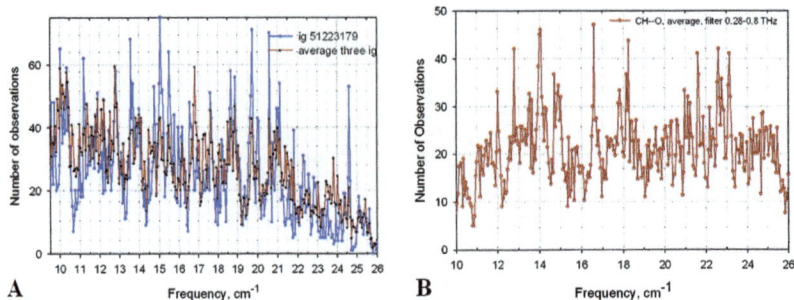

Fig. 9 A: Distribution of vibrational frequencies in sub-THz range for O···H−N bond using 0.28−0.8 THz filter, window 10ps, 10% threshold. B: The same as in A but for O···H−C bond. Simulation run with a random seed number ig 51223179.

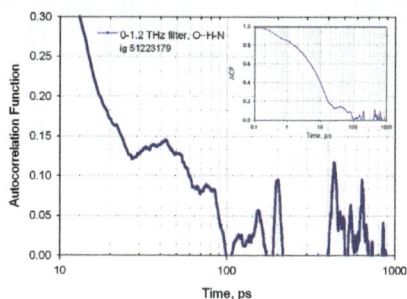

Fig. 10   ACF of oscillations after applying a 0–1.2 THz band-pass filter to the original trajectory of distances between O and N atoms of the O···H−N bond for time above 10 ps. The entire range is given in the inset.

$$W = \frac{\gamma}{\pi \cdot c},\tag{5}$$

where γ is in ps^{-1}, c is the speed of light (cm ps^{-1}), and W is in cm^{-1}. The line width and the dissipation factor differ only by a scaling factor πc. The procedure similar to finding γ was used to find the distribution function for relaxation time $\tau = \gamma^{-1}$ using equal intervals, $\delta\tau$, and for frequencies of vibrations using equal intervals, δ_F. The quality of data presentation for the distribution of spectral width values is sensitive to the choice of intervals, with too large of intervals resulting in a small number of possible observations, while too small interval gives noisy results. Several values of δW were used to find the best presentation of results. Fig. 5 demonstrates this effect for the spectral line widths distribution for the O···H−N bond.

The main result is, however, the same independent of δW: the distribution is not a smooth function of line width values, but shows sharp peaks. In addition to the existence of very narrow spectral lines (widths less than 0.1 cm^{-1}) having relatively high probability, almost equal distribution of line widths up to $W = 1$ cm^{-1}, and slow reducing of distribution of widths above ~ 1 cm^{-1}, are clearly demonstrated. Similar results are obtained for the second hydrogen bond, O···H−C, (Fig. 6). Line width distributions shown in Fig. 5 and 6 are not sensitive to the values of unexplained variations in the fitting procedure. This result is also

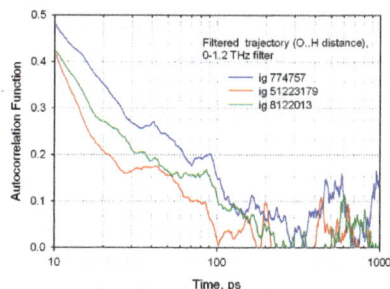

Fig. 11   Autocorrelation functions for the filtered trajectory of distances between O and N atoms in O···H−N bond calculated from three MD simulation runs.

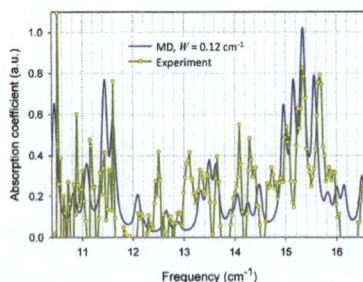

Fig. 12   Absorption spectrum of protein thioredoxin from *E. coli*: MD simulation and experimental results as measured using Vibratess spectroscopic sensor. Our work.[25]

demonstrated in Fig. 7, where the peak positions are independent of the cutting threshold.

Relaxation times can be recalculated as $1/\gamma$ from the line width distributions in Fig. 5–7 using eqn (5); some results are shown as numbers above curves in Fig. 7. Thus the atoms are involved in a number of collective oscillations, which are characterized by different relaxation time scales. However, Fig. 8 gives much more accurate results for the range of small relaxation times as calculated directly from the time distribution for two hydrogen bond vibrations using the filtered sub-range of 0.35–0.75 THz. The almost identical results for both bonds indicate that the distribution has a maximum at the relaxation time of 2–3 ps. These small values of relaxation time have been found in different experiments, and are often attributed to the rotational and translational relaxation of bulk water molecules.[7,10] However, relaxation time distributions show a very long tail that can not be described with only one relaxation process. There is not enough sensitivity for a more accurate analysis of longer relaxation time probabilities using this approach because of a small window used, $\Delta T = 5$ ps.

Fig. 9A shows the distribution of vibrational frequencies in the sub-THz range. Results are sensitive to the choice of initial Maxwell velocities determined by a random seed "ig" number. Results for one ig and averaging for 3 ig numbers however definitely indicate the presence of many modes in our relatively narrow sub-range and reduced modes density above 21 cm^{-1}. Vibrational frequencies are different for two bonds (Fig. 9A and B).

Autocorrelation function

We also analyzed the dissipation of THz oscillations using the autocorrelation function (ACF), $F(\Delta t)$, that was computed as

$$F(\Delta t) = \frac{1}{N-k} \sum_{i=1}^{N-k} [(Y_{t_i} - \overline{Y})(Y_{t_{i+k}} - \overline{Y})]/\sigma_Y^2, \qquad (6)$$

where Y_{t_i} and $Y_{t_{i+k}}$ are trajectory coordinates separated by $\Delta t = t_{i+k} - t_i = k \cdot \delta t$. In our case the time step $\delta t = 10$ fs and the total number of steps $N = 5 \times 10^5$ for 5 ns MD run. $\overline{Y}$ and σ_Y^2 are the sample mean and variance of Y.

Two types of time-domain data were analyzed using ACF. First, ACF was calculated for the original trajectory of distances between studied atoms. Second, it was computed for FFT filtered oscillations where upper and lower frequency limits were varied. ACF of oscillations after applying a 0–1.2 THz band-pass filter is shown in Fig. 10 for time above 10 ps, and for the entire range it is given in the inset.

Finally, Fig. 11 shows the variation in results for the O···H–N bond ACF calculated from three MD simulations. All results reveal overall a rather smooth reducing of ACF with time for t below ~30 ps.

However at higher time values, the peaks are reproducibly observed at ~40, 80–90, and ~170 ps from all three simulations. Above ~200 ps the results from different simulation runs are not consistent. This fact might indicate the possibility of molecule transition into a different conformational state. It can also explain the variability of other results from different simulation runs having different initial Maxwell velocities distribution.

4. Comparison with experimental data and discussion

We have recently measured the absorption spectra from *E.coli* protein thioredoxin using a new spectrometer with high spectral resolution.[4] The existence of intense and narrow spectral resonances was observed in transmission/absorption spectra with spectral line widths as narrow as $W \sim 0.1$ cm^{-1}.

In Fig. 12, the experimental spectrum is compared with MD simulation to verify both. Although the overall correlation between the theory and experimental data confirms again the existence of intense and narrow absorption lines, not all peaks are reproduced in the measured and simulated spectra. The most obvious reason for differences is that the same damping factor γ was used to calculate absorption for all vibrational modes. Our current results demonstrate a broad distribution of spectral line widths (Fig. 5–7) and of damping factor γ. Additional explanation has to be given to the fact that narrow width vibrations having small values of γ and large values of τ are well observed in absorption data and, at the same time, the highest probability is demonstrated for vibrations with very short τ ~ 2 ps (Fig. 8). Fig. 9A and B demonstrate vibrational frequencies for two hydrogen bonds as simulated with different spectral line widths. However, not all of these vibrations contribute equally to the absorption spectrum shown in Fig. 12. For each particular vibrational mode, the absorption coefficient is inversely proportional to W (eqn (1)), and absorption due to vibrations with small relaxation times (or with large γ and spectral line width) has a very low probability of being observed.

5. Conclusions

In this work, relaxation dynamics for two intra-molecular hydrogen bonds, O···H–N and O···H–C, were studied. Time domain trajectories (atomic displacements) demonstrate the complicated character of atomic movements. Fourier transform filtering was used to restrict the data into a frequency domain within the limit of 0.35–0.7 THz, which corresponds to the experimental conditions used to characterize vibrational spectra. This procedure permitted the selection and visualization of quasi-harmonic components of atomic movements. The fluctuations were further analyzed by fitting with a model of two non-interacting damped oscillators. A new fitting procedure permitted us to extract more information about relaxation parameters from atomic trajectories and justifies the application of a statistical approach for the quantitative analysis of oscillations. In the statistical analysis used, sample distributions for relaxation factors and spectral line widths reveal multiple peaks with an almost equally high probability in the range between 0.1 and 1 cm^{-1}. By studying atomic displacements, it was found that the atoms are involved in a number of collective oscillations, which are characterized by different relaxation time scales ranging from 2–3 ps to more than 150 ps. The relaxation time distribution gives a sharp peak for $\tau \sim 2$ ps, that was earlier observed in different experiments, and a very long tail that can not be described with only one relaxation process. A broad spectrum of vibrational frequencies is demonstrated. The results from analysis of autocorrelation functions for hydrogen bond distances confirm the existence of processes with long time characteristic scales.

The existence of long relaxation processes makes it possible to directly observe and study H-bond vibrational modes in sub-THz absorption spectra of biomolecules measured with the proper spectral resolution. The results obtained in this study are in general agreement with measurements using a recently developed time domain spectroscopic sensor with the spectral resolution of 1 GHz. These new results permit to better understand the dynamics of hydrogen bonds in thioredoxin and will be further used for more detailed analysis and prediction of experimental sub-THz absorption spectra from biological macromolecules.

Acknowledgements

We would like to express our great appreciation to Dr Ngai Wong, Senior Physical Scientist at DTRA, for his valuable support during the planning and development of this research work. This work is supported by the contract from the US Army Research Office #W911NF-12-C-0046.

References

1 D. F. Plusquellic and E. J. Heilweil, Terahertz Spectroscopy of Biomolecules, in: *THz Specrtoscopy: Principles and Applications*, Book CRC Press, 2007, pp. 269–297.

2 W. Zhang, E. R. Brown, M. Rahman and M. L. Norton, *Appl. Phys. Lett.*, 2013, **102**(2), 023701–023701.

3 T. Globus, T. Dorofeeva, I. Sizov, B. Gelmont, M. Lvovska, T. Khromova, O. Chertihin and Y. Koryakina, *American Journal of Biomedical Engineering*, 2012, **2**(4), 143–154.

4 T. Globus, A. M. Moyer, B. Gelmont, T. Khromova, M. I. Lvovska, I. Sizov and J. Ferrance, *IEEE Sens. J.*, 2013, **13**(1), 72–79.

5 T. Globus, I. Sizov and B. Gelmont, *Adv. Biosci. Biotechnol.*, 2013, **4**(3A), 493–503.

6 I. Sizov, M. Rahman, B. Gelmont, M. L. Norton and T. Globus, *Chem. Phys.*, 2013, **425**, 121–125.

7 T. Li, A. A. Hassanali, Y.-T. Kao, D. Zhong and S. J. Singer, *J. Am. Chem. Soc.*, 2007, **129**(11), 3376–3382.

8 K. E. Furse and S. A. Corcelli, *J. Phys. Chem. Lett.*, 2010, **1**(12), 1813–1820.

 9 W. Qiu, L. Wang, W. Lu, A. Boechler, D. A. R. Sanders and D. Zhong, *Proc. Natl. Acad. Sci. U. S. A.*, 2007, **104**(13), 5366–5371.

10 K. Moritsugu and J. C. Smith, *J. Phys. Chem. B*, 2005, **109**(24), 12182–12194.

11 J. Smith, T. Becker, S. Fischer, F. Noé, A. Tournier, G. Ullmann and V. Kurkal, Physical and Functional Aspects of Protein Dynamics, in: *Soft Condensed Matter Physics in Molecular and Cell Biology*, Taylor & Francis, 2006, pp. 225–241.

12 N. Q. Vinh, S. J. Allen and K. W. Plaxco, *J. Am. Chem. Soc.*, 2011, **133**(23), 8942–8947.

13 E. Balog, J. C. Smith and D. Perahia, *Phys. Chem. Chem. Phys.*, 2006, **8**(47), 5543–5548.

14 H. Frauenfelder, B. H. McMahon, R. H. Austin, K. Chu and J. T. Groves, *Proc. Natl. Acad. Sci. U. S. A.*, 2001, **98**(5), 2370–2374.

15 S. Cusack and W. Doster, *Biophys. J.*, 1990, **58**(1), 243–251.

16 M. D. Ediger, *Annu. Rev. Phys. Chem.*, 2000, **51**, 99–128.

17 A. A. Golosov and M. Karplus, *J. Phys. Chem. B*, 2007, **111**(6), 1482–1490.

18 N. Alijabbari, Y. Chen, I. Sizov, T. Globus and B. Gelmont, *J. Mol. Model.*, 2012, **18**(5), 2209–2218.

19 M. Karplus and J. N. Kushick, *Macromolecules*, 1981, **14**(2), 325–332.

20 R. M. Levy, M. Karplus, J. Kushick and D. Perahia, *Macromolecules*, 1984, **17**(7), 1370–1374.

21 T. Globus, D. Woolard, M. Bykhovskaia, B. Gelmont, L. Werbos and A. Samuels, *Int. J. High Speed Electron. Syst.*, 2003, **13**(04), 903–936.

22 E. F. Pettersen, T. D. Goddard, C. C. Huang, G. S. Couch, D. M. Greenblatt, E. C. Meng and T. E. Ferrin, *J. Comput. Chem.*, 2004, **25**(13), 1605–1612.

23 http://www.mathworks.com/matlabcentral/fileexchange/12377-interactive-fourier-filter-version-1-5/content/FourierFilter15/FouFilter.m.

24 R. M. Daniel, R. V. Dunn, J. L. Finney and J. C. Smith, *Annu. Rev. Biophys. Biomol. Struct.*, 2003, **32**(1), 69–92.

25 T. Globus, A. Moyer, B. Gelmont, I. Sizov and T. Khromova, Dissipation Time in Molecular Dynamics and Discriminative Capability of Sub-THz Spectroscopic Characterization of Biological Molecules and Cells, *Chemical and Biological Defense Science and Technology Conference*, DTRA, Las Vegas, Nov. 2011.

Faraday Discussions

ROYAL SOCIETY
OF CHEMISTRY

PAPER

Formation of coherent rotational wavepackets in small molecule-helium clusters using impulsive alignment†

Gediminas Galinis,[a] Luis G. Mendoza Luna,[a]
Mark J. Watkins,[a] Andrew M. Ellis,[b] Russell S. Minns,[c]
Mirjana Mladenović,[d] Marius Lewerenz,[d] Richard T. Chapman,[e]
I. C. Edmond Turcu,[e] Cephise Cacho,[e] Emma Springate,[e]
Lev Kazak,[f] Sebastian Göde,[f] Robert Irsig,[f] Slawomir Skruszewicz,[f]
Josef Tiggesbäumker,[f] Karl-Heinz Meiwes-Broer,[f]
Arnaud Rouzée,[g] Jonathan G. Underwood,[h] Marco Siano[i]
and Klaus von Haeften*[a]

Received 7th May 2014, Accepted 14th May 2014

DOI: 10.1039/c4fd00099d

We show that rotational line spectra of molecular clusters with near zero permanent dipole moments can be observed using impulsive alignment. Aligned rotational wavepackets were generated by non-resonant interaction with intense femtosecond laser pump pulses and then probed using Coulomb explosion by a second, time-delayed femtosecond laser pulse. By means of a Fourier transform a rich spectrum of rotational eigenstates was derived. For the smallest cluster, C_2H_2–He, we were able to establish essentially all rotational eigenstates up to the dissociation threshold on the basis of theoretical level predictions. The C_2H_2–He complex is found to exhibit distinct features of large amplitude motion and very early onset of free internal rotor energy level structure.

[a]University of Leicester, Department of Physics & Astronomy, Leicester, LE1 7RH, UK. E-mail: kvh6@le.ac.uk; Fax: +44 (0)116 252 2770; Tel: +44 (0)116 252 3525

[b]University of Leicester, Department of Chemistry, Leicester, LE1 7RH, UK

[c]University of Southampton, Department of Chemistry, Southampton, SO17 1BJ, UK

[d]Université Paris-Est, Laboratoire Modélisation et Simulation Multi Echelle, MSME UMR 8208 CNRS, 5 bd Descartes, 77454 Marne-la-Vallée, France

[e]Central Laser Facility, Rutherford Appleton Laboratory, UK

[f]Institute for Physics, University of Rostock, Germany

[g]Max Born Institute, Berlin, Germany

[h]Department of Physics & Astronomy, University College London, London WC1E 6BT, UK

[i]Blackett Laboratory, Imperial College London, London SW7 2BW, UK

† Electronic supplementary information (ESI) available. See DOI: 10.1039/c4fd00099d

1 Introduction

Laser-induced alignment is an emerging technology for preparing molecules in well-defined quantum states, thereby aligning them with respect to the laser field.[1-8] When used with a relatively long laser pulse, the technique is often referred to as adiabatic alignment,[3-5,9] whereas for pulse durations shorter than the rotational period of the molecule, impulsive or field-free alignment is a common term. The impulsive alignment technique is derived from the well-known rotational coherence spectroscopy method that has been applied to the investigation of the rotational structure of free molecules,[10-14] clusters[15-17] and liquids.[18] Laser-induced impulsive alignment can be applied to all molecules with an anisotropy of polarisability. While initial studies were performed on small linear molecules,[4,19,20] 1D and 3D impulsive alignment of complex, asymmetric top molecules has been demonstrated.[21-25] Control over molecular alignment has enabled the investigation of processes such as strong field ionisation,[26,27] high harmonic generation,[28-30] gas-phase X-ray[31] and (photo-)electron diffraction experiments.[31,32]

In impulsive alignment an intense laser pulse interacts with molecules non-resonantly and populates rotational eigenstates *via* virtual states corresponding to the respective laser wavelength. The excitation pathway *via* a virtual state means that in impulsive alignment excitation of rotational levels requires two photons. The excitation of rotational states is equivalent to a Raman process with the associated selection rules for the parity and total angular momentum of the excited system. Depending on the duration, shape and intensity of the laser field, the Raman process can induce sequential transitions leading, for example, to excitation of high rotational levels. Likewise, rotational levels can be de-excited, or remain unchanged. The final state constitutes a wavepacket of rotational eigenstates that periodically reassembles, representing alignment in space with respect to the laser polarisation axis. Following the propagation of the wavepacket in time provides dynamical information straightforwardly, such as the determination of coherence times. This is important for the investigation of time-dependent interactions, such as molecular collisions.[33] To illustrate the state of alignment in a classical picture, diatomics exposed to linearly polarised laser pulses will, for example, exhibit their internuclear axis aligned parallel and anti parallel with respect to the laser polarisation axis (z). In quantum mechanical terms, alignment is equivalent to a population shift towards the highest positive and negative M_z rotational states.[8] If the amplitude of the excitation laser pulse decreases much faster than the rotational period of the molecule, then the rotational states remain populated and the wavepacket propagates in time and space even after the laser field has vanished.[7]

This paper is motivated by the prospect of adopting the impulsive alignment technique for elucidating the properties of free clusters. The investigation of free atomic and molecular clusters is an important pillar for understanding size effects and complexity in condensed matter.[34-37] Furthermore, knowledge of intrinsic cluster properties is essential for the understanding of how clusters interact with their environment and is ultimately important for applications of clusters in new materials.[38,39]

To generate clusters free of external interactions, a supersonic expansion in vacuum is commonly employed: pressurised gas expanding through a nozzle into vacuum cools rapidly, thereby facilitating cluster growth, and can form a molecular beam, propagating through the vacuum apparatus.[40,41] A particular benefit of using molecular beams is that the cluster samples are continuously renewed, so that effects of any probing, such as radiation damage or fragmentation, does not affect subsequent measurements. An alternative possibility is to study clusters confined in ion traps.[42–45]

The propagating molecular beam can be probed by interaction with light, static electric or magnetic fields, by scattering from charged or neutral particles, or off space-fixed nanostructures. Thus, a variety of different techniques are currently available to study free clusters: laser spectroscopy,[46,47] ion depletion spectroscopy,[45] fluorescence spectroscopy,[48,49] mass spectrometry,[50–52] Stern–Gerlach deflection,[53–55] static electric field deflection,[56–58] electron beam scattering[59–61] and diffraction from space-fixed periodic nanostructures.[62,63] Very recently, time-resolved, single cluster diffraction using soft X-ray[64] and hard X-rays[65] has been employed.

Impulsive alignment is specifically beneficial for probing weakly bound clusters because spectral analysis is facilitated and enough information to benchmark physical models can be gained as we will illustrate here for small clusters of helium and a molecule. A characteristic feature of weakly bound molecular clusters is large amplitude motion. For such weakly bound clusters bending vibrations, internal rotations or other types of internal motion occur on similar time scales and therefore it is difficult, and often impossible, to separate the wavefunction into distinct rotational and vibrational parts. Hence, a conventional rigid rotor-type Hamiltonian fails to describe their rotational level structure.[66] While pure rotational spectroscopy provides insight into entangled internal motions of molecular complexes and clusters, the experimental data reported in the literature is comparatively sparse. Furthermore, to form weakly bound clusters, very low temperatures have to be provided. This implies that only the lowest rotational levels are populated with the consequence that a conventional rotational spectrum may contain only few lines, or perhaps even just one – usually not enough for a comprehensive analysis of the structure, as pointed out by Nesbitt and Naaman.[67] By exciting high lying rotational levels in a wavepacket it is therefore possible to generate a rich rotational line spectrum despite the limitation of a cold initial thermal population.

Small helium clusters also attract considerable attention because they establish model systems by which the effect of size and complexity on quantum phenomena, such as superfluidity, can be studied. In the past twenty years molecules have been embedded in large helium droplets as well as in small helium clusters and their pure rotational and rovibrational frequency spectra have been studied using conventional microwave (MW) and infrared (IR) spectroscopy. In broad terms these studies have established that in ^{4}He droplets the molecules rotate almost unhindered, but with an increased apparent moment of inertia compared to the gas phase. Also, their spectral lines are broadened with respect to the gas phase, reflecting the interaction with the helium (see review articles for further details[68–74]). Although much progress has been made the processes leading to the increase of apparent moment of inertia and line broadening are not entirely understood. The availability of the complete pure rotational spectrum of small molecule-helium clusters will be an important complement to the previous and ongoing infrared and microwave-based spectral analysis of these systems and pave the way towards a comprehensive under-standing of incipient superfluidity.

Here, we show the generation of rotational wavepackets of C_2H_2–He_n clusters using impulsive alignment. The phase of the wavepacket was detected as a function of time using Coulomb explosion. Fourier transformation produced a rich spectrum of discrete lines in the frequency domain that were attributed to rotational eigenstates of the clusters. A detailed analysis was carried out for the smallest complex, C_2H_2–He. This weakly bound complex is an important model system that has previously been investigated using *ab initio* calculations[75–78] and infrared spectroscopy in its deuterated form,[77] but whose *pure* rotational spectrum has not previously been reported. As part of the current study we also present an improved potential energy surface of this complex. A complete set of bound rovibrational levels derived from this surface was computed with a numerically exact discrete variable approach. The measured transitions of C_2H_2–He mapped practically all theoretically predicted transitions between bound rotational levels, showing excellent agreement between theory and experiment and demonstrating the successful application of the impulsive alignment method to obtain a complete rotational spectrum of a weakly bound complex. The knowledge of the complete level structure has prompted a more detailed theoretical analysis which revealed the importance of dynamical effects contributing to the pronounced free-internal rotor behaviour of this complex.

2   Experimental details

The experiments were conducted at the Rutherford Appleton Laboratory using the Artemis femtosecond laser beam line and the Atomic and Molecular Physics ultra high vacuum endstation, which was equipped with a source for the production of doped helium clusters in co-expansion. In all experimental runs the base pressure inside the vacuum chamber was lower than 5×10^{-9} mbar. Apart from the cluster production the experiment was conceptually similar to that of Pentlehner *et al.*,[79] particularly the generation of rotational wavepackets and their detection using a pump–probe scheme. Briefly, C_2H_2–He_n clusters were generated in a supersonic expansion of 0.01% C_2H_2 diluted in 9 MPa He through a cooled pulsed valve of conical shape (half opening angle, $\alpha = 20°$) with a throttle diameter of 100 μm (Even–Lavie nozzle[80]). The cluster beam propagated approximately 50 mm through the vacuum chamber into the focus of a velocity map imaging (VMI) detector[81] where it was excited by two pulsed laser beams, both originating from a 30 fs, 1 kHz Ti : sapphire laser operating at 800 nm (KM Labs Red Dragon). The laser system had two separate grating compressors, allowing the pulse durations of each laser beam to be independently controlled. The beams were co-linearly focused through a $f = 500$ mm lens into the supersonic expansion. The number of rotational levels in the wavepacket was controlled by adjusting the pulse length and intensity of the first laser pulse (the pump laser). The pump laser had a duration of 300 fs and was operated across a range of intensities from 2×10^{11} W cm^{-2} to 5×10^{12} W cm^{-2}, well below the level required to ionise the molecules. The probe pulse had a duration of 50 fs and an intensity of 1×10^{15} W cm^{-2}, sufficient to instantly break molecular bonds in a Coulomb explosion, thereby generating

C^+, C^{2+} and H^+ fragment ions. The velocity vectors of the fragment ions, which carried the molecular alignment information, were mass-selectively detected in a velocity map imaging (VMI) spectrometer[81] whose detector plane was parallel to the polarisation plane of both laser beams (xy plane in Fig. 1). Mass-selection was established by switching the gain of the microchannel plate detector so that it was only sensitive at the specific times when the ions of interest arrived at the detector. To extract alignment information the C^+ mass channel was chosen for detection.

The two-dimensional projection of the recoiling C^+ fragment directions and intensities was used to determine $I \times \cos^2 \theta$ for each position on the detector, where θ designates the angle between the polarisation of the pump laser and the projected velocity vector and I is the intensity. The origin of this vector was set to the centre of the image defined by the centre of mass of the ion distribution. The end of the vector pointed to a pixel representing the intensity of ions at this position. The average over the entire detector area, $\langle \cos^2 \theta_{2D} \rangle$, is proportional to the molecular alignment $\langle \cos^2 \theta_{3D} \rangle$.[79,82] The quantity $\langle \cos^2 \theta_{3D} \rangle$, indicating the phase of the rotational wavefunction, is a true quantum mechanical observable whose exact determination would require the projected velocity map images to be inverted, using an inverse Abel transformation.[83–85] Since the recorded ion distributions did not exhibit spherical symmetry the inversion procedure is technically challenging. The parameter $\langle \cos^2 \theta_{2D} \rangle$ was determined instead as a function of time by scanning the delay between the pump and probe laser pulses to reveal the rotational dynamics of the clusters. Because of the proportionality between $\langle \cos^2 \theta_{3D} \rangle$ and $\langle \cos^2 \theta_{2D} \rangle$ the lack of spherically symmetric images had no consequences for the derivation of spectroscopic information and the conclusions of our experimental results.

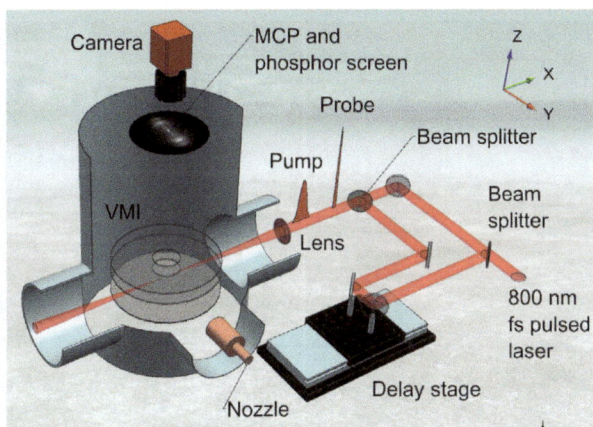

Fig. 1 Schematic of the experimental setup comprising a molecular beam machine, a laser system and a velocity map imaging detector coupled to a CCD camera. Clusters and molecules prepared in the molecular beam were impulsively aligned and probed using a laser pump–probe scheme, where the delay between the two laser pulses is generated with a movable stage. The state of alignment was probed for time delays of up to 600 ps, corresponding to the maximum displacement of the delay stage.

3 Results

3.1 Coulomb explosion

Fig. 2 shows raw velocity map images recorded selectively for the C^+ fragment under different expansion conditions. These raw images show ion fragment intensities with an angular and radial distribution that illustrate the Coulomb explosion process. Dashed rings have been introduced to indicate the centre position and to highlight differences in the radial distribution. Distinct changes in the Coulomb explosion pattern are observed for varying expansion conditions: panel (a) in Fig. 2 shows the ion image for expansions where the nozzle was held at a temperature of 293 K and panel (b) shows the image obtained when the nozzle was cooled to 203 K to facilitate the formation of clusters of C_2H_2 with helium. The ion image in (a) shows relatively sharp features, with evidence of more than one fragmentation channel leading to the formation of C^+. These features were attributed to unbound C_2H_2 as the dominant species. The angular distributions of the two images are identical and largely influenced by the anisotropic laser ionisation. However, the radial distributions differ: panel (b) shows considerably higher kinetic energies than (a). The increased kinetic energy can be taken as an indicator for the generation of further charges, even though the laser parameters themselves have not changed. The temperature variation from 293 to 203 K increases the gas number density in the interaction region by 30%, which corresponds to a reduction of the internuclear separation of particles within the jet of 11%. This small change means that space charge effects do not explain the increase of ion velocities observed in Fig. 2. It is more likely that the processes responsible occur within the clusters because of the much higher atomic number density than within the unclustered jet. Further charges can be generated within the C_2H_2 molecules, but also He atoms could be ionised.[86] A second possibility is that the higher kinetic energies originate from inelastic scattering of C^{2+}

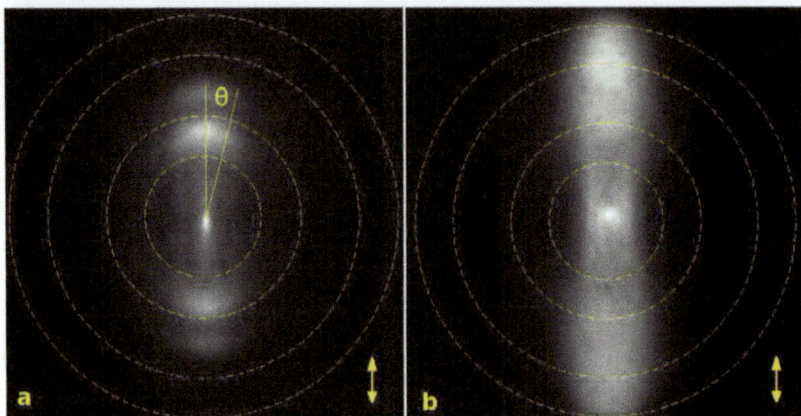

Fig. 2 Velocity map images of C^+ fragments generated *via* Coulomb explosion. (a) was recorded at a nozzle temperature of 293 K and shows fragments originating from free C_2H_2. In (b) the nozzle temperature was 203 K, facilitating cluster formation, but otherwise identical conditions. (b) shows increased kinetic energy compared to (a). The yellow double-headed arrow indicates the direction of the pump and probe laser polarisations, which are parallel to the *y*-axis in Fig. 1. The dashed circles are guides to the eye.

fragments or even higher charged fragments when leaving the cluster. Subsequently, these highly charged ions recombine with electrons, producing C^+ fragments.

3.2 Time-resolved alignment and Fourier transformation

The time-resolved molecular alignment parameter, $\langle\cos^2(\theta)_{2D}\rangle(t)$, is shown in Fig. 3(a) for a time period of 16 ps and for conditions favouring cluster formation similar to Fig. 2(b). The strongest features consist of a pattern of six oscillations which reappear every rotational period τ. This pattern matches the time-resolved alignment spectrum expected for free, cold C_2H_2 molecules. The features extend over the full range of the scan up to 600 ps, as shown in Fig. 3(b). No damping in amplitude is observed, so one can infer that the coherence of the rotational wavepacket is at least 600 ps. Revivals of small acetylene-helium clusters $(C_2H_2-He_n)$ are difficult to directly identify in the alignment scan because of the dominant signals from free acetylene rotation. A much clearer picture is obtained when a discrete Fourier transform is performed, producing a rotational frequency spectrum.[87]

The discrete Fourier transform, F_ω, is defined by

$$F_\omega = \sum_{t=0}^{N-1} f_t e^{-2\pi i\omega t/N} \tag{1}$$

where N designates the number of data points being transformed, f_t is the tth element in the time domain and ω is a variable in the frequency domain. The N complex numbers derived from eqn (1) were converted into a power spectrum by adding the squared real and imaginary parts of F_ω:

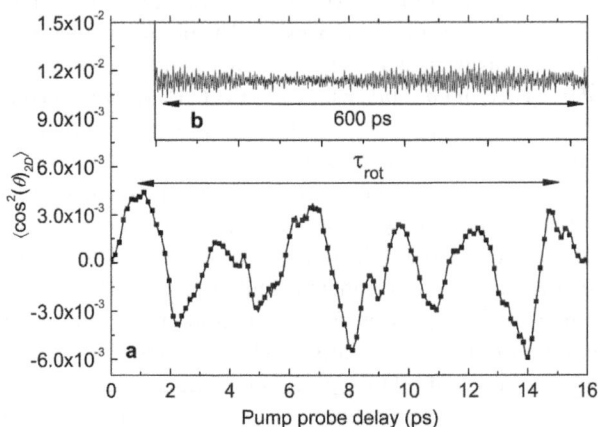

Fig. 3   Time-resolved alignment of C_2H_2 co-expanded with helium. (a) Rotational revivals obtained under conditions where the helium stagnation pressure and temperature were 9 MPa and 212 K, respectively. Note that the baseline has been subtracted and the data have been smoothed. The time difference between two full revivals, indicated by a double-headed arrow, is equal to the rotational period of acetylene and inversely proportional to 6b. (b) For time delays up to the maximum of 600 ps, no damping in the amplitude was observed. The coherence time of the wavepacket for free C_2H_2 molecules is therefore at least 600 ps.

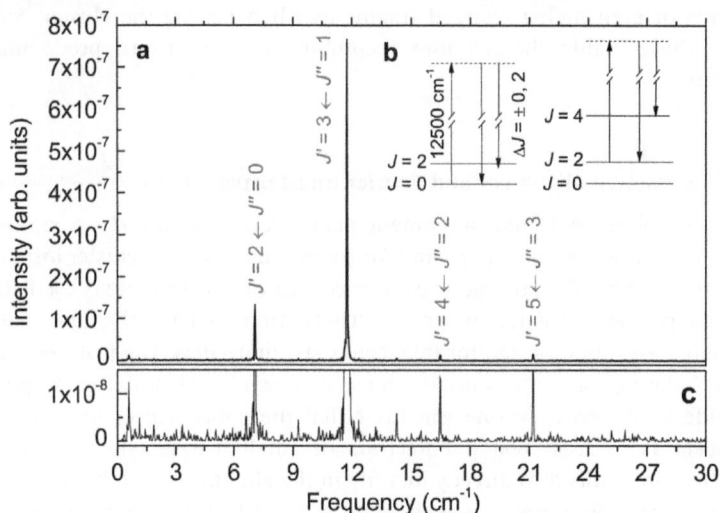

Fig. 4 Overview spectra. (a) Power spectrum of the time-resolved molecular alignment. The power spectra are dominated by strong peaks arising from beats between rotational levels of free C_2H_2 connected by $\Delta J = 0, \pm 2$. (b) The rotational levels in the wavepacket are excited through sequential Raman excitations with an 800 nm (12 500 cm^{-1}) laser pulse. This process sequentially populates higher levels via virtual states, as schematically illustrated in the inset. Panel (c) shows an expanded view of the power spectrum revealing numerous weak peaks.

$$Re(F_\omega)^2 + Im(F_\omega)^2 \qquad (2)$$

The resulting power spectrum is shown in Fig. 4.

A series of discrete lines in Fig. 4 corresponds to particular frequency contributions of C_2H_2 to the rotational wavepackets. The observed lines were recorded at a pump laser intensity of 2.5×10^{11} W cm^{-2} and coincide with the beat frequencies of free C_2H_2 at $6b + 4nb$, where b is the C_2H_2 rotational constant‡ ($b = 1.1767$ cm^{-1})[88] and $n = 0, 1, 2, 3$. The first line at $6b$ frequency is equivalent to the $J' = 2 \leftarrow J'' = 0$ transition, with successive lines at $10b$, $14b$, $18b$. The significantly higher intensity of the contribution from the $J = 1$ state compared with the $J = 0$ state is attributed to nuclear spin statistics describing the occupation ratio of even J(para) : odd J(ortho) levels, which in this case is 1 : 3.[89] Since Fig. 4 displays the power spectrum, the ratio in principle should be 1 : 9. However, when additional J-dependent coupling terms for two-photon transitions are taken into account a revised ratio of 6.25 is expected. This expected ratio is in satisfactory agreement with the observed relative peak intensities.

A much weaker series of discrete lines is seen at $4b$, $8b$ and $12b$, notably around 9.2 and 14.2 cm^{-1}. The position of these lines can be expressed through the series $2b + 2nb$, for $n \geq 1$, with every second line overlapping with the Raman allowed transitions at $6b + 4nb$. The occurrence of these lines does not comply with the

---

‡ B conventionally denotes the rotation constant of a linear molecule. However, in this context we want to differentiate between the internal rotation of the molecule within a complex, later in the text denoted as b, and the overall rotational constant of the complex, as B.

$\Delta J = 0, \pm 2$ selection rule for Raman transitions of diatomic molecules. We note that the $\Delta J = 0, \pm 2$ selection rule would be relaxed for symmetric top molecules and may indeed indicate the involvement of C_2H_2-$(He)_n$ clusters. However, the good match of these transitions with multiples of the b constant of free C_2H_2 is difficult to explain considering that in other molecule-helium complexes such lines are always shifted.[90]

4 Assignment of C_2H_2–He lines: statistical analysis and theoretical predictions

In the low frequency range in Fig. 4(c) numerous peaks are observed which have frequencies well below those of free C_2H_2. These lines are attributed to clusters because they correspond to a larger moment of inertia than is possible for free C_2H_2.[90,91] To fully assess our assignment in this frequency range and to account for the low intensities a careful statistical analysis of the noise was performed. A region of the spectrum between 100 to 240 cm^{-1} which was free of molecular frequencies was identified as suitable for a quantitative determination of the confidence level of our data and analysed to assess whether the intensity variation is stochastic. The signal height distribution was then investigated and the level at which a random event could be excluded with 99.9% certainty was determined. Details of the noise analysis are provided in the ESI.†

While this assessment of the noise level revealed a large number of lines exceeding the confidence level, focus was placed on the detailed analysis of the smallest complex, the C_2H_2–He complex. For this complex a potential energy surface and the energies of rotational states were calculated to guide the assignment. The predicted energy levels were then compared with the line positions observed in the spectrum.

4.1 Coherence times of C_2H_2–$(He)_n$

An enlarged view of the low frequency region, including a dashed line to indicate the 99.9% confidence level, is shown in Fig. 5. A large number of lines, attributed to C_2H_2–He$_n$ clusters and with intensities above the confidence level, are observed. The full widths at half maximum of these lines are 0.03 cm^{-1}, which matches the experimental limit in resolution set by the total time delay of 600 ps. Hence, the width of the lines assigned to C_2H_2–He$_n$ clusters is consistent with a coherence time for the rotational wavepacket of at least 600 ps. We note that such long coherence times are not unexpected for small molecular clusters, given that predissociation or other events that destroy coherence are very unlikely on a time scale of sub-nanoseconds. This finding further conforms with spectroscopic work in the frequency-domain, which has established that molecules residing inside superfluid ^{4}He droplets rotate almost freely.[92] Interestingly, a recent study by Pentlehner et al.[79] showed rapid loss of coherence for non-adiabatic alignment of CH_3I in helium droplets, i.e. within a single rotational period. This observation contradicts our results for C_2H_2–He and the many frequency-domain studies of molecules in helium droplets. This discrepancy is currently unexplained.

As detailed below we can assign some of the C_2H_2–He$_n$ features to the C_2H_2–He complex. The other peaks in Fig. 5 are attributed to larger C_2H_2–He$_n$ ($n \geq 2$) complexes. A detailed assignment is difficult due to a lack of models for weakly

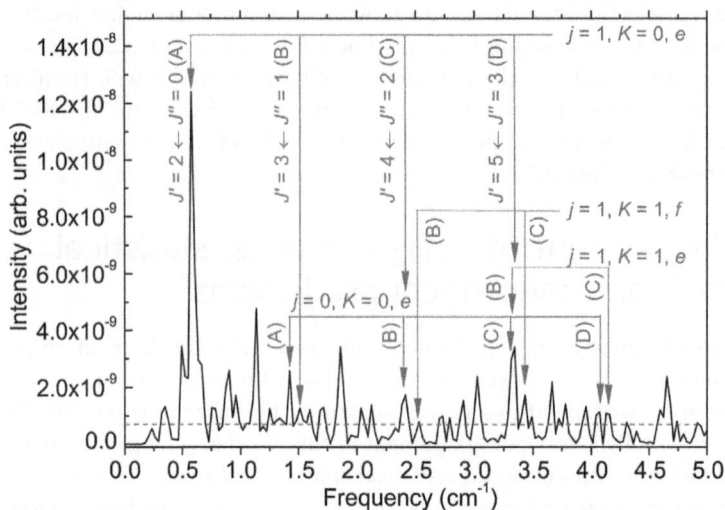

Fig. 5 C_2H_2–He spectrum in the frequency region below 5 cm^{-1}. The 99.9% confidence level, established through the second order χ^2 spectral distribution of the noise, is indicated by a horizontal dashed line. The spectral lines are labelled in alphabetical order with respect to the total angular momentum of the initial state. See text and the energy diagram in Fig. 8 for further information.

bound clusters consisting of more than one helium atom. Our assignment of the $n = 1$ case is guided by theory, as described in the next section.

4.2   Electronic structure calculations

Total electronic energies for C_2H_2 and C_2H_2–He were computed with the coupled cluster method with single and double substitutions and perturbative treatment of triples, CCSD(T), and all electrons were included in the correlation treatment. Core-optimised correlation consistent augmented (doubly augmented for He) basis sets, (d)aug-cc-pCVXZ, X = T,Q,5,6, as developed by Dunning and coworkers,[93,94] were employed as implemented in the MOLPRO electronic structure package.[95] Geometry optimisations carried out for the complex and isolated C_2H_2 with basis set sizes up to sextuple zeta quality show that complexation with He changes the geometry of C_2H_2 at most by 0.0001 Å. The C_2H_2 unit was consequently fixed at its experimental ground state expectation geometry,[88] $r_0(CC) = 1.20830$ Å and $r_0(CH) = 1.05756$ Å.

The interaction between C_2H_2 and a helium atom was explored in a Jacobi coordinate system with a Jacobi vector $\mathbf{R}$ pointing from the centre of mass of C_2H_2 to the helium atom and a Jacobi angle θ enclosed between $\mathbf{R}$ and the C_2H_2 molecular axis, as shown in Fig. 6. We used a grid of 300 points covering $0 \leq \theta \leq 90°$ in steps of 10° and radial grids which were optimised at each value of θ, typically ranging between 2.50 and 20 Å. Total energies for the complex obtained with basis sets from triple zeta, (d)aug-cc-pCVTZ, to quintuple zeta, (d)aug-cc-pCV5Z, level were extrapolated to the complete basis set (CBS) limit using the procedure of Peterson et al.,[96,97] which turned out to give more consistent results than the more common procedure of Helgaker and coworkers.[98] Interaction

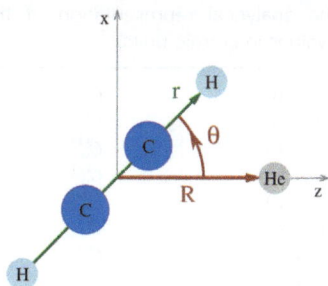

Fig. 6 Jacobi coordinates for the complex C_2H_2–He. The body-fixed z-axis is aligned with the Jacobi vector $\mathbf{R}$.

energies were determined from the estimated CBS energies by subtracting the result of an extrapolation for essentially separated monomers at $R = 100$ Å and $\theta = 0$. The grid point with the strongest interaction is at $R = 4.32$ Å and $\theta = 0$ with a potential value of $V_{int} = -25.100$ cm^{-1} relative to separate monomers at $V = 0$. This is noticeably below the previous best estimate of -24.21 cm^{-1}.[99] The rotational constant of the complex at its electronic equilibrium geometry is found to be $B_e = 0.2119$ cm^{-1} and the corresponding permanent dipole moment is 0.029 D. The spherically averaged dipole polarisability exhibits a very weak dependence on the complex geometry and has a value of $\bar{\alpha} = 24.0a_0^3$ which is essentially the sum of the monomer values of $22.6a_0^3$ and $1.4a_0^3$ for C_2H_2 and He, respectively, providing further indication of the very weak van der Waals interaction.

4.3 Analytical representation of the interaction potential

The interaction between C_2H_2 and a helium atom contains dispersion and induction contributions, where the latter arise from the quadrupole moment of C_2H_2. The asymptotic R-dependence should therefore contain only even inverse powers of R. We adopted an angle-dependent extended Tang–Toennies[100] form to represent the *ab initio* interaction energies:

$$V_{int}(R,\theta) = A(\theta)\exp\left\{-\left[b_1(\theta)R + b_2(\theta)R^2\right]\right\} - \sum_{k=3,8} f_{2k}(b(\theta),R)\frac{C_{2k}(\theta)}{R^{2k}} \qquad (3)$$

The radial parameters $X = A, b_1, b_2, C_6, C_8, C_{10}$, are expanded over even order Legendre polynomials according to

$$X(\theta) = \sum_{l=0,2,\ldots} X^{(l)}P_l(\theta) \qquad (4)$$

Higher order coefficients C_{2k}, $k > 5$, are defined by the standard recursion $C_{2k+2} = (C_{2k}/C_{2k-2})^3 C_{2k-4}$ and the functions f_{2k} are Tang–Toennies damping functions with $b(\theta) = b_1(\theta) + 2b_2(\theta)R$. The present fitting model includes terms up to $l = 6$ and contains 24 free parameters which were adjusted by a non-linear least squares procedure.[101] This reproduces the 235 *ab initio* interaction energies falling below +200 cm^{-1} with a root mean square error of 0.039 cm^{-1} and a largest deviation of 0.22 cm^{-1}. The coefficients for this analytical representation are

Table 1 Coefficients for the analytical representation of the C_2H_2–He interaction potential, eqn (3) and (4) (all values in atomic units)

$A^{(0)}$	0.1918812×10^2	$C_6^{(0)}$	0.1248242×10^2
$A^{(2)}$	0.2504029×10^2	$C_6^{(2)}$	-0.2662071×10^1
$A^{(4)}$	0.8700405×10^1	$C_6^{(4)}$	-0.4200783×10^1
$A^{(6)}$	0.1256556×10^1	$C_6^{(6)}$	-0.5564107×10^1
$b_1^{(0)}$	0.1327564×10^1	$C_8^{(0)}$	0.1318068×10^4
$b_1^{(2)}$	-0.3394156×10^0	$C_8^{(2)}$	0.1819554×10^4
$b_1^{(4)}$	-0.1268924×10^0	$C_8^{(4)}$	0.8531615×10^3
$b_1^{(6)}$	0.1446208×10^{-1}	$C_8^{(6)}$	0.1011623×10^4
$b_2^{(0)}$	0.4538917×10^{-1}	$C_{10}^{(0)}$	-0.2223242×10^5
$b_2^{(2)}$	0.4238423×10^{-1}	$C_{10}^{(2)}$	-0.2251989×10^5
$b_2^{(4)}$	0.1040915×10^{-1}	$C_{10}^{(4)}$	-0.1157502×10^5
$b_2^{(6)}$	$-0.1945577 \times 10^{-2}$	$C_{10}^{(6)}$	-0.3578250×10^5

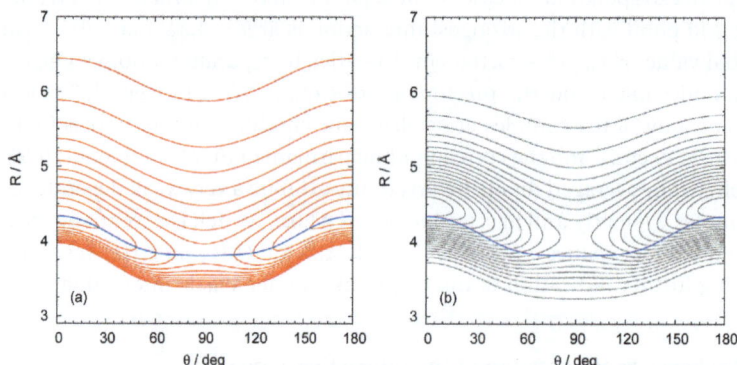

Fig. 7 (a) Contour plot of the C_2H_2–He interaction potential. The contour interval is 2 cm^{-1} and the lowest contour is at $V_{int} = -24$ cm^{-1}. (b) Contour plot of the $|\Psi_0|^2$ probability density of the ground state of C_2H_2–He. Contours are given at intervals of 5% of the maximum value. For both figures, the blue curve indicates the minimum energy path.

given in atomic units in Table. 1. Fig. 7(a) displays the interaction potential with its minimum at linear C_2H_2–He and a saddle point at the T-shaped arrangement.

4.4 Rovibrational calculations

Bound rovibrational levels for the fitted surface were calculated with the DVR-DGB method, which uses a discrete variable representation (DVR) for the angular coordinate and a distributed Gaussian basis (DGB) for the radial degree of freedom.[102] This method uses an exact kinetic energy operator and makes no approximations aside from the Born–Oppenheimer concept of a potential energy surface, whose accuracy is only limited by the quality of the electronic structure treatment. We used atomic masses in conjunction with 51 Gauss-Legendre DVR points in θ and an angle dependent radial basis composed of up to 85 non-evenly distributed Gaussians between $4a_0$ and $300a_0$. Energy eigenvalues and wave functions were computed for both parities ($p = 0, 1$ corresponding to even and odd parity states, respectively) and for total angular momentum $0 \leq J \leq 10$, but only levels up to $J = 5$ were found to be bound. Energy levels are converged to better than 0.001 cm^{-1}. The computed rovibrational ground state energy is -7.417 cm^{-1}. The square, $|\Psi_0|^2$, of the corresponding ground state $J = 0$ wave function is displayed in Fig. 7(b). Maxima are visible for the two equivalent linear minimum energy arrangements but the probability density is clearly spread over the entire angular domain and remains above 40% of its maximum even at the saddle point at $\theta = 90°$. At all angles the pronounced anharmonicity shifts the radial position of the $|\Psi_0|^2$ maximum significantly outward with respect to the minimum energy path. The impact of this strong delocalisation on the rovibrational level structure will be described in more detail in the Discussion section.

4.5 Assignment of experimentally observed lines

The assignment of the peaks in Fig. 5 was made using the transition frequencies predicted from the theoretical model described above. In particular, we looked for direct coincidence between theory and experiment within experimental error (0.03 cm^{-1}). The theoretical prediction for the first vibrationally excited state, the intermolecular stretching vibration, is 7.415 cm^{-1}. Our spectrum shows a signal at a matching energy but the marginal nature of this state, only 0.002 cm^{-1} below the dissociation threshold and at the limits of the theoretical treatment does not allow a firm assignment without further work. The expected Raman transitions for C$_2$H$_2$–He are marked in Fig. 8 by vertical arrows and text, showing the measured energy differences in cm^{-1}. The energy levels are labelled using the quantum numbers j, J, K and parity labels e and f.[77] Note that only the total angular momentum quantum number J and the parity of the wave functions are rigorous quantum numbers. The j quantum number refers to the internal

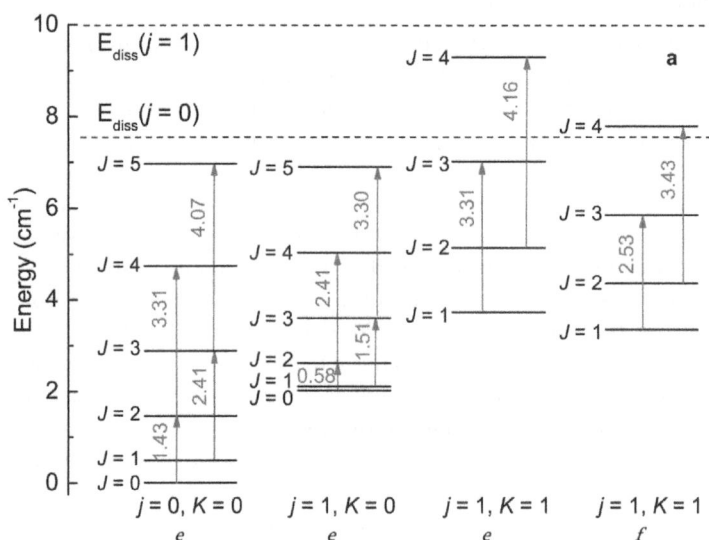

Fig. 8 Rotational energy level diagram of C$_2$H$_2$–He. The energy levels are grouped with respect to quantum numbers j, K and parity labels e, f. The vertical arrows designate the transitions observed experimentally. The numbers next to the arrows designate the corresponding transition frequencies in cm^{-1} (see Table 2). The horizontal dashed lines indicate the dissociation limits corresponding to C$_2$H$_2$ in $j = 0$ or $j = 1$.

Table 2 Energy level assignment of measured lines. Quantum numbers of the initial states are denoted by j'', K'', J'' and of the final states by j', K', J'. The experimental line positions are given in cm^{-1} and the estimated error margin in each case is ± 0.03 cm^{-1}

Sym.	j''	K''	J''	j'	K'	J'	Exp.	Calc.
e	0	0	0	0	0	2	1.43	1.46
e	0	0	1	0	0	3	2.41	2.40
e	0	0	2	0	0	4	3.31	3.28
e	0	0	3	0	0	5	4.07	4.08
e	1	0	0	1	0	2	0.58	0.60
e	1	0	1	1	0	3	1.51	1.49
e	1	0	2	1	0	4	2.41	2.41
e	1	0	3	1	0	5	3.30	3.31
e	1	1	1	1	1	3	3.31	3.29
e	1	1	2	1	1	4	4.16	4.15
f	1	1	1	1	1	3	2.53	2.50
f	1	1	2	1	1	4	3.43	3.43

rotation of the C_2H_2 unit and K is the projection of J onto the intermolecular axis (see Fig. 6). Although approximate, these alternative quantum numbers turn out to be useful to describe the energy level pattern. The present calculation provides a better match to the measured line positions than the best potential energy surface available prior to this work.[78] The assignment of the lines is also shown in Table 2.

5   Discussion

5.1   Rotational level pumping

The identified lines of C_2H_2–He constitute a rich spectrum showing the possibility of sequential rotational energy level pumping. For example, the lower levels in the $J' = 4 \leftarrow J'' = 2$ and the $J' = 5 \leftarrow J'' = 3$ transitions in the $j = 0$, $K = 0$, e manifold, the $J' = 4 \leftarrow J'' = 2$ and $J' = 5 \leftarrow J'' = 3$ transitions in the $j = 1$, $K = 0$, e manifold, as well as the $J' = 4 \leftarrow J'' = 2$ transitions in the $j = 1$, $K = 1$, e and f manifolds, have presumably gained population through Raman transitions from lower levels. We note that increased pump laser power does not increase the intensity of these lines in a straightforward fashion. At the highest pump intensity of 5×10^{12} W cm^{-2} the $J' = 3 \leftarrow J'' = 1$ transition in the $j = 1$, $K = 1$, f manifold becomes the strongest line. This transition also remains stronger than all other C_2H_2–He$_n$ lines when the expansion conditions are changed to promote clusters comprising more than one helium atom. More experimental work is needed to confirm this observation and to elucidate details of these trends.

The highest rotational level populated lies beyond the dissociation threshold for C_2H_2–He when internal rotation is not excited ($j = 0$). This level, which belongs to the $j = 1$, $K = 1$, e manifold, can predissociate provided a pathway for de-excitation of the internal rotation of the C_2H_2 molecule exists.

The observation of this energy level on the timescale of this experiment, where measurements up to 600 ps are possible, is not unreasonable. For example, the CO_2–He complex has a lower limit of 6 ns for the predissociation life time, as determined by laser spectroscopy measurements.[103]

5.2 Analysis of the rotational energy level pattern of C_2H_2–He

The computed level structure and the analysis of the wave functions shows that C_2H_2–He does not behave like a typical linear molecule in spite of its linear electronic minimum energy structure. Two limiting zero order models and corresponding labelling schemes can be used to describe the present situation, namely the rigid linear molecule picture in which He executes a bending motion, or the free internal rotor (potential-free) asymptote with a freely rotating C_2H_2 unit. The numerically exact wave functions obtained for given rigorous quantum numbers J and p in the present calculations can be projected onto either zero order representation in order to assign corresponding approximate quantum numbers and an additional analysis using an adiabatic projection is possible.

The bending vibration of linear molecules is two-dimensional and described by the quantum label ν_{lin}^{ℓ}, where the quantum number ℓ of the vibrational angular momentum $\hat{\ell}$ is given by $\ell = \nu_{\text{lin}}, \nu_{\text{lin}} - 2, \ldots, -\nu_{\text{lin}}$ leading to $\nu_{\text{lin}} + 1$ fold degeneracy of the level ν_{lin}. Due to angular momentum conservation, $\hat{\ell}_z = \hat{J}_z$ also holds for linear triatomic molecules, where $\hat{J}$ is the total rotation. This means that the first bending level for $J = 0$ is $2\nu_{\text{lin}}^0$ and that ν_{lin}^1 is accessible only for $J \geq 1$. Above the angular barrier, the bending energy pattern shows the onset of a two-dimensional rotor structure. The latter structure arises in the potential-free situation and is characterised by $\ell = \nu_{\text{lin}}, \nu_{\text{lin}} - 1, \ldots, -\nu_{\text{lin}}$, such that the degeneracy of the level ν_{lin} is $2\nu_{\text{lin}} + 1$. In the free-rotor limit, Coriolis interaction may prominently affect the overall molecular behaviour.[104]

A very useful quantitative analysis of the exact wavefunctions is possible through an adiabatic projection scheme designed for the DVR-DGB approach in the spirit of the method previously developed for tetratomic molecules.[105] The radial eigenvalues computed for each angular DVR point during the construction of the full problem provide adiabatic angular profiles for a given stretching quantum number ν_s. The lowest of these profiles corresponding to $\nu_s = 0$ defines the adiabatic ground state bending profile $^{\text{adi}}V^0(\theta)$, which is depicted in Fig. 9, together with the minimum energy path $V_{\text{MEP}}(\theta)$ along the Jacobi angle θ. The $^{\text{adi}}V^0(\theta)$ profile differs from $V_{\text{MEP}}(\theta)$ by the angle dependent ground-state energy of the intermolecular stretching vibration and provides a more useful rationalisation of the angular motion of the complex. Fig. 9 shows that the inclusion of the ground-state stretching energy lowers the effective barrier height at $\theta = 90°$ from 8.3 cm^{-1} for $V_{\text{MEP}}(\theta)$ to 3.4 cm^{-1} for $^{\text{adi}}V^0(\theta)$ due to strong stretch-bend coupling. Another interesting observation is the very weak θ dependence of $^{\text{adi}}V^0(\theta)$ for $\theta \in (0, 45°)$ and $\theta \in (135°, 180°)$, whereas $V_{\text{MEP}}(\theta)$ shows a somewhat parabolic θ shape in this region. Similar flat potentials over a wide angular range around linearity are typical for quasi-linear molecules.[106]

Solving the angular problem for an adiabatic profile with given ν_s provides bending functions with well defined adiabatic quantum numbers, ν_s and ν_b. The functions $|\nu_b, \nu_s; K\rangle$, where K is the body-fixed z-projection of $\hat{J}$, are used to construct adiabatic expansions of the exact wavefunctions, which allow assignments by identification of the dominant zero-order contribution(s). The adiabatic energies $^{\text{adi}}\varepsilon^{(K)}$ differ from the exact rovibrational energies $E^{(J,p)}$ due to missing Coriolis coupling and bend-stretch coupling contributions.

The bound rovibrational states for $J = 0$ and 1 in Fig. 9 are labelled by $(\nu_b; K)$ using the adiabatic quantum number ν_b and the quantum number K. The

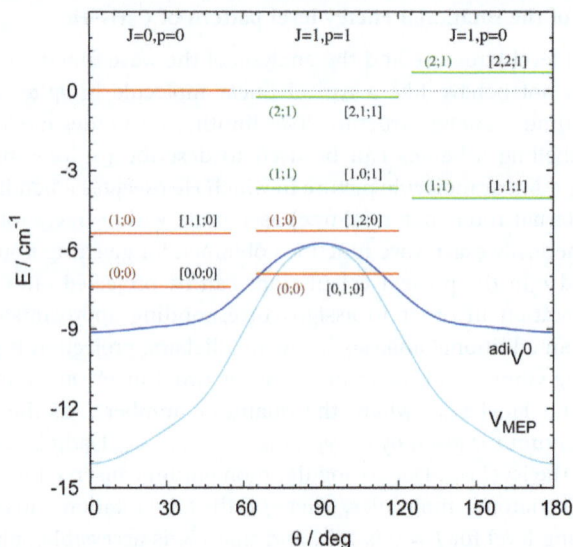

Fig. 9 Minimum energy path V_{MEP} and effective ground-state potential $^{adi}V^0$ along the Jacobi angle θ. The curve V_{MEP} is shifted upward by 10.99 cm^{-1} to coincide with $^{adi}V^0$ at θ = 90°. The quantum labels of the exact rovibrational states are given as both $(v_b;K)$ and $[j,\ell;K]$. The $J = 0$, $p = 0$ level without a quantum label is the excited stretching state. $K = 0$ and $K = 1$ levels are additionally coded in red and green, respectively.

complex has three bound $J = 0$ states. As seen in Fig. 9, only the vibrational ground state falls below the adiabatic isomerisation barrier (saddle point in T-shaped arrangement) at $^{adi}V^0(\theta = 90°) = -5.768$ cm^{-1}. The excited bending and stretching levels are 2.0 and 7.4 cm^{-1} above the ground state, respectively. Note that the stretching state at -0.002 cm^{-1} is very marginally bound and further work will be necessary to establish the nature of this state. Fig. 9 provides additional state labels in terms of $[j, \ell;K]$. This alternative scheme uses the angular momentum of C_2H_2, $\hat{j}$, and the end-over-end angular momentum of the complex, $\hat{\ell}$. Then $\hat{J} = \hat{j} + \hat{\ell}$, whereas the parity is $(-1)^{j+\ell}$. For the z-axis along the Jacobi **R** vector (see Fig. 6), $\ell_z = 0$ also holds. In terms of j and ℓ, the quantum label reads $[j, \ell;K]$ for a given J and parity p. The correlation between the labels $(v_b;K)$ and $[j,\ell;K]$ can be established by means of the angular momentum coupling rules and related Clebsch–Gordan coefficients.[107] For $j = 0$, we obtain $[0,J;0]$. For $j = 1$, there are three possible level groups: $[1,J+1;0]$ (e levels), $[1,J-1;1]$ (e levels), and $[1,J;1]$ (f levels). For the parity labelling in terms of e and f states, see ref. 108.

The energy level structure in Fig. 9 shows several features which are indicative of an early onset of a free-rotor energy pattern. The $J = 0$ energy of the bending vibration is below the energy of the complex rotating with $K = 1$. This is contrary to the situation in ordinary linear molecules, where the first bending $J = 0$ level, $2v_{lin}^0$, is above the first $K = 1$ level, v_{lin}^1. The splitting of the $K = 1$ levels of 0.37 cm^{-1} for $J = 1$ is large, in particular in relation to the very low energy scales in the present complex, and indicates a very strong rotation–vibration interaction. To clarify these points in more detail, we compare the exact energies $E^{(J,p)}$ with the adiabatic energies $^{adi}\varepsilon^{(K)}$ for $J = 1$ in Fig. 10. For $p = 1$, the adiabatic $K = 0$ and $K = 1$ components are prominently pushed apart by Coriolis interaction. The global

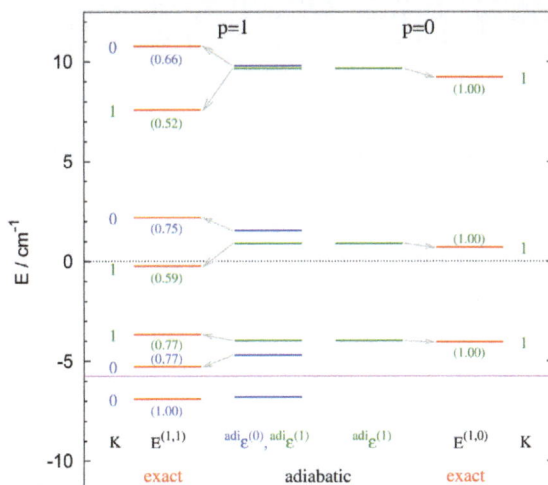

Fig. 10 Full-dimensional accurate level energies $E(J,p)$ and adiabatic level energies $^{adi}\varepsilon^{(K)}$ for $J = 1$. The horizontal line at -5.768 cm^{-1} indicates the position of the ground-state adiabatic isomerisation barrier (T-shape barrier). Arrows connect $^{adi}\varepsilon^{(K)}$ levels with the exact eigenstates $E(J,p)$ (red horizontal bars) originating from them. The quantity $^{(J)}P_K$ is given in parentheses. Adiabatic $K = 0$ and $K = 1$ levels are additionally coded in blue and green, respectively.

effect of this strong Coriolis coupling is clearly visible in Fig. 8(a): without this coupling the $j = 1, K = 1$, e and f levels are degenerate. The levels of the $j = 1, K = 1$, e stack are, however, systematically pushed upwards by Coriolis coupling with the $j = 1, K = 0$, e levels of the same J, such that the latter are compressed downward. As a result the level energies in the $j = 0, K = 0$ stack increase more rapidly leading to a cross-over at $J = 5$. Energy levels exhibiting Coriolis-type resonances are identified with the help of the probability $^{(J)}P_K$ that the wave function takes a certain K value. Coriolis mixing becomes more distinct as v_b increases. This is seen through the increase of the separations between the resulting exact levels and the decrease of the $^{(J)}P_K$ values with increasing v_b in Fig. 10.

The effect of Coriolis coupling appears marginal for the lowest $J = 1$ state in Fig. 10. This is consistent with Fig. 9, which shows that this level lies below the isomerisation barrier $^{adi}V^0(\theta = 90°)$ for $J = 0$ and $J = 1$. The vibrational ground state for $J > 2$ is, however, already above this barrier, such that it may also experience Coriolis mixing. This perturbation became visible when we fitted the rotational excitation of the vibrational ground state with the two-parameter and three-parameter effective term formulas

$$^{(2)}E_0(J) = {}^{(2)}B_0 J(J + 1) - {}^{(2)}D_0 J^2(J + 1)^2, \tag{5}$$

$$^{(3)}E_0(J) = {}^{(3)}B_0 J(J + 1) - {}^{(3)}D_0 J^2(J + 1)^2 + {}^{(3)}H_0 J^3(J + 1)^3. \tag{6}$$

Including rotational transitions up to $J = 5$, we obtain $[^{(2)}B_0, {}^{(2)}D] = [0.2466, 0.0005]$ and $[^{(3)}B_0, {}^{(3)}D] = [0.2451, 0.0003]$ with $^{(3)}H = 3 \times 10^{-6}$, where the standard deviation of the fit was $\sigma = 3 \times 10^{-3}$ and 2×10^{-4}, respectively (all values given

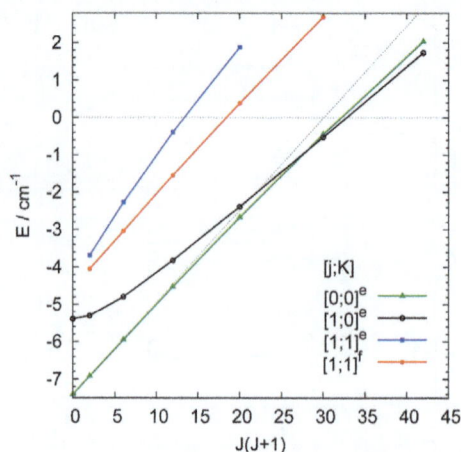

Fig. 11 Rotational energy of C_2H_2–He as a function of $J(J + 1)$. The dotted line stands for $E_0 = {}^{(2)}B_0J(J + 1)$.

in cm^{-1}). The sensitivity of B_0 on the number of fitting parameters and the large centrifugal D constant are indications of a rotational perturbation affecting the vibrational ground state. This can be seen in Fig. 11, which shows the importance of the centrifugal distortion constant D in the description of the rotational excitation in the vibrational ground state. The strong curvature seen for the levels $[1;0]^e$ (black line) and $[1;1]^e$ (blue line) results from Coriolis coupling between these two states.

In the free-rotor limit (potential-free situation), the level energy is approximated by

$$E_{j,\ell} = B\ell(\ell + 1) + bj(j + 1), \tag{7}$$

where B and b are the rotational constants of the complex and the monomer, respectively, and $b > B$. In this limit the bending $[1, 1;0]$ state and the $K = 1$ level $[1, 1;1]^f$ would have the same energy of $2(B + b)$. The $K = 1$ level $[1, 0;1]^e$ is then higher in energy by $2b$ than the ground vibrational state $[0, 0;0]$. The splitting of the $K = 1$ levels $[1, 0;1]^e$ and $[1, 1;1]^f$ would be $2B$, with the f state above the e state. In qualitative terms, the rovibrational structure of Fig. 9 features some of these properties. In quantitative terms, however, there are differences arising primarily from the very prominent Coriolis rotation–vibration interaction, such as the ordering of the e and f states with $K = 1$, which is reversed with respect to the potential-free situation.

6 Conclusions

In conclusion, we have applied impulsive alignment to build rotational wavepackets in complexes between acetylene and helium atoms. The phase of the wavepacket, represented by the degree of alignment, was measured as a function of time using Coulomb explosion. A Fourier transform revealed a rich spectrum of transitions between bound rovibrational eigenstates of the C_2H_2–He complex, which is in excellent agreement with theoretical predictions. The experimental verification of essentially all rotational eigenstates up to the dissociation threshold of the complex substantiated a detailed theoretical assessment of the level structure. Providing access to a large set of rotational states even for molecular systems with near zero dipole moment, as exemplified by the C_2H_2–He complex, is a particular strength of the impulsive alignment technique.

The C_2H_2–He complex is a highly delocalised system. This is the result of the small mass of helium in combination with a small angular barrier on the potential energy surface. The effective angular barrier is lowered further due to strong stretch-bend coupling, such that only two rovibrational eigenstates are below it. All states have a high probability in the barrier region. States above the barrier experience strong Coriolis coupling. In this respect the C_2H_2–He complex differs from linear complexes with a typical linear-molecule rovibrational structure, such as HCN–Ar,[102] and T-shaped complexes with a slightly asymmetric prolate top structure such as CO_2–He[103,109] and $C_2H_2^+$–Ar,[110] whose rotational levels can be well described in terms of effective rotational parameters. The present system exhibits some similarities with the complex CO^+–He,[111] which was found to exhibit an early change of rovibrational patterns due to a very low barrier to linearity.

These findings enable the extension of the impulsive alignment method to other molecular complexes and to clusters comprising multiple atoms. In the case of molecule-helium complexes, by increasing the number of helium atoms it will be possible to gain new insight into incipient superfluid effects. Because the impulsive alignment method does not require the molecule to possess a permanent electric dipole moment it can be applied to a wide range of molecules, including homonuclear diatomics and homonuclear atomic clusters. Incipient superfluid effects for these types of systems have not been explored previously. This methodology also offers the prospect of studying small rare gas clusters or homonuclear metal clusters, testing van der Waals forces and chemical interactions, respectively.

Provided the detection of alignment is successful for these systems impulsive alignment should then decisively facilitate the spectral analysis through the control of the constitution of the wavepacket by adjusting the pump laser power, making it possible to access information on rotational levels at higher energies. Equivalently, for low pump powers it is possible to selectively excite transitions from the lowest quantum states, thereby facilitating the interpretation of congested spectral data. All of this is possible in combination with a very cold molecular beam, enabling the formation of weakly bound species and an initial population of only the lowest quantum levels.

Acknowledgements

The authors wish to thank STFC for access to the Artemis facility at the Rutherford Appleton Laboratory and the University of Leicester for funding to support the studentship for GG associated with this project. We would like to thank S. Hook, P. Rice, N. Rodrigues and S. Thornton for technical support during the experiment. KVH kindly acknowledges funding by STFC (seed corn fund for experiments using 4th generation light sources) and the Leverhulme Trust (F/00212/AH). LGML acknowledges financial support from the Mexican Consejo Nacional de Ciencia y Tecnología (CONACYT) Scholarship number 310668, ID 215334. RSM would like to thank the Royal Society for a University Research Fellowship (UF100047), ML and MM acknowledge computational resources funded through ANR grant ANR-08-BLAN-0146-01. KHMB and JT acknowledge financial support by the Deutsche Forschungsgemeinschaft through SFB 652.

References

1 T. Seideman, *J. Chem. Phys.*, 1995, **103**, 7887–7896.

2 B. Friedrich and D. Herschbach, *Phys. Rev. Lett.*, 1995, **74**, 4623–4626.

3 J. J. Larsen, H. Sakai, C. P. Safvan, I. Wendt-Larsen and H. Stapelfeldt, *J. Phys. Chem.*, 1999, **111**, 7774–7781.

4 F. Rosca-Pruna and M. J. J. Vrakking, *Phys. Rev. Lett.*, 2001, **87**, 153902.

5 M. Tsubouchi, B. J. Whitaker, L. Wang, H. Kohguchi and T. Suzuki, *Phys. Rev. Lett.*, 2001, **86**, 4500.

6 H. Stapelfeldt and T. Seideman, *Rev. Mod. Phys.*, 2003, **75**, 543–557.

7 T. Seideman and E. Hamilton, *Adv. At., Mol., Opt. Phys.*, 2006, **52**, 289–329.

8 C. Vallance, *Phys. Chem. Chem. Phys.*, 2011, **13**, 14427–14441.

9 M. Machholm and N. Henriksen, *Phys. Rev. Lett.*, 2001, **87**, 193001.

10 J. P. Heritage, T. K. Gustafson and C. H. Lin, *Phys. Rev. Lett.*, 1975, **34**, 1299–1302.

11 P. M. Felker, *J. Chem. Phys.*, 1992, **96**, 7844–7857.

12 M. Morgen, W. Price, L. Hunziker, P. Ludowise, M. Blackwell and Y. Chen, *Chem. Phys. Lett.*, 1993, **209**, 1–9.

13 T. S. Rose, W. L. Wilson, G. Wäckerle and M. D. Fayer, *J. Chem. Phys.*, 1987, **86**, 5370–5391.

14 V. V. Matylitsky, W. Jarzeba, C. Riehn and B. Brutschy, *J. Raman Spectros.*, 2002, **33**, 877–883.

15 P. W. Joireman, L. L. Connell, S. M. Ohline and P. M. Felker, *J. Phys. Chem.*, 1991, **95**, 4935–4939.

16 V. A. Venturo, P. M. Maxton and P. M. Felker, *J. Phys. Chem.*, 1992, **96**, 5234–5237.

17 V. A. Venturo, P. M. Maxton, B. F. Henson and P. M. Felker, *J. Chem. Phys.*, 1992, **96**, 7855–7858.

18 M. Cho, M. Du, N. F. Scherer, G. R. Fleming and S. Mukamel, *J. Chem. Phys.*, 1993, **99**, 2410–2428.

19 P. W. Dooley, I. V. Litvinyuk, K. F. Lee, D. M. Rayner, M. Spanner, D. M. Villeneuve and P. B. Corkum, *Phys. Rev. A*, 2003, **68**, 023406.

20 V. Renard, M. Renard, S. Guérin, Y. T. Pashayan, B. Lavorel, O. Faucher and H.-R. Jauslin, *Phys. Rev. Lett.*, 2003, **90**, 153601.

21 E. Hamilton, T. Seideman, T. Ejdrup, M. D. Poulsen, C. Z. Bisgaard, S. S. Viftrup and H. Stapelfeldt, *Phys. Rev. A*, 2005, **72**, 043402.

22 E. Péronne, M. D. Poulsen, C. Z. Bisgaard, H. Stapelfeldt and T. Seideman, *Phys. Rev. Lett.*, 2003, **91**, 043003.

23 J. G. Underwood, B. J. Sussman and A. Stolow, *Phys. Rev. Lett.*, 2005, **94**, 143002.

24 K. F. Lee, D. M. Villeneuve, P. B. Corkum, A. Stolow and J. G. Underwood, *Phys. Rev. Lett.*, 2006, **97**, 173001.

25 A. Rouzée, E. Hertz, B. Lavorel and O. Faucher, *J. Phys. B: At., Mol. Opt. Phys.*, 2008, **41**, 074002.

26 V. Kumarappan, L. Holmegaard, C. Martiny, C. B. Madsen, T. K. Kjeldsen, S. S. Viftrup, L. B. Madsen and H. Stapelfeldt, *Phys. Rev. Lett.*, 2008, **100**, 093006.

27 L. Holmegaard, J. L. Hansen, L. Kalhøj, S. L. Kragh, H. Stapelfeldt, F. Filsinger, J. Küpper, G. Meijer, D. Dimitrovski, M. Abu-Samha, *et al.*, *Nat. Phys.*, 2010, **6**, 428–432.

28 J. Itatani, J. Levesque, D. Zeidler, H. Niikura, H. Pépin, J.-C. Kieffer, P. B. Corkum and D. M. Villeneuve, *Nature*, 2004, **432**, 867–871.

29 K. Miyazaki, M. Kaku, G. Miyaji, A. Abdurrouf and F. H. M. Faisal, *Phys. Rev. Lett.*, 2005, **95**, 243903.

30 T. Kanai, S. Minemoto and H. Sakai, *Nature*, 2005, **435**, 470–474.

31 M. Meckel, D. Comtois, D. Zeidler, A. Staudte, D. Pavičić, H. C. Bandulet, H. Pépin, J. C. Kieffer, R. Dörner, D. M. Villeneuve and P. B. Corkum, *Science*, 2008, **320**, 1478–1481.

32 J. Küpper, S. Stern, L. Holmegaard, F. Filsinger, A. Rouzée, A. Rudenko, P. Johnsson, A. V. Martin, M. Adolph, A. Aquila, S. C. V. Bajt, A. Barty, C. Bostedt, J. Bozek, C. Caleman, R. Coffee, N. Coppola, T. Delmas, S. Epp, B. Erk, L. Foucar, T. Gorkhover, L. Gumprecht, A. Hartmann, R. Hartmann, G. Hauser, P. Holl, A. Hömke, N. Kimmel, F. Krasniqi, K.-U. Kühnel, J. Maurer, M. Messerschmidt, R. Moshammer, C. Reich, B. Rudek, R. Santra, I. Schlichting, C. Schmidt, S. Schorb, J. Schulz, H. Soltau, H. Spence, J. C. D. Starodub, L. Strüder, J. Thøgersen, J. Vrakking, M. J. G. Weidenspointner, T. A. White, C. Wunderer, G. Meijer, J. Ullrich, H. Stapelfeldt, D. Rolles and H. N. Chapman, *Phys. Rev. Lett.*, 2014, **112**, 083002.

33 J.-M. Hartmann, C. Boulet, T. Vieillard, F. Chaussard, F. Billard, O. Faucher and B. Lavorel, *J. Chem. Phys.*, 2013, **139**, 024306.

34 W. A. de Heer, *Rev. Mod. Phys.*, 1993, **65**, 611–676.

35 T. P. Martin, *Phys. Rep.*, 1996, **273**, 199–241.

36 F. Baletto and R. Ferrando, *Rev. Mod. Phys.*, 2005, **77**, 371.

37 E. Roduner, *Chem. Soc. Rev.*, 2006, **35**, 583–592.

38 Y. Xia, Y. Xiong, B. Lim and S. Skrabalak, *Angew. Chem., Int. Ed.*, 2008, **48**, 60–103.

39 D. V. Talapin, J.-S. Lee, M. V. Kovalenko and E. V. Shevchenko, *Chem. Rev.*, 2009, **110**, 389–458.

40 E. Becker, *Laser Part. Beams*, 1989, **7**, 743–753.

41 R. Campargue, *Atomic and Molecular Beams: The State of the Art 2000*, Springer, 2001.

42 J. H. Parks, S. Pollack and W. Hill, *J. Chem. Phys.*, 1994, **101**, 6666–6685.

43 M. Maier-Borst, D. B. Cameron, M. Rokni and J. H. Parks, *Phys. Rev. A*, 1999, **59**, R3162.

44 J. W. Hager, *Rapid Commun. Mass Spectrom.*, 2002, **16**, 512–526.

45 J. T. Lau, J. Rittmann, V. Zamudio-Bayer, M. Vogel, K. Hirsch, P. Klar, F. Lofink, T. Möller and B. V. Issendorff, *Phys. Rev. Lett.*, 2008, **101**, 153401.

46 J. Higgins, C. Callegari, J. Reho, F. Stienkemeier, W. Ernst, K. Lehmann, M. Gutowski and G. Scoles, *Science*, 1996, **273**, 629.

47 U. Buck and F. Huisken, *Chem. Rev.*, 2000, **100**, 3863–3890.

48 F. Stienkemeier, J. Higgins, W. E. Ernst and G. Scoles, *Phys. Rev. Lett.*, 1995, **74**, 3592–3595.

49 K. von Haeften, T. Laarmann, H. Wabnitz, T. Möller and K. Fink, *J. Phys. Chem. A*, 2011, **25**, 7316–7326.

50 O. Echt, K. Sattler and E. Recknagel, *Phys. Rev. Lett.*, 1981, **47**, 1121–1124.

51 W. Knight, K. Clemenger, W. de Heer, W. Saunders, M. Chou and M. Cohen, *Phys. Rev. Lett.*, 1984, **52**, 2141–2143.

52 H. W. Kroto, J. R. Heath, S. C. O'Brien, R. F. Curl and R. E. Smalley, *Nature*, 1985, **318**, 162–163.

53 W. A. de Heer, P. Milani and A. Chatelain, *Phys. Rev. Lett.*, 1990, **65**, 488.

54 J. Bucher, D. Douglass and L. Bloomfield, *Phys. Rev. Lett.*, 1991, **66**, 3052.

55 M. B. Knickelbein, *Phys. Rev. Lett.*, 2001, **86**, 5255–5257.

56 V. Kresin, *Phys. Rev. B: Condens. Matter Mater. Phys.*, 1989, **39**, 3042.

57 R. Moro, R. Rabinovitch, C. Xia and V. V. Kresin, *Phys. Rev. Lett.*, 2006, **97**, 123401.

58 X. Xu, S. Yin, R. Moro, A. Liang, J. Bowlan and W. A. de Heer, *Phys. Rev. B: Condens. Matter Mater. Phys.*, 2007, **75**, 085429.

59 J. Farges, M. De Feraudy, B. Raoult and G. Torchet, *J. Chem. Phys.*, 1986, **84**, 3491.

60 S. Krückeberg, D. Schooss, M. Maier-Borst and J. H. Parks, *Phys. Rev. Lett.*, 2000, **85**, 4494.

61 M. N. Blom, D. Schooss, J. Stairs and M. M. Kappes, *J. Chem. Phys.*, 2006, **124**, 244308.

62 F. Luo, C. F. Giese and W. R. Gentry, *J. Chem. Phys.*, 1996, **104**, 1151–1154.

63 B. S. Zhao, G. Meijer and W. Schöllkopf, *Science*, 2011, **331**, 892–894.

64 C. Bostedt, E. Eremina, D. Rupp, M. Adolph, H. Thomas, M. Hoener, A. R. B. de Castro, J. Tiggesbäumker, K.-H. Meiwes-Broer, T. Laarmann, H. Wabnitz, E. Plönjes, R. Treusch, J. R. Schneider and T. Möller, *Phys. Rev. Lett.*, 2012, **108**, 093401.

65 T. Gorkhover, M. Adolph, D. Rupp, S. Schorb, S. Epp, B. Erk, L. Foucar, R. Hartmann, N. Kimmel, K.-U. Kühnel, *et al.*, *Phys. Rev. Lett.*, 2012, **108**, 245005.

66 Z. Bačić and R. E. Miller, *J. Phys. Chem.*, 1996, **100**, 12945–12959.

67 D. J. Nesbitt and R. Naaman, *J. Chem. Phys.*, 1989, **91**, 3801.

68 J. P. Toennies and A. F. Vilesov, *Annu. Rev. Phys. Chem.*, 1998, **49**, 1–41.

69 F. Stienkemeier and A. F. Vilesov, *J. Chem. Phys.*, 2001, **115**, 10119.

70 B. S. Dumesh and L. A. Surin, *Phys. Usp.*, 2006, **49**, 1113–1129.

71 F. Stienkemeier and K. K. Lehmann, *J. Phys. B: At., Mol. Opt. Phys.*, 2006, **39**, R127–R166.

72 J. P. Toennies and A. F. Vilesov, *Angew. Chem., Int. Ed.*, 2004, **43**, 2622–2648.

73 J. Küpper and J. Merritt, *Int. Rev. Phys. Chem.*, 2007, **26**, 249–287.

74 M. Barranco, R. Guardiola, S. Hernandez, R. Mayol, J. Navarro and M. Pi, *J. Low Temp. Phys.*, 2006, **142**, 1.

75 R. Moszynski, P. E. S. Wormer and A. van der Avoird, *J. Chem. Phys.*, 1995, **102**, 8385.

76 C. R. Munteanu and B. Fernández, *J. Chem. Phys.*, 2005, **123**, 014309.

77 M. Rezaei, N. Moazzen-Ahmadi, A. R. W. McKellar, B. Fernández and D. Farrelly, *Mol. Phys.*, 2012, **110**, 2743–2750.

78 B. Fernández, C. Henriksen and D. Farrelly, *Mol. Phys.*, 2013, **111**, 1173.

79 D. Pentlehner, J. H. Nielsen, A. Slenczka, K. Mølmer and H. Stapelfeldt, *Phys. Rev. Lett.*, 2013, **110**, 093002.

80 U. Even, J. Jortner, D. Noy, N. Lavie and C. Cossart-Magos, *J. Chem. Phys.*, 2000, **112**, 8068.

81 A. Eppink and D. Parker, *Rev. Sci. Instrum.*, 1997, **68**, 3477.

82 O. Ghafur, A. Rouzée, A. Gijsbertsen, W. K. Siu, S. Stolte and M. J. J. Vrakking, *Nat. Phys.*, 2009, **5**, 289–293.

83 G. Roberts, J. Nixon, J. Lecointre, E. Wrede and J. Verlet, *Rev. Sci. Instrum.*, 2009, **80**, 053104.

84 V. Dribinski, A. Ossadtchi, V. A. Mandelshtam and H. Reisler, *Rev. Sci. Instrum.*, 2002, **73**, 2634–2642.

85 G. Garcia, L. Nahon and I. Powis, *Rev. Sci. Instrum.*, 2004, **75**, 4989.

86 A. Mikaberidze, U. Saalmann and J. M. Rost, *Phys. Rev. Lett.*, 2009, **102**, 128102.

87 A. Przystawik, A. Kickermann, A. Al-Shemmary, S. Düsterer, A. M. Ellis, K. von Haeften, M. Harmand, S. Ramakrishna, H. Redlin, L. Schroedter, M. Schulz, T. Seideman, N. Stojanovic, J. Szekely, F. Tavella, S. Toleikis and T. Laarmann, *Phys. Rev. A*, 2012, **85**, 052503.

88 M. Herman, A. Campargue, M. I. El Idrissi and J. Van der Auwera, *J. Phys. Chem. Ref. Data*, 2003, **32**, 922–1361.

89 G. Herzberg, *Molecular Spectra and Molecular Structure. Volume II: Infrared and Raman Spectra of Polyatomic Molecules*, D. Van Nostrand, 16th edn, 1945.

90 L. Surin, A. Potapov, B. Dumesh, S. Schlemmer, Y. Xu, P. Raston and W. Jäger, *Phys. Rev. Lett.*, 2008, **101**, 233401.

91 J. Tang, Y. Xu, A. R. W. McKellar and W. Jäger, *Science*, 2002, **297**, 2030–2033.

92 M. Hartmann, R. E. Miller, J. P. Toennies and A. F. Vilesov, *Phys. Rev. Lett.*, 1995, **75**, 1566.

93 T. H. Dunning, Jr, *J. Chem. Phys.*, 1989, **90**, 1007.

94 R. A. Kendall, T. H. Dunning, Jr and R. J. Harrison, *J. Chem. Phys.*, 1992, **96**, 6796.

95 H.-J. Werner, P. J. Knowles, G. Knizia, F. R. Manby, M. Schütz, *et al.*, *MOLPRO, version 2012.1, a package of ab initio programs*, 2012, http://www.molpro.net.

96 K. A. Peterson, D. E. Woon and T. H. Dunning Jr, *J. Chem. Phys.*, 1994, **100**, 7410.

97 K. A. Peterson and G. C. McBane, *J. Chem. Phys.*, 2005, **123**, 084314.

98 T. Helgaker, W. Klopper, H. Koch and J. Noga, *J. Chem. Phys.*, 1997, **106**, 9639–9646.

99 B. Fernández, C. Henriksen and D. Farrelly, *Mol. Phys.*, 2013, **111**, 1173–1177.

100 K. T. Tang and J. P. Toennies, *J. Chem. Phys.*, 1984, **80**, 3726.

101 W. H. Press, S. A. Teukolsky, W. T. Vetterling and B. P. Flannery, *Numerical Recipes in FORTRAN. The Art of Scientific Computing*, Cambridge University Press, Cambridge, 2nd edn, 1992.

102 M. Mladenović and Z. Bačić, *J. Chem. Phys.*, 1991, **94**, 4988.

103 M. J. Weida, J. M. Sperhac, D. J. Nesbitt and J. M. Hutson, *J. Chem. Phys.*, 1994, **101**, 8351.

104 M. Mladenović, P. Botschwina, P. Sebald and S. Carter, *Theor. Chem. Acc.*, 1998, **100**, 134.

105 M. Mladenović, *Spectrochim. Acta, Part A*, 2002, **58**, 795.

106 B. P. Winnewisser, *Molecular Spectroscopy: Modern Research*, Academic Press, Orlando, 1985, vol. 3.

107 R. N. Zare, *Angular Momentum*, J. Wiley & Sons, New York, 1988.

108 M. Brown, J. T. Hougen, K.-P. Huber, J. W. C. Johns, I. Kopp, H. Lefebvre-Brion, A. J. Merer, D. A. Ramsay, J. Rostas and R. N. Zare, *J. Mol. Spectrosc.*, 1975, **55**, 500–504.

109 Y. J. Xu and W. Jäger, *J. Mol. Struct.*, 2001, **599**, 211–217.

110 O. Dopfer, R. V. Olkhov, M. Mladenović and P. Botschwina, *J. Chem. Phys.*, 2004, **121**, 1744.

111 M. Mladenović and M. Lewerenz, *Mol. Phys.*, 2013, **111**, 2068–2085.

Faraday Discussions

ROYAL SOCIETY OF CHEMISTRY

PAPER

Capturing interfacial photoelectrochemical dynamics with picosecond time-resolved X-ray photoelectron spectroscopy

Stefan Neppl,[*a] Andrey Shavorskiy,[a] Ioannis Zegkinoglou,[a] Matthew Fraund,[a] Daniel S. Slaughter,[a] Tyler Troy,[a] Michael P. Ziemkiewicz,[a] Musahid Ahmed,[a] Sheraz Gul,[bc] Bruce Rude,[a] Jin Z. Zhang,[c] Anton S. Tremsin,[d] Per-Anders Glans,[b] Yi-Sheng Liu,[b] Cheng Hao Wu,[ef] Jinghua Guo,[b] Miquel Salmeron,[e] Hendrik Bluhm[a] and Oliver Gessner[*a]

Received 8th March 2014, Accepted 2nd May 2014

DOI: 10.1039/c4fd00036f

Time-resolved core-level spectroscopy using laser pulses to initiate and short X-ray pulses to trace photoinduced processes has the unique potential to provide electronic state- and atomic site-specific insight into fundamental electron dynamics in complex systems. Time-domain studies using transient X-ray absorption and emission techniques have proven extremely valuable to investigate electronic and structural dynamics in isolated and solvated molecules. Here, we describe the implementation of a picosecond time-resolved X-ray photoelectron spectroscopy (TRXPS) technique at the Advanced Light Source (ALS) and its application to monitor photoinduced electron dynamics at the technologically pertinent interface formed by N3 dye molecules anchored to nanoporous ZnO. Indications for a dynamical chemical shift of the Ru3d photoemission line originating from the N3 metal centre are observed ~30 ps after resonant HOMO–LUMO excitation with a visible laser pump pulse. The transient changes in the TRXPS spectra are accompanied by a characteristic surface photovoltage (SPV) response of the ZnO substrate on a pico- to nanosecond time scale. The interplay between the two phenomena is discussed in the context of possible electronic relaxation and recombination pathways that lead to the neutralisation of the transiently oxidised dye after ultrafast electron injection. A detailed account of the experimental technique is given including an analysis of the chemical modification of the nano-structured ZnO

[a]Chemical Sciences Division, Lawrence Berkeley National Laboratory, Berkeley, California, USA. E-mail: sneppl@lbl.gov; ogessner@lbl.gov

[b]Advanced Light Source, Lawrence Berkeley National Laboratory, Berkeley, California, USA

[c]Department of Chemistry and Biochemistry, University of California Santa Cruz, Santa Cruz, California, USA

[d]Space Sciences Laboratory, University of California Berkeley, Berkeley, California, USA

[e]Materials Sciences Division, Lawrence Berkeley National Laboratory, Berkeley, California, USA

[f]Department of Chemistry, University of California Berkeley, Berkeley, California, USA

substate during extended periods of solution-based dye sensitisation and its relevance for studies using surface-sensitive spectroscopy techniques.

Introduction and motivation

A detailed understanding of charge transfer across molecule–semiconductor junctions is a prerequisite for the successful implementation of many envisioned renewable energy technologies. It is essential for electronic processes underlying novel solar photovoltaics, solar-to-fuel conversion schemes and energy storage technologies. It further plays a central role in the field of molecular electronics. In particular, dye-sensitised solar cells have attracted much attention as a promising low-cost alternative to silicon-based photovoltaic devices.[1] A key process in these photoelectrochemical cells is the optically induced electronic excitation of dye molecules, followed by charge separation through electron injection into a semiconductor. Significant ultrafast spectroscopy efforts have been devoted to gaining a detailed picture of the electron–hole separation and free charge carrier generation dynamics. The vast majority of time-domain studies are based on short optical, infrared, and near-ultraviolet (UV) laser pulses for both triggering and probing the time evolution of the excited electronic states by transient reflectivity, absorption and fluorescence.[2-8] Despite remarkable success in identifying the characteristic time scales that govern the interfacial processes, all-optical methods can only access valence electron orbitals, which are generally distributed over many different atomic centres. This delocalised character of the spectroscopic response often renders a detailed interpretation of pump–probe signals ambiguous, especially in the case of complex composite interfaces.

In principle, a deeper insight may be gained by using probe pulses with photon energies in the X-ray spectral range in order to access element-specific information through the involvement of inner-shell electrons. Extremely short X-ray pulses with durations of <100 attoseconds can nowadays be generated using table-top high-harmonic generation (HHG) in noble gases.[9] This technique enables time-domain studies with unparalleled temporal resolution and further benefits from the intrinsic synchronisation between the HHG driving laser pulses and the generated X-rays. However, for spectroscopic applications HHG currently provides sufficient photon flux only within a rather limited range of photon energies.[10,11]

Free-electron laser (FEL) facilities provide intense sub-100 fs X-ray pulses with tunable photon energies ranging up to 20 keV.[12-14] Currently operational X-ray FEL facilities, however, typically deliver X-rays with extremely high pulse energies but at comparably low repetition rates (<120 Hz), which can present challenges for photoemission spectroscopy experiments on non-replenishing condensed phase samples.[15]

Stationary energy-domain X-ray techniques, such as core-hole clock spectroscopy, have been applied with the goal to probe photoinduced femtosecond electron dynamics in molecule–semiconductor systems from the perspective of individual atomic sites.[16] However, these experiments are not equivalent to time-domain studies involving optical excitations since the charge transfer process is initiated by a resonant core-level to LUMO transition that perturbs the energy-level alignment at the interface more strongly than an optically induced

HOMO-to-LUMO transition.[17] In addition, the range of dynamic processes that can be addressed with this technique are dictated by the core-hole lifetime (typically a few femtoseconds) – which limits the general applicability of this approach to monitor interfacial electron dynamics that proceed on time scales spanning many orders of magnitude.

Many important catalytic and photochemical reactions include steps that proceed on pico-to-nanosecond time scales.[2,18-23] The intrinsically pulsed X-ray radiation produced at third-generation synchrotron radiation light sources provides an opportunity for real-time studies of the associated dynamics with pulse repetition rates in the MHz regime. These experiments benefit from very stable synchrotron beam characteristics in combination with a time resolution reaching tens of picoseconds – in principle only limited by the length of the electron bunches in the storage ring.

Over the past decade, picosecond time-resolved X-ray absorption spectroscopy has been implemented at a number of synchrotron radiation facilities.[21,22,24,25] The technique is now used extensively to uncover dynamics in isolated or solvated molecules, revealing important information about intra-molecular energy-, spin-, and charge-redistribution.[24] It has also been successfully employed to monitor laser-induced phase transitions of bulk materials.[26] Using electron-beam slicing techniques, the time resolution of these experiments can be further improved to ~150 fs[27,28]—albeit at the expense of a significantly reduced X-ray flux.

Compared to these achievements, the application of time-resolved X-ray spectroscopy to surfaces and interfaces is still in a very early stage. Some pioneering studies have employed transient X-ray absorption spectroscopy to monitor photoinduced dynamics at molecule–semiconductor interfaces in colloidal systems.[25,29] For applications using extended solid-state surfaces covered by molecular monolayers, time-resolved X-ray photoelectron spectroscopy (TRXPS) is a particularly suitable technique that promises to gain real-time insight into interfacial electron dynamics due to its intrinsically high surface sensitivity. Depth specificity and tunability on the order of single monolayers can be achieved by variation of the photoelectron kinetic energy and the measurement geometry.[30] So far, synchrotron- and HHG-based TRXPS has mostly been used to study transient surface photo-voltage (SPV) phenomena at clean or oxidised semiconductor surfaces.[31-36] The group of Kapteyn and Murnane applied TRXPS with photon energies below 50 eV to monitor optically induced changes in the valence electronic structure of oxygen molecules adsorbed on a Pt(111) surface.[37] Recently, Heinzmann and co-workers used femtosecond XUV pulses from a HHG source centred near ~95 eV to study light-induced dynamics in an iodo-phenylphenol self-assembled monolayer on Si(100).[38] The dynamics probed by inner-shell photoionisation of the iodine tail group are marked by a complex time evolution extending over many picoseconds.

Here we describe the implementation of TRXPS at the Advanced Light Source (ALS) with a time resolution of 70 ps (full-width-at-half-maximum, FWHM) and a usable photon energy range of ~70 eV to 1.5 keV, which grants access to the K-shells of all first row elements and the L and M core levels of many transition metals relevant for photocatalytic applications. The experiment combines the use of a high-repetition rate pump laser with time-stamping electron detection. The potential of the technique to provide atomic-site specific real-time access to complex interfacial electron dynamics is demonstrated using the prototypical molecule–semiconductor system of N3 dye molecules (bis(isothiocyanato)bis(2,2'-bipyridyl-4,4'-dicarboxylato)-ruthenium(II), Fig. 1A) adsorbed on a film of nano-structured ZnO. Photoinduced variations in the Ru3d inner-shell binding ener-gies are presented, which constitute a sensitive probe for the valence electron dynamics in the vicinity of the metal centre of the laser-excited dye molecules. Correlations between the intramolecular and interfacial electron dynamics and the macroscopic surface photovoltage response of the semiconductor support are discussed. The novel technique provides high quality picosecond TRXPS spectra across pump–probe time delays spanning hundreds of nanoseconds within less than one hour of data acquisition time.

Experimental details

Fig. 1 illustrates the key idea of the TRXPS experiment. A visible laser "pump" pulse promotes electrons from the highest occupied molecular orbital (HOMO) of the N3 chromophore to its lowest unoccupied molecular orbital (LUMO), which is aligned with the conduction band (CB) of the ZnO substrate. The dynamic evolution of the excited-state electronic structure is monitored by recording core-level photoelectron spectra of the N3 molecules as a function of the time delay between the laser excitation and the ionising X-ray "probe" pulse. In the following, we refer to positive (negative) time delays when the optical pump pulse arrives at the sample before (after) the X-ray probe pulse. Since the binding energies of the probed energy levels are sensitive to the local charge densities in the vicinity of the core holes, the local temporal variations of the excited valence charges due to electron injection and recombination processes may be reflected in corresponding transient chemical shifts of the core-level binding energies. The N3/ZnO system constitutes a particularly interesting target for TRXPS, since a comprehensive picture of the competition between different dynamic channels that enhance or limit the interfacial charge injection/recombination has not yet emerged.[6,20,23,39,40]

Sample preparation and characterisation

ZnO nanoparticles with an average diameter of 16 nm were synthesised according to a procedure described by Bauer *et al.* using zinc acetate dihydrate as a precursor.[23] About 0.20 ml of the resulting colloidal nanoparticle-ethanol

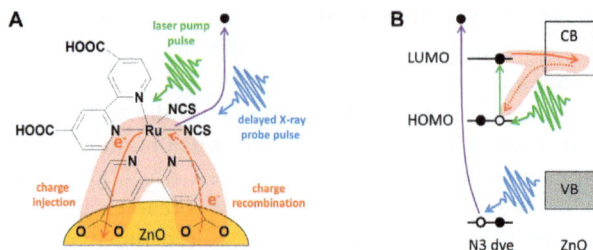

Fig. 1 Illustration of the application of TRXPS to study electron dynamics at the proto-typical dye-semiconductor interface N3/ZnO. The relevant processes are shown sche-matically (A) in real-space and (B) in the energy domain.

suspension is spin-coated on a 4 inch p-doped Si(100) wafer. Subsequently, the samples are dried at 60 °C and sintered in air at 380 °C for 30 minutes to generate a nanoporous film. Dye adsorption was performed by immersing the ozone/UV-cleaned ZnO films in a 0.2 mM solution of N3 (Solaronix) in absolute ethanol. Sensitisation times ranging from 2 hours to 6 days have been applied to test their impact on the film composition and the surface-sensitive TRXPS experiments. The resulting dye-sensitised ZnO films were thoroughly rinsed with ethanol to remove excess physisorbed molecules and blown dry in a stream of gaseous nitrogen. Samples were kept in the dark and under nitrogen atmosphere prior to loading them into the vacuum chamber (base pressure $\sim 10^{-5}$ mbar) for charac-terisation with stationary XPS and the TRXPS measurements. Non-sensitised ZnO films on Si and stainless steel supports have been prepared for reference purposes.

During all XPS and TRXPS experiments, the sample has been scanned with respect to the incident X-ray and laser beam to avoid beam damage of the adsorbed dye molecules. The scan speed was chosen in accordance with X-ray damage tests and optical fluorescence decay measurements.[41]

Whereas the interaction of N3 deposited from solution or by electrospray techniques onto single-crystalline and nanostructured TiO_2 has been extensively characterised with stationary XPS,[42-44] a similar comprehensive understanding of the N3 adsorption on ZnO substrates is still lacking. Compared to optical spec-troscopy, sample preparation for TRXPS experiments requires additional considerations. The XPS sampling depth in semiconductors is typically <4 nm for electron kinetic energies <600 eV,[45] which makes TRXPS an intrinsically surface-sensitive technique. However, most N3-sensitisation protocols have been devel-oped and tested using more established all-optical spectroscopies, which mainly sample bulk properties of the dye-sensitised semiconductor electrodes.

Hagfeldt and co-workers reported the formation of dye–Zn^{2+} complexes for ZnO films immersed in solutions of Ru-based chromophores.[46,47] Molecules bonded in these aggregates were found to be unable to efficiently inject electrons into the semiconductor conduction band.[48] The growth of dye–Zn^{2+} complexes involves dissolution of Zn–ions from the ZnO surface, which is driven by the protons in the carboxylic acid ligands.[47] This mechanism is supported by a recent XAS study, in which an exchange of zinc atoms between ZnO nanoparticles and H_2-protoporphyrins featuring carboxylic acid ligands has been observed, even-tually leading to metallisation of the molecules.[49]

Fig. 2A shows a stationary XPS spectrum of a nanoporous ZnO electrode that has been exposed to the N3 solution for 2 hours. Only peaks expected from the chemical composition of the N3 molecules and the ZnO substrate are observed. The energy axis was calibrated by setting the binding energy of the $Au4f_{7/2}$ line from a gold sample to 84.0 eV. The C1s region is dominated by a photoemission peak at $\sim$285 eV, which is ascribed to carbon atoms in the pyridine ligands of N3 with much smaller (unresolved) contributions from the carbon atoms in the two thiocyanate (NCS) groups and the $3d_{3/2}$ component of the Ru3d photoemission doublet.[43,50] The C1s emission from the carboxylic acid groups gives rise to an additional peak at $\sim$288.5 eV.

The $Ru3d_{5/2}$ photoelectrons from the N3 metal centre is observed at $\sim$280.6 eV binding energy. The complete Ru3d doublet with a spin-orbit splitting of 4.2 eV[51] is indicated as red shaded area in Fig. 2A. The shape and intensities of all peaks in

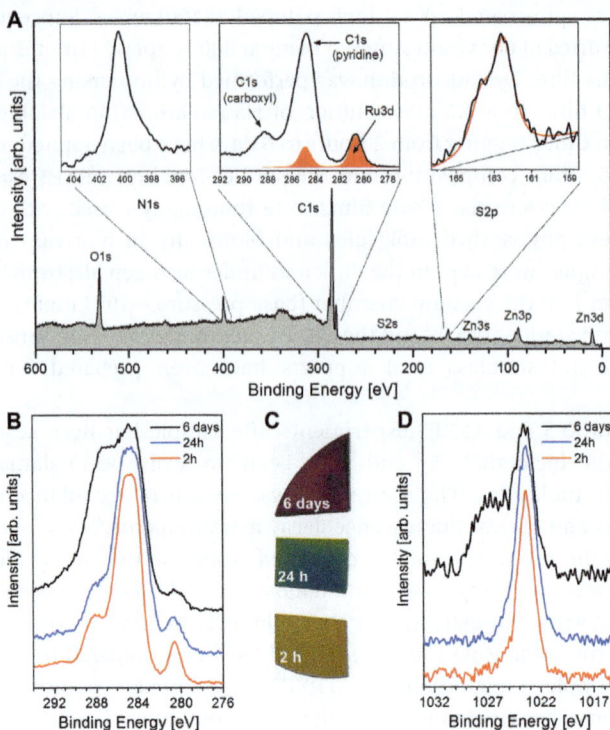

Fig. 2 Characterisation of N3-sensitised ZnO electrodes with stationary XPS. (A) XPS spectrum of an N3/ZnO electrode recorded with $h\nu = 850$ eV photon energy after 2 hours of sensitisation. Detailed views of the N1s, C1s and S2p core-level emission lines are shown as insets. (B) Comparison of XPS spectra in the region of the C1s and Ru3d photoemission lines obtained from ZnO electrodes after 2 hours, 24 hours and 6 days of N3 sensitisation. The spectra were recorded with $h\nu = 490$ eV to enhance the surface sensitivity. (C) Photographs of electrodes with different sensitisation times (from bottom to top): 2 hours, 24 hours, and 6 days. (D) Zn2p$_{3/2}$ photoemission lines of N3-sensitised electrodes produced with different sensitisation times ($h\nu = 1250$ eV).

the C1s region are very similar to spectra reported for N3 deposited *in situ* on TiO$_2$ using electrospray techniques in an ultra-high vacuum environment.[42,50] Furthermore, the N1s emission line has a pronounced tail towards lower binding energies, indicating the presence of nitrogen atoms in at least two different chemical environments – compatible with the molecular structure of N3 (Fig. 1A).[43,44,50]

It is generally accepted that chemisorption of N3 on metal–oxide surfaces involves one or more carboxylic acid groups, while the participation of the thio-cyanate ligands in the surface bonding is still a matter of debate and may be influenced by the presence of co-adsorbates.[52,53] An involvement of the N3 thio-cyanate groups in the N3–TiO$_2$ binding was proposed based on the observation of two chemically shifted components in the S2p photoemission spectrum.[44,50] In contrast, the S2p spectrum shown in Fig. 2A can be satisfactorily fitted by a single sulfur 2p doublet with the well-known spin-orbit-splitting of 1.2 eV[51] (red solid line in Fig. 2A). Moreover, its binding energy is rather close to the S2p line position previously assigned to thiocyanate groups not in contact with the substrate.[44,50] This favours a bonding motif for N3/ZnO in which only the carboxylic acid groups exhibit a strong interaction (deprotonation) with the surface, in agreement with a recent DFT study on this system.[41]

We find that samples prepared with 2 hours sensitisation time exhibit the highest homogeneity in surface chemical composition, with <8% variations in the total C1s XPS intensity and less than <5% variations in the relative peak ratios when scanning the X-ray beam across centimetre scale sample areas. These electrodes showed no obvious change in colour compared to the bare ZnO substrate, indicating that dye-adsorption was mainly restricted to the outer surface region of the ZnO. For longer sensitisation times, the increased uptake of N3 into the nanoporous film becomes visible as a characteristic coloration of the electrodes (Fig. 2C). However, as demonstrated in Fig. 2B and D, longer sensitisation times are also accompanied by notable changes in the C1s and $Zn2p_{3/2}$ XPS spectra, respectively: for the 6 day sample, a clear broadening of all peaks in the C1s/Ru3d region in combination with a significantly enhanced emission at ~288.5 eV and a reduction and shift/broadening of the $Ru3d_{5/2}$ emission line is observed. In addition, the $Zn2p_{3/2}$ substrate emission broadens and an additional broad peak appears at ~4 eV higher binding energy compared to the 2 hour sample, indicative of a significant chemical modification of the ZnO surface.[46,54] First indications for this chemical transformation are already discernable after 24 hours of sensitisation, which has been used in previous optical spectroscopy studies: the C1s, Ru3d and Zn2p photo-lines are broadened by ~20% compared to the 2 hour reference sample, and the $Ru3d_{5/2}$ peak exhibits a notable asymmetry and shift to higher binding energies.

Since etching and dissolution of ZnO is governed by the local dye concentration and effective exposure time,[48,54] $N3–Zn^{2+}$ agglomerates are expected to form more readily on the outer surface of the ZnO electrode which is in direct contact with the dye bath during sensitisation. In contrast, dye uptake into the bulk of the electrode is governed by slow diffusion of the molecules through the nanoporous ZnO network. Therefore, a sensitisation time of 24 hours might be sufficiently short to inhibit excessive $dye–Zn^{2+}$ complex formation within the bulk of the ZnO electrode, which is compatible with bulk sensitive optical probes. For the surface-sensitive TRXPS experiments reported herein, however, we limit the sensitisation times to 2 hours, which can be considered a reasonable compromise between sufficient dye chemisorption at the outer surface of the ZnO electrode and minimal chemical modification of the surface of the ZnO film.

Setup and X-ray/laser synchronisation

Generally, pump lasers used in synchrotron-based laser-pump/X-ray probe experiments operate at significantly lower pulse repetition rates than the X-ray light sources. This mismatch renders the traditional pump–probe approach that is marked by a single, tunable delay between the pump- and probe-pulses challenging. The laser is frequently synchronised with a specific electron bunch, employing special, low repetition rate storage ring fill patterns and/or mechanical and electronic gates to discriminate all signals but those associated with a single X-ray pulse at the desired pump–probe time delay.[55-57] This technique often results in significantly reduced data collection efficiency compared to static X-ray measurements since only a fraction of the available X-ray flux is used in the experiment. Additionally, it exposes the sample to significantly more, potentially damaging X-ray flux than required to acquire the X-ray spectra. However, the scheme provides the best possible time resolution, ultimately defined by the synchrotron pulse duration. In a different approach, fast time-resolving detectors are used to register and sort the data produced by all X-ray pulses in time, thereby 'passively' monitoring the pump–probe delay.[34,58] The time resolution in this time-tagging approach is usually limited to a few nanoseconds and dominated by the characteristics of the spectrometers and detectors.

The TRXPS setup presented here is schematically shown in Fig. 3. It combines aspects of both detection principles resulting in enhanced data acquisition efficiency, reduced target damage, and an X-ray pulse duration limited time resolution. The experiments are carried out at beamline 11.0.2 of the ALS using the ambient-pressure photoemission spectroscopy end-station.[59] A hemispherical electron analyser at this end-station is equipped with a two-dimensional delay line[60] (DLD) that records the arrival times and hit positions of all photoelectrons reaching the detector. The hit position along the dispersive plane of the analyser provides the kinetic energy of every electron while the time of arrival marks the pump–probe delay. The kinetic energy window that can be imaged in this snapshot mode is ~13% of the analyser pass energy. A pass energy of 150 eV was found to provide a good compromise between spectral resolution and total kinetic energy range for the TRXPS experiments described herein.

The overall accuracy for resolving different arrival times of electrons with the same kinetic energy is limited by their time-of-flight spread caused by different trajectories through the lens system and hemisphere of the XPS analyser. In particular, the latter is generally a function of the pass energy and lens mode.[34,58] For the setup shown in Fig. 3, we recently demonstrated a time resolution of 780 ps.[61]

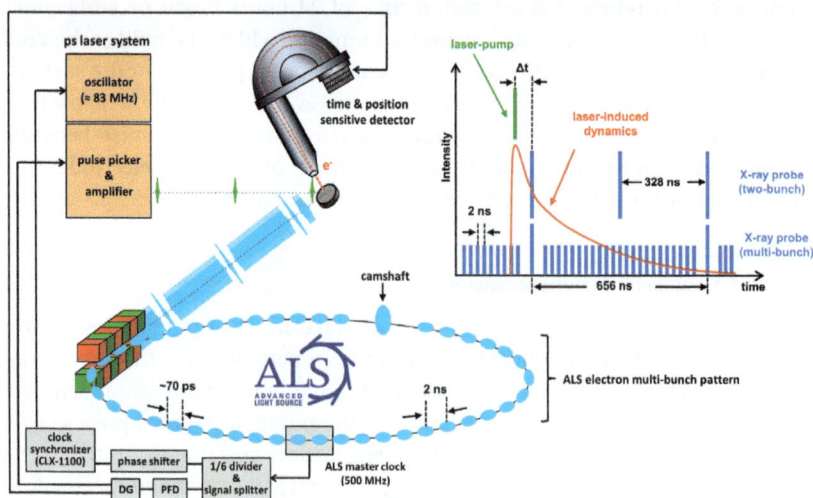

Fig. 3 Setup for picosecond time-resolved laser-pump/X-ray photoemission probe spectroscopy at the Advanced Light Source (ALS). Key components of the electronic laser/X-ray synchronisation scheme are indicated (DG = delay generator; PFD = programmable frequency divider). See text for details.

The laser system employed (Time Bandwidth Products DUETTO) produces 10 ps pulses with selectable wavelengths (1064 nm, 532 nm, 355 nm) and adjustable repetition rates ranging from single shots up to ~8 MHz. All experiments discussed in the following made use of the 532 nm output. Synchronisation between the laser pulses and the X-ray pulse train is achieved using the 500 MHz RF master clock of the ALS as a reference signal (Fig. 3). This master clock has a fixed phase relationship with the RF signal that defines the electron bunch pattern in the storage ring and is therefore intrinsically synchronised with the emitted X-ray pulse train. The laser oscillator operates at 83 MHz and can be locked to the reference signal with a phase-lock-loop provided by a commercial clock synchroniser unit (Time Bandwidth Products, CLX-1100). A 1/6 frequency divider is used to convert the 500 MHz master clock signal into an 83 MHz reference signal. A pulse picker routes a fraction of the oscillator pulses into the laser amplifier. The trigger signal for the pulse picker is derived from the same reference as the synchronisation signal for the oscillator and processed by a programmable frequency divider (PFD) to select pulses at a user-defined repetition rate. This signal also triggers the data acquisition of the DLD.

The time delay between the laser pulses and the ALS X-ray pulse train is coarsely adjusted in steps of 12 ns (oscillator pulse spacing) by using a delay generator (DG) installed in the pulse picker synchronisation branch. The laser/X-ray time delay is fine-tuned *via* an electronic phase shifter in steps of 5 ps (total range 5 ns). The optical pump pulse remains synchronised to individual X-ray pulses in the ALS pulse train with picosecond precision over periods up to several days.

During multi-bunch operation of the ALS, 296 of the available 328 storage ring fill pattern buckets contain electron bunches separated by ~2 ns and with a round-trip time of ~656 ns. Additionally, a more intense, isolated electron bunch ('camshaft') is placed in a *ca.* 60–70 ns wide gap of the fill pattern. As illustrated in Fig. 3, this enables efficient laser-pump multiple X-ray probe measurements that simultaneously monitor pico- to nanosecond time scales. In the ALS two-bunch mode, only one pair of intense electron bunches with a pulse separation of 328 ns is injected in the ring. The resultant time structure of the emitted X-rays is especially advantageous for the characterisation of dynamics proceeding on a picosecond time scale.

Laser and X-ray beam characterisation

The ability to achieve and monitor spatiotemporal overlap between the X-ray and laser beams at the location of the sample is an important prerequisite for any pump–probe experiment. In the TRXPS setup described herein, spatial overlap is accomplished by recording knife-edge scans of the X-ray and the laser beam using the sharply defined sensitive area of a photodiode. The diode is mounted on the same motorized *XYZ*-manipulator as the samples and scanned across both X-ray and laser beams. Typical results obtained for the horizontal and vertical scan directions in the sample plane are summarized in Fig. 4A and B, respectively. The laser and X-ray beams are combined using an annular mirror inside the vacuum chamber and directed to the sample surface at an angle of incidence of ~45°. The laser beam pointing is fine-tuned by a piezo-actuator (Newport Picomotor) controlled mirror mount outside the vacuum chamber. The centroids of the X-ray

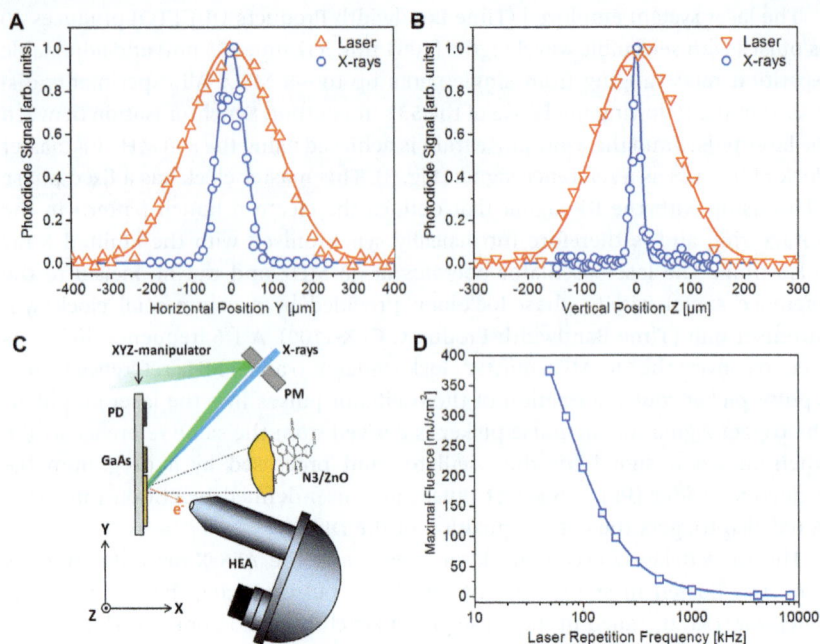

Fig. 4 Spatial profiles of the laser and X-ray beams, which are overlapped at the sample surface inside the vacuum chamber. Horizontal (A) and vertical (B) beam profiles are obtained by scanning a large-area photodiode mounted on the sample holder through the beams. Solid lines are Gaussian fits to the data. The beam paths employing a perforated mirror (PM) as a beam combiner and the measurement geometry with the photodiode (PD) and the hemispherical electron analyser (HEA) are illustrated in (C). The maximum fluence that can be delivered to the sample by the employed laser system based on the spot size determined in (A) and (B) is shown in (D) as a function of the laser repetition rate.

and laser beam profiles are overlapped in both dimensions by iterative scanning and tuning of the piezo-actuator. Spatial pump–probe overlap can be adjusted with a precision better than 20 μm within minutes.

The measurements in Fig. 4 indicate X-ray spot size diameters <100 μm (FWHM) in both the vertical and horizontal direction. In order to ensure uniform optical excitation density across the area probed by the X-ray beam, the laser spot size on the target was deliberately chosen to be significantly larger than the X-ray spot size, as illustrated by the $\sim$250 $\times$ 180 μm^2 (FWHM) optical beam profile at the sample surface. Note that all values include the projection-induced broadening in the horizontal beam dimensions caused by the $\sim$45° incidence angle. For the given spot size, the available laser output power at 532 nm is sufficiently high to enable pump–probe experiments with repetition rates of up to 1 MHz at pump fluences of several mJ cm^{-2} (Fig. 4D).

Temporal overlap and time resolution

The coarse relative timing of the laser pulses with respect to the ALS X-ray pulse train is adjusted to within ±0.3 ns by displaying the time traces of an avalanche photodiode mounted on the sample holder on a 1 GHz bandwidth oscilloscope. The timing stability is continuously monitored by observing the leakage through one of the laser steering mirrors on an oscilloscope that is triggered by the ALS bunch-marker clock. The pump–probe temporal overlap is tuned in the picosecond range by performing an X-ray/laser cross-correlation using the transient surface photo-voltage (SPV) effect of a semiconductor sample.

The position of the Fermi level at semiconductor surfaces can be pinned by surface- and impurity-states near the middle of the band gap.[62] The resultant charge transfer to the surface gives rise to a space charge region (typically ~100 nm deep) causing a characteristic downward (upward) bending of all energy levels for p-type (n-type) semiconductors toward the surface. Laser illumination with photon energies exceeding the band gap generates electron–hole pairs that partly compensate the electric field in the space charge layer. This leads to a transient flattening of all energy bands near the surface depending on the recombination dynamics of the electron–hole pairs inside the material. The SPV effect therefore manifests itself as a rigid, time-dependent shift of the entire photoelectron spectrum to lower (higher) binding energies for p-doped (n-doped) materials. The build-up of the transient SPV response is governed by the diffusion of photo-generated minority carriers in the space charge region to the semiconductor surface,[62] which usually proceeds on a sub-picosecond time scale.[32] The SPV effect therefore represents a suitable benchmark to measure the time resolution in TRXPS experiments.

We have investigated the SPV dynamics in thin (~3 μm) ZnO films grown on p-doped Si(100), which also served as substrates for the N3 dye molecules in the TRXPS experiments discussed later in this article. To identify the dominant time scale in the SPV response from the ZnO/Si samples, TRXPS in the time-tagging only mode, i.e., without any synchronisation between the laser and X-ray pulses, have been performed. As demonstrated in Fig. 5A, this allows the efficient measurement of transients in a time window only limited by the laser repetition rate (~102 kHz in this experiment). The measured deviation of the Zn3d binding energy relative to a laser-off reference spectrum is plotted as a function of the time delay between the optical pump pulse and the 850 eV X-ray probe pulse for different laser fluences. The sudden drop in the Zn3d binding energy observed around $\Delta t = 0$ signals the arrival of the laser pulse at the sample surface. The subsequent relaxation of the SPV shift lasts for several microseconds and is strongly influenced by the pump fluence. For laser fluences >20 μJ cm^{-2}, the Zn3d line position does not recover to the laser-off situation within the laser pump period (~10 μs). This accumulated SPV contribution is observed as a constant binding energy offset at negative pump–probe delays. We note that no SPV effect was observed for laser fluences up to 10 mJ cm^{-2} in ZnO films deposited on stainless steel substrates (green curve in Fig. 5A). Thus, the SPV dynamics probed by Zn3d TRXPS originates exclusively from laser-excited electron–hole pairs generated in the space charge region at the ZnO/Si interface. This also explains the transient drop (instead of an increase) in the observed Zn3d binding energy, which is not expected from the n-doped ZnO film but from the p-doped Si substrate.[31] The possibility to monitor electronic dynamics at buried interfaces by TRXPS has been reported before by the Wurth group for BaF$_2$/Si interfaces.[63] All traces shown in Fig. 5A have been acquired during ALS two-bunch operation with a data acquisition time of 15 min per trace.

The picosecond time-resolved drop of the effective Zn3d binding energy near $\Delta t = 0$ is presented in Fig. 5B. For this measurement, the laser was synchronised

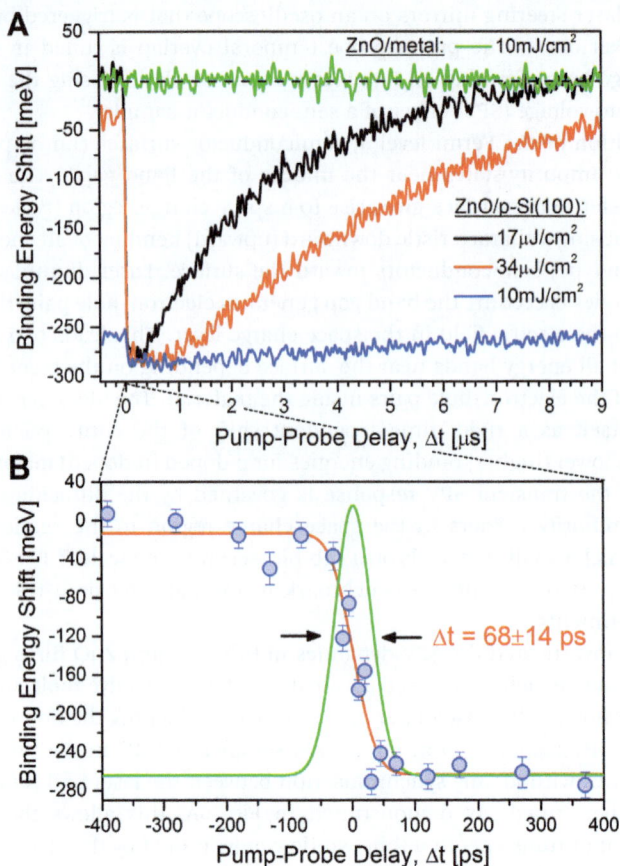

Fig. 5 Transient surface photo-voltage (SPV) dynamics at the ZnO/p-Si(100) interface probed by TRXPS. (A) Evolution of the Zn3d line position as a function of pump–probe delay for different laser pulse fluences. The data was acquired using the time tagging method without active laser/X-ray synchronisation. (B) SPV shifts extracted from pico-second time-resolved photoemission spectra near the temporal overlap between the X-ray-probe and the laser-pump pulse (17 μJ cm^{-2}) after laser/X-ray synchronisation. Fitting to a Gaussian error function (red solid line) yields the time resolution of the TRXPS experiment (~70 ps).

to the X-ray pulse train and the electronic phase shifter of the oscillator timing electronics (Fig. 3) was used to scan the delay of the laser pump pulse with respect to the ALS probe pulse. A fit of the SPV response to a Gaussian error function reveals a ~70 ps TRXPS time resolution (FWHM), which is in agreement with the previously reported ALS bunch length.[33]

To benchmark the performance of the TRXPS apparatus compared to existing setups, additional picosecond time-resolved SPV experiments on p-doped GaAs(100) have been carried out. The SPV effect in this material has been extensively studied and is more pronounced compared to other semiconductor materials with band gaps compatible with the pump laser photon energy (2.33 eV). Fig. 6 summarises the change in the Ga3d binding energy as a function of laser-pump/X-ray-probe (450 eV) delay. The data was recorded within 50 min acquisition time (integration time ~80 s per pump probe step) during ALS two-bunch operating mode with a laser repetition rate of ~500 kHz and a laser fluence of 10 nJ cm^{-2}, which is safely below the photo-chemical decomposition threshold of GaAs.[64]

The rapid build-up of the SPV effect is followed by a much slower decay of the SPV shift proceeding on a nanosecond time scale. The relaxation dynamics are governed by two different channels: a fast process leading to a reduction of the SPV shift by 40% within approximately 200 ps after laser excitation, and a second component responsible for a slower decrease over several nanoseconds. A bi-exponential decay model fitted to the data (red line in Fig. 6) yields time constants of 80 ± 30 ps and 2.3 ± 0.4 ns, respectively, for the fast and slowly decaying components. Such biphasic relaxation behaviour of the SPV effect has previously been observed in GaAs and was attributed to tunnelling and thermionic recombination of the photo-excited charge carriers on fast and slow time scales, respectively.[36,65,66] We note that photoelectron spectra recorded before the laser pulse reaches the surface are still shifted by ca. 50 meV to lower binding energies compared to spectra recorded without any laser illumination. The transient SPV effect has therefore not fully decayed within the time scale of the experiment, i.e. the time elapsed between two laser pulses (1.7 μs). This suggests the existence of (at least) a third relaxation channel with a time constant in the microsecond or even millisecond range.[66]

The time-resolved SPV measurements presented above highlight the capability of the TRXPS setup to capture electron–hole recombination dynamics on the pico- to microsecond time scale at semiconductor surfaces and buried interfaces by laser-pump/X-ray probe photoelectron spectroscopy with an overall time resolution only limited by the ALS electron bunch duration.

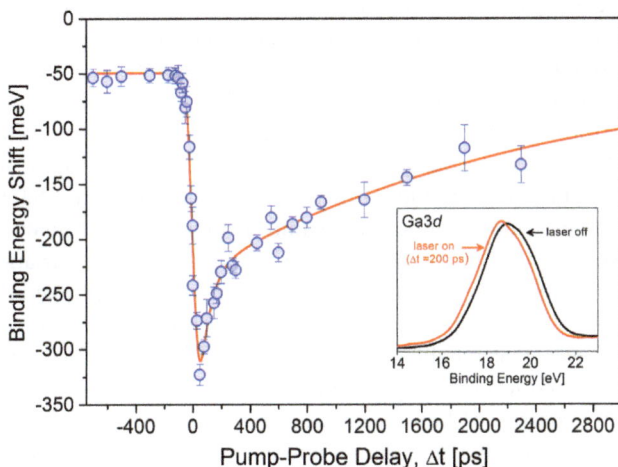

Fig. 6 TRXPS from p-doped GaAs(100). Blue circles show the SPV-induced reduction of the Ga3d core-level binding energy measured relative to the peak position in a laser-off spectrum (inset) as a function of pump–probe delay. The fit (red solid line) describes a step-like binding energy drop followed by a bi-exponential recovery convoluted with the Gaussian instrument response function.

Results

The TRXPS experiments presented here concentrate on the binding energy region comprising the C1s peaks and the Ru3d doublet. This provides, in principle, simultaneous sensitivity to charge dynamics affecting the metal centre and the aromatic ligands of the dye molecule (Fig. 1).

Typical C1s/Ru3d TRXPS data of N3-sensitised films of ZnO nanocrystals obtained during ALS multi- and two-bunch operating mode are summarised in Fig. 7. For these measurements, the laser system was operated at a repetition rate of ~102 kHz, which enables temporal overlap of the laser pulses with one out of 15 camshaft pulses in multi-bunch or one out of 30 camshaft pulses in two-bunch operating mode. The laser fluence was set to 10 mJ cm^{-2} and the sample was continuously scanned at a speed of 120 μm s^{-1} to minimise laser and X-ray induced beam damage.

Fig. 7A and B show false-colour representations of the acquired TRXPS spectra. Each slice of the false-colour map along the (horizontal) energy axis corresponds to an XPS spectrum recorded at a well-defined time delay after laser excitation. Projections onto the time axis reveal the characteristic multi- and two-bunch fill pattern of the ALS in Fig. 7A and B, respectively. Note that the individual pulses in the multi-bunch sections are not fully resolved due to the ~5 ns resolution of the time-tagging technique, which was determined by the pass energy and lens mode employed in this particular experiment. It is important to note, however, that this only represents the time resolution for the overlapping signals in the multi-bunch probe regions, whereas signals associated with the isolated camshaft pulses are still recorded with the bunch-length limited time-resolution of 70 ps.

For the two-bunch data set in Fig. 7B, the timing electronics of the setup were adjusted such that the laser pump pulse arrived at the sample surface 30 ps before the second camshaft pulse within the time span recorded by the detector. The relative timing of the laser and X-ray pulses was calibrated by performing an SPV-based cross-correlation immediately before the measurement (see Fig. 5B). XPS spectra recorded ~328 ns before (blue) and 30 ± 5 ps after (red) laser excitation are compared in Fig. 7C. The laser-affected spectrum exhibits a pronounced ~450 meV shift of both the C1s and the Ru3d$_{5/2}$ lines to higher binding energies. This rigid shift of the entire photoelectron spectrum is ascribed to an SPV effect in the substrate. Signatures of changes in the C1s and Ru line shapes, which are expected to contain information on the electron dynamics at the N3/ZnO inter-face, are obviously more subtle and require a more detailed analysis. In the following, we base our investigation of laser-induced line-shape variations in the TRXPS spectra exclusively on data recorded during two-bunch operation, since the multi-bunch TRXPS experiments were partly affected by detector saturation. We note, however, that reliable information on the SPV dynamics could never-theless be extracted from both types of data.

In Fig. 8 we contrast TRXPS spectra acquired at time delays of $\Delta t = 30$ ps (A) and $\Delta t = 30$ ps + 328 ns (C) to a laser-off reference spectrum ($\Delta t = 30$ ps − 328 ns). For accurate comparison of the line shapes, the SPV effect of the laser-on spectra has been corrected by shifting the spectra numerically along the binding energy axis to minimize the squared sum of residuals with the reference spectrum. Whereas the line shape of the spectrum taken at $\Delta t = 30$ ps + 328 ns and the laser-

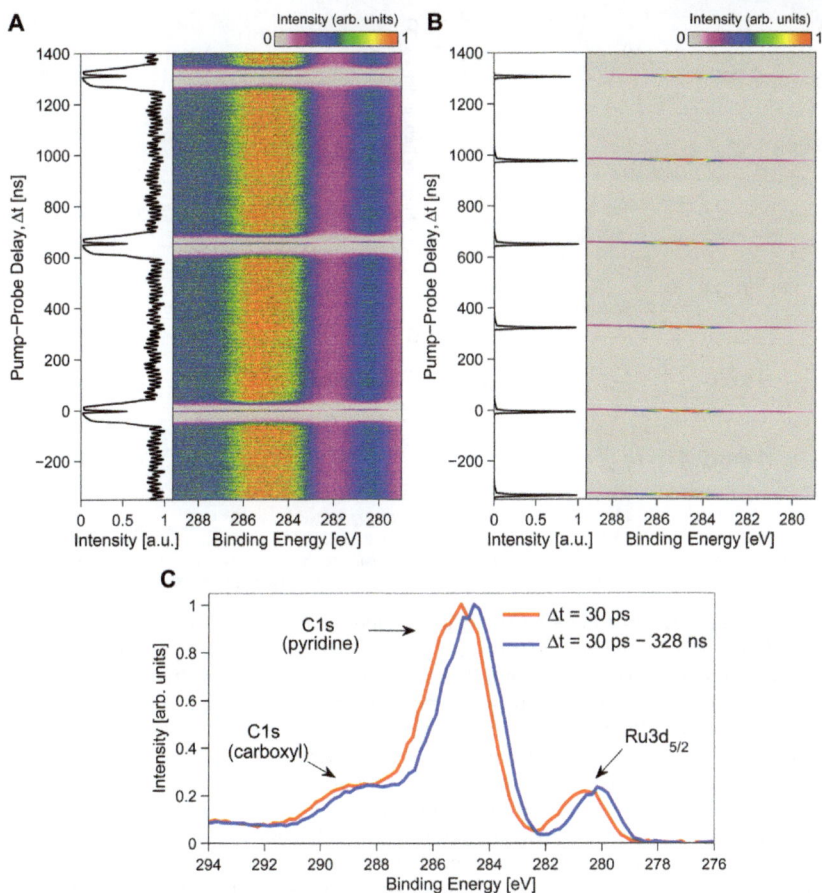

Fig. 7 C1s/Ru3d TRXPS maps acquired from N3/ZnO during the (A) multi- and (B) two-bunch operation modes of the ALS. Photoelectron spectra as a function of laser-pump/X-ray-probe time delay are displayed in a false colour representation – with a projection of the data onto the time axis shown on the left side. The laser pulse excites the sample surface 120 ps before the first camshaft X-ray pulse (A) and 30 ps before the second X-ray pulse (B) in the chosen detection time window. (C) Comparison of photoelectron spectra recorded during ALS two-bunch mode. The spectrum of the camshaft arriving $\Delta t = 30$ ps after the laser pulse at the sample (red) is shifted to higher binding energies with respect to the spectrum of the camshaft that arrives 328 ns before laser excitation (blue).

off reference are indistinguishable within the signal-to-noise of the measurement (Fig. 8C and D), distinct laser-induced changes are observed at $\Delta t = 30$ ps (Fig. 8A and B): intensities near the C1s pyridine and the $Ru3d_{5/2}$ peak centres are reduced, accompanied by intensity enhancements in the regions between peaks. This can be seen more clearly in the difference spectra shown in Fig. 8B and D. At $\Delta t = 30$ ps, there are minima at ~281 eV and ~285 eV with corresponding maxima near ~283 eV and ~287 eV. Compared to this, no pronounced correlations are observed in the spectral difference at $\Delta t = 30$ ps + 328 ns.

Fig. 9 summarises the overall temporal evolution of the SPV response of the N3/ZnO samples. The transient was obtained by combining data sets acquired

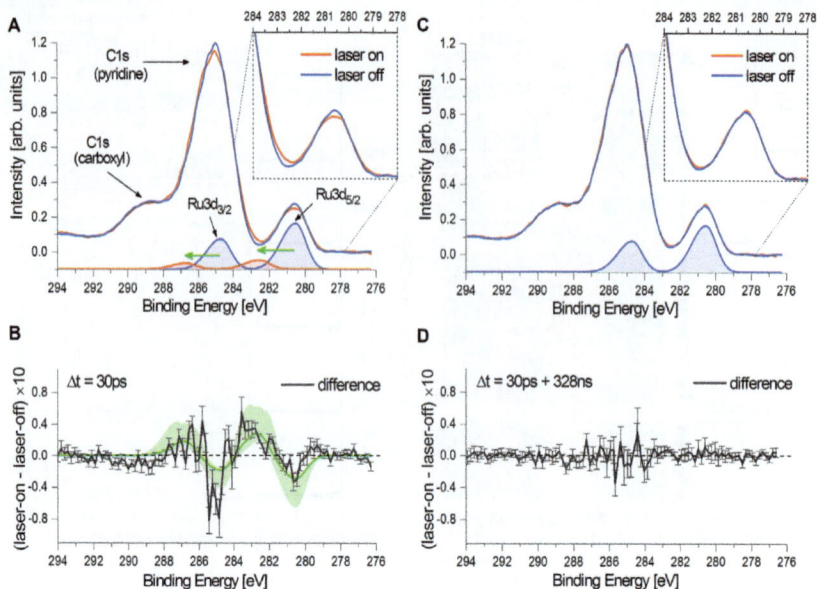

Fig. 8 Evidence for laser-induced TRXPS line-shape variations. (A) Comparison of a laser-off C1s/Ru3d photoemission spectrum ($\Delta t = 30$ ps − 328 ns, blue) to a spectrum recorded at a pump–probe delay of 30 ps (red). The laser-on spectrum was shifted numerically to compensate for the SPV effect. The insets highlight the spectral region near the Ru3d$_{5/2}$ emission line. (B) Difference spectrum obtained by subtracting the laser-off reference from the SPV-corrected laser-on spectrum. The fit result (green solid line) is based on a model that describes the entire laser-on/laser-off difference by a shift ΔBE of the Ru3d doublet in a fraction $F \approx 10\%$ of the probed dye molecules to ~2 eV higher binding energies (indicated by green arrows in (A)). The green shaded area highlights the array of difference spectra predicted by this model for the parameter range $\Delta BE = 1.5$–3 eV and $F = 10$–25%. See text for details. (C) and (D) Show the corresponding analysis of a photo-emission spectrum recorded at a pump–probe delay of $\Delta t = 30$ ps + 328 ns. No systematic differences between the laser-on and laser-off spectra are observed in this case.

during different ALS operating modes and with different data acquisition schemes. The amplitude of the SPV rises to almost 500 meV within the time resolution of the experiments and is followed by a rapid decay to 40% within only a few nanoseconds. However, the complete SPV relaxation involves several hundreds of nanoseconds. Note that the curves have a vertical offset of ~−400 meV due to the transient SPV effect at the ZnO/Si interface discussed above, which does not relax on the time scale of the laser pump period.

Discussion

In N3, photon absorption proceeds via excitation of a singlet metal-to-ligand charge transfer (^{1}MLCT) state. Subsequent depopulation of the initially excited state may occur, in principle, via direct (<100 fs) electron injection into the substrate or via intramolecular relaxation to the metastable ^{3}MLCT triplet state, followed by much slower electron injection (~10–100 ps).[4,7] The relative importance of both channels may depend on the specific dye–semiconductor system and has long been a matter of debate.[4,6,7,20,67] Since the HOMO of N3 is

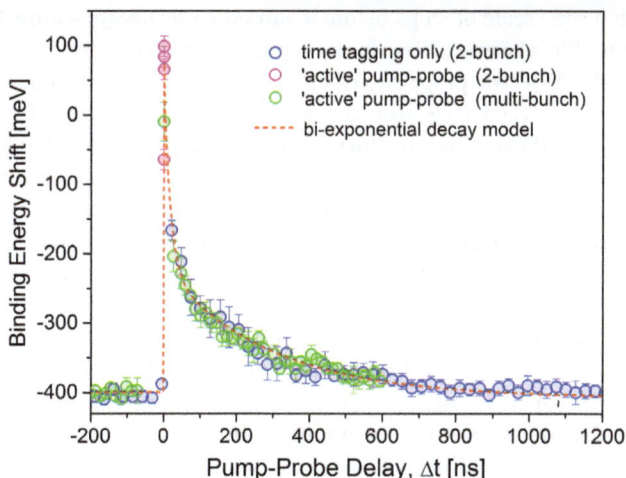

Fig. 9 Time evolution of the N3-induced SPV in the N3/ZnO/Si system. The SPV shift is quantified by minimizing the spectral difference with a laser-off reference spectrum. The change in binding energy is given relative to a spectrum recorded without any laser illumination. Colour coding of the circles refer to results obtained for different ALS operating modes and analyser configurations. The red dashed line is a bi-exponential decay approximation of the observed SPV dynamics.

predominantly located at the metal centre and the thiocyanate groups of the dye, whereas the LUMO resides mostly on the pyridine ligands that bond the molecule to the semiconductor, the ruthenium photoemission lines are expected to carry information of the transient charge redistribution within the dye.[52] In either relaxation channel, the light-induced charge redistribution is intuitively expected to manifest itself in an increase of binding energies of the ruthenium core levels because of the reduced electron density around the metal centre that screens the ruthenium core holes. This effect has indeed been observed for isolated N3 molecules in solution[21] and in a recent femtosecond TRXPS experiment on N3/ZnO at the Linac Coherent Light Source (LCLS).[41]

As illustrated in Fig. 8A and B, the difference spectrum at $\Delta t = 30$ ps can indeed be qualitatively rationalised in terms of a laser-induced shift of the Ru3d doublet by $\Delta BE \sim 2$ eV to higher binding energies for a fraction $F \sim 10\%$ of the N3 molecules. A corresponding fit is indicated as a green solid line. We note that F and ΔBE are the only free parameters in this model function, since the ground-state photoemission line shape can be extracted from the isolated $Ru3d_{5/2}$ emission line in the laser-off spectrum, and the well-known Ru3d spin-orbit splitting and intensity ratio.[51]

While ultrafast optical spectroscopy studies have shown that electron injection in N3/ZnO is dominated by "slow" electron transfer channels with time constants in the range of a few to hundreds of picoseconds (compared to, *e.g.*, $\leq$100 fs in N3/TiO$_2$), the microscopic origin for this behaviour is not fully understood.[8,40,67–69] Two competing descriptions for the charge injection process have been proposed: (i) an intermediate-interface-state model[40,69–72] and (ii) a two-state injection model.[8,39,67,73,74] In (i), optical excitation of the dye results in the formation of an electron–cation complex that temporarily binds the injected electron to the interface on a time scale of <5 ps before it subsequently decays within 10–150 ps to generate mobile charge carriers in the semiconductor conduction band.[8,67,68,74] In this case, the dielectric properties of the substrate are expected to have a major impact on the interfacial electron dynamics.[8] In (ii), unfavourable electronic coupling between the dye and the ZnO substrate is suggested, leading to reduced injection efficiency due to an increased competition with intramolecular relaxation. In this case, the majority of N3 molecules initially excited in the singlet metal-to-ligand charge transfer state ^{1}MLCT undergo relaxation to the metastable ^{3}MLCT triplet state that is expected to inject on a much slower time scale (10–100 ps).[40,69]

Evidently, the two scenarios are characterised by distinctly different spatial distributions of the excited electronic charge during the first tens of picoseconds after excitation. In (i) it would be transferred to the semiconductor but still localised at the interface at the beginning of the injection process, whereas it would initially reside on the dye in scenario (ii). The exact magnitude and time evolution of the TRXPS Ru3d binding energy shift may therefore provide valuable insight into the dominant injection mechanisms. Recent DFT calculations indicate that the two scenarios may lead to Ru3d chemical shifts that differ by approximately 1 eV.[75] Unfortunately, as indicated by the green shaded area in Fig. 8B, the signal-to-noise level in the current measurements does not permit an identification of the transient chemical shift at $\Delta t = 30$ ps within this level of precision. However, improvements of the experiment are expected to enable future investigations that will provide this information. Additional control experiments will also be performed to gauge the potential impact of line broadening effects that are not associated with a chemical shift within the dye molecules.

Apparently, the marked minimum at ~285 eV in the $\Delta t = 30$ ps difference spectrum is not well reproduced by the Ru3d model function described above. This might indicate that transient changes in one or more C1s photo-lines contribute to the differential laser-on/laser-off effect. In this respect it is noteworthy that a recent stationary high-resolution XPS study of N3 demonstrated that the most dominant C1s peak is composed of two different components corresponding to carbon atoms in two different chemical environments within the N3 pyridine groups.[43] The high binding energy component was assigned to the two carbon atoms bonded to the nitrogen atom, whereas the lower binding energy component was ascribed to the three carbon atoms in the pyridine ring that are in closer proximity to the carboxylic acid group. The binding energies of the two components differ by 1 eV. These findings are compatible with the slightly asymmetric C1s pyridine line shape in Fig. 8A and C. The additional differences observed at ~285 eV binding energy may therefore indicate a temporal increase of the C1s binding energies in carbon atoms that are located closer to the carboxylic acid groups than the nitrogen atoms (Fig. 1A). In addition, a potential contribution from the (spectrally unresolved) C1s emission originating from the thiocyanate groups has to be considered.

Even though the electron injection process in N3/ZnO is essentially completed after ~200 ps,[8] one may still expect signatures of the oxidised N3 dyes in the $\Delta t = 30$ ps + 328 ns TRXPS spectrum, unless the dye molecules have been completely reduced by charge recombination processes. The absence of evident correlations in the difference spectrum in Fig. 8D therefore suggests that ~300 ns after laser excitation, the majority of the initially excited N3 molecules have already returned to the neutral ground state by back-electron transfer from the ZnO substrate.

Reported time scales for relaxation processes at dye–semiconductor interfaces range from a few picoseconds to several milliseconds.[2,6,18,23] In this respect the question arises whether the N3/ZnO SPV transient in Fig. 9 contains information on the interfacial charge recombination dynamics. Thus, it is important to identify the exact origin of the SPV effect in this system. For laser fluences >1 mJ cm^{-2}, the SPV effect induced in the ZnO/p-Si support of the N3 molecules is completely saturated on a microsecond time scale (see Fig. 5A),[61] and can therefore not contribute to the nanosecond dynamics apparent in Fig. 9. The saturated SPV effect of the ZnO/p-Si is indeed observed as a constant ~400 meV shift of all TRXPS spectra to higher binding energy when the laser beam is blocked. Therefore, the n-type behaviour of the SPV transient in Fig. 9 must originate from the ZnO nanoparticles and/or the N3 molecules, and might be influenced by electron exchange across the interface.

The direct band gap of ZnO is ~3.3 eV (ref. 3) and significantly exceeds the laser photon energy in our experiment (2.33 eV). In order for direct ZnO photo-excitation to be the origin of the observed SPV dynamics, one would have to invoke excitation into or from band-gap states (a nonlinear response of the semiconductor is unlikely at the applied pump laser fluence). However, this effect can be ruled out since no SPV signature is observed in ZnO/stainless steel samples for laser fluences up to 10 mJ cm^{-2} (green trace in Fig. 5A). Since ZnO films are intrinsically n-doped (due to interstitial Zn atoms or oxygen vacancies)[76] the SPV transient in Fig. 9 may result from electron injection from the dye adsorbate into the ZnO substrate, which would have a similar effect as electron–hole pair creation directly in the substrate by super-bandgap excitation. For the n-type semiconductor ZnO, this would lead to transient lowering of the band edges near the surface (increased binding energies) during the electron injection process, followed by an SPV relaxation (decreasing binding energies) governed by the time scale of the charge recombination processes that neutralise the N3 molecules at the surface. This picture is consistent with the SPV behaviour in Fig. 9. We note that the N3/ZnO SPV relaxation dynamics can only be described satisfactorily using (at least) a bi-exponential decay model, which suggests contributions from recombination times of ~10 ns and ~100 ns (red dashed line in Fig. 9). These time scales are well within the range of recombination times reported in previous optical studies for N3/ZnO.[23]

Future studies will shed more light on the processes and mechanisms underlying the transient changes in the Ru3d/C1s line shapes and the substrate SPV response described herein. In particular, a more complete picture will be achieved by performing TRXPS experiments addressing the O1s, S2p and N1s core levels. Experiments involving the nitrogen atoms will be particularly interesting, since the thiocyanate- and pyridine-related emission lines can be distinguished in the N1s XPS spectrum and the N3 HOMO and LUMO have significantly different amplitudes on the corresponding atomic positions.[52]

Conclusion and outlook

An experimental setup for synchrotron-based picosecond time-resolved X-ray photoelectron spectroscopy (TRXPS) studies is presented. A unique combination of high temporal resolution and data collection efficiency is achieved by implementing both a picosecond laser/X-ray synchronisation scheme and a nanosecond time-resolved time-stamping technique. The potential of the method is demonstrated by monitoring light-induced electronic dynamics at the interface between N3 dye-molecules and a nanostructured ZnO semiconductor substrate. Transient changes in the C1s/Ru3d TRXPS line shapes are observed, which are indicative of a change in oxidation state of the Ru metal centre of the dye molecule 30 ps after laser excitation. Further studies are required to determine a comprehensive picture of the transient valence electron configurations that are reflected in the time-dependent XPS line shapes, and to explore their correlation with the macroscopic surface photovoltage response observed in this system.

Detailed static XPS studies suggest that the N3 sensitisation procedure of ZnO has to be adapted to the specific spectroscopic method by which the dynamics at the dye–semiconductor interface are probed in order to avoid a detrimental impact of chemical modifications.

Although the TRXPS experiments presented here have been performed under vacuum conditions, the ambient-pressure capability of the employed hemispherical analyser and beamline enables TRXPS studies with background pressures up to 10 Torr. We envision that this will enable *in situ* studies of the interplay between charge dynamics and solvent/electrolyte-induced modification of the dye–substrate interaction, which is believed to significantly influence the performance of dye-sensitised solar cells. The TRXPS experiments will be complemented by time-resolved X-ray absorption measurements based on Auger or partial electron yields recorded with the same experimental setup. These measurements will provide additional, site-specific information on the time evolution of the unoccupied valence states of the system. Combined, these methods have the potential to provide comprehensive, atomic site-specific real-time insight into interfacial photovoltaic and photocatalytic reaction dynamics approaching application-like conditions.

Acknowledgements

This work was supported by the U.S. Department of Energy, Office of Basic Energy Sciences, Chemical Sciences, Geosciences and Biosciences Division, through Contract no. DE-AC02-05CH11231. O. G. was supported by the Department of Energy Office of Science Early Career Research Program. J. Z. Z. is grateful for support by the Basic Energy Sciences Division of the US DOE (DE-FG02-ER46232). We appreciate the excellent support provided by the staff of the Advanced Light Source.

References

1 M. Grätzel, *J. Photochem. Photobiol., C*, 2003, **4**, 145–153.
2 Y. Tachibana, S. A. Haque, I. P. Mercer, J. R. Durrant and D. R. Klug, *J. Phys. Chem. B*, 2000, **104**, 1198–1205.
3 S. B. Rana, A. Singh and S. Singh, *Int. J. Nanoelectron. Mater.*, 2013, **6**, 45–57.
4 N. A. Anderson, X. Ai and T. Lian, *J. Phys. Chem. B*, 2003, **107**, 14414–14421.
5 G. Benkö, J. Kallioinen, J. E. I. Korppi-Tommola, A. P. Yartsev and V. Sundström, *J. Am. Chem. Soc.*, 2001, **124**, 489–493.

6 R. Katoh, A. Furube, A. V. Barzykin, H. Arakawa and M. Tachiya, *Coord. Chem. Rev.*, 2004, **248**, 1195–1213.

7 N. A. Anderson and T. Lian, *Annu. Rev. Phys. Chem.*, 2005, **56**, 491–519.

8 H. Němec, J. Rochford, O. Taratula, E. Galoppini, P. Kužel, T. Polívka, A. Yartsev and V. Sundström, *Phys. Rev. Lett.*, 2010, **104**, 197401.

9 E. Goulielmakis, M. Schultze, M. Hofstetter, V. S. Yakovlev, J. Gagnon, M. Uiberacker, A. L. Aquila, E. M. Gullikson, D. T. Attwood, R. Kienberger, F. Krausz and U. Kleineberg, *Science*, 2008, **320**, 1614–1617.

10 E. Magerl, S. Neppl, A. L. Cavalieri, E. M. Bothschafter, M. Stanislawski, T. Uphues, M. Hofstetter, U. Kleineberg, J. V. Barth, D. Menzel, F. Krausz, R. Ernstorfer, R. Kienberger and P. Feulner, *Rev. Sci. Instrum.*, 2011, **82**, 63104.

11 A. D. Shiner, C. Trallero-Herrero, N. Kajumba, H. C. Bandulet, D. Comtois, F. Légaré, M. Giguère, J. C. Kieffer, P. B. Corkum and D. M. Villeneuve, *Phys. Rev. Lett.*, 2009, **103**, 073902.

12 P. Emma, *et al.*, *Nat. Photonics*, 2010, **4**, 641–647.

13 W. Ackermann, *et al.*, *Nat. Photonics*, 2007, **1**, 336–342.

14 T. Ishikawa, *et al.*, *Nat. Photonics*, 2012, **6**, 540–544.

15 A. Pietzsch, A. Föhlisch, M. Beye, M. Deppe, F. Hennies, M. Nagasono, E. Suljoti, W. Wurth, C. Gahl, K. Döbrich and A. Melnikov, *New J. Phys.*, 2008, **10**, 033004.

16 L. Wang, W. Chen and A. T. S. Wee, *Surf. Sci. Rep.*, 2008, **63**, 465–486.

17 M. Weston, A. J. Britton and J. N. O'Shea, *J. Chem. Phys.*, 2011, **134**, 054705.

18 A. Listorti, B. O'Regan and J. R. Durrant, *Chem. Mater.*, 2011, **23**, 3381–3399.

19 J. J. H. Pijpers, R. Ulbricht, S. Derossi, J. N. H. Reek and M. Bonn, *J. Phys. Chem. C*, 2011, **115**, 2578–2584.

20 H. Němec, J. Rochford, O. Taratula, E. Galoppini, P. Kužel, T. Polívka, A. Yartsev and V. Sundström, *Phys. Rev. Lett.*, 2010, **104**, 197401.

21 B. E. Van Kuiken, N. Huse, H. Cho, M. L. Strader, M. S. Lynch, R. W. Schoenlein and M. Khalil, *J. Phys. Chem. Lett.*, 2012, **3**, 1695–1700.

22 M. Saes, C. Bressler, R. Abela, D. Grolimund, S. L. Johnson, P. A. Heimann and M. Chergui, *Phys. Rev. Lett.*, 2003, **90**, 047403.

23 C. Bauer, G. Boschloo, E. Mukhtar and A. Hagfeldt, *J. Phys. Chem. B*, 2001, **105**, 5585–5588.

24 C. Bressler and M. Chergui, *Annu. Rev. Phys. Chem.*, 2010, **61**, 263–282.

25 X. Zhang, G. Smolentsev, J. Guo, K. Attenkofer, C. Kurtz, G. Jennings, J. V. Lockard, A. B. Stickrath and L. X. Chen, *J. Phys. Chem. Lett.*, 2011, **2**, 628–632.

26 A. Cavalleri, H. H. W. Chong, S. Fourmaux, T. E. Glover, P. A. Heimann, J. C. Kieffer, B. S. Mun, H. A. Padmore and R. W. Schoenlein, *Phys. Rev. B: Condens. Matter Mater. Phys.*, 2004, **69**, 153106.

27 R. W. Schoenlein, S. Chattopadhyay, H. H. W. Chong, T. E. Glover, P. A. Heimann, C. V. Shank, A. A. Zholents and M. S. Zolotorev, *Science*, 2000, **287**, 2237–2240.

28 C. Stamm, T. Kachel, N. Pontius, R. Mitzner, T. Quast, K. Holldack, S. Khan, C. Lupulescu, E. F. Aziz, M. Wietstruk, H. A. Durr and W. Eberhardt, *Nat. Mater.*, 2007, **6**, 740–743.

29 M. H. Rittmann-Frank, C. J. Milne, J. Rittmann, M. Reinhard, T. J. Penfold and M. Chergui, *Angew. Chem. Int. Ed.*, 2014, **53**, 5858–5862.

30 S. Hüfner, *Photoelectron Spectroscopy: Principles and Applications*, Springer, 2010.

31 J. P. Long, H. R. Sadeghi, J. C. Rife and M. N. Kabler, *Phys. Rev. Lett.*, 1990, **64**, 1158–1161.

32 P. Siffalovic, M. Drescher and U. Heinzmann, *Europhys. Lett.*, 2002, **60**, 924.

33 T. E. Glover, G. D. Ackermann, A. Belkacem, B. Feinberg, P. A. Heimann, Z. Hussain, H. A. Padmore, C. Ray, R. W. Schoenlein and W. F. Steele, *Nucl. Instrum. Methods Phys. Res., Sect. A*, 2001, **467–468**(2), 1438–1440.

34 T. Gießel, D. Bröcker, P. Schmidt and W. Widdra, *Rev. Sci. Instrum.*, 2003, **74**, 4620–4624.

35 D. Bröcker, T. Gießel and W. Widdra, *Chem. Phys.*, 2004, **299**, 247–251.

36 T. Nakanishi, M. Kamada, T. Nishitani, K. Takahashi, S. D. More and S. Tanaka, *Surf. Rev. Lett.*, 2002, **09**, 1297–1301.

37 M. Bauer, C. Lei, K. Read, R. Tobey, J. Gland, M. M. Murnane and H. C. Kapteyn, *Phys. Rev. Lett.*, 2001, **87**, 025501.

38 H. Dachraoui, M. Michelswirth, P. Siffalovic, P. Bartz, C. Schäfer, B. Schnatwinkel, J. Mattay, W. Pfeiffer, M. Drescher and U. Heinzmann, *Phys. Rev. Lett.*, 2011, **106**, 107401.

39 P. Tiwana, P. Docampo, M. B. Johnston, H. J. Snaith and L. M. Herz, *Acc. Chem. Res.*, 2011, **5**, 5158–5166.

40 N. a. Anderson and T. Lian, *Coord. Chem. Rev.*, 2004, **248**, 1231–1246.

41 K. R. Siefermann, *et al.*, Atomic Scale Perspective of Ultrafast Charge Transfer at a Dye-Semiconductor Interface, submitted (2014).

42 M. Weston, A. J. Britton and J. N. O'Shea, *J. Chem. Phys.*, 2011, **134**, 054705.

43 L. C. Mayor, A. Saywell, G. Magnano, C. J. Satterley, J. Schnadt and J. N. O'Shea, *J. Chem. Phys.*, 2009, **130**, 164704.

44 E. M. J. Johansson, M. Hedlund, H. Siegbahn and H. Rensmo, *J. Phys. Chem. B*, 2005, **109**, 22256–22263.

45 S. Tanuma, C. J. Powell and D. R. Penn, *Surf. Interface Anal.*, 1991, **17**, 927–939.

46 K. Westermark, H. Rensmo, H. Siegbahn, K. Keis, A. Hagfeldt, L. Ojamäe and P. Persson, *J. Phys. Chem. B*, 2002, **106**, 10102–10107.

47 K. Keis, J. Lindgren, S.-E. Lindquist and A. Hagfeldt, *Langmuir*, 2000, **16**, 4688–4694.

48 H. Horiuchi, R. Katoh, K. Hara, M. Yanagida, S. Murata, H. Arakawa and M. Tachiya, *J. Phys. Chem. B*, 2003, **107**, 2570–2574.

49 R. N. González-Moreno, P. L. Cook, I. Zegkinoglou, X. Liu, P. S. Johnson, W. Yang, R. E. Ruther, R. J. Hamers, R. n. Tena-Zaera, F. J. Himpsel, J. E. Ortega and C. Rogero, *J. Phys. Chem. C*, 2011, **115**, 18195–18201.

50 L. C. Mayor, J. Ben Taylor, G. Magnano, A. Rienzo, C. J. Satterley, J. N. O'Shea and J. Schnadt, *J. Chem. Phys.*, 2008, **129**, 114701.

51 J. Chastain, J. F. Moulder and R. C. King, *Handbook of X-ray photoelectron spectroscopy: a reference book of standard spectra for identification and interpretation of XPS data*, Physical Electronics Division, Perkin-Elmer Corporation, 1995.

52 P. Persson and M. J. Lundqvist, *J. Phys. Chem. B*, 2005, **109**, 11918–11924.

53 M. Hahlin, E. M. J. Johansson, R. Schöin, H. Siegbahn and H. k. Rensmo, *J. Phys. Chem. C*, 2011, **115**, 11996–12004.

54 F. Yan, L. Huang, J. Zheng, J. Huang, Z. Lin, F. Huang and M. Wei, *Langmuir*, 2010, **26**, 7153–7156.

55 M. Saes, F. van Mourik, W. Gawelda, M. Kaiser, M. Chergui, C. Bressler, D. Grolimund, R. Abela, T. E. Glover, P. A. Heimann, R. W. Schoenlein, S. L. Johnson, A. M. Lindenberg and R. W. Falcone, *Rev. Sci. Instrum.*, 2004, **75**, 24–30.

56 F. A. Lima, C. J. Milne, D. C. V. Amarasinghe, M. H. Rittmann-Frank, R. M. van der Veen, M. Reinhard, V.-T. Pham, S. Karlsson, S. L. Johnson, D. Grolimund, C. Borca, T. Huthwelker, M. Janousch, F. van Mourik, R. Abela and M. Chergui, *Rev. Sci. Instrum.*, 2011, **82**, 063111.

57 A. M. March, A. Stickrath, G. Doumy, E. P. Kanter, B. Krässig, S. H. Southworth, K. Attenkofer, C. a. Kurtz, L. X. Chen and L. Young, *Rev. Sci. Instrum.*, 2011, **82**, 073110.

58 N. Bergeard, M. G. Silly, D. Krizmancic, C. Chauvet, M. Guzzo, J. P. Ricaud, M. Izquierdo, L. Stebel, P. Pittana, R. Sergo, G. Cautero, G. Dufour, F. Rochet and F. Sirotti, *J. Synchrotron Radiat.*, 2011, **18**, 245–250.

59 H. Bluhm, *J. Electron Spectrosc. Relat. Phenom.*, 2010, **177**, 71–84.

60 A. Oelsner, O. Schmidt, M. Schicketanz, M. Klais, G. Schönhense, V. Mergel, O. Jagutzki and H. Schmidt-Böcking, *Rev. Sci. Instrum.*, 2001, **72**, 3968–3974.

61 A. Shavorskiy, *et al.*, Sub-Nanosecond Time-Resolved Ambient-Pressure X-ray Photoelectron Spectroscopy Setup for Pulsed and Constant Wave X-ray Light Sources, *Rev. Sci. Instrum.*, to be submitted.

62 L. Kronik and Y. Shapira, *Surf. Sci. Rep.*, 1999, **37**, 1–206.

63 A. Pietzsch, A. Föhlisch, F. Hennies, S. Vijayalakshmi and W. Wurth, *Appl. Phys. A: Mater. Sci. Process.*, 2007, **88**, 587–592.

64 M. Kamada, J. Murakami, S. Tanaka, S. D. More, M. Itoh and Y. Fujii, *Surf. Sci.*, 2000, **454–456**, 525–528.

65 S. Tokudomi, J. Azuma, K. Takahashi and M. Kamada, *J. Phys. Soc. Jpn.*, 2007, **76**, 104710.

66 S. Tanaka, S. Dylan Moré, K. Takahashi and M. Kamada, *J. Phys. Soc. Jpn.*, 2003, **72**, 659–663.

67 A. Furube, R. Katoh, K. Hara, S. Murata, H. Arakawa and M. Tachiya, *J. Phys. Chem. B*, 2003, **107**, 4162–4166.

68 A. Furube, R. Katoh, T. Yoshihara, K. Hara, S. Murata, H. Arakawa and M. Tachiya, *J. Phys. Chem. B*, 2004, **108**, 12583–12592.

69 X. Ai, N. A. Anderson, J. Guo and T. Lian, *J. Phys. Chem. B*, 2005, **109**, 7088–7094.

70 G. Benkö, J. Kallioinen, P. Myllyperkiö, F. Trif, J. E. I. Korppi-Tommola, A. P. Yartsev and V. Sundström, *J. Phys. Chem. B*, 2004, **108**, 2862–2867.

71 J. Kallioinen, G. Benkö, V. Sundström, J. E. I. Korppi-Tommola and A. P. Yartsev, *J. Phys. Chem. B*, 2002, **106**, 4396–4404.

72 M. Pellnor, P. Myllyperkiö, J. Korppi-Tommola, A. Yartsev and V. Sundström, *Chem. Phys. Lett.*, 2008, **462**, 205–208.

73 R. Katoh, A. Furube, T. Yoshihara, K. Hara, G. Fujihashi, S. Takano, S. Murata, H. Arakawa and M. Tachiya, *J. Phys. Chem. B*, 2004, **108**, 4818–4822.

74 C. Strothkämper, A. Bartelt, P. Sippel, T. Hannappel, R. Schütz and R. Eichberger, *J. Phys. Chem. C*, 2013, **117**, 17901–17908.

75 C. D. Pemmaraju and D. Pendergast, private communication.

76 S. B. Zhang, S. H. Wei and A. Zunger, *Phys. Rev. B: Condens. Matter*, 2001, **63**, 075205.

Faraday Discussions

PAPER

How fast can a Peierls–Mott insulator be melted?

C. Sohrt, A. Stange, M. Bauer and K. Rossnagel*

Received 17th March 2014, Accepted 6th May 2014

DOI: 10.1039/c4fd00042k

Time- and angle-resolved extreme ultraviolet photoemission spectroscopy is used to directly determine the momentum-dependent electronic structure dynamics in the layered Peierls–Mott insulators $1T$-TaS$_2$ and $1T$-TaSe$_2$ on the sub-300 fs time scale. Extracted spectroscopic order parameters display a global two-time-scale dynamics indicating a quasi-instantaneous loss of the electronic orders and a subsequent coherent suppression of the lattice distortion on a time scale related to the frequency of the charge-density–wave amplitude mode. After one half-cycle of coherent amplitude-mode vibration, a crossover state between insulator and metal with partially filled-in and partially closed Mott and Peierls gaps is reached. The results are discussed within the wider context of electronic order quenching in complex materials.

I. Introduction

Complex materials, in which electron–electron and electron–lattice interactions are strong, exhibit some of the most intriguing phenomena in the quantum world, including (high-temperature) superconductivity, metal–insulator transitions, and ordering phenomena involving charge, orbital, spin, and lattice degrees of freedom.[1] Most of our knowledge and microscopic understanding of complex materials and their phenomena has been obtained from experimental and theoretical techniques that are applied in or near equilibrium. The non-equilibrium dynamics of complex materials, by contrast, is much less understood, yet powerful techniques and impressive results are emerging.[2–13]

The most important experimental approach to the non-equilibrium regime is by techniques that combine two ultrashort pulses: one for pump excitation of the material and the other as a time-delayed probe. For pumping, near-infrared laser pulses are most commonly used. For probing, photon pulses in the THz to hard X-ray spectral range and electron pulses with energies of a few 10 keV are now almost routinely applied. Thus, many powerful and complementary techniques have been implemented, including time-resolved optical and photoemission spectroscopy[14,15] as well as time-resolved X-ray and electron diffraction.[16]

Institut für Experimentelle und Angewandte Physik, Christian-Albrechts-Universität zu Kiel, D–24098 Kiel, Germany

With effective temporal resolutions of a few 10 fs, these techniques offer a new perspective on complex materials because they can resolve electron and lattice dynamics at or close to the fundamental time scales of electronic processes and ionic motion in solids. The direct dynamical information can be used to temporally dissect complex phases and identify—*via* temporal discrimination—the dominant degrees of freedom, order parameters, or interactions.[17] In particular, due to their distinct characteristic time scales, it is almost straightforward to separate electron–electron from electron–phonon interaction effects and coherent from incoherent phonon processes. These capabilities can provide novel insights not only into complex phases but also into thermally accessible phase transitions.[3–5,9,12] Moreover, pump–probe techniques can be used to investigate whether and to what extent phase transitions can be coherently controlled[2,6,11] and novel or "hidden" phases can be created that are not thermally accessible.[7,13]

While there are now various experimental techniques available and many different complex materials have been studied, a few central recurrent questions have emerged in the field, namely: how fast and how are symmetry-broken states melted and restored? What is the nature of transient non-equilibrium states? And what can non-equilibrium dynamics teach us about the nature and origin of complex equilibrium phases?

In this work, we address these questions for a paradigmatic example, the layered Peierls–Mott insulators $1T$-TaX_2 (X = S or Se), in which a large-amplitude charge-density wave (CDW) brings on a Mott transition, *i.e.*, electron–phonon and electron–electron interaction are simultaneously strong.[18–20] The materials serve as a good reference because the ground state is sufficiently complex and generic, rather well understood, and its experimental signatures are strong. $1T$-TaS_2, in particular, has been extensively studied with time-resolved methods.[5,13,17,21–27] Yet, the fundamental question about the speed and mechanism of the quenching of the combined Peierls–Mott-state after femtosecond laser excitation has not been investigated in depth.

Here, we use time- and angle-resolved photoemission spectroscopy (trARPES) with extreme ultraviolet (XUV) pulses from high-harmonic generation (HHG)[9] to track the ultrafast quenching dynamics in $1T$-TaX_2 (X = S or Se). Time-resolved XUV-ARPES is the ideal tool because it combines the high time resolution needed to resolve electron dynamics with the high momentum resolution and coverage needed to probe "spectroscopic order parameters" at different momenta in the electronic structure. Our focus is on the early stages (<300 fs) of the dynamics, *i.e.*, on the suppression rather than the reformation of electronic and lattice order.

Our results reveal a global two-time-scale dynamics of spectral weight and order-parameter transients indicating a quasi-instantaneous loss of Mott and CDW charge order and a subsequent coherent suppression of CDW lattice order on the time scale of the CDW amplitude mode. The resulting transient non-equilibrium state is characterized by filled-in and partially closed gaps at a relaxed periodic lattice distortion. The results are consistent with an explanation of the equilibrium Peierls–Mott state in $1T$-TaX_2 (X = S or Se) in terms of a linear cause-and-effect relationship where the Peierls distortion controls the critical parameters of the Mott transition. More generally, the present work provides evidence for a phenomenological hierarchy of the time scales of electronic order-parameter quenching.

II. Materials

$1T\text{-}TaS_2$ and $1T\text{-}TaSe_2$ are isostructural layered compounds consisting of X–Ta–X (X = S and Se) sandwiches.[28] Each atomic layer is hexagonally-packed, the coordination around each Ta atom is approximately octahedral, and the interaction between adjacent sandwiches is weak giving rise to highly anisotropic properties. Correspondingly, the electronic structure near the Fermi energy (E_F) is quasi-two-dimensional. In $1T\text{-}TaS_2$, the band structure is made up of completely filled S 3p valence bands and a partially (nominally d^1) filled Ta 5d band. The Fermi surface has elliptical electron pockets centered on the edges of the hexagonal Brillouin zone [Fig. 1(a)].[19,29] The band structure and Fermi surface of $1T\text{-}TaSe_2$ are similar to the ones of $1T\text{-}TaS_2$, except that the hybridization between the Se 4p and Ta 5d orbitals is significantly stronger, possibly even resulting in a Fermi-level crossing of the uppermost Se 4p valence band.[30,31]

The elliptical Fermi-surface pockets are prone to nesting [arrows in Fig. 1(a)] which in both compounds promotes the formation of an incommensurate CDW (ICCDW) considerably above room temperature. At lower temperatures, the

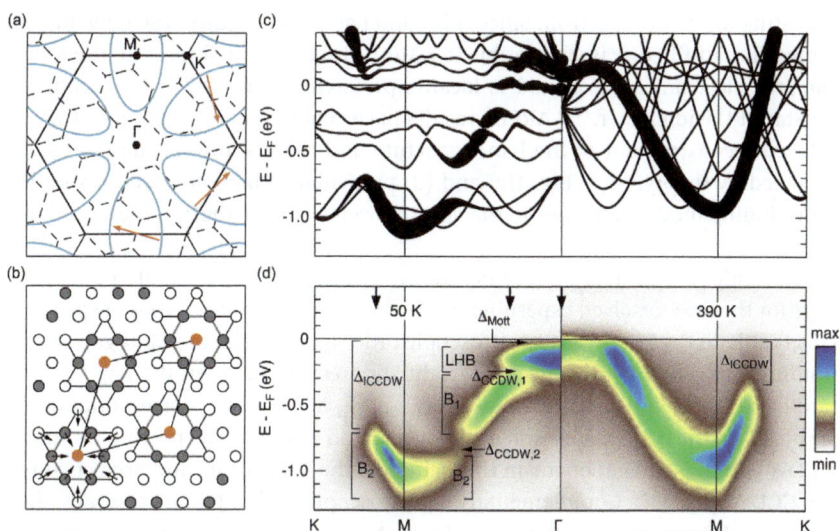

Fig. 1 Characteristic aspects of the lattice and electronic structure of $1T\text{-}TaS_2$ in connection with the commensurate $p(\sqrt{13} \times \sqrt{13})R13.9°$ charge-density wave. (a) Brillouin zones in the normal (thick solid lines) and reconstructed (thin dashed lines) phases. The unreconstructed Ta 5d Fermi surface has elliptical pockets. Possible nesting vectors are indicated. (b) Unit cell of the periodic lattice distortion (thick lines) and "Star-of-David" clusters (thin lines) in the Ta plane. Arrows indicate the displacements of the Ta atoms from their original positions. (c) Simulated reconstructed band structure (left) and folded-out band structure (right) with Umklapp bands generated by translation through the reciprocal lattice vectors of the reconstructed phase. The spectral weight carried by the electronic states is indicated by the line thickness. (d) Measured ARPES band structure in the commensurate (left) and incommensurate (right) charge-density–wave phase (data recorded with $h\nu = 96$ eV at the indicated temperatures). Characteristic spectral signatures are labeled. Downward pointing arrows along the top axis denote the momenta at which previous trARPES measurements were performed.[17,22,23,27]

ICCDW transforms into a commensurate one, in $1T\text{-}TaS_2$ *via* an intermediate nearly commensurate CDW phase. The corresponding transition temperatures are still high: ≈ 180–220 K for $1T\text{-}TaS_2$ and ≈ 470 K for $1T\text{-}TaSe_2$.[28] Intriguingly, the appearance of the commensurate CDW (CCDW) is accompanied by a Mott transition in both compounds. However, transport and ARPES results indicate that only $1T\text{-}TaS_2$ becomes an insulator, whereas $1T\text{-}TaSe_2$ remains metallic.[28,31] For $1T\text{-}TaSe_2$, it has been argued that the Mott transition is a surface effect.[32,33]

Our focus in this work is on the Peierls–Mott ground state. Fig. 1(b) depicts the characteristic periodic lattice distortion (PLD) in the Ta plane that comes along with the formation of the $p(\sqrt{13} \times \sqrt{13})R13.9°$ CCDW. The basic motif is a "Star-of-David" cluster consisting of a central Ta atom surrounded by two concentric 6-Ta-atom rings that are contracted towards the center.[34] The concomitant modulation of the conduction electron density involves electron transfer from the outer ring towards the inner atoms. The PLD and CDW amplitudes are remarkably large: $\approx 7\%$ of the in-plane lattice constant[34] and ≈ 0.5 electron per Ta atom,[35] respectively.

Accordingly, the Ta 5d electronic structure is strongly reconstructed in the CCDW phase. As shown in Fig. 1(c) for the case of $1T\text{-}TaS_2$, the occupied part of the band structure is split into submanifolds: there are two low-lying three-band submanifolds each filled with six electrons and one distinct and narrow band at E_F hosting the "13th" electron. Since this band has a small width <100 meV and is nominally half-filled, it is susceptible to a Mott–Hubbard transition.[19] In the Mott state, the electrons are expected to localize preferentially on the central atom of the Star-of-David cluster.[18,19] ARPES results have generally confirmed the CCDW-induced reconstruction of the band structure and the opening of the Mott gap predicted by theory.[20,36–38] Fig. 1(c) and (d) particularly show how well a simple tight-binding model can capture the spectral weight redistribution measured by ARPES.[20]

The static ARPES data of $1T\text{-}TaS_2$ shown in Fig. 1(d) (ref. 17 and 20) set the stage for the time-resolved experiments. There are three different types of spectral gaps that lend themselves as spectroscopic order parameters: the Mott gap Δ_{Mott} around Γ indicated by the spectral weight suppression between the lower Hubbard band (LHB) and E_F, the CCDW gaps $\Delta_{\text{CCDW,1}}$ and $\Delta_{\text{CCDW,2}}$ corresponding to the breaks in the band dispersion at about $\frac{1}{3}\overline{\Gamma M}$ and $\frac{2}{3}\overline{\Gamma M}$, and the nesting-induced ICCDW gap Δ_{ICCDW} that opens in the vicinity of the K–M–K line and projects into the CCDW phase.[20] Complementary spectroscopic order parameters are the amounts of "coherent" spectral weight that are transferred from the gapped regions to the three submanifolds denoted LHB, B_1, and B_2 in Fig. 1(d). We note that momentum resolution and full Brillouin-zone coverage are essential to probe the various spectral signatures.

Previous time-resolved UV- and XUV-ARPES studies on the ultrafast quenching dynamics of the Peierls–Mott state in $1T\text{-}TaS_2$ were performed under moderate to strong excitation, with absorbed pump fluences ranging from 0.135 to 2.2 mJ cm^{-2}.[17,22,23,27] Spectroscopic order-parameter transients with single rise times were recorded at selected momenta. The focus was mostly on the time constants and not so much on the magnitude of relative intensity changes. Specifically, it was shown that Δ_{Mott} and $\Delta_{\text{CCDW,1}}$ are filled in on an electronic (<30 fs) time scale, while simultaneously the intensity of the LHB is suppressed.[17,22,23,27] On the other hand, it was shown that a partial closing of

Δ_{ICCDW} and intensity changes in the B_1 manifold take place on a vibrational (≈ 200 fs) time scale related to the frequency of the CDW amplitude mode.[17,27] $1T$-TaSe$_2$ has so far not been investigated by trARPES. Yet, the material may be the better suited one because of a larger LHB distance to E_F (Δ_{Mott}) and reduced complexity due to the absence of the nearly commensurate CDW phase.

III. Experiment

The trARPES experiments were performed in a pump–probe scheme with an experimental setup sketched in Fig. 2.[39] The fundamental laser pulses were generated by a Ti:sapphire laser (KMLabs, Griffin) and amplified via a multipass configuration (KMLabs, Dragon) pumped by a Nd:YAG laser (Lee Laser, LDP-200MQG). The laser system was operated at a repetition rate of 8.2 kHz and delivered near-infrared (NIR) pulses at a photon energy of $h\nu = 1.57$ eV ($\lambda = 790$ nm) with 1 mJ pulse energy and 32 fs (FWHM) pulse duration. The laser output was split into pump and probe beams using a 10/90 beam splitter.

For photoexcitation of the samples, the fundamental of the laser was used in s-polarization. The pump intensity was adjusted using a variable neutral density filter in the pump line. The diameter of the pump beam at the sample was about 300 μm (FWHM).

The probe pulses were generated in two steps. The fundamental beam first passed a 200 μm thick beta-barium borate (BBO) crystal. The resulting second harmonic ($h\nu = 3.14$ eV, $\lambda = 395$ nm) was then focused into an argon-filled commercial high-harmonic-generation (HHG) source (KMLabs, XUUS).[40] The delivered XUV pulses had a near-Gaussian beam profile, as verified by a beam profile analysis using an image intensifier. A 300 nm thick aluminum filter was used to block the residual fundamental of the laser transmitted through the HHG source. The XUV radiation was focused with a gold-coated toroidal mirror to a diameter of about 170 μm (FWHM) at the sample position, as estimated from the results of ray-tracing simulations. The factor-of-two reduction of spot size compared to the pump beam ensured that a homogeneously excited surface region was probed.

The 14th harmonic ($h\nu = 22.1$ eV) of the fundamental was used for the time-resolved measurements.[41] The spectral width of the XUV pulses was measured to

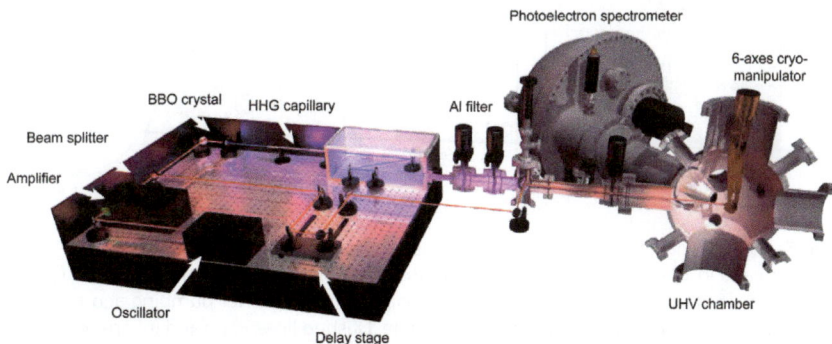

Fig. 2 Schematic illustration of the experimental setup for time-resolved high-harmonic-generation-based XUV-ARPES. Key components are labeled. For details, see text.

less than 170 meV employing a grating spectrometer. The temporal width was estimated from NIR–XUV cross-correlation measurements on the conduction-band population dynamics in a 1T-TiSe$_2$ sample. The quantitative analysis of the intensity transients yielded an upper limit for the XUV pulse duration of 13 fs. The probe pulses were p-polarized.

Photoemission spectra were recorded at a base pressure of 3×10^{-10} mbar with a commercial hemispherical electron spectrometer (SPECS, Phoibos 150) equipped with a 2D detection unit for parallel measurement of energy and momentum of the emitted electrons. The total energy resolution achieved in the present study was about 260 meV, determined from Fermi-edge spectra of a polycrystalline gold sample. The angular resolution of the experiment was esti-mated to 1.5°.

1T-TaS$_2$ and 1T-TaSe$_2$ single crystals were grown from high-purity elements by chemical vapor transport using iodine as transport agent.[42] Prior to the trARPES experiments, the sample quality was checked by electrical transport as well as high-resolution ARPES and XPS measurements. In the trARPES experiments, the samples were cleaved under ultrahigh vacuum conditions at room temperature using a scotch-tape method. During the trARPES measurements, the equilibrium sample temperature was 100 K.

IV.   Results

The central results of our comparative trARPES study of the Peierls–Mott dynamics in 1T-TaS$_2$ and 1T-TaSe$_2$ are summarized in Fig. 3 and 4. Fig. 3 displays energy–momentum maps (top row) and energy–distribution curves (bottom row) collected at selected pump–probe delays near the center of the Brillouin zone of

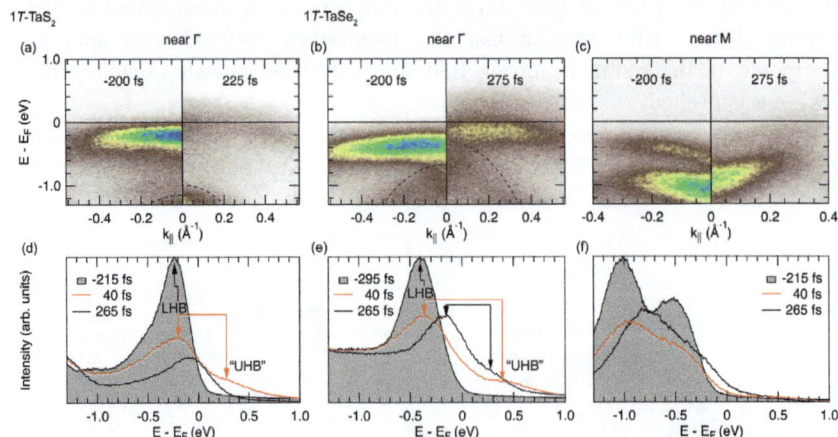

Fig. 3   ARPES snapshots of the electronic structure dynamics in 1T-TaS$_2$ near Γ (left column) and in 1T-TaSe$_2$ near Γ (middle column) and M (right column) ($T = 100$ K, $F_{abs} = 3.5$ mJ cm^{-2}, $h\nu_{probe} = 22.1$ eV). (a–c) Band maps before and after pumping acquired (a), (b) along the Γ–M line and (c) along the M–K line. Dashed lines in (a) and (b) are guides to the eye for the dispersion of the top of the chalcogen valence band. (d–e) Momentum-integrated energy–distribution curves at selected pump–probe delays, as obtained from band maps of the same data sets that the maps in (a–c) were taken from.

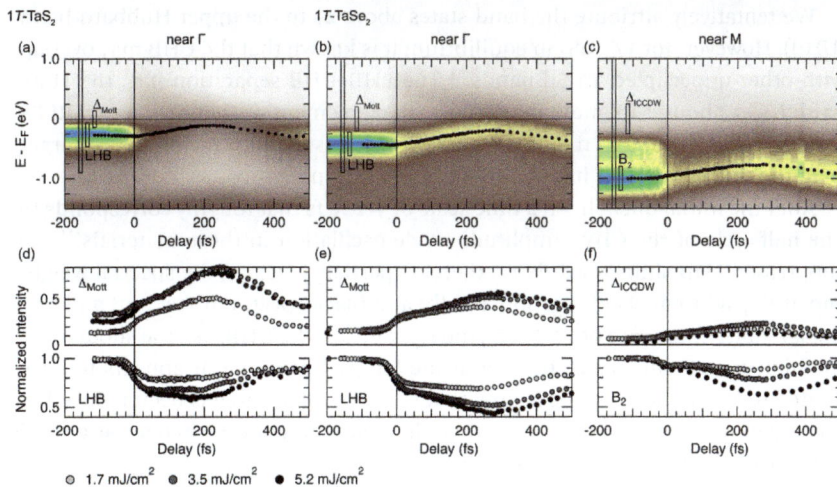

Fig. 4 Time-dependent spectral weight dynamics in 1T-TaS$_2$ near Γ (left column) and in 1T-TaSe$_2$ near Γ (middle column) and M (right column) ($T = 100$ K, $h\nu_{probe} = 22.1$ eV). (a–c) Momentum-integrated energy–distribution curves as a function of pump–probe delay ($F_{abs} = 3.5$ mJ cm^{-2}). Black dots track the time dependence of the center of mass of the spectral weight distribution. (d–e) Photoemission intensity averaged over energy intervals connected with the spectral gaps at E_F (upper panels) and the photoemission peaks (lower panels) ($F_{abs} = 1.7$–5.2 mJ cm^{-2}). For each spectrum, the averaged intensities were divided by the total intensity in the spectral region of the Ta 5d band and normalized to the intensity of the corresponding photoemission peak before time zero. The energy integration intervals used are indicated by boxes in (a–c).

1T-TaS$_2$ (left column) and near the zone center and the zone edge of 1T-TaSe$_2$ (right columns). Fig. 4 has the same column layout as Fig. 3 and shows the detailed time dependence of characteristic energy–distribution curves (top row) and order-parameter transients (bottom row). The electronic structure dynamics of 1T-TaS$_2$ near the Brillouin-zone edge has previously been reported in ref. 17. It is not shown here because it is almost identical to the one of 1T-TaSe$_2$.

The unpumped energy–momentum maps of 1T-TaS$_2$ and 1T-TaSe$_2$ acquired near Γ show the weakly dispersive LHB in excellent agreement with static ARPES results [left panels of Fig. 3(a) and (b), cf. Fig. 1(d)]. A notable difference between the two compounds is the larger LHB distance to E_F (by ≈ 200 meV) in 1T-TaSe$_2$. Another expected difference is the significantly reduced separation between the LHB and the (faintly visible) top of the chalcogen valence band: in 1T-TaS$_2$ there is a clear LHB-S 3p gap, while the LHB and the Se 4p band in 1T-TaSe$_2$ appear to overlap [dashed lines in Fig. 3(a) and (b)].

The time-selected Γ-point spectra depicted in Fig. 3(d) and (e) reveal the qualitatively similar dynamics of the LHB in the two compounds. At pump–probe delays ≤ 40 fs, the spectral weight carried by the LHB is quasi-instantaneously suppressed and transferred to higher energies; the spectral weight gap at E_F (Δ_{Mott}) is (incoherently) filled in and (coherent) band states above E_F become populated as indicated by a distinct shoulder in the high-energy spectral tail [red spectra in Fig. 3(d) and (e)]. We note that at the rather limited energy resolution of our experiment we find no evidence for the emergence of a mid-gap resonance as reported in ref. 23.

We tentatively attribute the band states above E_F to the upper Hubbard band (UHB). However, for 1T-TaS$_2$ in equilibrium it is known that the UHB may overlap with other unoccupied Ta 5d bands.[24] The LHB–UHB separation (*i.e.*, the Hubbard U) is about 510 meV in 1T-TaS$_2$, in agreement with previous trARPES results,[23] and about 790 meV in 1T-TaSe$_2$, which is about a factor of two larger than the value proposed in a scanning tunneling spectroscopy study.[33]

After the initial quench, on a time scale of ≈ 250 fs that roughly corresponds to one half-cycle of the CDW amplitude-mode oscillation in these materials,[43] the suppressed LHB shifts towards E_F, thereby (partially) closing the spectral weight gap at E_F [right panels of Fig. 3(a) and (b) and black spectra in Fig. 3(d) and (e)]. Remarkably, in the case of 1T-TaSe$_2$, the signature of the UHB, *i.e.*, the shoulder in the high-energy tail, is still present at the longer time scale. It appears to have shifted to lower energies, suggesting that the LHB–UHB gap has been reduced to about 430 meV. However, we note that electron temperature changes at a fixed Hubbard-U may lead to a similar effect.[24]

The unpumped energy–momentum map of 1T-TaSe$_2$ acquired along the Brillouin-zone edge shows the characteristic V-like dispersion of the lowest Ta 5d submanifold B$_2$ about the M point. It also shows a weak sign of dispersion away from E_F at the point where the gap opens [left panel of Fig. 3(c), *cf.* Fig. 1(c) and (d)]. The large (nesting-induced) spectral gap Δ_{ICCDW} is readily apparent. Inside the gap, a CDW-induced band belonging to the B$_1$ submanifold is observed. This "shadow" band is significantly more intense than the static ARPES data of 1T-TaS$_2$ in Fig. 1(d) would suggest. Possible reasons are electronic structure differences between the two compounds or a matrix-element effect due to the different photon energy used.

Upon photoexcitation, the spectral weight carried by the B$_1$ and B$_2$ bands is quasi-instantaneously suppressed and the previously split bands merge into one spectral feature [red spectrum in Fig. 3(f)]. On the longer time scale, this feature shifts towards E_F and its dispersion straightens out, but a clear spectral gap remains [right panel of Fig. 3(c) and black spectrum in Fig. 3(f)].

Fig. 4 presents the full temporal evolution of the spectral weight dynamics as a function of material and momentum. The data shown in Fig. 4(a)–(c) were obtained from time-resolved energy–momentum maps similar to those depicted in the top row of Fig. 3. The spectra were averaged over momentum intervals of ≈ 0.8 Å^{-1} to improve the statistics of the extracted photoemission intensity transients that are shown in Fig. 4(d)–(f). The photoemission intensity transients represent the temporal evolution of the spectral weight in energy intervals that correspond to the spectral (gap, peak) combinations (Δ_{Mott}, LHB) [Fig. 4(d) and (e)] and (Δ_{ICCDW}, B$_2$) [Fig. 4(f)].

The energy–time maps in Fig. 4(a)–(c) vividly illustrate the general, *i.e.*, material- and momentum-independent, two-step dynamics: a quasi-instantaneous suppression of spectral weight is followed by a continuous spectral weight shift towards E_F within 195–280 fs. The fast process occurs on a sub-vibrational (electronic) time scale on the order of the temporal resolution of the experiment; the time constant of the slower component agrees with measured half-cycle times of the CDW amplitude mode (203–224 fs in 1T-TaS$_2$ (ref. 21–23, 43 and 44) and 222–265 fs in 1T-TaSe$_2$ (ref. 43)), consistent with a displacive excitation of a coherent vibration of that mode.

All spectroscopic order-parameter transients shown in Fig. 4(d)–(f) also clearly display the two-time-scale dynamics. Initially, spectral weight is promptly transferred from the spectral peaks to the spectral gap regions resulting in the sharp (sub-40 fs) intensity drops and rises, respectively. Subsequently, the intensity changes reflect the shifting of the spectral peaks. The intensity changes become slower and the maximum intensity gains and losses are reached after one half-cycle oscillation of the amplitude mode. When the excitation density is increased, these half-cycle durations become longer, indicating a softening of the CDW amplitude mode. The fast component of the dynamics, on the other hand, remains within the resolution limit.

The peak intensity transients in the lower panels of Fig. 4(d)–(f) additionally display systematic excitation density-dependent amplitude changes. Upon stronger excitation, the amplitude of the amplitude-mode oscillation becomes larger resulting in larger peak shifts out of the fixed energy windows and thus larger spectral weight losses. This behavior is expected and consistent with the fact that the PLD is not completely suppressed, even for the high excitation densities applied here.

The amplitude of the fast, electronic component is more dependent on momentum than on fluence. Near M, where only CDW dynamics is probed, the intensity of the B_2 peak promptly drops by about 10%, independent of the fluence. By contrast, near Γ, where mostly Mott dynamics is probed, the initial relative intensity losses of the LHB are larger and wider spread, about 20–35%. Remarkably, the 20% intensity quench in $1T\text{-}TaS_2$ at the lowest fluence of 1.7 mJ cm^{-2} agrees with the value reported in ref. 22 and 23 where the absorbed fluence was more than one order of magnitude lower. We attribute both intensity quenches, the one of the LHB near Γ and the one of the B_2 band near M, to the loss of coherent spectral weight resulting from the suppression of charge order of the Mott component and the CDW component, respectively.

Our results thus confirm the general scenario that has emerged from previous trARPES work according to which the charge and lattice orders are suppressed on their own distinct time scales.[17,22,23,27] However, novel important aspects are added to that scenario here, namely, the global character of the two-time-scale dynamics and the possible change of the Hubbard U in response to the coherent amplitude-mode oscillation.

V. Discussion

Fig. 5 shows a simple density-of-states scheme that summarizes our trARPES results on the ultrafast electronic structure dynamics in $1T\text{-}TaX_2$ (X = S and Se). After femtosecond laser excitation, the strongly reconstructed band structure with clear Peierls and Mott gaps [Fig. 5(a)] is melted on two distinct time scales via two distinct non-equilibrium states. The photoexcited state at $t \lesssim 40$ fs is characterized by filled-in Mott and Peierls gaps and subbands carrying less coherent spectral weight [Fig. 5(b)]. The charge orders associated with Mott localization and CCDW formation are suppressed, whereas the PLD is still frozen. The state is highly non-thermal: neither the electron–phonon system nor the electron system itself will be thermalized. The second state, attained at $t \approx 200$ fs, also is a crossover state between insulator and metal [Fig. 5(c)]. However, now the gaps are not only filled-in, but they are partially closed. The lattice order had time to relax

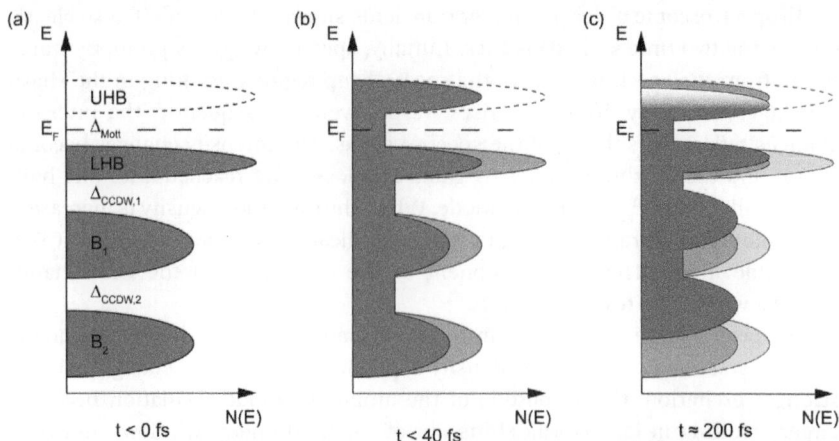

Fig. 5 Schematic representation of the ultrafast electronic structure dynamics in $1T\text{-}TaX_2$ (X = S or Se) as seen by trARPES. (a) Peierls–Mott ground state where the density of states below E_F exhibits three distinct peaks separated by CDW-induced gaps. The peaks directly below and above E_F are the lower and upper Hubbard band peak, respectively. The gap at E_F is the Mott gap. (b) Non-equilibrium state immediately following photoexcitation. Density-of-states peaks are suppressed and gaps are partially filled in. (c) Non-equilibrium state after about one half-cycle of a coherent amplitude-mode oscillation. Density-of-states peaks are shifted towards E_F and gaps are partially closed.

coherently so that in addition to the quenched charge orders the PLD will be partially suppressed. The electrons could be expected to have thermalized, but the electron temperature will be vastly higher than the lattice temperature.

These results readily provide answers to the three general questions raised in the introduction. First, regarding the speed limit and mechanism of the Peierls–Mott-state melting investigated here, we can conclude that the Peierls–Mott state as a whole—displaying strongly coupled charge and lattice orders—cannot be melted faster than the ions can move coherently. The speed limit for the quenching of this state is thus set by the duration of the relevant lattice vibrational mode. This mode is the CDW amplitude mode. In other words, the bottleneck of the melting process is the coherent breathing motion of the Star-of-David clusters [Fig. 1(b)] that is triggered by the femtosecond laser excitation.

Second, as to the nature of the transient non-equilibrium states, we find that from a spectroscopic viewpoint the crossover state at $t \approx 200$ fs appears similar to an equilibrium strong-coupling CDW state above the mean-field transition temperature: both states are characterized by smeared, partially closed, but persistent energy gaps ("pseudogaps"). However, the nature of these states is very different. In the non-equilibrium case, we have a hot, disordered electron system decoupled from a coherently fluctuating, but still long-range ordered lattice distortion. The equilibrium state, by contrast, exhibits strongly coupled, but fluctuating short-range electronic and lattice orders.[20,45]

Third, concerning the implications on the equilibrium ground state, we can conclude that the trARPES results are generally consistent with the model described in Sec. II. The Peierls–Mott transition in $1T\text{-}TaX_2$ (X = S or Se) is not a chicken-and-egg problem. It is the CCDW distortion that drives the Mott transition by controlling the degree of localization on the 13th Ta atom in the center of the Star-of-David cluster.[18,19] Our results may provide direct evidence for this linear cause-and-effect relationship in that they seem to indicate a reduction of the Hubbard U following the coherent reduction of the PLD amplitude. We note that a vibrational modulation of the Hubbard U has recently also been demonstrated in a time-resolved optical spectroscopy study on an organic Mott insulator.[46]

Finally, we place our results in the wider context of ultrafast photo-induced melting of electronic orders. Fig. 6 presents an overview of measured gap quenching times obtained by time-resolved optical and photoemission spectroscopy from a variety of materials comprising superconductors, CDW systems, and Mott insulators. The data are compiled from ref. 8–10, 17, 27 and 47–51 and plotted as a function of excitation density. The quenching times for 1T-TaSe$_2$ determined in this work are included. The higher sensitivity of time-resolved optical spectroscopy compared to trARPES is immediately seen in the generally lower excitation densities employed in the optical spectroscopy studies.

The plot reveals a phenomenological temporal hierarchy that reflects the distinct characteristic time scales of different processes dominating in the destruction of the three different forms of electronic order.

The quenching times of superconducting states are the slowest, in the range of 0.4–10 ps. This is because the energy transfer from the photo-excited hot carriers to the superconducting condensate is generally inefficient and incoherent. The hot carriers generate a dense population of high-energy ("2Δ") phonons and these phonons subsequently provide the dominant inelastic scattering channel for Cooper-pair breaking.[48] The quenching time is thus related to an empirical pair-breaking rate.

In CDW systems, as has also been shown in this work, there are two distinct quenching time scales: a slow, vibrational one of about 100–400 fs, and a fast, electronic one on the order of 20–100 fs. The two time scales reflect the fact that a

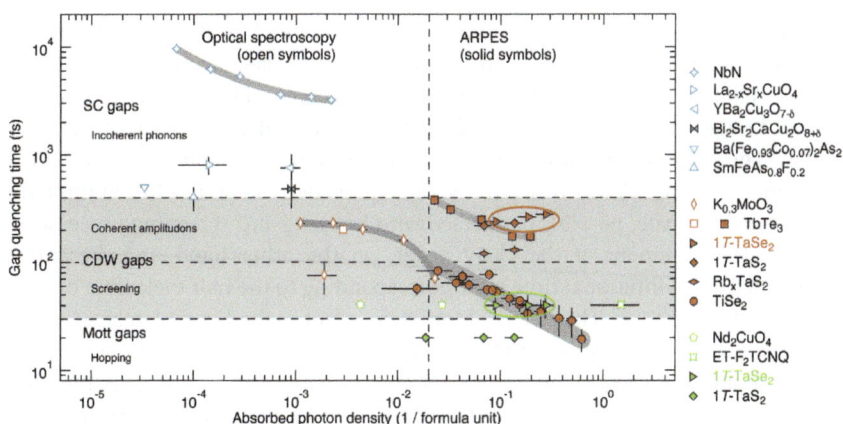

Fig. 6 Quenching times of electronic order parameters as a function of excitation density, as measured by time-resolved optical and photoemission spectroscopy for various superconductors, charge-density–wave systems, and Mott insulators. Data compiled from ref. 8–10, 17, 27 and 47–51. The results of this work are highlighted by ellipses.

CDW has two components: a lattice component, the PLD, and an electronic component, the periodic modulation of the conduction-electron density. The relevant collective mode mapping the PLD onto the undistorted lattice is the CDW amplitude mode, and indeed for all systems shown the measured lattice quenching time corresponds well to the half-cycle time of the amplitude mode. Collective electron behavior, on the other hand, is generally related to the plasma frequency. The plasma frequency may therefore be expected to set the time scale on which the electronic component of a CDW is suppressed.[9] Layered CDW materials of the dichalcogenide family display plasmon energies between 40–140 meV ($1T$-TiSe$_2$ (ref. 52)) and 1.05 eV ($2H$-TaSe$_2$ (ref. 53)) which indeed spans a range of screening times of 4–100 fs. Nonetheless, it is fair to say that the connection between screening and CDW quenching is not yet well established experimentally.

Lastly, Mott-gap quenching appears to be connected with the fastest rise times of about 20–40 fs. Effectively it will be even faster as typical experimental time resolutions are limited to a few 10 fs. The ultrafast collapse of the Mott gap proceeds *via* the buildup of coherent or incoherent mid-gap states[22–24] and the time scale is commonly associated with the fundamental electron hopping rate as given by the bandwidth W.[8,17,27] However, as suggested in Fig. 5(c), one can also imagine a scenario in which a coherent lattice vibration modulates the parameters U and W such that the system is driven into the metallic regime. The quenching dynamics would then exhibit a slower, vibrational component, similar to the case of a CDW.

Collectively, the data of Fig. 6 corroborate a central motivation for time-resolved studies: the possibility to discriminate fundamental electronic and phononic processes in the time domain. They also show that there is room at the bottom for improved time resolution. The true time scales of the fast electronic processes have not yet been measured.

VI.   Conclusion

In summary, time-resolved NIR pump-XUV probe ARPES has allowed us to study the non-equilibrium dynamics of the isoelectronic Peierls–Mott states of $1T$-TaS$_2$ and $1T$-TaSe$_2$ in the strong excitation regime. Femtosecond laser excitation in the mJ cm^{-2} range suppresses the coupled electron-lattice orders on two distinct time scales. The two charge orders due to Mott localization and the electronic component of the CDW, respectively, are quenched on a resolution-limited electronic time scale, possibly connected with hopping and screening processes. The lattice component of the CDW order, on the other hand, is coherently diminished on a vibrational time scale corresponding to the half-cycle time of the CDW amplitude mode. The Peierls–Mott state is not melted completely, even at the high excitation densities applied. The transient non-equilibrium state is a crossover state between insulator and metal. The Mott and Peierls gaps remain pseudo-gapped; they are filled-in and only partially closed. Overall, the results are consistent with an equilibrium phase-transition mechanism in which the CDW is the driver of the Mott transition. Moreover, the experimentally observed global two-time-scale dynamics fits into a general phenomenological hierarchy of the time scales involved in the quenching dynamics of electronically ordered states.

So, how fast can a Peierls–Mott insulator be melted? The bottleneck of the photo-induced melting process is the suppression of the PLD and the fastest way to get through this bottleneck is coherent lattice relaxation along the coordinate of the CDW amplitude mode. The period of the amplitude-mode oscillation (or a fraction thereof) thus defines the speed limit of the melting process. The bottleneck time scale for $1T$-TaS$_2$ and $1T$-TaSe$_2$ directly determined here is ≈ 200–300 fs.

Two immediate questions emerge from the presented trARPES results: can the Peierls–Mott state in $1T$-TaS$_2$ and $1T$-TaSe$_2$ be completely melted non-thermally? And does the (resolution-limited) quenching of the two different electronic orders happen on two different time scales? Both problems provide a challenge for trARPES experiments. The first one requires the use of even higher excitation densities than the ones applied here; the second one calls for sub-10 fs time resolution. In both directions, the price to be paid will be spectral resolution—because of pump-induced space–charge broadening and the time–bandwidth product, respectively. Yet, the strong spectroscopic signatures, specifically the large separation of the LHB from E_F in $1T$-TaSe$_2$, may make these experiments feasible.

Acknowledgements

This work was supported by the DFG *via* grant BA 2177/9-1 and by the BMBF *via* grant 05K10FK4.

References

1 E. Dagotto, *Science*, 2005, **309**, 257.

2 M. Rini, R. Tobey, N. Dean, J. Itatani, Y. Tomioka, Y. Tokura, R. W. Schoenlein and A. Cavalleri, *Nature*, 2007, **449**, 72.

3 F. Schmitt, P. S. Kirchmann, U. Bovensiepen, R. G. Moore, L. Rettig, M. Krenz, J.-H. Chu, N. Ru, L. Perfetti, D. H. Lu, M. Wolf, I. R. Fisher and Z.-X. Shen, *Science*, 2008, **321**, 1649.

4 R. Yusupov, T. Mertelj, V. V. Kabanov, S. Brazovskii, P. Kusar, J.-H. Chu, I. R. Fisher and D. Mihailovic, *Nat. Phys.*, 2010, **6**, 681.

5 M. Eichberger, H. Schäfer, M. Krumova, M. Beyer, J. Demsar, H. Berger, G. Moriena, G. Sciaini and R. J. D. Miller, *Nature*, 2010, **468**, 799.

6 D. Fausti, R. I. Tobey, N. Dean, S. Kaiser, A. Dienst, M. C. Hoffmann, S. Pyon, T. Takayama, H. Takagi and A. Cavalleri, *Science*, 2011, **331**, 189.

7 H. Ichikawa, S. Nozawa, T. Sato, A. Tomita, K. Ichiyanagi, M. Chollet, L. Guerin, N. Dean, A. Cavalleri, S. Adachi, T. Arima, H. Sawa, Y. Ogimoto, M. Nakamura, R. Tamaki, K. Miyano and S. Koshihara, *Nat. Mater.*, 2011, **10**, 101.

8 S. Wall, D. Brida, S. R. Clark, H. P. Ehrke, D. Jaksch, A. Ardavan, S. Bonora, H. Uemura, Y. Takahashi, T. Hasegawa, H. Okamoto, G. Cerullo and A. Cavalleri, *Nat. Phys.*, 2011, 7, 114.

9 T. Rohwer, S. Hellmann, M. Wiesenmayer, C. Sohrt, A. Stange, B. Slomski, A. Carr, Y. Liu, L. Miaja Avila, M. Kalläne, S. Mathias, L. Kipp, K. Rossnagel and M. Bauer, *Nature*, 2011, **471**, 490.

10 C. L. Smallwood, J. P. Hinton, C. Jozwiak, W. Zhang, J. D. Koralek, H. Eisaki, D.-H. Lee, J. Orenstein and A. Lanzara, *Science*, 2012, **336**, 1137.

11 K. W. Kim, A. Pashkin, H. Schäfer, M. Beyer, M. Porer, T. Wolf, C. Bernhard, J. Demsar, R. Huber and A. Leitenstorfer, *Nat. Mater.*, 2012, **11**, 497.

12 S. de Jong, R. Kukreja, C. Trabant, N. Pontius, C. F. Chang, T. Kachel, M. Beye, F. Sorgenfrei, C. H. Back, B. Bräuer, W. F. Schlotter, J. J. Turner, O. Krupin, M. Doehler, D. Zhu, M. A. Hossain, A. O. Scherz, D. Fausti, F. Novelli, M. Esposito, W. S. Lee, Y. D. Chuang, D. H. Lu, R. G. Moore, M. Yi, M. Trigo, P. Kirchmann, L. Pathey, M. S. Golden, M. Buchholz, P. Metcalf, F. Parmigiani, W. Wurth, A. Föhlisch, C. Schüßler-Langeheine and H. A. Dürr, *Nat. Mater.*, 2013, **12**, 882.

13 L. Stojchevska, I. Vaskivskyi, T. Mertelj, P. Kusar, D. Svetin, S. Brazovskii and D. Mihailovic, *Science*, 2014, **344**, 177.

14 D. J. Hilton, R. P. Prasankumar, S. A. Trugman, A. J. Taylor and A. D. Averitt, *J. Phys. Soc. Jpn.*, 2006, **75**, 011006.

15 U. Bovensiepen and P. S. Kirchmann, *Laser Photonics Rev.*, 2012, **6**, 589.

16 R. J. D. Miller, *Science*, 2014, **343**, 1108.

17 S. Hellmann, T. Rohwer, M. Kalläne, K. Hanff, C. Sohrt, A. Stange, A. Carr, M. M. Murnane, H. C. Kapteyn, L. Kipp, M. Bauer and K. Rossnagel, *Nat. Commun.*, 2012, **3**, 1069.

18 P. Fazekas and E. Tosatti, *Philos. Mag. B*, 1979, **39**, 229.

19 K. Rossnagel and N. V. Smith, *Phys. Rev. B: Condens. Matter Mater. Phys.*, 2006, **73**, 073106.

20 K. Rossnagel, *J. Phys.: Condens. Matter*, 2011, **23**, 213001.

21 J. Demsar, L. Forró, H. Berger and D. Mihailovic, *Phys. Rev. B: Condens. Matter*, 2002, **66**, 041101(R).

22 L. Perfetti, P. A. Loukakos, M. Lisowski, U. Bovensiepen, H. Berger, S. Biermann, P. S. Cornaglia, A. Georges and M. Wolf, *Phys. Rev. Lett.*, 2006, **97**, 067402.

23 L. Perfetti, P. A. Loukakos, M. Lisowski, U. Bovensiepen, M. Wolf, H. Berger, S. Biermann and A. Georges, *New J. Phys.*, 2008, **10**, 053019.

24 J. K. Freericks, H. R. Krishnamurthy, Y. Ge, A. Y. Liu and T. Pruschke, *Phys. Status Solidi B*, 2009, **246**, 948.

25 S. Hellmann, M. Beye, C. Sohrt, T. Rohwer, F. Sorgenfrei, H. Redlin, M. Kalläne, M. Marczynski-Bühlow, F. Hennies, M. Bauer, A. Föhlisch, L. Kipp, W. Wurth and K. Rossnagel, *Phys. Rev. Lett.*, 2010, **105**, 187401.

26 N. Dean, J. C. Petersen, D. Fausti, R. I. Tobey, S. Kaiser, L. V. Gasparov, H. Berger and A. Cavalleri, *Phys. Rev. Lett.*, 2011, **106**, 016401.

27 J. C. Petersen, S. Kaiser, N. Dean, A. Simoncig, H. Y. Liu, A. L. Cavalieri, C. Cacho, I. C. E. Turcu, E. Springate, F. Frassetto, L. Poletto, S. S. Dhesi, H. Berger and A. Cavalleri, *Phys. Rev. Lett.*, 2011, **107**, 177402.

28 J. A. Wilson, F. J. Di Salvo and S. Mahajan, *Adv. Phys.*, 1975, **24**, 117.

29 M. Bovet, D. Popović, F. Clerc, C. Koitzsch, U. Probst, E. Bucher, H. Berger, D. Naumović and P. Aebi, *Phys. Rev. B: Condens. Matter Mater. Phys.*, 2004, **69**, 125117.

30 F. Clerc, M. Bovet, H. Berger, L. Despont, C. Koitzsch, O. Gallus, L. Patthey, M. Shi, J. Krempasky, M. G. Garnier and P. Aebi, *J. Phys.: Condens. Matter*, 2004, **16**, 3271.

31 R. Ang, Y. Miyata, E. Ieki, K. Nakayama, T. Sato, Y. Liu, W. J. Lu, Y. P. Sun and T. Takahashi, *Phys. Rev. B: Condens. Matter Mater. Phys.*, 2013, **88**, 115145.

32 L. Perfetti, A. Georges, S. Florens, S. Biermann, S. Mitrovic, H. Berger, Y. Tomm, H. Höchst and M. Grioni, *Phys. Rev. Lett.*, 2003, **90**, 166401.

33 S. Colonna, F. Ronci, A. Cricenti, L. Perfetti, H. Berger and M. Grioni, *Phys. Rev. Lett.*, 2005, **94**, 036405.

34 R. Brouwer and F. Jellinek, *Phys. B*, 1980, **99**, 51.

35 M. Eibschutz, *Phys. Rev. B: Condens. Matter*, 1992, **45**, 10914.

36 B. Dardel, M. Grioni, D. Malterre, P. Weibel, Y. Baer and F. Lévy, *Phys. Rev. B: Condens. Matter*, 1992, **45**, 1462(R).

37 L. Perfetti, T. A. Gloor, F. Mila, H. Berger and M. Grioni, *Phys. Rev. B: Condens. Matter Mater. Phys.*, 2005, **71**, 153101.

38 R. Ang, Y. Tanaka, E. Ieki, K. Nakayama, T. Sato, L. J. Li, W. J. Lu, Y. P. Sun and T. Takahashi, *Phys. Rev. Lett.*, 2012, **109**, 176403.

39 S. Mathias, L. Miaja-Avila, M. M. Murnane, H. Kapteyn, M. Aeschlimann and M. Bauer, *Rev. Sci. Instrum.*, 2007, **78**, 083105.

40 A. Rundquist, C. G. Durfee III, Z. Chang, C. Herne, S. Backus, M. M. Murnane and H. C. Kapteyn, *Science*, 1998, **280**, 1412.

41 S. Eich, A. Stange, A. V. Carr, J. Urbancic, T. Popmintchev, M. Wiesenmayer, K. Jansen, A. Ruffing, S. Jakobs, T. Rohwer, S. Hellman, C. Chen, P. Matyba, L. Kipp, K. Rossnagel, M. Bauer, M. M. Murnane, H. C. Kapteyn, S. Mathias and M. Aeschlimann, *J. Electron Spectrosc. Relat. Phenom.*, 2014, DOI: 10.1016/j.elspec.2014.04.013.

42 F. J. Di Salvo, R. G. Maines, J. V. Waszczak and E. Schwall, *Solid State Commun.*, 1974, **14**, 497.

43 S. Sugai, *Phys. Status Solidi B*, 1985, **129**, 13.

44 T. Onozaki, Y. Toda, S. Tanda and R. Morita, *Jpn. J. Appl. Phys.*, 2007, **46**, 870.

45 W. L. McMillan, *Phys. Rev. B: Solid State*, 1977, **16**, 643.

46 S. Kaiser, S. R. Clark, D. Nicoletti, G. Cotugno, R. I. Tobey, N. Dean, S. Lupi, H. Okamoto, T. Hasegawa, D. Jaksch and A. Cavalleri, *Sci. Rep.*, 2014, **4**, 3823.

47 L. Stojchevska, P. Kusar, T. Mertelj, V. V. Kabanov, Y. Toda, X. Yao and D. Mihailovic, *Phys. Rev. B: Condens. Matter Mater. Phys.*, 2011, **84**, 180507(R).

48 M. Beck, M. Klammer, S. Lang, P. Leiderer, V. V. Kabanov, G. N. Gol'tsman and J. Demsar, *Phys. Rev. Lett.*, 2011, **107**, 177007.

49 A. Tomeljak, H. Schäfer, D. Städter, M. Beyer, K. Biljakovic and J. Demsar, *Phys. Rev. Lett.*, 2009, **102**, 066404.

50 F. Schmitt, P. S. Kirchmann, U. Bovensiepen, R. G. Moore, J.-H. Chu, D. H. Lu, L. Rettig, M. Wolf, I. R. Fisher and Z.-X. Shen, *New J. Phys.*, 2011, **13**, 063022.

51 H. Okamoto, T. Miyagoe, K. Kobayashi, H. Uemura, H. Nishioka, H. Matsuzaki, A. Sawa and Y. Tokura, *Phys. Rev. B: Condens. Matter Mater. Phys.*, 2011, **83**, 125102.

52 G. Li, W. Z. Hu, D. Qian, D. Hsieh, M. Z. Hasan, E. Morosan, R. J. Cava and N. L. Wang, *Phys. Rev. Lett.*, 2007, **99**, 027404.

53 A. König, R. Schuster, M. Knupfer, B. Büchner and H. Berger, *Phys. Rev. B: Condens. Matter Mater. Phys.*, 2013, **87**, 195119.

Faraday Discussions

ROYAL SOCIETY OF CHEMISTRY

PAPER

X-ray absorption spectroscopy with time-tagged photon counting: application to study the structure of a Co(i) intermediate of H₂ evolving photo-catalyst

Grigory Smolentsev,[*a] Alexander A. Guda,[b] Markus Janousch,[a] Cristophe Frieh,[a] Gaudenz Jud,[a] Flavio Zamponi,[c] Murielle Chavarot-Kerlidou,[d] Vincent Artero,[d] Jeroen A. van Bokhoven[ac] and Maarten Nachtegaal[a]

Received 7th March 2014, Accepted 14th April 2014

DOI: 10.1039/c4fd00035h

In order to probe the structure of reaction intermediates of photochemical reactions a new setup for laser-initiated time-resolved X-ray absorption (XAS) measurements has been developed. With this approach the arrival time of each photon in respect to the laser pulse is measured and therefore full kinetic information is obtained. All X-rays that reach the detector are used to measure this kinetic information and therefore the detection efficiency of this method is high. The newly developed setup is optimized for time-resolved experiments in the microsecond range for samples with relatively low metal concentration (~1mM). This setup has been applied to study a multicomponent photocatalytic system with a Co(dmgBF₂)₂ catalyst (dmg²⁻ = dimethylglyoximato dianion), [Ru(bpy)₃]²⁺ chromophore (bpy = 2,2′-bipyridine) and methyl viologen as the electron relay. On the basis of the analysis of hundreds of Co K-edge XAS spectra corresponding to different delay times after the laser excitation of the chromophore, the presence of a Co(i) intermediate is confirmed. The calculated X-ray transient signal for a model of Co(i) state with a 0.14 Å displacement of Co out of the dmg ligand plane and with the closest solvent molecule at a distance of 2.06 Å gives reasonable agreement with the experimental data.

1 Introduction

Development of catalysts for hydrogen evolution from water using sunlight is a challenging task of great practical importance.[1,2] Molecular catalysts based on

[a]Paul Scherrer Institut, WLGA 217, 5232, Villigen, Switzerland. E-mail: grigory.smolentsev@psi.ch; Tel: +41 56 310 5173
[b]Research Center for Nanoscale Structure of Matter, Southern Federal University, Sorge 5, 344090, Rostov-on-Don, Russia
[c]ETH Zurich, Wolfgang Pauli-Str. 10, 8093, Zurich, Switzerland
[d]Laboratory of Chemistry and Biology of Metals, Universite Grenoble Alpes, CEA, CNRS, France

coordination complexes of 3d metals,[3,4] and especially cobalt,[5] have high potential for large-scale applications since they are earth-abundant in contrast to more robust platinum-based catalysts, that are more expensive and less abundant. The catalyst under investigation in the current contribution is $Co(dmgBF_2)_2$ (dmg^{2-} = dimethylglyoximato dianion)[6,7] (Scheme 1). Its hydrogen evolving activity in multicomponent systems in combination with different chromophores[8-10] and in supramolecular systems[8,9,11] has been investigated with several hundreds of turn-overs achieved under homogeneous photocatalytic conditions in the presence of sacrificial electron donors. A few attempts have been made to improve the performance and stability by the modification of the Co ligands.[7,12,13] To fully exploit this approach and design ligands that stabilize the high energy interme-diates along the catalytic pathway, detailed knowledge about the reaction mechanism is essential. Insights in the reaction mechanism can be gained on the basis of electrochemical[5] and steady-state physicochemical (i.e. UV-visible or EPR)[12,14-16] characterization coupled with quantum chemical calculations,[14,17,18] but tracking of certain intermediates is possible only using time-resolved spec-troscopic techniques.

The key requirements to the techniques used to identify the photocatalytic intermediates are sensitivity to the structural and electronic changes of the catalytic center, selectivity that allows to distinguish the contribution of the catalyst from all other possible changes in the complicated chemical system and appropriate time range and resolution. Stopped flow and freeze quench methods are too slow to track the intermediates of H_2 evolving catalysts, especially for the most promising systems that produce hydrogen rapidly. Furthermore, since most of the reaction steps are intermolecular, the intermediates have lifetimes in the microsecond time range. Some initial reaction steps, for example, charge transfer from the chromophore to the catalytic center in the supramolecular systems can be intramolecular and occur within picoseconds,[19] but the charge-separated state has to be long lived to participate in the subsequent intermolecular reactions. Thus measurements in the microsecond range with sub-microsecond resolution are required for such applications.

Among time-resolved spectroscopic methods that are typically applied to study photocatalytic intermediates and reaction mechanism are transient optical absorption spectroscopy[20,21] and time-resolved IR.[22] Time-resolved X-ray absorp-tion spectroscopy (XAS) has a few advantages in comparison with optical and IR methods. First of all XAS is element-specific and therefore the contribution from the metal center of the catalyst is separated from all other changes in the system. Second, XAS spectra contain structural information about the local arrangement of the metal (in the fine structure of the near edge X-ray absorption spectrum

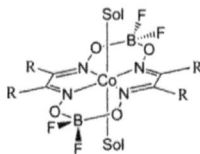

Scheme 1   Structure of $Co(dmgBF_2)_2$ (R = methyl) and $Co(dpgBF_2)_2$ (R = phenyl). Sol is a solvent molecule (CH_3CN).

known as XANES) and information about the oxidation state (mainly in the position of the absorption edge). Setups for time-resolved hard X-ray absorption spectroscopy measurements with ~100 ps resolution have been developed at a few beamlines worldwide.[23-28] Experiments in the microsecond range are quite rare and can be performed using spatially separated laser and X-ray beams (pump–flow–probe mode)[29,30] gating of the detector (pump–probe mode)[31] or recording the arrival time of all photons (pump–sequential-probes mode).[29,32-34]

Pump–flow–probe experiments use non-overlapping laser excitation and X-ray probing beams focused on the sample jet. The solution is circulating in the flow system and a cylindrical jet with a liquid speed of a few meters per second is formed. A continuous wave laser focused on the jet initiates the reaction, the distance between laser and X-ray beams and the velocity of the liquid flow determine the time delay, while the focusing of both beams defines the time resolution. The pump–flow–probe method is characterized by a high detection efficiency and the resolution is limited to ~40 µs with easily achievable flow parameters (flow speed ~5m s^{-1}, jet diameter ~1mm) and moderate X-ray spot size (~100 µm). Pump–probe experiments with a microsecond resolution can use the same data acquisition electronics as pump–probe experiments with pico-second resolution by increasing the width of the X-ray detector gate pulse. Since many X-ray bunches delivered by the synchrotron will fall within the time window defined by such a gate pulse, the time resolution is defined by the width of this pulse. The detection efficiency of pump–probe methods also depends on the gate, its shortening corresponds to higher time resolution but lower efficiency. Acquiring kinetic data with this approach requires scanning the delay between laser and gate pulses.

In the present manuscript we report a pump–sequential-probes setup to study photocatalytic intermediates that have lifetimes in the microsecond range. This setup is realized at the SuperXAS beamline of the Swiss Light Source and described in the experimental setup section. With this approach the arrival time of each photon in respect to the laser pulse is measured (method known as time-tagged photon counting) and therefore, in contrast to currently available pump–probe setups, full kinetic information is obtained without any delay scan. In the application section we demonstrate how this method is used to identify the intermediate states of a Co(dmgBF$_2$)$_2$ catalyst in the multicomponent system and to probe the structure of the Co(ii) and Co(i) states of the catalyst in solution.

2 Experimental setup

X-ray source

The pump–sequential-probes setup has been developed and implemented at the SuperXAS beamline of the Swiss Light Source (SLS), Villigen, Switzerland. The X-ray beam was delivered by the 2.9 T super-bend magnet. The SLS was run in the standard top-up mode with an average current of 400 mA. A collimated beam was formed by means of a Si mirror and the energy has been scanned by a channel-cut Si(111) monochromator. A toroidal mirror with Rh coating was employed after the monochromator to focus the incident X-rays with a spot size of 100 × 100 µm^2 on the sample. The photon flux obtained with this configuration at the sample was about 3.2 × 10^{11} photons/s.

Laser source

For time-resolved experiments the optimal repetition rate of the laser can be chosen maximizing the efficiency of X-ray flux usage. A useful X-ray signal of the excited state is obtained first of all during the time when intermediate species are formed in the detectable amount. Since the difference between excited and unexcited spectra is measured, the system has to fully return to the initial state between laser pulses. Therefore approximately half of the laser period can be used to measure the ground state contribution. As the result, the optimal repetition rate corresponds to a few lifetimes of the studied intermediate. Nevertheless if such species live too long or if the lifetime is not known *a priori* it is necessary to refresh the sample between two subsequent laser pulses. For samples in the liquid phase this can be achieved using a liquid jet flow system which is also beneficial to reduce X-ray and laser induced damage of the sample. In such a configuration, the period of the laser should not be much longer than the time required to refresh the sample. The second factor that has to be taken into account is the laser pulse energy that usually decreases with increasing repetition rate and must be sufficient to efficiently excite the sample.

For our source, the maximal focusing of the X-ray beam is $\sim$100 $\times$100 μm^2. The laser spot with diameter D has to be bigger in order to probe only the volume excited efficiently and taking into account that it is difficult to achieve a stable flow of the sample in the liquid jet with a speed v higher than $\sim$5m s^{-1}, the maximal repetition rate $f = v/D \sim$ 30 kHz is required. In our setup we used a Xiton IDOL laser with a repetition rate up to 40 kHz and maximal output power $\sim$2W at 15kHz. Available wavelengths for this laser are 447 nm and 671 nm. Maximal pulse energy is 125 μJ which is more than enough to excite samples of reasonable optical density (0.2 or higher) with a metal concentration of $\sim$1 mM in an irradiated volume of $\sim$200 μm diameter and $\sim$1 mm thickness.

Data acquisition system (DAQ)

The main advantage of our DAQ is the use of the so-called time-tagged single photon counting method to obtain kinetic and structural information from XAS in the range between tens of nanoseconds up to hundreds of microseconds. Such an approach allows us to use the flux of a bending-magnet beamline efficiently. The setup (see Fig. 1) is based on the multichannel digital X-ray processor XIA XMAP running in the so-called list mapping mode. In this regime information about each registered photon (arrival time, with respect to the trigger with a precision of 20ns, and energy) is buffered and then saved to file in HDF5 format. As detectors for incoming and fluorescence radiation we used avalanche photo diodes (APDs) coupled to the charge sensitive preamplifier. An additional APD that is sensitive to the laser light is connected to one of the channels of the digital X-ray processor and provides the reference signal of the timing of the laser pulse.

To achieve synchronization between X-rays, laser and DAQ we used the radio frequency (RF) signal from the storage ring (500 MHz), which is intrinsically synchronized with the X-ray pulses. It serves as the input for the event receiver (Micro Research, VME-EVR-230RF) that divides the frequency of the RF signal and generates square pulses corresponding to the required repetition rate of the experiment. These signals are further processed with the delay card (Micro Research, 4CHTIM-200) generating pulses of required duration (2 μs for the laser

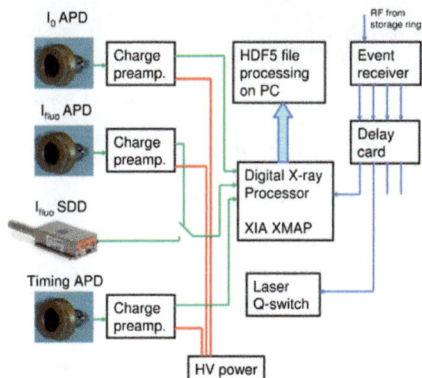

Fig. 1 Scheme of the data acquisition system for the pump–sequential-probes experiments.

trigger, 40ns for the digital X-ray processor trigger) and with required relative delay. The signal sent to the laser allows to control its acousto-optic modulator and thus to achieve synchronization between the actual laser pulse and DAQ better than 2 ns. HDF5 files generated by XMAP are processed online during the incoming X-ray energy scan with a short delay (time of measurement of 1–2 energy points). Histograms with the number of events as a function of time for selected energies of fluorescent photons are calculated. Further processing such us calculation of transient spectra is performed on the basis of these data.

Synchronization

The distribution of the current in synchrotron storage rings is not uniform and at the SLS a hybrid mode is often used. 390 bunches are separated by 2 ns from each other and they form the multibunch train while one additional more intense pulse, the so-called camshaft is located in the gap between multibunches (Fig. 2). While such a distribution is useful for experiments that require a 100 ps time resolution correlated to the width of the camshaft pulse,[35] a uniform average incoming intensity distribution is preferable for microsecond and nanosecond experiments. This has been achieved by keeping the synchronization between laser and DAQ and simultaneously shifting these two triggers relatively to the filling pattern of the synchrotron. This mode of our experimental setup we call asynchronous. Schematically it is shown in Fig. 2.

After each laser pulse intensities I_0 and I_f are measured as a function of time. A spectrum with good statistics can be obtained by averaging X-ray data following at least 10^5 laser pulses. However if the time position of the laser pulse is fixed to the camshaft position, the I_0 as a function of time would exhibit dips in intensity, as shown in Fig. 2 by I_0 synch. Instead, in the asynchronous mode the trigger pulses follow with the frequency so that the filling pattern of the storage ring is shifted relative to the laser pulse by 2 ns (1/480 of the storage ring cycle) after each laser period. As a result measuring each point of the spectrum for at least 10 s (corresponding to 10^4 laser cycles) one observes a uniform intensity distribution for the incoming beam, as shown in Fig. 2 by I_0 asynch. It allows to avoid "bad time points" that correspond to the measurements of the kinetic histogram during the

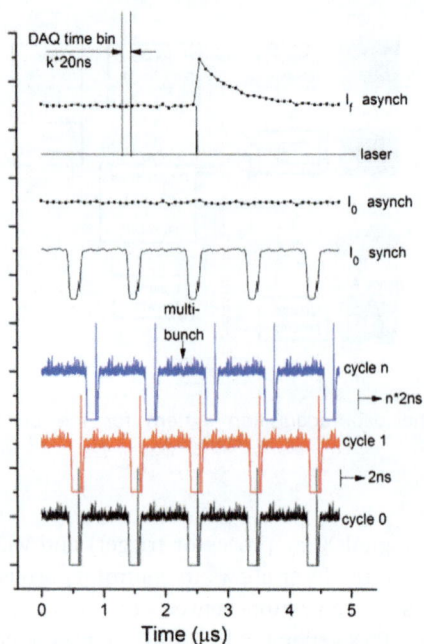

Fig. 2 Scheme of measurements using the synchronous and asynchronous modes of data acquisition system. Distribution of incoming X-ray intensity as a function of time is shown for a few individual cycles and for the averaged intensity in the synchronous and asynchronous mode. Position of the laser pulse and averaged intensity for the fluorescent detector are also shown.

time between multibunches of the storage ring. Additionally, we eliminated the need of incoming intensity measurements for each time point $I_0(t)$. Only the average value I_0 has to be measured. In the synchronized mode fluorescence intensity has to be normalized to the incoming intensity for each time point since bunches in the storage ring have a different current. Simplification of I_0 detection in the asynchronous mode has a positive impact on the signal to noise ratio of measured XAS spectra since the contribution of noise from I_0 measurements becomes negligible.

Detectors

Large area avalanche photo diodes (Advanced Photonix Inc, model 630-70-72-500) with a 200 mm^2 sensitive surface area and a ~400 μm sensor thickness were used as detectors. Rise time of signal from APD after the charge preamplifier (fast-ComTec, model CSP10) was measured and it is 30–40 ns (10–90%), depending on the device. The overall time resolution of the setup (that takes into account the contribution from the finite sampling frequency of the digital X-ray processor, APD and preamplifier contributions and all other synchronization jitters) is 30 ns (FWHM) as measured with only one bunch in the storage ring. The energy resolution of APDs is 700–800 eV at 5.9 keV which is not sufficient to discriminate efficiently the elastic scattering of X-rays and sample fluorescence. Therefore we used the combination of a Z-1 filter to reduce the unwanted background from the elastically scattered photons and Soller slits with conical geometry that partially protect the detector from re-fluorescence from the Z-1 filter. The distance between slit and the sample jet was fixed at 6 mm. The geometrical parameters of the slit were optimized numerically by finding a compromise between the efficiency of the slit and the minimization of the distance between sample and detector. Longer slits block re-fluorescence more efficiently, but the solid angle for the detector becomes smaller. As a result we have constructed a Soller slit with a 5.6 mm thickness that has an efficiency $\gamma = 0.962$ (only 3.8% of re-emitted photons will find a way to the detector) and with slit transmittance of 78% (22% of useful fluorescence photons will be blocked by the slit). Our DAQ is also compatible with silicon drift detectors that have lower time resolution and area, but at the same time have higher energy resolution and therefore can be used without Z-1 filter and Soller slits.

Samples

[Ru(bpy)$_3$]Cl$_2$, methyl viologen dichloride (MVCl$_2$), NH$_4$PF$_6$, NBu$_4$PF$_6$ and anhydrous acetonitrile were purchased from Sigma-Aldrich and used without further purification. As photo-sensitizer and electron-relay, [Ru(bpy)$_3$](PF$_6$)$_2$ and MV(PF$_6$)$_2$ were prepared from NH$_4$PF$_6$ and [Ru(bpy)$_3$]Cl$_2$ or MVCl$_2$, respectively, following a standard anion-exchange procedure. The catalyst [Co(dmgBF$_2$)$_2$(OH$_2$)$_2$] (dmg^{2-} = dimethylglyoximato dianion) was purchased from Strem Chemicals Inc. The sample consisted of an acetonitrile solution of [Ru(bpy)$_3$](PF$_6$)$_2$ (0.4 mM), MV(PF$_6$)$_2$ (8 mM), tetrabutylammonium hexafluorophosphate NBu$_4$PF$_6$ (0.1M) and [Co(dmgBF$_2$)$_2$(CH$_3$CN)$_2$] (0.8 mM), which forms spontaneously from [Co(dmgBF$_2$)$_2$(OH$_2$)$_2$] in CH$_3$CN solution. The solid-state metal complexes were handled in air. The freshly prepared sample solution was degassed with N$_2$ at least 30 min before the experiment and continuously purged and kept under N$_2$ during the experiment.

3 Theoretical method

Calculations of XANES spectra were performed using the full multiple scattering approach realized in the FEFF8 code.[36] The self-consistent potential within the muffin-tin approximation has been calculated for the cluster with a 4.9 Å radius. The energy-dependent exchange-correlation potential is obtained from the Hedin-Lundqvist approach. A cluster with radius 4.9 Å was also used for full multiple scattering calculations of spectra.

4 Application

The method has been applied for preliminary studies of the charge transfer in the multicomponent photo-catalytic system which consists of Co(dmgBF$_2$)$_2$ catalyst, Ru(bpy)$_3$$^{2+}$ chromophore and methyl viologen (MV^{2+}) that acts as the electron relay. This system was inspired from a similar study by Gray and coworkers using a Co(dpgBF$_2$)$_2$ catalyst (dpg^{2-} = diphenylglyoximato dianion) instead of Co(dmgBF$_2$)$_2$.[21] The initial Co species has a formal oxidation state of 2+. After photo-excitation of the chromophore the following reactions steps are hypothetically possible:

$$[\text{Ru(bpy)}_3]^{2+*} + \text{MV}^{2+} \rightarrow [\text{Ru(bpy)}_3]^{3+} + \text{MV}^+$$

$$[\text{Ru(bpy)}_3]^{2+*} + \text{Co(II)} \rightarrow [\text{Ru(bpy)}_3]^{3+} + \text{Co(I)}$$

$$\text{MV}^+ + \text{Co(II)} \rightarrow \text{Co(I)} + \text{MV}^{2+}$$

$$[\text{Ru(bpy)}_3]^{3+} + \text{Co(II)} \rightarrow [\text{Ru(bpy)}_3]^{2+} + \text{Co(III)}$$

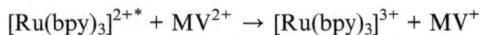

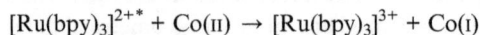

According to the time-resolved optical data reported for $\text{Co(dpgBF}_2)_2$,[21] the catalyst in the Co(III) state is formed in parallel with Co reduction at the conditions without any proton source. In most Co-based H_2-evolution systems,[5] Co(I) species play a crucial role in the catalytic mechanism: protonation of Co(I) yields Co-hydride intermediates which are the actual active species involved in H_2 evolution, either in the Co(III)-H state or in the Co(II)-H state obtained after a subsequent reducing event.

Our first goal has been to determine how many Co intermediates formed in the system. In the bottom panel of Fig. 3 we show the transient signal corresponding to different delay times between laser pump and registered X-ray photons. The width of the time window for the data corresponding to the long delay is significantly larger than for other two spectra to compensate the reduction of statistics due to the signal reduction, one can also notice that the data measured for the (0.25, 0.75) µs delay is noisier than the spectrum corresponding to a six times wider (−0.25, 2.75) µs time window. The amplitude of the transients has been re-normalized for easier comparison. The time-dependence of the signal monitored at a fixed X-ray energy corresponding to the maximum of the transient signal is shown in the top panel of Fig. 3. Since a liquid jet with continuous flow has been used to refresh the sample between two laser pulses the slow component of the decay can be caused by the movement of the excited volume out of the probed area, while the quick decay of the signal during the first 20 µs is related to the real kinetics of the system. Kinetic measurements utilizing the flow system are thus not as precise as those that one can imagine in the experiment with the full exchange of the sample during the pause before each laser pulse (the repetition rate of such measurements will be extremely low). Nevertheless, the experimental spectra (bottom panel of Fig. 3) measured at different time delays have very similar shapes indicating the formation of only one novel species upon irradiation of the photocatalytic system. The same result has been obtained from the analysis of the shape of the transient signals as a function of time over a set of 200 experimental spectra using the method reported in ref. 37, which indicated the presence of only 2 independent spectral components in our data corresponding to the initial species together with one intermediate species.

Fig. 4 shows the experimental XANES spectrum of the photocatalytic system measured before the laser pulse together with the transient spectrum that corresponds to the 10 µs time window after the laser pulse. The position of the absorption edge of the intermediate species is shifted towards lower energies with regards to the initial Co(II) (first peak of the transient signal is positive), indicative of the formation of a more reduced species, likely the Co(I) intermediate mentioned above and expected to be the entry point into a catalytic cycle for H_2 evolution.

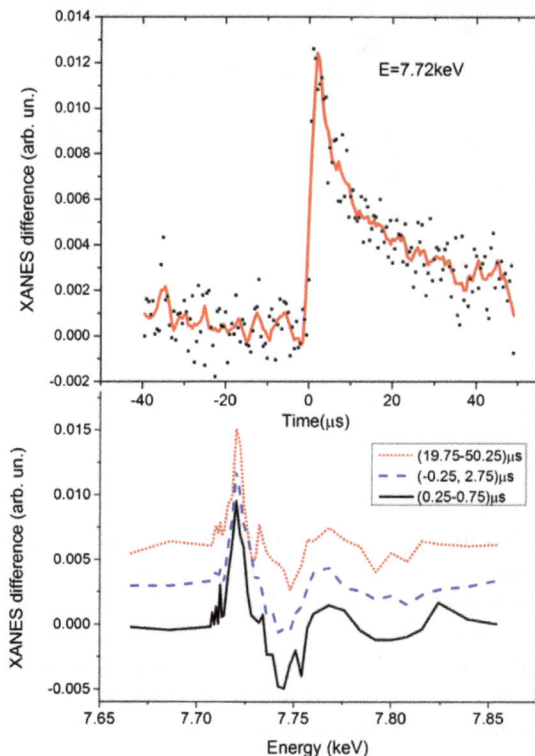

Fig. 3 Top Panel: Transient XANES signal measured for fixed energy of the incoming X-ray beam (E = 7.72keV) as a function of time between laser pump and X-ray probe. Bottom: Transient XANES spectra corresponding to different time windows after the laser pulse: (0.25, 0.75) μs, black line; (−0.25, 2.75) μs, blue line and (19.75, 50.25) μs, red line. Amplitudes were re-normalized for a better comparison.

The local structure around the metal center for molecules in solution can differ from that in the crystal phase measured using XRD and from that obtained by DFT optimization in vacuum. This is especially true for metal complexes possessing vacant coordination sites or labile ligands. $Co(dmgBF_2)_2$ belongs to this class of metal complexes and therefore we have started the simulations from the refinement of the structure of the initial Co(ii) state. Both XRD[38] and DFT[17] structures of $[Co(dmgBF_2)_2(CH_3CN)_2]$ have rather distant solvent molecules with bond lengths of 2.25 Å and 2.33 Å correspondingly, while some other Co(ii) complexes with distorted octahedral environment and acetonitrile in the first coordination shell have significantly shorter $Co-N_{Sol}$ bonds (2.08–2.09 Å) in the crystallized state.[39,40] Therefore in addition to XRD and DFT structures we have constructed models (3) and (4) with short $Co-N_{Sol}$ bonds (2.06 Å and 2.0 Å respectively). To take into account the expected disorder of the solvent molecules and avoid biases due to the strong multiple scattering along the Co–N–C path as found in the crystal and DFT-based structures, we have approximated the contribution of solvent to the XANES by taking into account the scattering from the nearest N atoms only.

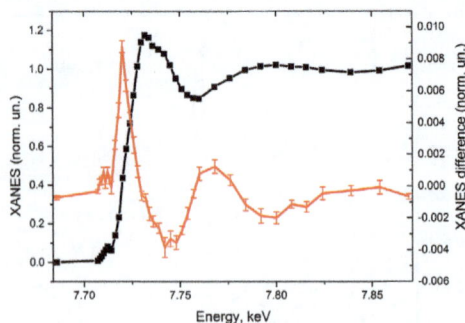

Fig. 4 Co K-edge XANES of the multicomponent photocatalytic system with Co(dmgBF$_2$)$_2$ catalyst in the ground state (black line, left scale) and transient X-ray absorption spectrum (red line, right scale) that corresponds to 10 μs time window after the laser pulse.

Fig. 5 shows that the agreement between the experimental spectrum and the spectra calculated for XRD (1) and DFT (2) models is not satisfactory. The relative intensity of peaks B and C is wrong. In addition, both calculated spectra display a shoulder between minimum D and maximum E that is not so well pronounced in the experimental data. Both models with short Co–N$_{Sol}$ bonds (3 and 4) agree reasonably well with the experiment. Variation of the Co–N$_{Sol}$ distance within these limits influences slightly the intensity of shoulder A which is observable, but rather weak in the experimental spectrum. As an additional small adjustment we have reduced the Co–N$_{eq}$ distances for models (3) and (4) by 0.03 Å relative to the XRD model. It has not changed the overall shape of the spectrum but has allowed to achieve a better agreement in the position of minimum D.

Experimentally determined reference structures for Co(ɪ) species with glyoxime ligands are very rare in the literature. The only available crystal structure of a Co(ɪ) complex with 2 diphenylglyoximato ligands and acetonitrile has been reported in ref. 12. Since the atomic cluster with radius 4.9 Å has been used to calculate X-ray absorption spectra, the substitution of methyl by phenyl group does not influence significantly the shape of XANES. According to this XRD model Co(ɪ) is five-coordinated in a distorted square-pyramidal environment with a short 1.97 Å Co–N$_{Sol}$ bond and 0.27 Å displacement of Co out of the plane formed by the equatorial nitrogen atoms. Acetonitrile is oriented almost perpendicularly to the equatorial plane with a Co–N–C angle of 174°. Hu et al. note that "solvent molecules were severely disordered and could not be refined accurately".[12] The DFT optimized structure for Co(ɪ) state reported by Muckerman and Fujita[17] has a significantly different orientation of the solvent molecules. Two of them were included in this model and the authors have found that both of them are oriented by the CH$_3$ group towards the Co atom with large Co–C distances of 3.7 Å and 4.8 Å, thus the Co atom is 4-coordinated.

Model (3) of the Co(ɪ) intermediate has been constructed starting from the model of the Co(ɪɪ) species with disordered solvents (curve (3) in Fig. 5) that gave a good agreement with the experiment. One of the solvent molecules has been moved away from the Co center to the distance 2.18 Å, while the second CH$_3$CN has been fixed at 2.06 Å. Additionally we have introduced a moderate out of plane

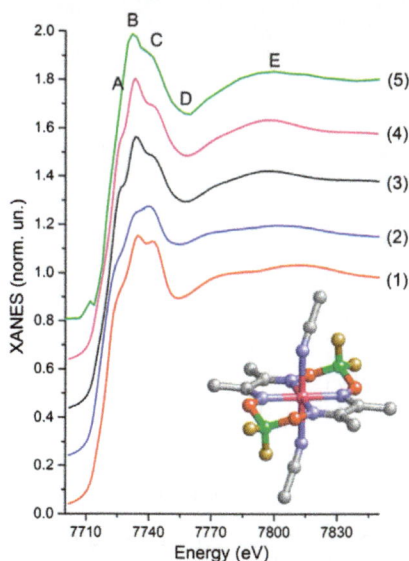

Fig. 5 Theoretical Co K-edge XANES spectra of [Co^{II}(dmgBF$_2$)$_2$(CH$_3$CN)$_2$] calculated for a model obtained using either X-ray diffraction[38] (red curve (1)), DFT optimization[17] (blue curve (2)), a model with disordered solvent molecules with Co–N$_{Sol}$ distances of 2.06 Å (black curve (3)) and 2.00 Å (magenta curve (4)). Curve (5) is the experimental spectrum of the multicomponent photocatalytic system with [Co(dmgBF$_2$)$_2$(CH$_3$CN)$_2$] catalyst measured before the laser pulse. The insert shows the structure of [Co(dmgBF$_2$)$_2$(CH$_3$CN)$_2$]. Grey atoms are C, red O, blue N, green B, yellow F, magenta Co. H atoms are omitted for simplicity.

displacement of the Co atom (0.14 Å) and a small contraction of Co–N$_{eq}$ (0.02 Å) as it has been suggested in both XRD and DFT models. Fig. 6 compares theoretical transient spectra calculated for these three models with the experimental pump–probe data. In addition to the structural changes we have taken into account 1 eV chemical shift of the absorption edge which is due to the movement of the core level as a result of changes of the screening by valence electrons. As one can see the shape of the transient signal is completely wrong for XRD model (1). The reason is similar to those observed also for the XRD structure of the ground state Co(II) species. Strong multiple scattering from the collinear Co–N–C chain of the solvent molecule influences the XANES significantly and therefore the model that ignores the disorder of solvent molecule fixing the solvent at idealized orthogonal orientation is not realistic. The DFT model is in better agreement with the experiment. A correct overall trend of the transient spectrum is seen, but all the main features are shifted in energy. Please note that even if the absolute scale of the energy calculations is not always correct we have performed already a few eV shift of the ground state spectrum and therefore there is no additional free parameter related with the independent shift of the transient spectrum. Among the 3 considered structures, model (3) gives the best agreement with the experiment. The theoretical spectrum is more structured between maxima A and B, which can be partially due to the fact that for the ground state peak A is more visible in the theory then in the experimental spectrum. Further improvement of

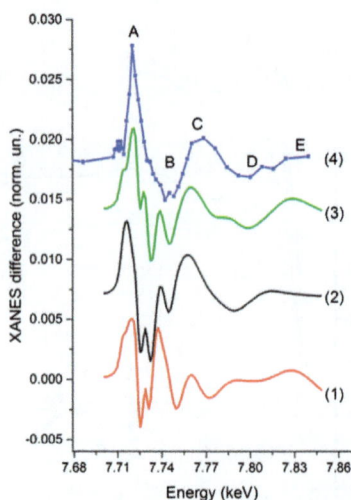

Fig. 6 Transient X-ray absorption spectrum corresponding to the transition from the Co(ii) to Co(i) state. Calculations for different models of Co(i) state based on XRD (red curve (1)), DFT (black curve (2)) and model with moderate out of plane displacement of Co and disordered solvent molecule (green curve (3)) are compared with the experimental data for the multicomponent photocatalytic system with Co(dmgBF$_2$)$_2$ catalyst (blue curve (4)).

the theoretical model can be based on more accurate modeling of disordered solvent around the metal complex by averaging many configurations of solvent molecules, for example using the approach previously proposed in ref. 41. Nevertheless, on the basis of our data we can conclude that the model with disordered solvent molecules with the shortest Co–N$_{Sol}$ distance ~2.06 Å and moderate displacement of Co out of the plane formed by N atoms of dmg ligands is the most realistic while the formation of a complex with perpendicular orientation of acetonitrile molecules has not been confirmed.

Discussing the experimental results shown in Fig. 4 that are plotted with the error bars we would like to mentioned that these measurements were taken during ~11 h and as one can see we have obtained a transient signal of very high quality. Please note that in comparison to the supramolecular systems,[42,43] or simple chromophorores,[44–46] the pump–probe signal for multicomponent systems is ~10 times weaker even if we are not limited by the pulse energy of the initial photoexcitation. Nevertheless, intermediates are relatively long-lived which somehow compensates the decrease of the statistics from the reduction of the signal amplitude. The incoming intensity of X-ray radiation at the bending magnet beamline of SLS is lower than, for example, at the undulator beamlines of the APS synchrotron. Paradoxically it can be seen also as an advantage since lower intensity allows to increase the detection efficiency using the single photon counting approach and at the same time to decrease the X-ray induced damage of the sample. In our previous work on the multicomponent system with similar Co-based catalyst that has been performed at 11ID-D beamline of the APS we observed a significant damage of the sample due to X-rays (incident X-ray flux ~6 × 10^{12} photons s^{-1}) while it was negligible at the SuperXAS beamline of the SLS (flux 3 × 10^{11} photons s^{-1}).[29] The number of photons registered by the fluorescence detector at the APS is around 3–4 per X-ray bunch for the samples with ~1mM concentration; therefore single photon counting is not possible and the data acquisition system works in the current mode. At SuperXAS of SLS for analogous sample we have a bit less than 1 photon μs^{-1} and therefore single photon counting is efficient. An additional advantage of our approach that has not been used for the present application of the setup is the possibility to use energy-resolving detectors, such as silicon drift detectors, that have a resolution of ~120 eV and therefore allow for a further increase of the signal to noise ratio for diluted samples.

Acknowledgements

L. X Chen, X. Zhang and V. Sundstrom are acknowledged for constructive discussion. This work was supported by the Swiss National Science Foundation (grant no. 200021-135226), the European Commission's Seventh Framework Programme (FP7/2007-2013) under grant agreement no. 290605 (COFUND: PSI-FELLOW), the French National Research Agency (ANR, Labex program ARCANE, ANR-11-LABX-0003-01) and the European Research Council (ERC grant agreement no. 306398). The COST Action CM1202 PERSPECT-H2O is also acknowledged. AG would like to thank Russian Foundation for Basic Research (project #14-02-31555).

References

1 A. Thapper, S. Styring, G. Saracco, A. W. Rutherford, B. Robert, A. Magnuson, W. Lubitz, A. Llobet, P. Kurz, A. Holzwarth, S. Fiechter, H. de Groot, S. Campagna, A. Braun, H. Bercegol and V. Artero, *Green*, 2013, **3**, 43–57.
2 N. S. Lewis and D. G. Nocera, *Proc. Natl. Acad. Sci. U. S. A.*, 2006, **103**, 15729–15735.
3 W. T. Eckenhoff and R. Eisenberg, *Dalton Trans.*, 2012, **41**, 13004–13021.
4 V. S. Thoi, Y. Sun, J. R. Long and C. J. Chang, *Chem. Soc. Rev.*, 2013, **42**, 2388–2400.
5 V. Artero, M. Chavarot-Kerlidou and M. Fontecave, *Angew. Chem., Int. Ed.*, 2011, **50**, 7238–7266.
6 C. Baffert, V. Artero and M. Fontecave, *Inorg. Chem.*, 2007, **46**, 1817–1824.
7 X. Hu, B. M. Cossairt, B. S. Brunschwig, N. S. Lewis and J. C. Peters, *Chem. Commun.*, 2005, 4723–4725.
8 A. Fihri, V. Artero, A. Pereira and M. Fontecave, *Dalton Trans.*, 2008, 5567–5569.
9 A. Fihri, V. Artero, M. Razavet, C. Baffert, W. Leibl and M. Fontecave, *Angew. Chem., Int. Ed.*, 2008, **47**, 564–567.
10 P. Du, J. Schneider, G. Luo, W. W. Brennessel and R. Eisenberg, *Inorg. Chem.*, 2009, **48**, 4952–4962.
11 C. Li, M. Wang, J. Pan, P. Zhang, R. Zhang and L. Sun, *J. Organomet. Chem.*, 2009, **694**, 2814–2819.
12 X. Hu, B. S. Brunschwig and J. C. Peters, *J. Am. Chem. Soc.*, 2007, **129**, 8988–8998.
13 P. Zhang, P.-A. Jacques, M. Chavarot-Kerlidou, M. Wang, L. Sun, M. Fontecave and V. Artero, *Inorg. Chem. Commun.*, 2012, **26**, 51.

14 A. Bhattacharjee, M. Chavarot-Kerlidou, E. S. Andreiadis, M. Fontecave, M. J. Field and V. Artero, *Inorg. Chem.*, 2012, **51**, 7087–7093.

15 E. Szajna-Fuller and A. Bakac, *Eur. J. Inorg. Chem.*, 2010, 2488–2494.

16 T. Lazarides, T. McCormick, P. Du, G. Luo, B. Lindley and R. Eisenberg, *J. Am. Chem. Soc.*, 2009, **131**, 9192–9194.

17 J. T. Muckerman and E. Fujita, *Chem. Commun.*, 2011, **47**, 12456–12458.

18 B. H. Solis and S. Hammes-Schiffer, *Inorg. Chem.*, 2011, **50**, 11252–11262.

19 B. S. Veldkamp, W.-S. Han, S. M. Dyar, S. W. Eaton, M. A. Ratner and M. R. Wasielewski, *Energy Environ. Sci.*, 2013, **6**, 1917–1928.

20 J. L. Dempsey, J. R. Winkler and H. B. Gray, *J. Am. Chem. Soc.*, 2010, **132**, 16774–16776.

21 J. L. Dempsey, J. R. Winkler and H. B. Gray, *J. Am. Chem. Soc.*, 2010, **132**, 1060–1065.

22 B. Probst, C. Kolano, P. Hamm and R. Alberto, *Inorg. Chem.*, 2009, **48**, 1836–1843.

23 L. X. Chen, X. Zhang, J. V. Lockard, A. B. Stickrath, K. Attenkofer, G. Jennings and D. J. Liu, *Acta Crystallogr., Sect. A: Found. Crystallogr.*, 2010, **66**, 240–251.

24 C. Bressler and M. Chergui, *Annu. Rev. Phys. Chem.*, 2010, **61**, 263–282.

25 T. Sato, S. Nozawa, K. Ichiyanagi, A. Tomita, M. Chollet, H. Ichikawa, H. Fujii, S. Adachi and S. Koshihara, *J. Synchrotron Radiat.*, 2009, **16**, 110–115.

26 A. M. March, A. Stickrath, G. Doumy, E. P. Kanter, B. Krässig, S. H. Southworth, K. Attenkofer, C. A. Kurtz, L. X. Chen and L. Young, *Rev. Sci. Instrum.*, 2011, **82**, 073110.

27 M. Tromp, A. J. Dent, J. Headspith, T. L. Easun, X.-Z. Sun, M. W. George, O. Mathon, G. Smolentsev, M. L. Hamilton and J. Evans, *J. Phys. Chem. B*, 2013, **117**, 7381–7387.

28 E. A. Stern, D. L. Brewe, K. M. Beck, S. M. Heald and Y. Feng, *Phys. Scr.*, 2005, 1044.

29 G. Smolentsev, A. Guda, X. Zhang, K. Haldrup, E. S. Andreiadis, M. Chavarot-Kerlidou, S. E. Canton, M. Nachtegaal, V. Artero and V. Sundstrom, *J. Phys. Chem. C*, 2013, **117**, 17367–17375.

30 D. J. Thiel, P. Līviņš, E. A. Stern and A. Lewis, *Nature*, 1993, **362**, 40–43.

31 H. Wang, G. Peng and S. P. Cramer, *J. Electron Spectrosc. Relat. Phenom.*, 2005, **143**, 1–7.

32 E. M. Scheuring, W. Clavin, M. D. Wirt, L. M. Miller, R. F. Fischetti, Y. Lu, N. Mahoney, A. Xie, J. Wu and M. R. Chance, *J. Phys. Chem.*, 1996, **100**, 3344–3348.

33 A. B. Stickrath, M. W. Mara, J. V. Lockard, M. R. Harpham, J. Huang, X. Zhang, K. Attenkofer and L. X. Chen, *J. Phys. Chem. B*, 2013, **117**, 4705–4712.

34 M. Haumann, P. Liebisch, C. Muller, M. Barra, M. Grabolle and H. Dau, *Science*, 2005, **310**, 1019–1021.

35 F. A. Lima, C. J. Milne, D. C. V. Amarasinghe, M. H. Rittmann-Frank, R. M. van der Veen, M. Reinhard, V.-T. Pham, S. Karlsson, S. L. Johnson, D. Grolimund, C. Borca, T. Huthwelker, M. Janousch, F. van Mourik, R. Abela and M. Chergui, *Rev. Sci. Instrum.*, 2011, **82**, 063111.

36 A. L. Ankudinov, B. Ravel, J. J. Rehr and S. D. Conradson, *Phys. Rev. B: Condens. Matter Mater. Phys.*, 1998, **58**, 7565.

37 G. Smolentsev, G. Guilera, M. Tromp, S. Pascarelli and A. V. Soldatov, *J. Chem. Phys.*, 2009, **130**, 174508.

38 J. Niklas, K. L. Mardis, R. R. Rakhimov, K. L. Mulfort, D. M. Tiede and O. G. Poluektov, *J. Phys. Chem. B*, 2012, **116**, 2943–2957.

39 J. P. Bigi, T. E. Hanna, W. H. Harman, A. Chang and C. J. Chang, *Chem. Commun.*, 2010, **46**, 958–960.

40 C. C. Lu, E. Bill, T. Weyhermüller, E. Bothe and K. Wieghardt, *Inorg. Chem.*, 2007, **46**, 7880–7889.

41 P. D'Angelo, O. M. Roscioni, G. Chillemi, S. Della Longa and M. Benfatto, *J. Am. Chem. Soc.*, 2006, **128**, 1853–1858.

42 S. E. Canton, X. Zhang, J. Zhang, T. B. van Driel, K. S. Kjaer, K. Haldrup, P. Chabera, T. Harlang, K. Suarez-Alcantara, Y. Liu, J. Pérez, A. Bordage, M. Pápai, G. Vankó, G. Jennings, C. A. Kurtz, M. Rovezzi, P. Glatzel, G. Smolentsev, J. Uhlig, A. O. Dohn, M. Christensen, A. Galler, W. Gawelda, C. Bressler, H. T. Lemke, K. B. Møller, M. M. Nielsen, R. Lomoth, K. Wärnmark and V. Sundström, *J. Phys. Chem. Lett.*, 2013, **4**, 1972–1976.

43 G. Smolentsev, S. E. Canton, J. V. Lockard, V. Sundstrom and L. X. Chen, *J. Electron Spectrosc. Relat. Phenom.*, 2011, **184**, 125–128.

44 W. Gawelda, M. Johnson, F. M. F. de Groot, R. Abela, C. Bressler and M. Chergui, *J. Am. Chem. Soc.*, 2006, **128**, 5001–5009.

45 G. Smolentsev, A. V. Soldatov and L. X. Chen, *J. Phys. Chem. A*, 2008, **112**, 5363–5367.

46 L. X. Chen, W. J. Jager, G. Jennings, D. J. Gosztola, A. Munkholm and J. P. Hessler, *Science*, 2001, **292**, 262–264.

Faraday Discussions

ROYAL SOCIETY OF CHEMISTRY

PAPER

Dynamics in next-generation solar cells: time-resolved surface photovoltage measurements of quantum dots chemically linked to ZnO (10$\bar{1}$0)

Ben F. Spencer,*[ab] Matthew J. Cliffe,[ab] Darren M. Graham,[a]
Samantha J. O. Hardman,[c] Elaine A. Seddon,[ab] Karen L. Syres,[d]
Andrew G. Thomas,[a] Fausto Sirotti,[e] Mathieu G. Silly,[e] Javeed Akhtar,[fg]
Paul O'Brien,[f] Simon M. Fairclough,[h] Jason M. Smith,[i]
Swapan Chattopadhyay[b] and Wendy R. Flavell[a]

Received 21st February 2014, Accepted 11th March 2014

DOI: 10.1039/c4fd00019f

The charge dynamics at the surface of the transparent conducting oxide and photoanode material ZnO are investigated in the presence and absence of light-harvesting colloidal quantum dots (QDs). The time-resolved change in surface potential upon photoexcitation has been measured in the *m*-plane ZnO (10$\bar{1}$0) using a laser pump-synchrotron X-ray probe methodology. By varying the oxygen annealing conditions, and hence the oxygen vacancy concentration of the sample, we find that dark carrier lifetimes at the ZnO surface vary from hundreds of μs to ms timescales, *i.e.* a persistent photoconductivity (PPC) is observed. The highly-controlled nature of our experiments under ultra-high vacuum (UHV), and the use of band-gap and sub-band-gap photoexcitation, allow us to demonstrate that defect states *ca.* 340 meV above the valence band edge are directly associated with the PPC, and that the PPC mediated by these defects dominates over the oxygen photodesorption mechanism. These observations are consistent with the hypothesis that ionized oxygen vacancy states are

[a]School of Physics and Astronomy and the Photon Science Institute, The University of Manchester, Manchester M13 9PL, United Kingdom. E-mail: ben.spencer@manchester.ac.uk

[b]The Cockcroft Institute, Sci-Tech Daresbury, Keckwick Lane, Daresbury, Warrington WA4 4AD, Cheshire, United Kingdom

[c]Manchester Institute of Biotechnology, Faculty of Life Sciences, University of Manchester, 131 Princess Street, Manchester M1 7DN, United Kingdom

[d]School of Chemistry, The University of Nottingham, University Park, Nottingham NG7 2RD, United Kingdom

[e]Synchrotron SOLEIL, BP 48, Saint-Aubin, F91192 Gif sur Yvette CEDEX, France

[f]Department of Chemistry, University of Manchester, Oxford Road, Manchester M13 9PL, United Kingdom

[g]Department of Physics, Nano-Science & Materials Synthesis Laboratory, COMSATS Institute of Information Technology, Chakshahzad Park Road, Islamabad 44000, Pakistan

[h]Department of Chemistry, University of Oxford, South Parks Road, Oxford OX1 3QR, United Kingdom

[i]Department of Materials, University of Oxford, Parks Road, Oxford OX1 3PH, United Kingdom

responsible for the PPC in ZnO. The effect of chemically linking two colloidal QD systems (type I PbS and type II CdS–ZnSe) to the surface has also been investigated. Upon deposition of the QDs onto the surface, the dark carrier lifetime and the surface photovoltage are reduced, suggesting a direct injection of charge carriers into the ZnO conduction band. The results are discussed in the context of the development of next-generation solar cells.

Introduction

The urgent need to reduce the cost of solar energy technology has led to an increasing interest in the transparent conducting oxide (TCO) ZnO as a potential photoanode,[1–3] particularly as part of next-generation solar cells utilising colloidal quantum dots (QDs) as the light harvesters.[4–6]

ZnO exhibits a persistent photoconductivity (PPC) after the excitation source has been switched off which, along with its large band gap of $\sim$3.4 eV,[1,2,7] makes it ideal as a photoanode material.[1] However, the origin of this PPC has been controversial:[1] for some years the capture of holes by chemisorbed oxygen was thought to be the primary mechanism,[8–14] and the surrounding oxygen environment has also been shown to influence the lifetime of PPC.[15,16] More recent theoretical work has instead suggested that oxygen vacancies play an important part in the PPC mechanism because band-gap states associated with metastable doubly-[17–19] or singly-charged[20] oxygen vacancies control the PPC, and recent experimental work has now strengthened this hypothesis.[21,22]

Colloidal QDs are nanometre-sized semiconductor crystals with sizes comparable to the Bohr radius of an exciton in the corresponding bulk material. This quantum confinement means that the effective band gap is tuned with the size of the nanocrystal.[23,24] QDs may also be especially useful for photovoltaic applications since many studies have shown that multiple carriers can be generated with increased efficiency over bulk materials by a single photon that is in excess of the band gap because the excess energy loss by phonon absorption is reduced; instead this excess energy creates additional carriers.[25–27] These quantum effects allow for the theoretical maximum efficiency of single junction solar cells (the Shockley–Queisser limit of approximately 30%)[28] to be overcome. Carrier multiplication has been identified in many QD materials including PbS,[5] PbSe,[26] InP,[29] InAs,[30] CdSe,[31] CdTe[32] and Si.[33] Photovoltaic devices using PbS QDs with carrier multiplication have already been implemented.[34]

A 'core–shell' structure may be introduced to passivate the QD or to engineer its electronic structure. So-called 'type I' QDs are those where on photoexcitation, the electrons and holes are both confined to either the core or shell of the nanocrystal. More recently, type II core–shell structures, which have a staggered band alignment, have become of greater interest because the photogenerated electron and hole can be separately localized in the core and shell of the QD, or *vice versa*. The core and shell thicknesses can effectively control the electron and hole wavefunction overlap and hence the recombination lifetime.[35–37] Type II core–shell QD structures studied so far include CdS–ZnSe,[35] CdSe–ZnTe,[38] CdTe–CdS,[39] CdTe–ZnTe,[40] CdTe–CdSe[41] and ZnTe–ZnSe.[42,43]

A proposed model for a next-generation solar cell includes the use of colloidal QDs as light harvesters and ZnO as the photoanode. Clearly the charge carrier dynamics at the interface must be well understood in order for such a solar cell to be designed. These may be probed through the surface potential change upon photoexcitation in the ZnO substrate. Upon photoexcitation, in the presence of a surface depletion layer (for an n-type semiconductor), electrons are promoted into the conduction band and migrate into the bulk, reducing the amount of band bending at the surface (Fig. 1). This change in the surface potential upon photoexcitation is known as the surface photovoltage (SPV) effect.[44,45] This effect can be observed using photoemission spectroscopy, or by illuminating the surface with laser radiation of energy larger than the band gap. Photoexcitation using a pulsed laser then allows for the time-dependent SPV to be probed using X-ray photoelectron spectroscopy (XPS).[21]

SPV measurements have been made on a variety of materials using both optical[46-56] and free-electron laser (FEL) pump beams[57,58] and a variety of experimental techniques, as recently reviewed by Yamamoto and Matsuda.[59] For example, Widdra et al.[50] and Bröcker et al.[51] utilized the BESSY synchrotron in single-bunch mode to provide a time window of 800 ns to study the SiO_2–Si (100) interface; a similar methodology was employed in measurements by us at the UK Synchrotron Radiation Source (SRS) in single-bunch mode to study the Si (111) 7 × 7 surface over a 320 ns time window.[21]

The PPC displayed by ZnO requires the use of transient SPV measurements on much longer timescales (µs to seconds). For these measurements we have used a modulated continuous-wave laser in conjunction with ns XPS at Synchrotron SOLEIL. The time period of the experiment is set by a signal generator that

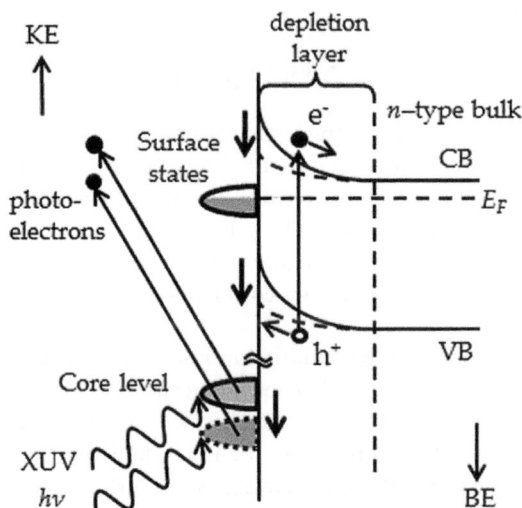

Fig. 1 Non-equilibrium SPV in a conventional n-type semiconductor. Laser illumination promotes electrons (e^-) across the band gap from the valence band (VB) to the conduction band (CB), which then migrate into the bulk due to the presence of a depletion layer. The corresponding holes (h^+) migrate to the surface. The electric field within the space charge region and hence the band bending are reduced (dashed lines). The binding energy (BE) of the core electron energy levels are thus increased, reducing the kinetic energy (KE) of the photoelectrons liberated upon X-ray absorption. Here, E_F denotes the Fermi level prior to photoexcitation, which is pinned by the surface states.

modulates the laser and triggers the XPS measurements, as detailed below. We study the effect of varying the oxygen vacancy content in the m-plane ZnO upon the transient SPV using band gap and sub-band gap laser radiation which demonstrates that sub-band gap states associated with oxygen vacancies are responsible for the PPC in ZnO. We show a consistent decrease in the SPV lifetimes upon photoexcitation when type I PbS QDs or type II CdS–ZnSe QDs are chemically linked to the ZnO substrate using 3-MPA ligands. The increase in lifetimes suggests that the attached QDs allow for direct injection of carriers into the conduction band of ZnO. Given the PPC in ZnO is controlled by oxygen vacancy concentration (*i.e.* sample preparation) this indicates that a QD–ZnO system could be a suitable basis for next-generation photovoltaics.

SPV theory

The amount of band bending at a semiconductor surface changes under photoexcitation. The total change in the band bending at the surface, or surface photovoltage ΔV_{SP}^{tot}, upon illumination is described by:[60]

$$\frac{\Delta V_{SP}^{tot}}{kT} \exp\left(\frac{\Delta V_{SP}^{tot}}{kT}\right) = \frac{n_P}{n_0} \exp\left(\frac{V_0}{kT}\right). \tag{1}$$

Here, n_0 is the doping carrier concentration, n_P is the photoexcited carrier concentration and V_0 is the equilibrium band bending. In our experiment, where we measure the change in SPV, ΔV_{SP}, induced by the laser illumination, the photoexcited carrier concentration is determined using the laser fluence, energy and absorption coefficient. A change in the surface potential also affects the photoexcited carrier lifetime, τ,[50]

$$\tau = \tau_\infty \exp\left(\frac{-\Delta V_{SP}}{\alpha kT}\right) \tag{2}$$

where α is a material parameter (typically with values ranging from 0.5 to 2)[51] and τ_∞ is the dark carrier lifetime (the lifetime of carriers in the absence of a surface photovoltage). The parameter α is likened to the ideality factor in a Schottky diode.[44] A theoretical study by Schulz *et al.* on p-type silicon (with a doping level of 10^{15} cm^{-3}) showed that values for α were consistently less than 1, and that the parameter also correlated with the equilibrium band bending V_0.[61]

After photoexcitation, the recombination rate is assumed to be limited by the process of overcoming the barrier induced by the band bending by thermionic emission across the depletion layer.[50] The SPV shift reduces in a dynamic way as recombination occurs (eqn (1)), and thus the photoexcited carrier lifetime increases with time (eqn (2)) as the surface potential returns to equilibrium. The decay of the SPV after the laser is switched off is thus modelled as a constant deceleration, as developed in Widdra *et al.*[50] and Bröcker *et al.*[51] For the $\Delta V_{SP}^{tot} > kT$ case, the decay of the SPV shift over time, $\Delta V_{SP}(t)$, can be described by:[50]

$$\Delta V_{SP}(t) = -\alpha kT \ln\left(\exp\left(\frac{-\Delta V_{SP}^{tot}}{\alpha kT}\right) + \frac{t}{\tau_\infty}\right). \tag{3}$$

For the case where $\Delta V_{SP}^{tot} \sim kT$, a more general form has been proposed:[51]

$$\Delta V_{SP}(t) = -\alpha kT \ln\left[1 - \exp\left(\frac{-t}{\tau_\infty}\right)\left(1 - \exp\left(\frac{-\Delta V_{SP}^{tot}}{\alpha kT}\right)\right)\right]. \tag{4}$$

The onset of the pump-induced SPV change when the laser is switched on may be modelled by a single exponential if the rate of carrier creation far exceeds recombination (*i.e.* at sufficiently high fluence).[44] Otherwise, a bi-exponential or a 'decelerated' exponential model analogous to eqn (4) may be appropriate. The latter reflects the dynamic increase in recombination rate as the surface band bending is reduced, which acts to counterbalance the rate of carrier creation.

In this model, the pump-induced change in the surface band bending is logarithmically dependent on the number of induced charge carriers (eqn (1)) and hence the photoexcitation fluence, Φ,

$$\Delta V_{SP}^{tot}(\Phi) = \alpha kT \ln(1 + \gamma\Phi), \tag{5}$$

where γ is another material parameter.[51]

Experimental

Time-resolved laser pump-synchrotron XPS probe experiments were carried out using a laser in combination with the TEMPO beamline at Synchrotron SOL-EIL.[21,62] A 10 mW CW laser (Coherent, CUBE) operating at 372 nm (3.33 eV) was modulated using a square-wave signal from a pulse generator, typically switching the laser on every 0.5 ms (*i.e.* a repetition rate of 2 kHz) to 2 ms (500 Hz), giving a fluence of approximately 25 µJ cm^{-2}. The pulse generator was also used to simultaneously trigger in-house software recording a narrow BE-range XPS spectrum (with a ~2 eV window) every 50 ns. These spectra were recorded using a SCIENTA SES 2002 analyser with a two-dimensional (2D) delay-line detector.[63,64] The time resolution was determined to be approximately 150 ns, which was

Fig. 2   Image of the laser beam overlapping the X-ray probe beam on the ZnO sample (measuring 5 × 10 mm) taken using a CCD camera at the TEMPO beamline, SOLEIL.

limited by the time difference in the signals from the delay-line detector and the speed of the electronics. Data were recorded over the time period of the pulse generator, which had a 50% duty cycle meaning the laser was illuminating the sample for half of the time period. In excess of 10 000 accumulations were required to achieve satisfactory signal-to-noise ratios. Spatial overlap of the X-ray probe beam (measuring 150 μm vertically by 100 μm horizontally) with the laser pump beam was achieved using a charge-coupled device (CCD) camera (Fig. 2). A photon energy of 200 eV was used to examine the Zn $3d$ core level, and the typical experimental resolution was 150 meV (monochromator + analyser). Care was taken to check for and eliminate sample charging: static measurements with and without laser illumination were measured repetitively to ensure that the spectrum returned exactly to its original BE position once illumination had stopped before transient measurements were started.

Materials

ZnO preparation

The m-plane ZnO ($10\bar{1}0$) surface was prepared using an established recipe.[65–69] The surface n-type conductivity of ZnO may be enhanced in UHV by sputtering, a process that creates donors such as oxygen vacancies at the surface[70–72] (possibly together with other defects and defect complexes with oxygen vacancies).[1,2,66] Careful sample preparation is required to enhance the surface concentration of these vacancies in order to avoid sample charging in photoemission. The sample underwent three cycles of argon ion sputtering and electron beam annealing, up to a temperature of 1023–1043 K. Following this, the sample was then annealed in 1.2–1.4 × 10^{-7} mbar oxygen at 703 K in order to heal some of the excess oxygen vacancies created by sputtering at the surface.[66] This step is important in controlling the final conductivity of the surface. In order to explore the effect of the oxygen vacancy concentration on the dynamics, the annealing period was varied between 10 and 20 min, creating different oxygen vacancy (and hence donor) concentrations in the near-surface layers.[73–75] The sample was then allowed to cool in the presence of oxygen, before a low temperature anneal (603 K, 20 min), followed by a short high temperature anneal *in vacuo* (1023–1043 K, 10 min),[76] completing the cleaning process. A final high temperature flash anneal *in vacuo* has previously been used to remove residual adsorbed oxygen;[76] here it was found to be necessary to eliminate charging during the pump–probe experiment.

The surface was diagnosed as uncontaminated using low energy electron diffraction, where a sharp ZnO ($10\bar{1}0$) 1 × 1 pattern was obtained,[68] and XPS showing no C $1s$ signal. Measurements were carried out at room temperature under UHV at pressures in the 1–2 × 10^{-10} mbar range.

Colloidal QD samples

The preparation and characterisation of the colloidal type I PbS QD sample has been described previously.[5,6] Briefly, the QDs were prepared using a novel environmentally-benign method that employed olive oil as both the solvent and capping agent.[6] The long-chain olive oil capping groups were found to be highly insulating, and so these were exchanged for 3-mercaptopropionic acid (3-MPA) ligands, which resulted in samples free from charging effects during photoemission measurements, indicating that charge transport into and out of the QDs was possible.[5] The QDs were characterised as having a mean diameter of 4.6 nm determined from a 1S absorption feature at 1.0 eV,[5,77] with well-defined excitonic features in the absorption spectrum which indicates a relatively small size dispersion.[5] These PbS QDs were also shown to exhibit a carrier multiplication for photon energies above ~2.5 times the effective (and tunable) band gap energy (*ca.* 1.0 eV here).[5] These measurements used a home-built transient absorption spectrometer described previously.[29]

The CdS core of the type II core–shell CdS–ZnSe QD sample was synthesized using a modified route previously used for synthesis of CdSe QDs where oleylamine was used as the sole surfactant.[78] The shelling technique used was that detailed by Blackman *et al.*;[79] precursors were injected into the CdS core solution under nitrogen (first the Zn precursor at 523 K, followed by the Se precursor five minutes later) before the temperature was raised to 553 K for twenty minutes.[79] Absorption and photoluminescence (PL) measurements showed a red shift in the first excitonic peak of 130 nm compared to the 5.1 nm diameter CdS core-only QDs (a PL peak shift from 470 to 600 nm as shown in Fig. 3). PL lifetime measurements (Fig. 4) confirm the type II behaviour of this sample, where electrons and holes are confined to the core and shell, respectively (unless photon energies greater than the shell effective band gap are used). The lifetime is fitted with a tri-exponential decay with the longest component at over 70 ns, indicating a type II structure with a significant reduction in the electron and hole wavefunction overlap caused by spatial separation.[37,42,80,81] PL lifetimes are consistently longer in type II than in type I QDs, for example, Kim *et al.* showed that PL lifetimes increased from 9.6 ns for CdTe QDs to 57 ns once a CdSe shell was added.[37] Type I CdS and CdSe QDs typically have PL lifetimes of ~10 ns.[82,83] Finally, transmission electron microscopy (TEM) images (Fig. 5) show peanut-shaped

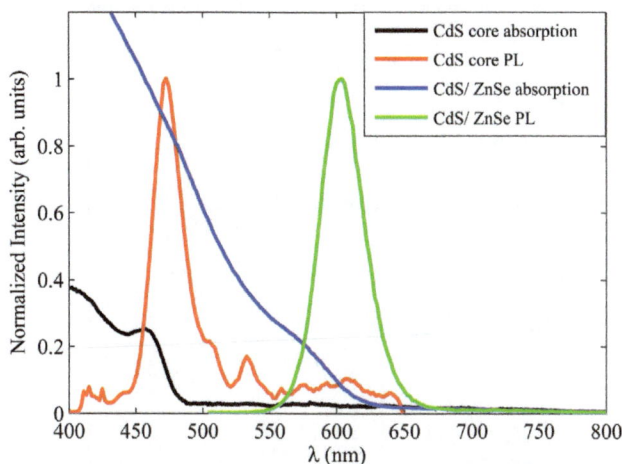

Fig. 3 Absorption and photoluminescence (PL) spectra for CdS core-only QDs and for CdS–ZnSe core–shell QDs, showing a red shift of the PL peak from 470 to 600 nm. The increased width (and the associated spreading seen in the absorption spectra) indicates an increase in size dispersion for the CdS–ZnSe QDs compared to the core-only CdS QDs.

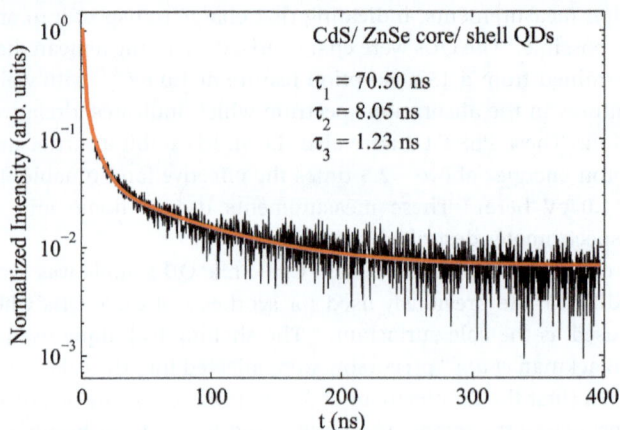

Fig. 4 PL lifetime of the CdS–ZnSe QDs which is fitted with a tri-exponential fit with lifetimes of 70.50, 8.05 and 1.23 ns.

CdS–ZnSe QDs of approximately 6.4 nm width and 8.4 nm in length with a size dispersion of ~9%.

Ligand exchange between oleylamine and 3-MPA was carried out using the methodology by Aldana *et al.*[84] and the samples were held in solution in chloroform. The QDs were deposited from solution onto the ZnO held *ex situ* outside the UHV chamber for less than one minute to minimize contamination,[5] and the substrate was washed with solvent to remove any QDs not chemically linked to the surface. The presence of the dots attached to the substrate was verified with XPS.

Once chemically attached to ZnO, the energy-level line-up is such that the lowest unoccupied molecular orbital (LUMO) of the QD lies at higher energy than

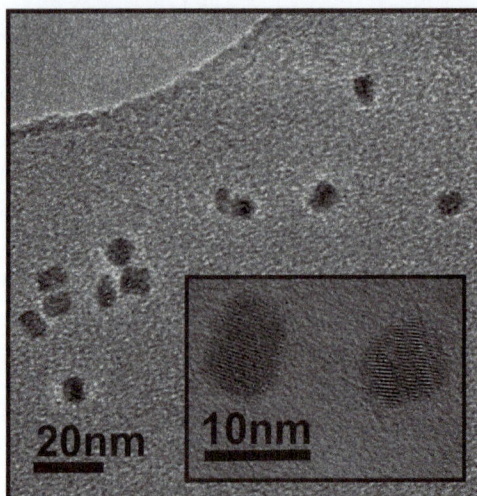

Fig. 5 Low and high (inset) resolution TEM image of the CdS–ZnSe core–shell QDs, showing peanut-shaped QDs with 6.4 nm width and 8.4 nm length with a size dispersion of approximately 9%.

the conduction band minimum in ZnO,[5,6,85,86] and carriers created in the PbS QD or in the shell of the CdS–ZnSe QD (promoted from the highest occupied molecular orbital, or HOMO) may be directly injected into the ZnO CB as shown in Fig. 6. The alignment obtained for the CdS–ZnSe QDs uses the work of Wang et al.,[85] Klimov et al.,[35] Ivanov and Achermann[86] and the absorption spectrum obtained for the sample (Fig. 3).

Results

ZnO (10$\bar{1}$0) with varied oxygen vacancy concentrations

A previous study of the ZnO (10$\bar{1}$0) surface at the SRS, Daresbury Laboratory, indicated no transient change in the SPV upon illumination over the 320 ns time window of these experiments, but a constant SPV shift of 115 meV at all pump–probe delay times, indicating PPC.[21] In order to study longer SPV decay times, we have used the time-resolved XPS facilities at the TEMPO beamline at SOLEIL. Fig. 7 shows XPS of the Zn 3d core level with and without 372 nm laser

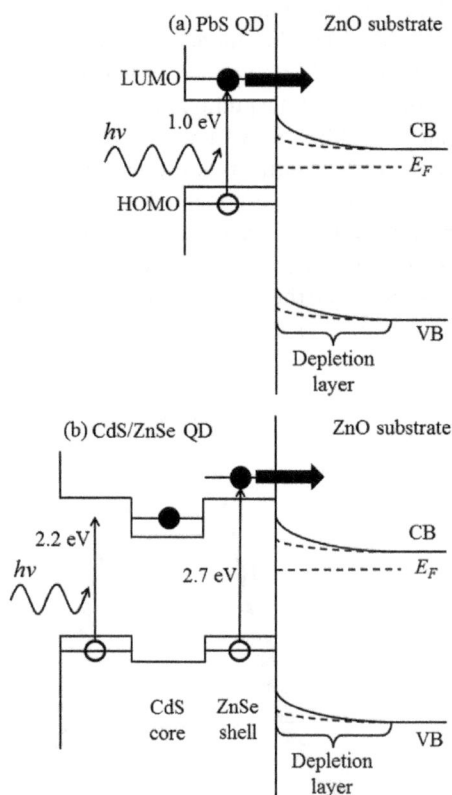

Fig. 6 Schematic energy-level line-up diagrams for (a) PbS QDs and (b) CdS–ZnSe QDs chemically linked to ZnO. The effective band gap energies were determined from the 1S absorption feature in the absorption spectra. In the CdS–ZnSe QD, for photon energies less than 2.7 eV, the type II structure of the QD will trap electrons (filled circles) in the core and holes (open circles) in the shell. For photon energies greater than 2.7 eV electrons in the shell may be photoexcited into the ZnSe CB and injected into the ZnO CB.

illumination; the second 'laser off' spectrum overlies the first, meaning the SPV shift was exactly removed in the absence of laser photoexcitation. The semicore Zn $3d$ level has a complex peak shape influenced by interactions with VB states[87] and is consistent with previous studies of this surface.[88]

In order to probe the PPC of the ZnO surface, time-resolved XPS measurements of the SPV decay following photoexcitation were carried out. The time-dependence of the laser-induced Zn $3d$ core level shift is shown in Fig. 8. As oxygen vacancies have been implicated in the PPC of ZnO,[19,20,89-91] the influence of a change in the concentration of oxygen vacancies (and hence donors), achieved by altering the length of the oxygen annealing cycle, was investigated. Fig. 8 shows the effect of changing the length of this part of the annealing cycle from 20 min to 10 min (Fig. 8(a) and 8(b), respectively). All other experimental conditions are identical. For the 20 min oxygen anneal case, the binding energy of the Zn $3d$ peak is plotted as the laser is switched on at 0 ms, and off after 0.5 ms (shown in the magnified section of Fig. 8(a)). When the laser is switched on, a total core-level shift of 48 meV to a higher binding energy is observed, but the rise time of the shift is very long. The SPV shift reaches its maximum value after approximately 0.1 ms. When the laser is switched off, the SPV shift decays back to equilibrium slowly, over almost half a millisecond.

We use self-decelerating relaxation models to obtain characteristic lifetimes for both the SPV onset and its decay. Eqn (4), rather than eqn (3), was found to provide the best fit of the decay of the pump-induced SPV. This is because in this experiment, the total SPV shift of 48 meV is comparable to kT ($\sim$26 meV), requiring use of the more general form of the expression for the SPV shift (in place of eqn (3)). Likewise, a decelerating exponential increase, analogous to eqn (4), fitted the onset of the SPV shift well. A dark carrier lifetime, τ_∞, of 150 μs is found for the 20 min oxygen anneal case (Fig. 8(a)). A much larger time constant for the SPV shift of 1.2 ms is found when the oxygen annealing time is reduced (Fig. 8(b)).

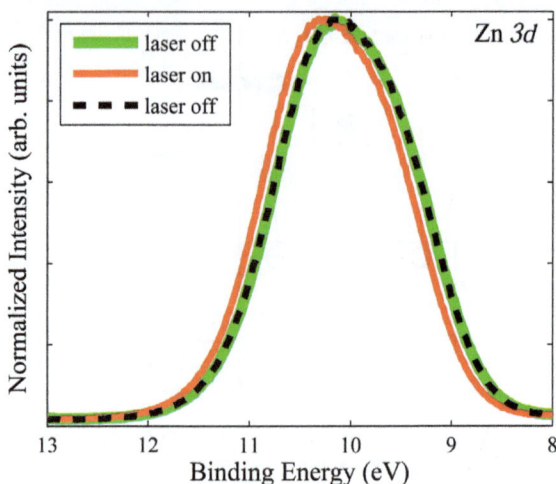

Fig. 7   The Zn $3d$ core level of the m-plane ZnO surface recorded using a photon energy of 200 eV with (red line) and without (green and dashed lines) laser photoexcitation with a CW laser (372 nm, 10 mW).

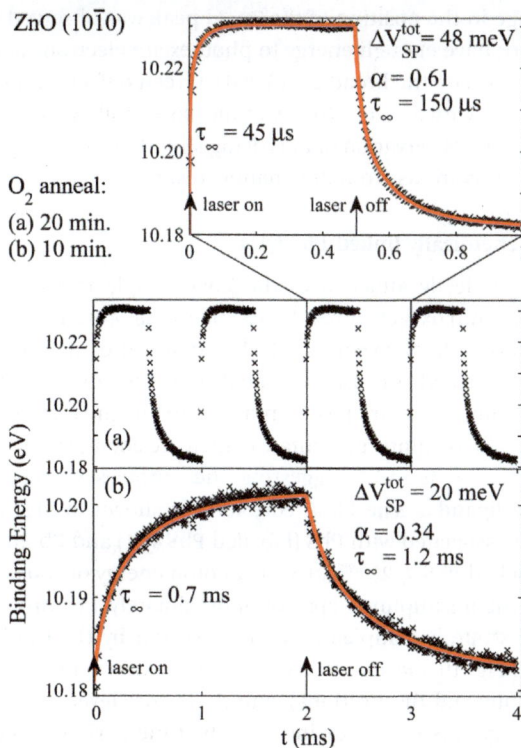

Fig. 8 Binding energy shift of the Zn $3d$ core level of the ZnO m-plane surface recorded using a photon energy of 200 eV, during modulation of 3.33 eV (372 nm) CW illumination for a minimum of 10 000 data accumulations. Laser modulation is indicated by arrows. The samples were prepared identically (as described in the Experimental section) except for annealing in 1.2–1.4 × 10^{-7} mbar oxygen at 703 K for (a) 20 min and (b) 10 min (the experimental data in (a) are shown repeated on the time axis in the middle panel for comparison with (b)). The decay and onset of the pump-induced SPV are fitted using eqn (4) and an analogous decelerating exponential increase respectively (red lines). Reducing the oxygen annealing time from 20 min to 10 min leads to an increase in the dark carrier lifetime, τ_∞, of approximately one order of magnitude.

The change in the oxygen annealing treatment of the substrate changes the timescales of the SPV decay by approximately an order of magnitude, with the slowest SPV decay time found in the less oxygenated surface. We also note that this surface shows a smaller SPV shift; an SPV shift of 20 meV is measured in Fig. 8(b) (10 min oxygen anneal), compared with 48 meV in Fig. 8(a) (20 min oxygen anneal). The material parameter, α, decreases from 0.61 to 0.34 as the oxygen annealing time is halved. The increase in τ_∞ from a few hundred μs to around 1 ms on decreasing the oxygen annealing time in this way was found to be reproducible in several different experiments conducted on successive synchrotron beam runs.

In order to explore the possible role of oxygen-vacancy-related band-gap states in PPC,[91] separate experiments were also conducted using sub-band-gap excitation to excite a sample prepared in a similar way to that in Fig. 8(b), *i.e.* annealed in oxygen for 10 min. A pulsed laser of significantly sub-band-gap energy (wavelength 590 nm, or 2.10 eV) was used to illuminate the sample every 2 ms, at a delay of 1 ms. No change in the position of the Zn *3d* peak was observed (Fig. 9); these laser pulses do not have enough energy to photoexcite electrons across the band gap of 3.4 eV, or to excite the broad 2.5 eV 'GD' (green defect) band.[89] However, a small photoresponse with a τ_∞ of 570 μs was measured following 405 nm (3.06 eV) excitation (Fig. 9). The observation of PPC using slightly sub-band-gap radiation is consistent with observations from ZnO nanowires.[22,91,92]

ZnO (10$\bar{1}$0) with chemically-linked PbS QDs

PbS QDs were chemically attached to the ZnO sample measured in Fig. 8(a) (prepared with a 20 min oxygen anneal). XPS shows the presence of the QDs at the surface. Fig. 10 shows the S *2p* core level XPS (analysed using CasaXPS[93]), where three species are identified as those associated with PbS (labelled S *2p* 1),[94] neutral S (labelled S *2p* 2)[94] and sulphur present in the 3-MPA ligand (labelled S *2p* 3).[95] No oxidized species such as sulfite and sulfate (which are chemically shifted by ~4 eV to higher BE)[96,97] were found,[5] suggesting the surfaces of the QDs are well passivated by the ligands.[5] The Pb *4f* core level is shown in Fig. 11, where two species are found associated with PbS (labelled PbS *4f* 1) and Pb directly attached to the ligand (labelled PbS *4f* 2).[5,98] An X-ray photon energy of 230 eV was used for both spectra (giving a sampling depth of approximately 1.5 nm). Fig. 11 shows that the ZnO substrate is sampled as well, as shown by the overlapping Zn *3s* signal.[99] All fits in Fig. 10 and 11 agree well with well-known literature parameters (doublet separations and BE positions), and these core levels were monitored to ensure laser damage was not occurring over the time period of the experiment. The valence band spectra for clean ZnO and with the PbS QDs attached are shown in Fig. 12, where the Zn *3d* core level signal is attenuated upon deposition of the QDs. Changes at the valence band energy (~1–3 eV) are attributed to the QDs,[5] where the valence band maximum corresponds to the highest occupied molecular orbit (HOMO) of the QD, at a lower BE than the clean ZnO, consistent with the

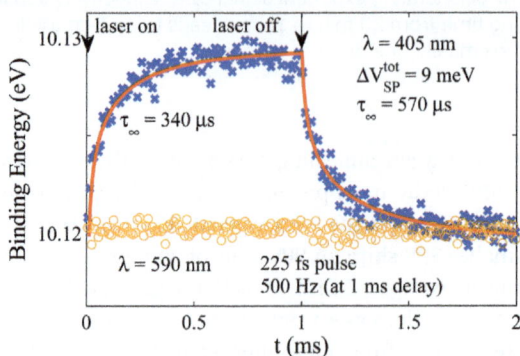

Fig. 9 Binding energy shift of the Zn *3d* core level of the ZnO *m*-plane surface recorded using a photon energy of 200 eV, during modulation of 3.06 eV (405 nm) CW laser illumination (blue crosses) for a minimum of 10 000 data accumulations. The sample was prepared as in Fig. 7(b). CW laser modulation is indicated by arrows. A dark carrier lifetime, τ_∞, of 570 μs is obtained by fitting with the self-deceleration model of eqn (4). Also shown is the binding energy of the Zn *3d* core level during illumination with 590 nm (2.10 eV) radiation (orange circles). Here a pulsed laser (225 fs pulse width) was modulated at 500 Hz, *i.e.* every 2 ms, with a delay of 1 ms.

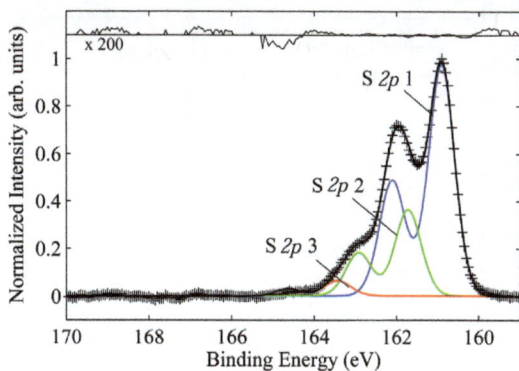

Fig. 10 XPS of the S *2p* core level measured using a photon energy of 230 eV. Three species are identified as being associated to PbS (1), neutral sulphur (2) and sulphur in the ligand (3). No features at higher BE due to oxidised species are observed. A magnified residual is shown above.

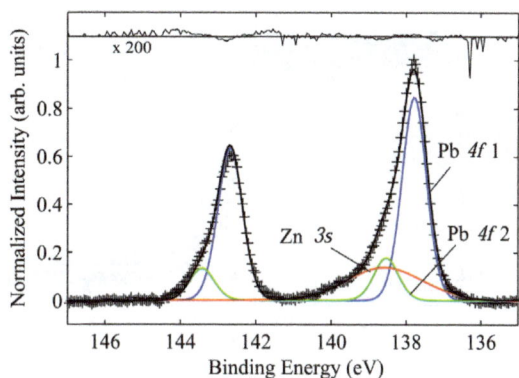

Fig. 11 XPS of the Pb *4f* core level measured using a photon energy of 230 eV. Two species are present which are due to PbS (1) and Pb atoms linked to the ligand (2). The Zn *3s* core level is also observed.

energy level line-up shown in Fig. 6(a). The increased signal in the BE range 4–8 eV after linking the QDs is due to the 3-MPA ligands.[6] A weak signal from the Pb *5d* core level is present at ~18.7 eV with a doublet separation of ~2.5 eV, consistent with literature values for PbS.[94,100]

The transient SPV measured with the Zn *3d* core level as in Fig. 8 was repeated, again using an exciting 372 nm laser. The ZnO sample was prepared and measured as shown in Fig. 8(a) (using a 20 min oxygen anneal) directly before the PbS QDs were linked to the surface. Fig. 13 shows that both the onset and decay times of the SPV decrease when QDs are attached to the surface. Indeed, the rising edge of the onset approaches the time resolution of the experiment (~150 ns), although the rise can be fitted with a rise time of 5 µs. The dark carrier lifetime, τ_∞, decreases from 150 µs to 65 µs. The size of the SPV shift, ΔV_{SP}^{tot}, also decreases from 48 meV to 15 meV and the material parameter, α, is decreased to 0.16 compared to 0.61 for the clean case shown in Fig. 8(a).

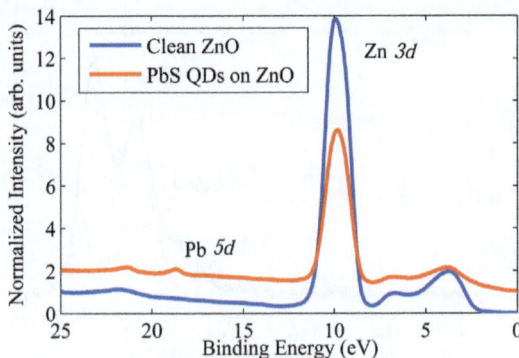

Fig. 12 XPS of the valence band of clean ZnO and PbS QDs attached to ZnO measured using a photon energy of 200 eV. The spectra have been normalized to the background at 25 eV BE and the PbS QDs spectrum has been offset vertically, showing changes in the valence band attributed to the QDs and 3-MPA ligands (see text). The Pb 5d core level is also observed at ~18.7 eV BE.

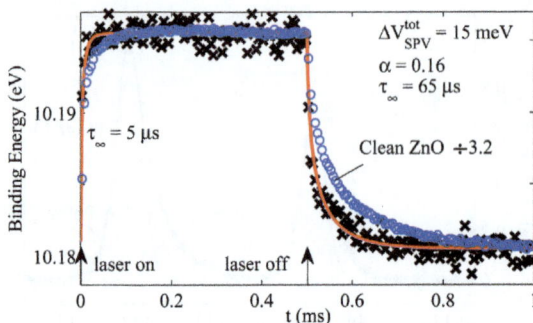

Fig. 13 Binding energy shift of the Zn 3d core level of the ZnO m-plane surface with PbS QDs chemically attached (black crosses), and of clean ZnO (blue circles, scaled down by a factor of 3.2 for comparison), recorded using a photon energy of 200 eV, during modulation of 3.33 eV (372 nm) CW illumination for a minimum of 10 000 data accumulations. Laser modulation is indicated by arrows. The decay and onset of the pump-induced SPV are fitted using eqn (4) and an analogous decelerating exponential increase respectively (red lines). The decay lifetime, SPV shift and material parameter, α, are reduced when QDs are attached to the surface compared to the clean ZnO case (see text).

ZnO (10$\bar{1}$0) with chemically-linked CdS–ZnSe QDs

Type II CdS–ZnSe core–shell QDs were chemically attached to ZnO in the same way as the PbS QD sample. The ZnO surface was prepared using a 10 min anneal as in Fig. 8(b). Fig. 14 shows the Zn 3p XPS measured using a photon energy of 800 eV. Three doublets are required to fit the spectrum, with the main doublet associated with ZnO at 88.3 eV BE (labelled Zn 3p 1),[99] a weak component associated with ZnSe in the shell of the QD at 2.0 eV lower BE (labelled Zn 3p 2),[101] and a doublet at 1.9 eV higher BE associated with attachment to the ligand (labelled Zn 3p 3). Doublets were fitted with a spin orbit splitting of 3.0 eV.[102] The fact that there are now additional Zn photoemission lines associated with the QDs and the attaching ligands means that the transient SPV fitting of the Zn 3d core level

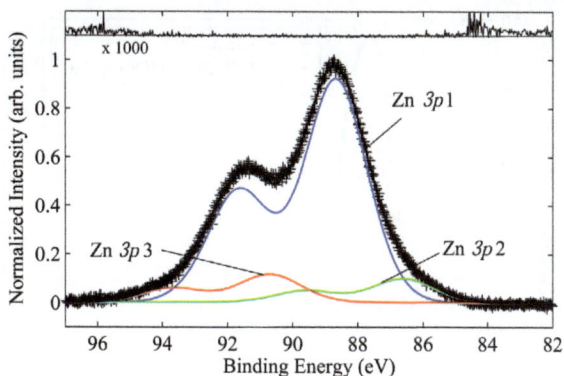

Fig. 14   Zn *3p* XPS measured using a photon energy of 800 eV, showing three species associated with ZnO (1), ZnSe in the shell of the QD (2), and Zn linked to the ligand (3).

should involve fitting with more than one doublet. Fig. 15 shows the XPS of the Se *3p* and S *2p* core levels measured with a photon energy of 600 eV. There are two species of S, one associated with CdS (S $2p_{3/2}$ at 161.7 eV BE) with a spin–orbit splitting of 1.2 eV (labelled S *2p* 1),[103] and a species chemically shifted by 1.9 eV to higher BE associated with S in the 3-MPA ligand (labelled S *2p* 2); the 'ligand' chemical shift is identical to that seen in the S *2p* spectra of the PbS QDs shown in Fig. 10. The Se $3p_{3/2}$ feature is present at 160.1 eV BE with a spin–orbit splitting of 5.6 eV,[104] and again all fitting parameters agree well with literature values. No oxidized S species in the range 168–172 eV BE are found. The Cd *3d* core level was recorded at a BE position of 405 eV in agreement with literature values for CdS.[103]

Fig. 16 shows the valence band for the clean ZnO and with the CdS–ZnSe QDs attached. As for the PbS QD–ZnO system, there is a shift of the valence band edge to slightly lower BE on linking the QDs to the surface in agreement with the energy level line-up diagram in Fig. 6(b), there is a small amount of additional

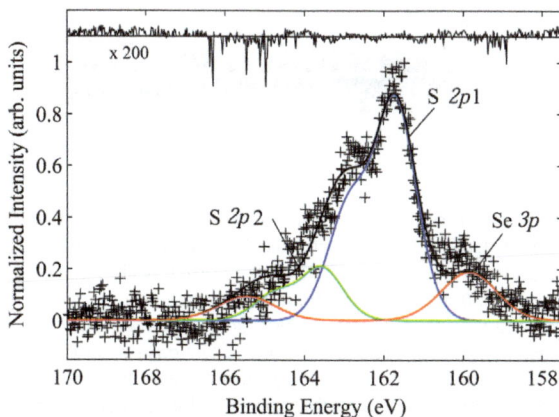

Fig. 15   XPS of the Se *3p* and S *2p* core levels measured with a photon energy of 600 eV. Two species of S *2p* are identified as that associated with the CdS core of the QD (1) and sulfur in the 3-MPA ligand (2).

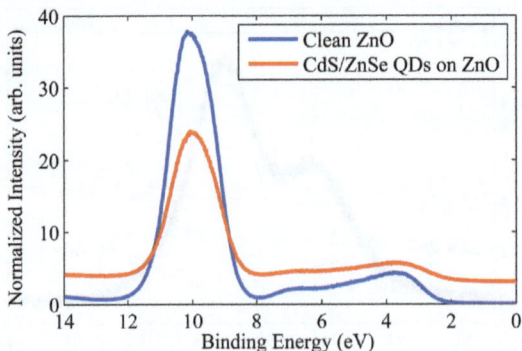

Fig. 16 Valence band XPS of clean ZnO and with CdS–ZnSe QDs chemically attached, measured using a photon energy of 200 eV. The spectra have been normalized to the background at 14 eV BE and the QD spectrum has been offset vertically to show changes in the valence band associated with the QDs and ligands (see text).

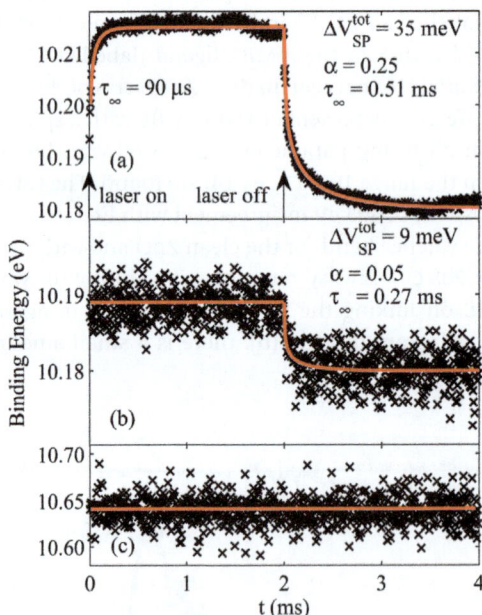

Fig. 17 Binding energy shift of the Zn 3d core level of (a) the ZnO m-plane surface, (b, c) the same surface with CdS–ZnSe QDs chemically attached, recorded using a photon energy of 200 eV, during modulation of 3.33 eV (372 nm) CW illumination for a minimum of 10 000 data accumulations. Laser modulation is indicated by arrows. The decay and onset of the pump-induced SPV are fitted using eqn (4) and an analogous decelerating exponential increase respectively (red lines). For the QD case where Zn is present in the shell, two components are fitted to the XPS spectra at each time interval. The lower BE component associated with the substrate (b) shows a reduced decay lifetime, SPV shift and material parameter, α, when QDs are attached to the surface (see text). (c) shows the transient signal of the higher BE component associated with Zn attached to the ligand which does not change upon photoexcitation.

intensity between 4 and 8 eV BE associated with the ligand,[6] and the Zn $3d$ core level intensity is attenuated by the covering of QDs.

The transient SPV of the Zn $3d$ core level of the ZnO surface when illuminated with 372 nm (3.33 eV) laser radiation was measured before and after chemically linking the CdS–ZnSe QDs, as shown in Fig. 17(a) and 17(b), respectively. A photon energy of 200 eV was used. The clean ZnO surface shows a dark carrier lifetime of 0.51 ms and the fitted material parameter is 0.25, similar to that found for the surface measured in Fig. 8(b), which was annealed in oxygen for a similar time. XPS taken before and after the time-resolved measurements verified that there was no significant laser damage during the experiment.

When the QDs are attached to the surface, the dark carrier lifetime decreases (here to 0.27 ms) as found in the PbS QD sample. The initial rising edge of the onset now appears as a step function, suggesting that the dynamics are happening on faster timescales than the time resolution of the experiment. The material parameter also decreases, here to 0.05, and the SPV shift is reduced from 35 meV to 9 meV. The 3.33 eV laser is energetic enough to photoexcite carriers into the ZnSe shell, meaning these two QD samples can directly inject carriers into the ZnO conduction band upon photoexcitation. The Zn $3d$ core level for this experiment required two components to be fitted to the XPS. The fitted secondary doublet associated with the QD does not shift upon photoexcitation as shown in Fig. 17(c), as anticipated, as only the ZnO substrate is subject to a change in band bending on photoexcitation. The secondary component is fitted at approximately 0.5 eV higher BE than the doublet associated with the ZnO substrate, indicating this component is associated with Zn attached to the ligand. Only the onset of this higher BE component is fit due to the BE window available.

Discussion

The change in the surface potential upon photoexcitation with 3.33 eV laser radiation has been observed for the m-plane ZnO ($10\bar{1}0$) surface, first with different oxygen vacancy concentrations (adjusted by different durations of annealing in oxygen during surface preparation), and with two quantum dot samples, type I PbS and type II CdS–ZnSe core–shell QDs, chemically linked to the surface with 3-MPA ligands.

The decay in the ZnO SPV shift following photoexcitation occurs on relatively long timescales, typically over several ms here, which is a persistent photocon-ductivity (PPC). Rival explanations for the origin of this PPC all rely on the availability of oxygen. In the oxygen photodesorption model,[8–14] oxygen is provided by the environment by chemisorption at the surface, whereas in the model based on metastable band-gap defect states, the concentration of lattice oxygen vacancies controls the PPC.[19,20,89–91] The highly-controlled nature of our experiments under UHV conditions (where the residual vacuum was $\sim 1.5 \times 10^{-10}$ mbar) allows for these two effects to be distinguished. Calculations of the initial oxygen vacancy concentration created by the 10 and 20 min O_2 annealing cycles and the estimated effect of these on the dynamics shown in Fig. 8 (a) and (b) lead us to conclude that the observed dynamics are at least 10^3–10^4 times faster than can be explained by the oxygen photodesorption model.[21] The observation of PPC using sub-band gap photoexcitation (Fig. 9), allows us to demonstrate that an alternative model for PPC in ZnO (involving lattice oxygen vacancies) is dominant in these experiments.[21] Our results show that defect states approximately 340 meV above the valence band edge are directly associated with the PPC, consistent with the hypothesis that ionized oxygen vacancy states are responsible for PPC in ZnO.[18,21] These results support similar measurements from ZnO nanowires.[22,90,91]

The onset of the SPV shown in Fig. 8 and 9 can be understood as a competition between the constant rate of photogeneration of carriers produced by the CW laser illumination and the recombination process, and as such the onset is expected to be faster than the decay times as observed throughout (Fig. 8, 9, 13 and 17). When the laser fluence is high, the onset of such SPV transients may be fitted using a single exponential that yields a time constant inversely proportional to the intensity of illumination.[44] However, fitting with a single exponential is not possible here, indicating the laser fluence is not sufficiently high for the optical generation to completely dominate the charge dynamics. Therefore, the onset of the SPV is instead fitted with a decelerated exponential model analogous to eqn (4), which reflects the dynamic increase in recombination rate as the surface band bending is reduced, which acts to counterbalance the rate of carrier photogeneration.

The 3.33 eV photoexcitation laser has sufficient energy to photoexcite electrons in both the PbS and the shell of the CdS–ZnSe QDs. In both cases, the effect of attachment of these QD samples to the surface of ZnO is to reduce the total change in the surface photovoltage, ΔV_{SPV}^{tot}, reduce the material parameter, α, and to reduce the dark carrier lifetime, τ_∞. The onset time of the SPV change was also reduced in both cases. Fig. 13 and 17 (b) show fast initial onsets, suggesting the dynamics here are occurring on timescales faster than the time resolution of the experiment (*ca.* 150 ns). This is anticipated if direct injection of charge from the QDs into the ZnO is occurring; charge injection from QDs into oxide surfaces may occur on fs timescales.[105,106]

Both samples were chemically attached with 3-MPA ligands. Valence band and core level XPS for both samples (Fig. 10–12, 14–16) show features associated with these ligands. The consistent decrease in the material parameter, α, which has been shown to scale with the equilibrium band bending,[61] along with the consistent decrease in the total SPV change measured, suggests that the adsorbed QDs at the surface lead to a decrease in the equilibrium band bending at the ZnO surface, through charge donation into the depletion layer at the surface of *n*-type ZnO. This decrease in the surface potential causes the photoexcited carrier life-times to change (according to eqn (2)), as well as the total SPV change measured upon photoexcitation (eqn (1)). Therefore, the data suggest that photoexcited carriers in the PbS QDs and in the shell of the CdS–ZnSe QDs are injected into the conduction band of the ZnO substrate instantaneously (relative to the intrinsic time resolution of the experiment), causing the band bending to immediately be reduced (observed as a shift to higher BE of the Zn *3d* core level). In the clean ZnO samples the onset can be understood to be due to photogeneration by the CW laser competing with the lifetime of the generated carriers, whereas upon depo-sition of the QDs, a new, faster route for photoinjection into the ZnO conduction band from the QDs is opened up. This fast injection of carriers from QD to substrate is consistent with the energy level line-up diagrams illustrated in Fig. 6(a) and (b), and with other literature investigating charge transfer in similar systems involving QDs chemically linked to TCOs,[107–113] which may be occurring on ps and fs timescales.[105,106,114]

The results presented here suggest the need for further experimentation into charge transfer between quantum dots and transparent conducting oxide substrates. Photoexcitation of the systems with laser photon energies greater than the effective band gap of the QDs, but insufficient energy to directly photoexcite the substrate is a logical next step. A transient SPV in the substrate when only the QDs have been photoexcited would confirm that carriers have been injected from the QD into the substrate.

There is also a clear need for laser pump-X-ray probe methodologies that allow for relatively long timescales to be measured (*e.g.* ms as required to measure the transient SPV for PPC in ZnO) with a much greater timing resolution (on sub ps timescales) in order to effectively resolve ultrafast onsets of transient SPV once QDs are attached to the surface. Clearly, the advent of 4th generation radiation from FELs and other low emittance electron beam sources, combined with time-of-flight electron energy analysis,[54] provides a route to addressing this issue, provided sample damage can be controlled.

These results demonstrate the potential for light-harvesting quantum dots attached to ZnO as a basis for next-generation solar cells. As the effective band gap energy of QDs can be tuned by varying the QD diameter, these light harvesters can be optimized to absorb the majority of the solar energy spectrum, and the use of type II QDs allows recombination rates to be reduced. The lifetime of carriers injected into the ZnO conduction band is by comparison very long, and can be tuned by varying the oxygen vacancy concentration in the sample, controlled by the preparation process. Thus charge can be extracted from the system efficiently before significant recombination has occurred.

Conclusions

Photoexcited carrier dynamics have been investigated in the potential next-generation photoanode material ZnO (using the nonpolar *m*-plane ($10\bar{1}0$) surface). A laser pump-X-ray probe methodology has allowed for the transient change in the surface potential upon photoexcitation to be measured. PPC is observed in ZnO where recombination of carriers occurs on ms timescales. The highly-controlled nature of the experiment under UHV, as well as photoexcitation with sub-band gap radiation, has allowed us to show that this PPC originates from defect band-gap states (340 meV above the valence band maximum) associated with lattice oxygen vacancies.[18,21] Light-harvesting colloidal quantum dots have been chemically linked to the surface with 3-MPA ligands. The total change in the surface photovoltage, the material parameter, α, and dark carrier lifetime consistently decrease when PbS and type II CdS–ZnSe core–shell QDs are attached to the ZnO surface. These changes are likely to be due to a decrease in the equilibrium band bending at the surface due to charge donation from the QD into the depletion layer at the surface of *n*-type ZnO. The initial onset times for the transient SPV are also decreased, suggesting that the onset of the SPV is occurring within the time resolution of the experiments. This suggests that direct injection of charge carriers from the QD into the conduction band of the substrate is occurring. The combination of fast electron injection with a very slow recombination rate due to PPC in the ZnO photoanode means that the system therefore shows great potential in next-generation solar cell technology.

Laser pump-X-ray probe measurements are powerful probes of photoexcited carrier dynamics over a wide range of timescales. The results presented here indicate a need for further studies in this area, and for development of methodologies capable of measuring relatively long (ms) timescales with increased (sub-ps) timing resolution.

Acknowledgements

The research leading to these results received funding from the European Community's Seventh Framework Programme (FP7/2007-2013) under grant agreement no 226716, allowing access to Synchrotron SOLEIL. Work was also supported by the Cockcroft Institute *via* its STFC core grant ST/G008248/1.

Notes and references

1 P. D. C. King and T. D. Veal, *J. Phys.: Condens. Matter*, 2011, **23**, 334214.

2 C. F. Klingshirn, *Zinc Oxide: from fundamental properties towards novel applications*, Springer, Heidelberg; London, 2010.

3 C. Klingshirn, J. Fallert, H. Zhou, J. Sartor, C. Thiele, F. Maier-Flaig, D. Schneider and H. Kalt, *Phys. Status Solidi B*, 2010, **247**, 1424.

4 B. Carlson, K. Leschkies, E. S. Aydil and X. Y. Zhu, *J. Phys. Chem. C*, 2008, **112**, 8419.

5 S. J. O. Hardman, D. M. Graham, S. K. Stubbs, B. F. Spencer, E. A. Seddon, H. T. Fung, S. Gardonio, F. Sirotti, M. G. Silly, J. Akhtar, P. O'Brien, D. J. Binks and W. R. Flavell, *Phys. Chem. Chem. Phys.*, 2011, **13**, 20275.

6 J. Akhtar, M. A. Malik, P. O'Brien, K. G. U. Wijayantha, R. Dharmadasa, S. J. O. Hardman, D. M. Graham, B. F. Spencer, S. K. Stubbs, W. R. Flavell, D. J. Binks, F. Sirotti, M. El Kazzi and M. Silly, *J. Mater. Chem.*, 2010, **20**, 2336.

7 J. A. McLeod, R. G. Wilks, N. A. Skorikov, L. D. Finkelstein, M. Abu-Samak, E. Z. Kurmaev and A. Moewes, *Phys. Rev. B: Condens. Matter Mater. Phys.*, 2010, **81**, 245123.

8 I. Tashiro, T. Kimura and K. Endo, *Appl. Opt.*, 1969, **8**, 180.

9 J. Lagowski, E. S. Sproles and H. C. Gatos, *J. Appl. Phys.*, 1977, **48**, 3566.

10 A. Rothschild, Y. Komem and N. Ashkenasy, *J. Appl. Phys.*, 2002, **92**, 7090.

11 X. G. Zheng, Q. S. Li, W. Hu, D. Chen, N. Zhang, M. J. Shi, J. J. Wang and L. C. Zhang, *J. Lumin.*, 2007, **122**, 198.

12 T. E. Murphy, K. Moazzami and J. D. Phillips, *J. Electron. Mater.*, 2006, **35**, 543.

13 R. J. Collins and D. G. Thomas, *Phys. Rev.*, 1958, **112**, 388.

14 M. J. Liu and H. K. Kim, *Appl. Phys. Lett.*, 2004, **84**, 173.

15 D. C. Hou, A. Dev, K. Frank, A. Rosenauer and T. Voss, *J. Phys. Chem. C*, 2012, **116**, 19604.

16 J. M. Bao, I. Shalish, Z. H. Su, R. Gurwitz, F. Capasso, X. W. Wang and Z. F. Ren, *Nanoscale Res. Lett.*, 2011, **6**, 404.

17 S. Lany and A. Zunger, *Phys. Rev. Lett.*, 2007, **98**, 045501.

18 S. Lany and A. Zunger, *Phys. Rev. B: Condens. Matter Mater. Phys.*, 2005, **72**, 035215.

19 S. B. Zhang, S. H. Wei and A. Zunger, *Phys. Rev. B: Condens. Matter*, 2001, **63**, 075205.

20 A. Janotti and C. G. Van de Walle, *Appl. Phys. Lett.*, 2005, **87**, 122102.

21 B. F. Spencer, D. M. Graham, S. J. O. Hardman, E. A. Seddon, M. J. Cliffe, K. L. Syres, A. G. Thomas, S. K. Stubbs, F. Sirotti, M. G. Silly, P. F. Kirkham, A. R. Kumarasinghe, G. J. Hirst, A. J. Moss, S. F. Hill, D. A. Shaw, S. Chattopadhyay and W. R. Flavell, *Phys. Rev. B: Condens. Matter Mater. Phys.*, 2013, **88**, 195301.

22 I. Beinik, M. Kratzer, A. Wachauer, L. Wang, Y. P. Piryatinski, G. Brauer, X. Y. Chen, Y. F. Hsu, A. B. Djurisic and C. Teichert, *Beilstein J. Nanotechnol.*, 2013, **4**, 208.

23 V. I. Klimov, *Semiconductor and metal nanocrystals: synthesis and electronic and optical properties*, Marcel Dekker, New York, 2004.

24 V. I. Klimov, *Nanocrystal quantum dots*, 2nd edn, CRC Press, Boca Raton, 2010.

25 R. D. Schaller and V. I. Klimov, *Abstr Pap Am Chem S*, 2005, **229**, U729.

26 R. D. Schaller and V. I. Klimov, *Phys. Rev. Lett.*, 2004, **92**, 186601.

27 R. D. Schaller, M. Sykora, S. Jeong and V. I. Klimov, *J. Phys. Chem. B*, 2006, **110**, 25332.

28 W. Shockley and H. J. Queisser, *J. Appl. Phys.*, 1961, **32**, 510.

29 S. K. Stubbs, S. J. O. Hardman, D. M. Graham, B. F. Spencer, W. R. Flavell, P. Glarvey, O. Masala, N. L. Pickett and D. J. Binks, *Phys. Rev. B: Condens. Matter Mater. Phys.*, 2010, **81**, 081303(R).

30 R. D. Schaller, J. M. Pietryga and V. I. Klimov, *Nano Lett.*, 2007, **7**, 3469.

31 R. D. Schaller, M. A. Petruska and V. I. Klimov, *Appl. Phys. Lett.*, 2005, **87**, 253102.

32 Y. Kobayashi, T. Udagawa and N. Tamai, *Chem. Lett.*, 2009, **38**, 830.

33 M. C. Beard, K. P. Knutsen, P. R. Yu, J. M. Luther, Q. Song, W. K. Metzger, R. J. Ellingson and A. J. Nozik, *Nano Lett.*, 2007, **7**, 2506.

34 J. B. Sambur, T. Novet and B. A. Parkinson, *Science*, 2010, **330**, 63.

35 V. I. Klimov, S. A. Ivanov, J. Nanda, M. Achermann, I. Bezel, J. A. McGuire and A. Piryatinski, *Nature*, 2007, **447**, 441.

36 A. Piryatinski, S. A. Ivanov, S. Tretiak and V. I. Klimov, *Nano Lett.*, 2007, **7**, 108.

37 S. Kim, B. Fisher, H. J. Eisler and M. Bawendi, *J. Am. Chem. Soc.*, 2003, **125**, 11466.

38 S. Kaniyankandy, S. Rawalekar, S. Verma and H. N. Ghosh, *J. Phys. Chem. C*, 2011, **115**, 1428.

39 S. Rawalekar, S. Kaniyankandy, S. Verma and H. N. Ghosh, *J. Phys. Chem. C*, 2010, **114**, 1460.

40 S. Rawalekar, S. Kaniyankandy, S. Verma and H. N. Ghosh, *J. Phys. Chem. C*, 2011, **115**, 12335.

41 D. Oron, M. Kazes and U. Banin, *Phys. Rev. B: Condens. Matter Mater. Phys.*, 2007, **75**, 035330.

42 S. M. Fairclough, E. J. Tyrrell, D. M. Graham, P. J. B. Lunt, S. J. O. Hardman, A. Pietzsch, F. Hennies, J. Moghal, W. R. Flavell, A. A. R. Watt and J. M. Smith, *J. Phys. Chem. C*, 2012, **116**, 26898.

43 M. Cadirci, S. K. Stubbs, S. M. Fairclough, E. J. Tyrrell, A. A. R. Watt, J. M. Smith and D. J. Binks, *Phys. Chem. Chem. Phys.*, 2012, **14**, 13638.

44 L. Kronik and Y. Shapira, *Surf. Sci. Rep.*, 1999, **37**, 1.

45 M. H. Hecht, *Phys. Rev. B: Condens. Matter*, 1991, **43**, 12102.

46 J. P. Long, H. R. Sadeghi, J. C. Rife and M. N. Kabler, *Phys. Rev. Lett.*, 1990, **64**, 1158.

47 S. Tanaka, S. D. More, J. Murakami, M. Itoh, Y. Fujii and M. Kamada, *Phys. Rev. B: Condens. Matter*, 2001, **64**, 155308.

48 S. Tanaka, T. Nishitani, T. Nakanishi, S. D. More, J. Azuma, K. Takahashi, O. Watanabe and M. Kamada, *J. Appl. Phys.*, 2004, **95**, 551.

49 S. Tanaka, T. Ichibayashi and K. Tanimura, *Phys. Rev. B: Condens. Matter Mater. Phys.*, 2009, **79**, 155313.

50 W. Widdra, D. Bröcker, T. Gießel, I. V. Hertel, W. Kruger, A. Liero, F. Noack, V. Petrov, D. Pop, P. M. Schmidt, R. Weber, I. Will and B. Winter, *Surf. Sci.*, 2003, **543**, 87.

51 D. Bröcker, T. Gießel and W. Widdra, *Chem. Phys.*, 2004, **299**, 247.

52 T. E. Glover, G. D. Ackermann, A. Belkacem, B. Feinberg, P. A. Heimann, Z. Hussain, H. A. Padmore, C. Ray, R. W. Schoenlein and W. F. Steele, *Nucl. Instrum. Methods Phys. Res., Sect. A*, 2001, **467**, 1438.

53 T. E. Glover, G. D. Ackermann, Z. Hussain and H. A. Padmore, *J. Mod. Opt.*, 2004, **51**, 2805.

54 M. Ogawa, S. Yamamoto, Y. Kousa, F. Nakamura, R. Yukawa, A. Fukushima, A. Harasawa, H. Kondoh, Y. Tanaka, A. Kakizaki and I. Matsuda, *Rev. Sci. Instrum.*, 2012, **83**, 023109.

55 A. Vollmer, R. Ovsyannikov, M. Gorgoi, S. Krause, M. Oehzelt, A. Lindblad, N. Martensson, S. Svensson, P. Karlsson, M. Lundvuist, T. Schmeiler, J. Pflaum and N. Koch, *J. Electron Spectrosc. Relat. Phenom.*, 2012, **185**, 55.

56 M. Hajlaoui, E. Papalazarou, J. Mauchain, G. Lantz, N. Moisan, D. Boschetto, Z. Jiang, I. Miotkowski, Y. P. Chen, A. Taleb-Ibrahimi, L. Perfetti and M. Marsi, *Nano Lett.*, 2012, **12**, 3532.

57 M. Marsi, M. E. Couprie, L. Nahon, D. Garzella, T. Hara, R. Bakker, M. Billardon, A. Delboulbe, G. Indlekofer and A. TalebIbrahimi, *Appl. Phys. Lett.*, 1997, **70**, 895.

58 M. Marsi, R. Belkhou, C. Grupp, G. Panaccione, A. Taleb-Ibrahimi, L. Nahon, D. Garzella, D. Nutarelli, E. Renault, R. Roux, M. E. Couprie and M. Billardon, *Phys. Rev. B: Condens. Matter*, 2000, **61**, R5070.

59 S. Yamamoto and I. Matsuda, *J Phys Soc Jpn*, 2013, **82**, 00213.

60 J. P. Long and V. M. Bermudez, *Phys. Rev. B: Condens. Matter*, 2002, **66**, 121308.

61 J. Schulz, P. Wurfel and W. Ruppel, *Phys. Status Solidi B*, 1991, **164**, 425.

62 F. Polack, M. Silly, C. Chauvet, B. Lagarde, N. Bergeard, M. Izquierdo, O. Chubar, D. Krizmancic, M. Ribbens, J. P. Duval, C. Basset, S. Kubsky and F. Sirotti, *AIP proceedings: The 10th International Conference on Synchrotron Radiation Instrumentation*, 2010, **1234**, 185.

63 N. Bergeard, M. G. Silly, D. Krizmancic, C. Chauvet, M. Guzzo, J. P. Ricaud, M. Izquierdo, L. Stebel, P. Pittana, R. Sergo, G. Cautero, G. Dufour, F. Rochet and F. Sirotti, *J. Synchrotron Radiat.*, 2011, **18**, 245.

64 G. Cautero, R. Sergo, L. Stebel, P. Lacovig, P. Pittana, M. Predonzani and S. Carrato, *Nucl. Instrum. Methods Phys. Res., Sect. A*, 2008, **595**, 447.

65 K. Ozawa and K. Edamoto, *Surf. Sci.*, 2003, **524**, 78.

66 W. Göpel and U. Lampe, *Phys. Rev. B*, 1980, **22**, 6447.

67 W. Göpel, *Surf. Sci.*, 1977, **62**, 165.

68 O. Dulub, L. A. Boatner and U. Diebold, *Surf. Sci.*, 2002, **519**, 201.

69 U. Diebold, L. V. Koplitz and O. Dulub, *Appl. Surf. Sci.*, 2004, **237**, 336.

70 S. J. Pearton, D. P. Norton, K. Ip, Y. W. Heo and T. Steiner, *Prog. Mater. Sci.*, 2005, **50**, 293.

71 J. M. Lee, K. K. Kim, S. J. Park and W. K. Choi, *Appl. Phys. Lett.*, 2001, **78**, 3842.

72 V. E. Henrich and P. A. Cox, *The surface science of metal oxides*, Cambridge University Press, Cambridge; New York, 1994.

73 G. W. Tomlins, J. L. Routbort and T. O. Mason, *J. Am. Ceram. Soc.*, 1998, **81**, 869.

74 A. C. S. Sabioni, A. M. J. M. Daniel, W. B. Ferraz, R. W. D. Pais, A. M. Huntz and F. Jomard, *Mater Res-Ibero-Am J*, 2008, **11**, 221.

75 I. S. Jeong, J. H. Kim and S. Im, *Appl. Phys. Lett.*, 2003, **83**, 2946.

76 W. A. Tisdale, M. Muntwiler, D. J. Norris, E. S. Aydil and X. Y. Zhu, *J. Phys. Chem. C*, 2008, **112**, 14682.

77 L. Cademartiri, E. Montanari, G. Calestani, A. Migliori, A. Guagliardi and G. A. Ozin, *J. Am. Chem. Soc.*, 2006, **128**, 10337.

78 X. H. Zhong, Y. Y. Feng and Y. L. Zhang, *J. Phys. Chem. C*, 2007, **111**, 526.

79 B. Blackman, D. Battaglia and X. G. Peng, *Chem. Mater.*, 2008, **20**, 4847.

80 J. Bang, J. Park, J. H. Lee, N. Won, J. Nam, J. Lim, B. Y. Chang, H. J. Lee, B. Chon, J. Shin, J. B. Park, J. H. Choi, K. Cho, S. M. Park, T. Joo and S. Kim, *Chem. Mater.*, 2010, **22**, 233.

81 P. T. K. Chin, C. D. M. Donega, S. S. Bavel, S. C. J. Meskers, N. A. J. M. Sommerdijk and R. A. J. Janssen, *J. Am. Chem. Soc.*, 2007, **129**, 14880.

82 J. I. Kim, J. Kim, J. Lee, D. R. Jung, H. Kim, H. Choi, S. Lee, S. Byun, S. Kang and B. Park, *Nanoscale Res. Lett.*, 2012, **7**, 1.

83 X. Y. Wang, L. H. Qu, J. Y. Zhang, X. G. Peng and M. Xiao, *Nano Lett.*, 2003, **3**, 1103.

84 J. Aldana, Y. A. Wang and X. G. Peng, *J. Am. Chem. Soc.*, 2001, **123**, 8844.

85 M. W. Wang, J. O. Mccaldin, J. F. Swenberg, T. C. Mcgill and R. J. Hauenstein, *Appl. Phys. Lett.*, 1995, **66**, 1974.

86 S. A. Ivanov and M. Achermann, *ACS Nano*, 2010, **4**, 5994.

87 P. D. C. King, T. D. Veal, A. Schleife, J. Zuniga-Perez, B. Martel, P. H. Jefferson, F. Fuchs, V. Munoz-Sanjose, F. Bechstedt and C. F. McConville, *Phys. Rev. B: Condens. Matter Mater. Phys.*, 2009, **79**, 205205.

88 W. Göpel, J. Pollmann, I. Ivanov and B. Reihl, *Phys. Rev. B*, 1982, **26**, 3144.

89 H. L. Mosbacker, Y. M. Strzhemechny, B. D. White, P. E. Smith, D. C. Look, D. C. Reynolds, C. W. Litton and L. J. Brillson, *Appl. Phys. Lett.*, 2005, **87**, 012102.

90 A. Kushwaha and M. Aslam, *J. Appl. Phys.*, 2012, **112**, 054316.

91 P. Liu, G. W. She, Z. L. Liao, Y. Wang, Z. Z. Wang, W. S. Shi, X. H. Zhang, S. T. Lee and D. M. Chen, *Appl. Phys. Lett.*, 2009, **94**, 063120.

92 K. Keem, H. Kim, G. T. Kim, J. S. Lee, B. Min, K. Cho, M. Y. Sung and S. Kim, *Appl. Phys. Lett.*, 2004, **84**, 4376.

93 Casa XPS, www.casaxps.com.

94 R. B. Shalvoy, G. B. Fisher and P. J. Stiles, *Phys. Rev. B: Solid State*, 1977, **15**, 1680.

95 G. Gonella, O. Cavalleri, S. Terreni, D. Cvetko, L. Floreano, A. Morgante, M. Canepa and R. Rolandi, *Surf. Sci.*, 2004, **566**, 638.

96 B. J. Lindberg, *Acta Chem. Scand.*, 1970, **24**, 2242.

97 V. I. Nefedov, *Surf. Interface Anal.*, 1981, **3**, 72.

98 A. Lobo, T. Moller, M. Nagel, H. Borchert, S. G. Hickey and H. Weller, *J. Phys. Chem. B*, 2005, **109**, 17422.

99 B. R. Strohmeier and D. M. Hercules, *J. Catal.*, 1984, **86**, 266.

100 F. R. McFeely, S. Kowalczy, L. Ley, R. A. Pollak and D. A. Shirley, *Phys. Rev. B: Solid State*, 1973, **7**, 5228.

101 F. Y. Yang, D. Y. Ban, R. C. Fang, S. H. Xu, P. S. Xu and S. X. Yuan, *J. Electron Spectrosc. Relat. Phenom.*, 1996, **80**, 193.

102 A. Lebugle, U. Axelsson, R. Nyholm and N. Martensson, *Phys. Scr.*, 1981, **23**, 825.

103 E. Agostinelli, C. Battistoni, D. Fiorani and G. Mattogno, *J. Phys. Chem. Solids*, 1989, **50**, 269.

104 D. W. Langer and C. J. Vesely, *Phys. Rev. B: Solid State*, 1970, **2**, 4885.

105 W. A. Tisdale, K. J. Williams, B. A. Timp, D. J. Norris, E. S. Aydil and X. Y. Zhu, *Science*, 2010, **328**, 1543.

106 Y. Yang, W. Rodriguez-Cordoba, X. Xiang and T. Q. Lian, *Nano Lett.*, 2012, **12**, 303.

107 V. Gonzalez-Pedro, Q. Shen, V. Jovanovski, S. Gimenez, R. Tena-Zaera, T. Toyoda and I. Mora-Sero, *Electrochim. Acta*, 2013, **100**, 35.

108 A. G. Pattantyus-Abraham, I. J. Kramer, A. R. Barkhouse, X. H. Wang, G. Konstantatos, R. Debnath, L. Levina, I. Raabe, M. K. Nazeeruddin, M. Gratzel and E. H. Sargent, *ACS Nano*, 2010, **4**, 3374.

109 I. Robel, V. Subramanian, M. Kuno and P. V. Kamat, *J. Am. Chem. Soc.*, 2006, **128**, 2385.

110 K. S. Leschkies, R. Divakar, J. Basu, E. Enache-Pommer, J. E. Boercker, C. B. Carter, U. R. Kortshagen, D. J. Norris and E. S. Aydil, *Nano Lett.*, 2007, **7**, 1793.

111 B. R. Hyun, Y. W. Zhong, A. C. Bartnik, L. F. Sun, H. D. Abruna, F. W. Wise, J. D. Goodreau, J. R. Matthews, T. M. Leslie and N. F. Borrelli, *ACS Nano*, 2008, **2**, 2206.

112 D. F. Watson, *J. Phys. Chem. Lett.*, 2010, **1**, 2299.

113 X. W. Sun, J. Chen, J. L. Song, D. W. Zhao, W. Q. Deng and W. Lei, *Opt. Express*, 2010, **18**, 1296.

114 K. Zidek, K. B. Zheng, C. S. Ponseca, M. E. Messing, L. R. Wallenberg, P. Chabera, M. Abdellah, V. Sundstrom and T. Pullerits, *J. Am. Chem. Soc.*, 2012, **134**, 12110.

Faraday Discussions

ROYAL SOCIETY OF CHEMISTRY

PAPER

Coherent dynamics of the charge density wave gap in tritellurides

L. Rettig,†[a] J.-H. Chu,‡[b] I. R. Fisher,[bc] U. Bovensiepen[a] and M. Wolf*[d]

Received 21st March 2014, Accepted 17th April 2014

DOI: 10.1039/c4fd00045e

The dynamics of the transient electronic structure in the charge density wave (CDW) system RTe_3 (R = rare-earth element) is studied using time- and angle-resolved photoemission spectroscopy (trARPES). Employing a three-pulse pump–probe scheme we investigate the effect of the amplitude mode oscillations on the electronic band structure and, in particular, on the CDW energy gap. We observe coherent oscillations in both lower and upper CDW band with opposite phases, whereby two dominating frequencies are modulating the CDW order parameter. This demonstrates the existence of more than one collective amplitude mode, in contrast to a simple Peierls model. Coherent control experiments of the two amplitude modes, which are strongly coupled in equilibrium, demonstrate independent control of the modes suggesting a decoupling of both modes in the transient photoexcited state.

1   Introduction

Cooperative effects in low-dimensional materials represent a fascinating topic of condensed matter research. Various interactions of the electronic, orbital, spin and lattice degrees of freedom can lead to instabilities and broken-symmetry ground states leading to new emergent properties, which are of both fundamental and technological interest. Examples for such low-temperature broken-symmetry phases are superconducting, charge density wave (CDW) or magnetically ordered states, which are connected to other typically lower symmetry states by phase transitions as function of temperature, pressure, external fields, or are induced by optical excitation.

[a] Fakultät für Physik, Universität Duisburg-Essen, Lotharstr. 1, 47048 Duisburg, Germany

[b] Geballe Laboratory for Advanced Materials and Department of Applied Physics, Stanford University, CA 94305, USA

[c] Stanford Institute for Materials and Energy Sciences, SLAC National Accelerator Laboratory, 2575 Sand Hill Road, Menlo Park, CA 94025, USA

[d] Abteilung Physikalische Chemie, Fritz-Haber-Institut der MPG, Faradayweg 4-6, 14195 Berlin, Germany. E-mail: wolf@fhi-berlin.mpg.de

† Current address: Swiss Light Source, Paul Scherrer Institut, 5232 Villigen PSI, Switzerland.

‡ Current address: Department of Physics, UC Berkeley, USA.

As a model system for emergent order, the formation of a CDW ground state found in many materials of reduced dimensionality is one of the well-established and intensely studied examples at the heart of quantum many-body physics.[1] CDW formation refers to a periodic modulation of the charge density (with wave vector $\mathbf{q}$) which is accompanied by a periodic lattice distortion. A predominant mechanism for CDW formation is the Peierls scenario,[2] which is most pronounced in a one-dimensional (1D) system or in systems where nesting ($\mathbf{q} = 2\mathbf{k_F}$) across the Fermi surface occurs. In this mechanism a key ingredient is a high density of states at the Fermi level together with strong, anisotropic electron–phonon coupling giving rise to the opening of an energy gap $\Delta(\mathbf{k})$ at the Fermi surface along the nesting direction.[1] The energy gap $\Delta(\mathbf{k})$ is a direct measure of the order parameter of the CDW phase and can be determined by appropriate spectroscopic tools from the occupied and unoccupied single-particle band structure.

Fluctuations or impulsive (e.g. optical) excitations can lead to collective excitations of the amplitude or phase of the complex order parameter $\Delta = |\Delta|\exp(i\phi)$. While for the phase mode the magnitude of Δ remains constant, the so-called amplitude mode directly modulates the magnitude of the gap leading to oscillations of $|\Delta|(t)$ in the free energy potential with a frequency associated to the phonon mode driving the Fermi surface nesting[1] (Fig. 1(a)). A sketch of a simple linear band model shown in Fig. 1(b) illustrates the effect of the amplitude mode on the electronic band structure: The linear dispersing band (dashed black line) is modified by the opening of a CDW gap of magnitude Δ_{CDW} at E_F, which is

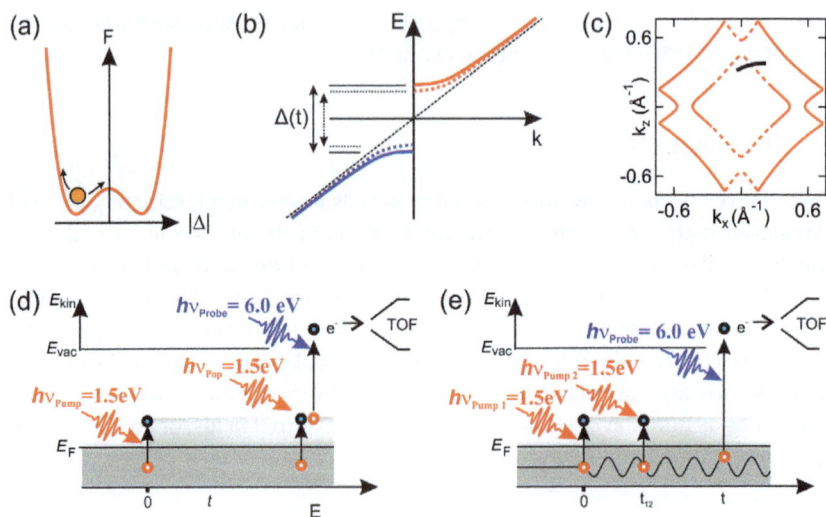

Fig. 1 (a) Schematic of the free energy surface as a function of the order parameter $|\Delta|$, and the motion of the amplitude mode. (b) Effect of the amplitude mode on the gapped dispersion of the CDW state, which transiently modulates the gap size $\Delta(t)$. (c) Normal state Fermi surface of RTe_3. The dashed lines indicate the gapped part of the Fermi surface. The black line marks the k-space cut shown in Fig. 2. (d) Sketch of the three-pulse experiment for the investigation of the unoccupied CDW band dynamics. The second population pulse $h\nu_{Pop}$ was kept at fixed delay before the probe pulse. (e) Sketch of the pulse sequence for the coherent control of the amplitude mode. Here, the separation of the two pump pulses, t_{12} was set fixed for each experiment.

proportional to the CDW order parameter, and is split into a lower and upper band (solid blue and red lines). The effect of the amplitude mode results in a modification of the magnitude of the gap, leading to a shift of both lower and upper CDW bands towards E_F (dashed lines) and to a transient modulation of the gap size $\Delta(t)$.

As the formation of broken symmetry ground states and emergent properties in general depends on a complex interplay of various elementary interactions, which occur on intrinsically different energy- and time scales, studies of the non-equilibrium dynamics of complex materials can reveal new insights into the underlying mechanisms and strengths of interaction between the various degrees of freedom.[3,4] Ultrafast time-resolved spectroscopy provides direct access to these dynamics because the relevant elementary scattering and relaxation processes in solids occur typically on femto- to picosecond time scales. In the case of CDW formation the most direct measurement would be the complete mapping of the band structure to extract the order parameter Δ and to study the interaction of the electronic system with other (phonon) degrees of freedom. In the experiment this is achieved by time- and angle-resolved photoemission spectroscopy (trARPES). Thereby, an ultrashort optical pump pulse excites the system inducing $e.g.$ the melting and subsequent recovery of the CDW phase and time evolution of the electronic structure changes are probed as snapshots by ARPES using a time-delayed ultrashort UV probe pulse (see $e.g.$ ref. 4–8).

The material family of Tritellurides, RTe$_3$ (R = rare-earth element), represents a quasi-1D model system, well suited to study the Fermi surface nesting driven CDW formation.[9-13] Angle-resolved photoemission spectroscopy (ARPES) has revealed a great deal of information about the electronic structure in RTe$_3$. The Fermi surface is modified by large energy gaps of several 100 meV in the CDW phase, making RTe$_3$ an ideal candidate to study the dynamics intrinsically linked to the CDW state. Substitution of various lanthanides allows direct tuning of the CDW transition temperature T_{CDW} by chemical pressure, whereby for the heavier elements ($e.g.$ DyTe$_3$, HoTe$_3$), a second, perpendicular CDW transition at a lower transition temperature occurs.[13,14]

In previous trARPES studies on TbTe$_3$ a transition into a transient metallic state together with a coherent oscillation at $\sim$2.3 THz of the CDW band was observed after ultrafast optical excitation.[6,15] The analysis was based on the changes of the transient dispersion of the gapped CDW band close to E_F and the coherent oscillation was attributed to the CDW amplitude mode due to its exclusive occurrence close to k_F and in the gapped region of the Brillouin zone of the CDW phase below T_{CDW}. In particular, the observed oscillation modulates the peak position of the occupied CDW band, as well as its amplitude and spectral width. In addition to the dynamics of the occupied band, a strong downshift of the unoccupied band was observed in the first $\sim$300 fs, leading to an initial suppression of the gap size.[15] However, this previous work was lacking a direct measurement of the order parameter $\Delta(t)$ and its temporal evolution over an extended range of time delays to extract detailed information about the coupling of various phonon modes to the band structure and, in particular, to the CDW gap. Therefore, a novel approach is necessary to investigate the coherent dynamics of the full CDW energy gap, as will be presented below.

In this work we study the complete dynamics of the CDW order parameter and investigate the occupied and unoccupied band structure in the gapped region as well as the influence of the amplitude mode oscillations on the position and size of the CDW gap in full detail. We employ a novel three-pulse excitation scheme to repopulate and probe the population in the upper (normally unoccupied) CDW band over extended periods of time (see Fig. 1(d)). From a detailed analysis of the frequencies of the modes modulating the gap size we obtain two dominating contributions at 2.2 THz and 1.75 THz, which demonstrates the existence of more than one amplitude mode, in contrast to a simple Peierls picture. Furthermore, by employing coherent control of the amplitude modes by a two pulse excitation scheme, we demonstrate a transient decoupling of the amplitude modes, which show strong coupling in thermal equilibrium.[16]

2   Experimental methods

The experimental setup for trARPES has been described in detail elsewhere.[15,17] Single crystals of RTe_3 (R = Tb, Dy) grown by slow cooling of a binary melt[14] were mounted on a 45° slanted sample holder. The samples where cleaved in ultrahigh vacuum ($p < 5E - 11$ mbar) at $T = 30$ K, where also the trARPES measurements were performed. For the three-pulse experiments, the output of a Ti:Sa regenerative amplifier (Coherent RegA) operating at a repetition rate of 300 kHz was frequency quadrupled using two successive β-bariumborate (BBO) crystals to yield $h\nu_{probe} = 6.0$ eV used as probe pulses. The rest of the fundamental beam at $h\nu_{pump} = 1.5$ eV was split into two parts and delayed separately using two individual optical delay stages in a Mach-Zehnder interferometer configuration, used as the two pump pulses. For the momentum- and energy-resolved detection of photoelectrons emitted by the probe pulses, the sample was rotated in front of a time-of-flight (TOF) spectrometer with an effective opening angle of ±3.5°. The overall time and momentum resolution was 100 fs and 50 meV, respectively.

3   Results

The occupied and unoccupied band structure of $TbTe_3$ in the gapped region probed 50 fs after excitation is shown in Fig. 2 in a false color representation, along the Brillouin zone (BZ) cut indicated in Fig. 1(c). Here, a single weak pump pulse with a fluence of $F_{Pop} \sim 90$ μJ cm^{-2} (population pulse), set at $t_{Pop} = 50$ fs before the probe pulse was used to transiently populate the normally unoccupied states at $E > E_F$. Both lower and upper CDW bands are visible below and above the Fermi level, respectively, separated by the CDW energy gap of $\Delta_{CDW} \sim 400$ meV. The delay t_{Pop} was carefully chosen such that the transient population of the unoccupied states was maximal.

In order to investigate the influence of the amplitude mode oscillation on the size and position of the gap, the three-pulse excitation scheme in Fig. 1(d) was employed. A first pump pulse ($h\nu_{Pump}$) with a fluence of $F_{Pump} = 240$ μJ cm^{-2} arriving at a variable time delay t with respect to the probe pulse is used to excite the coherent amplitude mode oscillations. The weaker population pulse is kept at the fixed delay t_{Pop} with respect to the probe pulse to transiently repopulate the unoccupied states. The subsequent probe pulse $h\nu_{Probe}$ then monitors both occupied and unoccupied bands. The much weaker fluence $F_{Pop} \ll F_{Pump}$ ensures minimizing the influence of the population pulse on the amplitude mode oscillation.

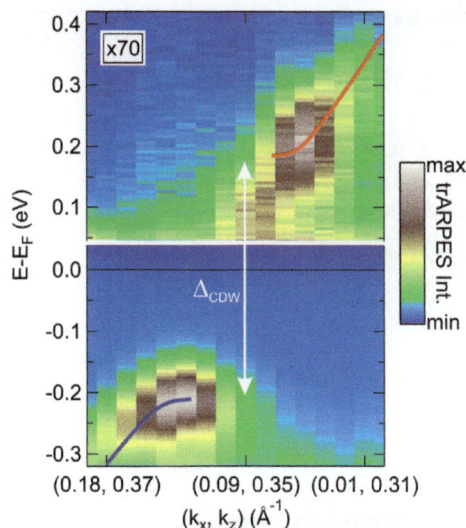

Fig. 2 trARPES intensity of TbTe$_3$ at an equilibrium temperature of 30 K recorded at 50 fs after excitation with a single pump pulse as a function of energy and momentum along the cut through the Brillouin zone shown in Fig. 1(c) in a false color representation. The upper color scale is enhanced by a factor of 70. Red and blue lines are guides to the eye to highlight the dispersion of the upper and lower CDW band, respectively, which are separated by the CDW energy gap Δ_{CDW} (see Fig. 1(b)).

3.1 Coherent modulation of the CDW gap size

Fig. 3(a) depicts the transient three-pulse trARPES intensity as a function of pump–probe delay and energy $E - E_F$, below and above E_F, at a k-position corresponding to the normal state Fermi momentum (white arrow in Fig. 2). For better visibility, the intensity scale for energies above E_F is enhanced by 30 times. After excitation at $t = 0$, we find the characteristic oscillations of the amplitude mode in position and intensity of the occupied peak,[6,15] where the dashed lines are guides to the eye to highlight the oscillation period. In the unoccupied band, also a weak intensity modulation by the amplitude mode is observed, which is highlighted by the contour lines. The transient trARPES spectra shown in Fig. 3(b) at a maximum (blue) and a minimum (green) of the oscillations in the position of the lower band reveal a shift in the band position also in the unoccupied band, which exhibits an opposite sign with respect to the occupied band.

For further analysis, the peak positions of the occupied and unoccupied band are determined by Lorentzian line fits and the results are depicted in the upper panel of Fig. 4. We clearly find oscillations of the same magnitude in the position of both bands with a frequency of $f \sim 2.2$ THz, which, however, show a beating pattern with a period of $t \sim 3$ ps, indicating the presence of more than one mode. Most remarkably, the oscillations are clearly anti-correlated, where a maximum in the lower band corresponds to a minimum in the upper band and *vice versa*. This behavior corresponds to a transient oscillation of the gap size $\Delta_{CDW} = E_{upper} - E_{lower}$, as depicted in the lower panel (green). In contrast, the gap center $E_{center} = (E_{upper} + E_{lower})/2$ (yellow) shows virtually no oscillations.

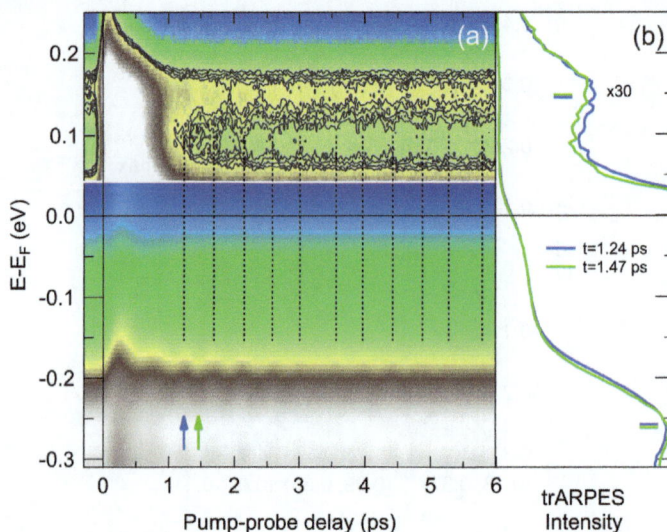

Fig. 3   (a) Three-pulse trARPES intensity of TbTe$_3$ as a function of pump–probe delay and energy $E - E_F$ taken at $(k_x, k_z) \approx (0.09, 0.35)\text{Å}^{-1}$ in a false color representation, using $F_{Pump} \sim 3F_{Pop}$. The upper color scale is enhanced by a factor of ×30. Dashed lines mark the positions of maxima of the lower peak position and contour lines highlight intensity oscillations in the upper band. (b) trARPES spectra at $t = 1.24$ ps (blue) and $t = 1.47$ ps (green) after excitation, corresponding to a maximum and minimum of the lower peak position, respectively. The blue and green ticks mark the peak positions obtained by Lorentzian line fits.

This is indeed the expected behavior for the influence of the amplitude mode of the CDW order parameter, which modulates the amplitude of the complex order parameter and hence the CDW gap size (see Fig. 1(b)). The observation of the coherent oscillations of the CDW gap size as demonstrated by our experiments thus provides an unambiguous and direct proof that the oscillations previously only observed in the lower CDW band[6,15] indeed correspond to amplitude mode excitations of the system.

The main frequencies of the oscillations are analyzed by fitting the transient gap size with two damped oscillators and an exponential background function, depicted as dark green line in Fig. 4. The fit, which shows a good agreement with the oscillations and in particular with the beating pattern, yields central frequencies of $f_1 = 2.230(6)$ THz (2.2 THz mode) and $f_2 = 1.77(2)$ THz (1.75 THz mode). The initial phases of both modes are found to be cosine-like, in agreement with a displacive excitation of the modes. A fast Fourier transformation (FFT) of the residual data show indications of additional frequencies around 2.6 THz and 3.5 THz, which are however too weak in amplitude to be unambiguously fitted. Both the observed frequencies and the ratio of amplitudes, $A_1/A_2 \sim 6$ agree very well with the coherent phonon modes observed for TbTe$_3$ with transient optical reflectivity.[18]

The observation of multiple frequencies that couple to the magnitude of the order parameter is remarkable, as in the mean-field description of the Peierls transition, only one specific phonon mode at the CDW wave vector **q** couples to the electronic degrees of freedom and shows a critical softening at

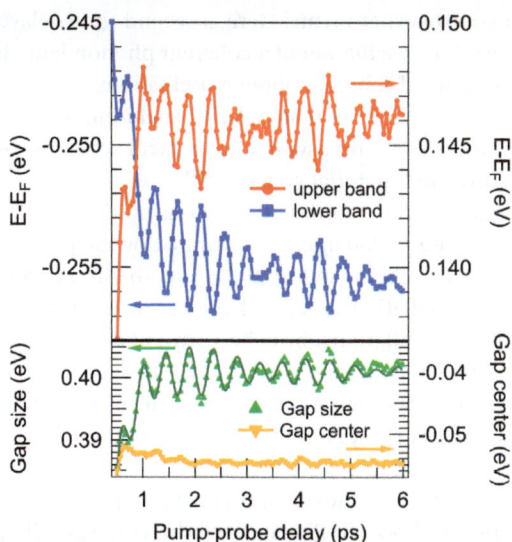

Fig. 4 Top: Peak positions of lower (blue) and upper (red) CDW band of TbTe$_3$ determined by fitting of the trARPES spectra. Both traces show an antipodal oscillation with the characteristic frequency of the amplitude mode. Bottom: Transient CDW gap size $\Delta = E_{upper} - E_{lower}$ and transient gap center $(E_{upper} + E_{lower})/2$. The solid line is a fit to the data (see text).

the phase transition.[1] In TbTe$_3$, this mode has been identified with the 2.2 THz mode, that shows sizable softening close to the phase transition.[18,19] However, this previous work also demonstrated that there exists a complex interplay between the 1.75 THz mode and 2.2 THz mode, suggesting a strong coupling between both modes§. Our results show that the amplitude of the CDW order parameter Δ_{CDW} is modulated by both modes, demonstrating the existence of more than one amplitude mode. This can be understood from the quasi-1D nature of RTe$_3$, which exhibits more than one phonon branch at the CDW wave vector. These additional phonon branches can also linearly couple to the modulated electronic charge density and exhibit the characteristic features of amplitude modes, as explained in the time-dependent Ginzburg-Landau model.[21,22]

3.2   Coherent control of the amplitude mode

The observation of both the 2.2 THz and the 1.75 THz mode in the transient oscillations of the order parameter raises the question, how far these modes are mutually coupled and influence each other or whether they independently modulate the CDW gap. The mode crossing of the two modes observed as a function of temperature in transient reflectivity measurements[18] and Raman scattering[19] indeed suggests a substantial coupling, especially near the crossing point of the modes. Additional insight into this coupling can be gained from time-

§ An assignment of the 1.75 THz and 2.2 THz modes to two different (perpendicular) CDW vectors appears unlikely as both modes are observed far above a potential second low temperature CDW transition[20] and modulate the previously observed k-dependent CDW gap in TbTe$_3$ along c^*.

resolved coherent control experiments. Here, a second time-delayed pump pulse is employed to control the oscillation of a coherent phonon launched by the first pump pulse. Such control of coherent phonon oscillations in the time domain has been demonstrated in a number of materials,[23–27] including selective excitation of certain vibrational modes[28,29] and asymmetric control schemes exploring collective dynamics in correlated material systems.[30,31]

For the coherent control experiments, two excitation pulses $h\nu_{Pump1}$ and $h\nu_{Pump2}$ of equal fluence $F \sim 250$ µJ cm^{-2} separated by the pump-pump delay t_{12} were used employing the scheme in Fig. 1(e). Here, the first pump pulse triggers the amplitude mode oscillation at t_0, and the second pump pulse is used for coherent control of the oscillations, by coherently enhancing or suppressing the oscillations depending on t_{12}. The resulting oscillations are observed by the probe pulse after the pump1-probe delay t. For each scan of t, t_{12} was held fix, and multiple scans for various t_{12} spanning a whole period of the amplitude mode were performed.

The resulting trARPES intensities of DyTe$_3$ for a pump-pump delay of $t_{12} = 0.67$ ps and $t_{12} = 0.93$ ps are shown in Fig. 5(a) and (b), corresponding to a coherent suppression or enhancement of the amplitude mode oscillation, respectively. The cross-correlation (XC) of the two pump pulses with the probe pulse are depicted in the lower panels, showing the arrival times and equal excitation amplitudes of both pump pulses. Clearly observed is the efficient suppression of the coherent oscillation for out-of-phase excitation in Fig. 5(a), whereas the oscillation is enhanced for in-phase excitation in Fig. 5(b).

The transient peak position of the lower CDW band determined by line fits is shown for various pump-pump delays t_{12} in Fig. 6(a), where the arrival of the control pulse is indicated by the red markers. After the initial shift of the peak position associated with the arrival of each pump pulse, the coherent oscillations

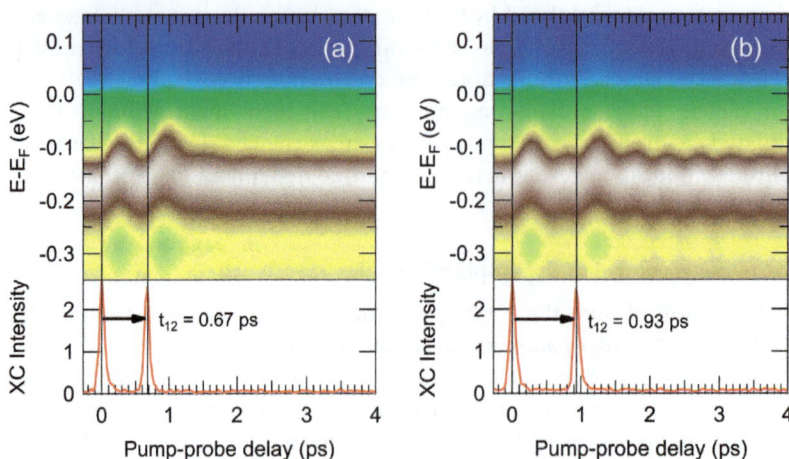

Fig. 5 (a) trARPES intensity of DyTe$_3$ at $(k_x, k_z) \approx (0.13, 0.33)$Å^{-1} for a pump-pulse separation $t_{12} = 0.67$ ps, corresponding to a coherent quench of the amplitude mode oscillation (out-of-phase). The lower panel shows the cross correlation (XC) trace of the two pump pulses with the probe pulse obtained from electrons with $E - E_F > 1.3$ eV. (b) trARPES intensity for a pump-pulse separation $t_{12} = 0.93$ ps, corresponding to a coherent enhancement of the amplitude mode oscillation (in-phase).

are well resolved in all cases. Traces of out-of-phase and in-phase excitation, corresponding to Fig. 5(a) and (b) are highlighted by thick orange and green lines, respectively.

In order to quantify the coherent control of the amplitude mode, fast FFT traces obtained from the transient peak position after background subtraction for $t > 2$ ps are shown in Fig. 6(b). The position of the two modes at $f = 2.2$ THz and $f = 1.75$ THz are marked by red and black dashed lines, respectively. While both peaks are present at $t_{12} = 0.49$ ps (bottommost curve), at $t_{12} = 0.68$ ps (orange curve), the intensity of the 2.2 THz mode is almost completely suppressed, and only the weak mode at 1.75 THz remains. Similarly, at $t_{12} = 0.93$ ps, only the enhanced oscillation of the 2.2 THz mode is observed and the 1.75 THz mode is absent. A fit of two Lorentzian line shapes centered at $f = 1.75$ THz and $f = 2.2$ THz to the FFT spectra is used to determine the oscillation amplitudes, which are

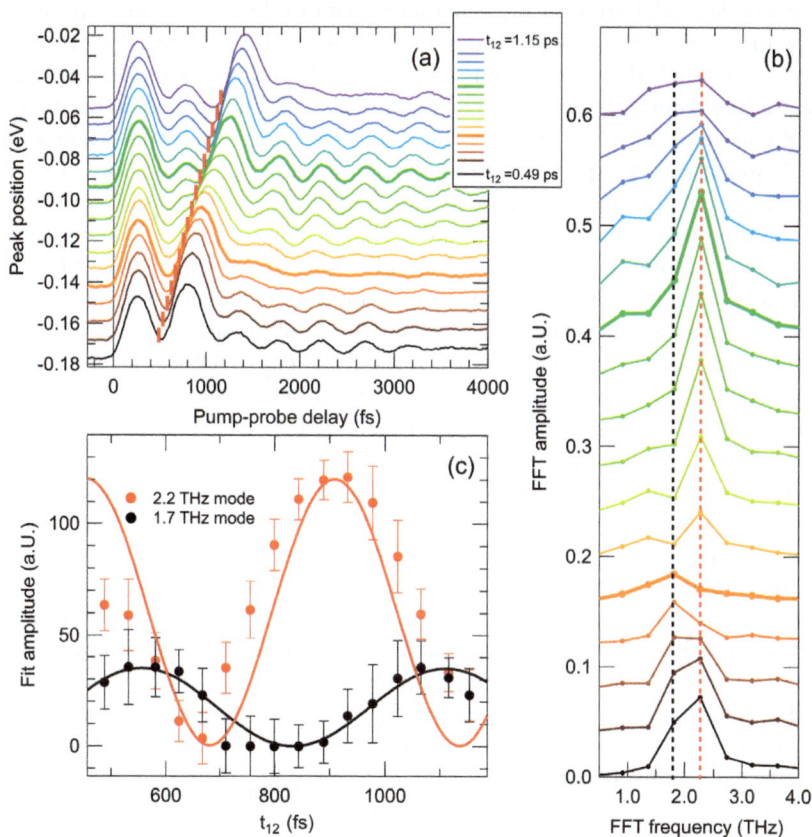

Fig. 6 (a) Transient peak position of the lower CDW peak in DyTe$_3$ as a function of pump–probe delay for various pump-pulse separations t_{12}, indicated by red markers. The data corresponding to the situation in Fig. 5(a) and (b) are highlighted by thick orange and green lines, respectively. Traces are vertically offset for clarity. (b) FFT amplitudes for various t_{12}. Positions of the 1.75 THz and 2.2 THz mode are marked with black and red vertical lines, respectively. (c) Fitted FFT amplitude of the 1.75 THz (black) and 2.2 THz (red) mode as a function of pump-pulse separation t_{12}. Solid lines are fits to eqn (1).

shown in Fig. 6(c) as a function of pump-pump delay t_{12} as black and red symbols, respectively.

From the response of the spectral weight of the two modes to the control pulse, information on the coupling of the two modes can be gained. Neglecting the decay of the coherent oscillations between the two pump pulses, the coherent oscillation amplitude $A(t_{12})$ of independent modes can be described by[25]

$$A(t_{12}) = \frac{A_0}{2}[1 + \cos(f \cdot t_{12} \cdot 2\pi)], \tag{1}$$

where f is the frequency of the oscillation and A_0 is the maximal amplitude for resonant excitation with both pump pulses. Fits to eqn (1) are shown as solid black and red lines in Fig. 6(c) for the 1.75 THz and 2.2 THz mode, respectively. The data are described nicely by a model with two independent modes.

4   Discussion

The simple model captures the general behavior of the oscillation amplitudes as a function of t_{12} of both modes remarkably well and especially perfectly reproduces their different periods with t_{12}. Slight deviations in the extracted amplitudes from the model could be explained due to the phase shift occurring to the oscillations when crossing the resonance condition, which leads to a deviation of the FFT spectra from Lorentzian line shapes, or nonlinearities in the fluence dependence of the oscillation amplitudes, which could result in a saturation of the oscillation amplitude. The good agreement of the coherent oscillation amplitude control to the simple model of independent oscillators indicates weak coupling of the different amplitude modes. Especially the fact that we can selectively control the amplitude of each mode, in particular that we can completely suppress the 2.2 THz mode with almost maximal amplitude of the 1.75 THz mode at $t_{12} = 0.68$ ps, and the absence of the 1.75 THz mode at $t_{12} \approx 0.8$ ps, underlines the independent nature of the two amplitude modes in the coherent control experiment.

The observation of weak coupling of the amplitude modes in the photoexcited phase is surprising, as both modes couple to the same electronic order parameter and modulate similar coordinates of the same atoms.[16] Our observation is in contrast to temperature-dependent studies of the amplitude modes in RTe$_3$ observed by time-resolved optical spectroscopy[18] and Raman spectroscopy.[19] These works demonstrated a mixing and avoided crossing of the two modes at 1.75 THz and 2.2 THz due to strong coupling during the softening of the modes with increasing temperature. In the time-dependent Ginzburg-Landau model this behavior arises due to the coupling of the lattice degrees of freedoms, described as structural order parameters, to the electronic part of the order parameter, which leads to an indirect coupling of the different phonon modes in the system.[22] Thus, the observation of weakly coupled amplitude modes in the coherent control experiments suggests a transient decoupling of the lattice degrees of freedom in the photoexcited state, which is not observed in the static case. A more complete description of the transient order parameter dynamics would require a proper description in the framework of the time-dependent Ginzburg-Landau model,[22] which is however beyond the scope of the current manuscript.

5 Summary

In summary, using a three-pulse pump–probe scheme in time- and angle-resolved photoemission experiments we demonstrated the effect of the amplitude mode oscillations on the electronic band structure in the prototypical charge-density wave compounds $RT3_3$. We observe coherent oscillations in both lower and upper CDW band with opposite phases, which result in a modulation of the CDW energy gap. The observation of two dominating frequencies modulating the CDW order parameter demonstrates the existence of more than one collective amplitude mode, which is in contrast to the simple Peierls model. This can be understood from the quasi-1D nature of RTe_3, which allows more than one phonon branch at the CDW wave vector $\mathbf{q}$ to couple to the order parameter. Coherent control experiments of the two amplitude modes, which are strongly coupled in equilibrium, allow independent control of the modes and demonstrate a transient decoupling of the amplitude modes in the photoexcited state.

Acknowledgements

The authors acknowledge financial support from the Deutsche For-schungsgemeinschaft through BO 1823/2 and FOR 1700, the Mercator Research Center Ruhr through Grant No. PR-2011-0003, and the US Department of Energy, Office of Basic Energy Sciences under contract DE-AC02-76SF00515.

References

1 G. Grüner, *Density Waves in Solids*, Addison-Wesley, 1994, vol. 89.
2 R. E. Peierls, *Quantum Theory of Solids*, Oxford University Press, New York, 1955.
3 *Dynamics at Solid State Surfaces and Interfaces*, Vol. 1, ed. U. Bovensiepen, H. Petek and M. Wolf, Wiley-VCH Weinheim, Germany, 2010.
4 S. Hellmann, T. Rohwer, M. Kalläne, K. Hanff, C. Sohrt, A. Stange, A. Carr, M. Murnane, H. Kapteyn, L. Kipp, M. Bauer and K. Rossnagel, *Nat. Commun.*, 2012, **3**, 1069.
5 L. Perfetti, P. A. Loukakos, M. Lisowski, U. Bovensiepen, H. Berger, S. Biermann, P. S. Cornaglia, A. Georges and M. Wolf, *Phys. Rev. Lett.*, 2006, **97**, 067402.
6 F. Schmitt, P. S. Kirchmann, U. Bovensiepen, R. G. Moore, L. Rettig, M. Krenz, J. H. Chu, N. Ru, L. Perfetti, D. H. Lu, M. Wolf, I. R. Fisher and Z. X. Shen, *Science*, 2008, **321**, 1649–1652.
7 T. Rohwer, S. Hellmann, M. Wiesenmayer, C. Sohrt, A. Stange, B. Slomski, A. Carr, Y. Liu, L. M. Avila, M. Kalläne, S. Mathias, L. Kipp, K. Rossnagel and M. Bauer, *Nature*, 2011, **471**, 490.
8 J. C. Petersen, S. Kaiser, N. Dean, A. Simoncig, H. Y. Liu, A. L. Cavalieri, C. Cacho, I. C. E. Turcu, E. Springate, F. Frassetto, L. Poletto, S. S. Dhesi, H. Berger and A. Cavalleri, *Phys. Rev. Lett.*, 2011, **107**, 177402.
9 E. DiMasi, M. C. Aronson, J. F. Mansfield, B. Foran and S. Lee, *Phys. Rev. B*, 1995, **52**, 14516–14525.
10 G.-H. Gweon, J. D. Denlinger, J. A. Clack, J. W. Allen, C. G. Olson, E. DiMasi, M. C. Aronson, B. Foran and S. Lee, *Phys. Rev. Lett.*, 1998, **81**, 886–889.

11 V. Brouet, W. L. Yang, X. J. Zhou, Z. Hussain, N. Ru, K. Y. Shin, I. R. Fisher and Z.-X. Shen, *Phys. Rev. Lett.*, 2004, **93**, 126405.

12 V. Brouet, W. L. Yang, X. J. Zhou, Z. Hussain, R. G. Moore, R. He, D. H. Lu, Z.-X. Shen, J. Laverock, S. B. Dugdale, N. Ru and I. R. Fisher, *Phys. Rev. B: Condens. Matter Mater. Phys.*, 2008, **77**, 235104.

13 R. G. Moore, V. Brouet, R. He, D. H. Lu, N. Ru, J.-H. Chu, I. R. Fisher and Z.-X. Shen, *Phys. Rev. B: Condens. Matter Mater. Phys.*, 2010, **81**, 073102.

14 N. Ru, C. L. Condron, G. Y. Margulis, K. Y. Shin, J. Laverock, S. B. Dugdale, M. F. Toney and I. R. Fisher, *Phys. Rev. B: Condens. Matter Mater. Phys.*, 2008, **77**, 035114.

15 F. Schmitt, P. S. Kirchmann, U. Bovensiepen, R. G. Moore, J.-H. Chu, D. H. Lu, L. Rettig, M. Wolf, I. R. Fisher and Z.-X. Shen, *New J. Phys.*, 2011, **13**, 063022.

16 M. Lavagnini, M. Baldini, A. Sacchetti, D. Di Castro, B. Delley, R. Monnier, J.-H. Chu, N. Ru, I. R. Fisher, P. Postorino and L. Degiorgi, *Phys. Rev. B: Condens. Matter Mater. Phys.*, 2008, **78**, 201101.

17 M. Lisowski, P. Loukakos, U. Bovensiepen, J. Stähler, C. Gahl and M. Wolf, *Appl. Phys. A: Mater. Sci. Process.*, 2004, **78**, 165–176.

18 R. V. Yusupov, T. Mertelj, J.-H. Chu, I. R. Fisher and D. Mihailovic, *Phys. Rev. Lett.*, 2008, **101**, 246402.

19 M. Lavagnini, H.-M. Eiter, L. Tassini, B. Muschler, R. Hackl, R. Monnier, J.-H. Chu, I. R. Fisher and L. Degiorgi, *Phys. Rev. B: Condens. Matter Mater. Phys.*, 2010, **81**, 081101.

20 A. Banerjee, Y. Feng, D. M. Silevitch, J. Wang, J. C. Lang, H.-H. Kuo, I. R. Fisher and T. F. Rosenbaum, *Phys. Rev. B: Condens. Matter Mater. Phys.*, 2013, **87**, 155131.

21 H. Schäfer, V. V. Kabanov, M. Beyer, K. Biljakovic and J. Demsar, *Phys. Rev. Lett.*, 2010, **105**, 066402.

22 H. Schäfer, V. V. Kabanov and J. Demsar, *Phys. Rev. B: Condens. Matter Mater. Phys.*, 2014, **89**, 045106.

23 T. Dekorsy, W. Kütt, T. Pfeifer and H. Kurz, *Europhys. Lett.*, 1993, **23**, 223.

24 M. Hase, K. Mizoguchi, H. Harima, S. Nakashima, M. Tani, K. Sakai and M. Hangyo, *Appl. Phys. Lett.*, 1996, **69**, 2474.

25 T. Onozaki, Y. Toda, S. Tanda and R. Morita, *Jpn. J. Appl. Phys.*, 2007, **46**, 870–872.

26 O. V. Misochko, R. Lu, M. Hase and M. Kitajima, *J. Exp. Theor. Phys.*, 2007, **104**, 245–253.

27 S. Wall, D. Wegkamp, L. Foglia, K. Appavoo, J. Nag, R. Haglund, J. Stähler and M. Wolf, *Nat. Commun.*, 2012, **3**, 721.

28 H. Takahashi, K. Kato, H. Nakano, M. Kitajima, K. Ohmori and K. G. Nakamura, *Solid State Commun.*, 2009, **149**, 1955–1957.

29 Y. Okano, H. Katsuki, Y. Nakagawa, H. Takahashi, K. G. Nakamura and K. Ohmori, *Faraday Discuss.*, 2011, **153**, 375–382.

30 R. Yusupov, T. Mertelj, V. V. Kabanov, S. Brazovskii, P. Kusar, J.-H. Chu, I. R. Fisher and D. Mihailovic, *Nat. Phys.*, 2010, **6**, 681–684.

31 P. Kusar, T. Mertelj, V. V. Kabanov, J.-H. Chu, I. R. Fisher, H. Berger, L. Forró and D. Mihailovic, *Phys. Rev. B: Condens. Matter Mater. Phys.*, 2011, **83**, 035104.

Faraday Discussions

Non-equilibrium Dirac carrier dynamics in graphene investigated with time- and angle-resolved photoemission spectroscopy

Isabella Gierz,*[a] Stefan Link,[b] Ulrich Starke[b] and Andrea Cavalleri[a]

Received 21st February 2014, Accepted 28th April 2014

DOI: 10.1039/c4fd00020j

We have used time- and angle-resolved photoemission spectroscopy (tr-ARPES) to assess the influence of many-body interactions on the Dirac carrier dynamics in graphene. From the energy-dependence of the measured scattering rates we directly determine the imaginary part of the self-energy, visualizing the existence of a relaxation bottleneck associated with electron–phonon coupling. A comparison with static line widths obtained by high-resolution ARPES indicates that the dynamics of photo-excited carriers in graphene are solely determined by the equilibrium self-energy. Furthermore, the subtle interplay of different many-body interactions in graphene may allow for carrier multiplication, where the absorption of a single photon generates more than one electron-hole pair *via* impact ionization. We find that, after photo-excitation, the number of carriers in the conduction band along the ΓK-direction keeps increasing for about 40 fs after the pump pulse is gone. A definite proof of carrier multiplication in graphene, however, requires a more systematic study, carefully taking into account the contribution of momentum relaxation on the measured rise time.

1   Introduction: quasiparticle dynamics in graphene

The peculiar electronic structure of graphene results from the conical intersection of two cosine-shaped bands at the K-point of the hexagonal Brillouin zone. While the single-particle dispersion is well described by a simple tight-binding model,[1] it has been shown that the linear band structure in the vicinity of the Fermi level is strongly renormalized by many-body interactions.[2,3] A careful line width analysis based on high-resolution ARPES measurements has revealed that the photo-hole decays into phonons, electron-hole pairs, and plasmons.[2] The influence of a

[a]Max Planck Institute for the Structure and Dynamics of Matter, Hamburg, Germany. E-mail: isabella.gierz@mpsd.mpg.de; Fax: +49 (0)40 8998 1958; Tel: +49 (0)40 8998 5362

[b]Max Planck Institute for Solid State Research, Stuttgart, Germany

particular decay channel depends strongly on energy. Decay *via* phonon emission becomes important for energies $|E| > \hbar\omega_{ph}$, while the emission of plasmons is restricted to the region around the Dirac point E_D, where valence and conduction band touch. Electron-hole pairs dominate the relaxation for energies $|E| > 2E_D$.[2] The decay of photo-excited carriers *via* many-body interactions can be directly observed in the time domain.[4]

From previous optical pump–probe experiments[5–15] supported by theoretical investigations[16–21] the following picture of hot carrier dynamics in graphene has emerged. The matrix element describing the coupling between electrons and light is anisotropic with nodes along the direction of light polarization, resulting in a correspondingly anisotropic initial distribution of electrons and holes peaked at $E = \pm\hbar\omega_{pump}/2$ away from the Dirac point.[19] Subsequent carrier-carrier scattering leads to an ultrafast thermalization of the electronic system resulting in a hot Fermi–Dirac (FD) distribution with a momentum-dependent electronic temperature within some tens of femtoseconds.[19] This distribution cools down and becomes isotropic *via* the emission of optical phonons within $\sim$200 fs,[5,6,19] resulting in a hot optical phonon population that has been directly observed with time-resolved Raman scattering.[8,11,15] These hot phonons subsequently decay into two acoustic modes on a picosecond time scale.[8,11,13,15] Furthermore, in the presence of lattice defects, the electronic system can couple directly to acoustic phonons resulting in supercollision cooling of the electronic system.[21–23]

The initial ultrafast thermalization resulting in an anisotropic FD distribution is beyond the time resolution of most pump–probe experiments. Measurements performed with sub 10 fs pulse duration, however, indicate the existence of separate electron and hole distributions in the conduction and valence band at early times.[7,14] These findings have been confirmed by time-resolved photoemission experiments[24,25] and the observation of a transient negative optical conductivity.[26] In ref. 17 the occurrence of population inversion after optical excitation has been attributed to the dominance of intraband electron–phonon scattering at early times. The decay of the population inversion on the femtosecond time scale is believed to be driven mainly by Auger recombination.[17,25]

Recently, the prediction of carrier multiplication by impact ionization[18–20] has motivated considerable experimental activity, revealing crucial misconceptions and resulting in conflicting interpretations. We would like to clarify that the observation of hot-carrier multiplication in ref. 27 and 28 is no proof for multiple electron-hole pair generation by a single absorbed photon as predicted by ref. 18-20. While ref. 29, based on the comparison of the measured transmission change after optical excitation with different theoretical models, argues in favor of carrier multiplication, tr-ARPES experiments find no indication for this effect.[25,30]

In the present work we show that scattering rates of photo-excited carriers in graphene measured with tr-ARPES indeed resemble the MDC line width determined by static ARPES as predicted by ref. 4. Furthermore, we find that the rise time of the number of carriers in the conduction band for excitation at 1.5 eV is slightly longer than the pump pulse duration. Before attributing this finding to the occurrence of carrier multiplication, alternative explanations for an increased rise time need to be excluded in a more systematic study.

2   Equilibrium properties of graphene

The present study is performed on quasi-freestanding hydrogen-intercalated graphene monolayers on a SiC(0001) substrate as first reported in ref. 31. The static band structure perpendicular to the ΓK-direction is displayed in Fig. 1a. The samples are slightly hole-doped with the Dirac point about 200 meV above the Fermi level. The influence of many-body interactions in similar samples has been addressed previously in a high-resolution ARPES measurement.[32] Fig. 1b shows the measured line width of momentum distribution curves (MDCs) as a function of energy. The steep increase in line width between the Fermi level and $E \approx -200$ meV is caused by the strong coupling of the electrons to the E_{2g} phonon mode at Γ and the A_1' phonon mode at K.[16,32] The continuous increase at lower energies is attributed to electron-hole scattering.[32]

3   Time- and angle-resolved photoemission spectroscopy

Photoemission spectroscopy is based on the photoelectric effect, where an incident photon in the extreme ultra-violet (EUV) range impinges on the surface of a crystal and releases a photoelectron. The resulting photocurrent as a function of kinetic energy and emission angle gives direct access to the electronic structure in momentum space. The photocurrent is directly proportional to the spectral function $A(\vec{k},E)$ of the sample that, using the sudden approximation, is given by[33]

$$A\left(\vec{k}, E\right) = \frac{1}{\pi} \frac{\operatorname{Im}\left(\Sigma\left(\vec{k}, E\right)\right)}{\left[E - E_{\vec{k}}^0 - \operatorname{Re}\left(\Sigma\left(\vec{k}, E\right)\right)\right]^2 + \left[\operatorname{Im}\left(\Sigma\left(\vec{k}, E\right)\right)\right]^2},$$

where $E_{\vec{k}}^0$ is the single particle dispersion and $\Sigma(\vec{k},E)$ is the complex self-energy due to many-body interactions. $A(\vec{k},E)$ is a Lorentzian function with the peak position at $E_{\vec{k}}^0 + \operatorname{Re}(\Sigma(\vec{k},E))$ and a peak width of $\operatorname{Im}(\Sigma(\vec{k},E))$. Thus, ARPES presents the ideal tool to investigate the influence of many-body interactions on the electronic structure. A careful analysis of peak positions and line widths has been used to identify many-body interactions in graphene.[2,32] More importantly, the imaginary part of the self-energy $\operatorname{Im}(\Sigma(\vec{k},E))$ can be obtained directly by transferring the technique to the time domain, where the energy- and momentum-dependent scattering rate is given by[4] $1/\tau(\vec{k},E) = -2\operatorname{Im}(\Sigma(\vec{k},E))$.

A schematic setup for tr-ARPES is shown in Fig. 2a. A pump pulse at $\hbar\omega_{\text{pump}}$ excites the sample and the response of the electronic structure is probed with ARPES, where a time-delayed EUV pulse releases photoelectrons. For the present experiment the pump wavelength was chosen to be $\hbar\omega_{\text{pump}} = 1.5$ eV, the fundamental wavelength of a femtosecond Ti:Sapphire laser system (800 nm, 60 fs, 1

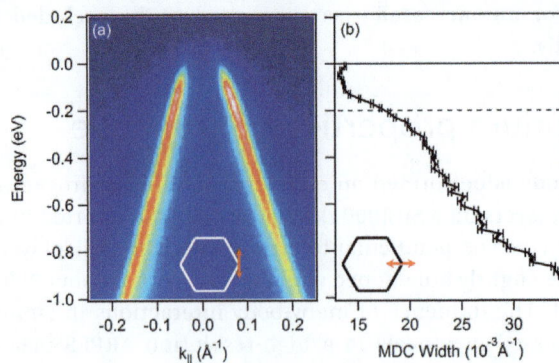

Fig. 1 (a) Static band structure of hydrogen-intercalated monolayer graphene around the K-point of the Brillouin zone measured along a cut perpendicular to the ΓK-direction (see inset). The samples are slightly hole-doped with the Dirac-point ~200 meV above the Fermi level. Panel (b) shows fit results of the Lorentzian line width for momentum distribution curves (MDCs) along the ΓK-direction (see inset) as a function of energy for a hydrogen-intercalated graphene monolayer. The horizontal dashed black line indicates the energy of the E_{2g} phonon mode.

kHz). For the EUV probe we use high harmonics of the fundamental wavelength generated in an argon plasma. A typical EUV spectrum obtained by high harmonics generation in argon is shown in Fig. 2b. The spectrum covers the energy range between 15 and 40 eV. A time-preserving grating monochromator[34] is used to select a single harmonic at $\hbar\omega_{probe} \approx 30$ eV for the present graphene measurements.

Fig. 3 shows snapshots of the electronic structure of a quasi-freestanding graphene monolayer in the vicinity of the K-point for different time delays after

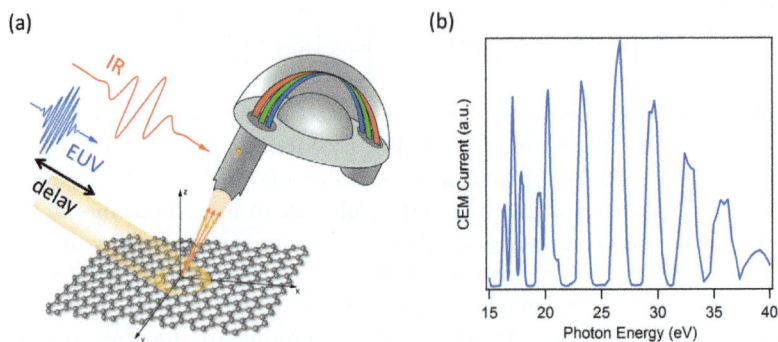

Fig. 2 (a) Schematic setup for time- and angle-resolved photoemission spectroscopy. The graphene samples are excited with femtosecond laser pulses at $\hbar\omega_{pump} = 1.5$ eV. The response of the electronic structure is probed with angle-resolved photoemission spectroscopy, where a time-delayed pulse at $\hbar\omega_{probe} \approx 30$ eV ejects photoelectrons from the sample. Both the emission angle and the kinetic energy of the photoelectrons are measured with a hemispherical analyzer, giving direct access to the distribution of electrons in momentum space as a function of energy and pump–probe delay. (b) Typical EUV spectrum from high harmonics generation in argon.

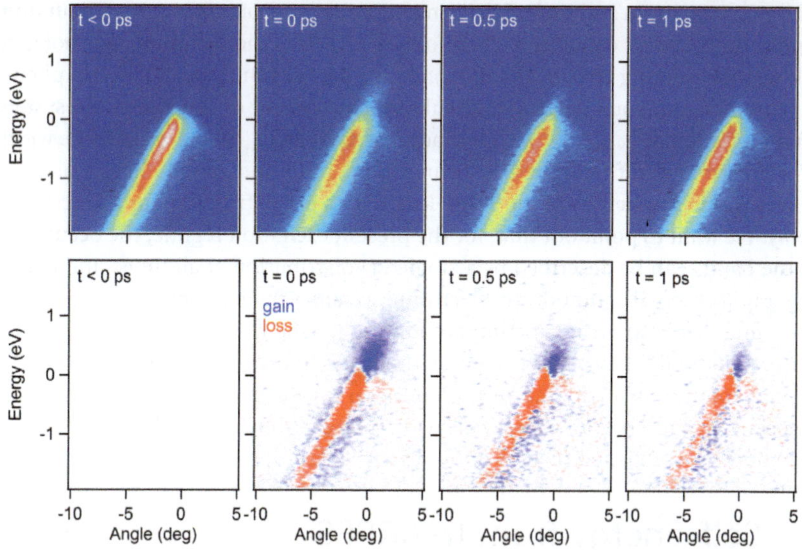

Fig. 3 Upper panel: Snapshots of the band structure along the ΓK-direction for a hydrogen-intercalated graphene monolayer for different pump–probe delays for excitation at $\hbar\omega_{pump} = 1.5$ eV with a fluence of 6.6 mJ cm^{-2} at room temperature. Lower panel: corresponding pump–probe signal.

excitation at $\hbar\omega_{pump} = 1.5$ eV (upper panel) together with the corresponding pump–probe signal (lower panel). The present experiments were performed with a pump fluence of 6.6 mJ cm^{-2} at room temperature. The pump–probe signal is dominated by an instantaneous heating of the electronic system accompanied by a momentum-broadening of the π-band and subsequent relaxation within ∼1 ps. Fig. 3 also illustrates the main advantage of tr-ARPES over all-optical pump–probe techniques: tr-ARPES can track the whereabouts of each electron as a function of

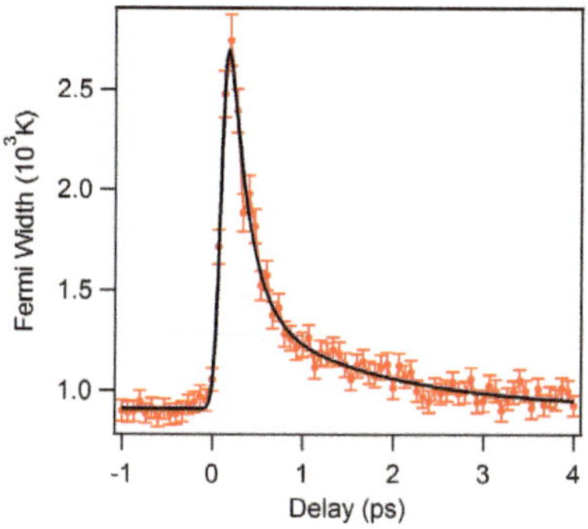

Fig. 4 Width of the Fermi level (data points) as a function of pump–probe delay obtained by fitting momentum-integrated energy distribution curves in the vicinity of the Fermi level with a Fermi–Dirac function. The continuous line represents a double exponential fit with $\tau_1 = 207 \pm 25$ fs typically attributed to electron-optical phonon scattering and $\tau_2 = 1.5 \pm 0.2$ ps due to emission of acoustic phonons.

energy, momentum and time, delivering a complete set of information about non-equilibrium carrier dynamics. Therefore, tr-ARPES is the technique of choice to address the existing controversies concerning carrier multiplication in graphene.

The instantaneous heating and subsequent cooling of the electronic system after excitation at $\hbar\omega_{pump} = 1.5$ eV is addressed in Fig. 4, where we plot the width of the Fermi cut-off (obtained by fitting momentum-integrated energy distribution curves in the vicinity of the Fermi level with a FD distribution) as a function of delay. We want to point out that, for the present excitation regime, the occupancy of the bands can be described by a single FD distribution at all times, indicating very rapid thermalization of the electronic system. The data points in Fig. 4 have been fitted with an error function accounting for the rise time of the signal and two exponential decays. The two decay times, $\tau_1 = 207 \pm 25$ fs and $\tau_2 = 1.5 \pm 0.2$ ps, are commonly attributed to the emission of optical and acoustic phonons, respectively. Our results agree well with values published in literature[5-21] and support the established picture of hot carrier relaxation in graphene.

4 Self-energy from tr-ARPES

According to ref. 4 non-equilibrium dynamics for a model photo-excited electron–phonon system measured with tr-ARPES are determined by the static self-energy alone. The phonon energy ω_{ph} is found to determine the size of an energy window around the Fermi level, where a long-lived pump–probe signal is expected. Electrons will decay rapidly for $|E| > \omega_{ph}$, while for $|E| < \omega_{ph}$ decay via phonon emission is no longer possible. Following ref. 4 we plot the momentum-integrated pump–probe signal as a function of delay (Fig. 5b) and determine the energy-dependence of the scattering time from our tr-ARPES data (Fig. 5c). Finally, we compare our results to the MDC line width obtained from static ARPES measurements on a similar sample (Fig. 6).

The data analysis is done as follows: in a first step, the snapshots in Fig. 3 are integrated over momentum and plotted as a function of delay (Fig. 5a). From the

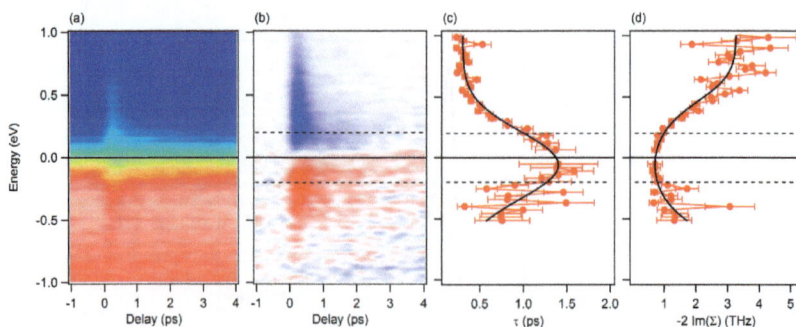

Fig. 5 (a) Momentum-integrated photocurrent as a function of pump–probe delay for excitation at $\hbar\omega_{pump} = 1.5$ eV with a fluence of 6.6 mJ cm^{-2} at room temperature. (b) Corresponding pump–probe signal. (c) Relaxation time τ as a function of energy obtained by fitting pump–probe time traces at different energies with an exponentially decaying function. (d) Inverse lifetime $1/\tau = -2\text{Im}(\Sigma)$ as a function of energy. Continuous black lines are guides to the eye. Horizontal dashed black lines indicate the energy of the in-plane bond-stretching phonon mode.

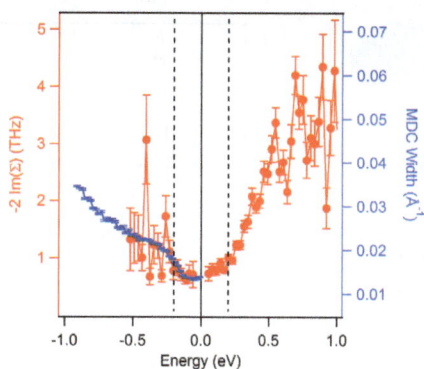

Fig. 6 Direct comparison between MDC width from Fig. 1c and $1/\tau$ from Fig. 5d. Vertical dashed black lines indicate the energy of the in-plane bond-stretching phonon mode.

momentum-integrated data we subtract the photocurrent at negative time delays to obtain Fig. 5b. Then we extract time traces at constant energy and fit them with an exponentially decaying function resulting in the scattering times plotted in Fig. 5c. In Fig. 5d we plot the scattering rate $1/\tau$ that is directly proportional to the imaginary part of the self-energy.[4]

In Fig. 5b, c, and d dashed black lines mark the size of the phonon window discussed above. As predicted,[4] decay times are much longer inside than outside of this phonon window. Hence, plots similar to the one in Fig. 5b offer a direct way to visualize the existence of relaxation bottlenecks in photo-excited systems associated with the coupling to a bosonic mode.

Both the scattering rate from tr-ARPES data and the MDC line width from static ARPES data are proportional to the imaginary part of the self-energy. In Fig. 6 we directly compare the two quantities to find out whether the non-equilibrium dynamics after photo-excitation are completely determined by the static self-energy as predicted in ref. 4. We indeed find that, within the experimental error bars, the scattering rate and the MDC line width follow the same trend for $E < 0$. In this context tr-ARPES results nicely complement the information gained by static ARPES investigations, as tr-ARPES directly measures the imaginary part of the self-energy via $1/\tau = -2\text{Im}(\Sigma)$ both below and above the equilibrium chemical potential.

5   Carrier multiplication

It has been proposed that, in graphene, the absorption of a single photon may generate several electron-hole pairs via impact ionization.[18-20] This process is illustrated in Fig. 7a. The prediction of carrier multiplication is based on the assumption that an ultrashort pump pulse generates a non-thermal carrier distribution with a narrow distribution of electrons in the conduction band and holes in the valence band centered at $E_D \pm \hbar\omega_{pump}/2$. Such a carrier distribution offers a large phase space for impact ionization (green arrows in Fig. 7) where part of the excess energy of a photo-excited electron is used to generate a second electron-hole pair. The inverse process of Auger heating is suppressed by Pauli blocking. According to ref. 18–20 carrier multiplication is expected to be robust

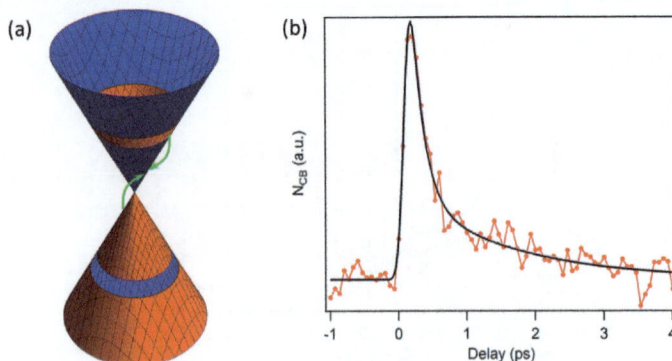

Fig. 7   (a) Schematic of carrier multiplication *via* impact ionization (green arrows) for the given non-equilibrium carrier distribution. Occupied (empty) states are shown in red (blue). (b) Number of carriers in the conduction band N_{CB} as a function of delay (data points) together with fit (continuous line).

even in the presence of electron–phonon coupling, where energy rapidly dissipates to the phonon bath. Optimum experimental conditions for the observation of carrier multiplication are a small pump fluence, a high excitation energy, a low initial temperature, and charge neutral graphene.[20,25]

In ref. 29 the authors have modeled experimental transmission changes at $\hbar\omega_{pump} = 2.25$ eV and $\hbar\omega_{probe} = 1.37$ eV, incorporating either dynamical screening that suppresses Auger scattering or static/regularized dynamical screening with a high impact of Auger scattering. The latter seems to result in a good agreement with the experimental data. The extracted carrier multiplication factor reaches a maximum value of ~1.1. Recent tr-ARPES results[25,30] using $\hbar\omega_{pump} = 950$ meV did not find any indication for carrier multiplication. In principle, tr-ARPES is the tool of choice for the search for carrier multiplication as it directly measures the number of electrons in the conduction band as a function of energy, momentum and time. However, the excitation energy for the tr-ARPES experiments is lower than the one in ref. 29, so that carrier multiplication may occur in ref. 29 but not in ref. 25 and 30. Clearly, the possible occurrence of carrier multiplication in graphene deserves further attention.

In Fig. 5b we plot the momentum-integrated pump–probe signal for excitation at $\hbar\omega_{pump} = 1.5$ eV as a function of delay. This graph directly mirrors the transient changes of the electron distribution across the Dirac cone along the ΓK-direction. The smoking gun evidence for the occurrence of carrier multiplication would be that the number of electron-hole pairs keeps increasing after the pump pulse is already gone.[18] In order to determine the number of carriers in the conduction band N_{CB} along ΓK, we integrate the data in Fig. 5b for $E > E_D = 200$ meV (data points in Fig. 7b). The data are fitted with an error function to describe the rise time and two exponential decays (continuous line in Fig. 5b). We find a rise time of 100 ± 10 fs that is slightly longer than the measured pulse duration of 60 fs.

This finding may be interpreted in terms of carrier multiplication. However, this is not the only possible explanation. As discussed in the introduction, interband transitions result in an anisotropic carrier distribution at earliest times with nodes along the direction of the pump polarization.[19] The pump polarization in the present experiment is along the ΓK-direction, *i.e.*, the transient band structure is measured along a node of the initially anisotropic carrier distribution. This means that carriers in the conduction band can be detected only after the initially anisotropic distribution has partially relaxed *via* phonon emission. Therefore, the increased rise time may also originate from the transition from an anisotropic to an isotropic distribution of carriers on the 200 fs time scale of electron–phonon coupling. Further, we want to point out that for the present excitation regime we observe a thermal FD distribution of carriers at all times (see Fig. 4) which is not expected to result in carrier multiplication.[25]

To conclude, a definite proof of carrier multiplication in graphene by tr-ARPES needs to be based on a complete mapping of the whole Dirac cone. A single cut along any particular direction in momentum space is not sufficient to exclude the contribution of an initially anisotropic carrier distribution to an increased rise time. Furthermore, an observation of carrier multiplication might require sub 10 fs time resolution to enable access to the initially non-thermal carrier distribution that lies at the heart of carrier multiplication.

Summary and outlook

In summary, we have used tr-ARPES to determine the imaginary part of the self-energy $Im(\Sigma)$ in graphene by measuring the scattering rate $1/\tau = -2Im(\Sigma)$ as a function of energy after photo-excitation at $\hbar\omega_{pump} = 1.5$ eV. A comparison with static MDC line widths indicates that the dynamics of photo-excited carriers in graphene are determined by the equilibrium self-energy as suggested in ref. 4. Furthermore, we have addressed the issue of carrier multiplication in graphene. While a definite proof for its existence is still missing, we have identified important steps towards a future, more systematic study of carrier multiplication in graphene using tr-ARPES.

Acknowledgements

We thank Haiyun Liu and Hubertus Bromberger for technical support with the tr-ARPES measurements and Matteo Mitrano for fruitful discussions and comments concerning the present manuscript. Stiven Forti, Camilla Coletti and Konstantin V. Emtsev helped with the static ARPES measurements, partially supported by the German Research Foundation (DFG) within the priority program 'graphene' SPP 1459 (Sta 315/8-1).

References

1 P. R. Wallace, *Phys. Rev.*, 1947, **71**, 622.

2 A. Bostwick, T. Ohta, T. Seyller, K. Horn and E. Rotenberg, *Nat. Phys.*, 2007, **3**, 36.

3 A. Bostwick, F. Speck, T. Seyller, K. Horn, M. Polini, R. Asgari, A. H. MacDonald and E. Rotenberg, *Science*, 2010, **328**, 999.

4 M. Sentef, A. F. Kemper, B. Moritz, J. K. Freericks, Z.-X. Shen and T. P. Devereaux, *Phys. Rev. X*, 2013, **3**, 041033.

5 T. Kampfrath, L. Perfetti, F. Schapper, C. Frischkorn and M. Wolf, *Phys. Rev. Lett.*, 2005, **95**, 187403.

6 S. Butscher, F. Milde, M. Hirtschulz, E. Malić and A. Knorr, *Appl. Phys. Lett.*, 2007, **91**, 203103.

7 M. Breusing, C. Ropers and T. Elsaesser, *Phys. Rev. Lett.*, 2009, **102**, 086809.

8 H. Yan, D. Song, K. F. Mak, I. Chatzakis, J. Maultzsch and T. F. Heinz, *Phys. Rev. B: Condens. Matter Mater. Phys.*, 2009, **80**, 121403(R).

9 H. Wang, J. H. Strait, P. A. George, S. Shivaraman, V. B. Shields, M. Chandrashekhar, J. Hwang, F. Rana, M. G. Spencer, C. S. Ruiz-Vargas and J. Park, *Appl. Phys. Lett.*, 2010, **96**, 081917.

10 D. Sun, C. Divin, C. Berger, W. A. de Heer, P. N. First and T. B. Norris, *Phys. Rev. Lett.*, 2010, **104**, 136802.

11 K. Kang, D. Abdula, D. G. Cahill and M. Shim, *Phys. Rev. B: Condens. Matter Mater. Phys.*, 2010, **81**, 165405.

12 C. H. Lui, K. F. Mak, J. Shan and T. F. Heinz, *Phys. Rev. Lett.*, 2010, **105**, 127404.

13 P. J. Hale, S. M. Hornett, J. Moger, D. W. Horsell and E. Hendry, *Phys. Rev. B: Condens. Matter Mater. Phys.*, 2011, **83**, 121404(R).

14 M. Breusing, S. Kuehn, T. Winzer, E. Malić, F. Milde, N. Severin, J. P. Rabe, C. Ropers, A. Knorr and T. Elsaesser, *Phys. Rev. B: Condens. Matter Mater. Phys.*, 2011, **83**, 153410.

15 I. Chatzakis, H. Yan, D. Song, S. Berciaud and T. F. Heinz, *Phys. Rev. B: Condens. Matter Mater. Phys.*, 2011, **83**, 205411.

16 N. Bonini, M. Lazzeri, N. Marzari and F. Mauri, *Phys. Rev. Lett.*, 2007, **99**, 176802.

17 T. Winzer, E. Malić and A. Knorr, *Phys. Rev. B: Condens. Matter Mater. Phys.*, 2013, **87**, 165413.

18 T. Winzer, A. Knorr and E. Malić, *Nano Lett.*, 2010, **10**, 4839.

19 E. Malić, T. Winzer, E. Bobkin and A. Knorr, *Phys. Rev. B: Condens. Matter Mater. Phys.*, 2011, **84**, 205406.

20 T. Winzer and E. Malić, *Phys. Rev. B: Condens. Matter Mater. Phys.*, 2012, **85**, 241404(R).

21 J. C. W. Song, M. Y. Reizer and L. S. Levitov, *Phys.Rev. Lett.*, 2012, **109**, 106602.

22 A. C. Betz, S. H. Jhang, E. Pallecchi, R. Ferreira, G. Fève, J.-M. Berroir and B. Plaçais, *Nat. Phys.*, 2013, **9**, 109.

23 M. W. Graham, S.-F. Shi, D. C. Ralph, J. Park and P. L. McEuen, *Nat. Phys.*, 2013, **9**, 103.

24 S. Gilbertson, T. Durakiewicz, J.-X. Zhu, A. D. Mohite, A. Dattelbaum and G. Rodriguez, *J. Vac. Sci. Technol., B: Microelectron. Nanometer Struct.*, 2012, **30**, 03D116.

25 I. Gierz, J. C. Petersen, M. Mitrano, C. Cacho, I. C. E. Turcu, E. Springate, A. Stöhr, A. Köhler, U. Starke and A. Cavalleri, *Nat. Mater.*, 2013, **12**, 1119.

26 T. Li, L. Luo, M. Hupalo, J. Zhang, M. C. Tringides, J. Schmalian and J. Wang, *Phys. Rev. Lett.*, 2012, **108**, 167401.

27 K. J. Tielrooij, J. C. W. Song, S. A. Jensen, A. Centeno, A. Pesquera, A. Zurutuza Elorza, M. Bonn, L. S. Levitov and F. H. L. Koppens, *Nat. Phys.*, 2013, **9**, 248.

28 Justin C. W. Song, Klaas J. Tielrooij, Frank H. L. Koppens and Leonid S. Levitov, *Phys. Rev. B: Condens. Matter Mater. Phys.*, 2013, **87**, 155429.

29 D. Brida, A. Tomadin, C. Manzoni, Y. J. Kim, A. Lombardo, S. Milana, R. R. Nair, K. S. Novoselov, A. C. Ferrari, G. Cerullo and M. Polini, *Nat. Commun.*, 2013, DOI: 10.1038/ncomms2987.

30 J. C. Johannsen, S. Ulstrup, F. Cilento, A. Crepaldi, M. Zacchigna, C. Cacho, I. C. E. Turcu, E. Springate, F. Fromm, C. Raidel, T. Seyller, F. Parmigiani, M. Grioni and P. Hofmann, *Phys. Rev. Lett.*, 2013, **111**, 027403.

31 C. Riedl, C. Coletti, T. Iwasaki, A. A. Zakharov and U. Starke, *Phys. Rev. Lett.*, 2009, **103**, 246804.

32 S. Forti, K. V. Emtsev, C. Coletti, A. A. Zakharov, C. Riedl and U. Starke, *Phys. Rev. B: Condens. Matter Mater. Phys.*, 2011, **84**, 125449.

33 S. Hüfner, "Photoelectron Spectroscopy, Principles and Applications," ISBN 978-3-662-09280-4.

34 F. Frassetto, C. Cacho, C. A. Froud, I. C. E. Turcu, P. Villoresi, W. A. Bryan, E. Springate and L. Poletto, *Opt. Express*, 2011, **19**, 19169.

Faraday Discussions

ROYAL SOCIETY
OF CHEMISTRY

DISCUSSIONS

Chemical reaction dynamics II and Correlated systems, surfaces and catalysis: general discussion

Henry Chapman, Majed Chergui, Jochen Küpper, Martin Wolf, Katharine Reid, Wendy Flavell, Gwyn Williams, Isabella Gierz, Elaine Seddon, Hans Jakob Wörner, Kai Rossnagel, Jonathan Underwood, Michael Woerner, Julia Weinstein, Christian Bressler, Grigory Smolentsev, Klaus von Haeften, Ben Spencer, Stefan Neppl and Nora Berrah

DOI: 10.1039/c4fd90015d

Gwyn Williams opened the discussion of the paper by **Christian Bressler**: How does the signal to noise at the rather fluctuating output of the LCLS compare with that of a steady synchrotron?

Christian Bressler replied: The short answer is: horribly. SASE XFEL sources (without self-seeding), especially if monochromatized, deliver fluctuations up to 100%, while synchrotrons deliver a photon number entirely governed by photon statistics (shot noise). For a physicist SR is the ideal probe with its extreme stability.

However, XFEL sources can deliver superior (scientific) information (due to their incredible single-pulse intensities), if you can reliably measure the incident intensity on a shot-by-shot basis. I am not sure about the level of I0 accuracy these days (probably on the percent level, thus improvements can be expected), but overall the data quality resembles what has been achieved earlier in TR X-ray experiments at SR sources with kHz laser systems, but here with femtosecond time resolution. This is a developing field, and improvements should be expected over time, similar as in the field of MHz-SR X-ray experiments.

Henry Chapman addressed **Christian Bressler** and **Stefan Neppl**: How heroic are the MHz rate experiments at synchrotrons, and are technologies that are being developed for those experiments useful for the European XFEL, which also has MHz pulse rates? What combinations of technologies of sources and detectors are ideally required for the types of experiments you are doing? For example, would

---

† Electronic supplementary information (ESI) available. See DOI: 10.1039/c4fd90015d

an energy recovery linac (ERL) be better than either a synchrotron or FEL for high-rep rate experiments? What about diffraction-limited storage rings?

Stefan Neppl responded: High repetition rate X-ray sources are generally desirable for time-resolved XPS experiments. They compensate efficiently for the relatively low pulse fluences that can be tolerated in these measurements (mainly due to space charge and sample damage limitations). There are, however, also limits to the pulse repetition rates that may be used in a particular experiment. These are usually defined by the recovery time of the system under investigation (it should have sufficient time to return to its unperturbed state between two consecutive pump pulses) and the speed by which the sample has to be renewed (*i.e.* scanned for large solid-state samples or flow for liquid/solvated samples). Dye-sensitized semiconductor systems such as the ones studied here usually have recovery times ranging from hundreds of nanoseconds up to milliseconds. Correspondingly, the most suitable pulse repetition rates often range from kHz to MHz regimes.

The use of time-resolving detectors and event tagging, as demonstrated in this paper for MHz synchrotron radiation (see *e.g.* Fig. 7A and B), is an interesting approach for experiments at future high-repetition rate XFEL light sources such as the European XFEL and LCLS II. The technique will resolve pump–probe signals corresponding to individual microbunches while recording events for all micro-bunches, simultaneously.

Diffraction-limited storage rings provide extremely small source sizes that result in correspondingly small X-ray focal spot sizes at the sample. This characteristic will potentially have several advantages for the technique presented here. Small X-ray focal spots enable the use of correspondingly small (optical) pump laser focal spots and, in turn, lower pump pulse energies to achieve the same excitation density. This relaxes the requirements for the pump lasers, which can be operated at higher repetition rates. Smaller spot sizes may also enable lower sample scan speeds and, therefore, the use of smaller samples.

Christian Bressler responded: Time-resolved X-ray experiments at MHz repetition rates are not "heroic" at all, at least no longer, as these can make use of the full X-ray flux at the larger class of synchrotrons: ESRF, APS, Petra-III, can deliver electron bunches in the 0.1–10 MHz range, and—if a MHz laser system is synchronized accordingly—one can collect up to around 1×10^{12} photons/s.

This has to be compared to the older TR experiments with only few kHz repetition rates: monochromatic beamlines typically deliver between 1×10^2 and 1×10^6 photons/pulse, thus one can do these experiments with only 1×10^5 – 1×10^9 photons/s, and this limits the class of systems, which can be studied, and certainly the information content due to the lower (by up to a few orders of magnitude!) signal-to-noise ratio.

Gwyn Williams asked: For absorption experiments one has to normalize to the incident flux. For a synchrotron this is steady, particularly if the beam current is constant, but for the LCLS the incident flux fluctuates, so do you have to use an incident flux monitor? If so, how is this implemented?

Christian Bressler answered: At LCLS this is implemented *via* a quadrant backscattering foil into four diodes. Prior studies established its general functionality down to the percent level, but in principle accuracies down to 1×10^{-6} should be possible. The need for better I0 normalization is merely user-driven, thus these should assess their data quality and, if needed, demand better I0 monitors.

Jonathan Underwood said: In the discussion you mentioned that, for the X-ray spectroscopy and scattering techniques described in your paper, as long as high dynamic range detection is available, the low repetition rate and high photon numbers per pulse characteristics of FEL sources are as effective as high (MHz) repetition rate sources with low photon numbers per pulse with digital detection. Do you think there are nonetheless still advantages to high repetition rate sources with low photon numbers per pulse with regard to new opportunities incorporating coincidence detection of scattered photons (for example, such as proposed by Mukamel and colleagues in *J. Phys. B: At. Mol. Opt. Phys.*, **47**, 124037)?

Christian Bressler responded: NB: A large dynamic range is always convenient, but not necessarily paramount. *E.g.*, for X-ray diffuse scattering a dynamic range of 100 (from single photons onwards) appears sufficient (in some cases only a dynamic range of ten is required). Mukamel and colleagues apply extreme intensities on the attosecond timescale, while chemical dynamics seeks to use the X-rays as a linear probe. In this spirit high-rep-rate laser and X-ray sources are preferred over sources with extreme peak brightness (and this includes XFELs with moderate focussing conditions), but Mukamel may prove me so wrong one day...

Katharine Reid addressed **Christian Bressler** and **Stefan Neppl**: There is a drive to keep reducing the pulse durations that can be achieved at FEL sources. Is there an optimum pulse duration to aim for, or do we need tools that provide a variety of pulse durations in order to probe dynamical processes on different timescales? Related to this, are there situations where the bandwidths associated with ultrashort pulse durations cause complications for measurements?

Stefan Neppl replied: The detection of transient chemical shifts with time-resolved XPS requires an experimental energy resolution that is at least comparable to the magnitude of the expected shifts. Since these shifts may vary between different systems, one would ideally want to select the best compromise of photon bandwidth and pulse duration for a given experiment. For example, transient line-shape changes—as described in our paper—would be difficult to reveal with sub-femtosecond X-ray pulses, which are inherently associated with bandwidths larger than 1 eV.

Christian Bressler replied: I sincerely believe that any answer to this question will be revisited, as newer experiments are being performed. For now, and in view of ease of use of optical laser beams, the sweet spot may be set into the 15–50 fs range. This would already help settle questions on reaction intermediates, possibly even some transition states, in photochemical reaction pathways. However, I think there is no hard limit for elucidating the elementary steps in chemical reactivity, and therefore I foresee a need for even shorter pulse durations in the future.

Grigory Smolentsev asked: You have used a few time-resolved X-ray techniques. Which one will be the most efficient to monitor the formation of first MLCT state in $Fe(bpy)_3$?

Christian Bressler replied: I still think that XANES would be well suited, although the first hard fact was delivered by K-beta XES.

Henry Chapman addressed **Christian Bressler** and **Gwyn Williams**: Gwyn Williams stated that there is a limit to the number of photons per unit time (*i.e.* peak X-ray power) you can generate from an electron bunch (*e.g.* in an FEL) which depends on the peak current that can be achieved, ultimately limited by space charge in the electron bunch. This is analogous to limitations in generating high pulse intensities in lasers, which is limited by damage to the gain medium by the high-power pulse. The problem is circumvented in that case by chirped pulse amplification, and the same should be true for FEL pulses. That is, use a long chirped electron bunch of limited peak current and then compress the generated chirped X-ray pulse with gratings. Such a scheme was originally proposed by Claudio Pellegrini, and efforts are underway in Hamburg to develop and test this method by Adrian Cavalieri and Sasa Bajt and their colleagues. For some experiments it might not actually be necessary to compress the X-ray pulse if your sample disperses the pulse and time-resolved information can be obtained, as proposed by Moffat in a previous Faraday Discussion (vol. 122, 65–77, 2003).

Gwyn Williams responded: In addition to this, it has been proposed that one can develop even shorter bunches by rotating the chirped bunch and passing it through a physical aperture. This is an area that would be of considerable interest in machine development. The calculations referred to above are here: 239. S. L. Hulbert and G. P. Williams, "Calculations of Synchrotron Radiation Emission in the Transverse Coherent Limit", *Rev. Sci. Instrum.*, 2009, **80**, 106103.

Christian Bressler responded: TR X-ray experiments on chemical and biologically relevant samples do not benefit from a further increased peak power, *au contraire*. The goal in chemical reaction dynamics is to resolve the ensuing reaction, and not to perturb it by the probe beam.

Jochen Küpper communicated: Christian, the "Gedanken" experiment you suggested regarding the measurement of the change from molecule to atoms was already performed (B. Erk, R. Boll, S. Trippel, D. Anielski, L. Foucar, B. Rudek *et al.*, "Imaging charge transfer in iodomethane upon X-ray photoabsorption", *Science*, 2014, **345**(6194), 288–291, DOI: 10.1126/science.1253607).

Christian Bressler communicated in reply: Thanks. This refers to one observable, the charge states of iodine up to 14+ or so. I could imagine that effects concerning the valence orbitals have been completely missed, but this is a cool experiment in the right direction!

Wendy Flavell opened the discussion of the paper by **Stefan Neppl**: This is a rather philosophical question, but how do you interpret the surface photovoltage shift in the case of your nanoparticle samples? How do you understand the large difference between clean ZnO and ZnO + N3 dye?

Stefan Neppl answered: This is an interesting question. According to theory, we would expect the band bending of an extended semiconductor crystal to be larger as compared to a nanoparticle of the same material. In our case, the substrate is a sintered film of ZnO nanoparticles and might therefore be considered as an intermediate between these two limiting cases. A quantitative description of the bend bending in such a nanoporous semiconductor network has—to the best of our knowledge—not been derived and clearly motivates comparative surface photovoltage studies on these systems.

The large difference in the SPV response of the bare ZnO substrate and the N3-sensitized ZnO sample can be explained by the fact that the photon energy of the 532 nm laser excitation is much lower than the ZnO band gap. Therefore, no electron-hole pairs are created and no SPV is observed for the bare ZnO substrate at this wavelength. For N3-ZnO, the 532 nm photons excite electrons in N3 from the HOMO to the LUMO, from which they are injected into the ZnO conduction band. These electrons partly compensate the positive space charges in the ZnO depletion layer which in turn gives rise to a transient SPV effect. The magnitude of this SPV shift (> 400 meV) is indeed surprisingly high. It might be related to an enhancement of the initial ZnO bend bending upon chemisorption of N3 molecules

Christian Bressler asked: How did you assess the excited state population? Is this assessment of general importance for any kind of TR experiment? If so, how important is this (prior or *in situ*) knowledge?

Stefan Neppl responded: In our analysis, we fit the Ru3d/C1s difference spectrum to a model that describes the entire difference spectrum as being due to a laser-induced shift of the Ru3d doublet. This model has only two fit parameters, the intensity of the shifted Ru3d component (= excited state population) and the shift in binding energy itself. Generally, these two parameters are correlated and may lead to ambiguous fit results when the energy shift is much smaller than the linewidth of the spectral features. This, however, is not the case for the 2–3 eV shifts observed in our TRXPS experiment and, therefore, both the population and the shift can be determined from a single fit.

Majed Chergui asked: You show that the XPS transient can be reproduced by shifting the entire spectrum by 2 eV and subtract it from the unshifted spectrum, but I wouldn't expect the Ru and C lines to be shifted by the same amount?

Stefan Neppl replied: We observe two different shifts in our time-resolved XPS experiment. The first one is a rigid shift of the entire C1s/Ru3d spectrum (see Fig. 7C of the manuscript) induced by the surface photovoltage (SPV) effect in the semiconductor substrate. If we correct for this SPV shift, and compare the spectrum to the laser-off reference as shown in Fig. 8, we also detect a characteristic laser-induced change in the C1s/Ru3d spectral shape, which can partly be explained as the result of a *ca.* 2–3 eV transient shift of the Ru3d doublet in the excited molecules to higher binding energies. This trend might be expected, since the oxidation state of the Ru-metal center increases upon charge injection. When modeling the difference spectra (green line in Fig. 8B), we implicitly assumed that a similar transient shift of the C1s levels—due to the dynamic charge reconfiguration within the molecule—is much less pronounced. This is of course an approximation, which has to be tested in future experiments with a better signal-to-noise ratio. Qualitatively, however, one might expect the effect on most of the carbon core levels to be smaller as compared to the Ru lines, since the laser excitation is associated with a HOMO–LUMO transition and subsequent charge transfer from the dye into the semiconductor substrate, leading to an electronic configuration that is well approximated by a hole in the HOMO of the dye and an additional electron in the substrate (ref. 41: Siefermann *et al.*, *J. Phys. Chem. Lett.*, 2014, 5(15), 2753–2759). The dye HOMO is highly localized on the Ru center and the thiocyanate groups. Thus, the net change in local electron density seen by an individual carbon atom in the pyridine groups is much smaller than for the ruthenium center and, therefore, the by far most intense C1s signal from the pyridine rings is expected to be much less affected by the photo-induced dynamics than the Ru3d lines. Detailed studies of core-level shifts associated with other atoms in the system, in particular the sulfur atoms in the thiocyanate ligands and the nitrogen atoms bound to the Ru center, will be performed in future experiments.

Martin Wolf addressed **Wendy Flavell** and **Stefan Neppl**: The electronic structure of ZnO surfaces depends very much on surface preparation. On several single crystal surfaces hydrogen acts as a donor and adsorption may lead to a metallic surface (charge accumulation layer). See, for example, *Phys Rev. B*, 2011, **83**, 125406. This leads to a pinning of the conduction band and prevents a (significant) surface voltage shift.

Wendy Flavell responded: Indeed that is the case. But what is observed at the surface of ZnO is, as you say, very dependent on surface preparation. In the work described in *Phys. Rev. B*, 2013, **88**, 195301 (and paper 13) the combination of vacuum and oxygen annealing cycles always yields an SPV shift to high binding energy, consistent with a surface depletion layer when excited with band gap (or even slightly sub-band-gap radiation.) The size of the shift and its dynamics change with oxygen content/donor concentration in a consistent way that implicates oxygen vacancies in the persistent photoconductivity (as suggested by Lany and Zunger, *Phys. Rev. B*, 2005, **72**, 035215). The involvement of hydrogen at these sites cannot be excluded.

Stefan Neppl replied: We would like to add that the ZnO nanoparticle films used in our experiments are prepared *ex situ*, *i.e.* not under UHV conditions. This kind of preparation will likely result in the coexistence of different adsorbate- and/or defect-induced electronic states, which all contribute to the bend bending. Future SPV experiments using both sub- and super-band gap illumination will provide more insight into the nature of the bend bending in this system.

Julia Weinstein asked: Dear Stefan, several questions:

(i) Is it possible to distinguish between the two variants of the attachment of the dye to the surface of a semiconductor: two carboxylic groups from one bpy-ligand, or one carboxylic group per bpy ligand, *e.g.*, from the electron density on the C atoms?

(ii) Does the experiment capture the redistribution (delocalisation) of electron density between several bpy-type ligands on the metal centre following MLCT excitation?

(iii) How do the timescales (and perhaps the yields?) of injection estimated by the method discussed compare with the results obtained by the optical methods?

Stefan Neppl replied: These are very interesting questions concerning the potential and limits of time-resolved XPS technique when applied to study dynamics in dye-semiconductor systems.

(i) Unfortunately, this information is challenging to derive from XPS spectra. The C1s photolines of all carboxylic acid groups form essentially one single XPS peak (see, *e.g.*, Mayor *et al.*, *J. Chem. Phys.*, 2008, **129**, 114701 and *J. Chem. Phys.*, 2009, **130**, 164704). The O1s line is sensitive to protonation/deprotonation of the carboxylic acid groups but this does not provide clear-cut information on the adsorption geometry either since both deprotonated as well as protonated carboxylic acid groups may participate in the attachment of the dye to the substrate (see, *e.g.*, Siefermann *et al. J. Phys. Chem. Lett.*, 2014, **5**, 2753).

(ii) In a recent femtosecond time-resolved XPS study, we have determined that the electron leaves the dye within less than 500 fs after the MLCT excitation (Siefermann et al., J. Phys. Chem. Lett., 2014, 5, 2753). Therefore, we expect that only experiments with a time resolution in the range of a few to, perhaps, a few tens of femtoseconds would be able to trace the electron delocalization step before electron injection into the substrate. Principally, carbon atoms in the individual bpy units which are located closer to the nitrogen or carboxylic acid groups, give rise to C1s emission lines which differ in binding energy by ca. 1 eV (see Mayor et al., J. Chem. Phys., 2009, 130, 164704). Given the above mentioned time resolution can be achieved, time-resolved XPS might therefore be used to trace intramolecular charge reconfiguration in N3 affecting these two parts in all the bpy-ligands differently, but cannot address the bpy-ligands individually.

(iii) We have not (yet) been able to derive a timescale for electron injection from the ALS-based picosecond time-resolved XPS experiments. In our recent femtosecond time-resolved XPS study at the LCLS (Siefermann et al., J. Chem. Phys. Lett., 2014, 5, 2735 we found evidence that, while the electron leaves the dye within less than 500 fs, it remains in an interfacial charge-transfer state for more than 1 ps. This result is in agreement with studies in the (near) optical domain that measured time constants of tens of ps for the appearance of free carriers inside the semiconductor, e.g., via time-resolved THz spectroscopy (Nemec et al., Phys. Rev. Lett., 2010, 104, 197401, Tiwana et al., ACS Nano, 2011, 5, 5158). A more detailed and general comparison (which indeed would be very interesting) between time-resolved XPS (TRXPS) and ultrafast optical spectroscopy studies on the electron dynamics and injection yields at dye-semiconductor interfaces will be possible as soon as more TRXPS results become available.

Majed Chergui asked: You show the nsec recovery kinetics of the dye, which has at least 2 components. What are the origins of these two timescales?

Injection is known to occur on several timescales, do you see a multicomponent growth of the cation signal in your measurements ?

Stefan Neppl answered: Multi-exponential recombination dynamics have been observed before for various dye-semiconductor systems, e.g., in all-optical transient absorption studies. For ruthenium-based dyes attached to ZnO and TiO_2 substrates, Bauer et al. (J. Phys. Chem. B, 2001, 105, 5585) found bi-exponential electron back transfer kinetics with time constants in the range of hundreds of nanoseconds to a few microseconds. The kinetics were the same for both substrates, which implies that the general electronic structure of the substrate is not the main factor governing the recombination dynamics. In general, the different timescales observed in the electron injection and electron back transfer process are at least partly ascribed to sample inhomogeneities, e.g. the co-existence of different binding configurations with different electronic coupling to the semiconductor substrate or electronic trap states induced by defects in the semiconductor crystal structure (see, e.g., Tachibana et al., J. Phys. Chem. B, 2000, 104, 1198). Another possibility is that the dye-electron recombination dynamics may be affected by transient surface/interfacial potentials in a similar fashion as electron-hole recombination dynamics in band-gap excited semiconductors (see the contribution of Wendy Flavell's group to this Faraday Discussion). The feedback between interfacial electronic dynamics and the potentials governing them generally leads to transient signals that cannot be described by single exponentials (see Bröcker *et al.*, *Chem. Phys.*, 2004, **299**, 247). We will study these effects in greater detail in the near future.

In our measurements, the transient shift of the Ru3d photoemission lines is the most direct probe for the cationic state of the dye molecule during charge injection. So far, we have measured this dynamical chemical shift only for two time delays (ref. 41 and Fig. 8 in the manuscript). The full time dependence of this shift will be studied in future experiments, which will also require access to few-femtosecond X-ray light sources (*e.g.* FELs) to resolve the possibly fast (sub-picosecond) dynamics of the involved electron injection pathways.

Wendy Flavell remarked: In the case of the decay transient for the ZnO SPV, why should the dynamics necessarily be bi-exponential with a fast and a slow part? The 'self-decelerating' model of Widdra and Broecker (*e.g.* W. Widdra *et al.*, *Surf. Sci.*, 2003, **543**, 87) allows for the increase in lifetime as the SPV decays, *i.e.* a variation in the lifetime during the decay could be an intrinsic part of the physics, allowing one function to be fitted to the decay transient.

Stefan Neppl replied: Indeed, the SPV decay in Fig. 9 can be fitted equally well using the self-decelerating model with a (dark) carrier lifetime of 390 ns and an α-parameter of 3. An α-parameter of 3, which is an additional fit parameter in this model, has so far not been observed in SPV experiments, and is significantly larger than α parameters reported, *e.g.*, for single crystalline ZnO ($\alpha < 1$). Generally, the α parameter is related to the ideality factor used in the theory of Schottky diode performance, but its interpretation in the context of SPV measurements, and how this parameter can be related to different recombination mechanisms, is still an open question.

Majed Chergui enquired: You excited the dye and you probe it by XPS, but evidence that you have injected the electron into the substrate, is to probe the lines of Zn (or O). We did such an experiment on RuN719 adsorbed at the surface of TiO_2 and observed the oxidation of Ru by Ru L edge spectroscopy and the reduction of Ti centres by Ti K edge absorption (M. Hannelore Rittmann-Frank, Chris J. Milne, Jochen Rittmann, Marco Reinhard, Thomas J. Penfold and Majed Chergui, *Angew. Chem. Int. Ed.*, 2014, **53**, 5858–5862). So in your case, what is the evidence that the electron is in the substrate.

Stefan Neppl answered: We think that the positive transient surface photo-voltage (SPV) response (see Fig. 9 of the manuscript) is the spectroscopic signature for the electrons being transferred to the semiconductor: photo-excited electrons from the dye can effectively screen the positive excess charge inside the n-type ZnO substrate only when they diffuse into the tens of nanometer wide depletion layer. This additional screening mediated by the injected electrons causes reduction of the bending near the surface of the ZnO substrate, which is observed as a sudden increase in binding energy of all energy levels that are referenced to the semiconductor Fermi level. Since the dye molecules are chemisorbed on the ZnO substrate, their core-level photoemission can be expected to be as sensitive to the SPV effect as photoemission lines originating from the semiconductor material itself.

Kai Rossnagel said: In your talk, you also presented time-resolved XPS data that were acquired at the LCLS. Since free-electron lasers have significantly lower repetition rates as compared to storage rings, the data collection efficiency of the presented photoelectron detection scheme is significantly reduced. It is limited by the fact that the delay-line detector can only detect one electron per photon probe pulse. Will you continue to perform time-resolved XPS measurements at the LCLS and, if so, do you have plans to make these measurements more efficient, for example, by combining a time-of-flight spectrometer with a multi-segment delay-line detector?

Stefan Neppl replied: The detector in the LCLS experiment was not a delay-line anode, but a phosphor screen combined with a fast CCD camera. Generally, the strongest constraints for time-resolved XPS experiments at FELs are imposed by space charge effects, which have been observed for X-ray pulse fluences of less than 10×10^7 photons/pulse, *i.e.*, several orders of magnitude lower than the maximum pulse fluences available (see, *e.g.*, Pietzsch *et al.*, *New J. Phys.*, 2008, **10**, 033004). For the N3/ZnO experiment at the LCLS (ref. 41: Siefermann *et al.*, *J. Phys. Chem. Lett.*, in press), the photon flux had to be reduced to *ca.* 10×10^6 photons/pulse to avoid space-charge induced peak distortions and shifts. Under these conditions, less than one photoelectron was detected per X-ray pulse. Thus, detector saturation was not an issue and multi-segmented anodes would not be required for this particular experiment to improve data collection efficiency. However, they may still be advantageous, for example, in combination with time-of-flight spectrometers to increase the angular acceptance and sensitivity of an XPS setup.

Cleary, time-resolved X-ray photoelectron spectroscopy will greatly benefit from high-repetition rate XFELs, where femtosecond time-resolved X-ray measurements that require moderate pulse energies can still be efficiently performed.

Wendy Flavell addressed **Stefan Neppl, Christian Bressler, Klaus von Haeften** and **Nora Berrah**: As time resolution improves, I'd like to make the comment that we must also build in the ability to measure over quite long timescales, with this good resolution. For condensed matter systems of the type we are discussing, dynamics can be happening over very long timescales (in the case of paper 13, we needed to measure at delays up to ms to capture the persistent photoconductivity of ZnO). The ArToF analyser coupled with hybrid mode operation at synchrotron, as implemented for example at SPring8 BL07LSU seems a productive direction for the future. Synchronising to successive 'rotations' of the large bunch allows the dynamics to be measured over many orders of magnitude of time delay.

Klaus von Haeften replied: I agree. In our paper on the 'Formation of coherent rotational wavepackets in molecule-helium clusters' we record the propagation of a rotational wavepacket over 600 ps. In our example we have chosen laser conditions that allow for the excitation of rotational level close and beyond the dissociation limit of the He–C_2H_2 complex. It would be very interesting to observe the propagation of the rotational wavepacket for longer, for example up to 10 ns because dissociation times of related systems (CO_2) have been reported to be 6 ns, M. J. Weida, J. M. Sperhac, D. J. Nesbitt and J. M. Hutson, *J. Chem. Phys.*, 1994, **101**, 8351. I would be particularly interested whether predissociation depends on the actual composition of all rotational states. A chemical reaction could then be triggered by shaping of the alignment laser field.

Stefan Neppl replied: We strongly support this statement. The combination of active pump–probe synchronization and simultaneous time-tagging as implemented in our experiment at the Advanced Light Sources enables both efficient synchrotron-bunch length limited time-resolved experiments, and the probing of dynamics proceeding on micro- and even millisecond timescales. In addition, this setup benefits from the adjustable repetition rate (single shot to MHz) of the used laser system, which allows to match the pump-pulse spacing to the timescale of the process to be investigated.

Christian Bressler answered: I would also agree, albeit the time resolution is no longer superimportant, when we enter the 100 ps range and much further above.

Nora Berrah responded: I agree and in fact the LCLS has extended its long timescales to 500 fs.

Jochen Küpper opened the discussion of the paper by **Klaus von Haeften**: You said that you had to use laser pulses for alignment and Coulomb explosion imaging that were parallel—why was that?

Then, it seemed as if you said that you would not see alignment along the alignment laser axis if using a perpendicularly polarized ionization laser—did I understand that correctly? If so, is it valid to speak of "alignment", in the sense of strong angular confinement of the molecular rotational distribution, or would it be more appropriate to refer to the experiment as "rotational coherence spectroscopy" (RCS) as introduced by Felker (P. Felker, Rotational Coherence Spectroscopy—Studies of the Geometries of Large Gas-Phase Species by Picosecond Time-Domain Methods, *J. Phys. Chem.*, 1992, **96**(20), 7844–7857). My next question refers to Fig. 3 in your paper, or, actually, to the alignment trace you showed during the conference, which seems to significantly deviate from the figure in the manuscript:

Why do your ¼ (and $^3/_4$) revivals show only alignment (or antialignment) but not the correspondingly pre- or post-anitalignment (alignment)?

See, for instance, Nielsen, *Phys. Chem. Chem. Phys.*, **13**, 18971 where these are clearly visible.

Samuel Leutwyler has recently performed a nice detailed analysis of this in his RCS experiments [(a) G. Brügger, H.-M. Frey, P. Steinegger, F. Balmer and S. Leutwyler, Accurate Determination of the Structure of Cyclohexane by Femto-second Rotational Coherence Spectroscopy and Ab Initio Calculations, *J. Phys. Chem. A*, 2011, **115**(34), 9567–9578, doi: 10.1021/jp2001546; (b) H. M. Frey, D. Kummli, S. Lobsiger and S. Leutwyler, High-resolution Rotational Raman Coherence Spectroscopy with Femtosecond Pulses, in *Handbook of High-resolution Spectroscopy*, Chichester, UK, John Wiley & Sons, Ltd., 2011, doi: 10.1002/9780470749593.hrs055]. He showed that accurate pump-probe delays are a very crucial effect and that is quite an effort to do this reasonably well. Therefore, it would be useful to learn what your statistical and systematic errors of the delay time were.

3. Since you suggest that you have no observable alignment with the laser polarizations perpendicular, is it correct to speak of "aligned molecules", or should we really consider your experiment a rotational coherence spectroscopy (RCS) experiment as introduced by Kim and Felker in the 80s.
2. How did you calibrate your temporal axis, what statistical and systematic errors do you get on the time-axis of the RCS trace, or, what precision?

What is the error propagation into the rotational constants? What are the precision and accuracy of the temporal axis in your RCS experiment, *i.e.*, what are the limits on its systematic errors?

How do these uncertainties—statistical (precision) and systematic (accuracy) estimates—propagate into the rotational constants determined? In how far can you use these numbers to determine "structures", as was done in other experiments (Ratzer, *Chem. Phys.*, **283**, 153 and references therein)?

Klaus von Haeften replied: 1. Indeed you must have misunderstood me. In all experiments in our paper the polarisation was parallel for consistency within the experimental run. Experiments with perpendicular polarisation were planned but the granted beamtime at the Rutherford Appleton Lab did not allow to begin such experiments. The orientation of polarisation had no consequences for our results.

2. I think it is fair to call our experiments RCS with the distinction to older work by Felker *et al.* that we use impulsive alignment to generate rotational wavepackets. This has specific, and very important advantages as outlined in the introduction of our paper 3. The temporal evolution of alignment depends on laser parameters and the target system. Our results shown are the first of a target comprising of clusters of acetylene and a variable number of helium atoms. Hence, it makes no sense to compare these with other work on other target systems, with different laser parameters, *etc.*
4. Calibration can be achieved by investigating a known system, for example carbon monoxide. The accuracy that can be achieved is nicely illustrated in A. Przystawik, A. Kickermann, A. Al-Shemmary, S. Dusterer, A. M. Ellis, K. von Haeften, M. Harmand, S. Ramakrishna, H. Redlin, L. Schroedter, M. Schulz, T. Seideman, N. Stojanovic, J. Szekely, F. Tavella, S. Toleikis and T. Laarmann, *Phys. Rev. A*, 2012, **85**, 052503.
5. Rotational constants only support structural models. I would therefore consider rotational spectroscopy not as a method to 'determine' structures. It is not uncommon, particularly for quantum mechanical systems that several observables add complementary information to the whole picture.

Jochen Küpper communicated: 1. Typically the Coulomb explosion laser is polarized perpendicular to the alignment laser to avoid overestimating the degree of alignment—why did you not do that?

What error, do you estimate, was introduced by that Sam Leutwyler has recently performed a nice detailed analysis of this in his RCS experiments and showed that this is a very crucial effect?

3. Since you suggest that you have no observable alignment with the laser polarizations perpendicular, is it correct to speak of "alignmed molecules", or should we really consider your experiment a rotational coherence spectroscopy (RCS) experiment as introduced by Kim and Felker in the 80s.

2. How did you calibrate your temporal axis, what statistical and systematic errors do you get on the time-axis of the RCS trace, or, what precision?

What is the error propagation into the rotational constants?

Klaus von Haeften communicated in reply: 1. It is correct that in many conceptually similar experiments perpendicular polarisation between pump and probe laser is chosen. A requirement to observe impulsive alignment is that polarisation of the pump laser is parallel to the detector plane (for Coulomb explosion detection). Choosing a probe laser polarisation perpendicular to the detector plane will remove any possible anisotropy in the images that may be caused by the excitation process, which is good to remove background signal for the determination of pump-laser-induced alignment. An advantage of a parallel probe-laser polarisation is that anisotropies in the excitation can be used to detect alignment induced in the pump process. For example, we observed a modulation of the ion yield.

2. I think these questions overlap with a previous one.

Katharine Reid remarked: The authors of paper 3 were able to achieve quite substantial alignment of a polyatomic molecule ($<\cos^2\theta> = 0.89$), but the alignments shown in your paper are substantially lower. Is there a prospect of attaining much higher alignment of molecules embedded in helium clusters? Following this, is there a prospect of probing photoinduced dynamics of aligned molecules embedded in helium clusters?

Klaus von Haeften replied: The degree of alignment observed in our work was comparatively low, but the target was unusual.

For a known molecule, *e.g.* with well-defined polarisability, in a laser field of known intensity, pulse length, *etc.*, the degree of alignment is defined and can be calculated. The degree of alignment is then reflected by the respective values of and the characteristic time-dependence. For example, a sine-wave-type time dependence indicates involvement of only two rotational levels. Correspondingly, the level of alignment is low. In our case, the target was different. It consisted of molecular clusters comprising a single HCCH molecule and a variable number of helium atoms. Consequently, 1. (t) shows the superposition of the alignment of all the individual clusters 2. Alignment information is washed out during the Coulomb explosion process (see another question by H-J Wörner) Our work clearly shows that it is possible to generate coherent rotational wavepackets in very small molecule-helium clusters. Our results do not justify commenting on the prospects of achieving high alignment of molecules in helium clusters or droplets, which would certainly be of great advantage for investigating photo-induced dynamics of embedded molecules and molecular complexes.

Jonathan Underwood asked: In the paper presented by von Haeften, the observed Coulomb explosion images are a convolution of the molecular axis alignment produced by the pump pulse and the strong orientational dependence of the Coulomb explosion probe process. Because the probability for Coulomb explosion has a strong orientational dependence, it is non-trivial to quantitatively extract the molecular axis alignment, and it has become the norm in this field to characterize the molecular axis alignment in terms of expectation values calculated for the resulting image. Recently, in collaboration with the Stapelfeldt group in Aarhus we have developed and demonstrated a technique for deconvoluting the orientational dependence of the Coulomb explosion process allowing for complete characterization of the molecular axis alignment through inversion of the images. This methodology requires an image to be recorded for the Coulomb explosion of isotropically oriented molecules ahead of doing the experiment with aligned molecules. This work will be published in the near future.

Hans Jakob Wörner commented: I have two questions. First, could you provide an estimate of the maximal cluster size or of the cluster-size distribution in your experiment? Second, have you obtained evidence for avalanche-like ionization of the helium clusters (*Phys. Rev. Lett.*, **102**, 128102) in other observables than the C^+ kinetic energies (Fig. 2b)?

Klaus von Haeften replied: We are unable to provide figures for the maximal cluster size or the cluster size distribution. The clusters are produced using a pulsed expansion of a gas mixture in an Even-Lavie valve. While the cluster sizes in continuous expansion can be fairly reliably predicted for a large range of gases using the formalism developed by Hagena, this cannot be said for gas mixtures. Secondly, the shape of conical nozzles, in particular the opening angle controls the cluster size. Scaling laws have not been published for the Even-Lavie nozzle, although we can confirm by our own unpublished work on argon cluster formation in a continuous expansion through a conical E-L-nozzle that Hagena's formula can be applied and that an effectve half-opening angle of $20°$ can be confirmed.

We have no firm evidence for avalanche-like ionization of the helium clusters, although we see indications that such an effect takes place. For expansion conditions favouring formation of larger clusters we observe higher kinetic energies of the Coulomb explosion carbon ion fragments. In our paper we suggest that an effect similar to that suggested by Mikaberidze, Saalmann and Rost, *Phys. Rev. Lett.*, 2009, **102**, 128102, may take place. We also explain in the paper that recombination between electrons and higher charged carbon ion fragment may contribute to the increased kinetic energies of C^+, an idea originally proposed by J. M. Rost in a personal communication.

The rather complicated fragmentation process following the Coulomb explosion has the consequence that angular information originally contained in the explosion process becomes washed out, leading to apparent lower levels of alignment.

Jochen Küpper remarked: You mentioned that you were using mixtures of gases. Which gases were you using, at what concentrations/partial pressures, and why was that necessary?

How did you make sure that you were not observing signals from mixed clusters, or simply from clusters with other atoms?

Klaus von Haeften responded: In the presented paper HCCH–He$_n$ clusters were produced by co-expansion of HCCH–He mixtures at stagnation pressures between 50 and 100 bar. The HCCH concentration was varied. Signatures of HCCH–He cluster formation were observed when the HCCH concentration was below 0.1%. Good results were achieved for even lower concentrations, for example, 0.03%. The concentrations used to produce the spectra shown in our paper are indicated.

The signals observed were from pure HCCH, and indeed from mixed HCCH–He clusters. Signals from pure He clusters were not expected and not observed.

The background pressure in the apparatus used was 5×10^{-9} mbar. Cluster formation with atoms other than helium can be safely excluded.

Michael Woerner opened the discussion of the paper by **Kai Rossnagel:** What is the energetic resolution of your experiments?

What determines the energetic resolution of your experiments?

Can you get energetic and spatial infomation simultaneously in the future ?

Kai Rossnagel answered: The effective energy resolution of our time-resolved ARPES experiments was determined from Fermi-edge spectra of a polycrystalline gold sample to be 260 meV. The two dominant contributions to the resolution are generally the spectral width of the XUV probe pulses and the resolution of the photoelectron spectrometer. For the presented experiments, an upper limit on the spectral width of the XUV pulses could be independently determined to be 170 meV, and the nominal spectrometer resolution was 200 meV. Note that we do not use a monochromator in our experimental setup, which would reduce the XUV pulse duration and photon flux. Monochromation of the XUV probe pulses is not necessary because by pumping with the second harmonic of a Ti:Sapphire laser at a fundamental wavelength of 790 nm we can generate an isolated, well-separated, and intense high harmonic pulse at about 22 eV. For experimental details, see DOI: 10.1016/j.elspec.2014.04.013.

To get spectral and spatial information at the same time would be possible with a combined femtosecond time-resolved ARPES and X-ray photoelectron diffraction (XPD) experiment. Time-resolved ARPES would provide direct information on the dynamics of the band structure and Fermi surface, while time-resolved XPD would provide complementary information on the changes of structural parameters. Our plan is to realize such an experiment at the European XFEL.

Majed Chergui remarked: Since you mention determining lattice structure by photoelectron diffraction combined with ARPES, how would you distinguish the electronic structure from the lattice structure in this case?

Kai Rossnagel replied: ARPES and XPD, when applied in the mode with fixed photon energy, measure the same thing: the angular distribution of photoelectrons at specific kinetic energies. The differences lie in the energies of the electronic states from which the photoemission occurs and in the information the photoemission intensity distribution contains. In the case of ARPES, valence electrons are knocked out and the recorded intensity maps directly reflect the dispersion of the electronic states as a function of crystal momentum. In the case of XPD, on the other hand, core electrons are photoemitted and the measured intensity distributions are interpreted as interference patterns from which structural parameters can be extracted more or less directly. Electronic structure determination by ARPES works best at XUV photon energies, whereas the interpretation of XPD patterns is the simplest and most intuitive when soft X-rays excite high-energy photoelectrons that are scattered by the atoms in the sample predominantly in the forward direction.

So when I dream of a combined femtosecond time-resolved ARPES and XPD experiment, I think about two separate sets of measurements to be performed on the same sample, one XUV-ARPES measurement and one soft-X-ray-XPD measurement. Presently, there is no photon source available where such experiments can be done. However, there is well-justified hope that, with up to 27000 pulses/s, the European XFEL, specifically its soft X-ray SASE3 beamline, will be such a source.

Martin Wolf enquired: You have pointed out that there are two timescales in the dynamics "melting" of the Mott and CDW gap in TaS_2 und $TaSe_2$. It is clear that the slow (200 fs) dynamics originates for ion motion by the amplitude mode, whereas you assign the ultrafast (<40 fs) dynamics to an electronic process leading to a filling of states in the electronic gap. Could you explain what mechanism you have in mind for this electronic process? Why is there similar behaviour for both the Mott and the CDW gap? Should they not behave differently?

Kai Rossnagel replied: The process leading to the ultrafast filling-in of the gap that I have in mind is the photoinduced creation of electronic disorder *via* excitation of electrons across the Mott or CDW gap or, in real space, *via* local disruptions of the underlying charge orders. Here, I particularly think of quasi-1D CDW systems for which it is well known that static random disorder as well as lattice fluctuations lead to a smearing of the gap edges and a filling-in of the gap.

Shouldn't (nonthermal) electronic disorder (fluctuations) have a similar effect on the gap?

The Mott and CDW gaps in the layered Ta compounds are similar in the sense that they both have an electronic and a lattice component. This is maybe clear for the CDW gap because a CDW generally consists of a periodic modulation of the conduction electron density and periodic lattice distortion. It is less clear for the Mott gap because a standard Mott gap is purely electronically driven. In the layered Ta compounds, however, the Mott gap acquires a lattice component because it is the CDW with its periodic lattice distortion that gives rise to the Mott transition: the amplitude of the periodic lattice distortion controls the degree of electron localization in the combined CDW-Mott phase. Thus, since both the CDW and the Mott gap now have electronic and lattice components, they both show a similar two-step melting dynamics with a fast quenching of the electronic order and a slower reduction of the periodic lattice distortion *via* coherent amplitude-mode vibration.

Isabella Gierz asked: 1T-TaS$_2$ exhibits different band gaps at different momenta. Some of them are attributed to a periodic lattice distortion and, hence, electron–phonon coupling, others are caused by a Mott transition of a half-filled band at the Fermi level mediated by electronic correlations. It seems surprising that all of these gaps exhibit similar dynamics, *i.e.*, a rapid initial filling within the experimental time resolution and a partial closing on the timescale of the amplitude mode. Can you comment on that?

Kai Rossnagel replied: As pointed out in my answer to the previous question from Martin Wolf, the layered Ta compounds we have studied are special in that the periodic lattice distortion is an essential part of both the CDW and the Mott physics. The periodic lattice distortion is part of the CDW and at the same time is a prerequisite for the Mott transition. It is thus underlying and directly connected to two forms of electronic order: the periodic electron density modulation of the CDW and the characteristic arrangement of Mott localized electrons. Thus, the CDW as well as the Mott gap are connected to combined electronic and lattice orders, and they should therefore display similar melting dynamics after photo-excitation, *i.e.*, fast incoherent suppression of electronic order reflected in the smearing and filling-in of the gap and slower coherent suppression of lattice order reflected in the closing of the gap.

Martin Wolf addressed **Kai Rossnagel** and **Isabella Gierz**: I think we should ask the question in how far photoexcitation should lead at all to a complete closing of the gap as observed in phase transition in thermal equlibrium. There is no reason why photodoping should induce the same process as thermal fluctuation of the electronic and lattice degrees of freedom. It might well be that we need other ways of photodoping to drive transtions , *e.g.* charge transfer processes leading to changes in band filling. There is growing evidence that in a number of systems (including the CDW transition in tri-tellurides) photoexcitaion results only in a partial closing of the gap and there remains a "pseudogap" even at high excitation densities. The origin is currently not well understood and may be related to inhomogenities in the system.

Kai Rossnagel replied: I agree. It is very reasonable to assume that spatio-temporal order parameter fluctuations may generally lead to a smeared gap rather than a fully closed gap in photoexcited systems that exhibit long-range electronic or lattice order before the excitation. This type of pseudogap may in fact be similar to the one observed in strong-coupling CDW systems in equilibrium. In strong-coupling CDW transitions there is a pseudogapped intermediate phase between the CDW ground state and the normal metallic state. This intermediate phase is characterized by short-range CDW order; the amplitude of the order parameter is finite, but its phase is fluctuating. However, in photoexcited CDW systems not only the amplitude and phase of the order parameter will behave differently, but also the electronic and lattice components of the order parameter will decouple and display different behavior.

Isabella Gierz addressed **Kai Rossnagel** and **Martin Wolf**: As mentioned in the talk and the paper, a charge density wave has two contributions: a periodic charge modulation and a periodic lattice distortion. This explains the appearance of two characteristic timescales after photoexcitation. The charge modulation disappears first, the lattice follows on its own vibrational timescale.

It is still surprising that the Mott gap at the Γ-point at the Fermi level also (partially) closes on the vibrational timescale of the amplitude mode. This is not expected from a band gap associated with strong electronic correlations.

Again, as mentioned in the talk and the paper, the Mott transition can only occur once the periodic lattice distortion has created a half-filled band at the Fermi level. The fact that the Mott gap also closes on the vibrational timescale indicates that the Mott gap only closes because the periodic lattice distortion relaxes. Thus, it seems as if photoexcitation only melts the charge density wave associated with strong electron–phonon coupling. The electronic correlations resulting in the Mott transition seem to be robust. However, this seems to contradict the fact that the intensity loss of the lower Hubbard band at Γ is significantly larger than the intensity loss near M. In the paper, the loss of coherent spectral weight is attributed to the suppression of charge order due to electronic correlations (at Γ) and electron–phonon coupling (near M), respectively. If this is true, the Mott gap at Γ closes as a direct result of photoexcitation, not mediated by a relaxation of the periodic lattice distortion. A direct melting of the Mott gap is expected to happen on electronic rather than vibrational timescales.

Can you comment on that?

Martin Wolf replied: As far as I understand Kai Rossnagel attributes the metal insulator transition in TaS_2 and $TaSe_2$ to a Mott–Peiers scenario, wherby the Mott physics is strongly affected by changes in the orbital overlap due to the periodic lattice distortion (PLD). The dynamics of the PLD is dictated by a the characteristic phonon timescale (aprox. 200 fs).

However, all previous studies show that there is an ultrafast filling of states in the gap on an electronic timescale, but that the gap and the peak of the LHB do not completely vanish. This seams to be a more general phenomenon occurring in various other systems with PLD. For example, we also observe a persistence of the CDW gap in photoexcited tri-tellurides, were the gap closes partially but not completely when raising the excitation density. This may be attributed to inhomogeneities in the system giving rise to spacially different excited regions (with *e.g.* different phases). The complete collapse of the gap would require thermally activated fluctuations to mediate the transition.

Kai Rossnagel answered: You have correctly summarized our results and interpretations related to the Mott gap. The Mott gap closes on a vibrational timescale because the periodic lattice distortion is relaxed and the lower Hubbard band peak loses coherent spectral weight on an electronic timescale because charge order is suppressed. On the same electronic timescale, the Mott gap is filled in. As I have argued before, there is no contradiction between the two gap melting processes because in the present special case the Mott gap is a consequence of both electron–electron and electron–lattice interaction. But even in the case of a purely electronically driven Mott gap, one may remind oneself that it develops on a lattice and one can then imagine how coherent phonons modify the critical parameters of the Mott transition—the bandwidth W and the intra-atomic Coulomb repulsion U—on a vibrational timescale. Thus, Mott gaps do not necessarily have to melt on an electronic timescale.

Gwyn Williams asked: Does the technology exist to pump a high Tc superconductor, with the lattice cold, *i.e.* in the superconducting state, to the normal state and observe the electronic structure in the same manner as you report in this paper? In other words, in the future can we go to higher resolution and higher photon energies to get to the edge of the Brillouin zone?

Kai Rossnagel responded: The spectroscopic investigation of superconductors generally requires high energy resolution, because the energy gaps are small (≤ 50 meV), and high sensitivity, because the pumping needs to be gentle, with absorbed photon densities of typically less than 0.001 ph/formula unit (see Fig. 6 of the paper). The limiting factor in time-resolved ARPES studies on superconductors is the repetition rate of the photon source. Time-resolved ARPES using 6 eV probe pulses has the required energy resolution and sensitivity to resolve the gap dynamics of high-TC superconductors, but due to the low photon energy the technique is restricted in k-space to the region around the center of the first Brillouin zone. With time-resolved XUV-ARPES based on high-harmonic generation, on the other hand, we can access the electron dynamics at Brillouin zone edges, but because of the rather low repetition rates (<10 kHz) practically achievable energy resolutions and sensitivities are too low to probe the gap dynamics. There are, however, efforts (a strong one in Martin Wolf's group) to push time-resolved HHG-based ARPES to higher repetition rates and thus higher resolutions and sensitivities. My feeling is that time-resolved XUV-ARPES studies of the gap dynamics (*i.e.*, the dynamics of the superconducting gap and the pseudogap) in high-TC superconductors will become feasible in a few years.

Majed Chergui asked: In a recent paper, Carbone *et al.* (*Proc. Natl Acad. Sci.*, 2012, **109**, 5603–5608, see ESI†) found in the case of a lutetium based crystal that the charge-density wave is preserved until the lattice is sufficiently distorted to induce the phase transition. So in your case, what exactly melts the CDW, the pump pulse of the phonon dynamics?

Kai Rossnagel responded: Again, a CDW is generally composed of two charge densities: an electronic charge density and a lattice charge density. What our results suggest is that the CDW in the 1T Ta dichalcogenides is quenched in a two-step process. First, photoinduced electronic disorder melts the electronic component. This is indicated by the filling-in of the CDW gap on the 40 fs timescale. Second, the coherent amplitude mode vibration that is launched by the photoexcitation leads to a reduction of the lattice component of the CDW within the first half-cycle of the coherent oscillation. This is seen in the closing of the CDW gap on the 200–300 fs timescale. Thus, the CDW is quenched through both (incoherent) electron dynamics and (coherent) lattice dynamics, and we propose that this is a general scenario.

Michael Woerner asked: In the self-energy both electron–phonon and electron–electron interactions play a significant role.

Can you consider both phenomena simultaneously in your analysis?

Or do you assume from the very beginning that the two interactions are relevant on very different timescales?

Kai Rossnagel answered: The combined CDW and Mott physics that we see in the layered compounds here is indeed the result of an intriguing interplay of electron–phonon and electron–electron interactions. Both types of interactions are simultaneously present and both contribute to the opening of the CDW gap and the Mott gap. A central theme of time-resolved spectroscopy is the idea that these contributions can be disentangled in the time domain. And this is indeed what we demonstrate in our time-resolved ARPES measurements. We identify the electronic and lattice contributions to the gaps by the characteristic and distinct timescales of electronic and lattice processes involved in the suppression of the gaps.

Christian Bressler opened the discussion of the paper by **Grigory Smolentsev**: How good are XANES calculations to experiment these days? How difficult are TR-XANES calculations to actually assess what is really going on, on a molecular level? Is there any hope that XANES (TR or not) will ever settle questions about coordination, charge state, possibly spin state, or even more?

Grigory Smolentsev responded: I think that clear progress has been made in recent years in the development of methods for quantitative analysis of XANES. There are still a lot of investigations in which a qualitative comparison of theoretical spectra with experiment is used to get important information about the coordination and charge state. There are two steps that have to be made in order to go from qualitative to quantitative XANES analysis. First, the structure of the compound (often in a solution phase) has to be determined. There are codes (for example, FitIt http://nano.sfedu.ru/fitit.html) that have been developed recently, which allow to refine the structure on the basis of quantitative XANES fitting. They are also applicable to time-resolved XANES spectra. These algorithms rely on the assumption that the discrepancy between theoretical and experimental spectra is minimal for the correct structure. Therefore, the second step is to verify this assumption or in more general terms to develop methods for calculations of XANES that give good quantitative agreement with the experiment for known structures. In our research, we always check this assumption. The structure of the ground state of the compound is usually known much more precisely than the structure of the excited or intermediate states. Therefore we use the ground state to check how precisely the theory can reproduce the experiment. We have found that in many cases the full-potential approach for XANES calculations (proposed in *Phys. Rev. B*, 2001, **63**, 125120) is important to get the quantitative agreement with the experiment. Self-consistent full-potential methods are highly demanded by the X-ray absorption society. They were announced, but they are not widely available yet. L-edge XANES calculations of 3d elements require the multi-electron approach. Different methods have been developed recently (*J. Chem. Phys.*, 2013, **138**, 204101; *J. Phys. Chem. Lett.*, 2012, **3**, 3565; *J. Phys.: Condens. Matter*, 2011, **23**, 145501; *Phys Rev. B*, 2010, **81**, 125121). They have been applied only in a few cases and broad testing is required. In general, I think that there is definitely progress in the development of a method for quantitative calculations of XANES for known structures, but it is not as fast as we wish.

Majed Chergui said: I agree with Ch. Bressler that the theoretical fitting of XANES spectra (above-edge) is still in its infancy and we can only fit them at a qualitative level, not a quantitative one. On the other hand, for the region below the edge, good progress has been made in recent years based on DFT and TDDFT methods. See Milne *et al.*, *Coord. Chem. Rev.*, DOI: 10.1016/j.ccr.2014.02.013 for a review of the recent theoretical developments.

Grigory Smolentsev answered: I agree with you that TDDFT and methods that allow to take into account the multi-electron effects for calculations of the pre-edges of X-ray absorption spectra have made some progress in recent years. Nevertheless, I don't think that the quantitative interpretation of the region below the edge is better developed than the XANES calculations. Even for well-known structures like Fe(bpy)$_3$ the agreement of the intensities and also the positions of the peaks is qualitative (see Fig 16 of the review DOI: 10.1016/j.ccr.2014.02.013). The methods of XANES analysis have made one step further from the calculations of the spectra for known structures. At the current stage, XANES calculations aim at determination of unknown structures, for example, photo-excited species or intermediates of photocatalytic reactions. There are examples of quantitative XANES fitting: *Phys. Rev. B*, 2007, **75**, 144106; *J. Phys. Chem. A*, 2008, **112**, 5363; *J. Phys. Chem. A*, 2010, **114**, 12780. In the manuscript presented here we did not have the aim to achieve quantitative agreement using structural refinement, but compared a few models obtained with other methods. From such a qualitative comparison an important information about the coordination of the catalyst's metal center has been obtained.

Katharine Reid addressed **Grigory Smolentsev** and **Ben Spencer**: It appears that charge recombination dynamics can occur on a variety of timescales. Can you comment on the factors that determine this?

Ben Spencer replied: Our work demonstrates the vast differences in timescales that charge recombination can occur on. In the case of $ZnO(10\bar{1}0)$ a persistent photoconductivity (PPC) over milliseconds is exhibited due to defect localized states associated with oxygen vacancies. Indeed, PPC in ZnO has been observed to last for hours and even days. In the case of quantum dot (QD) systems, type II QDs (core/shell QDs) allow for electrons and holes to be confined in the shell and core or *vice versa*, and a reduction in wavefunction overlap extends radiative recombination lifetimes. Nonradiative recombination is dependent upon the density of trap states, typically at the surface of QDs (Califano and Gómez-Campos, *Nano Lett.*, 2013, **13**, 2047). To generalise, charge recombination dynamics are heavily influenced by the carrier density, carrier separation (wavefunction overlap) and the density and nature of trap states within the band gap, which is highly material dependent (defects, surface states, *etc.*). The semiconductor materials used here demonstrate how the carrier recombination lifetimes can be engineered to cater to a specific application, which for us is photovoltaics *e.g.* by varying the oxygen vacancy concentration in ZnO or the core and shell properties of QDs.

Grigory Smolentsev replied: For photocatalytic systems the charge recombination rate depends significantly on the type of system. If the system is multi-component (in this case there is no chemical bond between the photosensitizer and the catalyst) the charge recombination is limited by the diffusion. Therefore the process is rather slow. It occurs in the microsecond time range. For supramolecular systems the charge transfer and recombination is an intramolecular process. Therefore it is much faster. For supramolecular systems charge recombination often occurs in the nanosecond range (and can be even faster if the system is not efficient).

Gwyn Williams opened the discussion of the paper by **Ben F. Spencer**: Can you give me an idea of where you intend to take this work in the near future?

Ben Spencer responded: We would like to have an experiment with improved time resolution than available at SOLEIL, since the onset time of the surface photovoltage appears to be less than the time resolution when quantum dots are linked to the surface. However, measuring dynamics with picosecond time resolution or better, but over many decades of time greater than this resolution, is a challenge. These criteria can be met at the beamline BL07LSU at SPring-8, where an angle-resolved time of flight (ARTOF) detector is used in combination with an amplified ultrafast laser pulse typically at 1 kHz repetition rate, while the synchrotron is operated in a mode with five single bunches separated by hundreds of nanoseconds, with measurements taken using several traverses of the synchrotron, *i.e.* over several microseconds (Ogawa *et al.*, *Rev. Sci. Instrum.*, 2012, **83**, 023109). The material used in this work, ZnO, is still not suitable for these experiments given the extremely long (millisecond) timescales involved, but we plan to continue work on transparent conducting oxide surfaces, in particular TiO_2, again with quantum dots chemically linked to the surface.

Majed Chergui asked: are there other methods to probe the photovoltage effect?

Ben Spencer answered: The most notable is the Kelvin probe experiment which measures illumination-induced changes in workfunction (Zisman, *Rev. Sci. Instrum.*, 1932, **3**, 367). The technique is typically limited by electrical noise and stray capacitance effects. Kelvin probes are typically used in ambient environments but are also successfully used in UHV, and the technique has been extensively developed, *e.g.* into scanning Kelvin probes with micron resolution (Nabhan *et al.*, *Rev. Sci. Instrum.*, 1997, **68**, 3108). High-resolution scanning techniques have also been developed using scanning tunnelling microscopy, where the SPV is measured by varying the sample bias with and without laser illumination (Hamers and Markert, *Phys. Rev. Lett.*, 1990, **64**, 1051). Other techniques include the use of an electron beam where the sample functions as the anode, and illumination-induced changes in the workfunction are measured as rigid shifts in the *I–V* curve (Steinrisser and Hetrick, *Rev. Sci. Instrum.*, 1971, **42**, 304). An excellent review of SPV techniques can be found in Kronik and Shapira, *Surf. Sci. Rep.*, 1999, **37**, 1. Time-resolved SPV experiments using XPS under controlled conditions have built on the seminal work of Long *et al.*, *Phys. Rev. Lett.*, 1990, **64**, 1158. The combination of pulsed lasers with synchrotron radiation allows for ps resolution of dynamics on ns timescales (*e.g.* Spencer *et al.*, *Phys. Rev. B*, 2013, **88**, 195301). Another key advantage of this technique is the chemical specificity it gives since the SPV is monitored using core-level photoelectrons, as also shown in Stefan Neppl's paper where both the N3 dye and ZnO nanoparticles are monitored. Our results show how both the QD and ZnO substrate can be monitored simultaneously using the Zn 3d core level since two chemical species can be fitted to the XPS spectra, one associated with the substrate and the other the QD (which does not shift upon laser illumination, as expected).

Martin Wolf remarked: We have discussed already the difficulty to have an atomically well defined surface structure of ZnO. I was a bit puzzled by the band bending diagrams which you show for n-doped ZnO and assume an upward band bending as in a conventional n-type semiconductor. The work by Ozawa and Mase (*Phys. Rev. B*, 2011, **83**, 125406) and also our work on ultrafast carrier dynamics on ZnO(10–10) (Deinert *et al.*, *Phys. Rev. Lett.*, 2014, in press) shows that already very small amounts of hydrogen adsorption lead to a pinning and a small downward band bending of the conduction band. In that case, no surface voltage shift is expected. I understand that you see no surface voltage shift on a sub-ns timescale but you observe it on a 5–10 micro-second timescale. This will make is invisible in our laser experiments performed with 250 kHz. But I am still puzzled about the sign and magnitude of the band bending.

How do you prepare your surface and how can you be sure that they are hydrogen free considering that there is always hydrogen present in the residual gas?

Ben Spencer replied: We consistently measured n-type behaviour, with surface photovoltage (SPV) shifts to higher binding energy. This is consistent with many previous studies on the nonpolar ZnO(10–10) surface (*e.g.*, King and Veal, *J. Phys.: Condens. Matter*, 2011, **23**, 334214; Tashiro *et al.*, *Appl. Opt.*, 1969, **8**, 180). The magnitudes of the SPV shifts we measured were consistent with calculations of the equilibrium band bending (Spencer *et al.*, *Phys. Rev. B*, 2013, **88**, 195301).

Regarding your experiments, at a repetition rate of 250 kHz, where subsequent photoexcitations occur every 4 μs, I would not expect you to see a transient SPV since the timescales involved are likely to be much longer than this as our results indicate. Indeed, we saw this when we performed measurements with ps resolution, at a repetition rate of 3.123 MHz, or a 320 ns time window. However, under static conditions (pulsed laser photoexcitation at a set pump–probe time delay), we measured a rigid shift to higher binding energy (Spencer *et al.*, *Phys. Rev. B*, 2013, **88**, 195301). Consecutive photoexcitations lead to a steady-state SPV shift measured at all time delays in our 320 ns time window. These initial measurements indicated that the ZnO was n-type and that we needed an experiment able to measure the SPV over much longer timescales, which lead to our work at SOLEIL. Therefore, I would expect that in your experiments at 250 kHz you should be able to measure a rigid shift in the binding energy of a ZnO core level under static conditions but not a transient change over the time window of your experiment. If this was not the case, then it may be the differences in the surface preparation that are responsible. Our surface preparation included an anneal in oxygen (at $p = {\sim}1 \times 10^{-7}$ mbar) for between 10 and 20 min. A high temperature vacuum anneal was also required after this to remove excess oxygen adsorbed to the surface, and this was required to prevent charging problems during our experiments (long-term shifts in the binding energy position of the Zn 3d core level). As the length of this oxygen anneal was increased, the timescales for onset and decay of the SPV decreased, which we attribute to a lower concentration of oxygen vacancies being present as you anneal in oxygen for longer. Defect states associated with these oxygen vacancies have been implicated in the persistent photoconductivity in ZnO (Lany and Zunger, *Phys. Rev. B*, 2005, **72**, 035215).

The role of hydrogen in ZnO is a valid and important point to consider—indeed, it is unlikely that the surface is hydrogen free due to hydrogen being present in the residual vacuum. I agree that there is evidence that H adsorption can lead to metallization at the surface, meaning no SPV shift would be measured upon photoexcitation. The work by Ozawa and Mase (*Phys. Rev. B*, 2011, **83**, 125406) adsorbed atomized H_2 onto the surface of ZnO under a partial pressure of 10^{-6} mbar, much higher than the base pressure of our experiment (10^{-10} mbar). Whilst this work indicates that this metallization could occur at much lower partial pressures, there is a lack of experimental evidence of this in the literature (that I am aware of). Clearly this metallization is not occurring in our experiments since we consistently measure SPV shifts to higher binding energy.

There is also substantial evidence that hydrogen acts as a donor in ZnO, giving n-type behaviour (including Van de Walle, *Phys. Rev. Lett.*, 2000, **85**, 1012; Hofmann *et al.*, *Phys. Rev. Lett.*, 2002, **88**, 045504; Janotti and Van de Walle, *Nat. Mater.*, 2007, **6**, 44; Lavrov, *Physica B*, 2009, **404**, 5075; Weber *et al.*, *Physica B*, 2012, **407**, 1456). Here the n-type conductivity is still dependent upon the oxygen vacancy concentration because hydrogen donors occupy oxygen vacancy sites (H_O).

Majed Chergui commented: In your paper, you show the decay kinetics of the QD's in solution (Fig. 4), which is tri-exponential. Do you know the origin of this behaviour? Actually, we studied the case of CdSe QDs and found that the multi-exponential decay is due to a distribution of Stark shifted electronic levels, due to the fact that these QDs are inhomogeneous and have large permanent dipole moments, which leads to a distribution of electronic level energies (see Al Salman *et al.*, *Chem. Phys.*, 2009, **357**, 96–101). Can this be the same in your case?

My second question is if you measured these kinetics for the QDs on the surface of the ZnO?

Ben Spencer responded: The long component of the kinetics is the radiative recombination component, whereas the fast components are nonradiative, attributed to Auger-mediated trapping (Califano and Gómez-Campos, *Nano Lett.*, 2013, **13**, 2047). Trap states within the band gap can arise from the surface of QDs, the core/shell interface, and core impurities. Trap states will vary due to inhomogeneity, meaning there is likely a complex distribution of trap states. Fitting with multiple exponentials is thus common practice, although it is not clear that the decay is actually multi-exponential in form, *e.g.* there has been recent work describing the nonradiative kinetics in terms of a power law (Sher *et al.*, *Appl. Phys. Lett.*, 2008, **92**, 101111). The work you mention (Salman *et al.*, *Chem. Phys.*, 2009, **357**, 96) does contradict the view of nonradiative decay channels being associated with trapping, and may be worth considering since it is well known that QDs can have large permanent dipole moments (*e.g.* Shim and Guyot-Sionnest, *J. Chem. Phys.*, 1999, **111**, 6955). The PL is used here to assess the type II nature of the core/shell QD sample, where the radiative decay component is extended by an order of magnitude when a shell is added to the core due to a reduction in wavefunction overlap of electron and hole due to separation in the core and shell (or *vice versa*). We did not measure these kinetics once they were linked to the ZnO surface. Indeed, the quenching of the PL upon attachment to ZnO would have been useful to assess the efficiency of charge transfer into the substrate. However, oxidation and other contamination may well have obscured this measurement.

Kai Rossnagel communicated: You say that there is a need for developing time-resolved X-ray spectroscopy techniques that are capable of measuring millisecond dynamics with sub-picosecond time resolution. What processes occurring at the surfaces of solids, other than the surface photovoltage effect, would be interesting to study with such techniques?

Ben Spencer communicated in reply: Time-resolved XPS techniques are well-suited to study a wide range of phenomena including oxidation reactions (*e.g.*, Baraldi *et al.*, *J. Elect. Spect. Rel. Phenom.*, 1995, **76**, 145; Höfert *et al.*, *Rev. Sci. Instrum.*, 2013, **84**, 093103), charging/ decharging dynamics (*e.g.*, Demirok *et al.*, *J. Phys. Chem. B*, 2004, **108**, 5179), and magnetization dynamics (*e.g.* Beaulieu, *J. Elect. Spect. Rel. Phenom.*, 2013, **189S**, 40). Being able to use fast XPS techniques over short and long timescales lends itself to investigating a range of reaction kinetics including photocatalysis, degradation, the influence of subsequent chemical reactions, *etc.* In our case we required millisecond timescales due the persistent photoconductivity exhibited by ZnO. Repeating these measurements with improved time resolution would allow to us better resolve additional fast processes such as charge injection into ZnO from light-harvesting quantum dots.

Kai Rossnagel opened the discussion of the paper by **Martin Wolf**: I find the observation of two CDW amplitude modes very intriguing. In the standard mean-field theory of the CDW transition, upon lowering the temperature across the transition temperature, a single phonon mode goes soft and a single CDW amplitude mode emerges. Could you explain the two-mode scenario in a little more detail?

Martin Wolf responded: In our ARPES work on TbTe3 we have developed a 3 pulse (pump-repopulate-probe) scheme which allows us to analyze the periodic oscillations of the CDW gap size (*i.e.* the order parameter) oven an extended the time range. For these measurements it is clear that at least two frequencies (associated with optical phonon modes) modulate the gap size. We believe the this directly shows that the lattice motion associated with the amplitude mode is more complex than in a simple (quasi 1D) Peierls picture. In fact any (normal state) phonon at the CDW wave vector which is strongly coupled to the electronic structure at the Fermi Level will modulate the band structure and in TbTe3 the two modes at 1.77 THz and 2.23 THz are most prominent. Demsar and coworkers have described this scenario of phonon softening in a time dependent Ginzburg–Landau model and could reproduce the observed phonon softening and anti-crossing of the 1.77 THz and 2.23 THz modes by time resolved optical reflectivity measurements (see *Phys. Rev. B*, **89**, 045106 (2014) and references therein). However, there is non complete softening observed. Note that in optical spectroscopy as well as in time resolved ARPES we excite $q = 0$ phonons, which do not exhibit complete softening, whereas the X-ray diffraction work by Frank Weber (KIT, Karlsruhe, Germany, to be published) shows that the phonon softening is most pronounced at non-zero q (CDW wave vector). Interestingly, we observe for the photoexcited state similar amplitudes for gap coupling for the gap modulation by the 1.77 THz and 2.23 THz modes, which is quite different from the findings for the thermal induced transition in *Phys. Rev. B*, **89**, 045106. This indicates that the photoinduced transition is different from the thermal equilibrium transition. In the latter case the CDW gap completely closes while in the photoinduced transition the gap remains finite and does not close even at higher excitation fluence.

Martin Wolf opened the discussion of the paper by **Isabella Gierz**: In your work you compare scattering rates obtained by time-resolved ARPES with the measured self energy obtained from the linewidth in static ARPES measurements. Both quantities show the same energy dependence (as has been predicted in the theoretical work by Tom Devereaux and coworkers at Stanford, *Phys. Rev. X*, 2013, **3**, 041033). However, quantitatively there is a large difference. What could be the origin? I would also like to remark that in a system like graphene, where you have cooling of photoexcited non equilibrium electrons by phonon emission and presumably even carrier multiplication, the relaxation rate which you measure is not the intrinsic lifetime of a carrier at that particular energy. In fact you are measuring the population dynamics in an energy interval which results from the balance of filling from higher lying states and the depletion by energy relaxation to lower lying states. As such it would be surprising if the data would agree quantitatively with the self energy in the static case.

Isabella Gierz answered: I agree that the equilibrium self-energy and the non-equilibrium scattering rate measure two different quantities. As indicated by Martin Wolf, the equilibrium self-energy is related to the lifetime of the photo-hole, whereas the non-equilibrium scattering rate is determined by a complex interplay of state filling and depletion. However, in *Phys. Rev. X*, **3**, 041033 (2013), Sentef *et al.* have shown that for their photoexcited electron–phonon model system there is a good agreement between non-equilibrium relaxation rates and the equilibrium self-energy. The purpose of Fig. 6 in our paper is to test this statement on an actual data set. The figure compares the energy-dependent scattering rate determined by tr-ARPES with the MDC line width from high-resolution static ARPES experiments. The scattering rate continuously increases from ~0.5 THz at the Fermi level ($E = 0$) to ~4 THz at $|E| = 1$ eV. The MDC line width rises from ~0.01 A^{-1} at $E = 0$ to ~0.035 A^{-1} at $|E| = 1$ eV in agreement with *Nat. Phys.*, **3**, 36 (2007). After multiplying the MDC line width with the Fermi velocity ($v_F \approx 6.6$ eV A) this corresponds to an increase from 16 to 56 THz. Overall, the scattering rate and the MDC line width show a qualitatively similar energy dependence. However, the values extracted from static ARPES measurements are about one order of magnitude larger than the non-equilibrium scattering rates. The most likely reason for this discrepancy is that static MDC line widths are broadened by the experimental resolution, suggesting higher scattering rates. This results in a considerable over-estimation of the electron–phonon coupling constant in graphene (see *Nat. Phys.*, **3**, 36). Therefore, Sentef *et al.* suggested the extraction of the electron–phonon coupling constant from tr-ARPES scattering rates instead. Nevertheless, as the equilibrium self-energy and the non-equilibrium scattering rate measure different quantities, the validity of this approach may be restricted to particular systems.

Kai Rossnagel asked: Could you comment a little bit more on why a system that has been excited rather strongly into a non-equilibrium state should relax back to equilibrium at scattering rates determined from the equilibrium state?

Isabella Gierz replied: Please also refer to my previous answer. It is quite counterintuitive that a system far from equilibrium relaxes at the equilibrium scattering rate. Nevertheless, this is the result of the model calculation in *Phys. Rev. X*, **3**, 041033 (2013). There, the authors argue that (in the framework of the double-time Green's function methodology employed in their work) changes in the occupation of the band structure occur in the first order for small deviations from equilibrium, while actual changes in the dispersion are in the second order. As long as the self-energy does not depend too strongly on the electronic distribution function, the non-equilibrium system is supposed to relax *via* the equilibrium scattering rate. We wanted to test this hypothesis on actual (not simulated) data. For photo-excited graphene, we found that the equilibrium self-energy and the non-equilibrium scattering rate are qualitatively similar, in the sense that they show a similar energy dependence. However, they differ quantitatively by about one order of magnitude. The static MDC line widths suggest much shorter lifetimes than those measured by tr-ARPES (see Fig. 6 of our paper). This discrepancy can be, at least partly, attributed to the limited momentum and energy resolution of the static ARPES results.

Katherine Reid asked: Have measurements similar to those that you have performed on graphene been performed on a graphite surface, and if so can you comment on the differences that are observed in the two cases?

Isabella Gierz responded: There is one time-resolved photoemission study on graphite from 2001 (*Phys. Rev. Lett.*, **87**, 267402, (2001)). In that work, the authors combined a visible pump at 2.3 eV with 4.6 eV probe pulses, just exceeding the work function of the sample. The small photon energy only allowed for photoemitted electrons from the vicinity of the Γ-point, and the spectra were recorded in angle-integrated mode. Despite these technical differences, the 2001 work obtained similar results for the energy dependence of the relaxation time and identified two different decay times that have been attributed to the internal equilibration of the electronic system (~250 fs) and the interaction with phonons (~1 ps). The interpretation of the two decay times has slightly changed since 2001, the main results of the graphite paper, however, remain valid. On the level of the observed decay channels and associated relaxation times graphite and graphene are very similar.

Gwyn Williams addressed Martin Wolf: What other tools would you like to have at your disposal to extend these experiments? Think about FELs in particular, with tunable THz and IR pumps, and higher photon energies, and 100 MHz repetition rates.

Martin Wolf responded: In almost all time-resolved pump–probe experiments we would like to have more specific tools for excitation. A promising avenue is IR or THz pumping to induce low energy excitations (*e.g.* just above the band gap of a superconductor or phonon pumping of solids). IR pumping is currently slowly penetrating in the field of tr-ARPES. THz pumping is yet to be applied, but we should keep in mind that combining THz pumping and ARPES will lead to the streaking of the outgoing electrons for light polarization with components normal to the surface. As Robert Moshammer has pointed out this may be useful to fully characterize the THz pulse. For pumping of the sample in plane polarization appears most promising (and could induce spin currents, dressing of Bloch states, *etc.*). I would propose that most ARPES work will be done with laboratory based sources and that there are currently important developments opening new areas by high rep-rate HHG, which may allow also for spin-resoved ARPES using new detectors. This will allow to study both ultrafast dynamics in ferromagnets as well as spin textures (*e.g.* in topological insulators) in great detail. In the field of X-ray science time resolved RIXS has great potential (see comment above)

Isabella Gierz answered: The biggest drawback of our lab-based tr-ARPES setup is the low repetition rate of 1 kHz (resulting in low signal-to-noise ratios) and the maximum achievable pump wavelength of 15 μm. We are currently considering the following upgrades:

(1) combine carrier envelope phase (CEP) stable mid-infrared pump pulses with few femtosecond HHG probe pulses to follow pump-induced coherent dynamics in real time

(2) go to higher repetition rates

(3) develop a narrow band THz pump source

Some of these aspects can be met at FELs, however, the use of FELs will also cause additional problems such as timing jitter.

Kai Rossnagel remarked: Regarding the persistence of the CDW gap even after rather strong photoexcitation, it could be interesting to compare the absorbed energy density to the electronic condensation energy, as given by equation (3.40) in Grüner's book (G. Grüner, *Density Waves in Solids*, (Perseus, Cambridge, MA, 2000)), and to the energy density required to heat up the excited sample volume to above the CDW transition temperature (see, *e.g.* A. Tomeljak *et al.*, *Phys. Rev. Lett.* , **102**, 066404 (2009)). If the absorbed energy density used in the experiment lied in between these two values, the persistent gap would reflect a non-equilibrium CDW state in which the electronic order is quenched, while the lattice order persists. Have you estimated these two values?

Martin Wolf responded: In our experiments on TbTe3 we used an absorbed laser fluence between 0.5 mJ cm^{-2} and 1.5 mJ cm^{-2}. For a fluence of 0.75 cm^2 the

CDW gap collapses to about half of its equilibrium value and for higher fluences no further decrease of the CDW gap is observed. For a fluence of 1.0 mJ cm^{-2} we estimate an absorbed energy density of approx. 1eV per unit cell or 250 meV per RTe3 unit. With a Debye temperature of ~180 K this energy corresponds to a lattice temperature of 350 K which is just above the transition temperature of the CDW transition. Note that the CDW gap does not collapse even at the 50% higher excitation fluence of 1.5 mJ cm^{-2}. The condensation energy according to equation (3.40) in Grüner's book is estimated to be 75 meV for $\Delta = 250$ meV. We are thus clearly exciting both electronic and lattice temperature to values exceeding the equilibrium CDW transition temperature, but still see only a partial collapse of the gap. However, one should keep in mind that a mean field model may not be correct for TbTe3. We speculate that other mechanisms (like spatial inhomogeneities) may play a role in stabilizing the CDW order. Without doubt the observations of "persisting pseudogaps" and "persisting order" needs further analysis.

Gwyn Williams asked: Do you consider that the system is not left in the ground state?

Martin Wolf replied: Your question addresses some fundamental issues of our understanding of the photoemission process. In photoemission and in particular in time resolved photoemission from a photoexcited sample we deal with an excited system and Koopman's theorem as well as the sudden approximation may not be valid. However, we might be in a favorable situation if we consider energy differences, *e.g* when we measure simultaneously the binding energy from a normally occupied an a transiently populated, but normally unocculied band, to evaluate the magnitude of a band gap. In this case we can assume that the energy difference is not affected by a breakdown of the sudden approximation. In general, fundamental studies of the photoemission process by time resolved methods are an important field and topic of attosecond spectroscopy from solids.

Katherine Reid addressed Martin Wolf, Isabella Gierz, Christian Bressler and Majed Chergui: Many talks have referred to time and angle-resolved photoemission. Can this be viewed as an "ultimate" dynamical probe, or are there other probes that would provide complementary or superior information in certain cases?

Martin Wolf replied: Time-resolved ARPES has its main advantage in studies of the valence electronic band structure of periodic solids and allows us in particular to access directly the momentum and energy resolved scattering rates of photoexcited electrons. In systems exhibiting photoinduced transitions or coherent lattice vibrations the transient evolution of the electronic band structure con be followed and we had two papers on the dynamics of gap closing and coherent modulation of binding energies mediated by coherent lattice vibrations (*e.g.* amplitude modes). When looking at the electronic structure near the Fermi level time-resolved ARPES is the method of choice, but it is limited by space charge effects to rather modest pump fluences (depending on the system and laser parameters like polarization). Most experiments on time-resolved ARPES will be done with lab based laser sources and currently different schemes (high-rep rate HHG, 80 MHz 6-7 eV ARPES and kHz HHG systems with IR/THz pumping) are implemented. There are of course many other time resolved probes which lead to complementary information on dynamics properties of the valence electronic structure. In the area of X-ray science XPE, XAS and XES provide element specific information about the chemical state and valence electronic structure. In the field of surface science and catalysis a first demonstration of element specific probing of a chemical reaction has been demonstrated using time-resolved RIXS (*Science*, **339**, 1302 (2013)). This provides information about chemical bond formation and breaking. As a core level spectroscopy tr-RIXS provides local information which is complementary/different to the electronic band structure in a well ordered periodic system. Important information about the electronic properties can be obtained also by time domain THz spectroscopy, which allows us to measure the complex conductivity as well as low energy modes (phonons, magnons) of solids after photoexcitation. This may provide important valuable information about the dynamics of the electric system. Also important are various structural probes. In particular, time resolved electron diffraction has made remarkable progress in the last few years, but is still limited in time resolution. X-ray diffraction is thus a promising tool.

Majed Chergui answered: I do not think there is an ultimate dynamical probe. The probe depends on the scientific question you want to answer. It is true that ARPES is a very powerful tool, but: (a) it is starting to be increasingly used and still has to show the full range of its capabilities; (b) it is valid only for solid systems, that are single crystals, therefore, you can immediately imagine that it cannot be used for any sample, least of all disordered ones; (c) it is not a structural tool in the sense that it does not provide you with geometric structure. Of course you can infer it indirectly, as with many other spectroscopic techniques. However, you do get an extremely fine and detailed electronic structure. In summary, just as with other methods, it is their combination (if you can afford it) that is the best strategy.

Isabella Gierz replied: Tr-ARPES is the technique of choice to measure the response of the momentum-resolved valence electronic structure in the time domain after excitation at various different wavelengths. There is, at present, no better technique for that purpose. However, tr-ARPES could be nicely complemented with time-resolved electron diffraction (to obtain structural information with similar surface sensitivity), time-resolved STM (for spatial resolution), or spin- and time-resolved ARPES (to address spin dynamics).

Christian Bressler answered: I am convinced there will never be one single probe which will answer all scientific questions around chemical reaction dynamics. X-ray tools are also quite advantageous, as they can likewise deliver electronic, spin and geometric structural information. Guest–host interactions are extremely difficult to study due to the many-bodies involved, but X-ray scattering may deliver useful information. Photoemission studies are limited in their penetration depth, while hard X-ray tools (X-ray absorption XAS, X-ray emission XES, X-ray diffuse scattering XDS) can look several (tens of) microns inside the sample.

Gwyn Williams addressed Martin Wolf and Isabella Gierz: Are there any other experiments, such as Auger resonant emission or inverse photoemission or spin-resolved photoemission that might shed insight into the phenomena you are studying?

Martin Wolf replied: Auger resonant emission may be an interesting probe which has not yet been explored in detail. Spin resolved photoemission will greatly benefit from new detector development with a 100 times better sensitivity compared to current schemes. These will be combined with spatial resolution and imaging to allow for multi parallel detection.

Isabella Gierz answered: Tr-ARPES is the tool of choice to follow the evolution of the valence electronic structure in the time domain. Time- and spin-resolved ARPES as well as time-resolved (photo-)electron diffraction would provide useful complementary information.

Elaine Seddon said: There have already been a couple of questions relating to future developments, especially in terms of techniques and observables. Electron spin is one such observable that has not been discussed in detail. I would therefore appreciate your views on potential future developments that involve electron spin.

Martin Wolf responded: There are currently important developments to open new areas by XUV pulse generation *via* high repetition rate high harmonic generation (100 kHz to several MHz OPCPA HHG) sources, which may allow also spin-resoved ARPES using new detectors with an enhanced sensitivity. This will allow us to study both ultrafast dynamics in ferromagnets as well as spin textures (*e.g.* in topological insulators) in great detail. In the field of X-ray science time resolved RIXS has a great potential.

Katherine Reid addressed Isabella Gierz and Martin Wolf :You mentioned time-resolved STM as a possible technique; what would be the achievable time-resolution?

Martin Wolf replied: There have been numerous attempts to achieve femto-second time resolutions using STM. A promising attempt is the electronic pump probe STM using short voltage pulses developed by the Almaden group (S. Loth *et al.*, *Science*, **329**, 1628 (2010)), which is currently improved to obtain a high sub-ns resolution. Another idea is to use THz pulses in an STM junction as an "ulrafast switch" of the bias voltage. There is also the ongoing development of time resolved low-electron electron diffraction (LEED) to study ultrafast structural phase transitions at surfaces. Thereby an ultrashort electron pulse is emitted from an STM tip. In transmission experiments (*e.g.* on freestanding graphs covered with adsorbed surfactants, *Science*, **345**, 200 (2014)) time-resolved LEED has been already demonstrated. In back reflection mode there is the challenge to control the electron imaging in LEED using an STM tip close to the surface.

Isabella Gierz responded: Time-resolved STM using voltage pulses from a pulse generator is limited to a time resolution of about 1 ns with atomic spatial resolution (*Science*, **329**, 1628 (2010)). Recently, the technique has been combined with THz light fields (*Nat. Photon.*, **7**, 620 (2013)), achieving subpicosecond time and 2 nm spatial resolution.

Michael Woerner asked Isabella Gierz: As you mentioned in your talk, one expects an anisotropic carrier distribution after photo-excitation. Did you see any dependence of the photoemission data on the laser polarization ?

Isabella Gierz replied: We have not checked yet. However, there is clear evidence for an initially anisotropic charge carrier distribution from all-optical pump–probe experiments (*Nano Lett.*, **14**, 1504 (2014)).

Majed Chergui asked: Regarding the carrier multiplication, have you done any fluence dependence studies?

Isabella Gierz answered: According to *Nano Lett.*, **10**, 4839 (2010) and *Phys. Rev. B*, **85**, 241404(R) (2012) carrier multiplication is supposed to occur at low pump fluences, high pump photon energies, and low sample temperatures. We systematically varied pump fluence (between 0.6 and 6 mJ cm^{-2}) and sample temperature (20 K, 77 K, and room temperature) and determined the number of carriers in the conduction band as a function of the pump–probe delay. We found that the rise time is always longer than the pulse duration. However, we also observed a thermal distribution of hot electrons at all times. As carrier multiplication is not expected to occur for a thermalized electronic distribution we rather attribute the increased rise time to the anisotropy of the initial carrier distribution in momentum space as predicted in *Phys. Rev. B*, **84**, 205406 (2011). In conclusion, we found no carrier multiplication in the experimentally accessible parameter range. However, at the lowest pump fluence of 0.6 mJ cm^{-2}, we are severely restricted by our signal-to-noise ratio due to the 1 kHz repetition rate. Therefore, carrier multiplication may occur at even lower fluences in agreement with *Phys. Rev. Lett.*, **111**, 027403 (2013) and *Nat. Commun.*, **4**, 1987 (2013).

Majed Chergui remarked: There is a debate about the occurrence of band gap renormalization in graphene, can you comment on it and whether it shows up in your studies?

Isabella Gierz replied: Monolayer graphene on SiC(0001) does not have a band gap at equilibrium. Nevertheless, it is conceivable that photoexcitation might induce a band gap or renormalize the electronic structure due to the screening of electronic correlations by the presence of photoexcited carriers. However, within our experimental resolution we do not find indications for a transient band gap renormalization. Our pump–probe signal is completely dominated by the effect of an elevated electronic temperature.

Katherine Reid asked: At the end of your talk you referred to an alternative explanation; would you be able to elaborate on this?

Isabella Gierz responded: The question refers to the fact that the observed rise time of the pump–probe signal is longer than the pulse duration, which might indicate the occurrence of carrier multiplication by impact ionization. However, we also observe a thermal Fermi–Dirac distribution of electrons at all times. This distribution does not allow for carrier multiplication to occur. Hence, the enhanced rise time of the signal must have a different origin. We propose, that the reason for the enhanced rise time is the gradual relaxation of the initially anisotropic carrier distribution. The matrix element for photo-excitation is anisotropic, resulting in an anisotropic charge carrier distribution in the conduction band with nodes along the direction of light polarization. This distribution becomes isotropic *via* the emission of optical phonons within ~200 fs. In the present experiment the photocurrent was measured along the nodes of the initially anisotropic contribution. Hence, the maximum pump–probe signal is reached only after the excited electrons have scattered into the field of view of the analyzer, which takes about ~200 fs.

Faraday Discussions

PAPER

Non-negative matrix analysis for effective feature extraction in X-ray spectromicroscopy

Rachel Mak,[a] Mirna Lerotic,[b] Holger Fleckenstein,[c] Stefan Vogt,[d] Stefan M. Wild,[e] Sven Leyffer,[e] Yefim Sheynkin[f] and Chris Jacobsen*[dag]

Received 27th February 2014, Accepted 28th April 2014

DOI: 10.1039/c4fd00023d

X-Ray absorption spectromicroscopy provides rich information on the chemical organization of materials down to the nanoscale. However, interpretation of this information in studies of "natural" materials such as biological or environmental science specimens can be complicated by the complex mixtures of spectroscopically complicated materials present. We describe here the shortcomings that sometimes arise in previously-employed approaches such as cluster analysis, and we present a new approach based on non-negative matrix approximation (NNMA) analysis with both sparseness and cluster-similarity regularizations. In a preliminary study of the large-scale biochemical organization of human spermatozoa, NNMA analysis delivers results that nicely show the major features of spermatozoa with no physically erroneous negative weightings or thicknesses in the calculated image.

1   Introduction

Images let us see what is present in a material, while spectra let us understand what we see. Combining the two in spectromicroscopy (also known as spectrum imaging, or hyperspectral imaging) provides rich data on the composition of complex materials, whether applied to electron energy loss in electron microscopy,[1,2] X-ray emission spectroscopy with X-ray excitation,[3] or electron excitation,[4] infrared microscopy,[5,6] or X-ray absorption microscopy.[7,8] The challenge we address here involves the interpretation of these data, which is required in order

[a]Department of Physics & Astronomy, Northwestern University, 2145 Sheridan Road, Evanston, IL 60208, USA. E-mail: rachel.mak@u.northwestern.edu

[b]2nd Look Consulting, Room 1702, 17/F, Tung Hip Commercial Building, 248 Des Voeux Road, Hong Kong

[c]Center for Free-Electron Laser Science (CFEL), DESY, Notkestrasse 85, 22607 Hamburg, Germany

[d]Advanced Photon Source, Argonne National Laboratory, 9700 South Cass Avenue, Argonne, IL 60439, USA

[e]Mathematics and Computer Science Division, Argonne National Laboratory, 9700 South Cass Avenue, Argonne, IL 60439, USA

[f]Department of Urology, Stony Brook Medicine, Stony Brook, NY 11794-8093, USA

[g]Chemistry of Life Processes Institute, Northwestern University, 2170 Campus Drive, Evanston, IL 60208, USA

to go from observation to understanding. With spectroscopy of pure, uniform substances there exists a long and rich tradition of understanding observed spectra based on calculations of various electron or phonon interactions in the substance (see, for example, Stöhr[9]). However, microscopy is used to study materials including heterogeneous mixtures and reactive phases on fine spatial scales, and in images of 10^5–10^7 pixels. It is clearly impractical to carry out a painstaking investigation of the spectrum of each pixel on its own. Instead, one can hope to find a reduced set S of spectra that, when combined, can reproduce the spectrum observed in any one pixel. One can then carry out analysis on this smaller set of spectra or compare them to spectral "standards" of materials expected to be present in the specimen. We describe here an approach to carrying out this analysis based on a non-negative matrix approximation (NNMA, also referred to in the literature as NMF),[10] comparing it with previous methods we have developed (e.g., cluster analysis), and showing its utility for imaging chemical states in complex materials such as human sperm.

In X-ray spectromicroscopy, one obtains transmission images $I(x, y, E)$ at a series of positions (x, y) and N different photon energies E. By knowing the incident flux $I_0(E)$, one can determine an optical density $D(x, y, E) = -\ln[I(x, y, E)/I_0(E)]$, which is linear in the thickness t of the absorbing material in the beam direction because of the Lambert-Beer law of $I = I_0 \exp[-\mu(E)t]$. In this expression, $\mu(E)$ is a photon-energy-dependent linear absorption coefficient, which in principle can be calculated from quantum mechanics, and which in practice is often obtained from tabulations of absorption per element and per energy.[11] Missing from these tabulations are the details of $\mu(E)$ in the vicinity of an X-ray absorption edge: rather than reaching the threshold energy to excite and remove a core-level electron from an isolated atom, one instead reaches an energy where an atom's electron can be promoted into a state with an energy only a few electron volts away from the Fermi energy. Since these near-vacuum energy states are strongly affected by the nature of the atom's chemical bonds,[9] spectromicroscopy using near-edge X-ray absorption fine structure (NEXAFS) or X-ray absorption near-edge structure (XANES) provides a way to image the element-specific chemical bond distributions in a complex material.

Our challenge is that what has been measured is simply the optical density $D(x, y, E)$, but we would like to interpret it as a product of absorption spectra $\mu(E, S)$ from a set S with S spectroscopically distinguishable components and a set of thickness maps or weighting images $t(S, x, y)$ that show how much of each spectrum is present at each pixel. If we do not seek to find any spatial correlation of spectral responses (i.e., we do not assume the spectral response of any given pixel to be correlated with that of its neighbors), we can flatten the two-dimensional (x, y) coordinates and use a one-dimensional coordinate p to represent the position of each pixel. This is also generalizable to 3D tomographic spectromicroscopy data.[12] We are therefore left with a matrix equation for our desired analysis of

$$\mathbf{D}_{N \times P} = \boldsymbol{\mu}_{N \times S}\, \mathbf{t}_{S \times P}, \tag{1}$$

where N denotes the number of photon energy indices and P the number of pixels.[13] Our goal is to find the set of spectra $\boldsymbol{\mu}_{N \times S}$ that describes all the significant variations in the data. The absorption spectra $\boldsymbol{\mu}_{N \times S}$ should be non-negative (since negative absorption would imply that the material is adding energy to the transmitted beam instead of absorbing energy from it); the thickness or weighting maps $\mathbf{t}_{S \times P}$ should likewise be non-negative because of the additive nature of the densities of the materials in the sample. Because $\mathbf{D}_{N \times P}$ measures the optical density $-\ln(I/I_0)$, which is always non-negative (barring errors in the incident flux I_0 normalization), it should be possible, in theory, to find non-negative $\boldsymbol{\mu}_{N \times S}$ and $\mathbf{t}_{S \times P}$ such that eqn (1) is satisfied.

The problem of analyzing the measured data $\mathbf{D}_{N \times P}$ in terms of a set of spectra $\boldsymbol{\mu}_{N \times S}$ has been the subject of numerous multivariate statistical analysis approaches in energy loss electron microscopy[14,15] and in infrared spectromicroscopy.[5,16] In X-ray spectromicroscopy, approaches using spectral standards or hand-defined regions assumed to be of uniform, pure composition have allowed one to obtain a set of S spectra $\tilde{\boldsymbol{\mu}}_{N \times S}$ from which thickness maps $\mathbf{t}_{S \times P}$ can be obtained[17,18] by using the singular value decomposition (SVD) for matrix inversion, yielding

$$\mathbf{t}_{S \times P} = \tilde{\boldsymbol{\mu}}^{\dagger}_{S \times N} \mathbf{D}_{N \times P}, \qquad (2)$$

where $\tilde{\boldsymbol{\mu}}^{\dagger}_{S \times N}$ is the pseudo-inverse of $\tilde{\boldsymbol{\mu}}_{S \times N}$.

Approaches for understanding more complex samples in X-ray microscopy have included the use of principal component analysis (PCA)[19,20] to identify a limited or significant basis set $\bar{S}$ of orthonormal spectral signatures. However, SVD inversion does not guarantee a non-negative thickness map $\mathbf{t}_{S \times P}$, and PCA can produce a basis set $\tilde{\boldsymbol{\mu}}_{N \times S}$ that includes both positive and negative spectral values. Therefore, these approaches do not satisfy the non-negative condition of our desired solution described in eqn (1).

2   Cluster analysis and negative values

Although PCA does not provide a set of spectra that are individually interpretable as positive absorption spectra of separate materials present in the specimen, it *does* provide a well-organized and reduced-dimensionality search space for cluster analysis[13,21] as a way of finding pixels with similar spectra. Once the clusters are found, the spectra calculated from each cluster center provide a set $\tilde{\boldsymbol{\mu}}_{N \times S}$ for calculation of thickness weightings $\mathbf{t}_{S \times P}$ according to eqn (2). Cluster analysis has proven useful for a variety of applications including soil and environmental analysis;[22,23] however, it is also observed to yield some regions with slightly negative values in the thickness maps $\mathbf{t}_{S \times P}$, which are unphysical and thus represent limitations in the analysis.

To understand the way in which non-negative errors can arise in cluster analysis, we consider a simple example of a specimen with uniform thickness and a continuum of composition starting with 100% of material A, which is strongly absorptive at energy E_1, and ending with 100% of material B, which is strongly absorptive at energy E_2 (see Fig. 1). A scatterplot of the location of individual pixels based on their responses at the energies $\{E_1, E_2\}$ is shown schematically in Fig. 2A. If these pixels are organized into two clusters, the groupings shown in Fig. 2B will be the result, where the vectors shown point to the center of the respective cluster centers; this will give rise to a set of spectra $\tilde{\boldsymbol{\mu}}_{N \times S}$ (with $S = 2$ in this example) from which one can calculate thickness maps according to eqn (2). However, consider

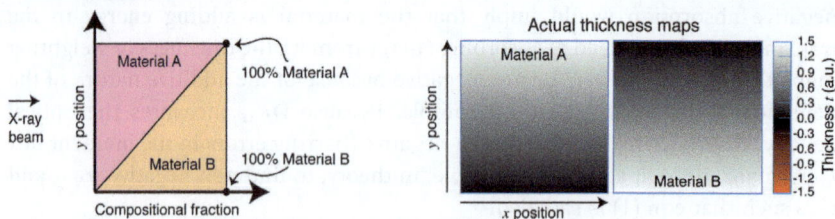

Fig. 1 Two-material test specimen to illustrate compositional mapping approaches. We assume that the specimen is a three-dimensional block comprises two separate materials A and B with a continuous variation between the two. The compositional variation is shown at left: the view is along the x direction (into the page), while the X-ray beam is traveling along the z direction. The thickness maps associated with each of the separate materials are shown at right.

the case of a pixel that is far from the median in composition, such as the upper left one in Fig. 2C. The only possible thickness map $\mathbf{t}_{S \times P}$ or weighting map of the cluster spectra $\boldsymbol{\mu}_{N \times S}$ able to reach that point is one that involves a negative weighting of one of the cluster spectra; that is, one that produces negative values in the thickness map $\mathbf{t}_{S \times P}$, which are unphysical in our desired interpretation of the measured positive optical densities of eqn (1). Of course, if the variation among spectral response of the pixels assigned to a cluster is small, these errors can be negligibly small; however, as Fig. 2 shows, there is no guarantee that cluster analysis will produce a thickness map with few negative pixels. Indeed, this negative thickness error is exactly what is observed in an actual cluster analysis of data of the form of Fig. 1, as shown in Fig. 3.

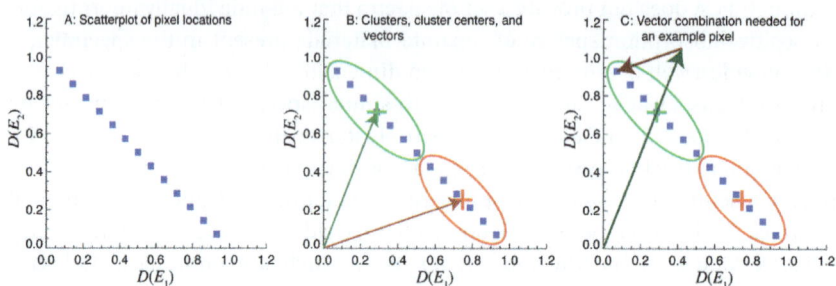

Fig. 2 Illustration of how cluster analysis can give rise to negative (and thus unphysical) values of $\mathbf{D}_{N \times P} = \boldsymbol{\mu}_{N \times S} \, \mathbf{t}_{S \times P}$. The figure at the left (A) shows a scatterplot of pixels from a continuously varying material combination as shown in Fig. 1, assuming that the two materials have opposing responses at energies E_1 and E_2. The middle figure (B) illustrates how these pixels will be grouped into two clusters; without other information, it would be natural to describe each cluster by the spectrum corresponding to the cluster center (marked with the red and green +). The figure at the right (C) shows how recreation of the spectrum of a pixel near one of the axes in this illustration would require a combination of positive weighting of one cluster spectrum (green in this case) but a *negative* weighting of the other cluster spectrum (red in this case); this would give rise to negative values in the thickness or weighting matrix $\mathbf{t}_{S \times P}$, implying negative absorption of the X-ray beam, which is unphysical.

Cluster segmentation | Cluster thickness maps | Cluster error map

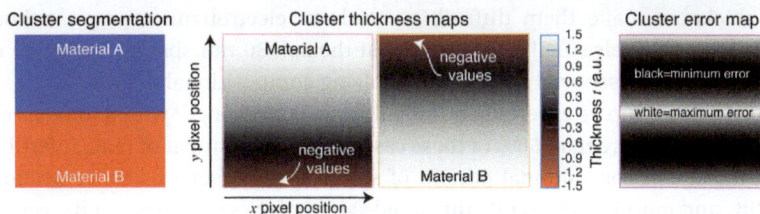

Fig. 3 Cluster analysis applied to the $\mathbf{D}_{N \times P}$ simulated data of Fig. 1 using two different spectra. The "Cluster segmentation" image at left shows how the simulated specimen is correctly classified as being made up of $S = 2$ distinct spectra $\boldsymbol{\mu}_{N \times S}$, while the "Cluster thickness maps" in the center show the weightings or thickness maps $\mathbf{t}_{S \times P}$ that result. Careful examination of these cluster thickness maps reveals that they include (unphysical) negative values, and the "Cluster error map" at right (which represents the error $|\mathbf{D}_{N \times P} = \boldsymbol{\mu}_{N \times S} \, \mathbf{t}_{S \times P}|$ as a grey-scale image) shows that both the extrema and midpoint mixtures are not reconstructed with full accuracy using cluster analysis.

A better solution to the simple example shown in Fig. 2 would locate the component vectors not in the center of clusters, but closer to extrema points so that the full range of spectroscopic variations can be represented; that is, in this example, the component vectors would point along (or close to) the respective axes rather than at the cluster centers. This is simple to arrive at for the deliberately constructed example of Fig. 2; but with nontrivial data distributions in multiple dimensions and in the presence of noise, the problem becomes much more challenging. We therefore present an alternative spectromicroscopy analysis approach in Sec. 4.

3 Application: biochemical analysis of spermatozoa

As an example application of X-ray spectromicroscopy analysis to a complicated, real-world specimen, we consider the case of human spermatozoa. Sperm are compact cells with tightly-packed and well-segregated materials in their head and a long flagellum that allows them to move through fluid (Fig. 4). Their density and

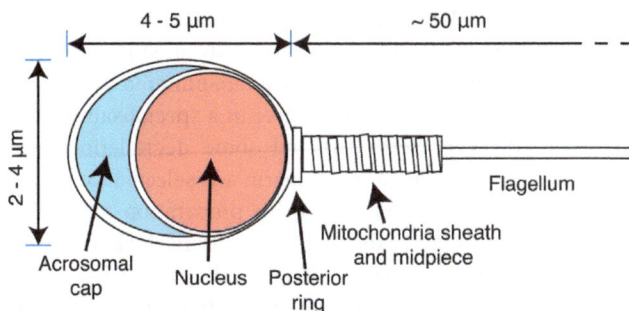

Fig. 4 Diagram of a human spermatozoa. The enzymes involved in penetrating the egg are in the acrosomal cap, while the nucleus contains the DNA tightly packed with histone proteins. The flagellar motor is within the posterior ring.

total thickness make them difficult to study in electron microscopy without sectioning, and their small size means that the sub-50 nm spatial resolution of X-ray spectromicroscopy is helpful for resolving compositional details.[24]

One in four couples experiences difficulties in conceiving,[25] with a male factor contributing in more than 50% of these cases.[26] The assessment of male infertility relies mostly on conventional criteria of sperm quality, such as concentration, motility, and morphology. While threshold values of these metrics can be used to classify men as subfertile, of indeterminate fertility, or fertile, none of the measures are reliable diagnostics of infertility.[27] This fact indicates that the sperm of infertile men may have hidden abnormalities in the composition of their nuclei. DNA damage above a certain threshold appears to impair fertilization and embryo development,[28,29] but little is known about the etiologies of sperm DNA damage and its full impact on human reproduction. Light microscopy does not deliver valuable information on sperm DNA or chromatin abnormalities, while bulk chemical measurements average over many morphologies and are not sensitive to individual spermatozoa. Flow cytometry can correlate sperm morphology with total DNA content,[30] but it is still useful to visualize overall biochemical organization at higher resolution and without relying on a single biochemical marker. X-ray spectromicroscopy insights into the correlation between sperm morphology and abnormal DNA or protein distributions could lead to a better understanding of the basis for light microscopy selection of one abnormality over another for *in vitro* fertilization in cases where no sperm are present with normal morphology.

Several investigators have carried out high-resolution soft X-ray microscopy studies of sperm.[31–34] Zhang *et al.* have used carbon near-absorption-edge X-ray spectromicroscopy for compositional mapping of hamster, rat, and bull sperm.[17] They acquired spectra of thin film standards of proteins protamine 1 and 2 and of calf thymus DNA; a species-weighted ratio of the protamine spectra was used along with the DNA spectrum to form a two-spectrum matrix μ, which was then inverted by using the SVD in order to yield thickness maps (eqn (2)) and estimate protein-to-DNA ratios. The results suggested that protamine content is independent of protamine 2 gene expression, but they did not allow one to discover other variations in biochemical organization because the analysis assumed a composition consisting of just the three targeted biochemicals. We therefore wish to consider analysis methods that are not based on such limiting assumptions.

We have used ejaculated sperm obtained *via* masturbation from randomly selected unidentified donors at the Andrology Lab at Stony Brook University Hospital. Fresh ejaculate was washed in phosphate-buffered saline to dilute the optically thick semen and then was imaged wet in a special sample holder,[35] air dried, or freeze dried. Wet sperm suffered some degradation during X-ray microscopy measurements, so an air-dried sperm was selected for the data shown here[24] since in images of dozens of sperm this preparation method seemed to preserve sperm morphology better than what we observed when plunge-freezing in liquid ethane followed by freeze drying (perhaps because of ice crystal formation in the dense sperm head during plunge-freezing). Images were taken with a scanning transmission X-ray microscope developed by us at Stony Brook University[36] and formerly operated on an undulator beamline X1A1 at the National Synchrotron Light Source at Brookhaven National Laboratory (the version of microscope used in these studies was upgraded to include laser interferometer control of the scanning stage[24, 37,38]). Images were acquired with 100 nm pixel size and at 133 photon energies across the carbon X-ray absorption near-edge spectroscopy region from 283.8 to 291.6 eV.

4   Non-negative matrix approximation methods

Cluster analysis based on data orthogonalized and reduced by using PCA is rapid and useful for analyzing complex data[13,21-23] although it can return negative values as described above. There is also a wide range of other productive approaches for spectromicroscopy analysis.[4,5,14,15,39] However, we restate our fundamental requirement, which is to find an approach that is constrained by the physics of X-ray absorption to yield only non-negative values for the matrices in the expression of eqn (1) of $\mathbf{D}_{N \times P} = \boldsymbol{\mu}_{N \times S}\, \mathbf{t}_{S \times P}$. This is precisely the requirement satisfied by non-negative matrix factorization, an analysis approach first explored by Paatero and Tapper[40] and later implemented with considerable notice by Lee and Seung.[10] We describe our implementation of NNMA analysis for X-ray absorption spectromicroscopy, realizing that the same approaches can be used for other types of spectral analysis,[41,42] spectrum imaging,[43] and hyperspectral imaging methods.[44,45]

The approach of Lee and Seung[10] for face recognition was to use a multiplicative update algorithm for non-negative matrix factorization of data in the form of eqn (1), with minimization of the basic data-matching cost function $F_0(\boldsymbol{\mu}, \mathbf{t})$ of (dropping matrix subscripts for simplicity):

$$F_0(\boldsymbol{\mu},\mathbf{t}) = \|\mathbf{D} - \boldsymbol{\mu}\mathbf{t}\|_2^2. \tag{3}$$

The Lee and Seung algorithm in our notation initializes with random non-negative values for the matrices $\boldsymbol{\mu}$ and $\mathbf{t}$ and then applies iterative updates[46] using multiplicative rules of the form

$$t \leftarrow t\, \frac{(\boldsymbol{\mu}^{\mathrm{T}}\mathbf{D})}{(\boldsymbol{\mu}^{\mathrm{T}}\boldsymbol{\mu}\mathbf{t})} \tag{4}$$

$$\boldsymbol{\mu} \leftarrow \boldsymbol{\mu}\, \frac{(\mathbf{D}\mathbf{t}^{\mathrm{T}})}{(\boldsymbol{\mu}\mathbf{t}\mathbf{t}^{\mathrm{T}})}, \tag{5}$$

where the multiplications and divisions not in parentheses are taken componentwise, until a minimum of the data-matching cost function F_0 of eqn (3) is reached (or, in practice, until F_0 falls below some predetermined threshold).

Minimizing the cost function to make the NNMA factorization of $\boldsymbol{\mu}\mathbf{t}$ as close as possible to the optical density data $\mathbf{D}$ is necessary but not sufficient for achieving a clear, easy-to-interpret analysis of X-ray spectromicroscopy data. With this basic cost function as the only consideration, one can miss several desired features of a useful solution; furthermore, the minimizer will not be unique (since any positive scaling between $\boldsymbol{\mu}$ and $\mathbf{t}$ would achieve the same cost function value). One approach is to introduce other considerations such as spectral smoothness as constraints, but this mixed strategy of optimization for some criteria, and constraints for others, can lead to very slow convergence.[24]

4.1 NNMA regularization

One method to narrow and refine the search space for NNMA is to introduce regularizations in addition to the basic cost function minimization. Regularization is one way to incorporate additional information we might have about the data into the overall cost function to be minimized. In this way, we find a balance between the error minimization from data-matching, and a good fit to data-modeling. Each regularization is controlled by a continuous regularization parameter λ in the cost function

$$F\left(\mu, t\right) = F_0\left(\mu, t\right) + \lambda_\mu J_\mu\left(\mu\right) + \lambda_t J_t(t), \tag{6}$$

where J_μ, J_t are the regularizers and λ_μ, λ_t are the regularization parameters applied to the spectral μ and thickness or weighting t matrices, respectively.

From a machine-learning perspective, λ represents the trade-off between errors in data-matching and the complexity of the model.[47] For small λ, the errors become smaller at the cost of not accurately modeling the data. For larger λ, the data-matching errors become reduced in importance relative to other desired characteristics of the data model. We consider here two regularization schemes for desired characteristics of our solution (our model): sparseness and similarity to cluster spectra.

Sparseness: In many X-ray spectromicroscopic datasets, the t matrix is expected to be sparse—each pixel would contain at most a few components (column sparseness), and each component would be favored to show up in only a small subset of pixels (row sparseness)—so that many entries in t would be zero or close to zero. The typical regularizer to model the sparseness of t is the one-norm:[48] $\|t\|_1 = \sum_{k,p} t_{k,p}$. The cost function to be minimized becomes

$$F(\mu, t) = \|D - \mu t\|_2^2 + \lambda_t \|t\|_1, \tag{7}$$

and the addition to the update rule for t is

$$\lambda_t \frac{\partial J_t}{\partial t} = \lambda_t \text{ ones}_{S \times P}, \tag{8}$$

so that eqn (4) becomes

$$t \leftarrow t \frac{(\mu^T D)}{(\mu^T \mu t + \lambda_t)}. \tag{9}$$

With spectral imaging, one of the consequences of incorporating sparseness into the model would be to create more clearly separable components, as seen in the reconstructed thickness maps t. Fig. 5 compares the results of applying NNMA without any regularization and one with sparseness regularization ($\lambda_t = 0.7$). Although we now have more distinguishable thickness maps, the reconstructed spectra μ do not resemble observed X-ray absorption spectra. This result is not surprising given that NNMA has no expectation of what the reconstructions should look like as long as the cost function is minimized.

Cluster similarity: Based on the above, we wish to also include a regularization to increase the similarity of solutions to observed X-ray absorption spectra. Since cluster analysis yields a set of spectra $\mu_{cluster}$ averaged from spectroscopically

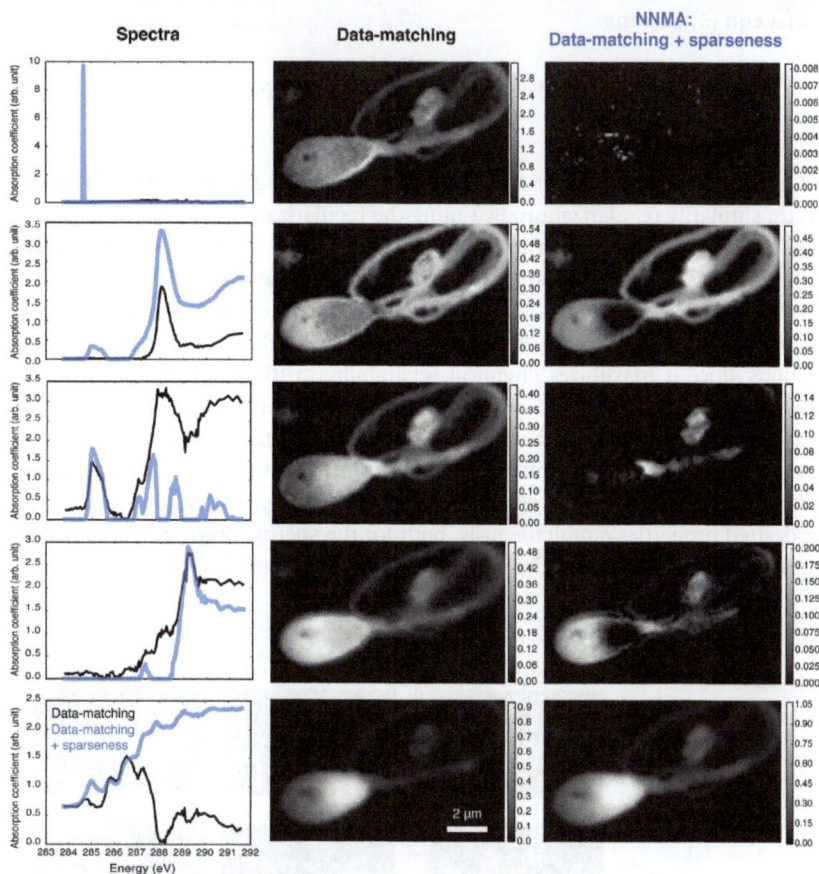

Fig. 5 Comparison between NNMA without and with sparseness regularization. The first column compares the reconstructed spectra μ. The second column shows the NNMA reconstructed thickness maps t of the sperm dataset without any sparseness regularization, while the third column shows NNMA reconstructed maps with sparseness regularization ($\lambda_t = 0.7$). In the last column, the components appear better separated than without sparseness regularization. However, some of the reconstructed spectra do not resemble observed X-ray absorption spectra; approaches to address this are illustrated in Figs. 6 and 7.

similar pixels, its spectra provide a good basis both for the starting solution of an optimization procedure and for a "similarity regularizer" $J_{\mu_{\mathrm{sim}}}$ to penalize reconstructions that deviate far from the input cluster spectra:

$$\lambda_{\mu_{\mathrm{sim}}} J_{\mu_{\mathrm{sim}}} = \lambda_{\mu_{\mathrm{sim}}} \| \mu - \mu_{\mathrm{cluster}} \|_2^2 \tag{10}$$

$$= \lambda_{\mu_{\mathrm{sim}}} \sum_{k=1}^{S} \sum_{n=1}^{N} \left(\mu_{n,k} - \mu_{\mathrm{cluster}\, n,k} \right)^2. \tag{11}$$

The addition to the update rule for μ is

$$\lambda_{\mu_{\mathrm{sim}}} \frac{\partial J_{\mu_{\mathrm{sim}}}}{\partial \mu} = \lambda_{\mu_{\mathrm{sim}}} 2(\mu - \mu_{\mathrm{cluster}}), \tag{12}$$

so that eqn (5) becomes

$$\mu \leftarrow \mu \, \frac{(\mathbf{D}\mathbf{t}^{\mathrm{T}})}{(\mu \mathbf{t}\mathbf{t}^{\mathrm{T}} + 2\lambda_{\mu_{\mathrm{sim}}}(\mu - \mu_{\mathrm{cluster}}))}. \tag{13}$$

Fig. 6 compares reconstructions from cluster analysis against NNMA with cluster spectra similarity regularization. By tuning the regularization parameter $\lambda_{\mu_{\mathrm{sim}}}$ to be high, we can obtain reconstructed spectra that match closely with those from cluster analysis, while eliminating the negative regions in the thickness maps.

While cluster similarity adds an important bias toward the properties of actual observed X-ray absorption spectra, sparseness is still a desirable property for our solutions since it maximizes chemical separability in the weighting or thickness maps. Fortunately, one can seek a simultaneous minimum of the three cost components together: the data-matching cost of eqn (3), the sparseness regularizer of eqn (7), and the cluster similarity regularizer of eqn (10) by using the combined updates of eqn (9) and (13). As shown in Fig. 7, this at last gives solutions that satisfy our desired properties simultaneously, both for the sperm spectromicroscopy data shown here and in other studies to be described elsewhere.

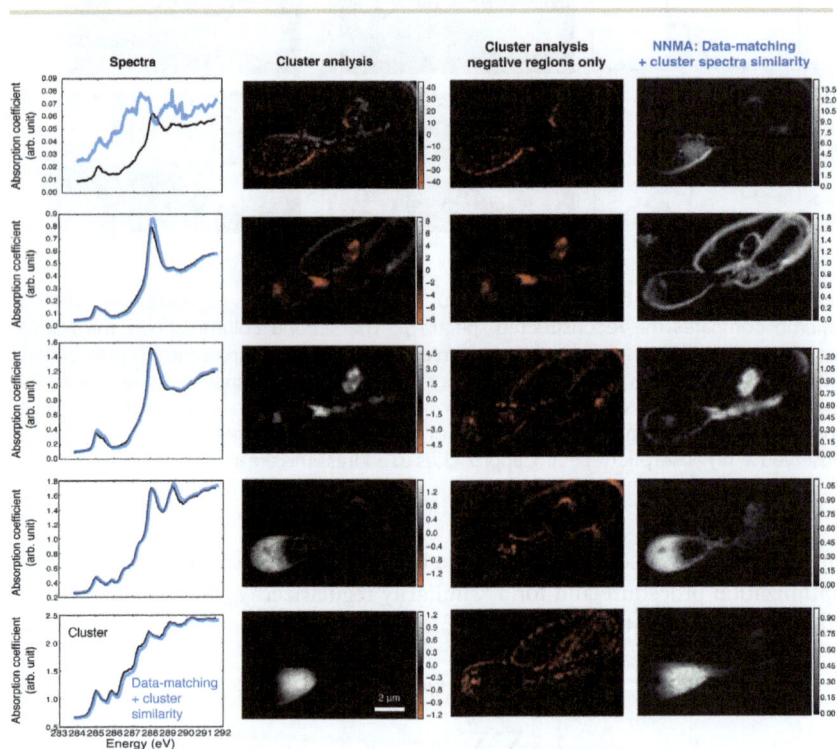

Fig. 6 Comparison between cluster analysis and NNMA with a cluster spectra similarity regularization scheme as described in Section 4.1. By tuning the regularization parameter $\lambda_{\mu_{\mathrm{sim}}}$ to be high (in this case, $\lambda_{\mu_{\mathrm{sim}}} = 100$), we can obtain NNMA reconstructed spectra that are similar to those from cluster analysis (first column). At the same time, the negative regions from the cluster thickness maps (second column) are eliminated, as seen in the NNMA reconstructed maps (fourth column). To highlight the negative regions in the cluster analysis thickness maps, only these regions are shown in the third column.

4.2   Selection of number of spectroscopically distinguishable components *S*

An important parameter in NNMA analysis (as well as in cluster analysis) is the selection of the number of spectroscopically distinguishable components S to seek. If S is too small, we will arrive at a basis set that is not able to reproduce all the important variations in the data; if S is too large, we may have simply analyzed variations due to noise from photon statistics or other sources. In PCA, the eigenvalues of the covariance matrix $\mathbf{Z} = \mathbf{DD}^T$ can provide a good

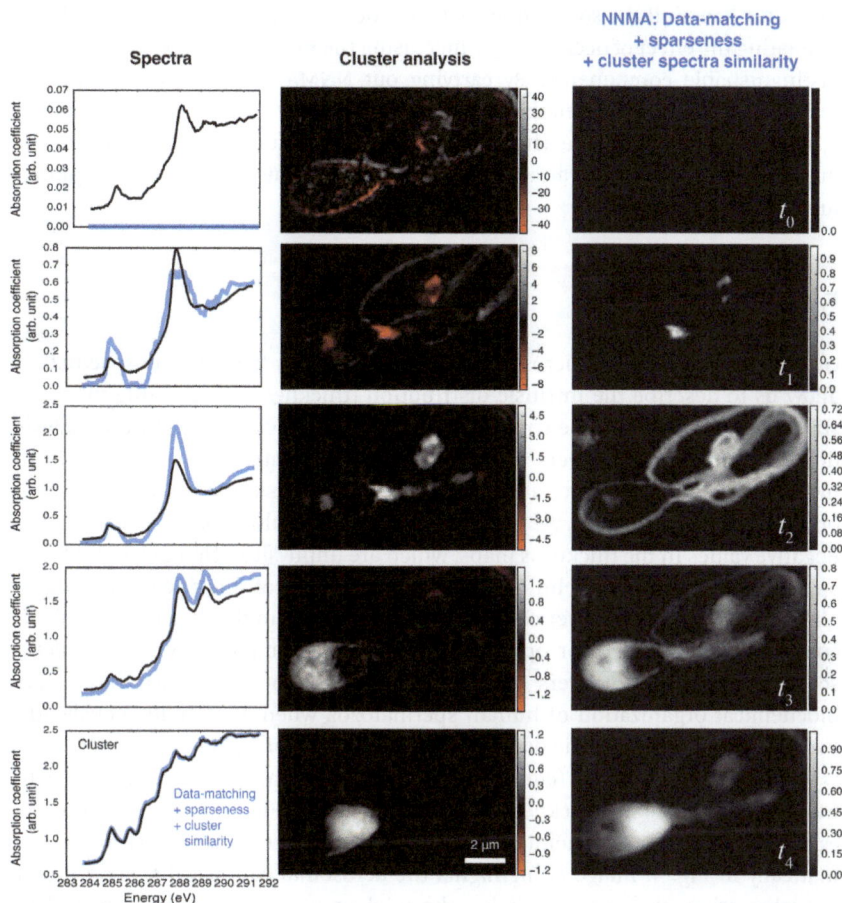

Fig. 7 X-ray spectromicroscopy analysis results obtained by the cluster analysis approach[21] and NNMA with data-matching (eqn (3)), sparseness (eqn (7)), and cluster similarity (eqn (10)) regularizers used in combination. As suggested in Fig. 2 and shown for this data set in Fig. 6, cluster analysis produces maps with negative weightings for some regions, which are not allowed by the physics of X-ray absorption, whereas the NNMA solution using data-matching, sparseness, and cluster similarity delivers an analysis result with recognizable X-ray absorption spectra and positive weightings or thickness maps, which nicely illustrate the large-scale biochemical organization of sperm. Image t_3 highlights the acrosomal cap, flagellar motor, and mitochondrion; image t_4 highlights the nucleus where histones are bound to DNA; image t_2 highlights the lipid membrane and flagellum; and image t_1 isolates a high-density area in the flagellar motor with some combination of chemical sensitivity and experimental absorption saturation limits. The regularization weightings used for the NNMA analysis were $\lambda_{\mu_{sim}} = 10$ and $\lambda_t = 0.5$.

estimate, since they often drop sharply from the first eigenvalue down to a point where subsequent eigenvalues decrease only slightly, indicating a transition from variations in significant signals to small variations due to different measures of the same noise factors.[49] In cluster analysis, we have used this "knee" in the eigenvalue curve to estimate the number of significant components $\bar{S}$ in the sample,[21] although in practice it has been found to be important to manually examine the analysis result. The number of clusters to seek would then be $\bar{S}$.

Since NNMA analysis involves the cost function F_0 ($\boldsymbol{\mu}$, $\mathbf{t}$) of eqn (3) that measures how well the solution $\boldsymbol{\mu}\mathbf{t}$ matches the data $\mathbf{D}$, we have a good basis for evaluating the effect of decreasing or increasing the number S of spectroscopically distinguishable components. By carrying out NNMA analysis with a range of values for S, we can see when the error F_0 ($\boldsymbol{\mu}$, $\mathbf{t}$) no longer decreases as a function of S; we can similarly examine when decreases to S are insufficient to capture all the important spectroscopic variations in the sample. This topic will be explored further in future work.

5   Discussion

Our goal in X-ray spectromicroscopy analysis is to find a "basis set" of spectra that allow us to describe the intrinsic distribution (thickness or weighting maps) of spectroscopically resolvable components in the specimen. While cluster analysis does this rapidly and delivers spectra that closely resemble those observed from individual pixels in the spectral image set (because they simply average a subset of observed spectra together), we have shown that the resulting basis set can lead to negative values in the thickness maps, which are unphysical. Non-negative matrix approximation analysis techniques provide a path out of this dilemma and also allow us to incorporate other characteristics desired from the data in a combined cost function approach for optimization. For our example X-ray spectromicroscopy data, NNMA delivers results consistent with the known large-scale biochemical organization of human spermatozoa when we simultaneously add two regularizers to the basic data-matching condition of eqn (3): sparseness (eqn (7)), and cluster similarity (eqn (10)). With this combined regularizer approach, we are able to obtain thickness maps and spectra (Fig. 7) that highlight the expected large-scale biochemical organization of spermatozoa as shown schematically in Fig. 4: image t_3 highlights the acrosomal cap, flagellar motor, and mitochondrion; image t_4 highlights the nucleus where histones are bound to DNA; image t_2 highlights the lipid membrane and flagellum; and image t_1 isolates a high-density area in the flagellar motor with some combination of chemical sensitivity and experimental absorption saturation limits. The spectroscopic peaks observed in Fig. 7 can be interpreted by careful comparison between theoretical calculations and experimental measurements of organic molecular assemblies such as amino acids[50] and manufactured polymers,[51] although detailed discussion is beyond the scope of the present work.

An exploratory version of the NNMA analysis approach described here is implemented in a Python open source code† called MANTiS[52] developed by 2nd

---

† http://code.google.com/p/spectromicroscopy

Look Consulting; a more refined interface to NNMA analysis is planned for an upcoming release of MANTiS. For the data shown here, the combined cost function converged to a minimum over about 10^4 iterations taking about 10 min on a laptop computer. These results show the potential of NNMA analysis on complicated data.

Acknowledgements

We thank the U.S. Department of Energy (DOE) Office of Science for support of this work under contract DE-AC02-06CH11357 to Argonne National Laboratory. Earlier exploration of NNMA as applied to sperm spectromicroscopy analysis was supported by the National Institutes of Health under grant R01 EB-000479; RM was supported in part by grant R01 GM-104530.

References

1 C. Jeanguillaume and C. Colliex, *Ultramicroscopy*, 1989, **28**, 252–257.

2 J. A. Hunt and D. B. Williams, *Ultramicroscopy*, 1991, **38**, 47–73.

3 C. J. Sparks, Jr, in *Synchrotron Radiation Research*, ed. H.Winick and S. Doniach, Plenum Press, New York, 1980, pp. 459–512.

4 P. G. Kotula, M. Keenan and J. R. Michael, *Microsc. Microanal.*, 2003, **9**, 1–17.

5 P. Lasch, W. Wäsche, W. J. McCarthy, G. J. Müller and D. Naumann, *Proc. SPIE–Int. Soc. Opt. Eng.*, 1998, **3257**, 187–198.

6 M. Diem, S. Boydston-White and L. Chiriboga, *Appl. Spectrosc.*, 1999, **53**, 148A–161A.

7 H. Ade, X. Zhang, S. Cameron, C. Costello, J. Kirz and S. Williams, *Science*, 1992, **258**, 972–975.

8 C. Jacobsen, G. Flynn, S. Wirick and C. Zimba, *J. Microsc.*, 2000, **197**, 173–184.

9 J. Stöhr, *NEXAFS Spectroscopy, Springer-Verlag*, 1992, **vol. 25**.

10 D. D. Lee and H. S. Seung, *Nature*, 1999, **401**, 788–791.

11 B. L. Henke, E. Gullikson and J. C. Davis, *At. Data Nucl. Data Tables*, 1993, **54**, 181–342.

12 G. A. Johansson, T. Tyliszczak, G. E. Mitchell, M. H. Keefe and A. P. Hitchcock, *J. Synchrotron Radiat.*, 2007, **14**, 395–402.

13 M. Lerotic, C. Jacobsen, J. Gillow, A. Francis, S. Wirick, S. Vogt and J. Maser, *J. Electron Spectrosc. Relat. Phenom.*, 2005, **144–147**, 1137–1143.

14 N. Bonnet, *J. Microsc.*, 1998, **190**, 2–18.

15 N. Bonnet, N. Brun and C. Colliex, *Ultramicroscopy*, 1999, **77**, 97–112.

16 P. Lasch, M. Boese, A. Pacifico and M. Diem, *Vib. Spectrosc.*, 2002, **28**, 147–157.

17 X. Zhang, R. Balhorn, J. Mazrimas and J. Kirz, *J. Struct. Biol.*, 1996, **116**, 335–344.

18 I. N. Koprinarov, A. P. Hitchcock, C. T. McCrory and R. F. Childs, *J. Phys. Chem. B*, 2002, **106**, 5358–5364.

19 P. L. King, R. Browning, P. Pianetta, I. Lindau, M. Keenlyside and G. Knapp, *J. Vac. Sci. Technol., A*, 1989, 7, 3301–3304.

20 A. Osanna and C. Jacobsen, *X-ray Microscopy: Proceedings of the Sixth International Conference (American Institute of Physics Conference Proceedings)*, 2000, pp. 350–357.

21 M. Lerotic, C. Jacobsen, T. Schäfer and S. Vogt, *Ultramicroscopy*, 2004, **100**, 35–57.

22 G. Mitrea, J. Thieme, P. Guttmann, S. Heim and S. Gleber, *J. Synchrotron Radiat.*, 2007, **15**, 26–35.

23 J. Lehmann, D. Solomon, J. Kinyangi, L. Dathe, S. Wirick and C. Jacobsen, *Nat. Geosci.*, 2008, **1**, 238–242.

24 H. Fleckenstein, Ph.D. thesis, Stony Brook University, 2008.

25 L. Schmidt, K. Munster and P. Helm, *BJOG*, 1995, **102**, 978–984.

26 D. S. Irvine, *Hum. Reprod.*, 1996, **13**, 33–44.

27 D. S. Guzick, J. W. Overstreet, P. Factor-Litvak, *et al.*, *N. Engl. J. Med.*, 2001, **345**, 1388–1393.

28 A. Ahmadi and S. C. Ng, *J. Exp. Zool.*, 1999, **284**, 696–704.

29 C. Cho, H. Jung-Ha, W. D. Willis, E. H. Goulding, P. Stein, Z. Xu, R. M. Schultz, N. B. Hecht and E. M. Eddy, *Biol. Reprod.*, 2003, **69**, 211–217.

30 U. B. Hacker-Klom, W. Göhde, E. Nieschlag and H. M. Behre, *Hum. Reprod.*, 1999, **14**, 2506–2512.

31 B. W. Loo, Jr., S. Williams, S. Meizel and S. S. Rothman, *J. Microscopy*, 1992, **166**, RP5–RP6.

32 R. Balhorn, R. E. Braun, B. Breed, J. T. Brown, D. Evenson, J. M. Heck, J. Kirz, I. McNulty, W. Meyer-Ilse and X. Zhang, *X-ray Microscopy and Spectromicroscopy*, Berlin, 1998, pp. II–29–46.

33 T. Vorup-Jensen, T. Hjort, J. V. Abraham-Peskir, P. Guttmann, J. C. Jensenius, E. Uggerhøj and R. Medenwaldt, *Hum. Reprod.*, 1998, **14**, 880–884.

34 J. Abraham-Peskir, E. Chantler, C. McCann, R. Medenwaldt and E. Ernst, *Med. Sci. Res.*, 1998, **26**, 663–667.

35 U. Neuhäusler, C. Jacobsen, D. Schulze, D. Stott and S. Abend, *J. Synchrotron Radiat.*, 2000, **7**, 110–112.

36 M. Feser, T. Beetz, C. Jacobsen, J. Kirz, S. Wirick, A. Stein and T. Schäfer, *Soft X-Ray and EUV Imaging Systems II*, Bellingham, Washington, 2001, pp. 146–153.

37 M. Lerotić, Ph.D. thesis, Department of Physics and Astronomy, Stony Brook University, 2005.

38 B. Hornberger, Ph.D. thesis, Department of Physics and Astronomy, Stony Brook University, 2007.

39 A. de Juan, R. Tauler, R. Dyson, C. Marcolli, M. Rault and M. Maeder, *TrAC, Trends Anal. Chem.*, 2004, **23**, 70–79.

40 P. Paatero and U. Tapper, *Environmetrics*, 1994, **5**, 111–126.

41 V. P. Pauca, J. Piper and R. J. Plemmons, *Linear Algebra Appl.*, 2006, **416**, 29–47.

42 G. Buchsbaum and O. Bloch, *Vision Res.*, 2002, **42**, 559–563.

43 J. S. Lee, D. D. Lee, S. Choi and D. S. Lee, *In 3rd International Conference on Independent Component Analysis and Blind Signal Separation*, 2001, pp. 629–632.

44 S. A. Robila and L. G. Maciak, *Proceedings SPIE*, 2006, p. 63840F.

45 C.-Y. Liou and K.-D. Ou Yang, *International Conference on Neural Information Processing*, 2005, pp. 280–285.

46 D. D. Lee and H. S. Seung, *Advances in Neural Information Processing Systems*, 2001, pp. 556–562.

47 E. Alpaydin, *Introduction to Machine Learning*, MIT Press, 2nd edn, 2010.

48 R. Tibshirani, *J. R. Stat. Soc. B*, 1996, **58**, 267–288.

49 E. R. Malinowski, *Factor Analysis in Chemistry*, John H. Wiley & Sons, New York, 2nd edn, 1991.

50 K. Kaznacheyev, A. Osanna, C. Jacobsen, O. Plashkevych, O. Vahtras, H. Ågren, V. Carravetta and A. Hitchcock, *J. Phys. Chem. A*, 2002, **106**, 3153–3168.

51 O. Dhez, H. Ade and S. Urquhart, *J. Electron Spectrosc. Relat. Phenom.*, 2003, **128**, 85–96.

52 M. Lerotic, R. Mak, S. Wirick, F. Meirer and C. Jacobsen, *J. Synchrotron Radiat.*, 2014, DOI: 10.1107/S1600577514013964.

Faraday Discussions

ROYAL SOCIETY OF CHEMISTRY

PAPER

Femtosecond X-ray diffraction maps field-driven charge dynamics in ionic crystals

Michael Woerner,[*a] Marcel Holtz,[a] Vincent Juvé,[a] Thomas Elsaesser[a] and Andreas Borgschulte[b]

Received 28th February 2014, Accepted 17th March 2014

DOI: 10.1039/c4fd00026a

X-Ray diffraction provides insight into the distribution of electronic charge in crystals. Equilibrium electron distributions have been determined with high spatial resolution by recording and analysing a large number of diffraction peaks under stationary conditions. In contrast, transient electron densities during and after structure-changing processes are mainly unknown. Recently, we have introduced femtosecond X-ray powder diffraction from polycrystalline samples to determine transient electron density maps with a spatial resolution of 0.03 nm and a temporal resolution of 100 fs. In a pump–probe approach with a laser-driven tabletop hard X-ray source, optically induced structure changes are resolved in time by diffracting the hard X-ray probe pulses at different time delays from the excited powder sample and recording up to several tens of reflections simultaneously. Time-dependent changes of the atomic arrangement in the crystal lattice as well as modified electron densities are derived from the diffraction data. As a prototypical field-driven process, we address here quasi-instantaneous changes of electron density in $LiBH_4$, LiH and $NaBH_4$ in response to a non-resonant strong optical field. The light-induced charge relocation in $LiBH_4$ and $NaBH_4$ exhibits an electron transfer from the anion (BH_4^-) to the respective cation. The distorted geometry of the BH_4 tetrahedron in $LiBH_4$ leads to different contributions of the H atoms to electron transfer. LiH displays a charge transfer from Li to H, i.e., an increase of the ionicity of LiH in the presence of the strong electric field. This unexpected behavior originates from strong electron correlations in LiH as is evident from a comparison with quasi-particle bandstructures calculated within the Coulomb-hole-plus-screened-exchange (COHSEX) formalism.

1 Introduction

Ultrashort optical pulses allow for applying peak electric fields to a crystal which are comparable to or even beyond the fields valence electrons experience under

[a]Max-Born-Institut für Nichtlineare Optik und Kurzzeitspektroskopie, 12489 Berlin, Germany. E-mail: woerner@mbi-berlin.de; Fax: +49 30 6392 1489; Tel: +49 30 6392 1470

[b]EMPA, Swiss Federal Laboratories for Materials Testing and Research, Laboratory for Hydrogen and Energy, CH-8600 Dübendorf, Switzerland

equilibrium conditions. In this regime of light-matter interaction, field-driven processes of charge transport and nonlinear optical phenomena under non-per-turb- ative conditions are made accessible to experiments with atto- to femto-second time resolution. Field-driven coherent ballistic electron motions[1] and electron emission,[2] interband tunneling of electrons,[3] and light-driven charge relocations[4-6] are prototype phenomena. Experiments based on a variety of optical and electro-optical methods have focused on the ultrafast dynamics of electric polarizations and currents while the spatial aspects of the electronic response have remained mainly unexplored. X-ray methods with a femtosecond time resolution, in particular X-ray diffraction, provide access to spatial electron distributions during and after ultrafast processes[7-13] and, thus, hold a strong potential for unraveling field-driven processes in *space and time*.

A basic field-driven process in the ionic material $NaBH_4$ is illustrated in Fig. 1 where the ionic potentials of Na^+ and BH_4^- are shown together with the (cell-periodic) electron wavefunctions in the highest valence band (vb) and the lowest conduction band (cb) state. Under equilibrium conditions, the vb wavefunction $|\Psi_{vb}\rangle$ shows a high amplitude on the BH_4^- and a small amplitude on the Na^+ ion. A strong external field of an amplitude comparable to the interionic field of approximately 10^9 V m^{-1} distorts the ionic potentials and leads to the admixture

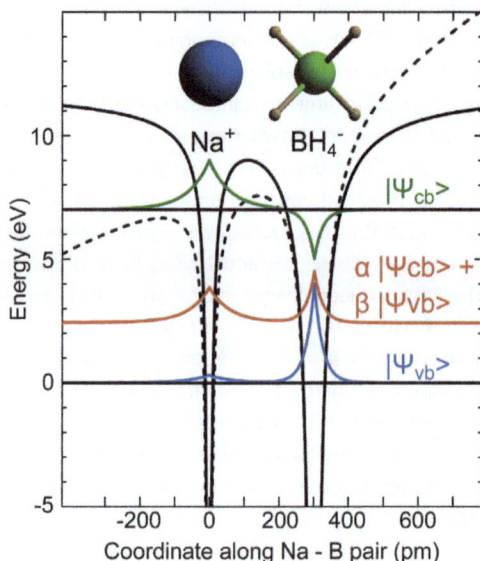

Fig. 1 Schematic illustrating the mechanism of electron transfer from a BH_4^- to a Na^+ ion in a $NaBH_4$ crystal. The unperturbed valence and conduction band wavefunctions, Ψ_{vb} and Ψ_{cb} (blue and green curves), and the ionic potentials (solid black lines) are plotted as a function of the inter-ionic distance. In the valence band, electrons are mainly localized on the BH_4^- ion while the conduction band wavefunction displays similar amplitudes on the two ions. Upon application of an external electric field with an amplitude comparable to the inter-ionic fields, the potentials are distorted (dashed black line) and the corresponding perturbed wavefunction $\alpha\Psi_{vb} + \beta\Psi_{cb}$ (red line) is a mixture of valence and conduction band states of the unperturbed Hamiltonian. Generation of this virtual state is connected with a partial electron transfer from the BH_4^- to the Na^+ ion and a strong electric polarization. In time, the electron transfer follows the external electric field.

of other ionic states, in particular conduction band states $|\Psi_{cb}\rangle$ with a similar amplitude on the two sites. The new wavefunction is given by $|\Psi_{virt}\rangle = \alpha|\Psi_{cb}\rangle + \beta |\Psi_{vb}\rangle$,[14] describing a state in which electronic charge is shifted from the BH_4^- to the Na^+ ion over the inter-ionic distance of some 300 pm. This mechanism generates a strong electric polarization of the material existing as long as the external field is present.

The elementary picture outlined in Fig. 1 neglects modifications of the charge distribution originating from the inherent long-range electron-electron interactions and their screening in the densely packed crystal lattice. Consequently, it remains open to what extent the picture of Fig. 1 exists and which modifications are introduced by the coulomb correlations among electrons. While such issues have been addressed for model systems in the theoretical literature,[15-26] experimental insight into field-driven electron relocations is scarce. Ultrafast time-resolved X-ray diffraction allows for testing the validity of the picture of Fig. 1 and, *vice versa*, should allow for probing the impact of correlations effects on transient electron distributions.

In this article, we demonstrate and discuss the potential of femtosecond X-ray powder diffraction[27] to measure transient electron density maps induced by a strong external optical field under nonresonant conditions. Our work is focused on the ionic materials LiH and $NaBH_4$, consisting of light elements and, thus, allowing for a mapping of valence electrons. While $NaBH_4$ displays a behavior expected for an ionic crystal, *i.e.*, a shift of electronic charge from the negative BH_4^- ion to the positive Na^+ ion, LiH behaves differently. Here, charge is shifted from Li to H, enhancing the ionicity of the material. We show that this unexpected behavior is a direct consequence of electron-electron corelations reproduced by theoretical bandstructure calculations within the Coulomb-hole-plus-screened-exchange (COHSEX) formalism.

The article is organized as follows. Section 2 gives a brief summary of the experimental method and data. The analysis of the diffraction data and extraction of electron density maps are discussed in detail in section 3, followed by a discussion of the results for $NaBH_4$ and LiH in section 4. Bandstructure theory and its application to the electron density maps are presented in section 5, followed by conclusions in section 6.

2   Femtosecond X-ray diffraction experiment

In a femtosecond pump–probe scheme, the electric field of a nonresonant 800 nm excitation pulse interacts with a powder sample and a synchronized hard X-ray probe pulse is diffracted from the sample at different fixed time delays. The angular positions and intensities of several Debye Scherrer rings which are of elliptic or hyperbolic shape on the two-dimensional X-ray detector, are recorded simultaneously and represent the primary data from which transient electron density maps are derived. Pump and probe pulses are derived from an amplified Ti:sapphire laser system delivering sub-50 fs pulses centered at 800 nm with an energy per pulse of 5 mJ and a repetition rate of 1 kHz. The excitation pulses have a peak amplitude of the electric field of approximately 1 GV m^{-1}. The main fraction of the laser output drives a plasma source[28] providing hard X-ray pulses of a photon energy of 8.06 keV (Cu-Kα) and a duration of approximately 100 fs. The X-ray diffraction patterns are detected with a large area detector (Pilatus Dectris

1M). Each diffraction ring corresponds to one (or several equivalent) set(s) of lattice planes hkl [Fig. 2(a)]. The exposure time per time delay step was 140 s and typically $\approx$ 2000 time delay steps were collected for several days with a fresh sample everyday. The all optical autocorrelation was measured repeatedly to assure a proper stacking of the different data sets.

The NaBH$_4$ and LiH samples consist of a 200 μm thick powder, pressed in-between and sealed by two 20 μm thick diamond windows. Assuming a spherical shape of ions in the crystal lattice, both NaBH$_4$ and LiH crystallize in a rock-salt structure (space group $Fm\bar{3}m$) with lattice constants of $a = 407.52$ pm (LiH)[29,30] and $a = 615.06$ pm (NaBH$_4$).[31,32] At $T = 300$ K, LiBH$_4$ crystallizes in the ortho-rhombic space group Pnma (No. 62). The unit cell dimensions are $a = 0.718$ nm, $b = 0.444$ nm, $c = 0.680$ nm[33] with four formula units per unit cell. Because of the high chemical reactivity, great care was taken to avoid air contamination of the sample which was prepared under an Ar atmosphere in a glovebox. During the measurements, the samples were continuously rotated.

In Fig. 2, we present diffraction patterns of NaBH$_4$ while similar data for LiH and LiBH$_4$ have been reported in ref. 13 and 12, respectively. Fig. 2(a) displays part of the detected ring pattern whereas Fig. 2(b) shows the intensity integrated over the stationary diffraction rings of NaBH$_4$ as a function of the diffraction angle 2θ. Two diamond reflections (blue) originate from the sample windows. When applying the field of the excitation pulse, we observe changes of diffracted intensity of the diffraction rings in all three samples with the angular positions remaining unchanged. In Fig. 3(a) and (b) the change of diffracted intensity on the (111) rings in NaBH$_4$ and LiH is plotted as a function of pump–probe delay. In panels (c–f) we show transient intensity changes for 4 selected allowed reflections of LiBH$_4$ as a function of pump–probe delay (symbols). Depending on the investigated material and the diffraction ring (hkl), one observes either an increase [(a) and (f)] or a decrease [panels ((b)–(e)] of diffracted intensity of a few percent around delay zero when excitation and probe pulses interact simulta-neously with the sample. For NaBH$_4$ and LiH the temporal behavior follows essentially the cross correlation between excitation and probe pulses. In LiBH$_4$ we

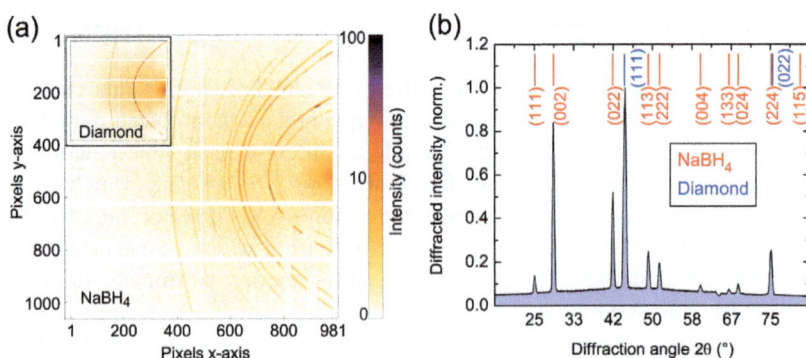

Fig. 2 (a) X-Ray diffraction pattern of the NaBH$_4$ powder sample as recorded with a large-area X-ray detector. The inset shows the diffraction pattern of the diamond windows only. (b) Diffracted X-ray intensity integrated over individual Debye-Scherrer rings from NaBH$_4$ as a function of the diffraction angle 2θ.

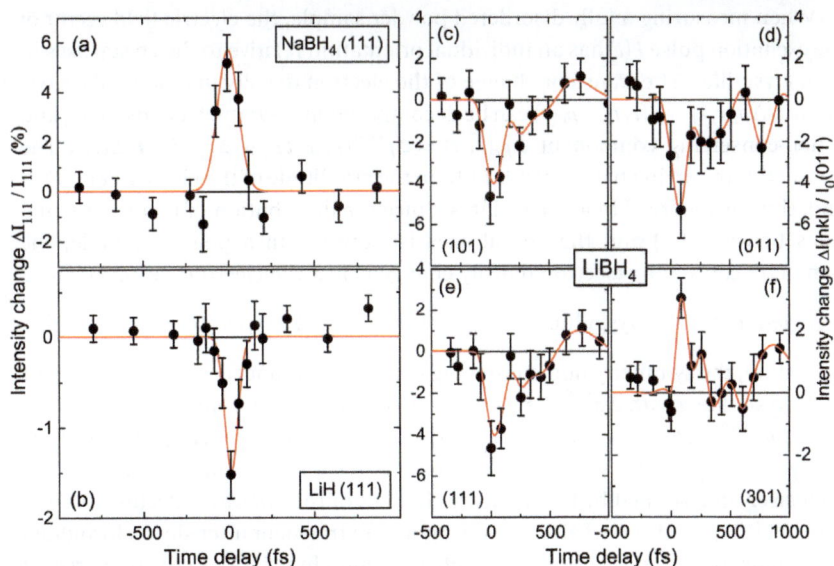

Fig. 3 (a) Relative change of diffracted intensity of $NaBH_4$ by the (111) plane *versus* the pump–probe time delay. (b) Relative change of the diffracted intensity of LiH by the (111) plane *versus* time delay. (c–f)Transient intensity changes for 4 selected allowed reflections measured on $LiBH_4$ powder as a function of pump–probe delay (symbols). The red lines are gaussian fits in panels (a) and (b) and B-splines in (c–f) as guides to the eye.

observe some small longer-lived intensity changes. This fact shows that the induced changes exist mainly with the excitation field present and that they are fully reversible.

The relative intensity change of the diffraction ring hkl is given by

$$\frac{\Delta I_{hkl}(t)}{I_{hkl}} = \frac{|F_{hkl}(t)|^2 - |F_{hkl}^0|^2}{|F_{hkl}^0|^2}, \qquad (1)$$

where $F_{hkl}(t)$ is the structure factor of the material modified by the external electric field and F_{hkl}^0 the known structure factor of the unexcited crystal. The structure factors are related to the electronic density $\rho(r)$ by a Fourier transform. The steady-state electronic density $\rho_0(r)$ is given by the Fourier transform of the structure factors F_{hkl}^0. At room temperature in all three materials $NaBH_4$, LiH, and $LiBH_4$ the structure factors are real due to the inversion symmetry of the rock-salt structure (space group $Fm\bar{3}m$) or the orthorhombic space group $Pnma$.

3 Reconstruction of the transient electron density $\rho(r,t)$

The equilibrium charge density $\rho_0(r)$ can be derived from the known structure and structure factors of the unexcited crystals.[29-33] To determine the transient $\rho(r,t)$ from the measured transient diffraction patterns, the phase factors of the structure factors $F_{hkl}(t)$ need to be known. The principal strategy for solving this problem is discussed next.

When measuring a fully disordered powder sample, the electric field vector of the excitation pulse $E(t)$ has an individual orientation relative to the crystal axes in each crystallite. The transient change of the electron density in a particular crystallite i, $\Delta\rho_i(r, t) = \rho_i(r, t) - \rho_0(r)$ can be decomposed into symmetry conserving and a non-conserving component $\Delta\rho_i(r, t) = \Delta\rho^{Sym}(r, |E(t)|) + \Delta\rho_i^{NoSym}(r, E(t))$. The symmetry conserving part depends only on the amplitude of the electric field $|E(t)|$ and, thus, is identical in all crystallites. Since the distribution of relative orientations between $E(t)$ and the crystal axes is isotropic in a powder sample, the symmetry conserving electron density change is just the electron density change averaged over all crystallites: $\Delta\rho_{Sym}(r, t) = N^{-1} \sum_{i=1}^{N} \Delta\rho_i(r, t)$. In this averaging procedure, the symmetry non-conserving components cancel each other since for each crystallite a with $\Delta\rho_a^{NoSym}(r, E(t))$ the ensemble contains another crystallite b with $\Delta\rho_b^{NoSym}(r, E(t)) = -\Delta\rho_a^{NoSym}(r, E(t))$. As a consequence, $\Delta\rho_{Sym}(r,t)$ exhibits the symmetry properties of the initial structure $\rho_0(r)$. Please note that $\Delta\rho_{Sym}(r,t)$ is exclusively determined by the intensity changes of the *allowed* reflections. In the case of $LiBH_4$ we observed also a small amount of transient intensity on forbidden reflections (see our discussion of this phenomenon in ref. 12). In the more recent studies on LiH and $NaBH_4$ any detectable intensity on forbidden reflections was absent and, thus, the symmetry non-conserving component is negligible compared to $\Delta\rho_{Sym}(r,t)$. The following discussion focuses exclusively on $\Delta\rho_{Sym}(r,t)$.

3.1 Maximum entropy method (MEM)

The MEM aims at finding the most likely charge-density distribution by maximizing the quantity $S = - \sum_{i=1}^{N_{pix}} \rho_i \log(\rho_i)$ (formally similar to the entropy, hence the name) under the constraints given by the information obtained from the experiment.[34,35] The quantities ρ_i are the electron density at a position i in space, the so-called pixel i. The total number of pixels is N_{pix}. The MEM effectively exploits all the information from the experiment without creating artifacts in the charge density distribution. The quantity S cannot be maximized analytically but is calculated recursively, e.g., by the Sakata-Sato algorithm[36] implemented in the BayMEM program.[37]

In general, an exact Fourier transform for determining the spatially resolved electron density requires an infinite number of structure factors F_{hkl}. In practice, any diffraction experiment only gives the structure factors for a limited number of reflections, typically restricted by an upper limit for the diffraction angle $2\Theta_{max}$. In such an incomplete data set, the scattering vector $q_{max} = \sin(\Theta_{max})/\lambda$ determines the spatial resolution at which the electron density map $\rho(r,t)$ can be reconstructed. For an unbiased analysis of the transient electron density $\rho(r,t)$, we prefer to avoid any refinement of the data, e.g., with least-squares methods starting from a model of the transient structure.

The MEM allows for a reconstruction of $\rho(r,t)$ from the experimentally observed reflections (F_{hkl}^{obs}) with $|q_{hkl}| < q_{max}$ without involving any model. A brief introduction to the MEM can be found in chapter 5.3 of Ref. 38. There has been quite some controversy about the application of the MEM, mainly caused by claims of a spatial 'super-resolution'[39] or the 'proof' of non-nuclear maxima of $\rho(r)$ in silicon.[36] In a proper application of the MEM, the amount of added unknown information, $i.e.$, the structure factors F_{hkl}^{MEM} with $|q_{hkl}| > q_{max}$, has to be definitely smaller than the experimentally known information F_{hkl}^{obs}. In contrast to least-squares methods the concept of the MEM allows for making the differences $|F_{hkl}^{MEM} - F_{hkl}^{obs}| \ll |F_{hkl}^{obs}|$ arbitrarily small.

Our previous reconstruction of $\rho(r,t)$ of $LiBH_4$ has been based on a published analysis of the equilibrium structure which has been determined with the least-squares method within the independent atom model.[33] The amplitude of the structure factors F_{hkl} are shown as black bars in Fig. 4. The corresponding charge density was used as the initial guess, the so-called PRIOR in the MEM.[37] Structure factors of overlapping Debye Scherrer rings were treated within the MEM using the so called "group amplitudes".[40]

Applying the MEM directly to the experimentally observed reflections (F_{hkl}^{obs}) with $|q_{hkl}| < q_{max}$ leads to unreliable predictions since the amount of forecast information ($i.e.$ F_{hkl}^{MEM} with $|q_{hkl}| > q_{max}$) is larger than that from the experiment covering the range of $|q_{hkl}| < q_{max}$. To circumvent this problem, we artificially reduced the spatial resolution in deriving both $\rho_0(r)$ and $\rho(r,t)$ by multiplying the corresponding structure factors with a gaussian profile $F_{hkl} \times \exp[-\ln(2)(q_{hkl}/q_{max})^2]$. The resulting structure factors are shown as red bars in Fig. 4(a).

The corresponding electron stationary electron density $\rho_0(x,0.25b,z)$ in the plane where the Li-, the B-, and two H-atoms of a $LiBH_4$ molecule reside is shown in Fig. 4(b). This spatial resolution corresponds to the experimental one and, thus, the MEM is not enhancing the spatial resolution, but just helps to avoid artifacts because of the abrupt end of the Fourier series, $e.g.$, the so-called Gibbs phenomenon (the appearance of ring-like structures around the atoms) in the charge-density distribution. Due to the low Z of all atoms in $LiBH_4$ one can clearly see the hydrogen atoms in the vicinity of the boron atom [Fig. 4(b)]. However, reducing the spatial resolution further, $F_{hkl} \times \exp[-\ln(2)(2q_{hkl}/q_{max})^2]$ [blue bars in Fig. 4(a)] leads to the situation that the MEM result shows an almost vanishing difference from the direct Fourier transform in the truncated reciprocal space, $i.e.$ $F_{hkl} = 0$ for $|q_{hkl}| > q_{max}$.

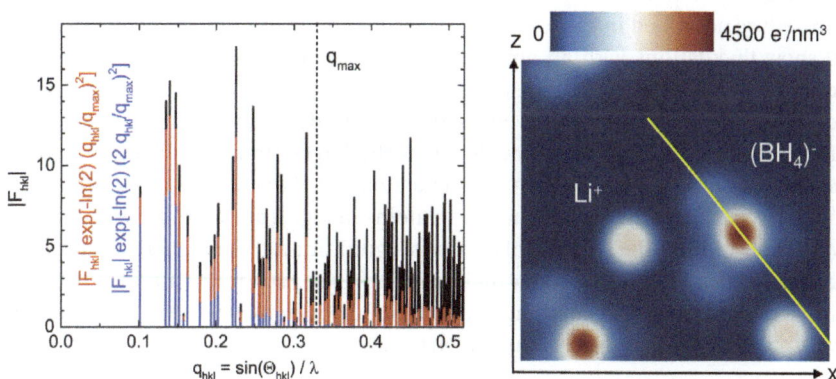

Fig. 4 (a) Amplitudes of structure factors of $LiBH_4$ as a function of $q = \sin(\Theta)/\lambda$ for different spatial resolutions. $q_{max} = \sin(\Theta_{max})/\lambda$ indicates the reflections measured in our femtosecond diffraction experiment. (b) Stationary total electron density distribution in the plane $Y = 0.25$), multiplied in the reciprocal space with a Gaussian [red columns in panel (a)] to fit our spatial resolution. The yellow line indicates the section along Fig. 7(a) is plotted.

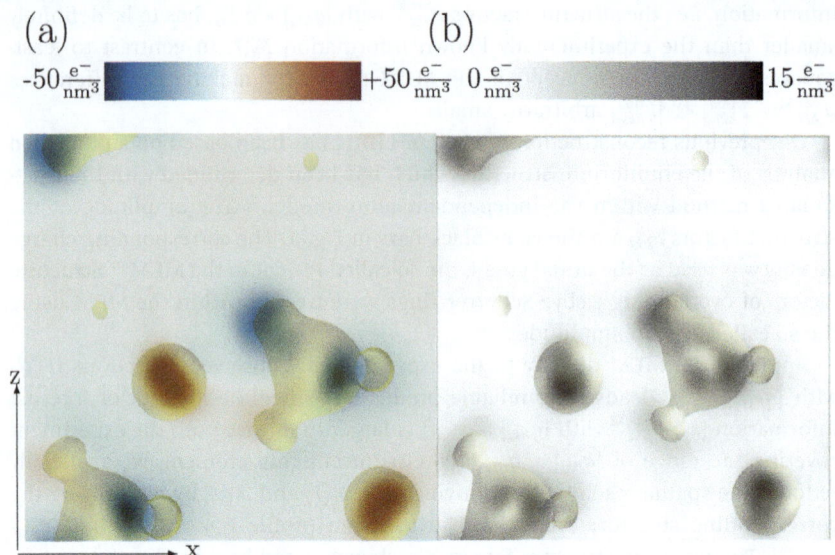

Fig. 5 Comparison between (a) the transient electron density map $\Delta\rho_{Sym}$ of LiBH$_4$ at $\tau = 0$ [Fig. 3(c) of ref. 12] together with (b) the spatially resolved error of such a map obtained from the fluctuations of maps measured at negative delay times $\tau < 0$.

To summarize, the MEM allows for a gradual transition from an exact reconstruction towards being increasingly speculative when increasing the (gaussian) spatial resolution from $q_{gauss} < q_{max}$ to $q_{gauss} > q_{max}$. Such an analysis avoids artefacts and allows for a proper assessment of the MEM results. This method has been applied for extracting the transient electron density maps of NaBH$_4$, LiH, and LiBH$_4$ from our femtosecond powder diffraction data.

3.2 Signal-to-noise–ratio of transient electron density maps

The signal-to-noise ratio of a transient electron density map is estimated by the procedure described in the following. Maps measured at negative delay times $\tau < 0$ show fluctuations with an amplitude strongly depending on the value of the electron density at a particular spatial position within the unit cell. The standard deviation of such fluctuations gives the amplitude of an error bar shown in Fig. 5(b) which is 3–4 times smaller than the largest transient changes $\Delta\rho(r,t)$ shown in Fig. 5(a). Typically, such fluctuations are large at spatial positions of high electron density, *e.g.*, on the boron atoms in Fig. 5(b). However, the position of an atom within the unit cell plays also a significant role for the size of the error bar. In particular, one gets a very high signal-to-noise ratio for $\Delta\rho(r,t)$ in situations if electronic charge transiently appears at spatial positions where initially no atom was present, *e.g.* hydrogen atoms in $(NH_4)_2SO_4$.[9]

4 Discussion and analysis of the transient charge density maps

Using the MEM described in the previous section we reconstructed from our femtoseond X-ray diffraction data (Fig. 3) the corresponding transient charge density maps $\Delta\rho(r, t = 0)$ which are shown for time delay zero in Fig. 6. In panel (a) we show the difference electron density map of $NaBH_4$. The difference maps $\Delta\rho(r, t = 0)$ of $LiBH_4$ and LiH are presented in panels (b) and (c). Another sectional view of $\Delta\rho(r, t = 0)$ of $LiBH_4$ has already been shown in Fig. 5(a). We now discuss the exchange of electronic charge between cations and anions, as well as details of transient charge changes on the hydrogen atoms which are clearly detected with our spatial resolution for $LiBH_4$.

4.1   Transient exchange of electronic charge between cations and anions

The initial electron density map $\rho_0(x,y,0)$ of LiH with our spatial resolution is shown in Fig. 2(a) of ref. 13 and exhibits a high electron density on the Li atom and a small density on the H atom. In Fig. 6(c), the change of electron density $\Delta\rho_{sym}(r,t)$ is plotted for zero delay time. This map shows a pronounced decrease of electron density on the Li atom and a corresponding increase on the hydrogen position, giving direct evidence for a quasi-instantaneous increase of the ionicity of LiH in the presence of the strong electric field. This surprising behavior is in

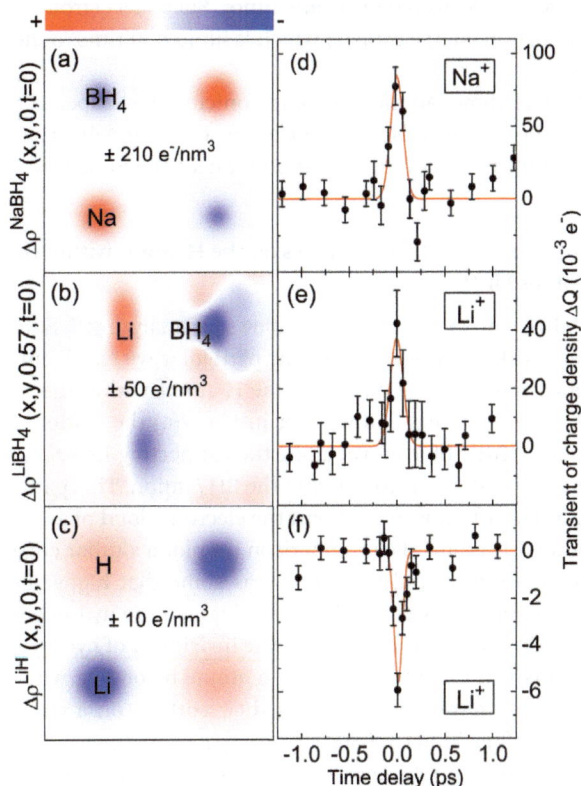

Fig. 6   (a) Difference electronic density map $\Delta\rho(r, t = 0)$ reconstructed by the MEM at zero delay time of $NaBH_4$. (b) and (c) $\Delta\rho(r,t = 0)$ of $LiBH_4$ and LiH, respectively. In contrast to the boron hydrides (i.e. charge transfer from anion to cation) LiH shows an increase of its ionicity when exposed to the strong electric field of the pump pulse. (d–f) Integrated charge changes on the respective cation Na or Li versus the delay time.

strong contrast to $\Delta\rho_{sym}(r,t)$ of NaBH$_4$ [Fig. 6(a)] and LiBH$_4$ [Fig. 6(b)], where the light-induced charge relocation exhibits an expected charge transfer from the anion (BH$_4^-$) to the respective cation.

To determine the amount of charge transferred, we divided the unit cell into sub-volumes, *i.e.*, each spatial position r within the unit cell is uniquely assigned to the sub-volume of the nearest atom. We then integrated the charge in the sub-volumes. For LiH this procedure leads to $(a/2)^3$ cubes around Li and H. In the case of BH$_4^-$ anions, the charge of the whole unit was put together. Time-dependent charge density changes are displayed in Fig. 6(d) to (f) where we plotted the integrated charge changes on the respective cation, *i.e.* Na$^+$ or Li$^+$, *versus* the delay time.

In LiH, the striking feature is a sharp drop of electronic charge on the Li atom [Fig. 6(f)] concomitant with the increase of the same amplitude on the hydrogen position due to charge conservation within each unit cell. The peaks of the transients in panels (d–f) have a width of ≈ 100 fs (FWHM) which agrees with the temporal cross-correlation function of the optical excitation and the hard X-ray probe pulses.[9,12] In the case of the boron hydrides [Fig. 6(d) and (e)] we see exactly the opposite behavior, *i.e.*, a light-induced charge transfer from the anion BH$_4^-$ to the respective cation Na$^+$ or Li$^+$. Outside the temporal overlap of pump and probe pulses, the changes of electron density are minor. Such facts strongly support our picture of a field-driven charge transfer which is limited in time to the presence of the driving field.

In summary, LiH shows an increase of its ionicity when exposed to the strong electric field of the pump pulse, in contrast to the boron hydrides where the charge transfer from the anion to the cation reduces the ionicity.

4.2 Asymmetric transient charge changes on the H-atoms within the distorted BH$_4$ tetrahedron of LiBH$_4$

In the case of LiBH$_4$ our spatial resolution allows for analyzing details of transient charge changes on the hydrogen atoms. In particular we can identify an asymmetric transient charge change on one of the H-atoms within the distorted BH$_4$ tetrahedron of LiBH$_4$. To this end we plot in Fig. 7(a) the stationary electronic density $\rho_0(r_{B-H})$ of LiBH$_4$ (blue curve) along the connecting line r_{B-H} between the boron atom and one hydrogen atom within the BH$_4^-$ anion. This path indicated by the yellow line in Fig. 4(b) traverses in part the electron cloud of a Li-atom but not those of the other 3 H-atoms. For comparison we plot a similar curve for NaBH$_4$ (red line). Although our spatial resolution was somewhat higher in the NaBH$_4$ experiment (*i.e.* larger q_{max}) we can clearly distinguish the hydrogen atoms from their boron atom for the BH$_4^-$ ion in LiBH$_4$. The invisibility of hydrogens in NaBH$_4$ is due to the statistical mixture of hydrogen atoms belonging with 50% probability to two different tetrahedrons having their corners on a cube around the boron atom (Wyckoff positions 32f of space group $Fm\bar{3}m$). In Fig. 7(b) we show the electron density changes $\Delta\rho(r_{B-H}, t = 0)$ for the two materials along the respective connecting lines. In the case of NaBH$_4$ the entire anion loses its charge without displaying an internal structure of $\Delta\rho(r_{B-H}, t = 0)$ within the BH$_4^-$ ion. In contrast, we observe a pronounced asymmetric transient charge change within the BH$_4$ tetrahedron of LiBH$_4$ [blue curve in Fig. 7(b) and difference charge density map in Fig. 5(a)]. Electronic charge is lost on one of the H-atoms (left minimum) and the

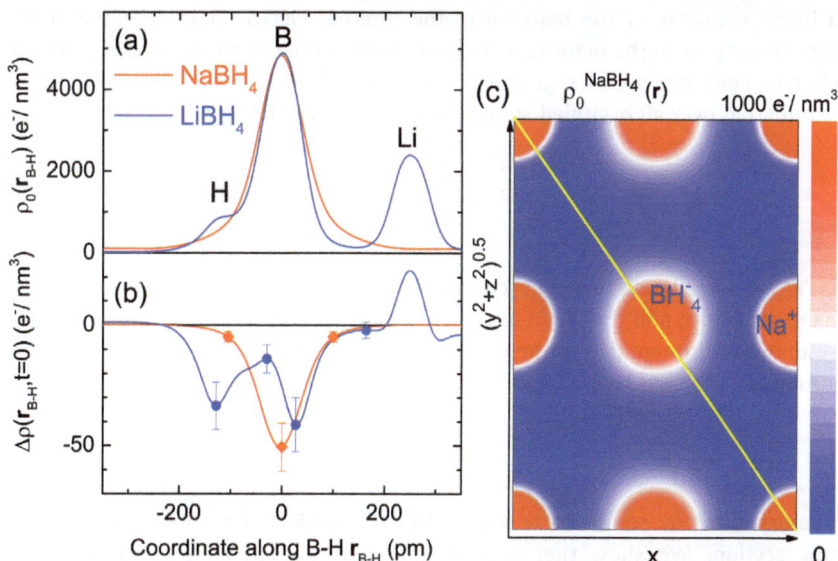

Fig. 7 (a) Stationary electronic density $\rho_0(r_{B-H})$ along the connecting line between the boron atom and one hydrogen atom [yellow lines in panel (c) and in Fig. 4(b)] within the BH_4^- anion for $LiBH_4$ (blue curve) and $NaBH_4$ (red curve). (b) Respective electron density change $\Delta\rho(r_{B-H},t=0)$ along the connecting lines. The error bars are calculated by means of the method described in section 3.2.

antipodal orbital of the B-atom (right minimum). In $LiBH_4$ the structure of the tetrahedrons is distorted with different lengths and angles of the B–H bonds.[33] Interestingly, the hydrogen H1 with the shortest bond length [$d_{B-H1} = 104$ pm, $cf.$ Table 2 of ref. 33] dominates in the charge loss of the anion. We consider this distortion the origin of the inequivalent charge transfer from the different hydrogens.

4.3 Microscopic mechanism of field-induced charge transfer between ions

We now discuss the microscopic physics underlying the material's polarization and the field-induced change of electron density $\Delta\rho_{sym}(r,t)$ in the different materials. Without external field, electrons in an insulator populate states up to the highest valence band. An external field of an amplitude comparable to the interionic field of the order of 1 GV m^{-1} distorts the ionic potentials and leads to the admixture of other ionic states, in particular conduction band states ($cf.$ Fig. 1). The wavefunction of the mixed state is given by

$$|\Psi_{b,k}(E)\rangle = \frac{1}{\sqrt{N}}\left[|\Psi_{b,k}\rangle + E\sum_{b'\neq b}\frac{\langle\Psi_{b',k}|er|\Psi_{b,k}\rangle}{\varepsilon_{b,k}-\varepsilon_{b',k}}|\Psi_{b',k}\rangle\right] \qquad (2)$$

at wave vector k in the (occupied) band b with the normalization constant N. The sum runs over all unoccupied bands b'. The perturbative approach in (2) is valid as the dipole interaction energy $|\langle\Psi_{b',k}|er|\Psi_{b,k}\rangle|\cdot|E| \simeq 0.2$ eV is much smaller than the smallest bandgap $\varepsilon_{b,k} - \varepsilon_{b',k} \simeq 5$ eV of the three materials, $i.e.$ LiH at the X point of the Brillouin zone. In other words, the experiments are in the regime of a linear response of the material to the external electric field. The symmetry conserving part of the deformed charge density is obtained by averaging over all electric field directions ($\hat{e}_\Omega$: unit vector in solid angle direction Ω) and a summation over all occupied states within the Brillouin zone

$$\rho_{Sym}(r,t) = \sum_{b,k}^{\text{occupied}} \frac{1}{4\pi} \int d\Omega \left| \Psi_{b,k}\left(\hat{e}_\Omega \cdot |E(t)|, r\right) \right|^2 . \tag{3}$$

The distorted wave function (2) shows that both the eigen-energies $\varepsilon_{b,k}$ (*i.e.*, the band structure) and the eigen-functions $\Psi_{b,k}(r)$ (*i.e.*, the Bloch functions) of the system Hamiltonian determine the exact shape of the electron density (3) in the strong electric field.

5 Theory

The experimental results were analyzed by the model calculations presented in this section. We show that a mean-field theory such as the Hartree–Fock approximation fails to account for the experimentally observed quasi-instantaneous increase of the ionicity of LiH in the presence of the strong electric field. However, calculations including Coulomb correlations on the most basic level, *i.e.*, quasi-particle bandstructures calculated within the Coulomb-hole-plus-screened-exchange (COHSEX) formalism,[20,23] predict the increase of the ionicity of LiH correctly. In the following, we use the mathematical nomenclature introduced in ref. 23 and 22.

The COHSEX formalism considers the dynamics of the one-particle Green's function, expands the self-energy operator in terms of a dynamically screened interaction (rather than a bare Coulomb interaction), and keeps the first term in such an expansion. The quasiparticle excitations in a closed shell many-electron system obey the integro-differential equation

$$\left[\frac{p^2}{2m} + V_N(r) + V_H(r)\right]\psi_{n,k}(r) + \int dr' \, \Sigma(r,r',E_{n,k})\psi_{n,k}(r') = E_{n,k}\psi_{n,k}(r) \tag{4}$$

where $V_N(r)$ is the nuclear Coulomb potential, $V_H(r)$ is the average Coulomb (Hartree) potential due to the electrons, and Σ is the electron self energy operator which includes all the exchange and correlation effects. It is in general a nonlocal, energy-dependent, non-Hermitian operator. The procedure for finding the quasi-particle energies $E_{n,k}$ and wave functions $\Psi_{n,k}(r)$ requires evaluating Σ and then solving eqn (4). In the following we use the static COHSEX approximation (first introduced by Hedin[20]) which constitutes one of the biggest steps in the theoretical basis for band structure calculations. In this approximation, the self energy is purely real (*i.e.* no finite quasi-particle lifetimes) and does not depend on the quasi-particle energy itself. As a consequence eqn (4) mathematically resembles a nonlinear "Schrödinger equation" with all its advantages for solving it. For our purpose it is convenient to expand the quasi-particle wave function in plane waves [eqn (B2) of ref. 23]

$$\psi_{n,k}(r) = \frac{1}{\sqrt{\Omega}} \sum_G \psi_{n,k}(G)e^{i(k+G)r} \tag{5}$$

where G are the reciprocal lattice vectors of the crystal and $\Omega = N\Omega_{uc}$ stands for the crystal volume determining the discretization in k space. $\Omega_{uc} = a^3/4$ is the volume of the primitive unit cell. The quasi-particle equation now has the form

$$\sum_{G'} \left[H^0_{G,G'}(k) + \Omega\, \Sigma_{G,G'}(k) \right] \psi_{n,k}(G') = E_{n,k} \psi_{n,k}(G). \qquad (6)$$

Both $H^0_{G,G'}(k)$ and $\Sigma_{G,G'}(k)$ contain purely local (i.e. k-independent) and nonlocal (i.e. k-dependent) contributions:

$$H^0_{G,G'}(k) = \frac{\hbar^2 |k + G|^2}{2m} \delta_{G,G'} + \frac{4\pi e^2}{\Omega_{uc} |G - G'|^2} [\rho_e(G - G') - \rho_N(G - G')] \qquad (7)$$

$$\rho_e(G) = \frac{2}{N} \sum_{n,k,G'}^{occ} \psi^*_{n,k}(G') \psi_{n,k}(G' + G), \qquad (8)$$

$$\rho_N(G) = Z_{Li}\, e^{i\, G\, R_{Li}} + Z_H\, e^{i\, G\, R_H}, \qquad (9)$$

The Fourier transform of the electron density $\rho_e(G)$ in eqn (8) is identical to the structure factor $F_{hkl} = \rho_e(G_{hkl})$ determined in an X-ray diffraction experiment. In the sum over occupied states with band index n and wave vector k (within the 1st BZ), only the states of the same spin as (n,k) contribute and the factor 2 accounts for the spin degeneracy. The self energy $\Sigma_{G,G'}(k)$ consists of a nonlocal screened exchange (SEX) and a purely local Coulomb hole (COH) contribution:

$$\Sigma^{SEX}_{G,G'}(k) = -\frac{1}{\Omega} \sum_{n,k'}^{occ} \sum_{G_1,G_2} \psi_{n,k'}(G - G_1) \psi^*_{n,k'}(G' - G_2)$$

$$\times\ \varepsilon^{-1}_{G_1,G_2}(|k - k'|, \omega = 0)\, \frac{4\pi e^2}{\Omega |k - k' + k_2|^2}, \qquad (10)$$

$$\Sigma^{COH}_{G,G'} = \frac{1}{\Omega} \sum_{n,q} \sum_{G_1,G_2} \psi_{n,k-q}(G - G_1) \psi^*_{n,k-q}(G' - G_2)$$

$$\times\ \frac{1}{2} \left[\varepsilon^{-1}_{G_1,G_2}(q, \omega = 0) - \delta_{G_1,G_2} \right] \frac{4\pi e^2}{\Omega |q + G_2|^2} \qquad (11)$$

$$= \frac{1}{2\Omega} \sum_{q,G_3} \left[\varepsilon^{-1}_{G+G_3,G'+G_3}(q, \omega = 0) - \delta_{G,G'} \right] \frac{4\pi e^2}{\Omega |q + G' + G_3|^2}.$$

The last line in eqn (11) can be derived from the completeness relation of the Bloch wave functions (5). The approximation of a statically screened Coulomb interaction (i.e. $\omega = 0$ in eqn (10) and (11)) makes these self energy contributions much simpler than their dynamically screened counterparts (B3a) and (B3b) in ref. 23. More importantly, the self energy is now essentially determined and controlled by the inverse dielectric matrix $\varepsilon^{-1}_{G,G'}(q,\omega = 0)$ which accounts correctly for inhomogeneous but static screening.[41,42] The static COHSEX formalism has been further approximated by Baroni et al.[22] who studied LiH with the dielectric function of a homogeneous electron gas, *i.e.*, a diagonal dielectric matrix $\varepsilon_{G,G'}^{-1}(\boldsymbol{q}, \omega = 0) = \varepsilon_{\text{hom}}^{-1}[|\boldsymbol{q} + \boldsymbol{G}|, \omega = 0]\,\delta_{G,G'}$. With such a dielectric matrix eqn (11) reduces to a much simpler integral [Eq. (5) of ref. 22]. The Coulomb hole around an electron in such a theory is independent of the quasi-particle state the electron occupies resulting in a constant energy shift of all bands[22] and corresponding to $\Sigma_{G,G'}^{\text{COH}} = E_{\text{CH}}\,\delta_{G,G'}$. A Hartree Fock calculation[43] goes even further and corresponds to $\varepsilon_{G,G'}^{-1}(\boldsymbol{q}, \omega = 0) = \delta_{G,G'}$. The latter case is a mean field theory which neglects any spatial correlations in the electronic system. The Coulomb hole self energy contribution vanishes in the Hartree Fock approximation, *i.e.*, $\Sigma_{G,G'}^{\text{COH}} = 0$. In the following we will see that *inhomogeneous* screening,[41,42] *i.e.* off diagonal elements in the dielectric matrix $\varepsilon_{G,G'}^{-1}(\boldsymbol{q}, \omega = 0)$, has a crucial influence on both the stationary $\rho_0(\boldsymbol{r})$ and transient $\rho(\boldsymbol{r},t)$ electron distributions of LiH exposed to strong electric fields, a phenomenon which could not be observed in the theory of Baroni *et al.*,[22] since such phenomena are excluded from the very beginning.

In order to be as close as possible to experimentally measured physical quantities of LiH we did not use *ab initio* static dielectric matrices [*cf.* section III.A of ref. 23] but (similar to section II.C of ref. 22) constructed a model dielectric matrix $\varepsilon_{G,G'}^{-1}(\boldsymbol{q}, \omega = 0)$ which coincides with various measured experimental values of LiH. The basic idea of our approach is that in the sub-volume around each type of atom we introduce an individual *homogeneous* dielectric function of the Baroni type [eqns (8) and (9) of ref. 22]

$$\varepsilon_{\text{Li}}^{-1}(q, \omega = 0) = \frac{1}{\varepsilon_S} + c_{1,\text{Li}}\frac{q^2}{q^2 + k_{1,\text{Li}}^2} + c_{2,\text{Li}}\frac{q^2}{q^2 + k_{2,\text{Li}}^2} \tag{12}$$

$$\varepsilon_{\text{Li}}^{-1}(q \to \infty, \omega = 0) = 1 - \frac{16\pi\,n_{e,\text{Li}}}{a_B}\frac{1}{q^4} + O\left[\frac{1}{q^6}\right] + \cdots \tag{13}$$

$$\varepsilon_{\text{H}}^{-1}(q, \omega = 0) = \frac{1}{\varepsilon_S} + c_{1,\text{H}}\frac{q^2}{q^2 + k_{1,\text{H}}^2} + c_{2,\text{H}}\frac{q^2}{q^2 + k_{2,\text{H}}^2} \tag{14}$$

$$\varepsilon_{\text{H}}^{-1}(q \to \infty, \omega = 0) = 1 - \frac{16\pi\,n_{e,\text{H}}}{a_B}\frac{1}{q^4} + O\left[\frac{1}{q^6}\right] + \cdots \tag{15}$$

From the experiment we take the static dielectric constant $\varepsilon_S = 3.61$ of LiH and the average electron densities around the Li-atom $n_{e,\text{Li}}$ and the H-atom $n_{e,\text{H}}$. As discussed in ref. 22, the term of type q^{-2} must be rigorously lacking in the aymptotic behavior $q \to \infty$. Its presence would give rise to unphysical divergences in the response to a point-charge perturbation.[44] The constraints determined by the asymptotic behaviors (13) and (15) leave the screening vector $k_{1,\text{Li}}$ (eqn (12)) and the screening vector $k_{1,\text{H}}$ (eqn (14)) as free (fitting) parameters in our calculation. The construction of the model dielectric matrix is now straightforward. We assume that the screening of the Coulomb potential between two point charges at positions $\boldsymbol{r}$ and are $\boldsymbol{r}'$ within the LiH crystal is determined by the sub-volume of the unit cell in which the center of gravity $\boldsymbol{r} = (\boldsymbol{r} + \boldsymbol{r}')/2$ falls. Putting the Li nucleus into the origin of the unit cell leads to the following dielectric matrix for LiH

$$\varepsilon_{G,G'}^{-1}(\boldsymbol{q}, \omega = 0) = \frac{\Omega_{Li}}{\Omega_{uc}} \exp\left(-\beta^2 |\boldsymbol{G} - \boldsymbol{G'}|^2\right) \varepsilon_{Li}^{-1}\left(\left|\boldsymbol{q} + \frac{\boldsymbol{G} + \boldsymbol{G'}}{2}\right|, \omega = 0\right)$$

$$+ \left[\delta_{\boldsymbol{G},\boldsymbol{G'}} - \frac{\Omega_{Li}}{\Omega_{uc}} \exp\left(-\beta^2 |\boldsymbol{G} - \boldsymbol{G'}|^2\right)\right] \times \varepsilon_H^{-1}\left(\left|\boldsymbol{q} + \frac{\boldsymbol{G} + \boldsymbol{G'}}{2}\right|, \omega = 0\right).$$

$$(16)$$

The length β in (16) determines the size of the bell-shaped screening volume Ω_{Li} around the Li-atom and in turn the partitioning of the average electron densities $n_{e,Li}\Omega_{Li} + n_{e,H}[\Omega_{uc} - \Omega_{Li}] = 4$ in eqn (13) and (15). For different $\varepsilon_{Li}^{-1}(q, \omega = 0) \neq \varepsilon_H^{-1}(q, \omega = 0)$ the dielectric matrix (16) gets off-diagonal elements which account for the *inhomogeneous* screening in LiH. Choosing a common dielectric function $\varepsilon_{Li}^{-1}(q, \omega = 0) = \varepsilon_H^{-1}(q, \omega = 0) = \varepsilon_{hom}^{-1}(q, \omega = 0)$ we can recover the *homogeneous* screening result of Baroni et al.[22]

In solving eqn (6), we expanded the quasi-particle band structure in plane waves using 339 reciprocal lattice vectors in a sphere around the Γ-point with radius $G_{hkl} < 7 \times 2\pi/a$. For the $\boldsymbol{k}$ summations in eqn (8), (10), and (11) we used the 10 special points (and their symmetry equivalents) in the Brillouin zone of ref. 45. In order to delete the singularity of the Coulomb potential in eqn (10) and (11) we used the method described with eqn (3.13) by Ohkoshi.[43]

As a start we repeated both the Hartree Fock (Fig. 1 of ref. 22) and the COHSEX calculation with *homogeneous* screening (Fig. 5 of ref. 22) of Baroni et al. and found a very good agreement of the band structures with our calculations [cf.

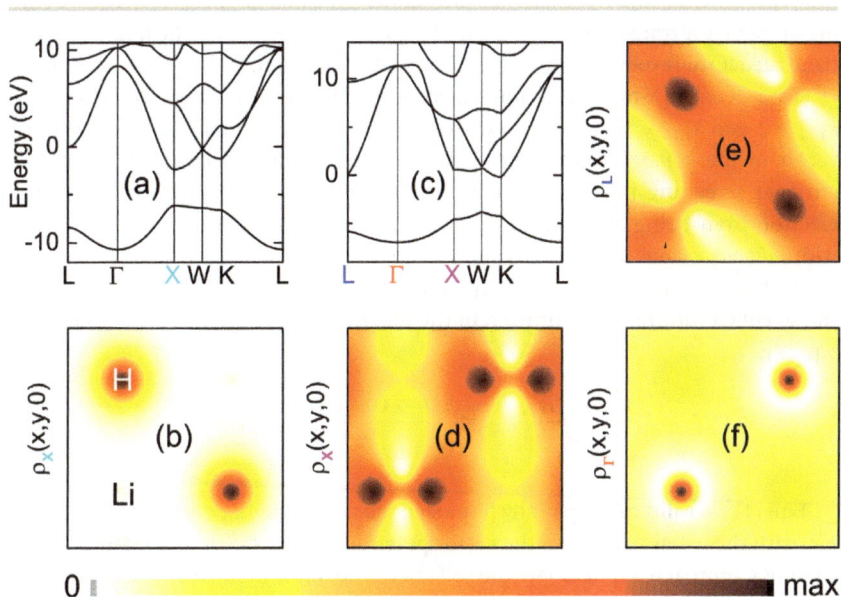

Fig. 8 (a) and (c) Quasi-particle band structures of LiH within the COHSEX approximation. (b) and (d) Stationary electron densities at the X-point, (e) L-point, and (f) Γ-point within the Brillouin zone (BZ). Panels (a) and (b) correspond to a calculation with homogeneous screening similar to that of ref. 22 leading to a 1S-like orbital on the H atom throughout the BZ. Panels (c,d,e,f) are calculated with a somewhat stronger screening on the proton resulting in 2P-like orbitals on both Li and H nuclei at the X-point (zone boundary).

Fig. 8(a)]. A significant discrepancy between experiment and theory is observed, however, when looking at the electron density map $\rho_0(r)$. In both the Hartree Fock and the COHSEX calculation with *homogeneous* screening, the valence band is dominated by 1S-like orbitals on the H atom throughout the Brilluoin zone.[26] As a consequence, the calculated stationary electron density correspond to the fully ionized Li^+ H^- situation. This is in strong contrast to the stationary X-ray diffraction experiments.[29,30] Counting the charges in LiH gives $Li^{0.5+}H^{0.5-}$, striking a happy medium between the ionic case Li^+H^- and the so-called covalent case $Li^{0+}H^{0-}$ in which electrons are shared between lithium and hydrogen. Obviously, a quasi-particle theory with *homogeneous* screening[22] misses key features of the electron correlations in LiH.

The COHSEX model discussed above allows for studying the influence of *inhomogeneous* screening on the electron correlations in LiH. To this end, we varied the screening vectors $k_{1,Li}$ and $k_{1,H}$ and the length β characterizing the size of the screening volume around Li in order to simultaneously fit the quasi-particle band structure [experimental values from Table V in ref. 22] and the stationary electron density [X-ray diffraction experiment from ref. 30]. The length β was chosen that the gaussian-shaped screening volume around the Li nucleus has the same width as the electron density on the Li-atom. It turned out that small variations of β around this starting value do not have a strong influence on both the band structure and the electron density $\rho_0(r)$.

Our calculations show that modifying the screening vectors $k_{1,Li}$ and $k_{1,H}$ in the COHSEX calculation of LiH allows for a gradual transition from $3e^-$ on Li^{3+} (metal-like wave functions) to $2e^-$ on Li^{3+} (insulator-like wave functions). Thus, the experimental value of $2.5e^-$ on Li^{3+} shows that LiH is at the very limit of an metal-insulator transition.[25,24] We found a good agreement with the quasi-particle band structure [experimental values from table V. in ref. 22] and the stationary electron density [X-ray diffraction experiment from ref. 30] when choosing $k_{1,H}/k_{1,Li} = 2$ and $k_{1,H} \times k_{1,Li} = k_{1,hom}^2$ with $k_{1,hom}$ being the Thomas Fermi screening vector of the *homogeneous* electron gas as used in ref. 22. In simple words, a stronger screening in the volume around the H-atom tries to repel the second electron from being attached to the H-atom. The corresponding band structure with *inhomogeneous* screening is shown in Fig. 8(c).

Next, we introduce light-matter interaction into our COHSEX model by making the so called minimal substitution in the kinetic energy term of the hamiltonian eqn (7):

$$\frac{\hbar^2|k+G|^2}{2m}\delta_{G,G'} \Rightarrow \frac{\hbar^2|k+G-eA(t)|^2}{2m}\delta_{G,G'}. \tag{17}$$

Eqn (17) is valid for a spatially homogeneous but time-dependent electric field $E(t)$ with the usual relation to the vector potential $A(t) = \int_{-\infty}^{t} dt' \, E(t')$. Since in the COHSEX approach applied here the self energy operator in (6) does not depend on the quasi-particle energy itself we can also substitute on the r.h.s. of (6) the eigenvalue $E_{n,k}\psi_{n,k}(G) \Rightarrow i\hbar\partial\psi_{n,k}(G)/\partial t$ to gain a nonlinear "Schrödinger equation" for the quasi-particles. In such a time-dependent COHSEX calculation all quantities including the self-energy became also time-dependent and had to be updated for each individual time step on the order of a few attoseconds. An advantage of time-dependent COHSEX is that it can also be used in the regime of non-perturbative light-matter interaction to investigate nonlinear electron correlations. On the other hand, such a theory is almost numerically intractable. Here, we study our system in linear response to an external electric field $E(t)$ oscillating at a frequency much lower than the lowest interband transition frequency. As a result, exclusively bound electrons interact with the external field. Such conditions are well matched with eqn 1.10 of ref. 16 from which we learn that the time-independent ground state electron correlations (i.e. the time-independent ground state self energy) determine the linear response to $E(t)$. As a consequence we can use time-independent perturbation theory as done with eqn (2) discussed above. Within our COHSEX formalism the optical interband dipole moments are calculated with help of[14]

$$\langle \Psi_{b',k}|er|\Psi_{b,k}\rangle = \frac{i\,e\,\hbar^2}{m}\sum_{G}\frac{\psi^*_{b',k}(G)[k+G]\psi_{b,k}(G)}{E_{b,k}-E_{b',k}}. \tag{18}$$

With this theoretical framework, we now investigate the deformation of the electron density in LiH in a strong electric field for two dielectric matrices: (i) using *homogeneous* screening with parameters of Baroni et al.[22] and (ii) using *inhomogeneous screening* with the parameters of our best fit for band structure [experimental values from table V. in ref. 22] and stationary electron density [X-ray diffraction experiment from ref. 30]. The calculations using eqn (2) and (3) are

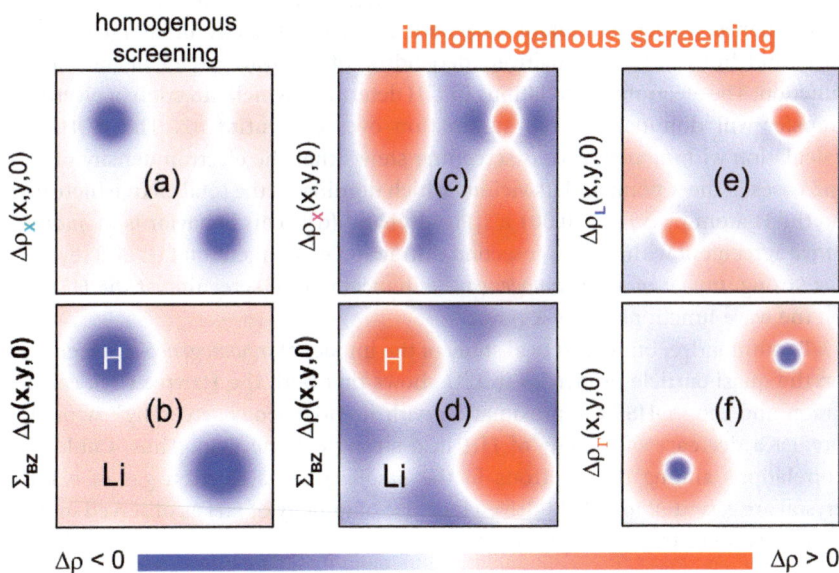

Fig. 9 (a) and (c) Deformed electron densities at the X point, (e) L-point, and (f) Γ-point within the BZ for an external field amplitude of $|E| = 10^9$ V m^{-1}. (b) and (d) Summation over the Brillouin zone gives the total density change $\Delta\rho_{Sym}(x,y,0,t=0)$ calculated with eqn (3). The respective charge integrations over $(a/2)^3$-boxes yield (b) $\Delta q_H = -0.1$ e^- and (d) $\Delta q_H = +0.01$ e^-. Panels (a) and (b) correspond to calculation with homogeneous screening similar to that of Ref. 22. Panels (c,d,e,f) are calculated with a somewhat stronger screening on the proton resulting in 2P-like orbitals on both Li and H nuclei at the X-point (zone boundary).

summarized in Fig. 8 and 9. In the left panels of the two figures, we show 8(a) the bandstructure, 8(b) the valence band electron density ρ_X at the X point where the smallest bandgap occurs, 9(a) the field-induced change of electron density at the X point, and 9(b) the total change of electron density calculated with eqn (3). This calculation assumes a homogeneous screening with the parameters of ref. 22. As in mean field calculations, i.e., Hartree Fock, the valence band is dominated by 1S-like orbitals on the H atom throughout the Brilluoin zone.[26] As a consequence, the stationary electron density correspond to the fully ionized Li^+ H^- situation. Applying an electric field would reduce the charge on the H^- anion by $\Delta q_H = -0.1$ e^- [Fig. 9(b)], a behavior in striking contrast to our experimental results for LiH [Fig. 6(c, f)] but close to the behavior of $NaBH_4$ [Fig. 6(a, d)]. In contrast, panels (c–f) of Fig. 8 and 9 are calculated using the COHSEX model with *inhomogeneous screening* and parameters of our best fit for the stationary electron density. In contrast to the homogeneous screening case the valence band Bloch functions depend now sensitively on their wave vector within the Brillouin zone (BZ). At the Γ-point [Fig. 9(f)] the valence band Bloch function has a strong contribution of 2S-like orbitals on the Li-atoms, whereas at the L-point (e) complex hybrid-orbitals on both the H- and Li-atom are formed. Most importantly, the Bloch functions develop into 2P-like orbitals on both Li and H nuclei when approaching the valence band at the X-point [Fig. 8(d)]. Under the external electric field, we observe a general trend of the deformed electron density at any point (wave vector) within the BZ. The electric field distributes the electronic charge more uniformly within the unit cell, causing in turn a reduction of amplitude of the highest peaks as observed for $\Delta\rho_X(x,y,0)$, $\Delta\rho_L(x,y,0)$, and $\Delta\rho_\Gamma(x,y,0)$ in panels (c,e,f) of Fig. 9. As a consequence, we have simultaneously both a transfer of electronic charge from H to Li and from Li to H, sensitively depending of the wave vector of the Bloch function. The electronic correlations in LiH determine which direction of electron transfer will dominate in the summation over the entire BZ. The COHSEX calculation with *inhomogeneous* screening shows that the electron density of H^+ increases at the X point and its vicinity which dominates the total charge increase on the H-atom by $\Delta q_H = +0.001$ e^- [Fig. 9(c) and (d)]. This behavior is in *quantitative* agreement with the femtosecond diffraction experiment on LiH and reveals the strong impact of an *inhomogeneous* enhancement of screening at the H^+ site on the wave functions at the X point.

To summarize, our theoretical study of the impact of *inhomogeneous* screening on the quasi-particle properties in LiH shows that both the Hartree Fock calculation and the COHSEX approximation with homogeneous screening[22] wrongly predict a decrease of ionicity of LiH in a strong electric field. Thus, Coulomb correlations among the electrons in the *inhomogeneous* electron gas of a LiH crystal are essential for a field-driven increase of ionicity of LiH as observed in the femtosecond diffraction experiments.

6   Conclusion

In conclusion, we studied in crystalline $NaBH_4$, $LiBH_4$, and LiH spatially resolved electron density maps determined by femtosecond X-ray powder diffraction in the response to a non-resonant strong electric field. Our experiments reveal the prominent role of electron correlations for the field-induced deformation of the electron density maps. In LiH, field-induced correlations between states in the valence and different conduction bands result in an enhancement of ionicity which is manifested in an electron transfer from Li to H and in agreement with quasi-particle calculations within the COHSEX formalism. In contrast, both $NaBH_4$ and $LiBH_4$ display an electron transfer from BH_4^- to the respective cation, as expected for an admixture of states in the lowest conduction band. Furthermore, the distorted geometry of the BH_4 tetrahedron in $LiBH_4$ leads to different contributions of the H atoms to electron transfer. Our results demonstrate the strong potential of femtosecond diffraction methods to uncover microscopic charge dynamics and determine electron transport mechanisms in crystalline matter.

Acknowledgements

This research has received funding from the European Research Council under the European Union's Seventh Framework Programme (FP7/2007-2012)/ERC Grant Agreement 247051 and from the Deutsche Forschungsgemeinschaft (Grant WO 558/13-1).

References

1 W. Kuehn, P. Gaal, K. Reimann, M. Woerner, T. Elsaesser and R. Hey, *Phys. Rev. Lett.*, 2010, **104**, 146602.
2 G. Herink, D. R. Solli, M. Gulde and C. Ropers, *Nature*, 2012, **483**, 190.
3 W. Kuehn, P. Gaal, K. Reimann, M. Woerner, T. Elsaesser and R. Hey, *Phys. Rev. B*, 2010, **82**, 075204.
4 T. E. Glover, D. M. Fritz, M. Cammarata, T. K. Allison, Sinisa Coh, J. M. Feldkamp, H. Lemke, D. Zhu, Y. Feng, R. N. Coffee, M. Fuchs, S. Ghimire, J. Chen, S. Shwartz, D. A. Reis, S. E. Harris and J. B. Hastings, *Nature*, 2012, **488**, 603.
5 M. Schultze, E. M. Bothschafter, A. Sommer, S. Holzner, W. Schweinberger, M. Fiess, M. Hofstetter, R. Kienberger, V. Apalkov, V. S. Yakovlev, M. I. Stockman and F. Krausz, *Nature*, 2013, **493**, 75.
6 A. Schiffrin, T. Paasch-Colberg, N. Karpowicz, V. Apalkov, D. Gerster, S. Mhlbrandt, M. Korbman, J. Reichert, M. Schultze, S. Holzner, J. V. Barth, R. Kienberger, R. Ernstorfer, V. S. Yakovlev, M. I. Stockman and F. Krausz, *Nature*, 2013, **493**, 70.
7 T. Elsaesser and M. Woerner, *J. Chem. Phys.*, 2014, **140**, 020901.
8 M. Bargheer, N. Zhavoronkov, Y. Gritsai, J. C. Woo, D. S. Kim, M. Woerner and T. Elsaesser, *Science*, 2004, **306**, 1771.
9 M. Woerner, F. Zamponi, Z. Ansari, J. Dreyer, B. Freyer, M. Prémont-Schwarz and T. Elsaesser, *J. Chem. Phys.*, 2010, **133**, 064509.
10 F. Zamponi, P. Rothhardt, J. Stingl, M. Woerner and T. Elsaesser, *Proc. Natl. Acad. Sci. U. S. A.*, 2012, **109**, 5207.
11 B. Freyer, F. Zamponi, V. Juv, J. Stingl, M. Woerner, T. Elsaesser and M. Chergui, *J. Chem. Phys.*, 2013, **138**, 144504.
12 J. Stingl, F. Zamponi, B. Freyer, M. Woerner, T. Elsaesser and A. Borgschulte, *Phys. Rev. Lett.*, 2012, **109**, 147402.
13 V. Juvé, M. Holtz, F. Zamponi, M. Woerner, T. Elsaesser and A. Borgschulte, *Phys. Rev. Lett.*, 2013, **111**, 217401.

14 C. Cohen-Tannoudji, B. Diu and F. Laloë, *Quantum Mechanics*, Wiley Interscience Publication, New York, 1977, Vol. 1–2.

15 D. H. Ewing and F. Seitz, *Phys. Rev.*, 1936, **50**, 760.

16 W. Kohn, *Phys. Rev.*, 1957, **105**, 509.

17 R. E. Behringer, *Phys. Rev.*, 1959, **113**, 787.

18 G. W. Pratt, Jr, *Phys. Rev.*, 1960, **118**, 462.

19 P. Hohenberg and W. Kohn, *Phys. Rev.*, 1964, **136**, B864.

20 L. Hedin, *Phys. Rev.*, 1965, **139**, A796.

21 K.-F. Bergren, *J. Phys. C: Solid State Phys.*, 1969, **2**, 802.

22 S. Baroni, G. Pastori Parravicini and G. Pezzica, *Phys. Rev. B*, 1985, **32**, 4077.

23 M. S. Hybertsen and S. G. Louie, *Phys. Rev. B*, 1986, **34**, 5390.

24 L. Bellaiche, J. M. Besson, K. Kunc and B. Lévy, *Phys. Rev. Lett.*, 1998, **80**, 5576.

25 S. Lebégue, M. Alouani, B. Arnaud and W. E. Pickett, *Europhys. Lett.*, 2003, **63**, 562.

26 N. Novakovic, I. Radisavljevic, D. Colognesi, S. Ostojic and N. Ivanovic, *J. Phys.: Condens. Matter*, 2007, **19**, 406211.

27 F. Zamponi, Z. Ansari, M. Woerner and T. Elsaesser, *Opt. Express*, 2010, **18**, 947.

28 N. Zhavoronkov, Y. Gritsai, M. Bargheer, M. Woerner, Th. Elsaesser, F. Zamponi, I. Uschmann and E. Förster, *Opt. Lett.*, 2005, **30**, 1737.

29 R. S. Calder, W. Cochran, D. Griffiths and R. D. Lowde, *J. Phys. Chem. Solids*, 1962, **23**, 621.

30 G. Vidal-Valat and J.-P. Vidal, *Acta Crystallogr., Sect. A: Found. Crystallogr.*, 1992, **48**, 46.

31 P. Fischer and A. Züttel, *Mater. Sci. Forum*, 2004, **Vols. 443–444**, 287.

32 R. S. Kumar and A. L. Cornelius, *Appl. Phys. Lett.*, 2005, **87**, 261916.

33 J. Soulié, G. Renaudin, R. Cerny and K. Yvon, *Journal of alloys and compounds*, 2002, **346**, 200.

34 E. Jaynes, *Phys. Rev.*, 1957, **108**, 171.

35 C. Gilmore, *Acta Crystallogr., Sect. A: Found. Crystallogr.*, 1996, **52**, 561.

36 M. Sakata and M. Sato, *Acta Crystallogr., Sect. A: Found. Crystallogr.*, 1990, **46**, 263.

37 S. Smaalen, L. Palatinus and M. Schneider, *Acta Crystallogr., Sect. A: Found. Crystallogr.*, 2003, **59**, 459.

38 P. Coppens, *X-Ray Charge Densities and Chemical Bonding*, Oxford University Press, 1997.

39 D. M. Collins, *Nature*, 1982, **298**, 49.

40 L. Palatinus, Maximum Entropy Method in Superspace Crystallography, PhD thesis, University of Bayreuth, Germany, 2003.

41 N. Wiser, *Phys. Rev.*, 1963, **129**, 62.

42 A. Baldereschi and E. Tosatti, *Solid State Commun.*, 1979, **29**, 131.

43 I. Ohkoshi, *J. Phys. C: Solid State Phys.*, 1985, **18**, 5415.

44 J. S. Langer and S. H. Vosko, *J. Phys. Chem. Solids*, 1959, **12**, 196.

45 D. J. Chadi and M. L. Cohen, *Phys. Rev. B: Solid State*, 1973, **8**, 5747.

Faraday Discussions

PAPER

Toward atomic resolution diffractive imaging of isolated molecules with X-ray free-electron lasers

S. Stern,†[ab] L. Holmegaard,[ae] F. Filsinger,‡[fg] A. Rouzée,[hi] A. Rudenko,[gjk] P. Johnsson,[l] A. V. Martin,§[a] A. Barty,[a] C. Bostedt,[m] J. Bozek,[m] R. Coffee,[m] S. Epp,[gj] B. Erk,[cgj] L. Foucar,[gn] R. Hartmann,[o] N. Kimmel,[p] K-U. Kühnel,[j] J. Maurer,[e] M. Messerschmidt,[m] B. Rudek,¶[gj] D. Starodub,‖[q] J. Thøgersen,[e] G. Weidenspointner,**[pr] T. A. White,[a] H. Stapelfeldt,[es] D. Rolles,[cgn] H. N. Chapman[abd] and J. Küpper*[abd]

Received 2nd March 2014, Accepted 21st March 2014

DOI: 10.1039/c4fd00028e

[a]Center for Free-Electron Laser Science (CFEL), Deutsches Elektronen-Synchrotron (DESY), Notkestrasse 85, 22607 Hamburg, Germany. E-mail: jochen.kuepper@cfel.de; stephan.stern@desy.de; Web: http://desy.cfel.de/cid/cmi

[b]Department of Physics, University of Hamburg, Luruper Chaussee 149, 22761 Hamburg, Germany

[c]Deutsches Elektronen-Synchrotron (DESY), Notkestrasse 85, 22607 Hamburg, Germany

[d]The Hamburg Center for Ultrafast Imaging, University of Hamburg, Luruper Chaussee 149, 22761 Hamburg, Germany

[e]Aarhus University, Department of Chemistry, 8000 Aarhus C, Denmark

[f]Fritz Haber Institute of the MPG, Faradayweg 4–6, 14195 Berlin, Germany

[g]Max Planck Advanced Study Group at CFEL, Notkestrasse 85, 22607 Hamburg, Germany

[h]FOM Institute AMOLF, Science Park 104, 1098 XG Amsterdam, The Netherlands

[i]Max-Born-Institute, Max Born Str. 2a, 12489 Berlin, Germany

[j]Max Planck Institute for Nuclear Physics, 69117 Heidelberg, Germany

[k]J. R. Macdonald Laboratory, Department of Physics, Kansas State University, Manhattan, KS 66506, USA

[l]Lund University, Department of Physics, P. O. Box 118, 22100 Lund, Sweden

[m]Linac Coherent Light Source, SLAC National Accelerator Laboratory, 2575 Sand Hill Road, Menlo Park, CA 94025, USA

[n]Max Planck Institute for Medical Research, 69120 Heidelberg, Germany

[o]PNSensor GmbH, 81739 Munich, Germany

[p]Max Planck Semiconductor Laboratory, 81739 Munich, Germany

[q]Department of Physics, Arizona State University, Tempe, AZ 85287, USA

[r]Max Planck Institute for Extraterrestrial Physics, 85741 Garching, Germany

[s]Aarhus University, Interdisciplinary Nanoscience Center (iNANO), 8000 Aarhus C, Denmark

† Presenting author at the Faraday Discussion.

‡ Present address: Bruker AXS GmbH, Karlsruhe, Germany

§ Present address: ARC Centre of Excellence for Coherent X-ray Science, School of Physics, The University of Melbourne, Australia

¶ Present address: Physikalisch-Technische Bundesanstalt, Bundesallee 100, 38116 Braunschweig, Germany

‖ Present address: Stanford PULSE Institute, SLAC National Accelerator Laboratory, 2575 Sand Hill Road, Menlo Park, California 94025, USA

** Present address: European X-ray Free Electron Laser (XFEL) GmbH, 22761 Hamburg, Germany

We give a detailed account of the theoretical analysis and the experimental results of an X-ray-diffraction experiment on quantum-state selected and strongly laser-aligned gas-phase ensembles of the prototypical large asymmetric rotor molecule 2,5-diiodobenzonitrile, performed at the Linac Coherent Light Source [*Phys. Rev. Lett.* **112**, 083002 (2014)]. This experiment is the first step toward coherent diffractive imaging of structures and structural dynamics of isolated molecules at atomic resolution, *i.e.*, picometers and femtoseconds, using X-ray free-electron lasers.

1   Introduction

The advent of X-ray Free-Electron Lasers (XFELs) opens up new and previously inaccessible research directions in physical and chemical sciences. One of the major scopes is the utilization of XFEL radiation in diffractive imaging experiments. Collecting single-shot X-ray diffraction patterns with the ultrashort, currently down to a few femtoseconds, X-ray pulses of extremely high brilliance at an XFEL allows the conventional damage limit in imaging of non-crystalline biological samples to be circumvented.[1] Experiments at the Linac Coherent Light Source (LCLS) confirmed the feasibility of utilizing XFELs for femtosecond single-shot imaging of non-crystalline biological specimens[2] as well as for femtosecond nanocrystallography of proteins.[3]

These results provide important steps on the path towards the paramount goal of atomically (picometer and femtoseconds) resolved diffractive imaging of structures and ultrafast structural dynamics during chemical reactions of even single molecules. However, the path toward this goal, often nicknamed as "recording of a molecular movie", is still long and many challenges have to be overcome in order to achieve the required spatio-temporal resolution.[4,5] The usually proposed experimental approach is to provide identical molecules, delivered in a liquid or gaseous stream to the focus of an XFEL.[6,7] Since the high single-shot XFEL intensity by far exceeds the damage threshold of single molecules, the molecules have to be replenished in each shot. Single-molecule diffraction data has to be collected for many shots with the molecule at many different orientations in order to fill up the three-dimensional diffraction volume. The relative orientation of single-molecule diffraction patterns from distinct shots could be determined computationally from the diffraction patterns themselves provided that the single-molecule diffraction signal is well above noise.[8–10] However, one of the main issues in single-molecule X-ray diffraction experiments is the weak scattering signal from single molecules, which, so far, is too weak to allow for orientation classification solely from the diffraction pattern, even at the high intensities of the novel XFELs. Therefore, diffraction data has to be recorded and averaged for many shots with the molecule at the same, pre-imposed alignment and/or orientation†† in space in order to obtain an interpretable diffraction pattern above noise. Strong molecular alignment in the laboratory frame can be achieved, for instance, through adiabatic laser alignment, while orientation requires additional dc electric fields.[11–14] Alignment and orientation can be varied easily by controlling the the alignment laser polarization and, in case orientation

---

†† Alignment refers to fixing one or more molecular axes in space, while orientation refers to breaking of the corresponding up-down symmetry.

is utilized as well, the direction of the dc field. Utilizing ensembles of such aligned molecules allows for averaging of many identical patterns, similar to recent experiments exploiting electron diffraction from CF_3I[15] or photoelectron imaging of 1-ethynyl-4-fluorobenzene[16,17] and dibromobenzene.[18]

An obstacle to this concept is that complex large molecules typically exist in various structural isomers, *e.g.*, conformers, which are often difficult to separate due to the small energy difference and low barriers between them. However, to achieve atomic-resolution in diffractive imaging experiments they have to be analyzed separately. We have proposed[7] to solve this by spatially separating shapes,[19] sizes,[20] or individual isomers[21–23] of the molecules before delivery to the interaction point of the experiment. These pre-selected ensembles can be efficiently, one- and three-dimensionally, aligned or oriented in the laboratory frame.[13,24,25]

Here, we give a detailed account of an X-ray diffraction experiment of ensembles of isolated gas-phase molecules at the Linac Coherent Light Source (LCLS).[26] Cold, state-selected, and aligned ensembles of the prototypical molecule 2,5-diiodobenzonitrile ($C_7H_3I_2N$, DIBN) were irradiated with XFEL pulses with a photon energy of 2 keV ($\lambda = 620$ pm) and X-ray diffraction data was recorded and analyzed. DIBN was utilized for this proof-of-principle experiment because it contains two heavy atoms (iodine) and it can be laser-aligned along an axis almost exactly coinciding with the iodine-iodine axis. Therefore, as the two-center iodine-iodine interference dominates the scattering signal, the experiment resembles Young's double slit on the atomic level. We achieved strong laser-alignment of the ensemble of DIBN molecules which allowed for averaging of many patterns from these weakly scattering molecules.

The outline of this paper is as follows: In section 2 the experimental setup is introduced. This includes details on the preparation of the molecular sample for the X-ray diffraction experiment: we present measurements of the molecular beam deflection profiles and two-dimensional ion-momentum distributions from which the molecular alignment of DIBN is quantified. In addition, the process of data acquisition, background subtraction, and spatial single-photon counting with the pnCCD photon detector[27,28] is outlined very briefly, while a comprehensive explanation of all the steps involved in the procedure of conditioning and correcting the X-ray diffraction data is given in Appendix A. The theory behind the numerical simulations of X-ray diffraction intensities to be compared with the experimental diffraction data is outlined in section 3. In section 4 the experimental results are presented and the manuscript concludes with a summary of the experimental findings and an outlook on future experiments in section 5.

2 Experimental

2.1 Experimental setup

The experiment was performed at the Atomic, Molecular, and Optical Physics (AMO) beamline[29,30] of LCLS,[31] using the CAMP (CFEL-ASG Multi-Purpose) experimental chamber.[27,32] The CAMP instrument was equipped with a state-of-the-art molecular beam setup providing gas-phase ensembles of cold and quantum-state selected target molecules.[20–24] For the X-ray diffraction experiment, a photon energy of 2 keV ($\lambda = 620$ pm) was used, which is the maximum photon energy available at AMO. The 2 keV X-ray pulses were focussed by a Kirkpatrick-

Baez (KB) mirror system into the CAMP experimental chamber, which was attached to the High Field Physics (HFP) chamber at the AMO beamline. The CAMP instrument contains multiple detectors to detect photons, electrons, and ions simultaneously and it is described in detail elsewhere.[27]

Fig. 1 shows a schematic view of the experimental setup inside CAMP. During the experiment, a pulsed molecular beam was formed by a supersonic expansion of a mixture of a few mbar of DIBN and 50 bar of helium (He) into vacuum through an Even-Lavie valve.[34] The target molecules were cooled to low rotational temperatures of ~1 K in the early stage of the expansion by collisions with the He seed gas.[35,36] Travelling through the electrostatic deflector, the molecules were dispersed along the vertical (y) axis according to their effective dipole moment, i. e., their quantum state. The deflector consists of two 24 cm-long electrodes, a cylindrical rod electrode at the top and a trough electrode at the bottom. The vertical distance between the two electrodes in the horizontal center of the deflector is 2.3 mm. By application of high static electric potentials of ±10 kV to the top and bottom electrodes, a strong inhomogeneous static electric field was created with an electric field strength of 120 kV cm^{-1} and an electric field gradient of 250 kV cm^{-2} in the center of the deflector as depicted in the inlet of Fig. 1. Quantum-state selection *via* the deflector is achieved due to the different Stark effect of distinct quantum states (*vide infra*). Furthermore, spatial separation of polar DIBN and non-polar He seed gas in the deflector was utilized to reduce the scattering background from the He in the X-ray diffraction experiment.

After passing through the deflector, the quantum-state dispersed molecular beam entered the detection chamber where it was crossed by three pulsed laser beams: Pulses from a Nd:YAG laser (YAG, 12 ns (FWHM), $\lambda = 1064$ nm, $E_I = 200$ mJ, $\omega_0 = 63$ μm, $I_0 \approx 2.5 \times 10^{11}$ W cm^{-2}) were used to align the ensemble of target molecules. The second laser, a Ti:Sapphire laser (TSL, 60 fs (FWHM), 800 nm, $E_I = 400$ μJ, $\omega_0 = 40$ μm, $I_0 \approx 2.5 \times 10^{14}$ W cm^{-2}) was used to ionize DIBN in order to optimize the molecular beam and the alignment without the LCLS beam. X-ray pulses from LCLS (100 fs, estimated from electron bunch length and pulse duration measurements,[37] $\lambda = 620$ pm, $E_I = 4$ mJ, $\omega = 30$ μm, $I_0 \approx 2 \times 10^{15}$ W cm^{-2}) were used to probe the ensemble of aligned DIBN. We deliberately worked out-of-focus of the X-ray beam at low fluence in order to mitigate electronic[38-40] and nuclear damage processes.[41] The X-ray photons diffracted from the ensemble were collected by the pnCCD photon detector at a distance (i. e., camera length) of 71 mm. 35% of the generated 1.25×10^{13} X-ray photons/pulse were estimated to be transported to the experiment.[42] The two panels of the pnCCD detector were opened by a significant amount in order to cover large scattering angles, i. e., the top pnCCD panel was moved by 44 mm (covering scattering angles of $31° \leq 2\Theta \leq 50°$) and the bottom panel to a distance of 17 mm ($13° \leq 2\Theta \leq 38°$) from the z-axis. All three laser beams were co-propagating, overlapped using dichroic (1064 nm and 800 nm) and holey (X-ray and infrared beams) mirrors. After intersecting the sample the lasers finally left the setup through a gap between the two panels of the pnCCD camera and another holey mirror in the back of the CAMP chamber to separate the laser beams again. Straylight from the optical lasers was reduced using a set of apertures mounted in a small tube directly in front of the interaction zone (named "light baffling tube" in Fig. 1). A similar light baffling tube was mounted downstream the interaction zone, reaching between the two pnCCD panels and containing a similar set of apertures in order to suppress straylight

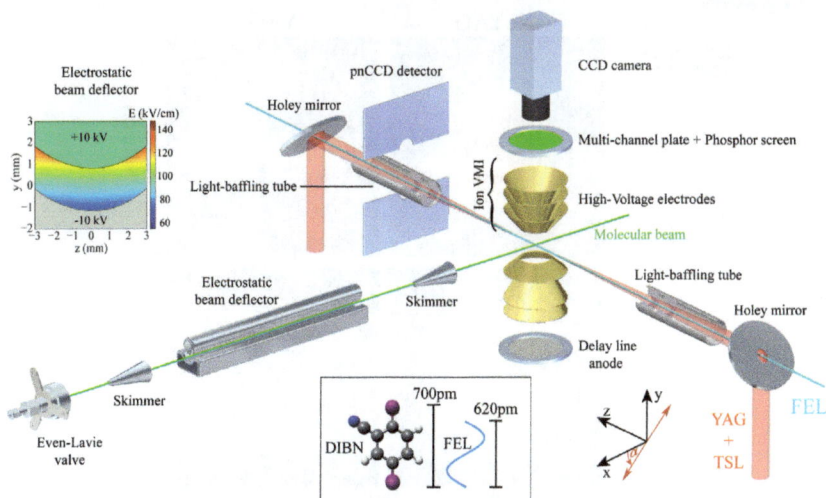

Fig. 1 Schematic view of the experimental setup inside the CAMP experimental chamber. The molecular beam, created by supersonic expansion of DIBN and He from the Even-Lavie valve on the left, enters the deflector and quantum-state selected molecules are delivered to the interaction point. In the center of the velocity map imaging spectrometer (VMI) the molecular beam is crossed by the laser beams copropagating from right to left. The direct laser beams pass through a gap in the pnCCD photon detectors that are used to record the X-ray diffraction pattern. The upper pnCCD panel is further away from the beam axis than the bottom panel in order to cover a wider range of scattering angles. The inlet on the upper left shows a cross section of the electrostatic beam deflector along the propagation direction of the molecular beam. The inlet on the lower edge illustrates the two significant lengthscales of the X-ray diffraction experiment, namely the molecular structure of DIBN with the iodine-iodine distance and the X-ray wavelength. The molecular structure of DIBN was obtained from *ab initio* calculations (GAMESS-US,[33] MP2/ 6-311G**), which predict a value of 700 pm for the iodine-iodine distance. Figure reproduced from ref. 26.

from optical or X-ray photons impinging from the back of the CAMP chamber onto the back of the pnCCD panels. In addition, the front side (*i. e.*, the side facing the interaction zone) of each pnCCD panel was covered using aluminum-coated filters in order to further suppress straylight from the optical lasers.

The CAMP chamber was equipped with a dual velocity-map-imaging (VMI) spectrometer in order to measure two-dimensional ion momentum distributions in the x-z plane, resulting from Coulomb explosion due to absorption of one or a few X-ray photons (or optical photons in case the TSL was utilized to probe the molecular alignment).[27] Operation of the VMI spectrometer as an ion time-of-flight (TOF) spectrometer in quasi-Wiley-McLaren configuration[43] allowed for mass selective detection of individual ionic fragments.

The X-ray diffraction experiment was performed with LCLS running at a repetition rate of 60 Hz while the YAG was running at 30 Hz. Hence, a dataset contains shots of aligned and randomly oriented molecules in an alternating manner. All diffraction measurements were conducted in the deflected part of the molecular beam, *i. e.*, at (nearly) optimal molecular alignment (*vide infra*). In the following, experimental results concerning preparation of the molecular

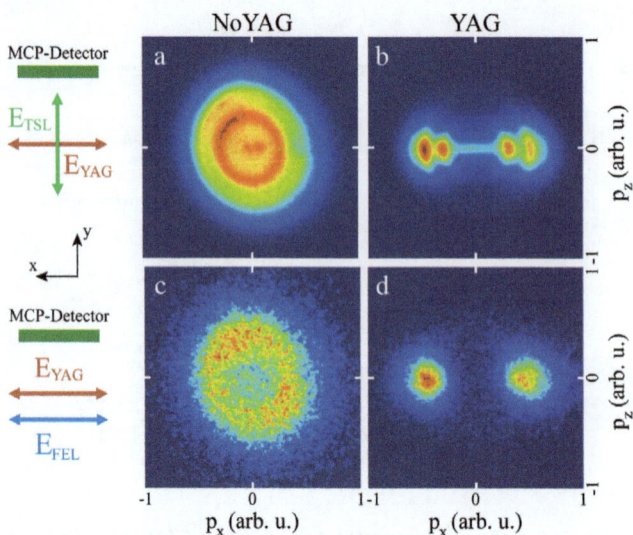

Fig. 2 I^+ ion momentum distributions recorded with the ion–VMI and MCP detector when the TSL (a,b) or the LCLS (c, d) was used to ionize and Coulomb explode the molecules. In (a, c) cylindrically symmetric distributions from isotropic ensembles are observed (the images are slightly distorted due to varying detector efficiencies). In (b, d) the horizontal alignment of the molecules, induced by the YAG, is clearly visible. In all measurements the YAG and the LCLS are linearly polarized along the x-axis, i. e., parallel to the detector plane, and the TSL is linearly polarized along the y-axis, i. e., perpendicular to the detector plane. Figure reproduced from ref. 26.

ensemble for the X-ray diffraction experiment are presented, namely quantum-state selection by deflection and laser-alignment.

2.2 Quantum-state selection and laser alignment

The benefit of quantum-state selection prior to laser alignment for cold ensembles of asymmetric top molecules[13] was exploited in our experiment in order to obtain strong alignment of the molecular sample for the X-ray diffraction experiment. For a large asymmetric top molecule such as DIBN, all populated rotational states in the molecular beam are so-called high-field-seeking (hfs) states. Molecules in these states are deflected towards increasing electric field strength, i. e., upwards along the y-axis.[24,44] The lowest states typically exhibit the largest Stark energy shift and, thus, the strongest deflection. Quantum-state selection is very beneficial for laser-alignment: As the lowest-lying states experience a stronger angular confinement in the electric field of a linearly polarized alignment laser, selection of the lowest-lying states prior to alignment significantly improves the degree of alignment.[13,24,25]

When the linearly polarized YAG was included, DIBN molecules aligned along their most-polarizable axis, which is nearly coincident with the iodine-iodine (I–I) axis. Utilizing Coulomb explosion imaging of aligned DIBN, induced by either the TSL or the FEL, strong alignment of DIBN ensembles was confirmed by two-dimensional momentum distributions of I^+ ions (which recoil along the iodine-iodine axis) recorded with the velocity-map imaging (VMI) spectrometer. Fig. 2

shows corresponding I^+ momentum distributions, recorded with (YAG) and without (NoYAG) the YAG alignment laser. In the NoYAG case, the I^+ images are circularly symmetric corresponding to an ensemble of isotropically-distributed molecules. The circularly symmetric image Fig. 2c, obtained following ionization with the horizontally polarized FEL also demonstrated that the interaction of the far-off resonant radiation with the molecule was independent of the angle between the molecular axis and the X-ray polarization direction: The x rays were a practically unbiased ideal probe of spatial orientation of molecules. Including the YAG laser, I^+ ions were strongly confined along the polarization axis of the YAG. The two distinct pairs of peaks in the TSL case correspond to two distinct ionization channels yielding I^+ ions from doubly and triply ionized molecules.[45] The degree of alignment is quantified by calculating $\langle \cos^2 \theta_{2D} \rangle$, where θ_{2D} is the angle with respect to the laser polarization axis in the projected, two-dimensional I^+ momentum distributions.

The deflector was utilized to improve the degree of alignment by quantum-state selection of the lowest states. Fig. 3 shows molecular beam density profiles obtained by measuring the I^+ signal probed at distinct positions along the y-axis when the deflector was off (blue) or on (green, 20 kV). Both graphs were normalized to the peak intensity. Only the upper part of the molecular beam was probed. The different deflection of distinct quantum states in the molecular beam leads to a shift of the beam profile as is shown in Fig. 3. The corresponding dispersion of quantum states can be illustrated by recording I^+ momentum distributions at distinct positions: as expected, the degree of alignment is significantly enhanced in the deflected part of the molecular beam. The resulting $\langle \cos^2 \theta_{2D} \rangle$ values are depicted by the red graph of Fig. 3 and the enhanced

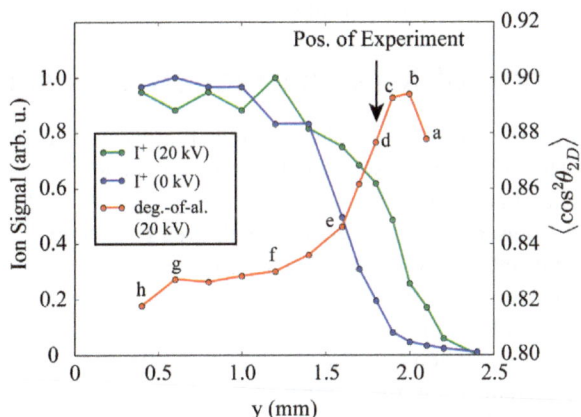

Fig. 3 The molecular beam density profiles, obtained by recording the I^+ signal (see left vertical axis) at different positions along the y-axis in the molecular beam for the undeflected (blue) and deflected (green) molecular beam. The different degree of alignment of DIBN in terms of $\langle \cos^2 \theta_{2D} \rangle$ (right vertical axis) at different positions in the deflected molecular beam illustrates the dispersion of quantum states (red). Considering the best compromise between degree of alignment and sufficient molecular beam density of target molecules, the X-ray diffraction experiment was performed at $y = 1.8$ mm ($\langle \cos^2 \theta_{2D} \rangle = 0.877$), not at the position where the highest degree of alignment was observed, i. e., $\langle \cos^2 \theta_{2D} \rangle = 0.894$ at $y = 2$ mm.

alignment in the deflected part of the molecular beam is obvious. The strongest alignment, quantified by $\langle \cos^2\theta_{2D}\rangle = 0.894$, was obtained at $y = 2$ mm. However, to utilize a higher beam density, the X-ray diffraction experiment was performed at $y = 1.8$ mm. At this position the degree of alignment was only slightly smaller ($\langle \cos^2\theta_{2D}\rangle = 0.877$), but the molecular beam density was still 60% of the unde-flected beam density, whereas it was only 20% at $y = 2$ mm.

During the X-ray diffraction experiment, the YAG polarization was rotated to $\alpha = -60°$ with respect to the horizontal axis. The alignment was probed repeatedly over the course of the X-ray diffraction measurement period of ~ 8 h and the average degree of alignment was $\langle \cos^2\theta_{2D}\rangle = 0.84$, mainly due to variations of the overlap of the YAG and FEL pulses. The degree of alignment is in agreement with measurements of adiabatic alignment of quantum-state selected ensembles of similar molecules[13,25] and matches requirements for diffraction experiments on aligned molecules.[7,15]

2.3 X-Ray diffraction data acquisition

A comprehensive description of the data conditioning procedure is given in Appendix A. In summary, single shot X-ray diffraction data was recorded by the pnCCD detectors and saved to file. Several sources of background signals (offset, gain, experimental background from the YAG alignment laser, *etc.*) and detector artifacts ("hot-pixels", *etc.*) were subtracted from the data by utilizing the CFEL-ASG Software Suite (CASS).[32] Eventually, single X-ray photon hits were extracted by application of a 3σ-threshold to these "clean" single-shot pnCCD data frames. This procedure yields 0.2 X-ray photons per shot (*i. e.*, on average only one scat-tered X-ray photon in five shots) that are scattered to the pnCCD detector. These photons are placed in a histogram which represents the molecular diffraction pattern obtained from aligned (labelled "YAG") and isotropically distributed ("NoYAG") ensembles of DIBN molecules.

3   Simulation of X-ray diffraction intensities

Diffraction intensities from ensembles of aligned and not-aligned DIBN mole-cules and the He seed gas were simulated for comparison with the experimental data. Unless stated otherwise, the underlying theory is either explicitly given by the book of Als-Nielsen & McMorrow[46] or was derived from there.[47]

X-ray scattering off ensembles of isolated molecules is very weak and hence the kinematical approximation (first Born approximation) is assumed to be valid, meaning that multiple scattering of a single photon is highly unlikely and can be neglected. For all calculations, the interaction point is regarded as the origin of the coordinate system. Then, the number of X-ray photons I_{sc} that are scattered from a single molecule to a certain pixel at position $\mathbf{R}$ can be calculated as

$$I_{sc} = \left| r_0 \cdot F_{mol}(\mathbf{q}) \cdot e^{i\mathbf{kR}} \right|^2 \cdot \Delta\Omega \cdot P \cdot \frac{I_0}{A_0} \qquad (1)$$

where r_0 is the Thomson scattering length of the electron which is given by

$$r_0 = \frac{e^2}{4\pi\varepsilon_0 mc^2} = 2.82\cdot 10^{-5}\ \text{Å} \qquad (2)$$

Utilizing conventional notation, $\mathbf{q} = \mathbf{k} - \mathbf{k}'$ is the scattering vector with $\mathbf{k}$ and $\mathbf{k}'$ being the wavevectors of the incident and scattered waves, respectively. $F_{mol}(\mathbf{q})$ is the molecular scattering factor (see below). $\Delta\Omega$ is the solid angle a certain pixel subtends to the incident XFEL beam, and P is the polarization factor depending on the X-ray source. Since LCLS is linearly polarized (along the x-axis), P takes the following form: $P(\mathbf{k}') = 1 - |\hat{\mathbf{u}} \cdot \hat{\mathbf{k}}'|^2$ with the unit vector $\hat{\mathbf{u}}$ pointing along the x-axis.[48] Finally, the number of incident photons is given by I_0 and the cross-sectional area of the incident X-ray beam is represented by A_0.

The scattering factor of a molecule $F_{mol}(\mathbf{q})$ is modeled as the sum of the atomic scattering factors $f_j(\mathbf{q})$ of the constituent j atoms (located at the positions $\mathbf{r}_j$ within the molecule) times the phase factor $e^{i\mathbf{q}\mathbf{r}_j}$, hence

$$F_{mol}(\mathbf{q}) = \sum_j f_j(\mathbf{q}) \, e^{i\mathbf{q}\mathbf{r}_j} \qquad (3)$$

A model of the atomic scattering factors f_j has been given by Waasmaier & Kirfel[49] by modelling atomic scattering factors in dependence of the scattering momentum transfer $s = \sin\Theta/\lambda$ as the sum of five gaussian functions and a constant.

$$f(s) = \sum_{i=1}^{5} a_i \, e^{-b_i \, s^2} + const. \qquad (4)$$

For the calculations presented here, the atomic scattering factors were modified by dispersion corrections given by Henke et al.,[50] thereby accounting for the dependence of the scattering strength from the photon energy.

(1) was used to calculate the diffraction pattern for a perfectly aligned molecule. However, the experimental diffraction pattern of an ensemble of DIBN molecules with a finite (i. e., non-perfect) degree of alignment is the incoherent superposition of single-molecule diffraction patterns at slightly different orientations with respect to the (linear) laser polarisation of the YAG. The relative weight of different orientations are described by an alignment-angular distribution function giving the relative population $n(\theta)$ where θ is the angle with respect to the YAG polarisation axis. The following approximation for strong alignment was applied in our model:[51]

$$n(\theta) = \exp\left(-\frac{\sin^2\theta}{2\sigma^2}\right) \qquad (5)$$

In practice, the blurred single-molecule diffraction pattern was obtained by averaging of single-molecule diffraction patterns calculated for 1000 distinct orientations of DIBN, weighted by (5). Then, this pattern is multiplied by the number of molecules N in the interaction volume V_0 in order to obtain the diffraction pattern of N molecules. However, as long as the X-ray beam is smaller than the molecular beam, the absolute number doesn't have to be known but rather the number density M of molecules: The number of molecules N can be written as $N = M \cdot V_0 = M \cdot A_0 \cdot l$ where the interaction volume is approximated as a cylindrical volume of lenght l in z-direction, and, in our case, l is the width of the

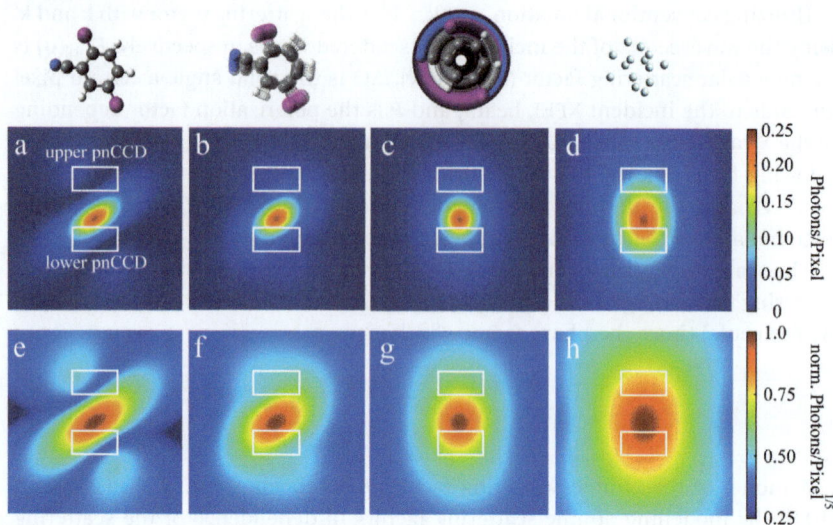

Fig. 4 Simulated scattering intensities for different degrees of alignment. a–d correspond to DIBN aligned with $\langle \cos^2\theta_{2D} \rangle = 0.99$ (a), 0.84 (b), 0.5 (isotropic,c). The signal for 5 580 He atoms (d) is the same as the DIBN signal at $q = 0$. To illustrate interference features (i. e., the weak first order diffraction maxima), the second row (e–h) shows the fifth root of the normalized intensities of the first row.

molecular beam which is ≈ 4 mm (determined by the last skimmer). Therefore, once (1) was multiplied by N, the factor N/A_0 in (1) could be replaced by $M \cdot l$.

Fig. 4 shows simulated diffraction patterns, i. e., the number of scattered photons on a plane detector at a camera length of 71 mm, for different degrees of alignment for 565 000 shots (4.375×10^{12} photons/shot), and a molecular beam density of $M = 1.2 \times 10^8$ cm^{-3}. The molecules were aligned at $\alpha = -60°$ with respect to the horizontal plane. White rectangles mark the position of the pnCCDs in the experiment. Images a–c correspond to DIBN aligned with $\langle \cos^2\theta_{2D} \rangle = 0.99$ (a), 0.83 (b), 0.5 (isotropic, c). The diffraction signal from 5 580 He atoms (d) at $q = 0$ is equal to the diffraction signal from a single DIBN molecule at $q = 0$. We do not exactly know the ratio of He atoms per DIBN molecule in our molecular beam, but it is in the 10^4–10^5 range. In order to illustrate interference features of the weak first order diffraction maxima, a different colorscale has been applied, enhancing the first-order iodine-iodine diffraction maxima: therefore, the second row (e–h) shows the fifth root of the normalized intensities. The main interference feature, originating in the interference of the two iodine atoms, is clearly visible for nearly perfect alignment, i. e., in Fig. 4 (e) while non-perfect alignment (f) significantly washes out the interference features at high angles.

4   Results and discussion

Diffraction patterns I_{NoYAG} and I_{YAG} were constructed independently for isotropic (NoYAG) and aligned (YAG) samples, respectively, by summing all photon hits in the energy range around 2 keV, corresponding to 1500–3200 ADU (analog-to-digital unit, see Appendix A) into a two-dimensional histogram. The resulting

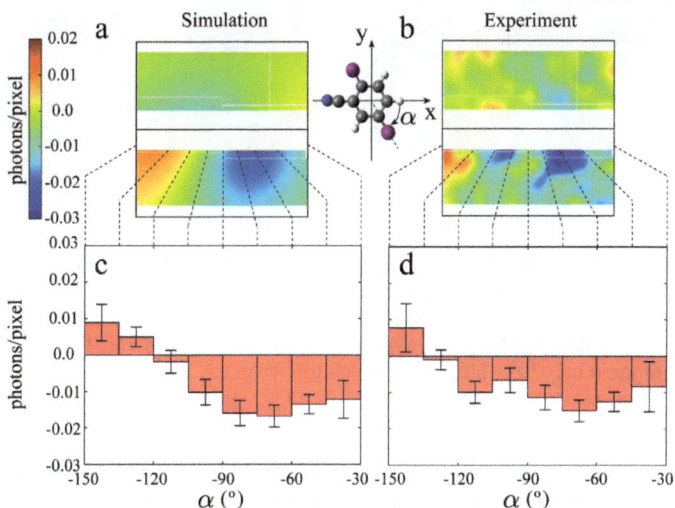

Fig. 5 Diffraction-difference $I_{YAG} - I_{NoYAG}$ of X-ray scattering in simulated (a) and experimental (b) X-ray-diffraction patterns. Histograms of the corresponding angular distributions on the bottom pnCCD (c, d) illustrate the angular anisotropy of the diffraction signal. Error bars correspond to 1σ statistical errors from Poisson noise. The molecular beam density in (a) is $M = 0.8 \times 10^8$ cm^{-3}. Figure reproduced from ref. 26.

images are shown in Fig. 15 c and d in Appendix A. In addition to the diffraction signal from aligned DIBN, the I_{NoYAG}- and I_{YAG}-data contain experimental background such as the isotropic atomic scattering from all individual atoms of DIBN, scattering from the helium seed gas, scattering from residual gas in the chamber, and scattering at apertures in the laser beam path. Since the scattering background from all these sources is the same under NoYAG and YAG conditions, it cancels out when calculating $I_{YAG} - I_{NoYAG}$.

Fig. 5 shows the diffraction-difference pattern $I_{YAG} - I_{NoYAG}$ for (a) simulated and (b) experimentally recorded X-ray diffraction data. The I_{NoYAG} data has been scaled to match the number of shots of the I_{YAG} data. The difference is almost entirely due to the iodine-iodine interference which dominates the anisotropic part of the scattering signal. The most notable diffraction features are the zeroth-order maximum and the first-order minimum appearing on the bottom pnCCD panel (i. e., at low resolution). The anisotropy of the diffraction signal of aligned DIBN is illustrated by the angular anisotropy with respect to the alignment angle α as shown in Fig. 5 c, d. This anisotropy is well beyond statistical uncertainties, thereby demonstrating X-ray diffraction signal from the aligned ensemble of isolated DIBN molecules.

Utilizing the iodine-iodine interference of the $I_{YAG} - I_{NoYAG}$ pattern, it was investigated whether the iodine-iodine distance could be estimated from the diffraction data. From *ab initio* calculations (GAMESS-US,[33] MP2/6-311G**), a value of 700 pm was predicted for the iodine-iodine distance. Taking into account the wavelength of 620 pm it is clear that the interference features extend to high scattering angles 2Θ, e. g., the first scattering maximum from the iodine-iodine interference appears at $2\Theta = 51°$ which was not covered by the detector; the outer corner of the top pnCCD panel corresponds to $2\Theta = 50°$, i. e., the resolution is

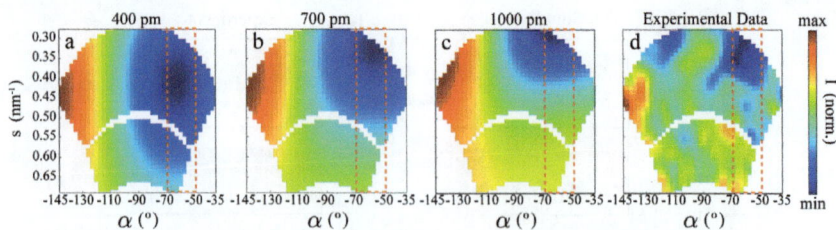

Fig. 6 Diffraction difference $I_{YAG} - I_{NoYAG}$ in the (s,α)-representation for simulated and experimental data. The simulated data shows the diffraction-difference $I_{YAG} - I_{NoYAG}$ for I–I distances of (a) 400 pm, (b) the theoretically expected I–I distance of 700 pm, and (c) 1000 pm. The experimental data is shown in (d). The dashed red frames mark the azimuthal range $\alpha \in [-70°, -50°]$ along which the $I(s)$ graph is obtained.

low. For this reason, direct methods such as phase-retrieval from the diffraction pattern alone were not applied. Instead, the data was compared to models of different iodine-iodine distances and the best fit of a particular model to the data was estimated as will be explained in the following.

Fig. 6 shows the diffraction-difference $I_{YAG} - I_{NoYAG}$ in a different representation. The (x,y)-coordinates were transformed to (s,α)-coordinates, where $s = \sin\Theta/\lambda$ is the scattering vector and α is the azimuthal angle. Due to the twofold symmetry of the diffraction pattern for rotations about the z-axis, the upper pnCCD was rotated by 180° and "connected" to the bottom edge of the lower

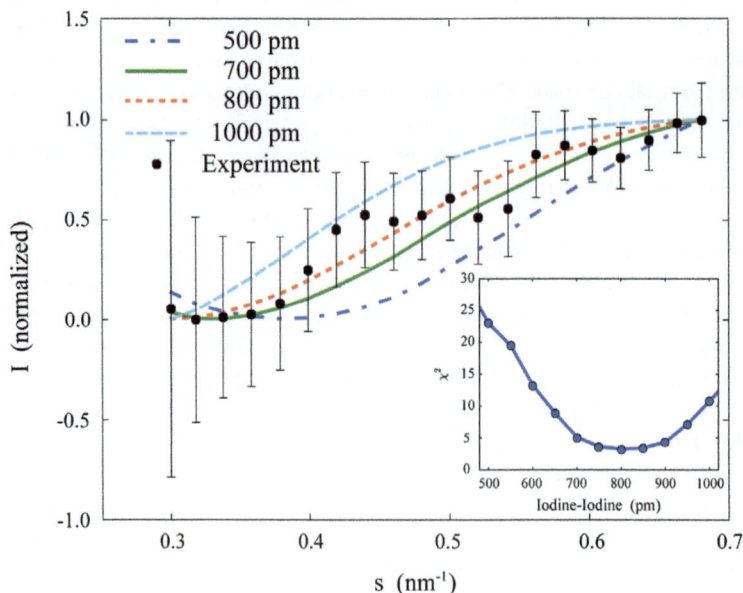

Fig. 7 Comparison of experimentally obtained intensity profiles $I(s)$ along the alignment direction of the diffraction-difference pattern $I_{YAG} - I_{NoYAG}$ with simulated profiles. The experimentally obtained $I(s)$ is best fitted (in terms of a χ^2 test) with the model for an iodine-iodine distance of 800 pm. In the inset the test-statistic χ^2 is shown in dependence of the iodine-iodine distance. Figure reproduced from ref. 26.

pnCCD, thereby extending the range of s-values. Due to the masking of pnCCD regions during the generation of photon hit lists (see Appendix A) the active regions of the two pnCCD panels do not overlap.

Varying the iodine-iodine distance d mainly results in squeezing/stretching of the diffraction minima/maxima in the diffraction pattern. This is most pronounced along the alignment direction $\alpha = -60°$ for the first diffraction minimum in our data. The intensity profile $I(s)$ of the $I_{YAG} - I_{NoYAG}$ data along with simulated $I(s)$ profiles for varying iodine-iodine distances is shown as a function of the scattering vector s in Fig. 7, averaged over $-70° \le \alpha \le -50°$. Each graph is normalized to be independent of the exact molecular beam density M of DIBN molecules, which merely changes the contrast, $i.$ $e.$, the depth of the minimum.

The agreement of the experimental data with a particular model is estimated in terms of a χ^2-test.[52] The best fit to the data, corresponding to the minimum χ^2-value, is obtained for an iodine-iodine distance of 800 pm, see Fig. 7. Due to the low resolution at the current experimental parameters, the fitting is not very accurate. Thus, in future experiments the use of shorter wavelengths will be crucial for an accurate determination of structural features with real atomic resolution. At LCLS, the shortest wavelength currently available is $\lambda \approx 130$ pm (photon energy $\sim$ 9.5 keV), while the European XFEL will be able to provide radiation at wavelengths down to $\lambda \approx 50$ pm (photon energy > 24 keV) from its start of operation in the near future.[53] In addition, the 27 000 pulses s^{-1} at European XFEL allows recordance of such diffraction patterns with better statistics in even shorter amounts of time than is currently possible.

Deviations from the equilibrium geometry could be explained by radiation damage effects, $i.$ $e.$, nuclear and/or electronic damage due to the intense XFEL radiation. However, we estimate that radiation damage effects could not be observed in our diffraction data. First, in contrast to previous experiments explicitly investigating the radiation damage induced by strongly focused XFEL beams,[41,42,54] we deliberately worked out of focus ($i.$ $e.$, at $\omega = 30$ μm), thereby avoiding significant electronic damage effects. Secondly, the wavelength of 620 pm and the range of recorded s-values is insufficient to resolve nuclear motion during the 100 fs X-ray pulses. The reasoning is given in the following.

Fig. 8 shows a time-of-flight spectrum, obtained by probing the molecular beam with the FEL. Iodine ions with increasing charge ($I^+...I^{+7}$) appear in the spectrum with decreasing intensity. In particular, singly-charged iodine I^+ is most abundant while fragments with charges higher than I^{+7} are virtually absent in the spectrum. When DIBN is ionized by 2 keV photons, predominantly the M-shell of iodine is accessed and the total photo-ionization cross-section of iodine of $\sigma_{abs} =$ 41.92 pm^2 (0.4192 Mbarn) is dominated by the cross-section of the $3p$ and $3d$ subshells. Considering the final charge states reached via Auger decay upon photoabsorption of a 2 keV photon in the $3p$ and $3d$ subshells of iodine, an Auger decay similar to xenon is expected, since the electronic decay processes do not strongly depend on the atomic number. For xenon, multiply charged Xe^{+n} ions are obtained from such a photoionization event,[55] $e.$ $g.$, an initial $3d$ vacancy in xenon yields Xe^{+4} as the most probable final charge state, while for a $3p$ vacancy, the charge-state distribution is shifted upwards and peaks around Xe^{+7}. The most-probable final charge state has, in both cases, a probability of $\approx$ 50%. Thus, by assuming similar ionization pathways for xenon and iodine, the absorption of a single 2 keV photon by DIBN is likely to result in a charge state distribution of DIBN peaking at $DIBN^{+4}$ or higher charges. Hence, iodine charge states of I^{+1} to I^{+7} could be entirely due to absorption of only a single photon. We conclude that typically one photon is absorbed per molecule. In the following, absorption of two or more photons is neglected.

For the moderate fluence conditions in our experiment, the probability p_{abs} for single-photon absorption of DIBN can be calculated based on the photo-absorption cross section of atomic iodine $\sigma_{abs} = 41.92$ pm^2 (0.4192 Mbarn).[56] Taking into account the number of photons $N_{photons} = 4.375 \times 10^{12}$ and the interaction area $A_0 = 7.068 \times 10^{-10}$ m^2 (706.8 µm^2), the probability for photo-absorption of a 2 keV photon by a single iodine atom is $p_{abs} = 0.25$, hence the probability for a DIBN molecule (i. e., two iodine atoms) is 0.5, i. e., half of the DIBN molecules absorb an X-ray photon, and, eventually, become multiply ionized by Auger relaxation and fragment due to Coulomb explosion.

We estimate the influence of scattering from fragmenting DIBN on the diffraction pattern in terms of a simple mechanical model concerning only nuclear damage, i. e., motion of ionic fragments happening during the 100 fs (FWHM) X-ray pulse due to Coulomb explosion. The effective spatial distribution of the two main scattering centers, i. e., the two iodine atoms, seen by the entire FEL during a single shot is estimated, taking into account the gradual ionization during the course of the FEL pulse, the total amount of ionization, and the velocity distribution obtained from the measured momentum distributions of I^+ ions.

A one-dimensional cut along the x-axis through the momentum distribution in Fig. 2 is shown in Fig. 9. It represents the measured I^+-velocity distribution v_I^+ in the laboratory frame. The data can be approximated by a Gaussian distribution with mean $\mu = 2700$ m s^{-1} and width $\sigma = 700$ m s^{-1}. Considering momentum conservation, the distribution of the relative velocities $v_{I\,-\,I}$ of the two iodine

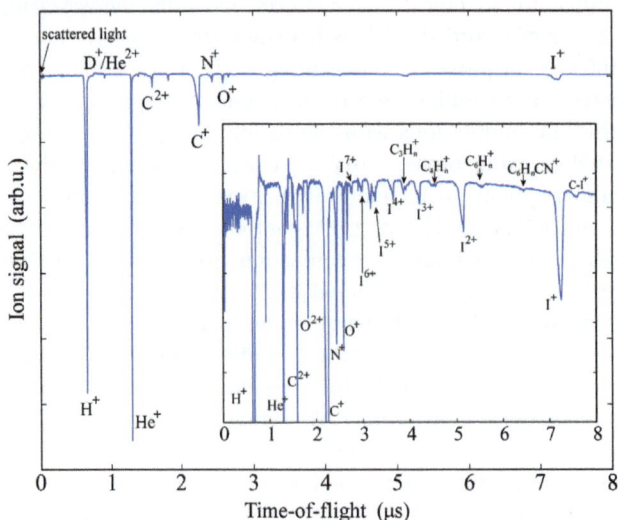

Fig. 8   Time-of-flight spectrum, obtained by probing the molecular beam with the FEL. The inlet is a zoom into the vertical axis in order to show the various iodine ions.

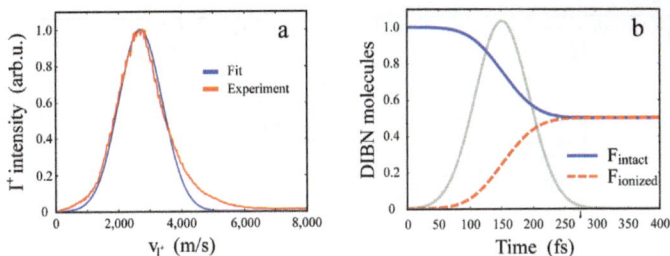

Fig. 9 (a) One-dimensional velocity distribution of I$^+$ ions, estimated from the momentum distributions as shown in Fig. 2. (b) The fractions of intact and ionized DIBN as a function of time for a 100 fs (FWHM) XFEL pulse indicated by the grey line.

atoms is then given by a Gaussian distribution with $\mu_v = 4200$ m s^{-1} and $\sigma_v = 1090$ m s^{-1}. Since a complete velocity distribution of all ions has not been determined experimentally, this model assumes fragmentation into I$^+$ and $[C_7H_3IN]^{+n}$.[‡‡] The resulting velocity distribution of I$^+$ fragments from ionized molecules is

$$v_{I-I} = C \cdot \exp\left(-\frac{(v - \mu_v)^2}{2\sigma_v^2}\right) \tag{6}$$

with the normalization constant C (such that $\int v_{I-I} dv = 1$). This translates into a spatial distribution of I–I distances $s(\Delta t, d)$ by the substitution $d = v\,\Delta t$, with the period $\Delta t = t - t_i$ between ionization time t_i and observation time t.

At each time t the probability for photoabsorption and ionization of molecules is $f_{\text{ionized}}(t) = I_{\text{FEL}}(t) \cdot \sigma_{\text{abs}} \cdot N/A_0$ with the FEL intensity $I_{\text{FEL}}(t)$, the photoabsorption cross section σ_{abs}, the number of molecules N, and the interaction area A_0. N/A_0 can be substituted by $M \cdot l$ with the molecular beam density M and the length of the interaction volume in z-direction l (see section 3), hence $f_{\text{ionized}}(t) = I_{\text{FEL}}(t) \cdot \sigma_{\text{abs}} \cdot M \cdot l$. For each time t, the distribution of I–I distances is given as the sum of intact-molecules with distances d_0 and the distributions of all previously ionized molecules

$$s(t,d) = F_{\text{intact}}(t) \cdot N \cdot s(0, d_0) + \sum_{t_i=0}^{t} s(t - t_i, d) \cdot f_{\text{ionized}}(t_i) \tag{7}$$

with the fraction $F_{\text{intact}}(t)$ of intact molecules at time t and the fraction $f_{\text{ionized}}(t_i)$ of molecules ionized at a certain particular time t_i with the property that

$$\sum_{t_i=0}^{t} f_{\text{ionized}}(t_i) = F_{\text{ionized}}(t), \text{ see Fig. 9 b.}$$

The spatial distribution of I–I distances as seen by the FEL pulse is the sum over $s(t,d)$ for all times, weighted by the instantaneous normalized FEL intensity $I_{\text{FEL}}^{\text{norm}}(t) = I_{\text{FEL}}(t)/\sum I_{\text{FEL}}(t)$:

---

‡‡ We note that this model contains two simplifications: First, the time for acceleration of the fragments as well as the time for ionization, i. e., the finite delay for Auger decay and subsequent charge rearrangement after photoabsorption was not taken into account (i. e., set to zero). Therefore, our model overestimates the atomic displacements. However, this is partly counteracted by the fact that higher charged I^{+n} fragments recoil faster than I$^+$ and hence lead to larger atomic displacements, which is not considered, because these momentum distributions of higher charged I^{+n} fragments were not measured.

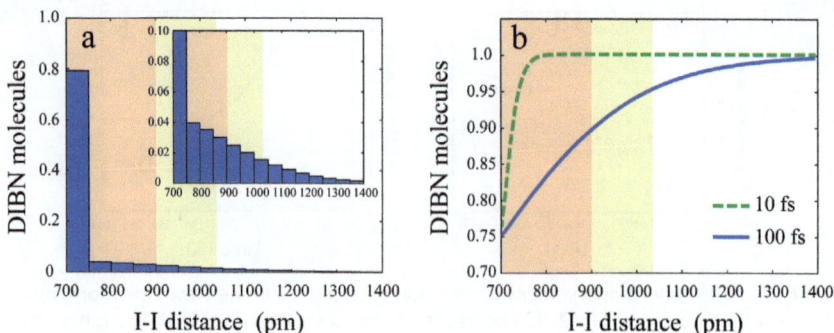

Fig. 10 (a) Histogram of $S(d)$, visualizing the fraction of molecules in different distance intervals, as seen by the 100 fs (FWHM) FEL pulse (blue). (b) The cumulative distribution of $S(d)$ (for smaller stepsize in I–I-distance). I–I distances that are less than 200 pm (330 pm) longer than the 700 pm equilibrium distance are marked by the red (yellow) shaded regions. The dashed green graph in (b) is for a theoretical case of a 10 fs pulse.

$$S(d) = \sum_t s(t, d) \cdot I_{\text{FEL}}^{\text{norm}}(t) \qquad (8)$$

This distribution is shown in Fig. 10 a. The corresponding cumulative distribution of I–I distances is illustrated by the solid blue line in Fig. 10 b. The latter gives the amount of molecules with I–I distances equal to or less than the given distance summed over the entire FEL pulse, e. g., 75% of the elastically scattered photons originate from scattering at intact molecules (i. e., I-I-distance at equilibrium distance of 700 pm) and another 15% (20%) of the diffraction signal originates from scattering of molecules corresponding to I–I distances that are less than 200 pm (330 pm) longer than the 700 pm equilibrium distance. These distances correspond to the red (yellow) shaded regions in Fig. 10. These damaged molecules might contribute to the experimentally determined elongated I–I distance of 800 pm in the minimum of the χ^2-fit. However, since the range of s-values (scattering vectors) covered is too small, these effects cannot be fully resolved in the current experiment with 620 pm wavelength radiation. Further suppressing such effects on the diffraction pattern could, for instance, be accomplished by using shorter pulses. For 10 fs practically no damage would be observed and even for the same pulse energy 95% of the molecules would be at equilibrium distance to within 40 pm.

5   Conclusion and outlook

We experimentally demonstrated coherent X-ray diffractive imaging of laser-aligned gas-phase samples of the prototypical complex molecule 2,5-diiodo-benzonitrile at the LCLS XFEL. This X-ray diffraction experiment resembles Young's double slit experiment on the atomic level due to the two-center interference of the two heavy iodine atoms. We implemented a state-of-the-art molecular beam setup in the CAMP experimental chamber at the AMO beamline of LCLS, utilized quantum-state selection of a cold molecular beam, and demonstrated the preparation of a strongly aligned ensemble of isolated gas-phase molecules. The controlled samples of DIBN were probed by the X-ray pulses in order to measure X-ray diffraction from these ensembles of aligned DIBN. Exploiting the high spectral resolution of the pnCCD detectors, we could successfully retrieve single scattered photons above noise and derive the molecular diffraction patterns from the weak and noisy signals. On average, 0.2 photons/shot were recorded on the camera. However, the angular structures contained in the diffraction patterns are well beyond experimental noise, *i. e.*, we succeeded to observe the two-center interference of the two heavy iodine atoms in the diffraction pattern which confirms the observation of a successful diffraction measurement from aligned DIBN. Even despite the limited resolution, *i. e.*, the long wavelength and the correspondingly limited range of scattering vectors s recorded, the heavy-atom distance was experimentally obtained and it is consistent with the computed molecular structure. Future experiments toward atomic resolution imaging will have to use shorter wavelength and collect diffraction data at higher resolution.

Our experiment confirms the feasibility of coherent X-ray diffractive imaging of small isolated gas-phase molecules and hence provides a first step towards single-molecule imaging at atomic resolution. Our controlled delivery approach is capable to provide three-dimensional alignment and orientation,[11,14,57] which would allow the determination of the 3D molecular structure using a tomographic approach similar to electron diffraction[15] or photoelectron tomography.[58]

Envisioned future experiments plan to make use of the unique short pulses of the XFELs in order to conduct fs pump–probe experiments in order to investigate ultrafast structural dynamics during, e. g., chemical reactions and open up a new field for experiments in femtochemistry and molecular dynamics. For the recording of molecular movies of ultrafast dynamics, x rays offer several advantages over electrons: X-ray pulses do not suffer from space-charge broadening of pulses nor from pump–probe velocity mismatch.[59,60] Hence, X-ray pulses from XFELs will permit better temporal resolution. Pulses as short as a 2–5 fs are already routinely created at XFELs,[61,62] and attosecond X-ray pulses are discussed.[63] These short pulses will allow the observation of the fastest nuclear motion and, moreover, the investigation of ultrafast electron dynamics, such as charge migration and charge transfer processes in molecular and chemical processes.[64,65]

We analyzed how damage effects can be avoided by using short pulses of low fluence at high repetition rates, which will be available at future XFELs, such as the upcoming European XFEL that will operate at 27 000 X-ray pulses/s. Our approach is suitable to study larger molecules provided moderately dense molecular beams of these samples can be generated. Hence, it should be applicable for coherent diffractive imaging of isolated biomolecules, as envisioned for a long time.[5,6,66,67]

Appendix A data acquisition and conditioning of X-ray diffraction data

X-ray diffraction data was recorded by the pnCCD photon detectors with the LCLS operating at 60 Hz. The YAG was operating at 30 Hz, hence single-shot YAG and NoYAG data was recorded in an alternating manner. The YAG and LCLS laser were

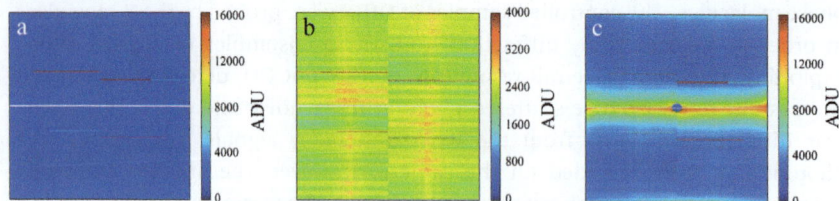

Fig. 11   Single shot raw data frames of an example dataset for the NoYAG case (a,b) and the YAG case (c).

propagating collinearly (see Fig. 1) which resulted in severe background levels from the YAG on the pnCCD despite the filters. This background as well as camera artifacts, known from dark frame measurements, were subtracted from the single shot data. The necessary single-photon counting required operation of the pnCCD cameras at the highest possible gain in order to give a good separation of 2 keV and optical and NIR photons (the latter from the YAG). Spectroscopic discrimination of rare events, i. e., single 2 keV X-ray photons, could be performed due to the high energy resolution of the pnCCD camera.[27] In this chapter we describe the steps necessary to correct the single-shot diffraction data for all artifacts and backgrounds. All processing of the data was performed using the CFEL-ASG Software Suite (CASS).[32]

Fig. 11 shows typical examples of single shot raw data frames for both panels of the pnCCD camera for (a, b) NoYAG and (c) YAG, which contain many artifacts. The measured pnCCD signals are given in ADU (analog-to-digital unit). The most significant difference between NoYAG and YAG data is the region of partly saturated signal at the inner edges of the two pnCCD panels in the YAG case. These signals are based on imperfect shielding of both pnCCD panels from near-infrared (NIR) photons especially at their respective edges.§§ This contribution to the experimental background is referred to as "YAG background". Furthermore, the single shot data contains pnCCD based artifacts such as offset- and gain variations, "hot pixels" or even "hot rows/channels", and time-dependent readout fluctuations called "common mode" (during read out of the pnCCDs, charges are shifted towards the ASIC along the horizontal direction). The pnCCD consist of 16 CAMEX modules.¶¶ The pnCCD-based artifacts and distinct offset within the 16 CAMEX modules become more obvious when zooming into the colorscale, see Fig. 11 (b).||||

In our experiment, scattering from ensembles of isolated molecules is very weak; in particular the probability for two or more X-ray photons scattered to the same pixel on the detector within the same single shot is negligible small. Therefore, single X-ray photon hits could be found by spectroscopic, i. e., energy-dependent discrimination of single-shot data which was corrected for all pnCCD

---

§§ The charge created by a single YAG photon is less than a 1/500th of the charge created by a single 2 keV X-ray photon. However, due to the high YAG intensity, many YAG photons pile up in a single pixel, especially in the regions not shielded thoroughly by the filters.

¶¶ For a description of the CAMEX modules, see ref. 27 and 28.

|||| Although there is a channel-specific offset and gain variation, all channels within the same CAMEX have similar gain and the difference of the channel-specific gain between distinct CAMEX is more pronounced than the gain variation within one CAMEX.

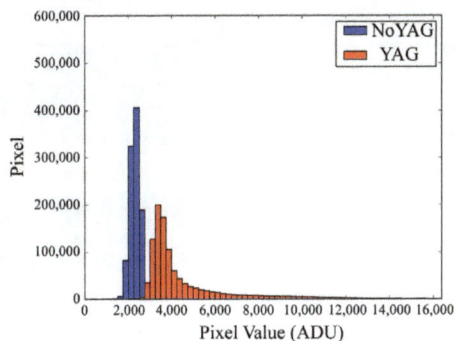

Fig. 12   Spectra for the single shot data frames given in Fig. 11 a, c; see text for details.

artifacts and the YAG background. The measured ADU value is proportional to the energy and a single 2 keV X-ray photon corresponds to a value of $\approx$ 2600 ADU.

Fig. 12 shows histograms for the 1024 × 1024 pnCCD values of the single shot data frames given in Fig. 11 (a–c). There is a constant offset of $\approx$ 2400 ADU in most pixels and in both cases (YAG and NoYAG). In the YAG case, in addition to the pnCCD-based offset, there is the huge background at the inner (and outer edges) of the two pnCCD panels, resulting in a shift of the spectrum towards higher values.

The YAG background was utilized to reliably distinguish single-shot YAG from single-shot NoYAG data. Fig. 13 shows a histogram of the integrated pnCCD signals for single YAG/NoYAG shots for an example dataset, illustrating the clear separation of YAG and NoYAG shots. The variation in the YAG case is due to the YAG intensity, varying on a shot-to-shot level.

First during the data conditioning process, single-shot YAG and NoYAG data was separated based on the integrated pnCCD signals. Then, the data was corrected for offset by subtracting an offset map, the latter obtained from averaging single shot pnCCD data under "dark" conditions. The common mode was corrected for by subtracting the median value along each vertical row from this row, separately for the upper and lower pnCCD panel. Fig. 14 shows the resulting frames for the YAG and NoYAG case. The channel-specific offset variation was

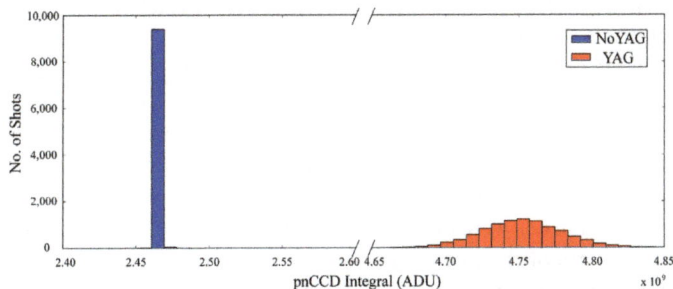

Fig. 13   Histogram of the total integrated value of individual YAG/NoYAG data for an example dataset containing 9451 shots with YAG off (NoYAG) and 9449 shots with YAG on (YAG).

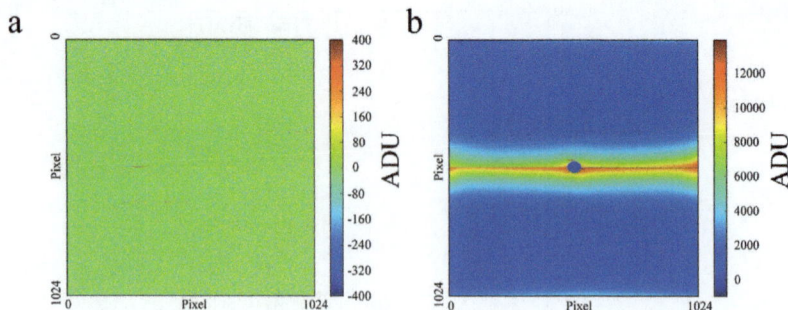

Fig. 14 Single shot pnCCD data frames, corrected for channel-dependent offset and common mode.

successfully corrected for. The NoYAG data is close to 0 value for almost every pixel while the YAG data still contains the severe background from the YAG.

The YAG background scattering was corrected for by subtracting a averaged YAG data frame, scaled to match the total intensity of the particular individual single-shot data frame, from the individual single-shot YAG frame. This method works reliable since the total YAG intensity is varying on a shot-to-shot level but the spatial distribution of the YAG on the pnCCD is independent of a certain shot (i. e., it can be scaled by a single number).

As a result from the steps mentioned above, the single-shot data frames were corrected for all backgrounds and artifacts except rare events such as single 2 keV X-ray photons scattered from the molecular sample. These photons, at ≈ 2600 ADU, were found by thresholding the background-corrected data frames and considering the charge spread of the X-ray photons: a 2 keV photon absorbed in the pnCCD creates a charge cloud which can cross the barrier of a single pixel and hence can give signals in two (or more) adjacent pixels. At 2 keV, almost all photon hits are single- or double-pixel hits (the latter is the case in which the charge cloud diffuses into a single neighboring pixel adjacent to the pixel where the photon is initially absorbed). This is justified by the experimental results, where 64% of all X-ray attributed hits are single-pixel hits, 35% are double-pixel hits while <1% make up for the rest. These photon hits were found by thresholding the background-corrected single-shot YAG and NoYAG data frames and combining adjacent pixels exceeding the threshold of 500 ADU. The X-ray hits found by this procedure were written to a list containing the coordinates, ADU value, and number of pixels the hit was combined from. By limiting the number of pixels a hit can be made of to six, rare events such as high energy particles impinging on the detector were neglected. Then, corrections for channel-dependent gain and charge-transfer-efficiency were applied to the energy values of the photon hits (although these corrections didn't affect the spatial distribution and also have almost no effect on the spectral distribution of the hits as well). Photon hits for certain regions of pixels were always neglected. This included "hot pixel" regions as well as the parts of the pnCCD that were (completely or nearly) saturated by YAG photons. The latter regions showed a high fluctuation of signal and, therefore, could not be discriminated successfully.

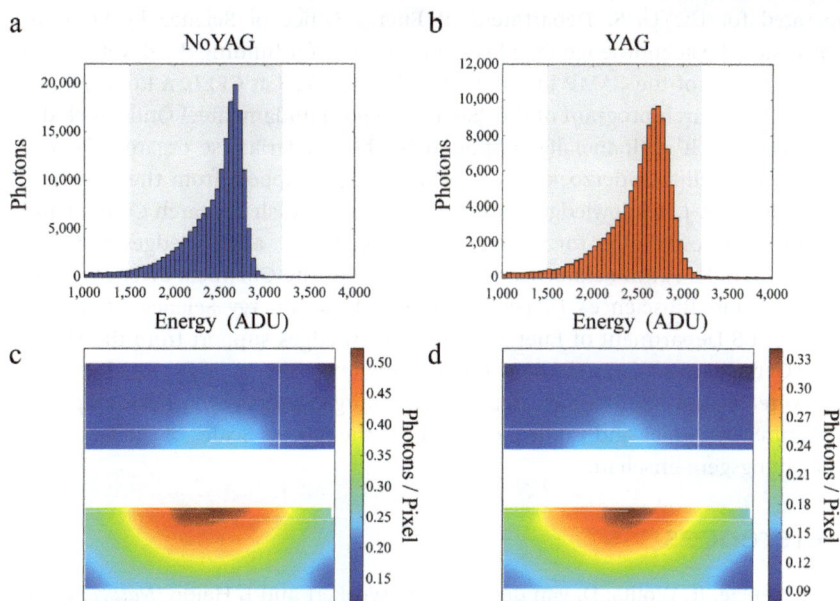

Fig. 15 Spectra of the hits for NoYAG (a) and YAG (b) for hits made up out of 1–2 pixels; spatial intensity distributions I_{YAG}, I_{NoYAG} of these hits in the energy interval 1500–3200 ADU (c, d), i. e., the "diffraction patterns". The raw data was convolved with a gaussian kernel.

Fig. 15 a, b show spectra of all photon hits from the NoYAG (a) and YAG (b) data. The spectrum is peaks at 2600 ADU, thereby matching expectations. The width of the peak can be attributed the energy resolution of the pnCCDs, the photon energy jitter of LCLS, and to the event recombination of double pixel hits (the latter being the major contribution to the broadening of the spectrum).

In the energy interval 1500–3200 ADU there are 172 499 photons for the NoYAG and 111 560 photons for the YAG data which are used for data analysis. The data was obtained from 842 722 shots (NoYAG) and 563 453 shots (YAG) respectively, hence the average hit rate on the whole pnCCD detector was 0.204 (0.197) photons/shot for the NoYAG (YAG) data. The spatial distribution of these photon hits, i. e., the diffraction patterns I_{NoYAG} and I_{YAG}, are shown in Fig. 15 c, d and are analyzed as described in section 4.

Acknowledgements

We thank Marcus Adolph, Andrew Aquila, Saša Bajt, Carl Caleman, Nicola Coppola, Tjark Delmas, Holger Fleckenstein, Tais Gorkhover, Lars Gumprecht, Andreas Hartmann, Günter Hauser, Peter Holl, Andre Hömke, Faton Krasniqi, Gerard Meijer, Robert Moshammer, Christian Reich, Robin Santra, Ilme Schlichting, Carlo Schmidt, Sebastian Schorb, Joachim Schulz, Heike Soltau, John C. H. Spence, Lothar Strüder, Joachim Ullrich, Marc J. J. Vrakking, and Cornelia Wunderer for help in preparing or performing the measurements. Parts of this research were carried out at the Linac Coherent Light Source (LCLS) at the SLAC National Accelerator Laboratory. LCLS is an Office of Science User Facility operated for the U. S. Department of Energy Office of Science by Stanford University. We acknowledge the Max Planck Society for funding the development and operation of the CAMP instrument within the ASG at CFEL. A.Ro. acknowledges the research program of the "Stichting voor Fundamenteel Onderzoek der Materie", which is financially supported by the "Nederlandse organisatie voor Wetenschappelijk Onderzoek". H. S. acknowledges support from the Carlsberg Foundation. P. J. acknowledges support from the Swedish Research Council and the Swedish Foundation for Strategic Research. H.N.C. acknowledges NSF STC award 1231306. A.Ru. acknowledges support from the Chemical Sciences, Geosciences, and Biosciences Division, Office of Basic Energy Sciences, Office of Science, US Department of Energy. D. R. acknowledges support from the Helmholtz Gemeinschaft through the Young Investigator Program. This work has been supported by the excellence cluster "The Hamburg Center for Ultrafast Imaging – Structure, Dynamics and Control of Matter at the Atomic Scale" of the Deutsche Forschungsgemeinschaft.

References

1 R. Neutze, R. Wouts, D. van der Spoel, E. Weckert and J. Hajdu, *Nature*, 2000, **406**, 752–757.

2 M. M. Seibert, T. Ekeberg, F. R. N. C. Maia, M. Svenda, J. Andreasson, O. Jönsson, D. Odić, B. Iwan, A. Rocker, D. Westphal, M. Hantke, D. P. Deponte, A. Barty, J. Schulz, L. Gumprecht, N. Coppola, A. Aquila, M. Liang, T. A. White, A. Martin, C. Caleman, S. Stern, C. Abergel, V. Seltzer, J.-M. Claverie, C. Bostedt, J. D. Bozek, S. Boutet, A. A. Miahnahri, M. Messerschmidt, J. Krzywinski, G. Williams, K. O. Hodgson, M. J. Bogan, C. Y. Hampton, R. G. Sierra, D. Starodub, I. Andersson, S. Bajt, M. Barthelmess, J. C. H. Spence, P. Fromme, U. Weierstall, R. Kirian, M. Hunter, R. B. Doak, S. Marchesini, S. P. Hau-Riege, M. Frank, R. L. Shoeman, L. Lomb, S. W. Epp, R. Hartmann, D. Rolles, A. Rudenko, C. Schmidt, L. Foucar, N. Kimmel, P. Holl, B. Rudek, B. Erk, A. Hömke, C. Reich, D. Pietschner, G. Weidenspointner, L. Strüder, G. Hauser, H. Gorke, J. Ullrich, I. Schlichting, S. Herrmann, G. Schaller, F. Schopper, H. Soltau, K.-U. Kühnel, R. Andritschke, C.-D. Schröter, F. Krasniqi, M. Bott, S. Schorb, D. Rupp, M. Adolph, T. Gorkhover, H. Hirsemann, G. Potdevin, H. Graafsma, B. Nilsson, H. N. Chapman and J. Hajdu, *Nature*, 2011, **470**, 78.

3 H. N. Chapman, P. Fromme, A. Barty, T. A. White, R. A. Kirian, A. Aquila, M. S. Hunter, J. Schulz, D. P. Deponte, U. Weierstall, R. B. Doak, F. R. N. C. Maia, A. V. Martin, I. Schlichting, L. Lomb, N. Coppola, R. L. Shoeman, S. W. Epp, R. Hartmann, D. Rolles, A. Rudenko, L. Foucar, N. Kimmel, G. Weidenspointner, P. Holl, M. Liang, M. Barthelmess, C. Caleman, S. Boutet, M. J. Bogan, J. Krzywinski, C. Bostedt, S. Bajt, L. Gumprecht, B. Rudek, B. Erk, C. Schmidt, A. Hömke, C. Reich, D. Pietschner, L. Strüder, G. Hauser, H. Gorke, J. Ullrich, S. Herrmann, G. Schaller, F. Schopper, H. Soltau, K.-U. Kühnel, M. Messerschmidt, J. D. Bozek, S. P. Hau-Riege, M. Frank, C. Y. Hampton, R. G. Sierra, D. Starodub, G. J. Williams, J. Hajdu, N. Timneanu, M. M. Seibert, J. Andreasson, A. Rocker, O. Jönsson, M. Svenda, S. Stern, K. Nass, R. Andritschke, C.-D. Schröter, F. Krasniqi, M. Bott, K. E. Schmidt, X. Wang,

I. Grotjohann, J. M. Holton, T. R. M. Barends, R. Neutze, S. Marchesini, R. Fromme, S. Schorb, D. Rupp, M. Adolph, T. Gorkhover, I. Andersson, H. Hirsemann, G. Potdevin, H. Graafsma, B. Nilsson and J. C. H. Spence, *Nature*, 2011, **470**, 73.

4 M. Chergui and A. H. Zewail, *ChemPhysChem*, 2009, **10**, 28–43.

5 A. Barty, J. Küpper and H. N. Chapman, *Annu. Rev. Phys. Chem.*, 2013, **64**, 415–435.

6 J. C. H. Spence and R. B. Doak, *Phys. Rev. Lett.*, 2004, **92**, 198102.

7 F. Filsinger, G. Meijer, H. Stapelfeldt, H. Chapman and J. Küpper, *Phys. Chem. Chem. Phys.*, 2011, **13**, 2076–2087.

8 N.-T. D. Loh and V. Elser, *Phys. Rev. E: Stat., Nonlinear, Soft Matter Phys.*, 2009, **80**, 026705.

9 R. Fung, V. Shneerson, D. Saldin and A. Ourmazd, *Nat. Phys.*, 2009, **5**, 64–67.

10 O. M. Yefanov and I. A. Vartanyants, *J. Phys. B: At., Mol. Opt. Phys.*, 2013, **46**, 164013.

11 J. J. Larsen, K. Hald, N. Bjerre, H. Stapelfeldt and T. Seideman, *Phys. Rev. Lett.*, 2000, **85**, 2470–2473.

12 H. Stapelfeldt and T. Seideman, *Rev. Mod. Phys.*, 2003, **75**, 543–557.

13 L. Holmegaard, J. H. Nielsen, I. Nevo, H. Stapelfeldt, F. Filsinger, J. Küpper and G. Meijer, *Phys. Rev. Lett.*, 2009, **102**, 023001.

14 I. Nevo, L. Holmegaard, J. H. Nielsen, J. L. Hansen, H. Stapelfeldt, F. Filsinger, G. Meijer and J. Küpper, *Phys. Chem. Chem. Phys.*, 2009, **11**, 9912–9918.

15 C. J. Hensley, J. Yang and M. Centurion, *Phys. Rev. Lett.*, 2012, **109**, 133202.

16 R. Boll, D. Anielski, C. Bostedt, J. D. Bozek, L. Christensen, R. Coffee, S. De, P. Decleva, S. W. Epp, B. Erk, L. Foucar, F. Krasniqi, J. Küpper, A. Rouzée, B. Rudek, A. Rudenko, S. Schorb, H. Stapelfeldt, M. Stener, S. Stern, S. Techert, S. Trippel, M. J. J. Vrakking, J. Ullrich and D. Rolles, *Phys. Rev. A: At., Mol., Opt. Phys.*, 2013, **88**, 061402.

17 R. Boll, D. Anielski, C. Bostedt, J. D. Bozek, L. Christensen, R. Coffee, S. De, P. Decleva, S. W. Epp, B. Erk, L. Foucar, F. Krasniqi, J. Küpper, A. Rouzée, B. Rudek, A. Rudenko, S. Schorb, H. Stapelfeldt, M. Stener, S. Stern, S. Techert, S. Trippel, M. Vrakking, J. Ullrich and D. Rolles, *Faraday Disc.*, 2014, **171**, DOI: 10.1039/C4FD00037D.

18 D. Rolles, R. Boll, M. Adolph, A. Aquila, C. Bostedt, J. Bozek, H. Chapman, R. Coffee, N. Coppola, P. Decleva, T. Delmas, S. Epp, B. Erk, F. Filsinger, L. Foucar, L. Gumprecht, A. Hömke, T. Gorkhover, L. Holmegaard, P. Johnsson, C. Kaiser, F. Krasniqi, K.-U. Kühnel, J. Maurer, M. Messerschmidt, R. Moshammer, W. Quevedo, I. Rajkovic, A. Rouzée, B. Rudek, I. Schlichting, C. Schmidt, S. Schorb, C. D. Schröter, J. Schulz, H. Stapelfeldt, M. Stener, S. Stern, S. Techert, J. Thøgersen, M. J. J. Vrakking, A. Rudenko, J. Küpper and J. Ullrich, *J. Phys. B*, 2014, **47**, 124035.

19 G. von Helden, T. Wyttenbach and M. T. Bowers, *Science*, 1995, **267**, 1483–1485.

20 S. Trippel, Y.-P. Chang, S. Stern, T. Mullins, L. Holmegaard and J. Küpper, *Phys. Rev. A: At., Mol., Opt. Phys.*, 2012, **86**, 033202.

21 F. Filsinger, U. Erlekam, G. von Helden, J. Küpper and G. Meijer, *Phys. Rev. Lett.*, 2008, **100**, 133003.

22 F. Filsinger, J. Küpper, G. Meijer, J. L. Hansen, J. Maurer, J. H. Nielsen, L. Holmegaard and H. Stapelfeldt, *Angew. Chem., Int. Ed.*, 2009, **48**, 6900–6902.

23 T. Kierspel, D. A. Horke, Y.-P. Chang and J. Küpper, *Chem. Phys. Lett.*, 2014, **591**, 130–132.

24 F. Filsinger, J. Küpper, G. Meijer, L. Holmegaard, J. H. Nielsen, I. Nevo, J. L. Hansen and H. Stapelfeldt, *J. Chem. Phys.*, 2009, **131**, 064309.

25 S. Trippel, T. Mullins, N. L. M. Müller, J. S. Kienitz, K. Długołecki and J. Küpper, *Mol. Phys.*, 2013, **111**, 1738.

26 J. Küpper, S. Stern, L. Holmegaard, F. Filsinger, A. Rouzée, A. Rudenko, P. Johnsson, A. V. Martin, M. Adolph, A. Aquila, S. Bajt, A. Barty, C. Bostedt, J. Bozek, C. Caleman, R. Coffee, N. Coppola, T. Delmas, S. Epp, B. Erk, L. Foucar, T. Gorkhover, L. Gumprecht, A. Hartmann, R. Hartmann, G. Hauser, P. Holl, A. Hömke, N. Kimmel, F. Krasniqi, K.-U. Kühnel, J. Maurer, M. Messerschmidt, R. Moshammer, C. Reich, B. Rudek, R. Santra, I. Schlichting, C. Schmidt, S. Schorb, J. Schulz, H. Soltau, J. C. H. Spence, D. Starodub, L. Strüder, J. Thøgersen, M. J. J. Vrakking, G. Weidenspointner, T. A. White, C. Wunderer, G. Meijer, J. Ullrich, H. Stapelfeldt, D. Rolles and H. N. Chapman, *Phys. Rev. Lett.*, 2014, **112**, 083002.

27 L. Strüder, S. Epp, D. Rolles, R. Hartmann, P. Holl, G. Lutz, H. Soltau, R. Eckart, C. Reich, K. Heinzinger, C. Thamm, A. Rudenko, F. Krasniqi, K. Kühnel, C. Bauer, C.-D. Schroeter, R. Moshammer, S. Techert, D. Miessner, M. Porro, O. Haelker, N. Meidinger, N. Kimmel, R. Andritschke, F. Schopper, G. Weidenspointner, A. Ziegler, D. Pietschner, S. Herrmann, U. Pietsch, A. Walenta, W. Leitenberger, C. Bostedt, T. Moeller, D. Rupp, M. Adolph, H. Graafsma, H. Hirsemann, K. Gaertner, R. Richter, L. Foucar, R. L. Shoeman, I. Schlichting and J. Ullrich, *Nucl. Instrum. Methods Phys. Res., Sect. A*, 2010, **614**, 483–496.

28 R. Hartmann, S. Epp, S. Herrmann, P. Holl, N. Meidinger, C. Reich, D. Rolles, H. Soltau, L. Strüder, J. Ullrich and G. Weidenspointner, *Nuclear Science Symposium Conference Record, 2008. NSS '08. IEEE*, 2008, pp. 2590–2595.

29 J. D. Bozek, *Eur. Phys. J. Spec. Top.*, 2009, **169**, 129–132.

30 C. Bostedt, J. D. Bozek, P. H. Bucksbaum, R. N. Coffee, J. B. Hastings, Z. Huang, R. W. Lee, S. Schorb, J. N. Corlett, P. Denes, P. Emma, R. W. Falcone, R. W. Schoenlein, G. Doumy, E. P. Kanter, B. Kraessig, S. Southworth, L. Young, L. Fang, M. Hoener, N. Berrah, C. Roedig and L. F. DiMauro, *J. Phys. B: At., Mol. Opt. Phys.*, 2013, **46**, 164003.

31 P. Emma, R. Akre, J. Arthur, R. Bionta, C. Bostedt, J. Bozek, A. Brachmann, P. Bucksbaum, R. Coffee, F. J. Decker, Y. Ding, D. Dowell, S. Edstrom, A. Fisher, J. Frisch, S. Gilevich, J. Hastings, G. Hays, P. Hering, Z. Huang, R. Iverson, H. Loos, M. Messerschmidt, A. Miahnahri, S. Moeller, H. D. Nuhn, G. Pile, D. Ratner, J. Rzepiela, D. Schultz, T. Smith, P. Stefan, H. Tompkins, J. Turner, J. Welch, W. White, J. Wu, G. Yocky and J. Galayda, *Nat. Photonics*, 2010, **4**, 641–647.

32 L. Foucar, A. Barty, N. Coppola, R. Hartmann, P. Holl, U. Hoppe, S. Kassemeyer, N. Kimmel, J. Küpper, M. Scholz, S. Techert, T. A. White, L. Strüder and J. Ullrich, *Comput. Phys. Commun.*, 2012, **183**, 2207–2213.

33 M. W. Schmidt, K. K. Baldridge, J. A. Boatz, S. T. Elbert, M. S. Gordon, J. H. Jensen, S. Koseki, N. Matsunaga, K. A. Nguyen, S. Su, T. L. Windus, M. Dupuis and J. A. Montgomery, *J. Comput. Chem.*, 1993, **14**, 1347–1363.

34 U. Even, J. Jortner, D. Noy, N. Lavie and N. Cossart-Magos, *J. Chem. Phys.*, 2000, **112**, 8068–8071.

35 *Atomic and Molecular Beam Methods*, ed. G. Scholes, Oxford University Press, New York, 1988.

36 K. Luria, W. Christen and U. Even, *J. Phys. Chem. A*, 2011, **115**, 7362–7367.

37 S. Düsterer, P. Radcliffe, C. Bostedt, J. Bozek, A. L. Cavalieri, R. Coffee, J. T. Costello, D. Cubaynes, L. F. DiMauro, Y. Ding, G. Doumy, F. Grüner, W. Helml, W. Schweinberger, R. Kienberger, A. R. Maier, M. Messerschmidt, V. Richardson, C. Roedig, T. Tschentscher and M. Meyer, *New J. Phys.*, 2011, **13**, 093024.

38 U. Lorenz, N. M. Kabachnik, E. Weckert and I. A. Vartanyants, *Phys. Rev. E: Stat., Nonlinear, Soft Matter Phys.*, 2012, **86**, 051911.

39 B. Ziaja, H. N. Chapman, R. Fäustlin, S. Hau-Riege, Z. Jurek, A. V. Martin, S. Toleikis, F. Wang, E. Weckert and R. Santra, *New J. Phys.*, 2012, **14**, 115015.

40 A. Fratalocchi and G. Ruocco, *Phys. Rev. Lett.*, 2011, **106**, 105504.

41 B. Erk, D. Rolles, L. Foucar, B. Rudek, S. W. Epp, M. Cryle, C. Bostedt, S. Schorb, J. Bozek, A. Rouzee, A. Hundertmark, T. Marchenko, M. Simon, F. Filsinger, L. Christensen, S. De, S. Trippel, J. Küpper, H. Stapelfeldt, S. Wada, K. Ueda, M. Swiggers, M. Messerschmidt, C. D. Schroter, R. Moshammer, I. Schlichting, J. Ullrich and A. Rudenko, *Phys. Rev. Lett.*, 2013, **110**, 053003.

42 B. Rudek, S.-K. Son, L. Foucar, S.-W. Epp, B. Erk, R. Hartmann, M. Adolph, R. Andritschke, A. Aquila, N. Berrah, C. Bostedt, N. Bozek, Johnand Coppola, F. Filsinger, H. Gorke, T. Gorkhover, H. Graafsma, L. Gumprecht, A. Hartmann, G. Hauser, S. Herrmann, H. Hirsemann, P. Holl, A. Hömke, L. Journel, C. Kaiser, N. Kimmel, F. Krasniqi, K.-U. Kühnel, M. Matysek, M. Messerschmidt, D. Miesner, T. Möller, R. Moshammer, K. Nagaya, B. Nilsson, G. Potdevin, D. Pietschner, C. Reich, D. Rupp, G. Schaller, I. Schlichting, C. Schmidt, F. Schopper, S. Schorb, C.-D. Schröter, J. Schulz, M. Simon, H. Soltau, L. Strüder, K. Ueda, G. Weidenspointner, R. Santra, J. Ullrich, A. Rudenko and D. Rolles, *Nat. Photonics*, 2012, **6**, 858–865.

43 W. Wiley and I. Mclaren, *Rev. Sci. Instrum.*, 1955, **26**, 1150–1157.

44 Y.-P. Chang, F. Filsinger, B. Sartakov and J. Küpper, *Comput. Phys. Commun.*, 2014, **185**, 339–49.

45 J. J. Larsen, H. Sakai, C. P. Safvan, I. Wendt-Larsen and H. Stapelfeldt, *J. Chem. Phys.*, 1999, **111**, 7774.

46 J. Als-Nielsen and D. McMorrow, *Elements of Modern X-ray Physics*, John Wiley & Sons, Chichester, West Sussex, United Kingdom, 2001.

47 S. Stern, *Dissertation*, Universität Hamburg, Hamburg, Germany, 2013.

48 R. A. Kirian, X. Wang, U. Weierstall, K. E. Schmidt, J. C. H. Spence, M. Hunter, P. Fromme, T. White, H. N. Chapman and J. Holton, *Opt. Express*, 2010, **18**, 5713–5723.

49 D. Waasmaier and A. Kirfel, *Acta Crystallogr., Sect. A: Found. Crystallogr.*, 1995, **51**, 416–431.

50 B. Henke, E. Gullikson and J. Davis, *At. Data Nucl. Data Tables*, 1993, **54**, 181–342.

51 B. Friedrich and D. Herschbach, *Phys. Rev. Lett.*, 1995, **74**, 4623–4626.

52 K. Nakamura, *J. Phys. G: Nucl. Part. Phys.*, 2010, **37**, 075021.

53 M. Altarelli, R. Brinkmann, M. Chergui, W. Decking, B. Dobson, S. Düsterer, G. Grübel, W. Graeff, H. Graafsma, J. Hajdu, J. Marangos, J. Pflüger,

H. Redlin, D. Riley, I. Robinson, J. Rossbach, A. Schwarz, K. Tiedtke, T. Tschentscher, I. Vartaniants, H. Wabnitz, H. Weise, R. Wichmann, K. Witte, A. Wolf, M. Wulff and M. Yurkov, The Technical Design Report of the European XFEL, *Desy technical report*, 2007.

54 L. Young, E. P. Kanter, B. Kraessig, Y. Li, A. M. March, S. T. Pratt, R. Santra, S. H. Southworth, N. Rohringer, L. F. DiMauro, G. Doumy, C. A. Roedig, N. Berrah, L. Fang, M. Hoener, P. H. Bucksbaum, J. P. Cryan, S. Ghimire, J. M. Glownia, D. A. Reis, J. D. Bozek, C. Bostedt and M. Messerschmidt, *Nature*, 2010, **466**, 56.

55 A. Kochur, A. Dudenko, V. Sukhorukov and I. Petrov, *J. Phys. B: At., Mol. Opt. Phys.*, 1994, **27**, 1709.

56 M. J. Berger, J. H. Hubbell, S. M. Seltzer, J. Chang, J. S. Coursey, R. Sukumar, D. S. Zucker and K. Olsen, *XCOM: Photon Cross Sections Database*, 2010, http://physics.nist.gov/xcom, available at http://physics.nist.gov/xcom, National Institute of Standards and Technology, Gaithersburg, MD.

57 J. L. Hansen, J. J. Omiste Romero, J. H. Nielsen, D. Pentlehner, J. Küpper, R. González-Férez and H. Stapelfeldt, *J. Chem. Phys.*, 2013, **139**, 234313.

58 J. Maurer, D. Dimitrovski, L. Christensen, L. B. Madsen and H. Stapelfeldt, *Phys. Rev. Lett.*, 2012, **109**, 123001.

59 J. C. Williamson and A. H. Zewail, *Chem. Phys. Lett.*, 1993, **209**, 10–16.

60 G. Sciaini and R. J. D. Miller, *Rep. Prog. Phys.*, 2011, **74**, 096101.

61 Y. Ding, A. Brachmann, F.-J. Decker, D. Dowell, P. Emma, J. Frisch, S. Gilevich, G. Hays, P. Hering, Z. Huang, R. Iverson, H. Loos, A. Miahnahri, H.-D. Nuhn, D. Ratner, J. Turner, J. Welch, W. White and J. Wu, *Phys. Rev. Lett.*, 2009, **102**, 254801.

62 Y. Ding, F.-J. Decker, P. Emma, C. Feng, C. Field, J. Frisch, Z. Huang, J. Krzywinski, H. Loos, J. Welch, J. Wu and F. Zhou, *Phys. Rev. Lett.*, 2012, **109**, 254802.

63 M. Dohlus, E. A. Schneidmiller and M. V. Yurkov, *Phys. Rev. Spec. Top.—Accel. Beams*, 2011, **14**, 090702.

64 R. Weinkauf, P. Schanen, A. Metsala, E. W. Schlag, M. Bürgle and H. Kessler, *J. Phys. Chem.*, 1996, **100**, 18567–18585.

65 A. H. Zewail, *J. Phys. Chem. A*, 2000, **104**, 5660–5694.

66 J. C. H. Spence, K. Schmidt, J. S. Wu, G. Hembree, U. Weierstall, R. B. Doak and P. Fromme, *Acta Crystallogr., Sect. A: Found. Crystallogr.*, 2005, **61**, 237–245.

67 D. Starodub, R. B. Doak, K. Schmidt, U. Weierstall, J. S. Wu, J. C. H. Spence, M. Howells, M. Marcus, D. Shapiro, A. Barty and H. N. Chapman, *J. Chem. Phys.*, 2005, **123**, 244304.

ROYAL SOCIETY OF CHEMISTRY

DISCUSSIONS

Nanoscale and bio imaging: general discussion

Henry Chapman, Majed Chergui, Martin McCoustra, Jasper van Thor, Jonathan Underwood, Michael Woerner, Christian Bressler, Grigory Smolentsev, Jochen Küpper and Chris Jacobsen

DOI: 10.1039/C4FD90016B

Martin McCoustra opened the discussion of the paper by **Chris Jacobsen** by asking: We have heard many comments on the idea of molecular movies in this Discussion. What likelihood do we have of what might be referred to as biomolecular movies showing the temporal evolution of biomolecules in a living organism?

Chris Jacobsen communicated in reply: The challenge here involves radiation damage. For repeatable phenomena in measurements that involve statistical measures of systems, one avoids radiation damage by measuring at a variety of time points after a stimulus (repeatability) with the irradiation hitting a number of biomolecules so that the dose to each individual biomolecule is reduced (statistical measures). This might apply for example to small angle scattering. When it comes to imaging larger and thus non-identical objects such as organelles and whole cells where there is considerable overall structural variation (no two cells look alike), direct real-space statistical averaging seems difficult to apply. Also, when one is doing even 20 nm resolution imaging, the radiation dose is in the 10^8 Gray range which is sufficient to break about one bond per molecule so the movie is far more likely to show the biomolecule being damaged rather than undergoing a "normal" temporal evolution. Of course, pump–probe crystallography can be used for stroboscopic (repeatable sample) and ensemble-averaged (crystal with many identical unit cells) measurements of the time evolution of a photostimulable biomolecule—which in the case of proteins represents a very small though important subset of all known proteins.

Henry Chapman asked: Would it be possible to use some sort of dynamic response in a pixel of an STXM scan as a contrast mechanism, for example to map the distribution of chemical activity in a sample?

Chris Jacobsen responded: Interesting thought. Of course X-ray photon correlation spectroscopy (XPCS) involves using the rearrangement of the coherently scattered intensity distribution to tell one about the timescale of rearrangements of optical phases of the material within the coherently illuminated beam spot. The challenge of course is one of radiation damage, in that when one goes from a large-beam-spot XPCS experiment to a nano focus scanning transmission X-ray microscope (STXM) beam spot, the number of photons per area goes up to very high levels and one is as likely to see a dynamic response caused by beam damage as a dynamic response caused by a more "natural" process in a pixel. The per pixel dose in 20–50 nm resolution STXMs is in the $1e^6$-$1e^8$ Gray range, and at around $1e^8$ Gray one has broken about one bond per molecule. At the same time, there are some indications that chemical damage takes on the order of a millisecond or so to be made manifest in diffraction patterns, so maybe there is some hope? Interesting question.

Grigory Smolentsev remarked: You use the regularization method to include *a priori* knowledge into your model. Are there any other methods that allows you to include *a priori* knowledge into the model, for example, those based on the linear combination of components obtained using principal component analysis? Could you compare your method with the alternative algorithms?

Chris Jacobsen replied: In fact the spectra arising from cluster analysis (which we include as the "cluster similarity" regularizer in eqn. 10–13 in our Faraday Discussion paper) are exactly what you suggest: a linear combination of components obtained from principal component analysis (PCA). As described in our earlier paper (Lerotic *et al.*, *Ultramicroscopy*, **100**, 35 (2004)), we use PCA to obtain a basis set with desirable properties like orthogonality and noise-supression. We then find weightings among these components by seeking clusters of pixels (independent spectral measurements) in this component space, yielding spectra that are linear combinations of PCA spectra. However, as explained in Section 2 of this Faraday Discussion paper, the resulting set of cluster spectra can be incomplete as a basis set and can lead to errors in analysis. What NNMA gives us is a way to use the cluster spectra as one part of the cost function (one regularizer). However, it is used in combination with the data-matching cost of eqn. 3, and the sparseness regularizer of eqn. 7–9, to yield an improved solution as shown by the lack of unphysical "negative absorption" values in the NNMA solutions shown in Fig. 6 of our paper.

Jonathan Underwood communicated: The regularization method presented in your paper bares some similarity to other widely used strategies such as Tikhonov regularization and particularly Projected Landweber regularization in terms of the non-negativity constraint. The convergence properties of these strategies are reasonably well studied and understood (*e.g. Num. Funct. Anal. Opt.*, **13**, 413 (1992)). Is anything known about the convergence properties of your regularization strategy?

Chris Jacobsen responded: To our understanding, the cited paper deals with the problem of determining A in $Af=g$ when f and g are known, subject to constraints; it shows good convergence properties. In our case we know neither A (the set of spectra) or f (the thickness maps) in advance; we only know the measured data g. We are not aware of analyses on convergence rates for global solutions, though the steepest-descent type approach we have used (based on derivatives of cost functions) at least aides in finding local solutions. We have carried out some work on the optimization of the regularizing parameter λ; we will report on that in a separate paper.

Henry Chapman addressed Chris Jacobsen and Jasper van Thor: I noticed that the update algorithm you present in your paper (eqn. 3–5) is essentially the algebraic reconstruction technique (ART) used in tomography (also called the Kaczmarz algorithm, used in Caterina Vozzi's tomographic reconstructions paper). Presumably regularisation has been applied in tomography too, or perhaps your developments could be of interest for that application. Your method is also immediately applicable to solving structural kinetics problems, and would be a big improvement upon the SVD method developed by Marius Schmidt and others for time-resolved WAXS and time-resolved crystallography,[1] just replacing energy (in your case) for time (in theirs). I have not seen a discussion in those applications how they impose the constraint that populations are positive. It seems to me that your regularisation technique would be a big improvement, especially sparsity which will enforce populations to occur more in a sequence than a jumble, and allowing one to avoid model bias.

1 M. Schmidt, S. Rajagopal, Z. Ren, and K. Moffat, Application of singular value decom- position to the analysis of time-resolved macromolecular X-ray data, *Biophys. J.*, 2003, **84**(3), 2112–2129.

Jasper van Thor communicated in reply: Dear Henry, in Singular Value Decomposition or other methods of matrix factorisation, a separation is made purely on the basis of orthogonality. The fundamental assumption is one of linearity. Thus, any individual measurement, or single column, of the input matrix must then be some linear combination of the left singular vectors. The basis spectra, or left singular vectors in themselves, have no physical meaning: these do not represent spectra for pure species. Therefore, it is always required to apply a model or connectivity scheme that relates to the populations found in the right singular vectors (see for example reference 1). By requiring amplitudes to be non-negative, as Chris has implemented, such a model is effectively being used.

1. L. J. G. W. van Wilderen, C. N. Lincoln, J. J. van Thor, *PLOS One*, **6**, 2011, e17373.

Chris Jacobsen answered: Good suggestions. I hasten to add that by no means did we invent regularization methods! We simply applied them to the case of spectromicroscopy analysis, where to our knowledge these methods had not been applied before (except for the PhD work of Holger Fleckenstein, who is a coauthor on the paper).

Jasper van Thor asked: Dear Chris, in the matrix decomposition did you perform a matrix inversion which presents the basis spectra μ_{SxN} which you show to have all the significant amplitude? Or did you perform a Singular Value Decomposition $D = \mu St^{T}$, where the singular values represent the amplitudes and also provide the statistical significance of the contributions of the basis spectra μ. The thickness maps as well as the basis spectra would need to be weighed by the respective singular value S. Does that correspond to the amplitudes which you have shown?

Chris Jacobsen replied: In the cluster analysis approach described in the first part of the paper, our procedure is as follows: (1) perform an eigenvector analysis from the covariance of the data matrix D, and then limit the vectors to those with statistical significance; and (2) carry out cluster analysis to find pixels with similar eigenspectra weightings. The spectra at those pixels are then averaged together to yield a matrix μ_{SxN} which is then inverted using singular value decomposition (SVD) as you describe. This is described in the 2004 Ultramicroscopy paper by Lerotic *et al.* cited in the manuscript. These spectral solutions are then used as just one of two regularizations in the non-negative matrix analysis (NNMA) approach described, where the main cost function is the data matching error (optical density matrix minus the product of the present guess of the set of spectra times the present guess of the weightings or thicknesses; that is, minimizing the square of $D - \mu t$). I hope this answers the question.

Michael Woerner communicated: I see some similarities of your method with the so called maximum entropy method (MEM). Is it possible to combine your method with the MEM?

Chris Jacobsen responded: Maximum entropy involves minimizing some norm that measures contrast or fluctuations in the solution. These can include norms such as L_1 or L_2. In our case, we used an L_1 norm on the thickness solution t, so in some sense we have included a regularization that is representative of what is used in some Maximum Entropy methods.

Majed Chergui opened the discussion of the paper by **Michael Woerner** by asking: Your set-up for fs powder diffraction is quite unique. How do you compare it with ultrafast electron diffraction where the electron beam can be focussed down to a single grain size, or also to the developments for nanodiffraction at synchrotrons, which may lead to ultrafast studies too?

Michael Woerner communicated in reply: First, I would like to mention, that table top femtosecond X-ray diffraction can be also performed using the rotation method as we have recently demonstrated (B. Freyer *et al.*, *Optics Express*, **19**, 15506 (2011)). However, the diffraction efficiency for X-rays is much smaller than that for electrons. Thus, femtosecond nanodiffraction of X-rays will be the playground of the big accelerator machines (at least in the near future). Recently, we started a collaboration with Andrius Baltuska of the TU Vienna to develop schemes for a table top femtoseond X-ray source using a long wavelength driver (*i.e.* an OPCPA system working at $\lambda = 4$ μm). Such experiments turned out to be very promising in the sense that we can expect for future table top systems a hundred times higher X-ray flux.

Jasper van Thor asked: Dear Michael, from your presentation it is clear that you apply a maximum entropy method to find the charge density distribution from your experimental data. Firstly, could you please reiterate the phasing method that was used, and the accuracy of the structure factors in terms of crystallographic quality indicators and R factors. Secondly, can you show that the pumped data are truly isomorphous with the unpumped data? From the strength of the charge density which you derive and have shown is surprising that no bond length modification is apparently observed, which would likely dominate the transient structure factor amplitude differences. Further, transient difference density is present only within the duration of the pump and is not seen after. For the field driven signal, would THz pumping then retrieve the time-domain response to the field?

Michael Woerner communicated in reply: First, I would like to mention that we use the maximum entropy method (MEM) (instead of using a direct Fourier transform in the truncated q-space) just as a tool to derive the electron density map. Thus, the MEM has nothing to do with phase problem. How we deal with the phasing problem is detailed at the beginning of section 3 of our article. In summary, our method of analysis determines the average unit cell of the entire powder sample which has per definition the symmetry properties of the initial structure. An analysis using the Patterson method (Stingl *et al.*, *Phys. Rev. Lett.*, **109**, 147402 (2012)) shows that the latter is the dominant contribution to the change of the electron density map. In general, an excitation using a polarized electric field might also change the intensity along a particular diffraction ring in powder diffraction. Our diffraction efficiency is too low to address such questions in more detail. Within an optical cycle of the 800 nm driving pulse the change of bond lengths are negligible. Impulsive Raman excitation, however, leads to changes of the bond lengths in $LiBH_4$ at later times as evident from Fig. 3(c–f) in our article. Similar to impulsive Raman excitation THz excitation might also lead to bond length changes on the longer times scales of vibrations.

Martin McCoustra commented: One wonders what might happen to the charge exchange as LiH is compressed? Would measurements in a diamond anvil cell be possible to experimentally address this question?

Michael Woerner responded: In the literature one finds stationary X-ray diffraction experiments on compressed LiH up to 250 GPa (Lazicki *et al.*, *Phys. Rev. B*, **85**, 054103 (2012)) using a diamond anvil cell. Thus, in principle femto-second X-ray diffraction under pressure should be possible. In practice, however, one expects additional problems in time-resolved experiments due to both X-ray absorption and dispersion of pump light in the anvil cell itself. So far, I did not find an analysis of stationary diffraction experiments concerning a pressure dependent ionicity of LiH. Depending on the ionicity of the pressure dependent ground state of LiH I expect a field induced change of the ionicity as discussed in our article.

Christian Bressler communicated: You have succeeded in extracting results *via* fs-powder diffraction in a laser-driven X-ray plasma source. Do you see the need to apply for beam-time at an XFEL source? What benefit from this source would you expect against your ample in-house beam-time? Or why did you not apply for XFEL beam-time anywhere so far?

Michael Woerner answered: I will apply for beam-time at an XFEL source once those devices provide >10 keV X-ray pulses at a repetition rate of 1 kHz or higher with a timing jitter between laser pump and X-ray probe pulses of less than 50 fs. So far the research we would like to do can be easily performed with our table top fs X-ray source (despite having much less X-ray photons per shot). The main reason I did not apply for XFEL beam-time is simply the lack of free personnel and working hours from our side. The experiments we are doing in our lab are very time consuming and personnel intensive. If this situation changes in the future (*i.e.* more personnel) we are happy to apply for beam-time at an XFEL.

Jonathan Underwood opened the discussion of the paper by **Jochen Küpper** by commenting: Prof Marangos asked what would be needed to extend the tech-nique presented by Küpper and colleagues for diffractive imaging of ground state (static) molecular structures to the measurement of excited state structural dynamics. If one considers an experiment where a resonant pump pulse elec-tronically excites molecules, and then the resulting structural dynamics are probed *via* diffraction with a time delayed X-ray pulse, then the optimal situation would be that the molecular alignment is maximal at the instance of probing. So, one can imagine an approach in which a strong non-resonant laser field is applied prior to the pump pulse to create a rotational wave packet in the ground state of the molecule. By careful design of the alignment laser field parameters (pulse shape, time-dependent polarization) and time-delay between the alignment and pump laser fields it may then be possible to transfer the rotational coherence to the excited electronic state where the subsequent wave packet evolution will produce the desired alignment at the instance of probing. However, this is clearly a very challenging proposition, as one needs to understand how the rotational wave packet will evolve in the excited state which may have very different (and time-dependent) rotational constants to the ground electronic state. Also, the alignment field parameters would need to be varied each time the pump and probe time delay was changed. There is very clearly a need to consider this problem in further detail as current understanding of what is and is not possible is currently poor.

Jochen Küpper communicated in reply: This experimental protocol suggested by Dr. Underwood is interesting and should, in principle, work. There are a few comments I wish to make below.

At the same time, I want to point out that we should really make big efforts to look at ground state chemistry. While there is very important electronically-excited state (photo)chemistry going on in chemistry and biology, most chemical dynamics does not happen in excited electronic states, but in the ground state. This involves almost all "classical", thermally activated chemistry and biology and comprises by far the largest part of all chemical and molecular dynamics. Therefore, we need to develop protocols to trigger and time chemical reactions on the ground state potential energy surface, and methods to investigate, image them with spatial and temporal atomic resolution. It might be possible to define the starting point of such ground-state reactions with (resonant or non-resonant) THz or IR photons. An alternative approach would be to use highly sensitive detection methods that are capable of simply observing the few, rare, reaction events in a thermally activated "soup" of reactants. While the latter is a dream, maybe solely a thought, the former approach of photo-triggering ground state reactions seems in reach. First experiments are already emerging[1,2] and need to be transferred to the ultrafast-time and structure resolved domain of imaging experiments.

Regarding the suggested transfer of coherence from the electronic ground state to the electronically excited state, a few points discussed at the meeting were related to the change of the inertial tensor upon excitation, *e.g.* the changes of rotational constants and the reorientation of the principal axes of inertia. These are generally important considerations, but we might be lucky in many cases: axis tilting effects are often very small (<1°) for valence excitations, rotational constants typically change by one to a few percent, and sometimes even the electronic properties, such as dipole moment and polarizability, change only weakly. Moreover, all these details can be derived from eigenstate resolved spectroscopies –for the prototypical molecule indole this has been shown in detail, including the details of strong vibronic cooling between different electronically excited states.[3–5] Regarding the alignment process, which is in the short pulse "impulsive" regime, I wish to point out that for the alignment to be visible in the coherent diffraction pattern, one has to obtain very strong degrees of alignment, *i.e.*, $\langle \cos^2\theta \rangle \gg 0.8$.[6] This will be very hard to achieve in an impulsive alignment approach, although multi-pulse alignment might be able to achieve it (Cryan *et al.*, *Phys. Rev. A*, 2009, **80**, 063412). Overall, it might be easier to implement a quasi-adiabatic alignment protocol, in which a moderately-long-pulse ac field angularly "permanently" confines the molecule—with significantly stronger degrees of alignment and independent of the molecule's electronic state. However, the presence of the field clearly has to be considered regarding the stability of the molecule in various states as well with respect to its influence onto the investigated dynamics (see also the next question for further details). Then, the truncated pulse method mentioned by Jonathan Underwood elsewhere might be a very viable compromise. Overall, I conclude by fully supporting the statements of Jonathan Underwood that "there is very clearly a need to consider this problem in further detail as current understanding of what is and is not possible is currently poor". Even the *sole* development of an understanding of the right approach would yield tremendous insight into the intricate workings of molecules.

1. B. Dian, A. Longarte and T. Zwier, *Science*, 2002, **296**(5577), 2369–2373.
2. B. C. Dian, G. G. Brown, K. O. Douglass and B. H. Pate, *Science*, 2008, **320**(5878), 924–928.
3. G. Berden, W. L. Meerts and E. Jalviste, *The Journal of Chemical Physics*, 1995, **103**(22), 9596–9606.
4. C. Brand, J. Küpper, D. W. Pratt, W. L. Meerts, D. Krügler, J. Tatchen and M. Schmitt, *Phys. Chem. Chem. Phys.*, 2010, **12**(19), 4968–4979.
5. J. Küpper, D. W. Pratt, W. L. Meerts, C. Brand, J. Tatchen and M. Schmitt, *Phys. Chem. Chem. Phys.*, 2010, **12**(19), 4980–4988.
6. F. Filsinger, H. N. Chapman, G. Meijer, H. Stapelfeldt and J. Küpper, *Phys. Chem. Chem. Phys.*, 2011, **13**(6), 2076.

Jonathan Underwood commented: Jochen Küpper makes very good points about the need to find ways to measure structural dynamics during ground state (chemically relevant) processes in molecules. My main concern about using the approach presented by Küpper and colleagues for structural imaging of ground state molecules undergoing chemical change is that the presence of the adiabatically applied alignment laser field could well be expected to change the observed chemical processes (beyond that expected from steric/geometric considerations). The laser field intensities used for producing molecular axis alignment typically lie in the region of 10^{12} W cm^{-2} corresponding to an electric field of about 10^{9} V m^{-1}. Such high fields strengths are expected to significantly distort the electronic structure of even ground state molecules – indeed the quantum mechanical expression for the molecular polarizability expresses the field-induced mixing of electronic states. As such, I am concerned that the alignment laser field would directly influence the distribution of charge in the molecules under study, affecting their reactivity.

Jochen Küpper responded: Again, I also consider this a very insightful and important comment by Jonathan Underwood. On the quantitative detail I wish to point out that the quantum-state-selection exploited in our work[1,2] not only allows for the selection of distinct molecular species,[3] but it allows the achievement of very strong alignment (and orientation) at moderated field strengths of only 10^{10} W cm^{-2}.[4] Nevertheless, while this strongly mitigates the problem, it does not solve it and one has to consider the (ac) Stark effect onto the chemical dynamics. Typical Stark energy shifts are of the order of 1 meV under the specified conditions. These shifts, clearly, could change the dynamics in a state-to-state resolved experiment where degeneracies might be lifted or induced, but it is likely not relevant in terms of the chemical energies in typical reactions, which could be of the order of 1 eV. Therefore, our current investigations of statistically averaged (*versus* quantum states) dynamics should not be much influenced.

At the same time, one should also relate these fields (~1×10^{8} V m^{-1}) to the "internal fields" from the reactions partners, which are of the order of the atomic unit of the electric field (~5 × 10^{11} V m^{-1}). Again, this is a difference of more than three orders of magnitude and it will take some effort and progress to be able to see these effects. In how far the more subtle effects hinted at, such as charge redistributions, *e.g.*, through the Stark-coupling of orbitals or states, are a different issue, is related to the discussion of the validity of the Born–Oppenheimer approximation, elsewhere in this Discussion.

Overall, the presence of an alignment laser will surely have an influence on the molecular system under investigation. Personally, I would be very happy to see this effect as it would show that we had finally arrived at the observation of intricate details of the dynamics of molecular systems. Under these circumstances we would first learn what this influence is and, subsequently, would surely be able to mitigate or avoid it altogether. The presented X-ray diffraction probe[1,2,5] should, at least, be a largely unaffected probe to observe these effects –once it is a routine tool with sufficient temporal and spatial resolution.

1. Stern *et al.*, *Faraday Discuss 171* (2014), doi: 10.1039/c4fd00028e.
2. J. Küpper, S. Stern, L. Holmegaard, F. Filsinger, A. Rouzée, A. Rudenko, *et al.*, *Physical Review Letters*, 2014, **112**, 083002.
3. F. Filsinger, J. Küpper and G. Meijer, *Faraday Discuss 142*, 155 (2009).
4. F. Filsinger, J. Küpper, G. Meijer, L. Holmegaard, J. H. Nielsen, I. Nevo, *et al.*, *The Journal of Chemical Physics*, 2009, **131**(6), 064309.
5. S. Trippel, T. Mullins, N. L. M. Mueller, J. S. Kienitz, J. J. Omiste, H. Stapelfeldt, R. González-Férez and J. Küpper, *Physical Review A*, 2014, **89**(5), 051401.

Faraday Discussions

ROYAL SOCIETY
OF CHEMISTRY

PAPER

Approaches to time-resolved diffraction using an XFEL

John C. H. Spence*

Received 27th February 2014, Accepted 25th March 2014

DOI: 10.1039/c4fd00025k

We describe several schemes for time-resolved imaging of molecular motion using a free-electron laser (XFEL), in response to the many challenges and opportunities which XFEL radiation has created for accurate time-resolved measurement of structure. For pump–probe experiments using crystals, the problem of recording full Bragg reflections (not partials) in each shot arises. Two solutions, the use of the large bandwith which necesarily results from using attosecond pulses, and the use the coherent convergent beam mode are suggested. We also show that with attosecond recording times shorter than the temporal coherence time, Bragg reflections excited by different wavelengths from different reflections can interfere, providing structure factor phase information. For slower processes, a mixing jet sample-delivery device is described to allow snapshot solution scattering during molecular reactions on the microsecond scale. For optically excited membrane proteins, we suggest the use of the lipid cubic phase sample delivery device operating at atmospheric pressure. The use of two-color and split-and-delay schemes is suggested for improved accuracy in the Monte-Carlo method of serial femtosecond crystallography (SFX).

1.   Introduction

Understanding of the function of biological macromolecules requires mapping of their three-dimensional (3D) structures at near-atomic resolution, together with observation of the dynamics of these structures during a chemical reaction or optical excitation. This has previously been achieved using "pump–probe" experiments based on time-resolved (TR) crystallography in the Laue mode,[1] over timescales from picoseconds to seconds.[2] Issues of reaction branching, the fraction of excited molecules in the crystal, and the periodic averaging over unit cells in the reconstructed density map are discussed elsewhere.[2] The conventional method of protein crystallography (MX), in which a sample is rotated slowly through the Bragg condition to provide angle-integrated reflections, cannot be used if a series of snapshots showing the molecular motions is required. (An angle integration over the crystal rocking curve is needed to obtain a structure factor). In the Laue method, fast integrated reflections are obtained by using broad-band

Department of Physics, Arizona State University, Tempe, Az., USA 85282

radiation in the transmission geometry, so that the corresponding range of Ewald sphere diameters spans the rocking curves of reflections near the Bragg condition, while also generating more reflections per exposure. This range of diameters is shown as the area AOB in Fig. 1.

Time-resolved measurements require a fast trigger to initiate the reaction in the crystal, followed after a delay by the recording of a diffraction pattern from the excited state of the crystal. Difference density maps are formed between the ground and excited states, with a different delay for each frame of a "movie". Many crystal orientations are needed for each delay for a 3D image reconstruction. Full, rather than partial, Bragg reflections must be recorded in each snapshot, which is then probed at different time points along the reaction coordinate, until complete data sets are obtained.

The invention of the hard X-ray laser[3] with its femtosecond pulses of intense radiation has created both opportunities and challenges for TR-MX. The first hard-X-ray XFEL, the Linac Coherent Light Source (LCLS) at SLAC, for example, provides about 1E12 photons of 8 kV radiation per pulse at 120 Hz repetition rate, into a focused spot as small as 0.1 micron diameter. Since the focused beam instantly vaporizes the sample and the shot-to-shot variations in beam intensity are large (eg 15%), while the beam is relatively monochromatic (eg 0.1% energy spread), Laue diffraction for TR-MX by the conventional method is impossible. However it has been found that these femotosecond pulses outrun radiation damage[4] so that atomic-resolution diffraction patterns are detected before the onset of serious damage from impact ionization due to the photoelectron cascade. (The limited direct ionization effects which occur during the pulse are unimportant for most MX applications, and detection ends well before the onset of atomic motion). The resulting need to maintain a fresh supply of protein nanocrystals for this "diffract-then-destroy" mode has been satisfied by using a

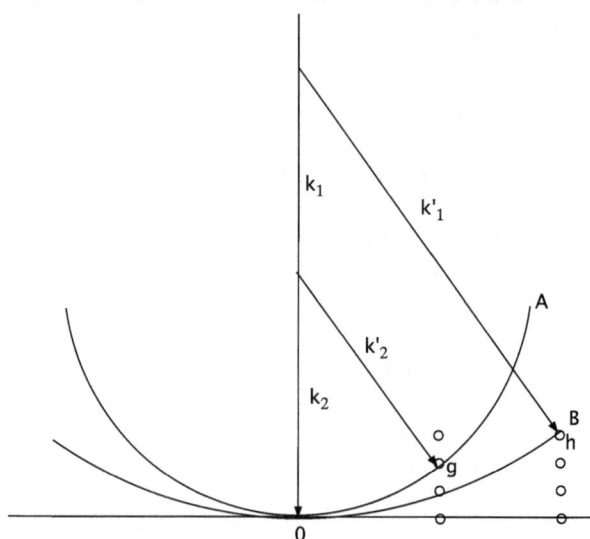

Fig. 1 Ewald sphere scattering geometry for two wavelengths scattering from different reflections into the same detector pixel. With a bandwidth spanning wavevectors from k_1 to k_2, all Bragg reflections falling within the area AOB are excited.

liquid jet a few microns in diameter, running across the pulsed X-ray beam in vacuum, carrying the appropriate concentration of protein nanocrystals. This has led to the development of the serial femtosecond crystallography (SFX) method (see[5] for a review). The first pump–probe experiments using a liquid jet for time-resolved SFX have recently been reported, applied to a stream of optically excited Photosystem 1 - ferredoxin nanocrystals, important for photosynthesis.[6] These showed changes in structure factors due to illumination which well exceeded the noise level. In this method, a pump laser crosses the liquid jet immediately prior to their transit of the X-ray beam. Diffraction patterns from excited and ground-state nanocrystals are interleaved, however normalization is difficult, since each crystal is a different size and each snapshot provides only partial reflections. One relies on the chance occurrence among millions of patterns to provide a range of orientations across the rocking curve (from nanocrystals of different size).

While this pump–probe-in-a-jet approach holds promise, the Monte Carlo process used to merge data (which averages over all stochastic experimental variables, such as beam intensity variations, nanocrystal size and orientation) converges slowly as 1/sqrt(N) for N shots, and is very wasteful of protein, since all protein between shots runs to waste. In addition, the time-resolution of the EXFEL is far better than needed for biological studies, where molecular motions occur on a time-scale of microseconds to milliseconds. Nevertheless we wish to take advantage of the absence of radiation damage made possible by the use of femosecond snapshots with destructive readout. In this paper we review some alternative schemes for TR-SFX under development. Among many challenges when attempting to use the poorly characterized experimental conditions of an XFEL for the highly accurate quantification of data needed in time-resolved work, we focus here mainly on the problem of obtaining full, rather than partial, Bragg reflections.

2.   Attosecond Laue diffraction

A proposal has recently been made to construct an attosecond (as) hard-X-ray laser, based on a coherent inverse Compton scattering (C-ICS) source.[7] This machine is currently under construction at DESY in Germany, and is much smaller than an XFEL. Taking pulses to be bandlimited and to have full temporal coherence, the pulse duration Δt and beam energy spread ΔE are then related by the uncertainty principle as

$$\Delta t \text{ (fs)} = 4.14/\Delta E \text{ (eV)} \qquad (1)$$

the necessary energy spread for Laue diffraction of, say, 3% at 10 kV can then be obtained using 14 as pulses, giving a bandwidth of 300 eV. These pulses will certainly also outrun radiation damage. The crystalline sample then acts as a monochromator, picking out wavelengths which satisfy Bragg's law for many reflections, and spanning their rocking curves.[1] In this way both radiation damage avoidance and Laue snapshot diffraction with full reflections becomes possible when using sufficiently brief pulses.

For bandlimited pulses, two different reflections may scatter at different wavelengths λ_1 and λ_2 into the same direction $\mathbf{k}_1$ (or $\mathbf{k}_2$) and detector pixel, as shown in Fig. 1. These will interfere coherently if the pulse duration is sufficiently brief. Then the resulting intensity will contain information on structure factor phase. Interference occurs if the recording time τ (the pulse duration) is less than the beat period T between the frequencies of the two wavelengths $\omega_1 = c\,k_1 = 2\pi\,c/\lambda_1$ and $\omega_2 = c\,k_2 = 2\pi\,c/\lambda_2$, satisfying the two different Bragg conditions, and this will be so if the pulses are bandlimited. For a crystal wider than the focused beam and thicker than an extinction distance, aside from unimportant constants, the scattered intensity in a single pulse in direction $\mathbf{k}_2'$ is

$$I_T = \int_0^\tau \left| F(g)\exp\left(ik_1'\cdot r + i\varpi_1 t\right) + F(h)\exp\left(ik_2'\cdot r + i\varpi_2 t\right) \right|^2 dt \qquad (2)$$

If the recording time $\tau < T = 2\pi/(\omega_1 - \omega_2)$ and the complex structure factors are $F(\mathbf{g}) = f(\mathbf{g})\exp(i\,\theta_g)$, this becomes

$$I_T = f(g)^2 + f(h)^2 + 2|f(g)||f(h)|\cos(\theta_g - \theta_h) \qquad (3)$$

which is sensitive to the phase difference between structure factors F_g and F_h. The result depends on choice of origin for $\mathbf{r}$. Phase sums which are independent of origin can be formed if the corresponding reciprocal lattice vectors sum to zero (forming a closed loop in reciprocal space), and may be found in this case if absorption is neglected and Freidel symmetry assumed. Then such loops can be formed within the region AOB of Fig. 1.

3. Coherent convergent-beam diffraction

A second method for obtaining full reflections in a single shot is provided by convergent-beam diffraction. The geometry is shown in Fig. 2. We see that since

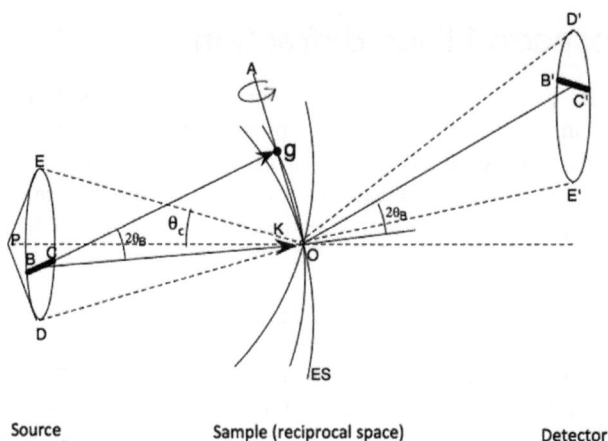

Source   Sample (reciprocal space)   Detector

Fig. 2 Coherent convergent-beam diffraction geometry, showing source DE, Ewald sphere ES and detector E'D'. Point source P fills illumination aperture DE coherently. Limiting Ewald sphere orientations (eg ES) are defined by marginal rays DO and EO. Plane-wave component K of converging spherical wave (with beam divergence θ_x) from midpoint of BC arrives at detector at midpoint of B'C'. Rotation of arrowed wavevectors originating along line BC about OA preserves the Bragg condition for reciprocal lattice vector g, defining line C'B' at detector, along which the Bragg condition is satisfied, producing Bragg lines instead of spots. The disk E'D' is an inverted image of ED.

the Bragg condition for reflection **g** is preserved when a crystal is rotated about **g**, it follows that for a stationary crystal, wavevectors within the illumination cone of semiangle θ_χ which rotate about **g** also preserve this condition, producing a line of intensity at the detector instead of a Bragg spot. The intensity profile across this line is then the crystal rocking curve, which may be measured, given sufficiently fine pixelation. In this way, the covergent beam method provides an "angular window" across the crystal rocking curve in each reflection simultaneously which lies near a Bragg condition. Integration across the line of intensity then gives full reflections, proportional to structure factors, as required for "single-shot" Laue diffraction.

If the illumination is coherent, as from an XFEL, an extremely small beam diameter can be produced, of diffraction-limited diameter $d_p \sim 1.2\ \lambda/\theta_\chi$ between first minima. Such small beams become smaller than one mosaic block, so that that the traditional theory of mosaicity with its smooth modelled rocking curve does not apply, and the shape transforms seen in experimental XFEL data must be considered when merging data.[13] By combining this expression for d_p (which ignores aberrations in the probe-forming optics) with Bragg's law $\lambda = 2\ d_{hkl} \sin (\Theta_B)$, putting $d_{hkl} = d_p/2$, we see that when the beam radius is just equal to the unit-cell dimension (for a first-order reflection) then the coherent convergent-beam disks will overlap and interfere by about 22%. (The current LCLS micro-diffraction beam diameter of 100 nm is not much bigger than the unit cell of some macromolecular crystals or viruses). This interference between coherent overlapping orders provides information on structure factor phases (reckoned about an origin at the center of the beam).[8] It has been demonstrated both in electron[8] and soft X-ray diffraction.[14] For protein crystallography, where autoindexing of a large number of reflections in three-dimensions is required, the requirement on beam divergence is that it be somewhat less than the Bragg angle, in order that individual reflections can be indexed without overlap in three dimensions on the curved Ewald sphere. Using simulations, we have shown the more rapid convergence of SFX data which results when beam divergence increases. Here the standard deviation R_{split} in the sum of the same reflection from N different nanocrystals of different size in slightly different orientations is plotted against the number of nanocrystals, and the results found to fall more rapidly as N increases due to the decrease in "partiality" which results from larger beam divergence. This experimental requirement for larger beam divergence of a few milliradians is entirely practical for protein crystals.

4.   Mixing jets

The liquid injector system used to spray submicron nanocrystals across the XFEL beam in a liquid stream a few microns in diameter, flowing in vacuum at about 10 m s^{-1} has been described in several papers.[9] The stream is surrounded by a coaxial jacket of high-pressure gas, which, by speeding up the liquid, focuses the stream. In this way, clogging is avoided by allowing the use of a larger nozzle but much finer focused liquid stream, which breaks up by Rayleigh instability into a stream of micron-sized droplets.

The arrangement consists of a hollow fiber-optic line (carrying the buffer and nanocrystal solution) running inside a glass capillary tube, with high pressure helium gas between the two tubes, emerging at the nozzle to focus the liquid.

A new version has now been made for the study of time-resolved chemical reactions in solution at an EXFEL.[10] As shown in Fig. 3, a third tube is used between the inner fiber and outer capillary. The space between this tube and the fiber supplies a second liquid, while gas outside it (and within the capillary tube) provides the focusing effect. The inner hollow fiber-optic tube slides telescopically along inside the intermediate tube to vary the reaction time of two fluids. In a typical experiment, a substrate may be supplied in the intermediate space, and an enzyme in solution in the fiber-optic line along the axis. With the fiber withdrawn, the two fluids, catalyst and substrate, mix at the end of the fiber, then react as they travel to the nozzle leading into vacuum. EXFEL shots then provide snap-shot SAXS data for different reaction times as the inner fiber is slowly withdrawn. The theory of these fluctuation SAXS patterns (FSAXS), originally given by Z. Kam, in which the recording time is much shorter than the rotational diffusion times of the molecules, has been reviewed elsewhere.[11] In particular, the patterns are two-dimensional, unlike conventional isotropic one-dimensional SAXS patterns, and hence contain more information to assist a three-dimensional reconstruction. (Two-dimensional scattering due to coherent interference between X-rays scattered from different particles is not used.) Experiments with this new mixing jet are planned for the near future and the performance of the jet has been evaluated using fluorescent dyes. The range of reaction times possible is 10–200 milliseconds, appropriate to many biological systems, while the mixing time at the end of the inner fiber, before reactions begin, is about 200 microseconds, and this sets the error in reaction timing measurement for each movie frame. In this way we hope to track the molecular motions involved in enzyme catalysis.

5. Pump–probe LCP jet experiments

In order to eliminate the protein which flows to waste between XFEL shots (at 120 Hz), a new sample injector has been developed using a more viscous medium, lipidic cubic phase (LCP).[12] Only about 1 out of 10,000 nanocrystals produce X-ray patterns in the earlier liquid jet, running at 10 m s^{-1} with a flow rate 10 microliters/min, requiring perhaps 10–100 mg of pure protein for 5 h of data collection.

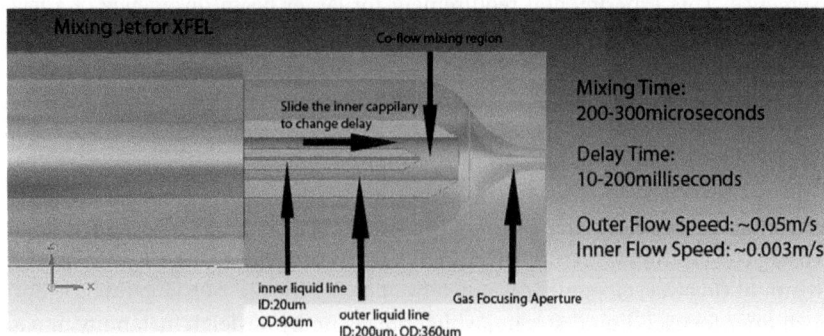

Fig. 3 Mixing jet, showing three concentric tubes and nozzle. Two fluids are mixed in a short mixing time, then allowed to react for a delay time before emerging from the nozzle in a jet a few microns in diameter. This is intercepted by the pulsed X-ray beam. By sliding the inner tube, the delay time can be adjusted.

The slower, more viscous LCP jet, with flow rate between 1–300 nl min^{-1}, then greatly reduces wasted protein by a factor of about 20, and also provides both a growth medium (for both membrane and some soluble proteins) in which to grow nanocrystals, while at the same time running slowly enough to produce a high hit rate (the nanoxtals emerge from this "toothpaste jet" at about the same rate as the X-ray pulses arrive). The jet has proven particulary useful for study of the important GPCR class of proteins, which may form nanocrystals but do not grow large enough crystals for conventional protein crystallography (MX). Fig. 4 shows recent results for XFEL diffraction from human serotonin receptor[12] and an image of the LCP jet running. Using photosensitive membrane proteins in this sample delivery system, it should also be possible to undertake pump–probe experiments for the microsecond time-scales (and longer) involved in biology. Experiments are planned to determine the maximum repetition rate of the XFEL which allows debris to be cleared due to the vaporization of one sample, before the next moves into place on the X-ray optic axis. The possibility is also being explored of doing this work at atmospheric pressure (or in a helium environment), since the LCP jet is found to work satisfactorily in air at STP.

6. Other approaches, discussion

For the TR-SFX experiments undertaken to date,[6] the Monte Carlo method of data analysis has been used.[13] This means that X-ray snapshots for the ground and excited state in a pump–probe experiment are taken from different nanocrystals of different size (each of which is destroyed after recording these patterns). In addition, one has shot-to-shot intensity variations of perhaps 15% when using an XFEL, much larger than the effect on structure factors of the pump laser. While scaling methods have been developed to deal with this, greater accuracy could be

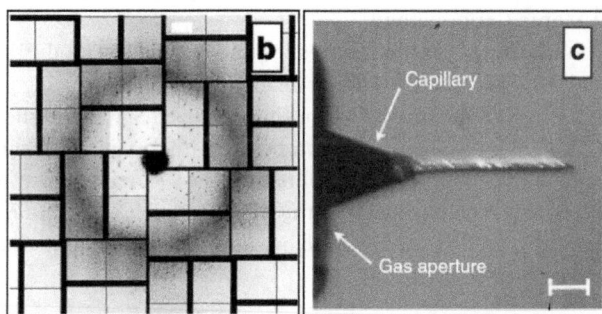

Fig. 4 (b) Diffraction from serotonin receptor 5-HT$_{2B}$ in cholesterol-doped 9.9 MAG + 7.9 MAG LCP. No sharp rings are visible suggesting that formation of L$_c$ phase has been avoided (X-ray intensity attenuated to 3.1% due to strong Bragg diffraction from 5 × 5 × 5 μm^3 sized crystals, 1.5 μm X-ray beam diameter, 50 fs pulse length, 9.5 keV, 50 μm LCP jet diameter, 190 nL min^{-1} flow rate, 120 Hz pulse rate). The resolution at the detector edge in both panels is 2.5 Å.[12] c) 9.9 MAG LCP extrusion in vacuum viewed between crossed polarizers. The tapered end of the capillary nozzle is seen protruding out of the gas aperture. Capillary inner diameter: 30 μm. With He as co-flowing gas. Birefringence (bright flecks) is an indication of a transition of the cubic phase to a lamellar crystalline phase due to evaporative cooling.

obtained if two snapshots, for ground and excited state, could be obtained from the same crystal. In this case the first shot must be below the damage threshold for the protein nanocrystal. These experimental conditions will become available with the development of "two-color" experiments, in which, using seeded beams or two undulators and gain modulation, it is possible to produce a pair of X-ray pulses of slightly different wavelength, with a delay between them. At present these delays are limited to a few hundred femtoseconds. Between these pulses must be inserted the pump laser pulse. It is not possible to read out two diffraction patterns separated by such a brief time interval, however their different wavelengths will displace them slightly on the detector, so that they may be read out together as a single frame. A serious challenge with this arrangement is the reduction of timing jitter on the pump laser, which must be less than the time between pulses.

Alternatively, schemes have been developed for "split-and-delay" of the X-ray pulses, using either mirrors or Bragg beam splitters. The beam is split into two beams of equal intensity, and one pulse sent on a longer optical path to introduce delay (the speed of light is 3.3 ns m^{-1}, so that a very long detour is needed to obtain the microsecond delays important for biology). Timing jitter on the pump laser is no longer an issue, but the overlap of the two beams in space then becomes challenging if submicron beam diameters are involved. The two beams have the same energy. One may then arrange for the two beams to strike the same nanocrystal (perhaps in LCP) at slightly different angles, so that the immediate and delayed diffraction patterns are displaced again on the detector, for common readout. Since only differences between the intensities of scattering from the same crystals are accumulated, this method can be shown to eliminate most error due to variations in crystal size, orientation and shot-to-shot beam intensity variation. The weaker pulses needed to avoid damage to the crystals do increase noise in high angle scattering, however detailed simulations show a much more rapid reduction (with number of shots) in the error associated with measurement of a difference in structure factor, due to pumping, than when using the conventional Monte Carlo method.

In conclusion, several new approaches to time-resolved XFEL diffraction have been presented. Of these, the convergent beam approach to reducing partiality in Bragg beams, the use of the mixing jet for time-resolved SAXS, and the use of the LCP jet in air for time-resolved pump–probe experiments may be implemented immediately. The development of split-and-delay schemes has undergone preliminary tests, while construction of the first attosecond hard X-ray laser may be some years away, in this rapidly evolving and exciting field of time-resolved atomic-resolution X-ray imaging.

Acknowledgements

I am grateful to Prof K. Schmidt and Prof U. Weierstall for many useful discussions. Supported by NSF STC award 1231306.

References

1 *Theory of X-ray diffraction in crystals*. W. H. Zachariasen. Dover 1967 Toronto.

2 T. Graber, S. Anderson, H. Brewer, Y. S. Chen, H. S. Cho, N. Dashdorj, R. W. Henning, I. Kosheleva, G. Macha, M. Meron, R. Pahl, Z. Ren, S. Ruan, F. Schotte, V. Srajer, P. J. Viccaro, F. Westferro, P. Anfinrud and K. Moffat, BioCARS: a synchrotron resource for time-resolved X-ray science, *J. Synchrotron Radiat.*, 2011, **18**(4), 658–670.

3 P. Emma, R. Akre, J. Arthur, R. Bionta, C. Bostedt, J. Bozek, A. Brachmann, P. Bucksbaum, R. Coffee, F.-J. Decker, Y. Ding, D. Dowell, S. Edstrom, A. Fisher, J. Frisch, S. Gilevich, J. Hastings, G. Hays, Ph. Hering, Z. Huang, R. Iverson, H. Loos, M. Messerschmidt, A. Miahnahri, S. Moeller, H.-D. Nuhn, G. Pile, D. Ratner, J. Rzepiela, D. Schultz, T. Smith, P. Stefan, H. Tompkins, J. Turner, J. Welch, W. White, J. Wu, G. Yocky and J. Galayda, First lasing and operation of an Ångstrom wavelength free-electron laser, *Nat. Photonics*, 2010, **4**(9), 641–647.

4 H. N. Chapman, P. Fromme, A. Barty, T. A. White, R. A. Kirian, A. Aquila, M. S. Hunter, J. Schulz, D. P. DePonte, U. Weierstall, R. B. Doak, F. R. N. C. Maia, A. V. Martin, I. Schlichting, L. Lomb, N. Coppola, R. L. Shoeman, S. W. Epp, R. Hartmann, D. Rolles, A. Rudenko, L. Foucar, N. Kimmel, G. Weidenspointner, P. Holl, M. Liang, M. Barthelmess, C. Caleman, S. Boutet, M. J. Bogan, J. Krzywinski, C. Bostedt, S. Bajt, L. Gumprecht, B. Rudek, B. Erk, C. Schmidt, A. Hömke, C. Reich, D. Pietschner, L. Strüder, G. Hauser, H. Gorke, J. Ullrich, S. Herrmann, G. Schaller, F. Schopper, H. Soltau, K. U. Kühnel, M. Messerschmidt, J. D. Bozek, S. P. Hau-Riege, M. Frank, C. Y. Hampton, R. G. Sierra, D. Starodub, G. J. Williams, J. Hajdu, N. Timneanu, M. M. Seibert, J. Andreasson, A. Rocker, O. Jönsson, M. Svenda, S. Stern, K. Nass, R. Andritschke, C. D. Schröter, F. Krasniqi, M. Bott, K. E. Schmidt, X. Wang, I. Grotjohann, J. M. Holton, T. R. M. Barends, R. Neutze, S. Marchesini, R. Fromme, S. Schorb, D. Rupp, M. Adolph, T. Gorkhover, I. Andersson, H. Hirsemann, G. Potdevin, H. Graafsma, B. Nilsson and J. C. H. Spence, Femtosecond X-ray protein nanocrystallography, *Nature*, 2011, **470**(7332), 73–77.

5 J. C. H. Spence, U. Weierstall and H. N. Chapman, X-ray lasers for structural and dynamic biology, *Rep. Prog. Phys.*, 2012, **75**, 102601.

6 Andrew Aquila, Mark S. Hunter, R. Bruce Doak, Richard A. Kirian, Petra Fromme, Thomas A. White, Jakob Andreasson, David Arnlund, Sa! a Bajt, Thomas R. M. Barends, Miriam Barthelmess, Michael J. Bogan, Christoph Bostedt, Hervé Bottin, John D. Bozek, Carl Caleman, Nicola Coppola, Jan Davidsson, Daniel P. DePonte, Veit Elser, Sascha W. Epp, Benjamin Erk, Holger Fleckenstein, Lutz Foucar, Matthias Frank, Raimund Fromme, Heinz Graafsma, Ingo Grotjohann, Lars Gumprecht, Janos Hajdu, Christina Y. Hampton, Andreas Hartmann, Robert Hartmann, Stefan Hau-Riege, Günter Hauser, Helmut Hirsemann, Peter Holl, James M. Holton, André Hömke, Linda Johansson, Nils Kimmel, Stephan Kassemeyer, Faton Krasniqi, Kai-Uwe Kühnel, Mengning Liang, Lukas Lomb, Erik Malmerberg, Stefano Marchesini, Andrew V. Martin, Filipe R. N. C. Maia, Marc Messerschmidt, Karol Nass, Christian Reich, Richard Neutze, Daniel Rolles, Benedikt Rudek, Artem Rudenko, Ilme Schlichting, Carlo Schmidt, Kevin E. Schmidt, Joachim Schulz, M. Marvin Seibert, Robert L. Shoeman, Raymond Sierra, Heike Soltau,

Dmitri Starodub, Francesco Stellato, Stephan Stern, Lothar Strüder, Nicusor Timneanu, Joachim Ullrich, Xiaoyu Wang, Garth J. Williams, Georg Weidenspointner, Uwe Weierstall, Cornelia Wunderer, Anton Barty, John C. H. Spence and Henry N. Chapman, Time-resolved protein nanocrystallography using an X-ray free-electron laser, *Optics Express*, 2012, **20**, 2706.

7 W. S. Graves, *et al.*, *Phys. Rev. Lett.*, 2012, **108**, 263904.

8 J. C. H. Spence, N. Zatsepin and C. Li, Coherent convergent-beam time-resolved X-ray diffraction, *Philos. Trans. R. Soc., B*, 2014, **369**, 20130325.

9 U. Weierstall, J. C. H. Spence and R. B. Doak, Injector for scattering measurements on fully solvated species, *Rev. Sci Instr.*, 2012, **83**, 035108.

10 D. Wang, J. C. H. Spence and U. Weierstall, A mixing jet for solution scattering at XFELs, *J. Synchrotron Radiat.*, 2014 Submitted.

11 R. Kirian, Structure determination through correlated fluctuations in X-ray scattering, *J. Phys. B*, 2012, **45**, 223001.

12 Uwe Weierstall, Daniel James, Dingjie Wang, Wei Liu, John C. H. Spence, R. Bruce Doak, Garrett Nelson, Petra Fromme, Raimund Fromme, Ingo Grotjohann, Christopher Kupitz, Nadia A. Zatsepin, Shibom Basu, Daniel Wacker, Chong Wang, Sébastien Boutet, Marc Messerschmidt, Garth J. Williams, Jason E. Koglin, M. Marvin Seibert, Cornelius Gati, Robert L. Shoeman, Anton Barty, Henry N. Chapman, Richard A. Kirian, Kenneth R. Beyerlein, Raymond C. Stevens, Dianfan Li, Syed T. A. Shah, Nicole Howe and Martin Caffrey, Lipidic cubic phase injector facilitates membrane protein serial femtosecond crystallography, *Vadim Cherezov Nature Communications*, 2014, **5**, 3309.

13 Richard A. Kirian, Thomas White, James Holton, Henry N. Chapman, Petra Fromme, Anton Barty, Lukas Lomb, Andrew Aquila, Filipe Maia, Andrew Martin, Raimund Fromme, Xiaoyu Wang, Mark Hunter, Kevin Schmidt and John C. H. Spence, Structure factor analysis of femtosecond microdiffraction patterns from protein nanocrystals, *Acta Cryst A*, 2011, **67**, 131–140.

14 H. N. Chapman, Phase-retrieval X-ray microscopy by Wigner-distribution deconvolution, *Ultramicroscopy*, 1996, **66**, 153–172.

Faraday Discussions

ROYAL SOCIETY
OF CHEMISTRY

PAPER

Signal to noise considerations for single crystal femtosecond time resolved crystallography of the Photoactive Yellow Protein

Jasper J. van Thor,[*a] Mark M. Warren,[a] Craig N. Lincoln,[a] Matthieu Chollet,[b] Henrik Till Lemke,[b] David M. Fritz,[b] Marius Schmidt,[c] Jason Tenboer,[c] Zhong Ren,[e] Vukica Srajer,[e] Keith Moffat[de] and Tim Graber†[e]

Received 13th February 2014, Accepted 2nd April 2014

DOI: 10.1039/c4fd00011k

Femtosecond time resolved pump–probe protein X-ray crystallography requires highly accurate measurements of the photoinduced structure factor amplitude differences. In the case of femtosecond photolysis of single $P6_3$ crystals of the Photoactive Yellow Protein, it is shown that photochemical dynamics place a considerable restraint on the achievable time resolution due to the requirement to stretch and add second order dispersion in order to generate threshold concentration levels in the interaction region. Here, we report on using a 'quasi-cw' approach to use the rotation method with monochromatic radiation and 2 eV bandwidth at 9.465 keV at the Linac Coherent Light Source operated in SASE mode. A source of significant Bragg reflection intensity noise is identified from the combination of mode structure and jitter with very small mosaic spread of the crystals and very low convergence of the XFEL source. The accuracy with which the three dimensional reflection is approximated by the 'quasi-cw' rotation method with the pulsed source is modelled from the experimentally collected X-ray pulse intensities together with the measured rocking curves. This model is extended to predict merging statistics for recently demonstrated self seeded mode generated pulse train with improved stability, in addition to extrapolating to single crystal experiments with increased mosaic spread. The results show that the noise level can be adequately

[a]Imperial College London, Division of Molecular Biosciences, South Kensington Campus, London SW7 2AZ, UK. E-mail: j.vanthor@imperial.ac.uk; Tel: +44(0)207 594 5071

[b]LCLS, SLAC National Accelerator Laboratory, Menlo Park, California 94025, USA

[c]Department of Physics, University of Wisconsin-Milwaukee, 1900 E. Kenwood Blvd, Milwaukee, WI 53211, USA

[d]Department of Biochemistry and Molecular Biology, and Institute for Biophysical Dynamics, University of Chicago, 920 East 58th Street, Chicago, Illinois 60637, United States

[e]Center for Advanced Radiation Sources, The University of Chicago, 5640 South Ellis Avenue, Chicago, Illinois 60637, USA

† Current address: Washington State University c/o Argonne National Laboratory Bld. 438F 9700 S. Cass AveArgonne, IL 60439, USA.

modelled in this manner, indicating that the large intensity fluctuations dominate the merged signal-to-noise ($I/\sigma I$) value. Furthermore, these results predict that using the self seeded mode together with more mosaic crystals, sufficient accuracy may be obtained in order to resolve typical photoinduced structure factor amplitude differences, as taken from representative synchrotron results.

1   Introduction

Structural reaction dynamics of proteins are fundamental to their biological function.

Femtosecond time resolved protein X-ray crystallography would provide a direct structural probe of these events, providing compelling motivation for exploiting novel XFEL sources for such studies.[1] Traditionally, time resolved pump–probe spectroscopic techniques provide access to the early time processes of activation in biological molecules. One particular example of photo-induced biological activation concerns the photoisomerisation of the biological p-coumaric acid chromophore of the Photoactive Yellow protein photoreceptor. Focusing on the sub-picosecond excited state dynamics, photoisomerisation couples the initial absorption event to trigger protein structural changes, which cause a cascading series of events and formation of reaction intermediates ranging from femtosecond to second time scales.[2-20] The primary photoproduct, 'I0', which is formed after excited state decay with a typical time constant of ~0.7 ps and approximately 30% primary quantum yield[2] is characterized by visible absorption at 510 nm, red-shifted from the ground state absorption at 446nm.[21,22] Ultrafast infrared spectroscopy has shown the *trans–cis* isomerisation is accompanied by a disruption of the hydrogen bonding between the chromophore C=O group and the Cys69 backbone, and includes the formation of a short lived ground state intermediate with distorted *trans–cis* configuration.[17] Further rearrangements of the chromophore and the surrounding protein environment occur in the electronic ground state configuration, and launch a photocycle which is thermally reversible on the ms-second time scale. The following intermediates I1, I2' and I2 occur on ns, ms and ms time-scale, followed by reformation of the ground state, and involve global protein structural changes that persist for ms-s, and are considered to form the biological signaling state of the photoreceptor.[7]

Using synchrotron radiation, pump–probe time resolved crystallography experiments of crystals of the Photoactive Yellow Protein use the Laue X-ray diffraction method to capture the photoinduced structure factor amplitude differences.[23-26] The PYP is one of relatively small number of light sensitive protein crystals that have been amenable to time resolved pump–probe studies using this technique, with other examples including heme proteins and photosynthetic reaction center.[27-30] Since their first demonstration, these experiments have continued to generate considerable interest and activity, particularly to exploit the technique to probe in detail the various reaction mechanisms and assignments from extensive time resolved measurements.[24-26,31]

With the emergence of novel X-FEL femtosecond pulsed hard X-ray sources, new capabilities for pump–probe crystallography now exist.[1] It has been demonstrated that XFELs can be used for protein X-ray crystallography, particularly using the stream of micron-sized crystals injected into the XFEL beam using a microjet.[32-36] Considering the crystallographic data quality, compared to that obtained at synchrotron sources, it is noted that good quality measurements of the structure factor amplitudes can be obtained by employing a 'Monte-Carlo' method of data integration of many observations from a large number of micro-crystals. This approach has now also allowed the observation of anomalous scattering using the serial femtosecond crystallography approach, demonstrating the possibility to use XFELs also for phasing of protein structures.[33,34] This result indicates that also time resolved pump–probe serial femtosecond crystallography may be possible, which requires a very precise measurement of the small photo-induced structure factor amplitude differences, with precision that exceeds the noise levels of the collected and processed data. Here, we consider the alternative possibility of using a single, macro-crystal for pump–probe studies with XFEL radiation, particularly in view of the noise levels which are considerably higher than those for datasets collected at synchrotron sources using either the rotation method or the Laue method. A distinct advantage of the Serial Femtosecond Crystallography approach is that radiation damage should not contribute to the photoinduced difference measurements, whereas single crystal experiments collected using a stroboscopic approach do experience the effects of radiation damage.

Achieving sufficient signal-to-noise of the difference measurements depends on both the concentration of the photointermediate as well as the crystallo-graphic data quality. Femtosecond photolysis is thus a critical parameter, which has been considered in detail on the basis of extensive power-, pulse duration- and wavelength dependence as well as the addition of second order dispersion using passive pulse shaping techniques.[2] In the case of PYP it was found that photochemical dynamics fundamentally limit the achievable time resolution to a few hundred femtoseconds, requiring stretching and addition of second order dispersion to suppress the stimulated emission path. Numerical modelling extracted the non-linear cross sections and projected photoproduct yields which were also confirmed using conventional model-free Z-scan measurements. Perhaps counter-intuitively,[37] optical penetration depth is not critical in this regime, in contrast to nanosecond excitation. Since the non-linear cross sections rapidly dominate the population transfer with high peak power for femtosecond excitation, little or no additional gain will be achieved for reduction of the pho-tolysed depth from large single crystals to micro-crystals. From the ultrafast spectroscopy studies, a compromise could be formulated which results in an instrumental cross correlation of ~300 fs with approximately 10% photolysed yield in the interaction region.[2] In this contribution we consider the impact of the X-ray source parameters on the crystallographic data quality collected on single crystals, firstly to evaluate the ability to scale and merge ground state data sets with sufficient accuracy.

2 Experiment

X-Ray crystallography experiments were conducted at the X-ray Pump–Probe station (XPP) at the Linac Coherent Light Source (LCLS) at Stanford, USA during experiments xpp23410 and xpp44112. For the xpp23410 experiments the X-ray slits were set to 80×40 μm or to 200×200 μm, the wavelength was 1.37757 Å, and a monochromator was used with a 1.6 eV band pass. The beam intensity was adjusted by inserting attenuators such that the strongest Bragg reflections did not saturate the detector and was focused to give a convergence of 0.007°, and a spot size of 80 × 40 µm. The distance to the detector was 68.5 mm, as refined from the data processing, using a MAR 165 detector. The repetition rate was 1 Hz and either 5 or 50 exposures were used to collect a single image with a stationary crystal or with continuous rotation, respectively. For the xpp44112 experiments, the wavelength was 1.3099 Å and a monochromator was used with 1.6 eV band-pass. The pulse duration was ~70 fs the X-ray slits were set to 40 × 40 µm. The beam was attenuated to below detector saturation with 5 X-ray pulses, with the aim to approach typical synchrotron conditions with regard to incident flux and pump–probe cycles. The distance to the detector was 59.1 mm and was calibrated from the powder pattern from SeO2. The detector was a mar 165 CCD. P6$_3$ crystals of PYP were mounted in capillaries at room temperature and were typically ~50 × 50 × 400 µm dimension. The beam was centered on the crystals and scanned across the needles to collect data from multiple positions. Data were processed using MOSFLM[38] and reduced with Scala[39] and CCP4,[40] and also with HKL-2000.[41] Rigid body refinement was performed with REFMAC[42] using coordinates from 2PHY.[43]

3   Results

3.1   X-ray beam mode structure and jitter in this structure are the origins of Bragg reflection intensity noise with static crystal orientation

With a detector distance of 68.5 mm, a wavelength of 1.3776 Å and a 1.6 eV bandpass, P6$_3$ crystals of PYP shows diffraction spots to the edge of the detector with up to 1.5 Å resolution. A test of the intrinsic noise of the Bragg diffraction intensity, as well as a characterization of the radiation damage under the atten-uated conditions, included a data collection of a single crystal volume with static crystal orientation. Setting the X-ray slits to 200 × 200 µm, a series of 59 images, with 5 exposures each, followed the general decay and reproducibility of the diffraction intensities. Fig. 1 shows an example of three high resolution Bragg reflections and their integrated intensities on the 59 subsequent images (panels A and C). Fig. 1B shows the total integrated intensities of all Bragg spots on each image.

The total diffraction intensities show large fluctuations which are primarily caused by the intensity fluctuations of the X-ray source. The intrinsic bandwidth of the SASE generated beam is ~30 eV, with a center wavelength of 1.38 Å. Therefore, the inclusion of a monochromator with a 1.6 eV bandpass results in essentially 100% pulse intensity fluctuations, such that also the average of the sums of five pulses still produces several-fold diffraction intensity fluctuations as seen in Fig. 1B. Furthermore, there is a general decay caused by radiation damage of the protein crystal, which limit the number of exposures of each crystal volume. Importantly, it is seen in Fig. 1C that there are also large fluctuations in the relative diffraction intensities of individual Bragg reflections, in spite of the stationary orientation and energy selection by the monochromator. Whereas these three reflections are presented as examples, this observation was also true for all other reflections measured. The possibility of crystal 'twitching' in response to X-ray exposure can very likely be discounted under these attenuated conditions. In contrast, un-attenuated conditions were seen to ablate protein

A

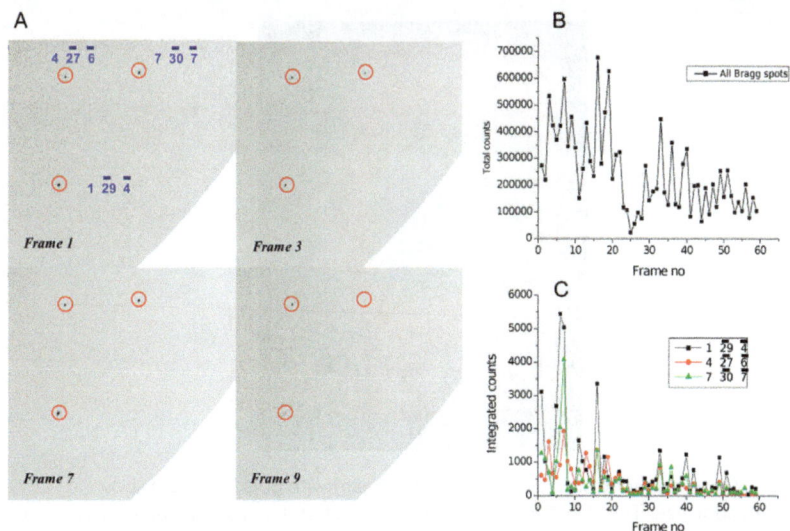

B

C

Fig. 1 Bragg diffraction intensity fluctuations, and decay due to radiation damage, with a static crystal orientation. A). At the detector edge, reflections (1,−29,−4), (4,−27,−6) and (7,−30,−7) are shown for representative frame numbers 1,3,7, and 9 in the data collection series of 59 images. B) The total integrated counts for all Bragg spots for frames 1 through to 59. C) Integrated intensities for (1,−29,−4) (squares), (4,−27,−6) (circles) and (7,−30,−7) (triangles) on each frame.

crystals and generate acoustic shock sufficient to disorder adjacent unexposed crystal volume to the extent that diffraction was lost. Under attenuated condition used here, the absorbed X-ray dose is on the same order of magnitude as that under conditions of pulsed Laue X-ray crystallography at synchrotron sources such as BioCARS 14ID.[44] Since both the 100 ps synchrotron source pulse duration and the 70 fs XFEL pulse duration are within the acoustic propagation time across the crystal, differences in pulse duration should not account for twitching occurring under XFEL radiation exposure. Instead, the differences in relative intensities of reflections are likely due to stimulation of spatially separated mosaic blocks due to strong structure of the X-ray beam mode and jitter of this structure (Fig. 2).

In comparison, a study of tetragonal Thaumatin crystals at the 19-BM synchrotron station at the Advanced Photon Source established the repeatability of retrieving the relative Bragg diffraction intensity after correction for radiation damage decay and additionally a small storage ring current drop, to be on the order of 0.7–7% r.m.s.d. for low and high resolution diffraction bins, respectively.[45]

The feasibility of small scale mosaic block microstructure is taken from phase contrast X-ray diffraction imaging of lysozyme crystals, which together with observation of anisotropic rocking curves indicate a micro-structure of crystal defects on length scales which would be expected to affect the measurements made with a 200 × 200 μm area shown in Fig. 1.[46] Reducing the X-ray slits to smaller size did not appreciably change the noise levels.

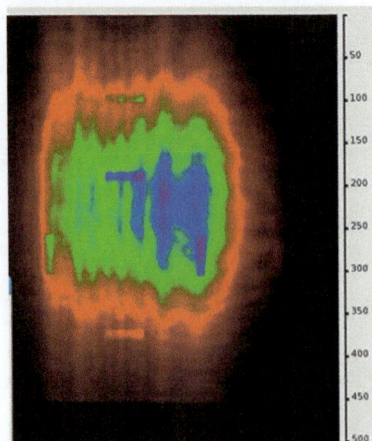

Fig. 2 Measured mode structure of the X-ray beam with open aperture showing ~250 × 250 μm spot size. The structure is caused by diffraction from an upstream mirror, which has smaller active length than the footprint of the beam. The structure shows additional shot-to-shot jitter.

3.2 Rocking curves of PYP crystals measured with non-diverging XFEL beam

The reported convergence of the LCLS X-ray source is 0.007°, which is considerably smaller than the mosaic spread of PYP crystals. A rocking curve was measured over a 0.33° range, with incrementing the goniometer in 0.01° steps, and each image read after five pulses, as used in Fig. 1. Each image contained between 150 and 200 Bragg spots with $I/\sigma I$ of 3.0 or above. Fig. 3 shows representative rocking curves collected for spots in low and medium resolution bins. The full width half max ($\theta_{1/2}$) of a Bragg diffraction of a perfect crystal is given by[46]

$$\theta_{1/2} = 2d\xi^{-1} \tag{1}$$

where d is the interplanar spacing and ξ is the extinction length which in the symmetric Laue case is given by

$$\xi = \frac{\pi \cos\theta_B}{Cr_e\lambda(F_g/V)} \tag{2}$$

where r_e is the classical electron radius ($r_e = 2.82 \times 10^{-5}$ Å), F_g is the structure factor, V is the volume of the unit cell, C is the polarisation factor, λ is the wavelength and θ_B is the Bragg angle. With the addition of crystal mosaicity the phenomenological appearance of the angular intensity spread was however better represented by Gaussian functions rather than cosine functions,[47] and the fitted f.w.h.m. values are shown in Fig. 3 for four representative reflections at various resolution.

The histogram of a set of selected reflections shows a considerable spread in the f.w.h.m. values between 0.002° and 0.075° (Fig. 4). These values should be taken only as estimates given the fluctuations of the source intensity as well as the additional structure factor amplitude noise demonstrated in Fig. 1. However, the results resembles for instance rocking curves determined for lysozyme crystals in

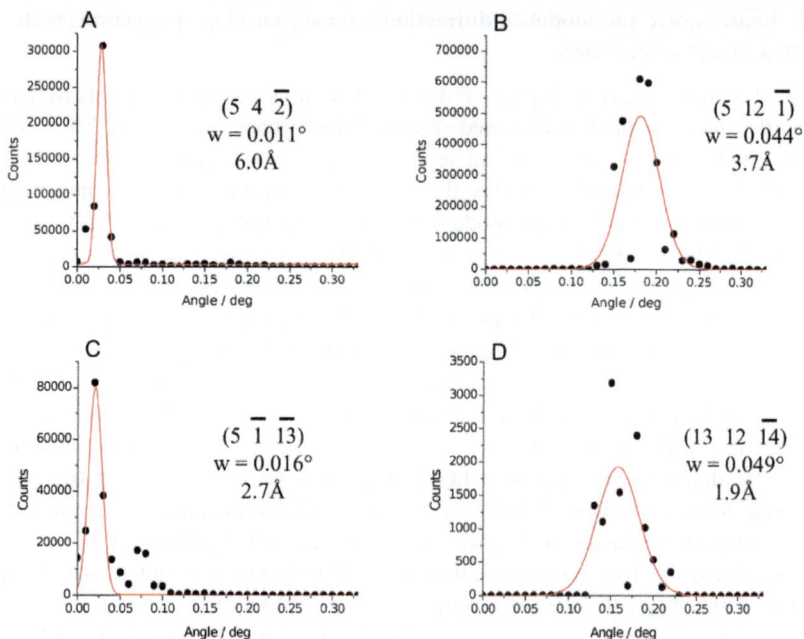

Fig. 3 Representative measured rocking curves for A) (5,4,−2) (f.w.h.m. 0.011°), B) (5,12,−1) (f.w.h.m. 0.044°), C), (5,−1,−13) (f.w.h.m. 0.016°) and D) (13, 12, −14) (f.w.h.m. 0.049°).

their anisotropic distribution but also absolute values as well as line shape defects and splitting (Fig. 3C).[47] The results indicate that the combination of the low mosaic spread of PYP crystals and the very low convergence angle result in very sparse observation when using the rotation method. Below, we show that increased mosaic spread would significantly improve the resulting signal-to-noise of data collected under such conditions.

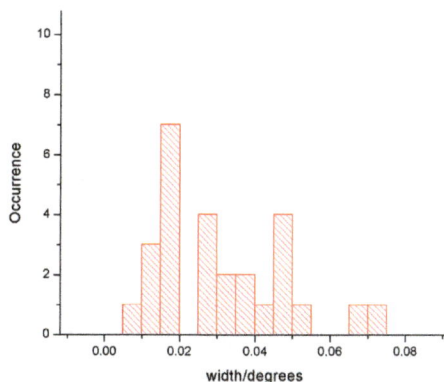

Fig. 4 Histogram of 27 individually fitted rocking curves of a single PYP crystal rotated through a 0.33° range, for reflections selected at various resolution and $I/\sigma I$ exceeding 10 at the peak.

3.3   Source noise and modelled diffraction intensity envelope projections with SASE and self-seeded mode

In the following, we consider the source intensity noise of SASE mode as well as recently demonstrated self-seeded mode, which have energy selection by including the monochromator and seeding, respectively. A photodiode detector recorded the intensity of each shot during data collection using multiple single crystals of PYP, using the SASE mode with energy selection at 1.3099 Å and 1.6 eV bandpass. Within a single period of data collection, a series of 8,475 pulses were measured (Fig. 5A). Recently, LCLS was operated in self-seeded mode, which leads to stabilization of the central wavelength, which a typical single shot f.w.h.m. of 0.5 eV and time averaged 1 eV bandwidth, at 8–9 keV.[48] The reported parameters included a 50% r.m.s. stability fluctuation, in part due to lack of FEL saturation in the seeded half of the undulator. Experimental results presented in Amann *et al.* (2012) Nature Photonics come from the very first self-seeding runs[48] (January 2012). At that time, the last four LCLS undulators were not used for the self-seeding. Recently, it was reported that all the undulators can be used for self-seeding operation having an impact both on the intensity and the stability of the self-seeded beam (personal communication, Alberto Lutman and Paul Emma, LCLS, SLAC). Fig. 5B shows a recent set of 500 pulses at 8.4 keV, generously provided by Alberto Lutman and Paul Emma, which may be used to model the resulting crystallographic noise in comparison with the SASE mode.

Together with the rocking curve information it is possible to estimate the accuracy of the estimate for the integrated structure factor amplitudes, on the basis of these intensity distributions alone, neglecting the additional noise resulting from mode structure and jitter. The rotation method was experimentally implemented by *continuous* rotation over a 1° range for each integrated image, with 50 X-ray probes, thus probing at 0.02° intervals. In this 'quasi cw' manner, the XFEL pulse train probes the envelope of the Bragg reflection up to 6 times, assuming an average value for the f.w.h.m. of 0.04° (Fig. 4). Fig. 6A shows the simulated histogram of the total resulting integrated intensity using the data shown in Fig. 5A, which correspond to the actual experimental conditions of data

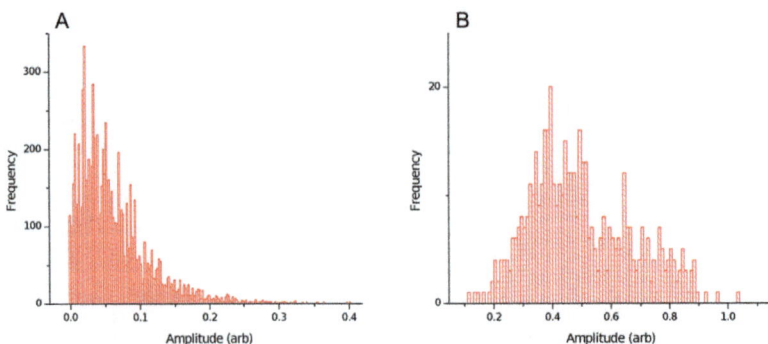

Fig. 5   A) Histogram of 8475 pulse intensities from SASE mode at 9.47 keV with a monochromator. The $I/\sigma I$ from this series is 1.21. B) Histogram of 500 pulses from self-seeded mode at 8.4 keV using the entire length of the undulator, as recently demonstrated. The $I/\sigma I$ from this series is 2.84.

collection (see below section 3.4). This distribution thus corresponds to the integrated 2D projection of a single observation of a Bragg spot, assuming only X-ray intensity noise contributions. Fig. 6B shows the modelled distribution expected under conditions of using self-seeded mode, using data from Fig. 5B. Fig. 6C simulates the resulting distribution if the mosaic spread would increase 10-fold to a value of f.w.h.m. of 0.4°, but maintaining the 0.02° rotation between pulses, resulting in an average number of 76 probes for each reflection. Finally, Fig. 6D simulates a mosaic spread of f.w.h.m. of 0.4° and a 0.02° step under conditions of self-seeding.

3.4 – Applying the 'quasi-cw' rotation method to PYP crystals using the SASE mode

Data were collected using the rotation method at 1.3099 Å and 1.6 eV bandpass, using a rotation of 0.1° and 5 probes at 0.02° steps for each image, corresponding to conditions shown in Fig. 5A and Fig. 6A. The beam was 80 × 40 μm and was translated by 100 μm to collect data on multiple crystal volumes on a single

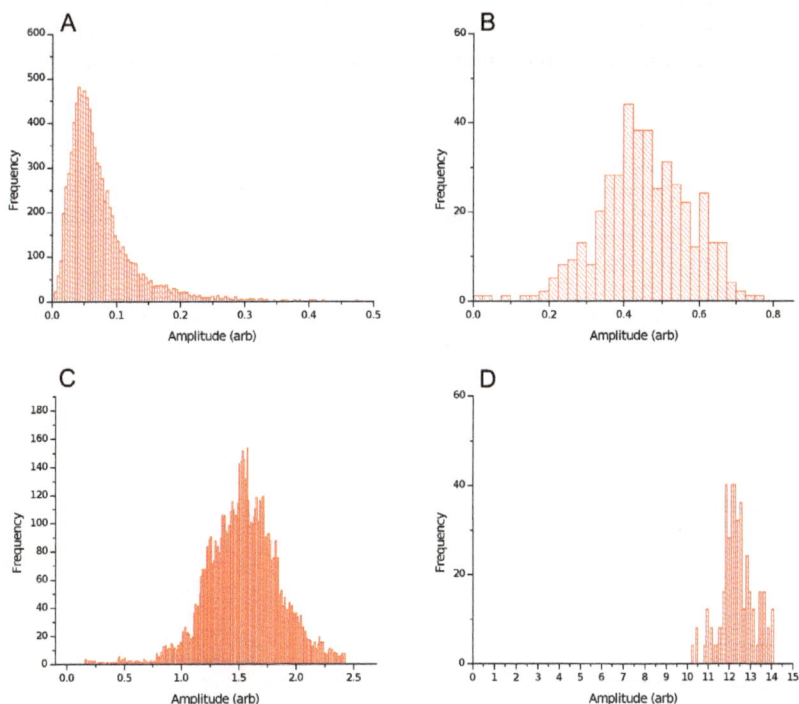

Fig. 6   Modelled distributions of the total integrated Bragg reflection intensity for SASE mode (A,C) and self-seeded mode (B,D), using data from Fig. 5A,B. (A) Expected distribution for a mosaic spread of f.w.h.m. of 0.04° and a step of 0.02° using SASE mode. The $I/\sigma I$ for this is expected to be 1.47. (B) Expected distribution for a mosaic spread of f.w.h.m. of 0.04° and a step of 0.02° using self-seeded mode. The $I/\sigma I$ for this is expected to be 3.81. (C) Expected distribution for a mosaic spread of f.w.h.m. of 0.4° and a step of 0.02° using SASE mode. The $I/\sigma I$ for this is expected to be 5.05. (D) Expected distribution for a mosaic spread of f.w.h.m. of 0.4° and a step of 0.02° using self-seeded mode. The $I/\sigma I$ for this is expected to be 16.4.

crystal. Six crystals in total were used, with each changeover starting at the equivalent indexed orientation to ensure continuous coverage of reciprocal space between crystals. Data were processed using both MOSFLM[38] and HKL-2000,[41] exploring conditions for outlier rejections. Table 1 presents statistics for a multi-crystal averaged dataset and a single crystal dataset.

The multi-crystal averaged dataset was processed in order to maximize the completeness, with an average multiplicity of 4.8. The relatively low completeness is caused by overlap of missing reflections from similar physical orientation of all mounted crystals, with the crystallographic c-axis corresponding to the long axis. Processing of data from a single crystal with increased rejection criteria appears to improve the R_{merge} and I/σ_I value of the data at the expense of 4.9% rejection, but having only low completeness. The R_{merge} and $I/\sigma I$ values are however meaningless in this regime with an average multiplicity of 1.2, thus lacking most of the intensity noise except for a small portion of reflections which had a multiplicity of 2 and were distributed throughout all resolution bins. For a Gaussian distributed error, it is expected that the PCV(I) value is a factor of sqrt(π/2) $\sim$ 1.25-fold higher than the R_{meas} value.[49] This ratio is 1.43 and 1.34 for the average and highest resolution shell, indicating that with the number of observations available some deviation from Gaussian distribution is present.

A simple relationship may be used to consider the dominating experimental error arising from the source fluctuations in terms of an expected R-value[49]

Table 1 Data reduction statistics

	Multi-crystal dataset	Single-crystal dataset
# of frames/# of crystals	96/6	23/1
Space group	P6$_3$	P6$_3$
Unit cell (a,b,c/Å;α,β,γ (°))	66.934, 66.934, 40.986 90.000 90.000, 120.000	66.867, 66.867, 40.985 90.000 90.000, 120.000
Resolution range (Å)	57.97–1.60	100–1.50
Reflections (Total/Unique)	137,459/13,877	20,145/16,947
Rejected outlier (#/%)	9,986/7.3	1,286/6.4
Completeness[a]	84.3(63.6)	51.4(21.6)
$I/\sigma I$	3.1(1.4)	7.1(1.5)
Average Multiplicity	4.8	1.2
R_{merge}[b]	0.400 (0.534)	0.239 (0.470)
R_{meas}[c]	0.445 (0.614)	N.D.
$R_{p.i.m.}$[d]	0.190 (0.289)	N.D.
PCV[e]	0.635 (0.822)	N.D.
R_{crys}/R_{free} (%)	0.26 (0.23)	N.D.

[a] Numbers appearing in parenthesis are for the high resolution shell.

[b] $R_{merge}(I) = \sum_{hkl} \sum_i |I_i(hkl) - \langle I(hkl)\rangle| / \sum_{hkl} \sum_i I_i(hkl)$.

[c] $R_{meas}(I) = \sum_{hkl} \sum_i \left(\frac{n}{n-1}\right)^{1/2} |I_i(hkl) - \langle I(hkl)\rangle| / \sum_{hkl} \sum_i I_i(hkl)$.

[d] $R_{p.i.m.}(I) = \sum_{hkl} \sum_i \left(\frac{1}{n-1}\right)^{1/2} |I_i(hkl) - \langle I(hkl)\rangle| / \sum_{hkl} \sum_i I_i(hkl)$.

[e] $PCV(I) = \sum_{hkl} \left(\frac{1}{n-1} \sum_i (I_i(hkl) - \langle I(hkl)\rangle)^2\right)^{1/2} / \sum_{hkl} \langle I(hkl)\rangle$.

$$R_{meas} = \left(\frac{2}{\pi}\right)^{1/2} \frac{\langle \sigma_I \rangle}{\langle I_o \rangle} \cong \frac{0.7979}{\langle I_o \rangle / \langle \sigma_I \rangle} \qquad (3)$$

Furthermore, by approximating the simulated distributions seen in Fig. 6 as purely statistical variation, the $I/\sigma I$ value may be estimated

$$\left(\frac{\langle I_o \rangle}{\langle \sigma_I \rangle}\right)_{merged} = n^{1/2} \left(\frac{\langle I_o \rangle}{\langle \sigma_I \rangle}\right)_{unmerged} \qquad (4)$$

where n is the multiplicity.

Following this model, based on the modelled $I/\sigma I$ value of 1.47 for a single observation under precisely the conditions of data collection (Fig. 5A, 6A, Table 1), multiplication with the square root of the average multiplicity yields a value of 3.22 for the expected average value of $(I/\sigma I)_{merged}$ for the multi-crystal dataset (Table 1). The close correspondence suggest that the source intensity fluctuations dominate the experimental error of this dataset, and that the used (average) value of 0.04° for the f.w.h.m. of the mosaic spread in order to simulate the distribution in Fig. 6A which provides $(I/\sigma I)_{unmerged}$ as a single observation appears to be appropriate. An analysis of this model for each resolution bin shows deviation between predicted and experimental $(I/\sigma I)_{merged}$ values in opposite directions at low and high resolution respectively (Fig. 7).

This deviation could arise from the error-normalisation procedure that is carried out by Scala, which adjusts the errors according to[50]

$$\sigma'(I_{hl}) = Sdfac[\sigma^2(I_{hl}) + SdB \cdot \langle I_h \rangle + (Sdadd \cdot \langle I'_h \rangle)^2]^{1/2} \qquad (5)$$

Sdfac, *SdB* and *Sdadd* are fitted parameters used in the scaling of the errors for all intensity bins, in order to normalize the deviations δ_{hl}.

$$\delta_{hl} = \left(\frac{n-1}{n}\right)^{1/2} (I_{hl} - \langle I_h \rangle)/\sigma'(I_{hl}) \qquad (6)$$

Sdadd is adjusted to model errors that are proportional to the intensity, and thus includes source fluctuations. Scala uses a default value of $Sdadd = 0.02$,

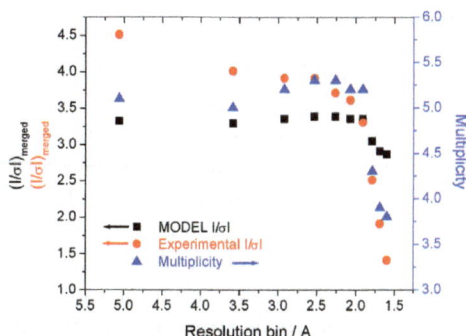

Fig. 7 Modelled $(I/\sigma I)_{merged}$ values (squares) and experimental $(I/\sigma I)_{merged}$ values (dots) as determined by Scala (Evans 2006), for separate resolution bins. The Multiplicity (triangles), together with a fit of $(I/\sigma I)_{unmerged}$ taken from Fig. 6A are used with eqn (4) to model the plotted values (squares).

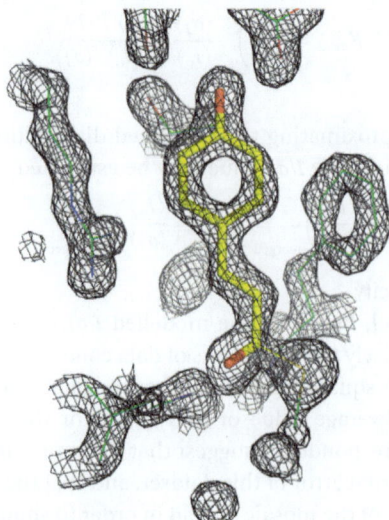

Fig. 8 2Fo-Fc electron density of the p-coumaric acid chromophore and direct environment contoured at 2σ level using the multi-crystal dataset (Table 1, center column).

whereas the multi-crystal dataset (Table 1) resulted in a values of 0.091 and 0.084 for full and partials reflections, respectively. The normalized deviation resulted in values of ~0.7 and ~0.5 for the two highest resolution bins, which also correspond with the lowest reported $(I/\sigma I)_{merged}$ values from Scala results (Fig. 7; Red dots), indicating that the standard error is overestimated for the weakest reflections.

The crystallographic R-factor at 26%, and R-free of 29%, indicate that the scaling and averaging has led to generally reasonable amplitudes. Fig. 8 shows resulting 2Fo-Fc electron density. It is noted however that at this level of accuracy the shape of the electron density is almost entirely determined by the synchrotron derived phase angles used in the Fourier synthesis.

4 Discussion

This work shows that crystallographic data quality collected under 'quasi-cw' and monochromatic conditions using the rotation method, in terms of its signal to noise parameter $(I/\sigma I)_{merged}$, may be understood from the dominating pulse intensity fluctuations. From data with a multiplicity of 4.8 there is correspondence with the square root law (eqn (4)) which is sufficiently to predict the average $(I/\sigma I)_{merged}$ vale, but showing some deviations in opposite directions for high and low resolution bins as seen in Fig. 7. We have identified an additional source of Bragg reflection intensity noise which is likely to be caused by the X-ray mode structure (Fig. 2) and jitter of this mode structure. Nevertheless, with up to five 'probes' of the XFEL pulse train, assuming an average mosaic spread and using a Gaussian distribution model of the source noise fluctuations, a reasonable estimate of the signal to noise for a single reflection can be made, when fully rotated through its rocking curve.

The analysis that is presented here is in essence a 'Monte Carlo' application as no knowledge or measurement of the source is used in the data processing. In principle, post-refinement of partiality from a fine-slicing method might take advantage of the pulse intensity recorded for each image. In this work that would still require the refinement of a single integral with five pulses and rotation through a known portion of the rocking curve. This could still result in increased signal to noise of the merged and scaled observations. However, the presence of the noise caused by the mode and mode jitter strongly suggests that this will not improve the data in the case of rocking curves with small f.w.h.m. as for the PYP crystals. Such an approach may be reconsidered if beamline optics could be improved to remove the diffraction and mode structure.

Taking these results together with recent progress at LCLS which shows considerably improved stability, it may be predicted that in particular for single crystals with larger mosaic spread, signal to noise may be attained that could be comparable with that collected at synchrotron sources. For example, assuming a mosaic spread of f.w.h.m. = 0.4°, a $(I/\sigma I)$ value of 16.4 was predicted for the self-seeded mode at LCLS (Fig. 6D). Assuming that eqn (3) remains valid in this regime, the $(I/\sigma I)_{\text{merged}}$ would become 32.8 with a multiplicity of four. Depending on the space group symmetry, this could be achieved with collecting ~100 images, which is a similar amount as for the dataset presented here (Table 1). The $(I/\sigma I)_{\text{merged}}$ value is presumably the most appropriate parameter in order to make a comparison with published synchrotron derived datasets. For example, time resolved pump–probe Laue X-ray crystallography data sets of the PYP which are sufficiently sensitive to detect the small photoinduced differences reproducibly, typically show an average value of $(I/\sigma I)_{\text{merged}} \sim 35$ for data with a characteristic resolution of 1.6 Å.[23,24,31,51,52]

Interestingly, these results also suggest that XFELs are a particularly well placed complementary sources to synchrotron stations, also for picosecond and longer time studies. This argument considers mainly the mosaic spread of protein crystals. Fast time-resolved synchrotron measurements rely on the Laue method, and a 'quasi-cw' scheme as used here is usually not feasible because of the lower intrinsic pulse power of spontaneous radiation. Even with a very bright source such as 14-ID beamline at BioCARS which uses two in vacuum undulators,[44] a monochromatic pulsed application would require too many pump–probe cycles in order to collect complete datasets. Instead, Laue crystallography is used. The Laue technique however requires very high quality crystals particularly those that have a very small mosaic spread, typically below f.w.h.m = 0.1°. Since the multiple frequencies stimulate mosaic blocks at different Bragg angles, Laue diffraction of crystals with higher mosaic spread show elongation of the Bragg spots which presents difficulties for the data integration due to close overlap and additionally the energy gradient that is present along the stretched reflection. This work suggests that for crystals which are not well suitable for Laue crystallography, pump–probe time resolved monochromatic experiments at XFELs are especially promising as shown in this contribution.

Acknowledgements

Portions of this research were carried out at the Linac Coherent Light Source (LCLS) at the SLAC National Accelerator Laboratory. LCLS is an Office of Science

User Facility operated for the U.S. Department of Energy Office of Science by Stanford University. The Linac Coherent Light Source is acknowledged for beam time access under experiment numbers xpp23410 and xpp44112. We thank Alberto Lutman and Paul Emma (LCLS, SLAC) for generously sharing stability data prior to publication. J.J.v.T acknowledges support from EPSRC (*via* award EP/I003304/1) and ERC (Grant Agreement No. 208650). M.S. is supported by NSF-STC 1231306 and NSF-0952643 (Career).

References

1 J. C. J. M. Glownia, J. Andreasson, A. Belkacem, N. Berrah, C. I. Blaga, C. Bostedt, J. Bozek, L. F. DiMauro, L. Fang, J. Frisch, O. Gessner, M. Gühr, J. Hajdu, M. P. Hertlein, M. Hoener, G. Huang, O. Kornilov, J. P. Marangos, A. M. March, B. K. McFarland, H. Merdji, V. S. Petrovic, C. Raman, D. Ray, D. A. Reis, M. Trigo, J. L. White, W. White, R. Wilcox, L. Young, R. N. Coffee and P. H. Bucksbaum, *Opt. Express*, 2010, **18**, 17620–17630.

2 C. N. Lincoln, A. E. Fitzpatrick and J. J. van Thor, *Phys. Chem. Chem. Phys.*, 2012, **14**, 15752–15764.

3 Y. Imamoto, M. Kataoka, F. Tokunaga, T. Asahi and H. Masuhara, *Biochemistry*, 2001, **40**, 6047–6052.

4 Y. Imamoto, K. Mihara, F. Tokunaga and M. Kataoka, *Biochemistry*, 2001, **40**, 14336–14343.

5 Y. Imamoto, H. Kamikubo, M. Harigai, N. Shimizu and M. Kataoka, *Biochemistry*, 2002, **41**, 13595–13601.

6 S. Devanathan, A. Pacheco, L. Ujj, M. Cusanovich, G. Tollin, S. Lin and N. Woodbury, *Biophys. J.*, 1999, **77**, 1017–1023.

7 K. J. Hellingwerf, *Antonie van Leeuwenhoek*, 2002, **81**, 51–59.

8 J. Hendriks, T. Gensch, L. Hviid, M. A. van Der Horst, K. J. Hellingwerf and J. J. van Thor, *Biophys. J.*, 2002, **82**, 1632–1643.

9 W. D. Hoff, A. Xie, I. H. Van Stokkum, X. J. Tang, J. Gural, A. R. Kroon and K. J. Hellingwerf, *Biochemistry*, 1999, **38**, 1009–1017.

10 T. E. Meyer, G. Tollin, T. P. Causgrove, P. Cheng and R. E. Blankenship, *Biophys. J.*, 1991, **59**, 988–991.

11 M. Unno, M. Kumauchi, N. Hamada, F. Tokunaga and S. Yamauchi, *J. Biol. Chem.*, 2004, **279**, 23855–23858.

12 M. A. van der Horst, W. Laan, S. Yeremenko, A. Wende, P. Palm, D. Oesterhelt and K. J. Hellingwerf, *Photochem. Photobiol. Sci.*, 2005, **4**, 688–693.

13 B. Borucki, C. P. Joshi, H. Otto, M. A. Cusanovich and M. P. Heyn, *Biophys. J.*, 2006, **91**, 2991–3001.

14 J. T. Kennis and M. L. Groot, *Curr. Opin. Struct. Biol.*, 2007, **17**, 623–630.

15 D. S. Larsen and R. van Grondelle, *ChemPhysChem*, 2005, **6**, 828–837.

16 I. H. van Stokkum, B. Gobets, T. Gensch, F. Mourik, K. J. Hellingwerf, R. Grondelle and J. T. Kennis, *Photochem. Photobiol.*, 2006, **82**, 380–388.

17 L. J. van Wilderen, M. A. van der Horst, I. H. van Stokkum, K. J. Hellingwerf, R. van Grondelle and M. L. Groot, *Proc. Natl. Acad. Sci. U. S. A.*, 2006, **103**, 15050–15055.

18 M. L. Groot, L. J. van Wilderen, D. S. Larsen, M. A. van der Horst, I. H. van Stokkum, K. J. Hellingwerf and R. van Grondelle, *Biochemistry*, 2003, **42**, 10054–10059.

19 M. L. Groot, L. J. van Wilderen and M. Di Donato, *Photochem. Photobiol. Sci.*, 2007, **6**, 501–507.

20 C. P. Joshi, B. Borucki, H. Otto, T. E. Meyer, M. A. Cusanovich and M. P. Heyn, *Biochemistry*, 2006, **45**, 7057–7068.

21 D. S. Larsen, M. Vengris, I. H. van Stokkum, M. A. van der Horst, F. L. de Weerd, K. J. Hellingwerf and R. van Grondelle, *Biophys. J.*, 2004, **86**, 2538–2550.

22 A. D. Stahl, M. Hospes, K. Singhal, I. van Stokkum, R. van Grondelle, M. L. Groot and K. J. Hellingwerf, *Biophys. J.*, 2011, **101**, 1184–1192.

23 B. Perman, V. Srajer, Z. Ren, T. Teng, C. Pradervand, T. Ursby, D. Bourgeois, F. Schotte, M. Wulff, R. Kort, K. Hellingwerf and K. Moffat, *Science*, 1998, **279**, 1946–1950.

24 H. Ihee, S. Rajagopal, V. Srajer, R. Pahl, S. Anderson, M. Schmidt, F. Schotte, P. A. Anfinrud, M. Wulff and K. Moffat, *Proc. Natl. Acad. Sci. U. S. A.*, 2005, **102**, 7145–7150.

25 Y. O. Jung, J. H. Lee, J. Kim, M. Schmidt, K. Moffat, V. Srajer and H. Ihee, *Nat. Chem.*, 2013, **5**, 212–220.

26 M. Schmidt, V. Srajer, R. Henning, H. Ihee, N. Purwar, J. Tenboer and S. Tripathi, *Acta Crystallogr., Sect. D: Biol. Crystallogr.*, 2013, **69**, 2534–2542.

27 V. Srajer, T. Teng, T. Ursby, C. Pradervand, Z. Ren, S. Adachi, W. Schildkamp, D. Bourgeois, M. Wulff and K. Moffat, *Science*, 1996, **274**, 1726–1729.

28 F. Schotte, M. Lim, T. A. Jackson, A. V. Smirnov, J. Soman, J. S. Olson, G. N. Phillips, Jr., M. Wulff and P. A. Anfinrud, *Science*, 2003, **300**, 1944–1947.

29 J. E. Knapp, R. Pahl, V. Srajer and W. E. Royer, Jr., *Proc. Natl. Acad. Sci. U. S. A.*, 2006, **103**, 7649–7654.

30 A. B. Wohri, G. Katona, L. C. Johansson, E. Fritz, E. Malmerberg, M. Andersson, J. Vincent, M. Eklund, M. Cammarata, M. Wulff, J. Davidsson, G. Groenhof and R. Neutze, *Science*, 2010, **328**, 630–633.

31 F. Schotte, H. S. Cho, V. R. Kaila, H. Kamikubo, N. Dashdorj, E. R. Henry, T. J. Graber, R. Henning, M. Wulff, G. Hummer, M. Kataoka and P. A. Anfinrud, *Proc. Natl. Acad. Sci. U. S. A.*, 2012, **109**, 19256–19261.

32 H. N. Chapman, P. Fromme, A. Barty, T. A. White, R. A. Kirian, A. Aquila, M. S. Hunter, J. Schulz, D. P. DePonte, U. Weierstall, R. B. Doak, F. Maia, A. V. Martin, I. Schlichting, L. Lomb, N. Coppola, R. L. Shoeman, S. W. Epp, R. Hartmann, D. Rolles, A. Rudenko, L. Foucar, N. Kimmel, G. Weidenspointner, P. Holl, M. N. Liang, M. Barthelmess, C. Caleman, S. Boutet, M. J. Bogan, J. Krzywinski, C. Bostedt, S. Bajt, L. Gumprecht, B. Rudek, B. Erk, C. Schmidt, A. Homke, C. Reich, D. Pietschner, L. Struder, G. Hauser, H. Gorke, J. Ullrich, S. Herrmann, G. Schaller, F. Schopper, H. Soltau, K. U. Kuhnel, M. Messerschmidt, J. D. Bozek, S. P. Hau-Riege, M. Frank, C. Y. Hampton, R. G. Sierra, D. Starodub, G. J. Williams, J. Hajdu, N. Timneanu, M. M. Seibert, J. Andreasson, A. Rocker, O. Jonsson, M. Svenda, S. Stern, K. Nass, R. Andritschke, C. D. Schroter, F. Krasniqi, M. Bott, K. E. Schmidt, X. Y. Wang, I. Grotjohann, J. M. Holton, T. R. M. Barends, R. Neutze, S. Marchesini, R. Fromme, S. Schorb, D. Rupp, M. Adolph, T. Gorkhover, I. Andersson, H. Hirsemann, G. Potdevin, H. Graafsma, B. Nilsson and J. C. H. Spence, *Nature*, 2011, **470**, 73–U81.

33 T. R. Barends, L. Foucar, S. Botha, R. B. Doak, R. L. Shoeman, K. Nass, J. E. Koglin, G. J. Williams, S. Boutet, M. Messerschmidt and I. Schlichting, *Nature*, 2013.

34 T. R. Barends, L. Foucar, R. L. Shoeman, S. Bari, S. W. Epp, R. Hartmann, G. Hauser, M. Huth, C. Kieser, L. Lomb, K. Motomura, K. Nagaya, C. Schmidt, R. Strecker, D. Anielski, R. Boll, B. Erk, H. Fukuzawa, E. Hartmann, T. Hatsui, P. Holl, Y. Inubushi, T. Ishikawa, S. Kassemeyer, C. Kaiser, F. Koeck, N. Kunishima, M. Kurka, D. Rolles, B. Rudek, A. Rudenko, T. Sato, C. D. Schroeter, H. Soltau, L. Strueder, T. Tanaka, T. Togashi, K. Tono, J. Ullrich, S. Yase, S. I. Wada, M. Yao, M. Yabashi, K. Ueda and I. Schlichting, *Acta Crystallogr., Sect. D: Biol. Crystallogr.*, 2013, **69**, 838–842.

35 L. C. Johansson, D. Arnlund, G. Katona, T. A. White, A. Barty, D. P. Deponte, R. L. Shoeman, C. Wickstrand, A. Sharma, G. J. Williams, A. Aquila, M. J. Bogan, C. Caleman, J. Davidsson, R. B. Doak, M. Frank, R. Fromme, L. Galli, I. Grotjohann, M. S. Hunter, S. Kassemeyer, R. A. Kirian, C. Kupitz, M. Liang, L. Lomb, E. Malmerberg, A. V. Martin, M. Messerschmidt, K. Nass, L. Redecke, M. M. Seibert, J. Sjohamn, J. Steinbrener, F. Stellato, D. Wang, W. Y. Wahlgren, U. Weierstall, S. Westenhoff, N. A. Zatsepin, S. Boutet, J. C. Spence, I. Schlichting, H. N. Chapman, P. Fromme and R. Neutze, *Nat Commun*, 2013, **4**, 2911.

36 S. Boutet, L. Lomb, G. J. Williams, T. R. Barends, A. Aquila, R. B. Doak, U. Weierstall, D. P. DePonte, J. Steinbrener, R. L. Shoeman, M. Messerschmidt, A. Barty, T. A. White, S. Kassemeyer, R. A. Kirian, M. M. Seibert, P. A. Montanez, C. Kenney, R. Herbst, P. Hart, J. Pines, G. Haller, S. M. Gruner, H. T. Philipp, M. W. Tate, M. Hromalik, L. J. Koerner, N. van Bakel, J. Morse, W. Ghonsalves, D. Arnlund, M. J. Bogan, C. Caleman, R. Fromme, C. Y. Hampton, M. S. Hunter, L. C. Johansson, G. Katona, C. Kupitz, M. Liang, A. V. Martin, K. Nass, L. Redecke, F. Stellato, N. Timneanu, D. Wang, N. A. Zatsepin, D. Schafer, J. Defever, R. Neutze, P. Fromme, J. C. Spence, H. N. Chapman and I. Schlichting, *Science*, 2012, **337**, 362–364.

37 R. Neutze and K. Moffat, *Curr. Opin. Struct. Biol.*, 2012, **22**, 651–659.

38 A. G. Leslie, *Acta Crystallogr., Sect. D: Biol. Crystallogr.*, 2006, **62**, 48–57.

39 W. Kabsch, *J. Appl. Crystallogr.*, 1988, **21**, 916–924.

40 M. D. Winn, C. C. Ballard, K. D. Cowtan, E. J. Dodson, P. Emsley, P. R. Evans, R. M. Keegan, E. B. Krissinel, A. G. Leslie, A. McCoy, S. J. McNicholas, G. N. Murshudov, N. S. Pannu, E. A. Potterton, H. R. Powell, R. J. Read, A. Vagin and K. S. Wilson, *Acta Crystallogr., Sect. D: Biol. Crystallogr.*, 2011, **67**, 235–242.

41 Z. Otwinowski and W. Minor, *Methods Enzymol.*, 1997, **276**, 307–326.

42 G. N. Murshudov, P. Skubak, A. A. Lebedev, N. S. Pannu, R. A. Steiner, R. A. Nicholls, M. D. Winn, F. Long and A. A. Vagin, *Acta Crystallogr., Sect. D: Biol. Crystallogr.*, 2011, **67**, 355–367.

43 G. E. Borgstahl, D. R. Williams and E. D. Getzoff, *Biochemistry*, 1995, **34**, 6278–6287.

44 T. Graber, S. Anderson, H. Brewer, Y. S. Chen, H. S. Cho, N. Dashdorj, R. W. Henning, I. Kosheleva, G. Macha, M. Meron, R. Pahl, Z. Ren, S. Ruan, F. Schotte, V. Srajer, P. J. Viccaro, F. Westferro, P. Anfinrud and K. Moffat, *J. Synchrotron Radiat.*, 2011, **18**, 658–670.

45 D. Liebschner, M. Dauter, G. Rosenbaum and Z. Dauter, *Acta Crystallogr., Sect. D: Biol. Crystallogr.*, 2012, **68**, 1430–1436.

46 Z. W. Hu, Y. S. Chu, B. Lai, B. R. Thomas and A. A. Chernov, *Acta Crystallogr., Sect. D: Biol. Crystallogr.*, 2004, **60**, 621–629.

47 D. Lubbert, A. Meents and E. Weckert, *Acta Crystallogr., Sect. D: Biol. Crystallogr.*, 2004, **60**, 987–998.

48 J. Amann, W. Berg, V. Blank, F. J. Decker, Y. Ding, P. Emma, Y. Feng, J. Frisch, D. Fritz, J. Hastings, Z. Huang, J. Krzywinski, R. Lindberg, H. Loos, A. Lutman, H. D. Nuhn, D. Ratner, J. Rzepiela, D. Shu, Y. Shvyd'ko, S. Spampinati, S. Stoupin, S. Terentyev, E. Trakhtenberg, D. Walz, J. Welch, J. Wu, A. Zholents and D. Zhu, *Nat. Photonics*, 2012, **6**, 693–698.

49 K. Diederichs and P. A. Karplus, *Nat. Struct. Biol.*, 1997, **4**, 269–275.

50 P. Evans, *Acta Crystallogr., Sect. D: Biol. Crystallogr.*, 2006, **62**, 72–82.

51 K. Moffat, *Nat. Struct. Biol.*, 1998, **5**, 641–643.

52 S. Rajagopal, K. S. Kostov and K. Moffat, *J. Struct. Biol.*, 2004, **147**, 211–222.

Faraday Discussions

PAPER

Core-level transient absorption spectroscopy as a probe of electron hole relaxation in photoionized $H^+(H_2O)_n$

Zheng Li,[ab] Mohamed El-Amine Madjet,†[a] Oriol Vendrell*[ac] and Robin Santra[abc]

Received 23rd April 2014, Accepted 29th April 2014

DOI: 10.1039/c4fd00078a

There is fundamental interest in understanding the coupled nuclear and electronic dynamics associated with charge transfer processes in complex molecules and materials, which are often mediated by electron, electron hole or proton motion. With dramatic improvements in the techniques to generate extreme ultraviolet (XUV) and X-ray femtosecond pulses, it now becomes possible to trigger and probe these kinds of processes in real time. Here we study the dynamics of an electron hole created by photoionization in the valence shell of protonated water clusters $H^+(H_2O)_n$. We demonstrate that the electron hole is strongly correlated with the protons forming the hydrogen bond network. We show that it is possible to probe key aspects of the valence electron hole dynamics and the coupled nuclear motion with femtosecond time resolution by resonantly exciting K-shell 1s electrons to fill the electron hole. This represents an opportunity for X-ray transient absorption spectroscopy.

1   Introduction

The rapid development of ultrafast and intense photon sources in extreme ultraviolet (XUV) and X-ray spectral ranges makes it possible to trigger and monitor electronic and nuclear dynamics and to probe them in real time.[1-3] Of special interest in the context of chemically relevant processes are the dynamics of valence electrons coupled to atomic displacements. These can be triggered by direct absorption of photons in the visible, ultraviolet or XUV spectral range or, at least in principle, by Raman-type core-valence transitions caused by X-rays.[4] In this context, theoretical proposals to pump and probe pure electronic valence dynamics in molecules by non-linear XUV and X-ray techniques on femtosecond

[a]Center for Free-Electron Laser Science, DESY, Notkestraße 85, D-22607 Hamburg, Germany. E-mail: oriol.vendrell@cfel.de

[b]Department of Physics, University of Hamburg, D-20355 Hamburg, Germany

[c]The Hamburg Centre for Ultrafast Imaging, Luruper Chaussee 149, D-22761 Hamburg, Germany

† Present address: Qatar Environment and Energy Research Institute (QEERI), 5825 Doha, Qatar.

to attosecond timescales have been put forward.[5-9] Pure (in the sense of pure state) electronic dynamics occurs when a superposition of valence electronic states is created upon interaction with a pulse of enough coherent bandwidth. The (potentially) coherent electronic dynamics that may be triggered with a pulse of enough bandwidth will dephase and become incoherent as soon as the overlap between nuclear wavepackets on different electronic states vanishes. Thereafter there is no further unitary dynamics of the electronic subsystem alone and it becomes driven by nuclear motion through non-adiabatic coupling. Non-adiabatic dynamics beyond the Born–Oppenheimer approximation is key in a wide scope of biological, chemical and physical processes.[10-21] In non-Born–Oppenheimer quantum molecular dynamics, nuclear wavepackets corresponding to the solution of the time-dependent Schrödinger equation can evolve within multiple vibronically coupled electronic states leading to an ultrafast and often irreversible energy transfer from the excited electrons to the nuclear degrees of freedom. Alternatively one can approximate the wavepacket dynamics of the nuclear degrees of freedom as a swarm of classical trajectories that are able to switch between the different potential energy surfaces (PES).

Nuclear dynamics naturally start upon photoionization, which can be accomplished *e.g.* by single XUV photon absorption, due to the sudden change of the potential experienced by the nuclei after an electron has been removed.[22] After electronic decoherence, which we assume to be fast compared to the time-scale of nuclear motion, the kind of rearrangements of *both* nuclei and valence electrons that follow occur unavoidably on timescales that correspond to nuclear dynamics, namely tens to hundreds of femtoseconds. This is the scenario for photoionization that we consider in this work, where we theoretically investigate the applicability of core-level transient absorption spectroscopy (TAS)[2,7,23-27] as a tool to probe valence electron hole dynamics and relaxation driven by non-adiabatic effects. In particular we study the photoionization dynamics of protonated water clusters $H^+(H_2O)_n$, which have been the subject of previous investigations using a free-electron laser.[28] We envision the situation in which a time-delayed X-ray probe pulse projects the evolving nuclear and electronic wave-packet of the previously photoionized system with an electron hole in the valence shell onto core-hole states, and the transient X-ray absorption is recorded as a function of pump–probe delay. The core-hole states therefore act as final states onto which electronic and nuclear components of the wave-packet are projected. A feature of core-level spectroscopy to probe valence shell dynamics in the described scenario is that the transition energy to promote a core electron to the valence hole is mostly determined by changes in the valence shell and not by the more inert and tightly bound core electrons. Another advantage is atom specificity through the very different core electron binding energies of different atomic species such as *e.g.* carbon, nitrogen and oxygen atoms. Moreover, transient absorption spectroscopy is a general technique that can be applied to a wide range of atomic and molecular systems in either gas or condensed phase. Either free-electron laser[29] or high-order harmonics sources[24] can be utilized to probe such ultrafast dynamics.

By correlating the features of the transient absorption spectra with the nuclear dynamics obtained from wavepacket or quantum-classical approaches, we are able to gain a detailed understanding of the non-adiabatic processes related to valence hole relaxation in molecules and clusters. Protonated water clusters $H^+(H_2O)_n$ offer a convenient example in which the correlations between nuclear displacements and hole dynamics are particularly strong.[22,30] In this work we focus our attention on the core-level spectroscopy and dynamics of protonated clusters $H^+(H_2O)_2$ and $H^+(H_2O)_{21}$ upon valence photoionization. The former corresponds to small molecular system with well separated PES except for few crossings at conical intersections. The latter example corresponds to a medium sized cluster with a large number of energetically close valence electronic states and multiple crossings, which may resemble in some aspects the situation encountered in solid state samples.

2 Theory

2.1 Electronic structure and dynamics

The self-consistent solution of the Hartree–Fock (HF) equations for an N-electron closed shell system

$$\hat{f}|\phi_j\rangle = \varepsilon_j|\phi_j\rangle \tag{1}$$

provides the set of occupied orbitals $|\phi_j\rangle$ of the anti-symmetrized wavefunction $|\Psi_0\rangle$ corresponding to the HF ground electronic state as well as their orbital energies ε_j. $\hat{f}$ is the Fock operator consisting of the one-body terms of the electronic Hamiltonian and the mean-field potential, which depends on all occupied orbitals. The HF energy for the ground state of the N-electron system is

$$E_{HF} = \langle\Psi_0|\hat{H}_e|\Psi_0\rangle \tag{2}$$

where $\hat{H}_e$ is the exact non-relativistic electronic Hamiltonian for a fixed configuration of the nuclei of the system. Within this context, Koopmans' theorem shows that

$$-\varepsilon_j = \langle\Psi_j|\hat{H}_e|\Psi_j\rangle - E_{HF} \tag{3}$$

$$= E_j - E_{HF} \tag{4}$$

where $|\Psi_j\rangle = \hat{c}_j|\Psi_0\rangle$ is an $(N-1)$-electron state constructed from the HF solution by removing an electron from the j-th orbital, $\hat{c}_j$ is the annihilation operator for the j-th orbital and E_j is an approximate eigenenergy through first order in many-body perturbation theory. If one accepts $|\Psi_j\rangle$ as an approximation to the j-th electronic eigenstate of the $(N-1)$-electron system, then the negative energies of the occupied orbitals obtain the meaning of ionization potentials, which constitutes Koopmans' approximation. Within this approximation, the energy of the j-th ionic state is simply $E_j = E_{HF} - \varepsilon_j$, which requires for its evaluation only one ground state HF calculation. With the self-consistent orbitals obtained from eqn (1), $\langle\Psi_i|\hat{H}_e|\Psi_j\rangle = E_j\delta_{ij}$. Therefore, in a configuration interaction sense and within the one-hole space, the $|\Psi_j\rangle$ configuration represents the best possible approximation to the j-th $(N-1)$-electron state. Of course, this is still an approximation to the true eigenstate. For example, orbital relaxation effects caused by the increased total charge and the contribution of two-hole one-particle configurations, which becomes crucial to describe inner-valence holes, are neglected. Nonetheless, the adoption of Koopmans' approximation provides a qualitative approach and a feasible alternative to describe the electronic structure and nuclear dynamics evolving within the large manifold of tens or even hundreds of $(N-1)$-electron outer-valence states accessible upon photoionization of medium sized molecules and clusters.

In order to perform on-the-fly dynamics (forces are computed as trajectories evolve) one needs an efficient evaluation of the gradient of the electronic energy with respect to nuclear displacements. For non-adiabatic dynamics the coupling terms between electronic states are needed as well. HF energy gradients are implemented in the majority of quantum chemistry packages and are calculated from the derivatives of one- and two-body integrals and the molecular orbital coefficients.[31] All that is needed to calculate $\frac{\partial E_j}{\partial \lambda}$, where λ represents an atomic displacement, is the derivative of the corresponding orbital energy $\frac{\partial \varepsilon_j}{\partial \lambda}$, which is calculated from the same integral derivatives as the total $\frac{\partial E_{HF}}{\partial \lambda}$. In our calculations, we use the multiconfigurational capabilities of the MOLCAS[32] package to generate the electronic wavefunctions $|\Psi_j\rangle$ with the orbitals from a previous HF calculation and then obtain the energy gradient for this single configuration. The coupled nuclear and electronic dynamics of $H^+(H_2O)_n$ clusters after photoionization have been simulated using the quantum-classical surface hopping scheme of Tully,[33,34] where nuclei are treated classically and electrons quantum mechanically. The non-adiabatic couplings were obtained with the usual finite differences approach requiring the calculation of overlaps of electronic wavefunctions at adjacent times.[34]

2.2 Transient core-level absorption spectrum

The time-domain expression for the transient absorption spectrum (TAS) of a non-stationary wavepacket $\chi_1(\mathbf{R}, \tau)$ evolving on the PES $E_1(\mathbf{R})$ and probed by an electronic transition to electronic state 2 is given by[35]

$$\sigma(\omega, \tau) = \frac{4\pi\omega}{c} \int d\mathbf{R} |\mu_{12}(\mathbf{R})|^2 |\chi_1(\mathbf{R}, \tau)|^2$$
$$\times \frac{\Gamma/2}{\Gamma^2/4 + [E_1(\mathbf{R}) - E_2(\mathbf{R}) - \omega]^2}, \tag{5}$$

where

$$\mu_{12}(\mathbf{R}) = \langle \Psi_1(\mathbf{r}; \mathbf{R}) | \hat{\mu}_{12} | \Psi_2(\mathbf{r}; \mathbf{R}) \rangle \tag{6}$$

is the transition dipole matrix element between the electronic states involved. Eqn (5) corresponds to the so-called Lorentzian limit in which dephasing of the nuclear wavepackets evolving on PES $E_1(\mathbf{R})$ and $E_2(\mathbf{R})$ is assumed to be fast in comparison to the time-scale of atomic motions. A short probe pulse is also assumed in obtaining this expression.[35] Basically, the cross-section corresponds to the transition dipole matrix element squared between two electronic states, averaged by the probability density in nuclear coordinate space and broadened by the lifetime of the final electronic state. This limit is meaningful in the present situation of core-level TAS due to large topological differences between PES of valence- and core-hole states, leading to fast dephasing, as well as the presence of purely electronic decay mechanisms, like Auger decay, for the core-hole final states.

Eqn (5) can be readily used within a surface-hopping framework. At every time τ the integral over configurational space is substituted by a sum over the swarm of classical trajectories. Each trajectory is found at one of the valence electronic states of the system whose dynamics are being probed, and we denote the index of the electronic state of trajectory J at time τ as $j(J, \tau)$. In addition, instead of a single final electronic state for the probe step, one needs to consider all possible transitions from the valence-hole state to core-hole states accessible in the energy range of interest. With these considerations our final working equation for the absorption cross-section reads

$$
\sigma(\omega, \tau) = \frac{4\pi\omega}{c} \frac{1}{M} \sum_J^M \sum_\alpha \left\{ \left| \mu_{j(J,\tau),\alpha}(\mathbf{R}_J) \right|^2 \right.
$$

$$
\times \left. \frac{\Gamma/2}{\Gamma^2/4 + \left[E_\alpha(\mathbf{R}_J) - E_{j(J,\tau)}(\mathbf{R}_J) - \omega \right]^2} \right\}. \tag{7}
$$

α denotes the final core-hole electronic state, $\mathbf{R}_J \equiv \mathbf{R}_J(\tau)$ and M is the total number of trajectories. For core-level TAS Γ can be chosen to be of the order of the linewidth of the relevant core-hole state, e.g. $O1s^{-1}$ in the present case. The linewidth of $O1s^{-1}$ has been measured to be 160 ± 5 meV corresponding to a lifetime of about 4.1 fs.[36] In our calculations we set $\Gamma = 400$ meV, which is of the order of the energy span between the lowest and highest energy transitions at any time delay and which accounts for the fact that finer structures are most probably related to a finite number of averaged classical trajectories.

3   Results

Protonated water clusters $H^+(H_2O)_n$ are closed-shell, stable ionic clusters and have been extensively studied in the gas phase as model systems due to the relevance of the so-called excess proton in acidic chemistry and positive charge transport in biological systems.[37] In recent works, the correlated dynamics of the excess proton and an electron hole created by photoionization were studied in detail in the smallest cluster of the family containing a hydrogen bond, namely $H_5O_2^{2+}$, for which it was shown that the breakdown of the Born–Oppenheimer approximation plays a crucial role in its electronic relaxation mechanism.[22,30] After valence photoionization, the proton and hole correlate their relative positions in the cluster within 5 fs involving strong non-adiabatic effects followed by Coulomb explosion, which constitutes an example of the kind of valence electronic and nuclear dynamics that one would like to address with time-resolved spectroscopic approaches. In the following, we discuss the correlated proton-hole dynamics in protonated water clusters $H^+(H_2O)_2$ and $H^+(H_2O)_{21}$ after photoionization and provide a connection between the valence hole relaxation dynamics in these systems and the features of simulated core-level TAS spectra. A schematic view of the concept of core-level TAS to probe valence hole dynamics is shown in Fig. 1. For the Zundel dication, we had previously performed fully quantum wavepacket calculations[22] using the multiconfigurational time-dependent Hartree (MCTDH) method.[38,39] We also performed comparative studies between fully quantum and

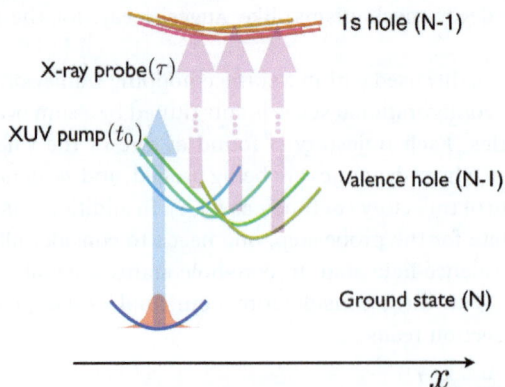

Fig. 1 Schematic representation of valence hole dynamics probed by core-level transient absorption spectroscopy. The curves represent potential energy surfaces and the vertical direction corresponds to energy. N represents total number of electrons in the system before photoionization. x represents an arbitrary nuclear coordinate. Arrow lengths schematically depict photon energy.

quantum-classical calculations[30] showing that the quantum-classical approach can reliably describe the correlated proton-hole dynamics and hole relaxation of relevance in the present study. The results in this contribution are all based on a quantum-classical treatment with the nuclei following classical trajectories.

3.1 Core-level transient absorption spectrum of photoionized Zundel cation

In the molecular orbital picture the water molecule features three filled outer-valence orbitals with binding energies $-\varepsilon_j$ ranging from roughly 14 to 20 eV. The filled molecular orbital with large atomic O_{2s} character has a binding energy of roughly 35 eV. At this energy the excited electronic states of the water cation have substantial two-hole one-particle character and the orbital picture breaks down.[41] As more water molecules are added to form clusters, the outer-valence band becomes denser and the occupied orbitals extend over several water molecules.[41] Therefore, the outer-valence shell of water clusters can be thought of as being composed of $3n$ molecular orbitals, where n is the number of water molecules in the cluster. Fig. 2 shows the six lowest PES of the dicationic (singly ionized) $H_5O_2^{2+}$ along the displacement coordinate of the central proton for fixed geometry of the rest of the system. The complex topology of these PES, being characterized by multiple crossings as a function of the displacement of the central proton coordinate, is apparent. This is the main reason why electronic non-adiabaticity plays a prominent role in the dynamics of water clusters upon ionization.

The core-level TAS of the Zundel cation obtained with eqn (7) is shown in Fig. 3. The spectrum can be decomposed into contributions corresponding to different binding energies of the photoelectron, which are accessible if the kinetic energy of the photoelectron is measured and the photon energy of the XUV pump pulse is known. The contribution from binding energies in the 20–25 eV range, which corresponds to ionization into the three lowest PES of the photoionized system in Fig. 2, is shown in Fig. 3a. In this case the electron hole is produced at or close to the ground electronic state of the ionized cluster and the core-to-valence

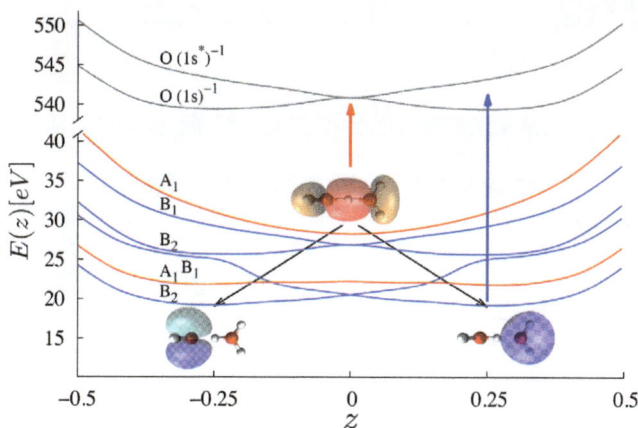

Fig. 2 Potential energy surfaces of the six lowest energy electronic states of the Zundel dication corresponding to outer-valence ionization. z indicates the position of the shared proton along the oxygen-oxygen (O–O) axis relative to the O–O distance and therefore has no units. A precise definition of z is found in Ref. 40. In the inset structures the symmetry point group of the cluster is D_{2d} with the central proton located at $z = 0$, which is energetically very close to the absolute PES minimum energy configuration in the ground electronic state.[40] The molecular orbitals correspond to the removed electron from each of the three lowest energy electronic states of the $(N - 1)$-electron system for $z = 0$. The symmetry labels correspond to the C_{2v} point group, which is reached when the central proton is displaced from $z = 0$ in either direction. Each pair of $B_{(1,2)}$ electronic states becomes a doubly degenerate state of E character in D_{2d} for $z = 0$. This degeneracy is lifted as the proton moves towards either water molecule in the form of an E ⊗ b conical intersection.[22]

transition energy is rather constant in time. Only an oscillatory component with a frequency of ∼3300 cm^{-1} (period of 10 fs) can be seen. This corresponds to a coherent O–H oscillation in the outgoing H_2O^+ water cation after the Coulomb explosion. Photoionization of electrons with binding energy in the 20–25 eV range leads to fragmentation into $(H_3O^+ + H_2O^+)$.[30]

The spectrum obtained for binding energies in the range 25–30 eV is presented in Fig. 3b. This corresponds to ionization into the group of the three upper PES of outer-valence character. Now, a relaxation of the outer-valence hole of about 6 eV on a timescale of about 20 fs can be seen. This is the time it takes for $H_5O_2^{2+}$ to reach the ground electronic state due to non-adiabatic transitions as a consequence of nuclear rearrangements.[22,30] The effect of such a relaxation is mirrored by the increase of the gap between the O_{1s} and valence hole electronic states, which is directly tracked by the probe pulse. The outer-valence hole relaxation is in the range of 5 to 10 eV and therefore has a spectral signature much wider than the broadening resulting in the absorption signal from electronic decay of the O_{1s} hole. The intensity of the TAS provides information on the $2p$ non-bonding character of the valence hole. This orbital character leads to a larger transition dipole than orbitals participating in the hydrogen bonding because of a larger overlap with oxygen 1s orbitals. The ground electronic state of the water cation features the electron hole in the 2p orbital perpendicular to the molecular plane. Indeed, as time goes on, the TAS intensity increases due to localization of the electron hole in this molecular orbital.

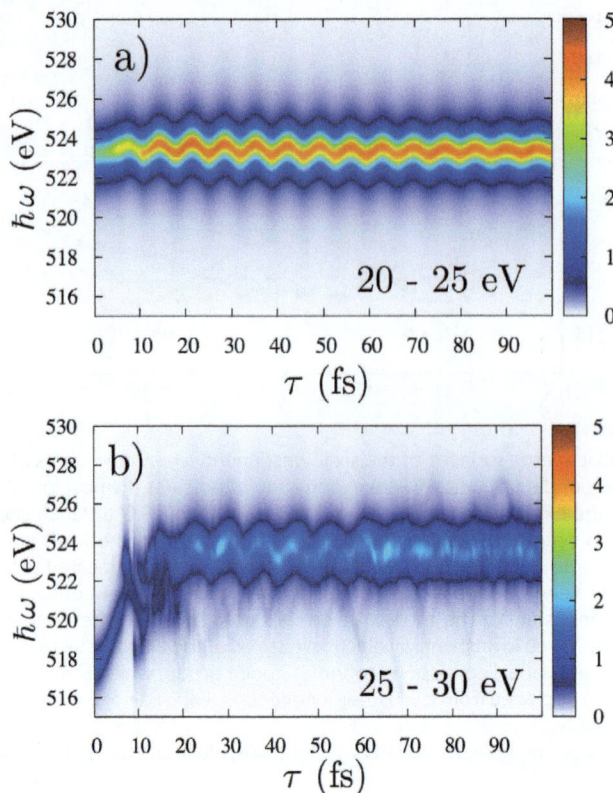

Fig. 3 Transient absorption spectra of the Zundel dication after photoionization of an electron with binding energy in the 20–25 and 25–30 eV ranges. Intensity is in Mb.

3.2 Transient core-level spectrum of photoionized $H^+(H_2O)_{21}$

The protonated water cluster $H^+(H_2O)_{21}$, firstly characterized by electrospray mass spectrometry,[42,43] is produced in special abundance in comparison to clusters with a similar number of water molecules but different than 21, which is referred to as a magic number.[44–46] As a consequence, the structure and other ground state properties of this system are well characterized. Contrary to what one might have expected, the symmetric structure of an enclosed hydronium H_3O^+ cation inside a cage of water molecules is not the most stable structure of $H^+(H_2O)_{21}$. Instead, calculations show that the cluster favors a surface-protonated cage with a neutral water in the center,[47–49] which is confirmed experimentally.[50] Such a structure was used as a starting point for sampling initial conditions for the molecular dynamics calculations. Due to its well characterized structure, this system represents a convenient example on which to study the applicability of TAS as a tool to disentangle valence hole dynamics in a manifold of tens of accessible electronic states combined with complicated nuclear dynamics.

As previously indicated, the outer-valence shell of every water molecule contributes three filled molecular orbitals, which in the case of $H^+(H_2O)_{21}$ hybridize to generate 63 outer-valence orbitals of the cluster. In the one hole approximation, ionization from each of these orbitals gives rise to an electronically excited state of $H(H_2O)_{21}^{2+}$. The electronic binding energies in this band span a range from about 12 to 25 eV. The lowest binding energies correspond to orbitals far from the location of the hydronium fragment on the surface of the cluster. An example of this is given by the molecular orbital related to the ground electronic state, D_0, of the photoionized cluster in Fig. 4. In contrast, the largest binding energy in the outer-valence corresponds to removing an electron from the close vicinity of the hydronium, which features already a positive charge associated to the excess proton before photoionization. An example is the electronic state with the highest energy within the outer-valence band, D_{62}, where the hole is seen to overlap with the region of the cluster holding the initially present positive charge, Fig. 4. In contrast to photoionized $H^+(H_2O)_2$, the electronic states of the photoionized $H(H_2O)_{21}^{2+}$ are tightly packed together forming a quasi-continuous band of valence-ionized states and the cluster does not Coulomb-explode at least during the time it takes for the electron hole to relax. The electron hole corresponding to each of the electronic states is delocalized among several of the water molecules and the relative stability of different hole states depends sensitively on the configuration of the hydrogen bond network of the cluster. The inner-valence hole states have binding energies $\sim$ 14 eV larger than the largest outer-valence binding energy and are excluded in our considerations.

The interaction of $H^+(H_2O)_{21}$ with a photon of energy larger than about 25 eV can lead to photoionization and production of a dication with a hole in any of the orbitals of the outer-valence band. For the calculation of the spectra in Fig. 5, 6 to 10 trajectories per electronic state were integrated up to 150 fs with a time-step of 0.24 fs (10 au). Starting nuclear configurations were sampled from a Boltzmann distribution at 300 K in the ground electronic state of $H^+(H_2O)_{21}$. The spectra of the different trajectories were averaged with equal weight independently of the initial electronic state, which assumes constant photoionization cross-section across the range of energies of the outer-valence. Immediately after photoionization, just within the first few femtoseconds, the hydrogen bond network starts to rearrange in response to the newly created positive charge. For electron binding energies in the 12 to 15 eV range the electron hole is produced such that the dication is close to its electronic ground state and the hole is on average far away from the excess proton. The corresponding TAS is shown in Fig. 5a. In this case, the core-to-valence transition remains relatively constant in energy since the

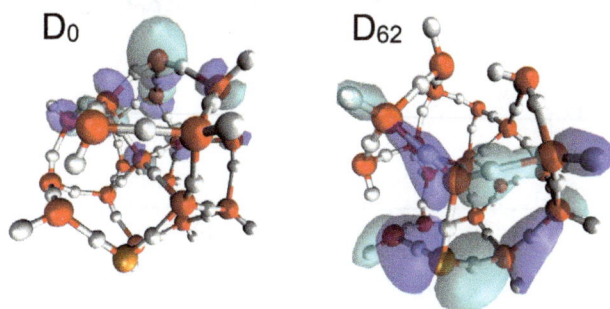

Fig. 4 Molecular orbitals corresponding to the electron hole in the ground (D_0) and highest energy (D_{62}) electronic states of the outer-valence band in $H(H_2O)_{21}^{2+}$. The orange oxygen atom represents the center with the largest hydronium (H_3O^+) character.

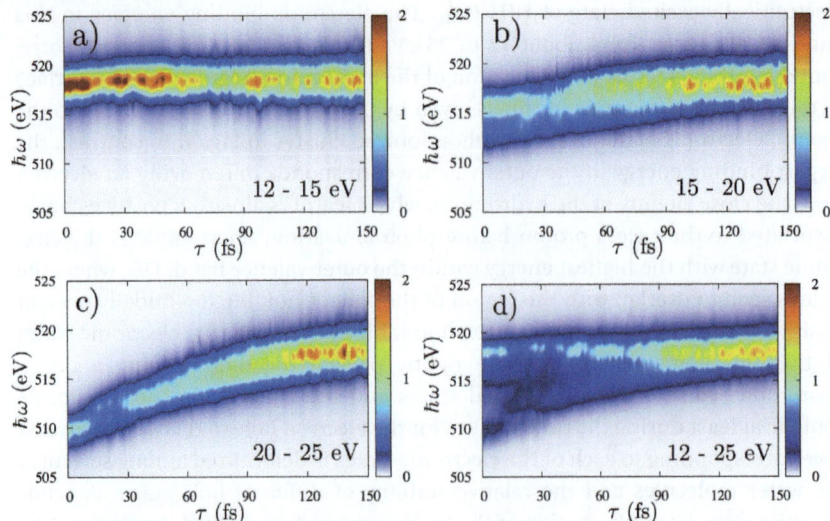

Fig. 5 Transient absorption spectra of the $H(H_2O)_{21}^{2+}$ dication of $H^+(H_2O)_{21}$. τ corresponds to the pump–probe delay time. The spectra are shown separately according to the binding energy range of the valence photoelectron (a), (b) and (c). The integrated spectrum over the whole range of outer-valence binding binding energies is shown in (d). Intensity is in Mb.

valence hole can no further relax. As the hole is produced deeper in the valence shell the hole relaxation dynamics start to manifest themselves in the TAS. This can be seen already for binding energies in the 15–20 eV range and more dramatically in the energy range 20–25 eV in Fig. 5b and 5c, respectively.

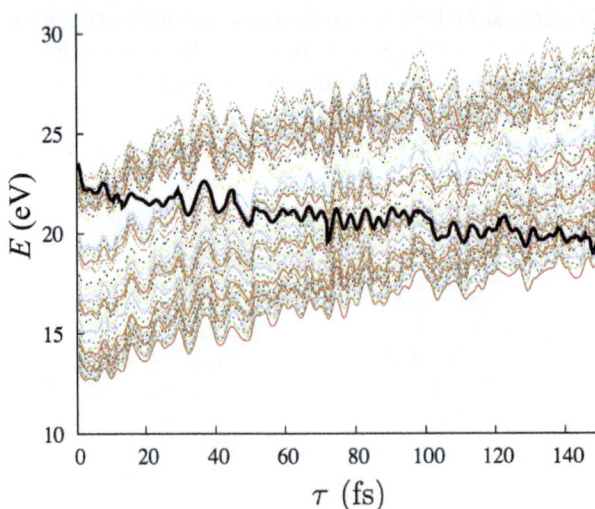

Fig. 6 Potential energy curves of the whole band of 63 valence hole states of the $H(H_2O)_{21}^{2+}$ dication along a single trajectory started from the 63rd electronic state. The black solid curve shows the non-adiabatic descending path in the outer-valence shell of the dicationic system.[36]

Interestingly, the TAS trace in Fig. 5c has the shape of an inverse exponential reminiscent of first order kinetics for the energy transfer from electronic to nuclear degrees of freedom and is completely irreversible, which is indicative of a large increase of available vibrational states as electronic excitation energy is transferred to vibrations. In contrast, the $H^+(H_2O)_2$ case presents an abrupt relaxation occurring at the delay time at which most trajectories reach the region of configurational space with a small potential energy gap. The hole relaxation by more than ~ 10 eV within the first 100 fs is a consequence of strong non-adiabatic effects. Similarly as in the transient absorption spectra of the Zundel cation, the increase of spectral intensity at long times of about ~ 100 fs corresponds to electron hole localization in orbitals close to oxygen centers in contrast to those related to hydrogen bond formation, which are more delocalized between the individual water molecules.

Fig. 6 shows the electronic energy of the dicationic states along one of the trajectories started from D_{62}, the highest electronic excitation within the outer-valence band. It takes the system about 150 fs to reach the ground electronic state of the dication. The band of electronic energies of the ionized cluster continuously shifts to larger values as a consequence of energy transfer to the vibrational degrees of freedom and the consequent molecular dynamics in higher energy regions of the corresponding PES. Again we emphasize the underlying main mechanism dominating the dynamics in the $H(H_2O)_{21}^{2+}$ dication, the Coulomb repulsion between the valence electron hole and the excess positive charge in the hydrogen bond network. In contrast to the Zundel dication though, the excess proton character in $H(H_2O)_{21}^{2+}$ is shared by many of the protons involved in the hydrogen bonds in the cluster. In this case it is the whole network of hydrogen bonds that responds to the positive hole charge by concerted motions of hydrogen atoms between the corresponding donor and acceptor oxygen atoms. This makes the charge separation process in the $H(H_2O)_{21}^{2+}$ dication a more complicated process than in the Zundel dication that operates also on a longer time-scale due to the large number of electronic states involved.

We note here that time and frequency axes are in principle not related by a time-frequency uncertainty relation in TAS. The time axis corresponds to the delay between pump and probe pulses and their pulse lengths determine the time resolution. The frequency axis is the Fourier pair of the time argument for propagation in the set of final states and the spectrum is proportional to the Fourier transform of the corresponding time correlation function.[35] This is equivalent to saying that the transmitted field can be sent through some grating and decomposed into its Fourier components, independently of the pump–probe delay time. A coherent source providing a photon energy centered at about 515 eV and pulse length of about 100 attoseconds, corresponding to a bandwidth of roughly 10 eV, would be required to probe the features in Fig. 3 and 5. Alternatively, the measurement could be performed by indirect means, e.g. total fluorescence at a given probe photon energy, which then would be scanned through the range of interest. In this case, the time and energy resolutions are not independent anymore because the knowledge about the incoming photon energy depends on the length of the probe pulse. One would wish to know the incoming photon energy with an uncertainty of at most 2 to 3 eV, which is achieved by (Fourier limited) pulses of a duration 2 fs or longer, still perfectly valid in order to resolve the features along the time axis.

4   Conclusions

The outer-valence photoionization of protonated water clusters triggers electron hole and nuclear rearrangements on the femtosecond time-scale. These dynamics are connected to strong non-adiabatic effects in which the electronic subsystem, usually in an excited state after photoionization, relaxes by coupling to atomic motion. In the case of the Zundel cation $H^+(H_2O)_2$ the hole relaxation can be as fast as about 10 to 20 fs, after which the system has relaxed to the ground electronic state of the ionized and fragmented cluster after a Coulomb explosion. In the case of $H^+(H_2O)_{21}$, a dense band of 63 outer-valence electronic states (three per water molecule) of the photoionized $H(H_2O)_{21}^{2+}$ can be accessed corresponding to electron binding energies in the range of 12 to 25 eV. The hole relaxation occurs in this case on a time-scale of the order of 100 fs. The valence hole relaxation affects the energy gap between outer-valence and core excited states substantially. Precisely this varying energy gap can be probed by core-level transient absorption spectroscopy providing a time-resolved picture of electron hole relaxation with femtosecond time-resolution. Information on the type of valence orbital accommodating the valence hole is encoded in the intensity of the transitions. The absorption intensity is larger for valence hole localization in molecular orbitals with strong O_{2p} character due to their good overlap with O_{1s} orbitals, which provides some positional information. An advantage of core-level transient absorption as a tool to probe hole relaxation dynamics is that it is mostly background free since the core-to-valence transition is only resonant in the presence of the valence hole. Finally, in a molecular system or cluster system with different kinds of atoms with well separated core-level absorption edges, these can be conveniently used to follow the location of the valence hole in time.

Acknowledgements

The authors acknowledge the Hamburg Centre for Ultrafast Imaging (CUI) for financial support.

References

1 V. S. Petrović, M. Siano, J. L. White, N. Berrah, C. Bostedt, J. D. Bozek, D. Broege, M. Chalfin, R. N. Coffee, J. Cryan, L. Fang, J. P. Farrell, L. J. Frasinski, J. M. Glownia, M. Gühr, M. Hoener, D. M. P. Holland, J. Kim, J. P. Marangos, T. Martinez, B. K. McFarland, R. S. Minns, S. Miyabe, S. Schorb, R. J. Sension, L. S. Spector, R. Squibb, H. Tao, J. G. Underwood and P. H. Bucksbaum, *Phys. Rev. Lett.*, 2012, **108**, 253006.

2 L. Fang, T. Osipov, B. Murphy, F. Tarantelli, E. Kukk, J. P. Cryan, M. Glownia, P. H. Bucksbaum, R. N. Coffee, M. Chen, C. Buth and N. Berrah, *Phys. Rev. Lett.*, 2012, **109**, 263001.

3 B. K. McFarland, J. P. Farrell, S. Miyabe, F. Tarantelli, A. Aguilar, N. Berrah, C. Bostedt, J. Bozek, P. H. Bucksbaum, J. C. Castagna, R. Coffee, J. Cryan, L. Fang, R. Feifel, K. Gaffney, J. Glownia, T. Martinez, M. Mucke, B. Murphy, A. Natan, T. Osipov, V. Petrovic, S. Schorb, T. Schultz, L. Spector, M. Swiggers, I. Tenney, S. Wang, W. White, J. White and M. Gühr, arXiv:1301.3104, 2013.

4 J. D. Biggs, Y. Zhang, D. Healion and S. Mukamel, *J. Chem. Phys.*, 2012, **136**, 174117.

5 S. Mukamel, *Phys. Rev. B*, 2005, **72**, 235110.

6 A. D. Dutoi, L. S. Cederbaum, M. Wormit, J. H. Starcke and A. Dreuw, *J. Chem. Phys.*, 2010, **132**, 144302.

7 E. Goulielmakis, Z.-H. Loh, A. Wirth, R. Santra, N. Rohringer, V. S. Yakovlev, S. Zherebtsov, T. Pfeifer, A. M. Azzeer, M. F. Kling, S. R. Leone and F. Krausz, *Nature*, 2010, **466**, 739.

8 B. Cooper and V. Averbukh, *Phys. Rev. Lett.*, 2013, **111**, 083004.

9 J. Leeuwenburgh, B. Cooper, V. Averbukh, J. P. Marangos and M. Ivanov, *Phys. Rev. Lett.*, 2013, **111**, 123002.

10 G. A. Worth and C. L. S., *Annu. Rev. Phys. Chem.*, 2004, **55**, 127.

11 D. R. Yarkony, *Rev. Mod. Phys.*, 1996, **68**, 985.

12 D. R. Yarkony, *Chem. Rev.*, 2012, **112**, 481.

13 S. Pisana, M. Lazzeri, C. Casiraghi, K. Novoselov, A. Geim, A. Ferrari and F. Mauri, *Nature materials*, 2007, **6**, 198.

14 H. Satzger, D. Townsend, M. Z. Zgierski, S. Patchkovskii, S. Ullrich and A. Stolow, *Proc. Natl. Acad. Sci.*, 2006, **103**, 10196.

15 M. Barbatti, A. J. A. Aquino, J. J. Szymczak, D. Nachtigallov, P. Hobza and H. Lischka, *Proc. Natl. Acad. Sci.*, 2010, **107**, 21453.

16 H. J. Wörner, J. B. Bertrand, D. V. Kartashov, P. B. Corkum and D. M. Villeneuve, *Nature*, 2010, **466**, 604.

17 D. Polli, P. Altoe, O. Weingart, K. M. Spillane, C. Manzoni, D. Brida, G. Tomasello, G. Orlandi, P. Kukura, R. A. Mathies, M. Garavelli and G. Cerullo, *Nature*, 2010, **467**, 440.

18 T. J. Martinez, *Nature*, 2010, **467**, 412.

19 V. I. Prokhorenko, A. M. Nagy, S. A. Waschuk, L. S. Brown, R. R. Birge and R. J. D. Miller, *Science*, 2006, **313**, 1257–1261.

20 T. Schultz, E. Samoylova, W. Radloff, I. Hertel, A. Sobolewski and W. Domcke, *Science*, 2004, **306**, 1765.

21 J. D. Savee, V. A. Mozhayskiy, J. E. Mann, A. I. Krylov and R. E. Continetti, *Science*, 2008, **321**, 826.

22 Z. Li, M. E.-A. Madjet, O. Vendrell and R. Santra, *Phys. Rev. Lett.*, 2013, **110**, 038302.

23 M. Dantus, M. J. Rosker and A. H. Zewail, *J. Chem. Phys.*, 1987, **87**, 2395.

24 Z.-H. Loh and S. R. Leone, *J. Chem. Phys.*, 2008, **128**, 204302.

25 R. Santra, V. S. Yakovlev, T. Pfeifer and Z.-H. Loh, *Phys. Rev. A*, 2011, **83**, 033405.

26 A. Wirth, M. T. Hassan, I. Grguraš, J. Gagnon, A. Moulet, T. T. Luu, S. Pabst, R. Santra, Z. A. Alahmed, A. M. Azzeer, V. S. Yakovlev, V. Pervak, F. Krausz and E. Goulielmakis, *Science*, 2011, **334**, 195–200.

27 S. Pabst, A. Sytcheva, A. Moulet, A. Wirth, E. Goulielmakis and R. Santra, *Phys. Rev. A*, 2012, **86**, 063411.

28 L. Lammich, C. Domesle, B. Jordon-Thaden, M. Förstel, T. Arion, T. Lischke, O. Heber, S. Klumpp, M. Martins, N. Guerassimova, R. Treusch, J. Ullrich, U. Hergenhahn, H. B. Pedersen and A. Wolf, *Phys. Rev. Lett.*, 2010, **105**, 253003.

29 H. T. Lemke, C. Bressler, L. X. Chen, D. M. Fritz, K. J. Gaffney, A. Galler, W. Gawelda, K. Haldrup, R. W. Hartsock, H. Ihee, J. Kim, K. H. Kim,

J. H. Lee, M. M. Nielsen, A. B. Stickrath, W. Zhang, D. Zhu and M. Cammarata, *The Journal of Physical Chemistry A*, 2013, **117**, 735–740.

30 Z. Li, M. E.-A. Madjet and O. Vendrell, *J. Chem. Phys.*, 2013, **138**, 094313.

31 P. Pulay, G. Fogarasi, F. Pang and J. E. Boggs, *J. Am. Chem. Soc.*, 1979, **101**, 2550–2560.

32 V. Veryazov, P.-O. Widmark, L. Serrano-Andres, R. Lindh and B. Roos, *Int. J. Quantum Chem.*, 2004, **100**, 626.

33 J. C. Tully, *J. Chem. Phys.*, 1990, **93**, 1061.

34 S. Hammes-Schiffer and J. C. Tully, *J. Chem. Phys.*, 1994, **101**, 4657.

35 S.-Y. Lee, W. T. Pollard and R. A. Mathies, *Chem. Phys. Lett.*, 1989, **163**, 11.

36 R. Sankari, M. Ehara, H. Nakatsuji, Y. Senba, K. Hosokawa, H. Yoshida, A. De Fanis, Y. Tamenori, S. Aksela and K. Ueda, *Chem. Phys. Lett.*, 2003, **380**, 647–653.

37 J. M. Headrick, E. G. Diken, R. S. Walters, N. I. Hammer, R. A. Christie, J. Cui, E. M. Myshakin, M. A. Duncan, M. A. Johnson and K. D. Jordan, *Science*, 2005, **308**, 1765–1769.

38 M. H. Beck, A. Jäckle, G. A. Worth and H.-D. Meyer, *Phys. Rep.*, 2000, **324**, 1–105.

39 G. A. Worth, M. H. Beck, A. Jäckle and H.-D. Meyer, The MCTDH Package, Version 8.2, (2000). H.-D. Meyer, Version 8.3(2002), Version 8.4 (2007). See http://mctdh.uni-hd.de.

40 O. Vendrell, M. Brill, F. Gatti, D. Lauvergnat and H.-D. Meyer, *J. Chem. Phys.*, 2009, **130**, year.

41 I. B. Müller and L. S. Cederbaum, *J. Chem. Phys.*, 2006, **125**, 204305.

42 J. Q. Searcy and J. B. Fenn, *J. Chem. Phys.*, 1974, **61**, 5282–5288.

43 G. Hulthe, G. Stenhagen, O. Wennerström and C.-H. Ottosson, *J. Chromatogr. A*, 1997, **777**, 155–165.

44 M. Miyazaki, F. A., T. Ebata and N. Mikami, *Science*, 2004, **304**, 1134.

45 J.-W. Shin, N. I. Hammer, E. G. Diken, M. A. Johnson, R. S. Walters, T. D. Jaeger, M. A. Duncan, R. A. Christie and K. D. Jordan, *Science*, 2004, **304**, 1137.

46 T. Zwier, *Science*, 2004, **304**, 1119.

47 A. Khan, *Chem. Phys. Lett.*, 2000, **319**, 440.

48 S. S. Iyengar, M. K. Petersen, T. J. F. Day, C. J. Burnham, V. E. Teige and G. A. Voth, *J. Chem. Phys.*, 2005, **123**, 084309.

49 M. P. Hodges and D. J. Wales, *Chem. Phys. Lett.*, 2000, **324**, 279.

50 C. Lin, H. Chang, J. Jiang, J. Kuo and M. Klein, *J. Chem. Phys*, 2005, **122**, 074315.

Faraday Discussions

ROYAL SOCIETY
OF CHEMISTRY

PAPER

Emerging photon technologies for probing ultrafast molecular dynamics

N. Berrah,[a] L. Fang,[ab] T. Osipov,[ab] Z. Jurek,[cd] B. F. Murphy[b] and R. Santra[cde]

Received 18th February 2014, Accepted 3rd April 2014

DOI: 10.1039/c4fd00015c

The understanding of physical and chemical changes at an atomic spatial scale and on the time scale of atomic motion is essential for a broad range of scientific fields. A new class of femtosecond, intense, short wavelength lasers, the free electron lasers, has opened up new opportunities to investigate dynamics in many areas of science. For chemical dynamics to advance however, a rigorous, quantitative understanding of dynamical effects due to intense X-ray exposure is also required. We illustrate this point by reporting here an experimental and theoretical investigation of the interaction of C_{60} molecules with intense X-ray pulses, in the multiphoton regime. We also describe the potential of new available instrumentation and explore their potential impact in physical, chemical and biological sciences when they are coupled with emerging photon technologies.

1 Introduction

The understanding of how nuclear and electron dynamics inside molecules can influence chemical reactions presents important implications in Physics, Chemistry and Biology with unforeseeable impacts. Nuclear motion occurs on femtosecond time scales and Angstrom spatial scales while electron motion is often much faster, particularly for tightly bound inner-valence and core electrons. New tools developed in the past two decades have made it possible to develop probes that can match many of these spatio-temporal requirements. Ultrafast infrared and optical wavelength lasers led to the detection of molecular vibrational, rotational and dissociative motion of molecules,[1–6] as well as some degree of quantum control of valence electrons. The more recent advancements in attosecond lasers promise direct control of the electronic motion as well.[7] All of these investigations contribute to the fundamental understanding of dynamic

[a]Department of Physics, University of Connecticut, Storrs, CT 06269, USA

[b]Department of Physics, Western Michigan University, Kalamazoo, MI 49008, USA

[c]Center for Free-Electron Laser Science, DESY, 22607 Hamburg, Germany

[d]The Hamburg Centre for Ultrafast Imaging, 22761 Hamburg, Germany

[e]Department of Physics, University of Hamburg, 20355 Hamburg, Germany

phenomena which could ultimately lead to the *control* of chemical reactions with an unprecedented temporal resolution.

Technological advances in building short pulse lasers in all wavelength regimes coupled with advanced instrumentation can also contribute, when paired with theoretical calculations and modeling, to answer the key questions about energy flow and transient processes in molecular dynamics. Emerging photon technologies have enabled a new class of femtosecond lasers to join the ultrafast laser family, namely the vacuum ultraviolet (vuv)[8] and X-ray free electron lasers (FELs).[9] They are new powerful femtosecond photonic tools, spanning a wide photon energy range from the infra-red (IR) to the hard X-rays. The FELs are tunable, covering a wide photon energy range and they are intense enabling a wide class of experiments, from non-linear science to time-resolved dynamics in physics, chemistry and biology, including chemical dynamics, the topic of the Faraday Discussion 171.

Ultrafast X-rays from FELs have photon energies sufficient to access core and inner-shell electrons, and like synchrotrons but unlike visible optical lasers, they enable inside-out ionization. The element-specificity of X-ray absorption, *i.e.* the ability to target specific atoms within molecules and select specific shells in those atoms (by tuning with high resolution the photon energy to specific spectral regions)[10,14] can be used to chart photochemical reactions and bioprocesses with atomic spatial resolution and femtosecond temporal resolution. Furthermore, the core-shell ionization and Auger decay processes, which are dominant in FEL-based work, lead to multiply charged fragments that are compared to strong-field optical and infrared laser cases. Thus FEL-based findings are relevant to the general optical laser community. Femtosecond optical laser pulses have led to the development of transition state spectroscopy and femtosecond chemistry,[15] and have been applied in pump–probe experiments to map out time-dependent nuclear motion in molecules.[16] Similar schemes are being used with accelerator-based FELs[17,18] which are complementary to table-top optical lasers offering the opportunity to interrogate molecular dynamics.

The understanding of dynamics depends upon investigating the intertwined electronic and nuclear motion which may require theoretical models beyond the Born–Oppenheimer approximation and including electron correlation. We need to understand the electronic structure because it determines the potential energy surfaces along which the nuclear motion evolves. This is very difficult, however, due to the different interactions and the large number of degrees of freedom that must be considered in order to completely describe even the smallest molecule. Here, we argue that judicious Molecular Dynamics (MD) modeling can result in advancing our understanding of molecular femtosecond dynamics as we show in this joint experimental and theoretical work.

Atomic, molecular and cluster physics experiments have been carried out with both vuv and X-ray FELs for investigating non-linear physics but also for revealing nuclear dynamics when using multiphoton ionization as a clock to determine the average time interval between the photoabsorption events,[19] for molecular transformation during isomerisation,[17] and for uncovering the time-scale of nucleobase ultraviolet photo-protection[20] relevant to chemical dynamics. Fundamental atomic experiments have provided insight into the nature of the interaction of light atoms such as Ne[21,22] and heavy atoms such as Xe[23] with X-ray FELs when tuning the intensity and/or pulse duration attributes of the X-ray FELs.

Even with the first experiments that started in the fall of 2009 when the LCLS was still being commissioned, the ability to tune the pulse duration from sub-10fs to 250fs enabled molecular investigations to observe transparency effects[21,24] and multiple core hole formation as a rebirth to electron spectroscopy for chemical analysis (ESCA).[12,25-30]

We report in this paper, which is a follow-up to an earlier work,[31] on the nature of the interaction and response of a large molecule, C_{60}, with intense femtosecond photon pulses. Here we focus our investigation on the wavelength, pulse duration and pulse energy photoionization dependence of a special molecule, C_{60}. We argue with this paper that femtosecond dynamic experimental investigation in C_{60} under intense X-ray exposure of different wavelengths and pulse duration paired with classical mechanics-based Molecular Dynamics (MD) modeling can provide useful quantitative insights to the scientific community. We also explore here a future dynamical investigation with recently emerging instrument technologies paired with ultrafast X-ray photons.

Buckminsterfullerene (C_{60}) is a system that keeps being studied because it connects to many fields of research. It is a highly symmetric compound consisting entirely of C–C bonds forming many novel materials like graphene or carbon nanotubes. We were motivated to investigate and understand the interaction of a molecule like C_{60} with short and intense radiation because the interaction of X-ray FEL with matter is still terra-incognita since these lasers are less than five years old and there is no published experimental data yet on large system like C_{60}. Furthermore, theoretical models of large molecular femtosecond dynamics under ultrafast and intense X-ray laser exposure are available and need to be systematically tested since the wavelength, pulse duration and fluence can be varied. Although our primary interest is of fundamental nature, our results impact matter under extreme conditions because this community interprets their data using fundamental atomic and molecular physics results. Our findings also bear on radiation damage which happens *within* 10–100 femtosecond X-ray pulse duration. This issue is very important because it impacts X-ray imaging in all of the sciences and our contribution is at the fundamental molecular level. Neutze *et al.*[32] predicted the potential radiation damage for biomolecular imaging with femtosecond X-ray pulses. More recently, Chapman *et al.*,[33] Nugent *et al.*[34] and Barty *et al.*[35] provided data on this topic with the current FELs. Our experimental and theoretical study of the C_{60} fragmentation dynamics under high X-ray fluence provides key insight into molecular dynamics in carbon-bonded molecules. The reported work enables a thorough understanding of the influence of intra-pulse radiation damage on high resolution X-ray diffraction imaging just like our previous work[21,24] uncovered electronic damage under intense radiation.

2   Experimental and theoretical methodology

The experiment was conducted at the atomic, molecular and optical physics (AMO) hutch of the Linac coherent light source (LCLS) at SLAC National Accelerator Laboratory using the high field physics instrument.[24,36] X-Ray optics focused the incoming X-ray pulses to a peak focal intensity of 10^{16}–10^{18} W cm^{-2}. A collimated molecular beam of C_{60} molecules from a resistively heated oven crossed the X-ray path at the focus. A magnetic bottle spectrometer allowed kinetic energy (KE) resolved ion time-of-flight spectroscopy with high collection efficiency, even at several hundred eV ion KE. A 2 meter long ion drift path provided high ion mass-to-charge and KE resolution while avoiding overlap of different fragment ion species.[22,31]

The last decade has seen several theoretical models developed to study the evolution of samples irradiated by X-ray FEL pulses.[31,37–45] In this work we utilized the XMDYN[42] tool to model finite samples irradiated by high intensity X-ray pulses. The approach uses an atomistic description of C_{60} combined with a molecular dynamics treatment of the real space dynamics. The electronic configurations of the individual atoms/ions were tracked and utilized. Cross sections and rates for photoionization, fluorescent and Auger relaxation processes were calculated by the XATOM toolkit[43,44] and the Monte Carlo algorithm was implemented to describe these stochastic events. At an ionization event a new classical (free) electron was 'created' within the model and launched with the proper velocity. XMDYN treats these electrons together with the atoms and atomic ions within a sample as classical particles. The fullerene-specific classical Brenner force field[45] accounted for the chemical bonds between atoms, and the Coulomb forces for the interaction between the charged particles. The Newton equations were solved numerically to evolve the system in real space.

Free electrons can ionize the sample when colliding with atoms (electron impact ionization) and this effect is included[37] in our model. Impact ionization is responsible for the generation of many low energy free electrons, which is significant even for the sixty atom large fullerene. Further important molecular effects were also introduced in XMDYN: (a) molecular Auger effect, when one of the two involved electrons is from an atom next to the atom with the initial core hole; (b) bond breaking due to the changes of the atomic electronic configurations driven by the ionizations. The experiment had significant impact on the development of the model by motivating the inclusion of many of the processes above, while modeling helped to interpret the measured data and reveal further details not accessible in present experiments.

3   Results and discussion

3A   Molecular fragmentation

The ionization of C_{60} with focused X-rays of 485, 600 and 800 eV was investigated. One of the challenges with the SASE FELs[8,9] is that the interaction region covers a volume around the focus with an inhomogeneous X-ray spatial fluence distribution. Thus the recorded data include contributions from the wide range of fluence with the peak X-ray fluence localized at the center of the focus. In addition, the photon and pulse energy vary for each X-ray pulse. This is both a challenge for the non-linear optics community but it is also an opportunity for the chemical dynamics community to observe and compare trends at different fluences since each laser shot represents an experiment which is recorded along with all the experimental parameters. The data can then be binned accordingly. It should be noted that the FEL pulse energy can be focused for non-linear studies[21–31] but it can also be defocused for other time-resolved studies that do not require intense pulses.[20]

The wavelength and intensity of the photon pulses lead to K-shell multiphoton ionization of the parent molecule which highly charges-up and then fragments into smaller observable molecular and atomic ionic fragments. We believe that the molecular fragments are generated dominantly in the wings of the fluence distribution while the atomic fragments are generated in the central high-

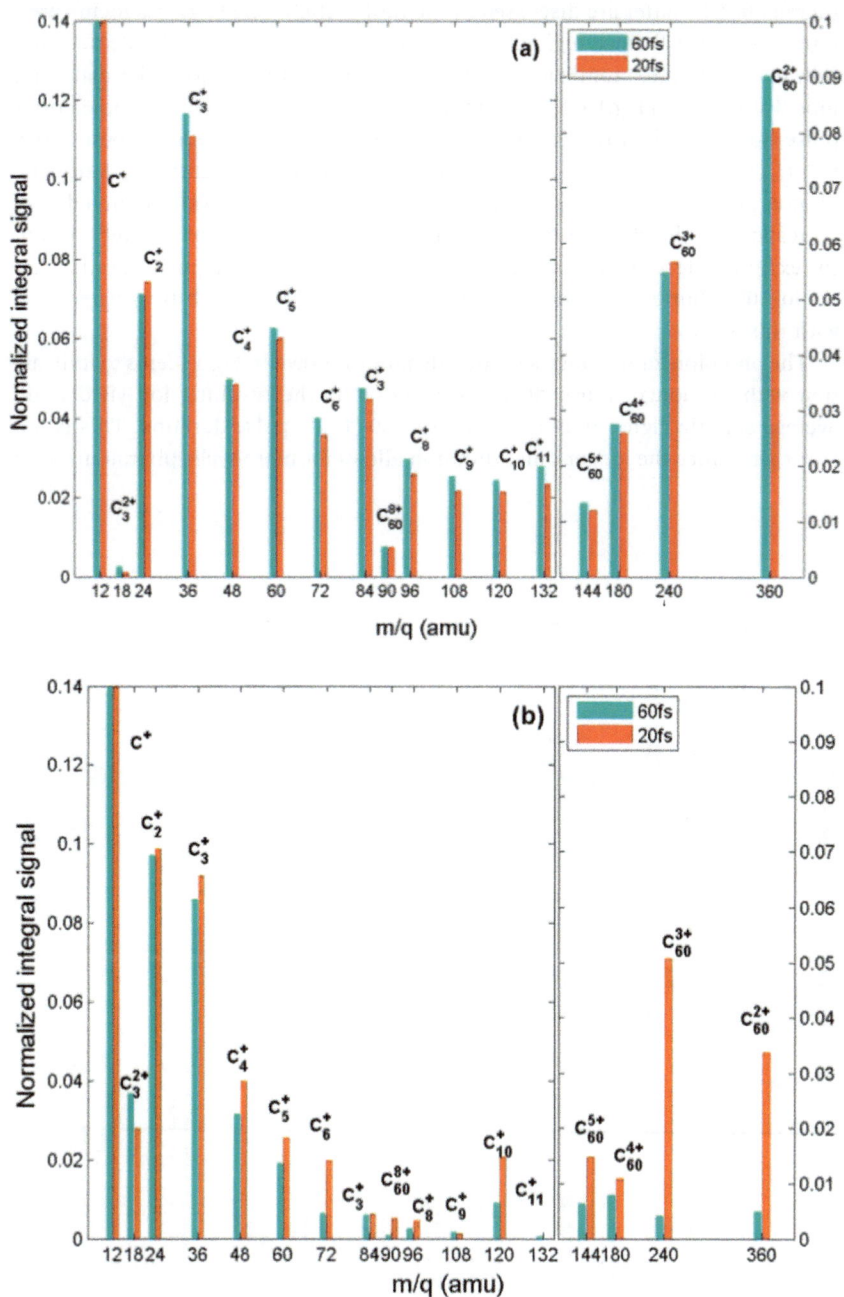

Fig. 1 (a) Experimental integral signals of molecular peaks at two pulse duration. The photon energy is 600 eV; the pulse energy is 0.61 mJ. The C^+ peak is out of scale in the current plots. (b) Calculations using parameters that model best the measurements at 60 fs (30fs) and 30 fs (13 fs).

fluence region of the focus. Fig. 1 shows the ionization with 600 eV photons pulses at 60 and 20 fs with the same pulse energy of 0.61 mJ. The spectra display the molecular ionic C_{60} charge states from C_{60}^{2+} to C_{60}^{8+} as well as the observed molecular ionic fragments ranging from C_{11}^{+} to C_2^{+} and C^{+} the start of the C ionic fragments. The latter are discussed in the next section. The large fragments were formed in lower fluence regions of the X-ray beam where the molecule does not fully dissociate into atomic ions. This trend agrees with the molecular dynamics model as shown in Fig.1b. Note that the calculations were carried out with shorter pulse durations than the quoted experimental values to obtain the best agreement. The used values in the calculations are given in parenthesis. Since the photon pulse durations are not measured, the experimental values quoted here correspond to the electron pulse durations which is the only parameter given to the experimenters. It has been shown however that the photon pulse duration can be 40–50% shorter than the electron pulse duration[21,24] which is in agreement with this work.

The photoionization fragmentation dynamics shown in Fig.1 seems to indicate that with the longer pulse duration of 60 (30)fs the resulting ion yield is on average slightly higher than at the shorter 20 (13) fs pulse duration. This is not surprising since the longer pulse duration allows for more cyclic photoionization

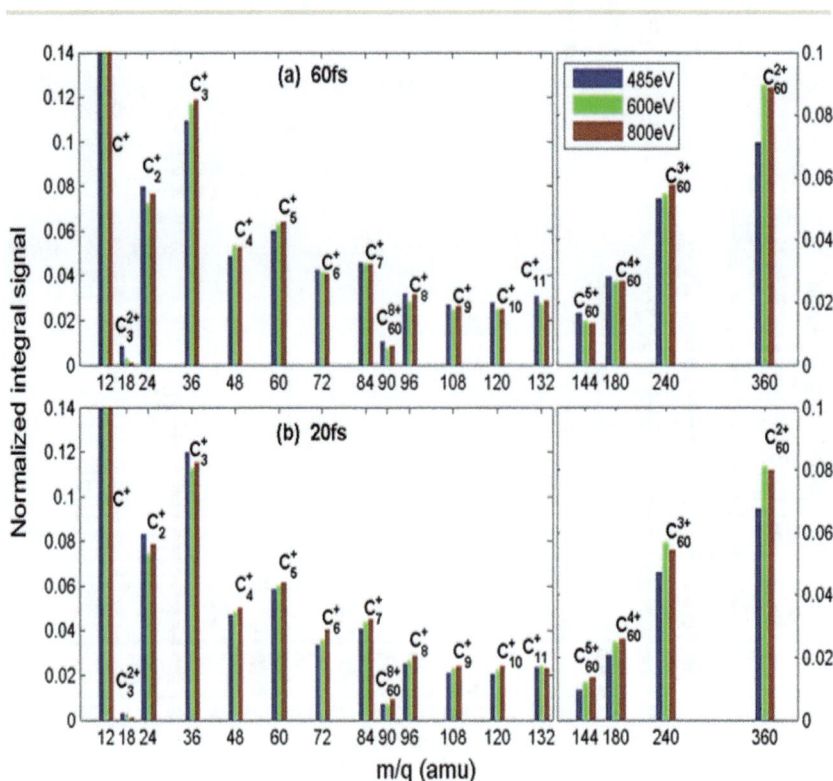

Fig. 2 Integral signals of molecular peaks at three different photon energies. (a) The pulse duration at 60 fs has pulse energy of 0.9 mJ. (b) The pulse duration at 20 fs has pulse energy of 0.5–0.55 mJ. The C^{+} peak is out of scale in the current plots.

and Auger decay during the same pulse, as observed in atoms[21] and smaller molecules.[24]

Fig. 2 shows the comparison of the molecular fragmentation at two pulse durations, 60 and 20 fs, for three photon energies, 485, 600 and 800 eV and for the same pulse energy. Although the K-shell photoionization cross section decreases from 485 to 800 eV (the K-edge for C is ∼280 eV), the parent and molecular fragments ion charge states decrease systematically from 800 eV to 485 eV. This trend is not understood and could be due to one of the challenges we need to overcome, namely better understanding of the X-ray focal width at different wavelengths.

3B Non linear physics

In our investigation, an average of 180 X-ray photons are absorbed per molecule in the X-ray focus, corresponding to 87 keV total energy transfer to the fullerene leading to the Coulomb explosion of the molecule generating highly charged atomic fragments ranging from C^{1+} to fully stripped C^{6+} at 150 fs as shown in Fig.3. The measurements were taken with 0.91 mJ pulse energy and 485 eV. We also show the carbon charge state distribution at 60 fs and despite the same pulse energy the distribution is slightly different, but consistent with measurements carried out in Ne atoms[21] and N_2 molecule.[24] Namely, the long pulse duration allow for more highly charged states, in this case, C^{4+} and C^{5+}. We do not observe fully stripped C at 60 fs since the pulse is not long enough to allow for the cyclic photoionization and Auger processes[21,24] to remove all six electrons from the C fragments. We also show in Fig. 4a,b the experimental and theoretical comparison of the C charge state distribution produced at 600 eV photon energy with 60 fs and 20 fs and with the same pulse energy of 0.61 mJ. The agreement between

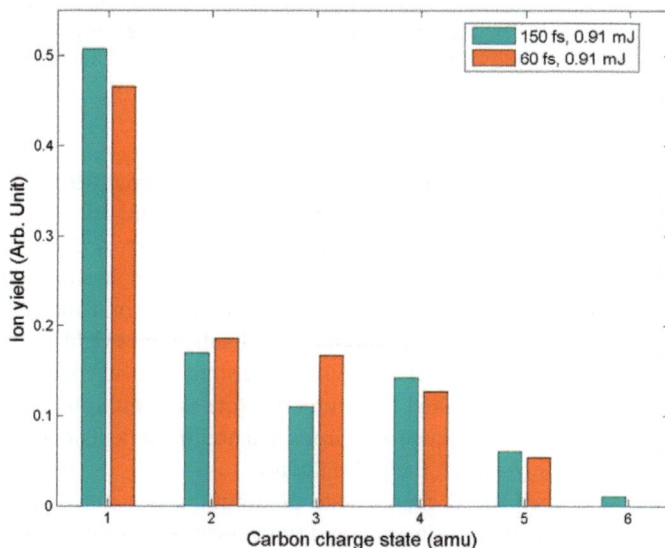

Fig. 3 Charge state distribution of atomic C ion fragments at pulse durations of 150 fs and 60 fs and the same pulse energy. The transmission of the spectrometer is included. The photon energy is 485 eV. Signals are normalized to the total C atomic ion yields.

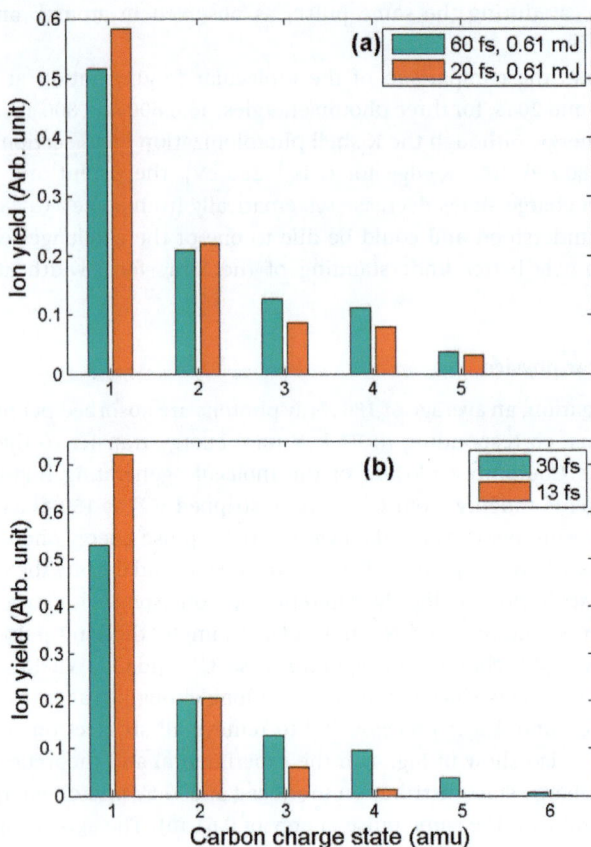

Fig. 4 (a) Experimental charge state distribution of atomic C ion fragments at pulse durations of 60 fs and 20 fs with the same pulse energies. The transmission of the spectrometer is included. The photon energy is 600 eV. The pulse energy is 0.61 mJ. (b) Calculated charge state distribution. Signals are normalized to the total C atomic ion yields.

theory and experiment is good and the same trend is observed with the higher photon energy. That is if one keeps the pulse energy constant, one will generate less highly charged states at the shorter pulse duration, close to or far from the C K-edge. This result is by far the *most nonlinear* sequential multiphoton process (during one X-ray laser pulse, C_{60} absorbs many photons), in a molecule ever reported in the X-ray spectrum. The production of substantial yields of C^{5+} and C^{6+} as shown in Fig.3 for both 150 and 60 fs with similar pulse energy, has not been seen before in the published literature.[46,47] Note that multiphoton absorption—be it sequential or nonsequential—leads to a nonlinear dependence, in this case of C^{5+} and C^{6+} ion yield, on the pulse fluence.

We also investigated the C charge state distribution as a function of pulse energy. Fig. 5 displays our experimental and theoretical results for C_{60} ionized at 485 eV, at 150 fs and for pulse energy varying between 1.2–0.6 mJ. We observe that C^{1+} is the most intense at the lowest pulse energy. This trend however inverts starting at C^{3+} through C^{6+} were the highest ion yield is obtained at the highest

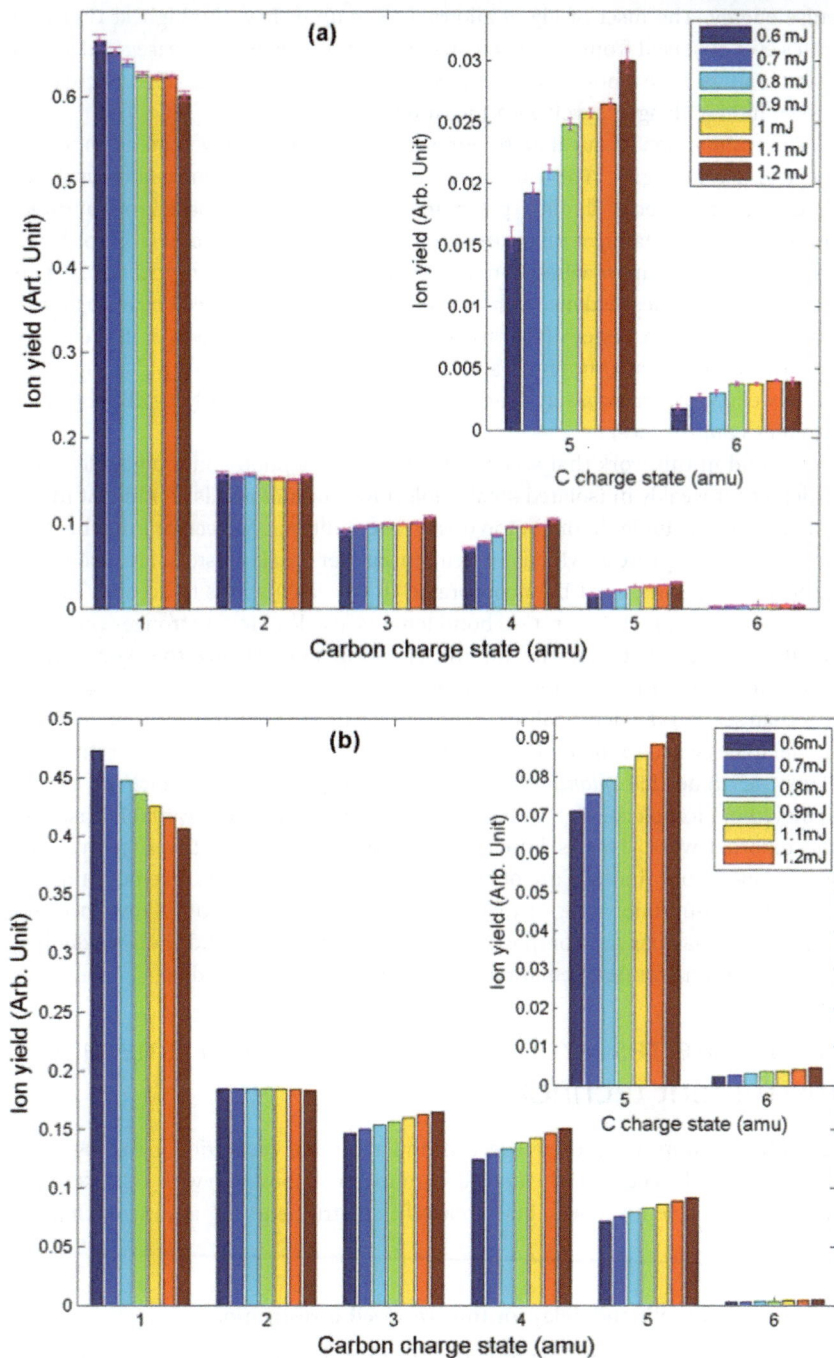

Fig. 5   (a) Experimental charge state distribution of atomic C ion fragments at various pulse energies and the same pulse duration. The photon energy is 485 eV and the pulse duration is 150 fs. (b) Calculated charge state distribution.

pulse energy. The inset of Fig. 5 allows a close-up of how the highest C charge states ever observed from C_{60} are created as the pulse energy increases and shows that 0.6mJ is high enough to produce fully stripped carbon. The calculation shown in Fig. 5b agrees well with measurements.

The comparison of our data to our MD calculations is very good as shown in the Fig. 4 and Fig. 5. The validation and refinement of molecular dynamics modeling at extreme fluence presented in this work is based on a direct comparison to dynamical quantities from experiment and is crucial in order to show that these approaches can make *quantitatively correct predictions*. Our experiment provides evidence that the charged particles produced by exposing an extended quantum system (C_{60}) to high-fluence x rays behave as if they were classical particles at different wavelength, pulse duration and pulse energies. Thus, we have *validated* a very fundamental assumption underlying all molecular dynamics approaches.

We find in this work that secondary ionization by photo- and Auger electrons which occur weakly in isolated small molecules[24] or atoms,[21] is significant in the dynamics of C_{60} under high photon dose rate conditions. Molecular influences on the Auger decay process, which are weak in van der Waals clusters, are also much stronger in C_{60} and must be incorporated in the calculations to account for the experimental data. The short C–C bond lengths also lead to far stronger Coulomb repulsion forces between the rapidly charging ions during the X-ray pulse, producing ion-ion forces more representative of biomolecules under high X-ray flux than seen in van der Waals clusters. Our findings allow the successfully tested model to be scaled to the higher photon energy typical of bio-molecular imaging. Thus, we argue that *quantitative radiation damage dynamics predictions* can be achieved in biomolecules at far higher intensity than currently available based on the reported work. This is important for high resolution diffraction imaging. Furthermore, our findings are of interest to the matter under extreme condition (MEC)[48,49] community using any source of photons (IR, UV, VUV, X-rays) because the formed plasmas are composed of highly charged ions and their modelling, often based on atomic calculation can now use experimental data.

4 Future research opportunities with emerging instrument technologies

As FELs are improving their spectral and temporal resolution by using laser seeding[50] or self-seeding,[51] the Berrah's group in collaboration with LCLS staff[52–55] and CAMP[56] scientists has built powerful instruments to advance scientific discoveries.

4A The X-ray split and delay for time-resolved investigations

We built an X-ray pump-X-ray probe instrument, called the X-ray split and delay (XRSD)[52,55] to carry out time-resolved experiments at the LCLS. The tool commissioned in June 2013 is now available for experiments. It consists of a compact two mirror device whose principle of operation is illustrated schematically in Fig.6. Briefly, two mirrors are located along the beam path such that the first mirror intercepts a portion of the beam and deflects it towards the interaction region. The second mirror intercepts the remaining portion of the beam

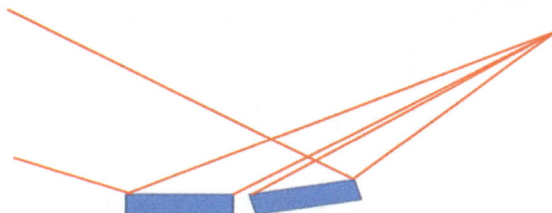

Fig. 6   Schematic of the two mirror soft X-ray split and delay.

downstream of the first mirror and deflects that portion through a slightly larger angle to intersect the beam from the first mirror in the interaction region of the experiment. In order to split the beam without significant losses, the first mirror was polished to its trailing edge where a short chamfer allows the beam not reflected by the mirror to pass over the edge to the second mirror. This design incorporates an extended second mirror requiring only vertical motion and pitch to properly intercept the delayed portion of the beam and reflect it towards the interaction region. This system was successfully commissioned in May 2013[52,55] and has already been used in two experiments.

4B   The LAMP instrument

We built a modular system called LAMP that consists of: **i)** a chamber that houses two velocity map imaging (VMI) detectors to measure electrons and ion-ion coincidences resulting from the photo-absorption of fs X-ray photons,[54] as well as **2i)** a chamber that houses two imaging detectors of the pn-CCD type that record scattering images or fluorescence resulting from the interaction of the photons with the samples.[53] This instrument was commissioned in November of 2013, was used successfully during an experiment that recorded the scattering of a biological system and is now available for experiments at the LCLS.[53,54] LAMP was designed similarly to the CAMP instrument[56] and it benefited from the first design and collaboration[54,56] to make significant improvements in its design and workmanship.

The LAMP-VMI system allows time-resolved nuclear dynamics experiments to be monitored *via* the detection of electron and ionic fragments resulting from photo-absorption of X-ray photons, as a function of pump probe delay and probe pulse intensity using the XRSD or using optical lasers synchronized to the FEL beam. With these new tools, we record the different fragmentation pathways by measuring multi-particles ion-ion coincidences/multi-particle correlations as a function of pump–probe delay to monitor the evolution of the kinetic energy release (KER) associated with different break-up patterns. We also simultaneously image the electrons momenta to capture the most detailed X-ray induced reaction in molecules and nano-systems. This instrument, available to any users, has the possibility to uncover new mechanisms in physics, chemistry and biology. The interpretation of the measurements can be done with molecular modeling and *ab initio* electronic structure calculations by comparing measured KER electron and ion spectra with classical and quantum-mechanical simulations. The identification of the various dynamic mechanisms can lead to improved understanding of

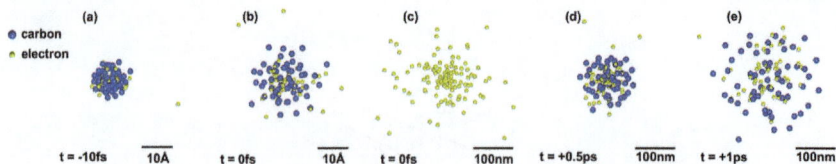

Fig. 7 Real space snapshots of the time evolution of a C_{60} molecule irradiated at the center of the focus of the X-ray free electron laser pulse, based on theoretical modelling. Carbon atoms (blue) and electrons (yellow) are shown as a function of time at fine spatial scale (a, b) and expanded spatial scale (c, d,e). The centre of the pulse is at t = 0 fs and the photon energy is 600 eV with 30 fs pulse duration and 0.6 mJ pulse energy.

the inner-working of molecular systems and to ultimately control chemical dynamics, the topic of this Faraday Discussion 171.

4C   Tracking the ionization dynamics of C_{60} by time-resolved measurements

We explore briefly here a possible future study that can utilize the new tools described above to carry out a time-resolved experiment of the absorption and dissociation dynamics of C_{60} under various pulse durations and pulse energies. Our calculations led by the Santra's group predict, as shown in Fig. 7, that the ionization of C_{60} at high fluence occurs within the first few fs. The fast electrons will be ejected, followed within the first 20 fs–130 fs of the expansion dynamics of a nanoplasma formed of trapped electrons and C ions. We plan to track the evolution of the ionization mechanisms by time-resolved measurement using X-ray pump-X-ray probe spectroscopy with the XRSD. We will take advantage of the *site specificity* of X-ray ionization by creating a hole (pump) on the K-shell site in C_{60}. We will then use a second, delayed X-ray pulse (probe) to interrogate the ionization/expansion of the system at various pump–probe delay times. We will record the photoelectron spectra and ion spectra simultaneously using the double VMI spectrometers in the LAMP end station. The varying delay will reveal the time-evolution of the ionized system.

5   Conclusions

The presented work reports on the fragmentation dynamics of C_{60} subsequent to absorption of X-ray FEL pulses of different photon energies (485 eV, 600 eV, 800 eV), pulse duration (20 fs, 60 fs, 150 fs) and pulse energy (0.6mJ–1.2mJ). Our findings seem to indicate that once a molecule, whether it is a small diatomic molecule like N_2 or a large molecule like C_{60}, absorbs photons from intense short X-ray pulses, it undergoes multi-photon ionization, followed by Coulomb explosion and dissociation. In addition, the data indicate that the fragment ions generated from a small or a large molecule behave the same way with respect to wavelength, pulse duration and pulse energy. However, from the radiation damage perspective, our modeling has shown that secondary ionization of C_{60} by photo-and Auger electrons, weak in isolated small molecules and absent in atoms, is significant in its dynamics under high dose rate conditions. Furthermore, our work revealed that molecular influences on the Auger decay process, which are weak in van der Waals clusters[57,58] are much stronger in C_{60} and must be incorporated in the calculations to account for the data. The short C–C bond lengths lead to far stronger repulsion forces between the rapidly charging ions during the X-ray pulse producing ion-ion forces more representative of chemically bonded systems than those seen in van der Waals clusters.

Our results on the fundamental interaction of C_{60} with intense femtosecond X-ray photons are interesting in their own right in chemical dynamics and atomic and molecular physics but they are also relevant to many fields of research, in particular to matter under extreme conditions as well as high resolution X-ray imaging and scattering experiments relevant to biology.

The world is on the threshold of a dramatic increase in the number of FEL photon sources which are planned to be even more intense—despite present severe world economic constraints—because of their importance to uncovering new science. In fact, advances in FEL technologies are pushing the pulse duration down to the attosecond realm.[59] It is therefore of utmost importance to understand, from a fundamental point of view, the interaction of atoms, molecules, nano-systems and bio-systems with short X-ray radiation with the present and future generation of FEL photon sources.

Acknowledgements

This work was funded by the Department of Energy office of Science, Basic Energy Sciences, Division of Chemical Sciences, Geosciences, and Biosciences under grant No. DE-FG02-92ER14299.A002. We would like to thank E. Kukk, M. Mucke, J. H. D. Eland, V. Zhaunerchyk, R. Feifel, L. Avaldi, P. Bolognesi, C. Bostedt, J. D. Bozek, J. Grilj, M. Guehr, L. J. Frasinski, D. T. Ha, K. Hoffmann, B. K. McFarland, C. Miron, K. Ueda, E. Sistrunk, R. J. Squibb and J. Glownia for their contribution to the earlier work.[31]

References

1 T. Ergler, A. Rudenko, B. Feuerstein, K. Zrost, C. D. Schroter, R. Moshammer and J. Ullrich, *Phys. Rev. Lett.*, 2006, **97**, 193001.

2 H. J. Worner, J. B. Bertrand, D. V. Kartashov, P. B. Corkum and D. M. Villeneuve, *Nature*, 2010, **466**, 604.

3 M. Meckel, D. Comtois, D. Zeidler, A. Staudte, D. Pavicic, H. C. Bandulet, H. Pepin, J. C. Kieffer, R. Dorner, D. M. Villeneuve, *et al.*, *Science*, 2008, **320**, 1478.

4 S. Baker, J. S. Robinson, A. Haworth, H. Teng, R. A. Smith, C. C. Chirila, M. Lein, J. W. G. Tisch and J. P. Marangos, *Science*, 2006, **312**, 424.

5 M. Magrakvelidze, *et al.*, *Phys. Rev. A: At., Mol., Opt. Phys.*, 2012, **86**, 013415.

6 L. Fang and G. Gibson, *Phys. Rev. A: At., Mol., Opt. Phys.*, 2008, **78**, 051402.

7 S. Haessler, J. Caillat, W. Boutu, *et al.*, *Nat. Phys.*, 2010, **6**, 200.

8 W. Ackermann, *et al.*, *Nat. Photonics*, 2007, **1**, 336.

9 P. Emma, *et al.*, *Nat. Photonics*, 2010, **4**, 641.

10 N. Berrah, *et al.*, *J. Mod. Opt.*, 2010, **57**, 1015.

11 A. Rudenko, J. Ullrich and R. Moshammer, *Annu. Rev. Phys. Chem.*, 2012, **63**, 635.

12 J. Cryan, *et al.*, *Phys. Rev. Lett.*, 2010, **105**, 083004.

13 T. Osipov, *et al.*, *J. Phys. B: At., Mol. Opt. Phys.*, 2013, **46**, 164032.

14 C. Bostedt, *et al.*, *J. Phys. B: At., Mol. Opt. Phys.*, 2013, **46**, 164003.

15 A. H. Zewail, *Angew. Chem., Int. Ed.*, 2000, **39**, 2586.

16 G. Sansone, F. Kelkensberg, F. Morales, J. F. Perez-Torres, F. Martin and M. J. J. Vrakking, *IEEE J. Sel. Top. Quantum Electron.*, 2012, **18**, 520.

17 V. Petrovic, *et al.*, *Phys. Rev. Lett.*, 2012, **108**, 253006.

18 J. M. Glownia, *et al.*, *Opt. Express*, 2010, **14**, 17620.

19 L. Fang, *et al.*, *Phys. Rev. Lett.*, 2012, **109**, 78.

20 B. K. McFarland *et al.*, (Submitted to Nature Comm.).

21 L. Young, *et al.*, *Nature*, 2010, **466**, 56.

22 L. J. Frasinski, *et al.*, *Phys. Rev. Lett.*, 2013, **111**, 073002.

23 B. Rudek, *et al.*, *Nat. Photonics*, 2012, **6**, 858.

24 M. Hoener, *et al.*, *Phys. Rev. Lett.*, 2010, **104**, 253002.

25 L. Fang, *et al.*, *Phys. Rev. Lett.*, 2010, **105**, 083004.

26 N. Berrah, *et al.*, *Proc. Natl. Acad. Sci. U. S. A.*, 2011, **108**(41), 16912.

27 P. Salen, *et al.*, *Phys. Rev. Lett.*, 2012, **108**, 153003.

28 H. Fukuzawa, *et al.*, *Phys. Rev. Lett.*, 2013, **110**, 173005.

29 K. Tamasaku, *et al.*, *Phys. Rev. Lett.*, 2013, **111**, 043001.

30 B. F. Murphy, *et al.*, *Phys. Rev. A: At., Mol., Opt. Phys.*, 2012, **86**, 053423.

31 B. Murphy *et al.*, (Submitted to Nature Comm.).

32 Neutze, *et al.*, *Nature*, 2000, **406**, 752.

33 Chapman, *et al.*, *Nature*, 2011, **470**, 73.

34 Nugent, *et al.*, *Nat. Phys.*, 2011, **7**, 142.

35 A. Barty, *et al.*, *Nat. Photonics*, 2012, **6**, 35.

36 J. D. Bozek, *Eur. Phys. J. Spec. Top.*, 2009, **169**, 2013.

37 Z. Jurek, G. Faigel and M. Tegze, *Eur. Phys. J. D*, 2004, **29**, 217; Z. Jurek, and R. Santra, *R. XMDYN*. CFEL, DESY, Hamburg, Germany (2013).

38 S. P. Hau-Riege, *Phys. Rev. Lett.*, 2012, **108**, 238101.

39 M. Bergh, N. Tîmeanu and D. van der Spoel, *Phys. Rev. E: Stat., Nonlinear, Soft Matter Phys.*, 2004, **70**, 051904.

40 C. Caleman, *et al.*, *J. Mod. Opt.*, 2011, **58**, 1486.

41 B. Ziaja, A. R. B. de Castro, E. Weckert and T. Möller, *Eur. Phys. J. D*, 2006, **40**, 465.

42 Z. Jurek, B. Ziaja & R. SantraXMDYN Rev. 1.0360. (CFEL, DESY, Hamburg, Germany, 2013).

43 S.-K. Son & R. Santra *XATOM –; an integrated toolkit for X-ray and atomic physics.* (CFEL, DESY, Hamburg, Germany, 2011).

44 S.-K. Son, L. Young and R. Santra, *Phys. Rev. A: At., Mol., Opt. Phys.*, 2011, **83**, 033402.

45 D. W. Brenner, *Phys. Rev. B*, 1990, **42**, 9458.

46 N. Hay, *et al.*, *J. Phys. B: At. Mol. Opt. Phys.*, 1999, **32**, L17.

47 R. C. Constantinescu, *et al.*, *Phys. Rev. A*, 1998, **58**, 4637.

48 B. Nagler, *et al.*, *Nature Physics*, 2009, **5**, 693.

49 S. M. Vinko, *et al.*, *Nature*, 2012, **482**, 59.

50 E. Allaria, *et al.*, *Nature Photonics*, 2012, **6**, 699.

51 J. Amann, *Nature Photonics*, 2012, **6**, 693; D. Cocco, *et al.*, *Proc. SPIE*, 2013, **8849**, 88490A, DOI: 10.1117/12.2024402.

52 J. C. Castagna, B. M. Murphy, J. D. Bozek and N. Berrah, *SPIE Proceedings*, 2013, **8504**, 9.

53 C. Bostedt, T. Osipov, J. C. Castagna, M. L. Swiggers, N. Berrah *et al.*, (to be submitted to Nuc. Inst. and Meth.).

54 T. Osipov, C. Bostedt, J. C. Castagna, M. L. Swiggers, A. Rudenko, D. Rolles, N. Berrah *et al.*, (to be submitted to Nuc. Inst. and Meth.).

55 B. Murphy, J. C. Castagna, M. L. Swigegrs, J. D. Bozek, N. Berrah *et al.* (to be submitted to Nuc. Inst. and Meth.).

56 L. Struder, *et al.*, *Nucl. Inst. Meth.*, 2012, **614**, 483.

57 H. Thomas, *et al.*, *Phys. Rev. Lett.*, 2012, **108**, 133401.

58 C. Bostedt, *et al.*, *Phys. Rev. Lett.*, 2012, **108**, 093401.

59 P. Emma *et al.*, *Proceedings of the 2004 FEL Conference*, 2014, 333–338; V. Wacker *et al.*, *Proceedings of FEL*, 2012, Nara, Japan, ISBN 978-3-95450-123-6.

Faraday Discussions

ROYAL SOCIETY OF CHEMISTRY

PAPER

Multi-colour pulses from seeded free-electron-lasers: towards the development of non-linear core-level coherent spectroscopies

Filippo Bencivenga,*[a] Flavio Capotondi,[a] Francesco Casolari,[a] Francesco Dallari,[a] Miltcho B. Danailov,[a] Giovanni De Ninno,[ab] Daniele Fausti,[a] Maya Kiskinova,[a] Michele Manfredda,[a] Claudio Masciovecchio*[a] and Emanuele Pedersoli[a]

Received 8th May 2014, Accepted 9th May 2014

DOI: 10.1039/c4fd00100a

We report on new opportunities for ultrafast science thanks to the use of two-colour extreme ultraviolet (XUV) pulses at the FERMI free electron laser (FEL) facility. The two pulses have been employed to carry out a pioneering FEL-pump/FEL-probe diffraction experiment using a Ti target and tuning the FEL pulses to the $M_{2/3}$-edge in order to explore the dependence of the dielectric constant on the excitation fluence. The future impact that the use of such a two-colour FEL emission will have on the development of ultrafast wave-mixing methods in the XUV/soft X-ray range is addressed and discussed.

1 Introduction

The last decades have witnessed the advent of two new classes of photon sources: optical lasers and synchrotrons, that are presently used in several fields of physics, chemistry and biology. Synchrotrons can provide high brightness photon pulses of 10–100 ps time duration ranging from the extreme ultraviolet (XUV) to the hard X-rays. A distinct feature of synchrotrons is the broad emission spectrum which permits to achieve continuous tunability in the photon energy while retaining a sufficient photon flux at the sample. This has allowed the development of core-level spectroscopies, *e.g.*: X-ray absorption and emission, anomalous diffraction, resonant inelastic X-ray scattering, *etc.* In all these methods, the source tunability across core transitions of selected elements permits to obtain information on the electronic, molecular and lattice structures of the sample with element and chemical state sensitivity. On the other hand, the pulse duration

[a]Elettra-Sincrotrone Trieste S.C.p.A., S.S. 14 km 163,5 in AREA Science Park, I-34149, Basovizza, Trieste, Italy. E-mail: filippo.bencivenga@elettra.eu; claudio.masciovecchio@elettra.eu; Tel: +39 040 375 8202; +39 040 375 8093

[b]Laboratory of Quantum Optics, University of Nova Gorica, Nova Gorica, SI-5000, Slovenia

hampers the study of ultrafast (<ps) processes. Only recently pioneering experiments succeeded in using synchrotrons to probe ultrafast dynamics induced by optical laser pulses ('pump').[1-3]

Optical lasers are able to provide coherent photon pulses as short as few fs with brightness larger than synchrotrons. Furthermore, the spatial and temporal coherence of laser radiation ensures the occurrence of fixed phase relationships between the values of the electric field in different positions and times. This feature is at the base of applications such as interferometry or holography and also allowed the prompt discovery of non-linear wave-mixing phenomena.[4] Methods based on optical wave-mixing[5-7] (e.g., transient grating,[8] photon echo,[9] coherent Raman scattering,[10] etc.) can probe a large timescale range and can provide selectivity in the probed excitations. To date, these methods are exploited in several fields of physics, chemistry and biology to study a large array of dynamical processes, ranging from ms diffusions[11] and μs–ns protein folding[12] to sub-ps reconstruction of wave-function in reacting molecules,[13] electronic relaxations and charge transfers.[9] The major limitation of these experiments is the long wavelength of optical radiation produced by table top lasers. This prevents the excitation of core-level transitions, that can be exploited to obtain atomic selectivity.

Such limitation can be overcome by XUV/X-ray wave-mixing methods, as already theoretically discussed in details.[14-17] As compared to (linear) pump-probe core-level spectroscopies, XUV/X-ray wave-mixing methods are potentially able to probe the non-local nature of the excitations as well as coherences between different atoms within a molecule,[14-17] thus potentially allowing to study, e.g., charge and energy transfer processes at the molecular scale with elemental sensitivity.

However, the requirements in terms of coherence, pulse duration and field intensity are beyond the capabilities of the most advanced synchrotron sources. On the other hand, sources based on the high-harmonic-generation process are able to deliver ultrafast coherent XUV/X-ray pulses[18] but, at the present stage of their development, cannot provide the required field intensity and have limited wavelength tunability. In this respect the advent of free electron laser (FEL) sources, providing ultrafast XUV/X-ray photon pulses with high field intensity, represents an opportunity for extending wave-mixing methods in the XUV/X-ray range. Recently, the FEL photons from LCLS (Stanford, USA) have been used to demonstrate a wave-mixing process involving X-ray photons, still with the assistance of an optical laser.[19] Furthermore, *"with an eye toward extending optical wave-mixing techniques to the X-ray regime"*,[20] the LCLS team recently demonstrated the first two-colour FEL emission, still in the self amplified spontaneous emission (SASE) regime. This is characterized by an emission spectrum consisting in several longitudinal modes with random phase relations among them, which results in a poor longitudinal (time) coherence. A two-colour SASE FEL emission has also been reported from SACLA facility (Koto, Japan).[21]

Highly coherent ("laser-like") FEL pulses can be obtained through a seeding process, which also reduces significantly the shot-to-shot fluctuations of the FEL output.[22] Seeding is adopted at LCLS (for the single-colour operation) and at FERMI (Elettra, Trieste, Italy). The scheme developed at FERMI is based on an ultrafast UV laser that triggers the FEL emission.[22] This makes it possible to control the FEL output by tuning the seed laser. For instance, the FEL photon energy is given by $\hbar\omega_{FEL} = N\hbar\omega_{UV}$, where $\hbar\omega_{UV}$ and N are the photon energy of the seeding pulse (tunable in the 4–5.5 eV range†) and an integer (harmonic) number, respectively. The latter is presently limited to the 3–15 range, but in a near future N values up to 60–70 will be possible by exploiting a double stage FEL scheme.[23] The possibility to chose both $\hbar\omega_{UV}$ and N allows for a continuous tunability in $\hbar\omega_{FEL}$.[24] In summary, seeded FELs, as FERMI, combine the key advantages of synchrotron and laser sources. Furthermore, we recently demonstrated the possibility to achieve a two-colour (easily upgradable at 'multi-colour') seeded FEL emission with independent control on relative intensity, time delay (Δt) and photon energy separation ($\Delta\omega$) between the two FEL pulses.[25]

Such two-colour FEL emission has been employed to study the XUV optical response of Ti 0.5 ps after the FEL irradiation.[25] The obtained results suggest the occurrence of a FEL-induced XUV transparency for fluences (F) larger than 2 J cm^{-2}, which can be qualitatively accounted for by a frequency shift (>2.2 eV) of the optical constants of the Ti sample. However, the limited set of data did not allowed us to quantify the amount of the shift nor to study its dependence on F. The latter is often the key variable for this class of experiments, since it ultimately determines the amount of energy transfer to the sample. In order to fill this gap of knowledge, we hereby report the results of new two-colour FEL-pump/FEL-probe measurements on Ti as a function of F.

In the F-range exploited in the present study a tangible photoionization occurs within the FEL pulse duration ($\sim$50 fs). This brings the system into a non-equilibrium state consisting in a hot dense electron plasma and a cold ionic lattice. Such an "exotic" state evolves in the sub-ps timescale towards a thermodynamic equilibrium state, termed warm dense matter (WDM), that survives until the hydrodynamic expansion (typically after a few 10's of ps) ineluctably destroys the sample. The transient nature of WDM samples imposes experimental approaches based on ultrafast methods, even for the determination of basic quantities such as, $e.g.$, temperature and density.[26–30] WDM is featured by extreme conditions of these thermodynamic parameters,[31] typical in astrophysical context.[32] The theoretical description of these extreme states of matter is a $grand$ $challenge$ for fundamental science. Indeed, on one hand, WDM is "too hot" to be described by classical condensed matter theories and, on the other hand, it is 'too dense' to be described in the frame of classical plasma. In these conditions the thermal energy compares to the Fermi energy and to the interaction energy of ions. This turns into a partial degeneracy of free electrons, strong couplings between ions and a significant population of excited states. These conditions lead to correlations between particles in the plasma and to the persistence of a short-range order. The understanding of these general aspects of interparticle interactions in solid density plasma is expected to have a deep impact, $e.g.$, in the development of inertial confinement fusion technology.[33]

Since XUV/X-ray FELs can overcome the issue of optical opacity of the induced electron plasma, they have already been exploited to study solid density plasma.[31] The unique capability to probe core electron states has enabled the discovery of new phenomena, as saturation of core transitions at the L-edge of Al.[34] Such pioneering experiment, carried out without time resolution, reveals how the inner-shell absorbers can be severely depleted when the photoionization rate surpasses the

---

† For the sake of simplicity we hereafter drop the symbol $\hbar$ and indicate frequencies in energy units.

recombination one. This turns into a reduced screening of the electrostatic atomic potential, which shifts the absorption edge towards higher energy.[34–37] The lack of time resolution limits this kind of study to the non-equilibrium state reached within the time duration of the FEL pulse (*i.e.*, ~15 fs for the experiment discussed in ref. 34). An FEL-pump/FEL-probe approach is thus required to obtain direct information on the time evolution of the system towards the WDM regime. The changes in the diffraction pattern from grating-shaped samples irradiated with two-colour FEL pulses allowed us to determine the *F*-dependence of both the absorption (β) and dispersion coefficients (δ) of Ti at the expected time (~0.5 ps) for the transition towards the WDM state. With respect to absorption measurements,[34] in which only β can be determined, the present approach allows to obtain both δ and β and, therefore, to determine the *F*-dependence of the dielectric constant (ε). The latter quantity embodies relevant information on the electronic properties of the sample, that can be used, *e.g.*, to map fast phase transitions induced by FEL radiation,[38,39] such as interatomic bond softening.

The present study reports on the first determination of ε close to a core atomic resonance in FEL-excited matter undergoing transition towards the WDM regime.

After the description of the two-colour FEL scheme developed at FERMI in section 2 and the discussion of the results obtained from Ti in section 3, section 4 is dedicated on discussing the use of such a two-colour coherent emission in wave-mixing applications, that will be developed at FERMI in the near future.

2 Double seed pulse operation at FERMI

The basic idea to achieve a multi-pulse FEL emission at FERMI is very simple and consists in seeding the electron bunch with multiple seed laser pulses.[40] Fig. 1a

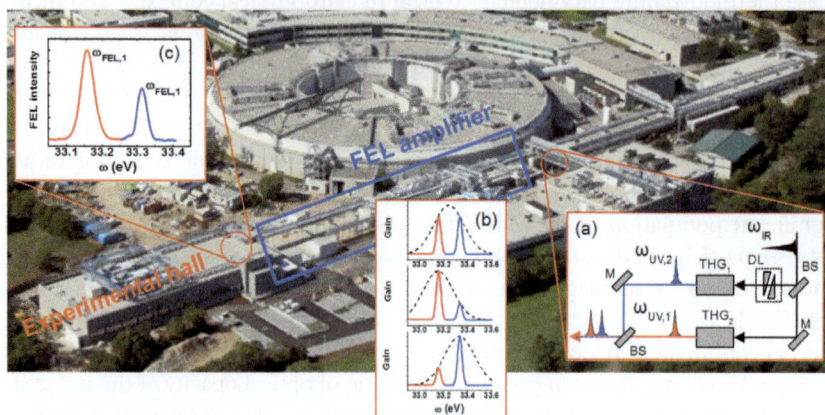

Fig. 1 (Main panel) FERMI facility; the blue hatched area is the FEL amplifier. (a) Optical system used to obtain the two seed laser pulses: BS, M, THG$_{1,2}$ and DL are 50 : 50 beamsplitters, reflective mirrors, third harmonic generation crystals and the delay line, respectively. (b) Changes in the relative intensity between the two-colour FEL pulses as a function of the central frequency of the FEL gain curve (dashed lines); the blue/red double peak structure sketches the spectrum of the FEL output. (c) Spectrum of the FEL output measured during the experiment on Ti; $\omega_{FEL,1} = 33.16$ eV and $\omega_{FEL,2} = 33.32$ eV are the pump and probe pulses, respectively.

shows the two seed pulses obtained by splitting the incoming pulse from a Ti:sapphire laser (photon energy 1.58 eV, time duration 120 fs) in two pulses travelling through separate arms. One of them can be time delayed by a pair of insertable wedges; the actual time delay is monitored by optical cross-correlation. Each beam path is equipped with independent filters, shutters and crystals for third harmonic generation, the latter being able to provide independent tunability in the photon energy of both pulses ($\omega_{UV,1}$ and $\omega_{UV,2}$) in the 4.77–4.73 eV range. A much wider range can be achieved by using an ultrafast UV optical parametric amplifier, also available at FERMI. The two UV seed pulses are finally collinearly recombined and sent into the FEL amplifier (main panel of Fig. 1). As long as $\omega_{UV,1}$ and $\omega_{UV,2}$ are within the gain bandwidth of the FEL amplifier (of about 0.5–1%), these seed pulses stimulate the two-colour FEL emission at $\omega_{FEL,1} = N\omega_{UV,1}$ and $\omega_{FEL,2} = N\omega_{UV,2}$.

The relative intensity between the two FEL pulses can be varied either by acting on the pulse energy of the seeds or by shifting in ω the gain bandwidth of the FEL, as sketched in Fig. 1b. Each of the two FEL pulses can be blocked by simply closing the shutter in the corresponding seed laser beamline. The FEL output is monitored shot-by-shot by an "on-line" XUV/soft X-ray spectrometer.[41] A representative two-colour FEL spectrum is reported in Fig. 1c. The jitters in intensity and photon energy were of about 15 and 0.005%, respectively, while $\delta\omega/\omega_{FEL} \sim 5 \times 10^{-4}$, where $\delta\omega$ is the FEL pulse bandwidth. Such a very narrow (almost transform limited) bandwidth and the extremely low jitters are practical manifestations of the "laser-like" features of the seeded FEL emission. The two-colour pulses are also expected to be coherent (*i.e.*, to have fixed phase relationship among them), since they arose from the same (coherent) laser pulse and the scheme adopted at FERMI allows to preserve the coherence properties of the seed laser in the FEL emission.[42] The time jitter between the two pulses is expected to be negligible (<1 fs), since it may only be due to fluctuations in the optical path difference accumulated by the two pulses along the beam paths shown in Fig. 1a, where they are separated. These paths are quite short ($\sim$0.5 m) and are situated in stable environmental conditions. In support of this statement we refer to recent timing measurements, where we found that jitters introduced by a very complex and $\sim$150 m long beam transport of the seed laser down to the FERMI experimental hall are in the sub-10 fs range.[43]

The maximum separation in $\Delta\omega = \omega_{FEL,1} - \omega_{FEL,2}$ is presently limited to $\sim$0.01 $\times$ ω_{FEL} by the bandwidth of the FEL amplifier; larger $\Delta\omega$-values can be obtained at the expense of intensity. Strategies to extend the exploitable $\Delta\omega$-range up to a few % of ω_{FEL} are under evaluation. The Δt range is also limited to values shorter than the length of the electron bunch ($\sim$1 ps), while interference effects occurring when the two seed pulses overlap in time set the lower limit for Δt at about 0.2–0.3 ps. The latter can be reduced by exploiting seed laser pulses shorter than the ones used here. Alternatively, a single chirped seed pulse with very high intensity can be used to achieve a two-colour FEL emission with Δt-values in the sub-100 fs range[44] or, eventually, exploiting seeding schemes with light pulses with orthogonal polarization (both 45° with respect to the axis of the modulator).

3   Behavior of photoexcited Ti in the sub-ps timescale

Measurements have been carried out at the DiProI experimental end-station.[45,46] The value of Δt was set to 0.5 ps, a value much shorter than the characteristic timescale for hydrodynamic expansion. The photon energy of the pump pulse ($\omega_{FEL,1}$) was set to 33.16 eV, *i.e.* across the Ti $M_{2,3}$-edge, while the photon energy of the probe one ($\omega_{FEL,2}$) was set to 33.32 eV, still within the Ti $M_{2,3}$-edge slope. The collinearity of the two FEL pulses hitting the target was checked by means of *in situ* and *ex situ* methods, that also allowed to estimate the focal spot area (~1000 μm²) provided by a Kirkpatrick-Baez active optical system.[47]‡ The total FEL intensity was measured shot-by-shot by two I_0 monitors placed after and before a gas attenuator cell.[41] The latter was used in combination with solid state filters to modify the intensity of the FEL pulses. The residual UV radiation from the seed laser was blocked by an Al filter. The area of each spectral line (see Fig. 1c) can also be used as a measure of the FEL pulse intensity, both in single or double pulse mode. This parameter showed a good correlation with both the I_0 monitor and a thermopile sensor, placed in the experimental end-station before the experiment for calibration purposes. The maximum FEL intensity at the sample was ~10 μJ, corresponding to a maximum F-value of ~1 J cm^{-2}; the damage threshold of the sample was found to be ~0.1 J cm^{-2}. The estimated time duration of the FEL pulses was 75 fs.[48,49]

The samples were gratings of 400 nm pitch (L) made out of parallel rectangular Ti bars (50 μm long, 165 nm wide and 70 nm thick) deposited on a 20 nm thick Si$_3$N$_4$ window; the total area of each sample was 50×50 μm². Since at the highest F-values the sample is destroyed by a single FEL shot, a matrix consisting in 169 (nominally identical) samples is used. The FEL pulses impinge onto the sample at normal incidence (see Fig. 2a). The intensity distribution of the diffracted intensity is reported in Fig. 2b, as calculated from the Huygens–Fresnel principle in the far field approximation, which is appropriate in the present case. Such a lineshape is given by $I(\theta)/I_0 = R(\theta)F(\theta)$, where I_0 is the intensity of incident radiation while $R(\theta)$ and $F(\theta, \delta, \beta)$ are, respectively, the N-slits interference pattern and a modulating function that depends on δ, β and sample geometry (dashed line in Fig. 2b). $R(\theta)$ is featured by a set of peaks located in correspondence of given directions, defined by the angles $\theta_{N,i} = \sin^{-1}(2\pi Nc/\omega_{FEL,i}L)$ with respect the incoming beam direction in the plane orthogonal to the grating lines; here N is an integer number (diffraction order) while c is the speed of light. The grating sample thus works as a spectral analyser, since it is able to separate in angle the signal from the $\omega_{FEL,1}$ (pump) and $\omega_{FEL,2}$ (probe) radiation. However, since $L > 2\pi c/\omega_{FEL,i}$ while $\omega_{FEL,1} \approx \omega_{FEL,2}$, the angular separation between the $\omega_{FEL,1}$ and $\omega_{FEL,2}$ radiation diffracted in the same N^{th} order ($\delta\theta_N = \theta_{N,1} - \theta_{N,2}$) is much smaller than the angular separation between adjacent orders ($\theta_{N,i} - \theta_{N+1,i}$). $\delta\theta_N$ actually compares to the observed intrinsic angular spread of the diffraction peaks, which is mainly due to the finite number of illuminated lines of the grating and to the divergence (~0.2°) of the incoming (focused) beams. The lineshape of the Nth

---

‡ Such a quite large value was found during this particular experiment; focal spots down of ~100 μm² are routinely achieved.

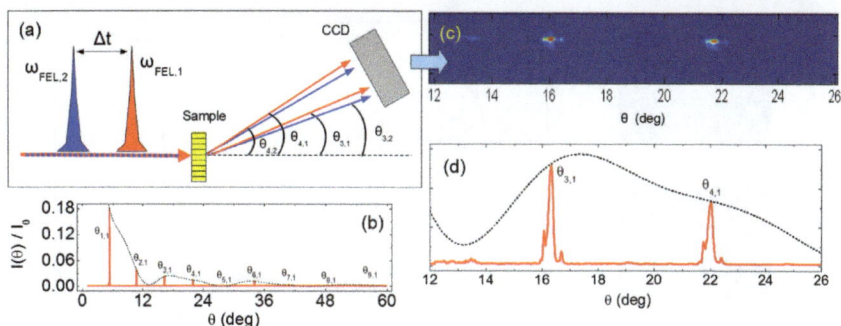

Fig. 2 (a) Sketch of the experimental setup: $\theta_{3,1}$, $\theta_{3,2}$, $\theta_{4,1}$ and $\theta_{4,2}$ are the diffraction angles (enlarged for clarity) collected by the CCD detector, $\Delta t = 0.5$ ps is the pump–probe delay while $\omega_{FEL,1} = 33.16$ eV and $\omega_{FEL,2} = 33.32$ eV are the FEL photon frequencies. (b) Full red and dashed black lines are the expected lineshape of $I(\theta)/I_0$ at $\omega_{FEL,1}$ and the modulating function $F(\theta, \delta, \beta)$, respectively. The angular positions ($\theta_{N,1}$) of the first nine diffraction spots ($N = 1$–9) are indicated. (c) Diffraction pattern from the Ti grating illuminated by a single FEL pulse at $\omega_{FEL,1}$, the bright spots located at $\theta \approx 16.4°$ and $\theta \approx 22°$ are the third and fourth order diffraction spots, respectively. (d) Full red and dashed black lines are the projected lineshape along θ of the image shown in panel (c) and the $F(\theta, \delta, \beta)$ profile scaled by an arbitrary factor, respectively.

diffraction spot from the two-colour beam thus consists in a broadening of the peak with respect to the single colour diffraction spot, rather than in two well defined peaks. This situation makes more difficult to separately determine the amount of pump and probe signals, thus complicating the data analysis and interpretation. Such an undesired angular overlap between the pump and probe diffraction spots is mitigated at large N-values, since $\delta\theta_N$ increases on increasing N. For this reason we placed the CCD detector (Princeton Instrument MTE-2048B) to monitor the highest diffraction orders (*i.e.*, $N = 3$ and $N = 4$; located in the 12–26° θ-range), compatibly with the geometrical constraints of the experimental setup. An example of the measured diffraction pattern is shown in Fig. 2c, while in Fig. 2d we report the corresponding projected lineshape along the θ-axis. In this case $\delta\theta_N \sim 0.2°$ while the angular separation between the third and fourth diffraction orders is $\sim 5°$.

In the low-F regime ($\ll 0.1$ mJ cm^{-2}), where the samples are not damaged by the FEL irradiation, three set of measurements were carried out: (i) only with the pump pulse, (ii) only with the probe one and (iii) with both pulses. These measurements consist in the acquisition of at least 100 FEL shots. After that, we performed a single-shot measurement with both pulses in the high-F regime (>0.1 mJ cm^{-2}). Fig. 3a–c report the images of the fourth diffraction order for cases (i), (ii) and (iii), respectively, while Fig. 3d is the corresponding high-F pump–probe single-shot measurement. By inspecting these raw data one can appreciate how the diffraction spot of the low-F pump–probe measurement (Fig. 3c) is reasonably similar to the sum of the low-F signal from pump (Fig. 3a) and probe pulses (Fig. 3b), while the profile of the high-F pump–probe measurement (Fig. 3d) resembles that of the pump pulse alone (Fig. 3a). This qualitative observation can be quantified by extracting the projected lineshape along θ, which is shown in the right hand side of the respective images (Fig. 3e–h). Fig. 3g

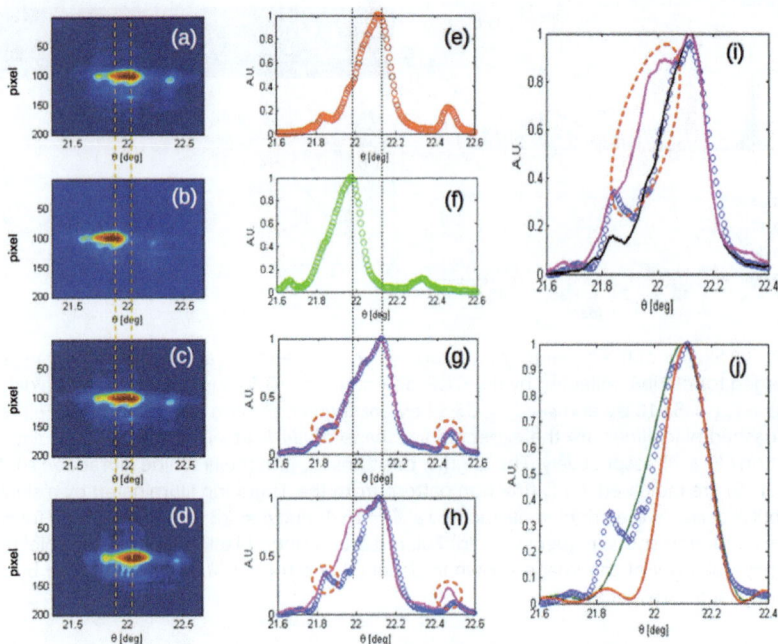

Fig. 3 Panels (a) and (b) are images of the fourth diffraction spot of the pump and probe pulses in the low-F regime ($F < 0.01$ J cm^{-2}), respectively. Panels (c) and (d) are pump-probe profiles in the low-F ($F < 0.01$ J cm^{-2}) and high-F ($F = 0.85$ J cm^{-2}) regime, respectively. Data are scaled to the peak intensity; the vertical dashed lines mark the barycenter of pump and probe peaks. Dots in panels (e)–(h) are the projected lineshape along θ of the images in panels (a)–(d); pump, probe and pump-probe data are reported as red, green and blue dots, respectively, while magenta lines in panels (g) and (h) are the expected pump-probe lineshape; see text for further details. Dashed circles and vertical dashed lines in panels (g) and (h) indicate the satellite peaks and the barycenter of pump and probe peaks, respectively. Panel (i) compares the pump-probe lineshape at high-F (blue dots; $F = 0.85$ J cm^{-2}) with the expected one (magenta line); the dashed ellipse highlights the decrease in the diffraction intensity in the probe's side. Black line is the result of the best fit procedure used to extract $C_{\text{pump}}^{\text{HF}}$ and $C_{\text{probe}}^{\text{HF}}$; see text for further detail. Panel (j) compares the high-F pump-probe lineshape (blue dots) with the "starting" (green line) and "target" (red line) profiles used in the fitting procedure to determine δ and β at the probe's photon energy; see text for further details.

shows how the diffraction lineshape of the low-F pump-probe measurements (blue circles) can be accounted for by a weighed sum of pump and probe profiles (solid red line), in which the weighting coefficients (C_{pump} and C_{probe}) are given by the intensity of pump and probe pulses, determined shot-by-shot by the 'on-line' spectrometer. This is not the case for the pump-probe lineshape at high-F (blue circles in Fig. 3h–i). Here a decrease in the scattering intensity in the probe's side with respect to the expected weighted sum (magenta lines in Fig. 3h–i) can be appreciated.

Such a decrease in the probe scattering at high-F is our pump-probe signal, which can be ascribed to an F-induced change in the β and δ coefficients of Ti at the photon frequency of the probe. In order to confirm this hypothesis and quantify the eventual F-dependencies of δ and β we carried out a data analysis based on Huygens–Fresnel calculations. Specifically, since the sample parameters and the values of δ and β at $\omega_{FEL,1}$ and $\omega_{FEL,2}$ are known (δ and β data are from ref. 50), we used the Huygens–Fresnel principle to calculate the expected diffraction lineshapes at $\omega_{FEL,1}$ and $\omega_{FEL,2}$. These profiles are then weighted by the measured values of C_{pump} and C_{probe} to obtain the expected diffraction pattern if no changes are induced by the pump. Such a lineshape is used as "starting profile" for a best fit procedure based on a standard χ^2 minimization in which the values of δ and β for the probe pulse are left free to vary in the Huygens–Fresnel calculations in order to match a "target profile" (see Fig. 3j). The latter is also calculated from the Huygens–Fresnel principle using as weighting coefficients (C_{pump}^{HF} and C_{probe}^{HF}) those obtained from a best fit procedure of the high-F pump-probe lineshape to a weighted sum of the low-F profiles of pump and probe signals (see Fig. 3i). The analysis also assumes that the FEL does not induce changes in the Si_3N_4 substrate, which is essentially transparent at the employed FEL photon frequencies. The need to use such "artificial" starting and target profiles instead of the experimental data is due to imperfections in the sample morphology, that randomly vary from sample to sample. The only feature systematically observed is a pair satellite peaks, of unknown origin, whose relative intensity seems to systematically increase with F (see Fig. 3g–h). These peaks were not taken into account in the data analysis.

The best fit results for δ and β are reported in Fig. 4a as a function of F. The quite large scattering of the data is mainly due to the aforementioned variability in the sample morphology and can be likely reduced by increasing the statistics (*i.e.*, the number of samples probed in single-shot mode), as we are planning to do. However, in spite of the poor accuracy in the determination of δ and β, at low-F both quantities are consistent with the expected values while, on increasing F, their absolute values show a marked decrease.

On general grounds, the latter behavior is quite striking since in the XUV range materials do not behave as a dispersiveless ($\delta \rightarrow 0$) and absorptionless ($\beta \rightarrow 0$) medium when probed close to atomic resonances, as in the present experiment. This result can be cast in the frame depicted in ref. 34, where saturable absorption of the Al L-transition reflects in an abrupt increase in the XUV transmission at fixed photon frequency with increasing F (FEL-induced transparency *i.e.* $\beta \rightarrow 0$ at high-F). Saturable absorption of the M-transition in Ti induced by the pump is expected to induces a shift of the absorption edge towards higher energy, due to the depletion of inner-shell absorbers.[34–37] Therefore, the subsequent probe radiation is no longer in resonance with this core transition and its optical response is similar to the one of the unperturbed sample irradiated with lower energy (out of resonance) photons. This picture is consistent with the observed trend, as highlighted in Fig. 4a, where we show the ω-dependence of δ and β for an unperturbed Ti sample (from ref. 50) along with their F-dependence at fixed photon frequency ($\omega_{FEL,2} = 33.32$ eV), as determined by the best fit procedure. It is worth noting that the FEL-induced transparency in Al was observed at a much faster timescale (<15 fs), lying within the FEL pulse duration.[34] In fact, the results shown in Fig. 4 represent the first quantitative evidence of the persistence of a highly excited state featured by "XUV-transparency" at time scales ($\sim$0.5 ps) exceeding significantly the time duration of FEL excitation pulse ($\sim$0.05 ps) and, most importantly, the core–hole recombination time (<0.01 ps)[51] as well. Within 0.5 ps the sample most likely undergoes a tangible relaxation of the electronic and

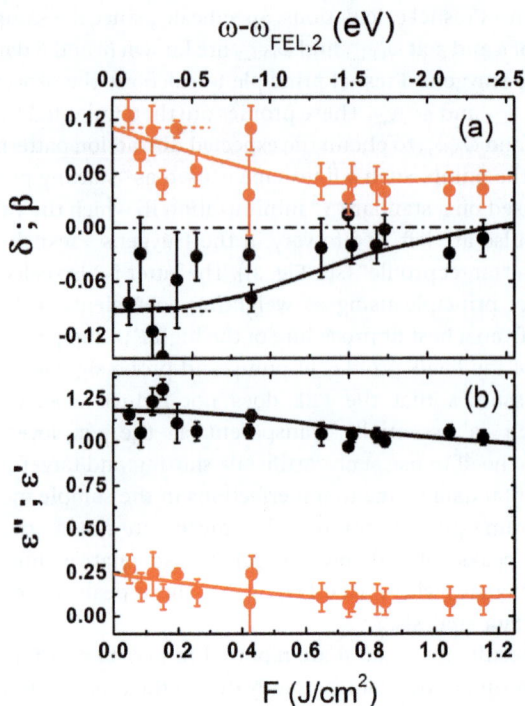

Fig. 4 (a) Black and red dots are the experimental values of δ and β, respectively, as obtained through the fitting process outlined in the text. Dashed black and red horizontal lines indicate, respectively, the expected δ and β values for an unexcited Ti sample;[50] dash-dotted line corresponds to $\delta = \beta = 0$. Full black and red lines are the ω dependence (upper horizontal scale) of δ and β, respectively, as obtained from ref. 42. (b) Corresponding F-dependence (bottom horizontal scale) of ε' (black dots) and ε'' (red dots); lines are the ω dependence of ε' and ε'' (upper horizontal scale).

ion sub-systems and may have reached (or be close to) an extreme thermodynamic equilibrium state, as WDM. Consequently, our results indicate that in these conditions the system still shows qualitative similarities with the "exotic" (non-equilibrium) state occurring at much faster timescales. In particular, it may be featured by a non negligible amount of ionized atoms, likely in dynamic equilibrium with a warm free electron gas. In order to understand such dynamics we plan to systematically study the Δt-dependence of such phenomenology, though in a time window presently limited to 0.3–1 ps, as well as to enlarge the F-range (1–2 orders of magnitude more in F can be exploited in terms of FEL pulse energy and smaller focal spot size).

We finally recall that the proposed approach has the advantage to allow for the determination of both δ and β, from which it is straightforward to compute the dielectric function: $\varepsilon = \varepsilon_0(\varepsilon' + i\varepsilon'') = \varepsilon_0(1 - \delta + i\beta)^2$, where ε_0 is the free space permittivity. Results are shown in Fig. 4b. These represent the first time-resolved determination of ε in highly FEL-excited matter in correspondence of a core atomic resonance. More generally, the possibility to determine the time and fluence dependence of ε would be of great relevance for several applications, not necessarily limited to the study of matter under extreme conditions, since ε

basically embodies all relevant information on the electronic properties of the sample.

The above described experiment has been carried out on a permanent ("static") grating. In the following section we describe how such a two-colour seeded FEL emission could be used to perform XUV/soft X-ray wave-mixing experiments, that are essentially based on "dynamic" gratings induced by the interference between coherent pulses.

4 Wave-mixing experiments with two-colour seeded FEL pulses

The experiment discussed in the previous section exploits the time and photon energy separation between the two FEL pulses and it undoubtedly points out the robustness of the two-color seeded FEL emission developed at FERMI. However, it does not directly use the full coherence provided by the seeding scheme, which would be exploited in the near future in XUV/soft X-ray wave-mixing experiments. The wave-mixing process results from the combined action of two or more input fields ($E_i(t)$) which may have different frequencies (ω_i) and wavevectors (k_i). The non-linear response of materials to these interacting fields can be accounted for by expressing the sample polarization in powers of $E_i(t)$, i.e.:[5]

$$P(t) = \varepsilon_0[\Sigma_i \chi^{(0)} E_i(t) + \Sigma_{i,j} \chi^{(2)} E_i(t) E_j(t) + \Sigma_{i,j,k} \chi^{(3)} E_i(t) E_j(t) E_k(t) + ...] = P^{L}(t) + P^{NL}(t) \qquad (1)$$

where $\chi^{(n)}$ are tensors of rank $n + 1$ termed n^{th}-order susceptibilities. The non linear polarization ($P^{NL}(t) = \varepsilon_0 [\Sigma_{i,j} \chi^{(2)} E_i(t) E_j(t) + \Sigma_{i,j,k} \chi^{(3)} E_i(t) E_j(t) E_k(t) + ...]$) acts as a driving force in the wave equation.[5] A nth-order non-linear process can then lead to the radiation of electric fields at frequencies $\omega_{(n+1)\text{wm}} = \omega_1 \pm \omega_2 \pm ... \pm \omega_n$, that are not necessarily present in the input fields. Such a non-linear mechanism is also referred to as $(n + 1)$-wave-mixing process. In this frame the coherence of the input fields plays a crucial role, as it defines the phase relationships of $P(t)$ in different locations and times within the sample. The $\omega_{(n+1)\text{wm}}$ radiation can thus add in amplitude rather than in intensity along a well defined, "phase-matched", direction $k_{(n+1)\text{wm}} = k_1 \pm k_2 \pm ... \pm k_n$.[52,53] This leads to an N^2 increase in the intensity of the wave-mixing signal (where N is the number of elementary emitters in the sample falling within the coherence volume of the interacting fields), rather than the N scaling typical of any linear process obeying the superposition principle. Coherence may then result in an enormous increase in the non-linear signal from condensed matter, which can even turns into a well defined coherent beam propagating after the sample. It is worth stressing how such a phase matching argument does not hold for isolated atoms/molecules, in this case the non-linear signal will be emitted in all directions. Furthermore, in extended samples with inversion symmetry all the elements of the $\chi^{(2)}$ tensor vanish. In many practical cases the lowest order non-linear term is hence the $\chi^{(3)}$ one, which is associated to the four-wave-mixing (FWM) process. This is probably the non-linear mechanism more thoroughly exploited in non-linear optical spectroscopy in condensed matter. A particular FWM method, also referred to as coherent Raman scattering (CRS), is reported in Fig. 5a. In this case two coherent beams ((ω_1, k_1) and (ω_2, k_2)) are crossed into the sample in time coincident conditions. The interference between these two pulses originates a (transient) electromagnetic wave with spatial periodicity $L = 2\pi/|k_1 - k_2|$ and 'beatings' at $\omega_1 + \omega_2$ and $\omega_1 - \omega_2$. The latter can be used, *e.g.*, to stimulate a given excitation having characteristic energy $\omega_{ex} = \omega_1 - \omega_2$, as illustrated in Fig. 5b. If a third coherent beam (ω_3, k_3) is sent into the sample in phase matching conditions (see Fig. 5c), then the time evolution of the selected excitation can be tracked back by looking at the intensity of the FWM signal at frequency $\omega_s = \omega_1 - \omega_2 + \omega_3$, that propagates along the 'phase matched' direction $k_s = k_1 - k_2 + k_3$.

In optical CRS $\omega_{1,2,3}$ are limited to below ~3–4 eV, so that $\omega_1 - \omega_2$ values are typically in the sub-eV range. This allows studying ultrafast dynamics of molecular vibrations and low-energy electronic excitations. In light of the larger field frequency of XUV/X-ray radiation, $\omega_1 - \omega_2$ values as large as several eV's can be achieved. This would allow studying the ultrafast dynamics of high-energy ('optical') excitations, such as valence band excitons. In the XUV/X-ray range atomic selectivity of the FWM process can be also attained, since ω_1 and ω_3 can be tuned to core transitions of selected atoms in condensed samples. In this case the virtual states depicted in Fig. 5b are substituted by real core–hole resonances. The localization of core shells ensures that the stimulated excitation is initially centered on the atom resonant at ω_1 (atom A), while the occurrence of a FWM signal at Δt means that the selected excitation propagated towards the atom resonant at ω_3 (atom B). The unique capability of XUV/X-ray FWM to stimulate high-energy excitations, as valence-band excitons, selecting the atomic specie in

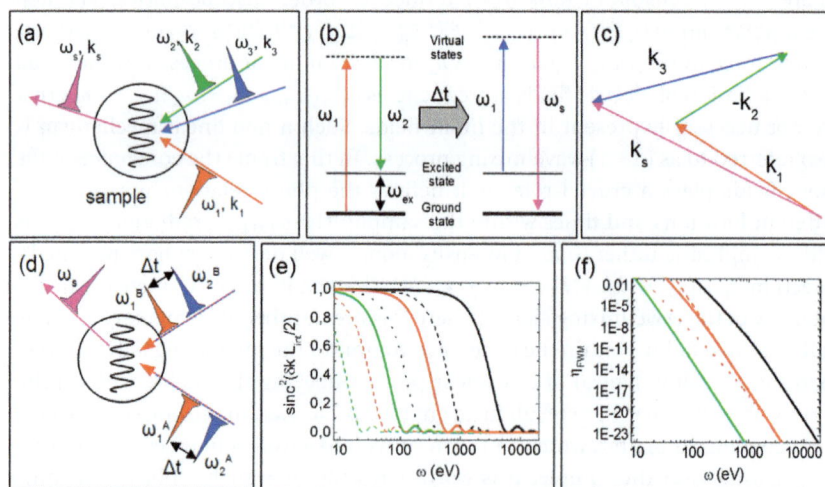

Fig. 5 Sketch of a FWM experiment: (b) input/output beams; (b) level scheme; (c) phase matching. (d) Sketch of a XUV CRS experiment that can be carried out at the EIS-TIMER beamline exploiting the two-colour operation of the FERMI FEL; note that the FWM signal (ω_s) propagates in a direction different from those of the input beams. (e) Phase matching factor for: $\delta k/k = 1 \times 10^{-4}$ and $L_{int} = 1$, 10 and 50 μm (full, dashed and dotted black lines), $\delta k/k = 1 \times 10^{-3}$ and $L_{int} = 1$ and 10 μm (full and dashed red lines) and $\delta k/k = 5 \times 10^{-3}$ and $L_{int} = 1$ and 10 μm (full and dashed green lines). (f) ω-dependence of η_{FWM} for resonant XUV/X-ray CRS (see text for further details) in the cases: $\gamma_0 = 0.01 \times \omega_0$, $\omega_{ex} = 0.1$, 1 and 10 eV (dashed, full and dotted red lines) and $\omega_{ex} = 1$ eV, $\gamma_0 = 0.001 \times \omega_0$, $0.01 \times \omega_0$ and $0.1 \times \omega_0$ (full black, red and green lines).

correspondence of which it is created as well as when and close to which atom it is eventually probed would allow, e.g., to study ultrafast charge transfers between different atomic species in molecular solids.[15–17,54,55]

The FWM process also provides wavevector selectivity through the k_s vector. For instance, in transient grating (TG) experiments $\omega_1 = \omega_2$ and, consequently, only excitations with energy falling within the bandwidth of the excitation pulses can be stimulated. This is the case, e.g., of acoustic phonons, whose frequency continuously varies from zero to ~10–100 meV following a k-dispersion relation. Among all phonon modes that are excitable within the bandwidth of the excitation pulses, only the one with $k_s = k_1 - k_2$ leads to a detectable (phase matched) FWM signal. The k_s-selectivity is of particular relevance in the XUV/X-ray range. Here the short wavelength of the radiation permits to probe k_s^{-1}-values that may compare with the characteristic lengthscales of inter- and intramolecular structures. The study of phonon modes in the so-called 'mesoscopic' k_s-range (0.1–1 nm^{-1}), nowadays inaccessible by any experimental method, is the main aim of the EIS-TIMER beamline.[56] The latter is a XUV/soft X-ray TG instrument that is going to be realized at FERMI in 2015 (a temporary setup with limited capabilities is ready to be hosted in the DiProI end-station for the first tests). Indeed, such a "mesoscopic" k_s-range is of the highest relevance for the study of collective atomic dynamics in disordered systems and nanostructures, since it matches the characteristic lengthscale of the topological disorder in the former class of samples and the "artificial" periodicity of the latter ones.

To date, the theoretical grounds for XUV/X-ray FWM methods have been already presented[14–17] and evidences of XUV wave-mixing are also available.[19,57,58] On such grounds we evaluated the possible impact of XUV/soft X-ray FWM, as well as the upgrades of the FERMI source and photon transport system to practically realize an 'ideal' XUV/soft X-ray FWM experiment, able to provide fs time resolution, polarization and atomic selectivity.[54,55]

In a short term perspective, the development of the two-colour seeded FEL emission opened up the way for additional wave-mixing applications (other than TG) at the EIS-TIMER end station. For instance, the photon transport system of EIS-TIMER will allow to split the FEL beam in two directions and then focus the two FEL beams at the sample with given crossing angles and variable time delays (up to some ps) among them.[56,59] This capability, in combination with the two-color operation, would permit, e.g., to realize the CRS experiment shown in Fig. 5d. In this case the first FEL pulse (ω_1^B) impinging into the sample is a "dummy pulse". Then two FEL pulses at ω_1 and ω_2 (labeled as ω_2^B and ω_1^A in Fig. 5d) are used to coherently stimulate excitations at $\omega_{ex} = \omega_1 - \omega_2$, that are finally probed by the time delayed ω_2^A pulse. Atomic selectivity in the excitation process can be achieved by tuning ω_1 to a core transition of a given atom within the sample. In order to estimate the amount of the FWM signal expected in this kind of experiments (on condensed matter), we may consider the following equation:[54]

$$\eta_{FWM} = |E_s|^2/|E_2^A|^2 = |\chi^{(3)}|^2|E_1^A|^2|E_2^B|^2 \exp^{-\alpha L} \sin c^2(\delta k L_{int}/2), \qquad (2)$$

where η_{FWM} is the efficiency of the FWM process, E_s and E_2^A are the signal and probe fields, respectively, while the $\exp^{-\alpha L}$ factor accounts for the signal loss due to absorption, being L and α the length of the sample crosses by the probe beam and the absorption coefficient at the probe's frequency, respectively. Finally, the $\sin c^2(\delta k L_{int}/2)$ term in eqn (2) is the phase matching factor, where L_{int} is the characteristic dimension of the overlapping volume of the input fields and δk is the wavevector mismatch. The latter defines the coherence length of the FWM process ($L_{FWM} = \pi/\delta k$). When $L_{FWM} > L_{int}$ the whole portion of the sample illuminated by the input beams gives rise to the FWM signal. Conversely, if $L_{FWM} < L_{int}$, then only an effective fraction of the interaction region coherently contributes to the FWM process. This situation may reflect in a substantial decrease of the signal; we recall that detectable XUV/X-ray FWM signals are expected from condensed samples owing to the coherent addition of scattering amplitudes occurring along the phase matching direction. In a non-collinear geometry δk is determined by both the pointing and wavelength stability of the source. The impact of the latter in a two-color wave-mixing experiment with $\omega_1 \sim \omega_2 = \omega$ and $\delta k_1 \sim \delta k_2$ is quantified in Fig. 5e, where we report the ω-dependence of the phase matching factor for some values of L_{int} and $\delta k/k$. In light of the excellent wavelength stability provided by seeding ($\delta k/k < 10^{-4}$),[22] a quite large value of L_{int} (of about 10 μm, compatible with the focal spot size routinely achieved at FERMI) can be attained. This is enough to keep the phase matching factor above an acceptable level throughout the entire ω-range (12–310 eV) exploitable by FERMI. On the other hand, the phase matching constraint becomes stricter on going deeper into the X-ray domain. Indeed, a given wavelength stability of the source (say $\delta k/k \sim 10^{-4}$) corresponds to much smaller L_{int}-values ($\sim$1 μm). Smaller focal spot sizes are hence necessary to keep the phase matching factor at the same level (see Fig. 5e). An estimate of the magnitude of $|\chi^{(3)}|$ can be gained by the following equation:[5]

$$\chi^{(3)} = (3ne^4\omega_0^2)/[8\varepsilon_0 m_e{}^3 r_a{}^2|\omega_1{}^2 - \omega_0{}^2 - 2i\gamma_0\omega_1|^2(\omega_2{}^2 - \omega_0{}^2 - 2i\gamma_0\omega_1)^2], \quad (3)$$

where n, r_a, e and m_e are the number density, the atomic radius, the electron charge and mass, respectively, while ω_0 and γ_0 are the characteristic energy and linewidth of the involved core transition. Fig. 5f reports the corresponding ω_1-dependence of η_{FWM} for different values of ω_{ex} and γ_0, as obtained from eqn (2) and (3) by assuming that one of the input fields is resonant with the core transition (i.e. $\omega_1 = \omega_0$), $\omega_1 \sim \omega_2 \gg \omega_{ex}$, $n = 3.5 \times 10^{28}$ m^{-3}, $r_a = 0.1$ nm, $\sin c^2(\delta k L_{int}/2) = 0.8$, $L = \alpha^{-1}$ and $|E_1^A|^2 = |E_1^B|^2 = 10^{20}$ V^2 m^{-2}. The choice of the latter value relies upon the consideration that the power series reported in eqn (1) does not necessarily converge for arbitrarily large values of $|E|$. In particular, when $|E|$ approaches the atomic field strength ($E_{at} \sim e/(4\pi\varepsilon_0 a_0^2) \sim 5 \times 10^{11}$ V m^{-1}; being a_0 is the Bohr radius) the non-linear response of materials is very different and eqn (1) loses its validity.[5] Again, in the X-ray regime the efficiency of the FWM process significantly drops down, even for equal conditions in terms of phase matching factor and input field strength. This points out how the XUV coherent FEL pulses provided by FERMI are likely the more suited ones to make the first step in bringing coherent wave-mixing spectroscopy beyond the optical range. However, we have to stress that these estimates were done for a particular experiment based on the FEL two-colour emission developed of FERMI and sketched in Fig. 5d; other FWM processes in which both pump and probe pulses are tuned to core resonances may have substantially larger $|\chi^{(3)}|$ values, even in the X-ray range.[14–17]

We finally recall that the ω_{ex} range exploitable in the experiment proposed in Fig. 5d is limited to below 1 eV (for ω's $\sim$100 eV) by the bandwidth of the FEL amplifier. This limits the range of detectable excitations to vibrational modes or low-energy electronic excitations, similarly to optical CRS, still with the option of atomic selectivity. However, the forthcoming development of novel multi-colour emission schemes can substantially enhance the potentiality of wave-mixing experiments that would be carried out at FERMI. In particular we mention an alternative approach (presently under testing) to generate a single FEL pulse that contains two photon frequencies, corresponding to the Nth and M^{th} harmonics of the seed laser. In this case values of $\Delta\omega = \omega_{ex} = (N - M)\omega_{UV}$ can be exploited, with the straightforward option to set ω_{ex} either by changing ω_{UV} and/or N and M.

5   Conclusions

We exploited the novel two-color seeded FEL emission scheme developed at the FERMI facility to carry out the first two-color FEL-pump/FEL-probe experiment. Data allowed us to determine the optical response of XUV photoexcited Ti close to the $M_{2/3}$-edge as a function of the excitation fluence on an ultrafast timescale. As a result, 0.5 ps after the action of the exciting (pump) pulse we observed the insurgence of an XUV transparency on increasing the excitation level of the sample, already reported in literature for much shorter (<15 fs) timescales. The persistence of such a 'FEL-induced transparency' at times much longer than the characteristic core–hole recombination time can be interpreted as an indication that the highly ionized state induced by the pump partially survives when the electronic and ionic subsystems are thermalized (or during their thermal-ization). Furthermore, we also observed a FEL-induced dispersionless behavior, which develops with fluence paralleling the absorptionless one. This trend can be cast in the framework of a FEL-induced shift of the absorption edge, since we found a quantitative (though poorly accurate) agreement between the fluence evolution of both absorption and dispersion coefficients and the dependencies of such quantities on the probing photon frequency in the unexcited sample. We also stress that the measuring scheme discussed here allows for the determination of the time-evolution of the dielectric function in FEL-excited samples.

The discussed experiment demonstrated the reliability of the two-colour operation mode developed at FERMI and, though it did not directly exploited the coherence properties of the seeded FEL source, it represents a milestone in our efforts towards bringing the most advanced optical methods into the realm of XUV/soft X-rays. We hereby discussed on the possibility to exploit the coherence of such a two-colour FEL source to perform non-linear wave-mixing experiments at the EIS-TIMER beamline, a used-dedicated instrument devoted to the development of XUV/soft X-rays wave-mixing experiments. These coherent methods are nowadays limited to the optical domain, also in light of the lack of bright, ultrafast (possibly multi-colour) sources of coherent XUV/X-ray radiation. Such a lacking can be definitely filled by seeded FELs, which may also provide the unique option of a fully controlled multi-pulse/multi-colour FEL emission.

Acknowledgements

The authors acknowledge support from the European Research Council through grant 202804-TIMER, the Italian Ministry of University and Research through grants FIRB-RBAP045JF2 and FIRB-RBAP06AWK3 and the Regional Government of Friuli Venezia Giulia through grant Nanotox 0060-2009. The authors gratefully thank all members of the FERMI Commissioning Team for their invaluable assistance in preparation of the experiment and during the measurements.

References

1 M. Cammarata, *et al.*, *Nat. Methods*, 2008, **5**, 881–886.

2 C. Bressler, *et al.*, *Science*, 2009, **323**, 489–492.

3 G. Vankó, *et al.*, *Angew. Chem., Int. Ed.*, 2010, **49**, 5910–5912.

4 P. A. Frenken, A. E. Hill, C. W. Peters and G. Weinreich, *Phys. Rev. Lett.*, 1971, 7, 118–119.

5 R. W. Boyd, *Nonlinear Optics*, Elsevier, Oxford, 2008.

6 P. D. Maker and R. W. Terhune, *Phys. Rev.*, 1965, **137**, A801–A818.

7 N. Bloembergen, Recent Progress in Four-Wave Mixing Spectroscopy, in *Laser Spectroscopy IV*, ed H. Walther and K. W. Rothe, Springer, Berlin, 1979.

8 L. Dhar, J. A. Rogers and K. A. Nelson, *Chem. Rev.*, 1994, **94**, 157–193.

9 E. T. J. Nibbering, D. A. Wiersma and K. Duppen, *Phys. Rev. Lett.*, 1991, **66**, 2464–2467.

10 S. Mukamel, *Principles of Nonlinear Optical Spectroscopy*, Oxford University, New York, 1995.

11 M. Terazima, K. Okamoto and N. Hirota, *J. Phys. Chem.*, 1993, **97**, 5188–5192.

12 G. Dadusc, *et al.*, *Proc. Natl. Acad. Sci. U. S. A.*, 2001, **98**, 6110–6115.

13 D. Avisar and D. J. Tannor, *Phys. Rev. Lett.*, 2011, **106**, 170405.

14 S. Tanaka, *et al.*, *Phys. Rev. A: At., Mol., Opt. Phys.*, 2001, **63**, 063405.

15 S. Tanaka and S. Mukamel, *J. Chem. Phys.*, 2002, **116**, 1877–1891.

16 S. Tanaka and S. Mukamel, *Phys. Rev. Lett.*, 2002, **89**, 043001.

17 O. Berman and S. Mukamel, *Phys. Rev. B: Condens. Matter Mater. Phys.*, 2004, **69**, 155104.

18 L. Gallmann, C. Cirelli and U. Keller, *Annu. Rev. Phys. Chem.*, 2012, **63**, 447–469.

19 T. E. Glover, *et al.*, *Nature*, 2012, **488**, 603–608.

20 A. A. Lutman, *et al.*, *Phys. Rev. Lett.*, 2013, **110**, 134801.

21 T. Hara, *et al.*, *Nat. Commun.*, 2013, **4**, 2919.

22 E. Allaria, *et al.*, *Nat. Photonics*, 2012, **6**, 699–704.

23 E. Allaria, *et al.*, *Nat. Photonics*, 2013, **7**, 913–918.

24 E. Allaria, *et al.*, *New J. Phys.*, 2012, **14**, 113009.

25 E. Allaria, *et al.*, *Nat. Commun.*, 2013, **4**, 2476.

26 G. Gregori, S. H. Glenzer, W. Rozmus, R. W. Lee and O. L. Landen, *Phys. Rev. E: Stat. Phys., Plasmas, Fluids, Relat. Interdisc. Top.*, 2003, **67**, 026412.

27 O. Gahl, *et al.*, *Nat. Photonics*, 2008, **2**, 165–169.

28 B. B. Pollock, *et al.*, *Rev. Sci. Instrum.*, 2012, **83**, 10E348.

29 E. Principi, *et al.*, *Phys. Rev. Lett.*, 2012, **109**, 025005.

30 F. Bencivenga, *et al.*, *Sci. Rep.*, 2014, **4**, 4952.

31 R. W. Lee, *et al.*, *J. Opt. Soc. Am. B*, 2003, **20**, 770–778.

32 R. J. Taylor, *The Stars: Their Structure and Evolution*, Cambridge University Press, Cambridge, 1994.

33 S. X. Hu, B. Militzer, N. N. Goncharov and S. Skupsky, *Phys. Rev. Lett.*, 2010, **104**, 235003.

34 B. Nagler, *et al.*, *Nat. Phys.*, 2009, **5**, 693–696.

35 G. Zimmerman and R. More, *J. Quant. Spectrosc. Radiat. Transfer*, 1980, **23**, 517.

36 S.-K. Son, H. N. Chapman and R. Santra, *Phys. Rev. Lett.*, 2011, **107**, 218102.

37 O. Ciricosta, *et al.*, *Phys. Rev. Lett.*, 2012, **109**, 065002.

38 J. Gaudin, *et al.*, *Phys. Rev. B: Condens. Matter Mater. Phys.*, 2013, **88**, 060101(R).

39 N. Medvedev, *et al.*, *New J. Phys.*, 2013, **15**, 015016.

40 H. P. Freund and P. G. O'Shea, *Phys. Rev. Lett.*, 2000, **84**, 2861–2864.

41 M. Zangrando, *et al.*, *Rev. Sci. Instrum.*, 2009, **80**, 113110.

42 E. Allaria, *et al.*, *New J. Phys.*, 2010, **12**, 075002.

43 M. B. Danailov, *et al.*, *Opt. Express*, 2014, **22**, 12869–12879.

44 G. De Ninno, *et al.*, *Phys. Rev. Lett.*, 2013, **110**, 064801.

45 E. Pedersoli, *et al.*, *Rev. Sci. Instrum.*, 2011, **82**, 04371.

46 F. Capotondi, *et al.*, *Rev. Sci. Instrum.*, 2013, **85**, 051301.

47 L. Raimondi, *et al.*, *Nucl. Instrum. Methods Phys. Res., Sect. A*, 2013, **710**, 131–138.

48 Fermi Commissioning Team, private communication.

49 D. Ratner, *et al.*, *Phys. Rev. Spec. Top.—Accel. Beams*, 2012, **15**, 030702.

50 Centre for X-Ray Optics, database freely available at,http://henke.lbl.gov/optical_constants/.

51 M. Ohno and G. A. van Riessen, *J. Electron Spectrosc. Relat. Phenom.*, 2003, **128**, 1–31.

52 J. A. Giordmaine, *Phys. Rev. Lett.*, 1962, **8**, 19–20.

53 N. Bloembergen, *J. Opt. Soc. Am.*, 1980, **70**, 1429–1436.

54 F. Bencivenga, *et al.*, *Proc. SPIE–Int. Soc. Opt. Eng.*, 2013, **8778**, 877807.

55 F. Bencivenga, *et al.*, *New J. Phys.*, 2013, **15**, 123023.

56 F. Bencivenga and C. Masciovecchio, *Nucl. Instrum. Methods Phys. Res., Sect. A*, 2009, **606**, 785–789.

57 L. Misoguti, *et al.*, *Phys. Rev. A: At., Mol., Opt. Phys.*, 2005, **72**, 063803.

58 K. Tamasaku, K. Sawada, E. Nishibori and T. Ishikawa, *Nat. Phys.*, 2011, **7**, 705–708.

59 F. Bencivenga, *Nuovo Cimento C*, accepted.

ROYAL SOCIETY
OF CHEMISTRY

DISCUSSIONS

Instrumentation and methods: general discussion

Jochen Küpper, Gwyn Williams, Henry Chapman, Katharine Reid, Jeppe Christensen, Hans Jakob Wörner, Elaine Seddon, Gopal Dixit, Eleanor Campbell, Jonathan Underwood, Michael Woerner, John Spence, Jasper van Thor, Martin Wolf, Oriol Vendrell, De Ninno Giovanni, Christian Bressler and Nora Berrah

DOI: 10.1039/C4FD90017K

Jochen Küpper opened the discussion of the paper by John Spence by commenting: Dear Speakers, following the discussions of relevant time-scales in your papers and especially based on the suggestion of John Spence, that biology is made up of "slow processes", I would like to hear your opinion on what are the relevant timescales to understand nature. What range of time-scales are relevant for the understanding of complex chemical and biological systems? What is the most important timescale (maybe in terms of SI prefix) to understand? And which is the most important timescale to investigate over the next years?

John Spence communicated in reply: For biological systems we can take the problem of protein folding as an example. Experiments must be undertaken on hydrated samples to be meaningful. Measurements of folding time have been simulated using atomic potentials and molecular dynamics (for shorter times and small proteins only, due to computer limitations). A large protein contains tens of thousands of atoms, almost entirely C, O, N, H, in the form of a series of any of the 20 small amino acid molecules (residues), each of known structure, in some sequence defined by DNA. Proteins at RT in solution differ from small molecules in that entropy is a large term in their free energy, and hydrogen bonding and van der Waals forces play important roles in folding, as does hydrophobic interactions (residues that hate water hide in the middle). A recent study found the folding time to be approximately N/100 microseconds for N amino acids. These times are long because of the time needed for the structure to explore stochastically the huge configuration space (Levinthal's paradox) in order to find a local total energy minimum. The study of shorter times (sub picosecond) would certainly be important for studies of the chemistry of individual amino acids, where entropy is not the dominant term in the free energy.

Oriol Vendrell responded: In chemical dynamics, the time-scale in which measurable changes can be detected by some probe scheme after a reactive event is triggered lies in the order of tens to hundreds of femtoseconds. This is the natural time-scale for bond vibrations and nuclear rearrangements, which is the consequence of the usual energy differences between vibrational levels in molecules. This comes as a natural thing after the great developments in the field of femtochemistry over more than two decades.[1] Most femtochemistry studies though are related to photochemical reactions, which can be triggered by ultrashort laser pulses and probed at well defined time-delays. However, the dynamics of ground-state chemical processes, meaning the time it takes for a individual reactive events to connect reactants to products *via* a transition state, is also of the order of tens to hundreds of femtoseconds.[2] Such thermally activated events, even if individually fast, are often very rare depending on the energy barrier height between reactants and products and cannot be triggered easily. This makes femtochemistry investigations of thermally activated reactions, the vast majority of chemistry, scarce with only a few exceptions.[3] A great challenge and opportunity for the next years is in my opinion the extension of usual femtosecond spectroscopy studies, which constitute an indirect probe of structural and electronic rearrangements, towards femtosecond structural determination in complex environments and for complex structures, where accurate theory predictions are often beyond reach. In this respect, we have been investigating opportunities for transferring large amounts of energy to liquid phases with short and intense THz pulses as a possible way to trigger thermal chemical reactions of dissolved molecular species.[4]

1. A. H. Zewail, *Science*, 1988, **242**, 1645.
2. B. J. Gertner, R. M. Whitnell, K. R. Wilson and J. T. Hynes, *JACS*, 1991, **113**, 74–87.
3. D.M. Newmark, *Acc. Chem. Res*, 1993, **26**, 33–40.
4. P. K. Mishra, O. Vendrell and R. Santra, *Angew. Chem. Int. Ed.*, 2013, **52**, 13685.

Jasper van Thor communicated in reply: The femtosecond time domain is key to studying activation processes in biological materials. The fundamental chemistry of bond rearrangement and dynamics occurs on ultrafast time scales, which subsequently trigger slower processes. From an experimental point of view both the ultrafast as well as the slow processes are of biological interest, and resolving the full cascading interconversion processes structurally is one important goal for XFEL science. It is of interest to note that XFELs have already shown particular utility also for resolving slow processes, beyond the nanosecond regime, taking advantage of the ability to conduct radiation damage free experiments as well as, in principle, detect single turnover mechanisms. For femtosecond time resolved studies, discussed in our paper, the key issue is the detection sensitivity in light of fundamental limits to femtosecond population transfer and intrinsic noise characteristics of the XFEL source. In order to develop the necessary signal-to-noise to detect the small structure factor amplitude differences technical details for a three-pulse probe–pump–probe scheme are considered and previously presented.[1] The paper by John Spence also references parts of this previous discussion.[1] In this scheme I proposed an internally referenced measurement of photoinduced femtosecond dynamics, which requires several geometrical and timing characteristics to be implemented. A pulse replica is generated from a monochromatic source with a small angle of incidence, in addition to focusing both beams to integrate over the rocking curve and the mosaic block which we have shown may lead to additional structure factor amplitude noise if the source otherwise has a small convergence (as discussed in our paper). By introducing also a time delay in between the pulse replicas, adding an optical pump will record both an un-pumped and a pumped diffraction pattern on the area detector in a single frame. A ratiometric measurement will thus provide the photoinduced differences in principle within the detector dynamic range. A Serial Femtosecond Crystallography application would need substantial attenuation to achieve non-destructive sampling of the first interaction, while attenuated defocused measurements of large crystals are shown to be non-destructive. Critical parameters of a split and delay unit include the stability of the intensity ratio and energies of the pulse replicas, also with pre-monochromation of either a self-seeded source or in SASE mode. A geometrical splitter may be insufficiently stable while also optical splitters based on thin crystals give rise to noise. The performance of the split and delay instrument will therefore likely dominate the sensitivity with which the photo induced structure factor amplitude differences are determined in a ratiometric manner with a probe–pump–probe scheme.

1. J.J. van Thor, (21 Feb 2014) 1st Ringberg Workshop on Structural Biology with FELs, "Considerations for ultrafast pump–probe X-ray crystallography: Non-linear cross sections dominate femtosecond time resolution and the rotation method for large crystals", Ringberg Castle, Germany.

Jeppe Christensen asked: In your paper you say that the time resolution of FELs are far better that needed for biological studies where processes happen on a micro- to millisecond time scale. At the same time you commented on the problems with shot-to-shot stability of the FELs. My question is, why go through the hassle of performing an FEL experiment, when synchrotrons work at the desired time scale and are much more stable. One could work at room temperature and just do one shot per crystal to avoid radiation damage.

John Spence replied: We use the XFEL in biology to outrun radiation damage, not to obtain high time resolution. Previously , samples were frozen to minimize damage at synchrotrons, which prevents us from studying dynamics. Radiation damage has always limited the quality of diffraction data from biomolecules, and in particular the resolution. Thus the XFEL opens the way to the study of dynamics at room temperature in a native environment, at atomic resolution, without damage.

Michael Woerner communicated: If you are only interested in the time-dependent relative positions of the nuclei I think the concept of "diffract before destroy" might work. However, X-ray diffraction gives information about the electron density map. Thus, the strong X-ray pulse might modify the electron density during diffraction. Do you consider such phenomena in your analysis ?

John Spence answered: Radiation damage is due to impact ionization by photoelectrons, which takes time to develop. It is found that with pulses shorter than about 70 fs , the atomic structures determined using an XFEL are the same as those obtained on a synchrotron, so the "diffract before destroy" method does indeed work, as shown in many papers. Elastic scattering commences instantaneously, and for short pulses, some of the atom images will be ionized, however this has little effect on a density map at a 3 Angstrom resolution, especially if the phases are obtained by molecular replacement from models in the protein data base. By avoiding the need to freeze samples on a synchrotron (to avoid damage), it therefore opens the way to the analysis of dynamics at room temperature. The comparison of XFEL and synchrotron structure determination for the same sample is given in *Science*, 2012, **337**, 362, the evolution of the damage in time can be understood by studying the intensity of Bragg beams as a function of pulse duration (see *Nature Photonics*, 2011, **6**, 35) and a new protein structure is determined by XFEL in Weierstall, *Nature Comms.*[1] It is important to understand that the effects of radiation damage depend on resolution – fine detail is destroyed first, and high order Bragg beams fade first with increasing dose.

1. Uwe Weierstall, Daniel James, Dingjie Wang, Wei Liu, John C.H. Spence, R. Bruce Doak, Garrett Nelson, Petra Fromme, Raimund Fromme, Ingo Grotjohann, Christopher Kupitz, Nadia A. Zatsepin, Shibom Basu, Daniel Wacker, Chong Wang, Sébastien Boutet, Marc Messerschmidt, Garth J. Williams, Jason E.Koglin, M. Marvin Seibert, Cornelius Gati, Robert L. Shoeman, Anton Barty, Henry N. Chapman, Richard A. Kirian Kenneth R. Beyerlein, Raymond C. Stevens , Dianfan Li, Syed T.A. Shah, Nicole Howe, Martin Caffrey, Vadim Cherezov, Lipidic cubic phase injector facilitates membrane protein serial femtosecond crystallography, *Nature Commun.*, 2014, 5, 3309.

Jonathan Underwood addressed John Spence and Jasper van Thor: There has been much discussion during this conference on the opportunities presented by X-ray FEL technology for structural (diffractive) imaging of static and dynamic molecular structures, and the results presented in this session show that this technique holds great promise. A complementary and proven technique for measuring structural dynamics is offered by electron diffraction. In comparison with X-ray diffraction, electron diffraction has several appealing features: (1) scattering cross sections for electrons are typically 4–6 orders of magnitude larger due to the Coulombic interaction with both the electrons and nuclei in the target; (2) the inelastic/elastic scattering cross section ratio for electrons is lower for electrons than for X-rays; and (3) the energy deposited into the target per inelastic collision is lower for electrons than X-ray photons. The net result of these factors is that 3 orders of magnitude less energy is deposited per useful scattering event for electrons than for X-rays, significantly reducing the problem of sample damage. Historically, when employing electrons in the 30–200 keV range, the temporal resolution in such experiments has typically been limited to *ca.* 0.5 ps by several factors: (1) the velocity mismatch between the laser and electron pulses as they traverse the sample; (2) the space-charge repulsion which acts to temporally broaden the electron bunch (and also may stochastically blur the observed image leading to reduced spatial resolution); and (3) the spread of initial electron velocities (corresponding to the energy spread of the electrons produced from the photocathode) which leads to broadening of the electron bunch as it travels to the sample. In addition, the space–charge repulsion also places an upper limit on the electron bunch charge requiring many electron bunches to be scattered in order to build up a diffraction pattern.

More recently, Hastings and co-workers (*Appl. Phys. Lett.*, 2006, **89**, 184109) demonstrated that electron diffraction is possible with relativistic electrons in the few MeV energy range. At this energy, the limitations described above are removed, and so this brings the possibility of electron diffraction with sub-100 fs time resolutions with high bunch charge, potentially giving single shot images. Do you think this approach holds promise for the sorts of problems currently being targeted by X-ray diffraction at FELs? Where do you see the role, if any, of ultrafast relativistic electron diffraction in the study of structural dynamics?

John Spence responded: Many points need to be made within the context of Henderson's *Quart Rev Biophysics*, 1995, **28**, 171, comparison of X-rays, neutrons and electrons as probes for structural biology, the considerable volume of literature on the use of MeV TEM in materials science and biology in the 1970's, and work published by the few groups now operating either fast electron diffraction cameras or fast electron microscopes (which solve the phase problem by direct imaging). In addition, it is important to specify if one is imaging single particles in ice, gas diffraction from small molecules (not viruses or large proteins), 2D crystals in ice, or solution scattering. Further important distinctions must be made between single-shot and stroboscopic methods, and between 2D projections and 3D images, which required data to be merged, perhaps from shots from identical objects in different orientations. Crystalline redundancy reduces sensitivity to damage by periodic averaging, so that if large enough crystals can be made, it is very difficult to compete with X-ray crystallography. Note that for biological significance, samples must be wet, frozen or otherwise hydrated.

1. We use the XFEL to outrun radiation damage, not to obtain high time resolution. Consider a pulse which is a delta-function in time. The elastic diffraction pattern would be recorded before the onset of the damaging photoelectron cascade, so that damage-free diffraction would result. Historically, radiation damage has always limited the resolution and data quality in most biological imaging methods at high resolution.
2. An important difference between X-rays and electron beams is that, after losing energy in the sample, an electron continues to the detector to create inelastic background (unless an Omega filter is used for diffraction; see Spence and Zuo, *Electron Microdiffraction*, Plenum Press, New York, 1992). No MeV Omega filter has been built, and these do not exclude phonon-scattering losses. X-rays are annihilated (in the creation of photoelectrons) during the most probable inelastic interactions, so that inelastic background is then not created.
3. While Henderson shows that the ratio of image-forming elastic scattering to damaging inelastic scattering (and the amount of energy dumped in the sample) are favorable to electron beams over X-rays, the XFEL is capable of outrunning damage altogether (see Barty *et al.*, *Nature Photonics*, 2012, **6**, 35), so these considerations do not apply. Using this "diffract-and-destroy" capability it becomes possible to study dynamics at room temperature and high resolution, without the need for freezing, as in cryo-EM, which prevents the study of dynamics (unless "quenching" methods are used). Although the elastic cross section for electrons is relatively much greater than that of X-rays, a 0.1 micron diameter XFEL hard X-ray beam contains about 1E12 photons in 50 fs, whereas an electron field emission gun produces about 20 electrons per picosecond. Larger photocathodes for electron beams degrade spatial coherence, and the chromatic stability of the multi-MeV electron beams used in accelerators is far worse (as seen in Hasting's paper) than those used in *e.g.* a 1MeV electron microscope, where an energy spread of 1 part in 1E6 or better is obtained using high stability voltage doublers.
4. Can femtosecond electron beams outrun radiation damage? Under the spatial and chromatic coherence conditions needed for single-particle imaging one has much less

than one beam electron in each 50 fs pulse from a field-emission electron source. Using the periodic averaging available in a 2D organic crystal in ice it may be possible, but would not have obvious advantages over existing 2D crystal cryo-EM methods, for which frozen samples normally do not allow study of dynamics. Zewail's group have shown how stroboscopic methods can be used to build up an electron diffraction pattern or image from the repeated excitation of a reversible process in a sample for which a sharp optical trigger exists. Space charge effects in the beam can in principle be eliminated by working with one electron per pulse and MHz repetition rates.

5. The effects of coherent multiple elastic electron scattering in thin samples have been extensively studied (see Spence, *High Resolution Electron Microscopy*, OUP, 4th edn, 2014). For protein nanocrystals we find (Subramanian and Spence, *Ultramic.*, 2014, submitted) a maximum tolerable thickness of about 70 nm thickness at 1 MeV. Since most of the information in a density map comes from phases, this limit may be increased by modeling the PDB to get phases (molecular replacement method). At high energies, where ionization damage decreases, damage due to ballistic "knock-on" processes increases. (This factor, plus cost, lead to the demise of HVEM microscopy in materials science in the 1980's.)

6. Cryo-electron microscope imaging of two-dimensional protein crystals in a thin (*e.g.* 50 nm thick) film of ice, combining Bragg diffraction (to measure structure factors) and TEM imaging (to solve the phase problem) has been highly successful, and offers the highest resolution of any cryo-EM method. These crystals are typically less than 10 nm thick, while 0.5 MeV electron beams are commonly used. These conditions avoid multiple scattering and allow the study of dynamics by quenching the crystal in different intermediate states. The samples are hydrated, as required. I believe it would be very difficult to compete with this approach using the beam from an accelerator, which does not allow imaging for phasing, and causes knock-on damage.

Henry Chapman asked: Regarding the issue of reducing crystal size to the point that electron penetration is not an issue, it is quite easy to make a 20 nm crystal (of one unit cell), it is called a single molecule. How does electron diffraction from single molecules in the gas phase compare with X-ray FEL single-molecule diffraction?

John Spence communicated in reply: Gas-phase electron diffraction has a long history, largely restricted to the small molecules which can easily be vaporized, unlike proteins. Using electrospray or similar methods, it is now possible to create a vapor of large hydrated molecules such as proteins in the vacuum conditions needed for electron diffraction. The possibility of undertaking serial electron diffraction from a stream of molecules was discussed in Spence and Doak, *Phys Rev Lett.*, 2004, **92**, 198102. Any water jacket will add significantly to the thickness of the molecule, which needs to be less than about 20 nm to avoid multiple scattering perturbations to the data. My comments elsewhere on background due to inelastic scattering also apply (electrons which loose energy in the sample continue on to the detector). Fast electron diffraction from gas-phase small organic molecules has been developed extensively in Zewail's laboratory at Caltec. A field emission electron source produces about 40 electrons per picosecond, but may be readily focussed down to nanometer dimensions. Since protein unfolding times are long, if a method could be found for launching proteins from liquids into vacuum without a thick water jacket, and provided an Omega type parallel-detection energy filtering device were used to reduce inelastic background, then electron diffraction at perhaps 400 keV of gas-phase proteins would be worthwhile.

Jonathan Underwood raised the question: Are the data you presented the results of calculations or experiments? How do you expect the results to scale to say 6–7 MeV electron energies?

John Spence answered: Our paper (Subramanian and Spence, *Ultramic*, submitted, 2014) gives multiple scattering electron diffraction simulations for protein nano crystals up to 1 MeV beam energy. Beyond that the strength of the interaction does not change significantly (see Spence, *High Resolution Electron Microscopy*, OUP, 4th edn, 2014, Figure 6.8 gives the dependence of phase-contrast images of an atom with beam energy, not diffracted Bragg beam intensities). The appropriate theoretical form for scattering at several MeV is the Moliere High Energy Approximation (see T.-Y. Wu and T. Ohmura, *Quantum theory of scattering*, Prentice Hall, 1st edn, 1962, p. 50). The design of electron microscopes for energies up to 3MeV has been described in detail in the literature, and several 1 MeV machines are currently operating. The difficulties in designing a diffraction camera to operate above 1 MeV are likely to be: 1. The design of the required area detector. Those currently used in 1MeV TEMs should be carefully studied, based on 45 degree mirrors. 2. The very small Bragg angles, and correspondingly powerful very high current lenses needed to magnify these patterns up to the pixel size of the detectors. (Quadrupoles might be used instead, to reduce current.) 3. The stability of the accelerating potential, which causes chromatic aberration. This effect must be less than the Bragg angle. 4. The effects of knock-on ballistic damage. 5. The construction of transfer stages for hydrated or cryo-cooled samples in a suitable goniometer, if crystals must be used to obtain a sufficient intensity of high angle scattering. 6. The effects of inelastic scattering, causing background.

Certainly before embarking on a large construction project for a multi-MeV diffraction camera for biology, the samples of interest should be studied in an existing 1 MeV machine, fitted with cryo-EM sample handling facilities. This exists in Japan. In the USA, a time-resolved high energy machine is being considered for materials science. My answer to the question on whether electron beams can outrun radiation damage is also relevant.

Jochen Küpper asked: Dear John, thank you very much for the detailed explanation of the issues with high-energy-electron diffraction. Now, with your calculations, what are the prospects of the investigation of ultrafast chemical dynamics in relatively small gas-phase molecules, let us say isolated molecules with a size up to 10 nm, using coherent electron diffractive imaging with few-femtosecond few-MeV electron beams from accelerator-based electron sources? Will it be possible to obtain images of intact molecules, maybe in a diffract-only-from-intact-molecules approach as described for small-molecule X-ray diffraction (Küpper *et al.*, *Phys. Rev. Lett.*, 2014, **112**, 083002)?

John Spence replied: Much of my answer to Henry Chapman's question applies, but the need for hydration is removed for small molecules. The Zewail group has pioneered this type of fast gas-phase electron diffraction at lower beam energies. In principle, the use of MeV beams eliminates the "space–charge" problem, since for relativistic reasons the electric and magnetic components of the Lorentz force between charges in the beam cancel at infinite energy. In practice, one has the challenges of high voltage engineering to obtain a small energy spread (sufficient chromatic coherence), and the problems of building an efficient area detector which is not damaged by the beam (one design uses a phosphor on a pellicle screen, viewed by a CCD at 90 degrees through a 45 degree mirror with a hole in it for the central beam. The screen is replaced as it damages, and one must consider the X-ray background from the beam striking anything, getting into the detector). In addition the de Broglie wavelength of the electron is so small that scattering angles become extremely small, requiring strong magnetic lenses to magnify the diffraction pattern, and very high collimation (much less than a Bragg angle for any crystalline sample), which reduces fluence. Finally, one has the problem of radiation damage to the sample, in the form of knock-on ballistic damage, which probably cannot be out-run by an electron beam. If the plan is for the beam to span many randomly oriented molecules per pulse, then data analysis becomes a headache, although in principle the method of angular correlations due to Z. Kam can disentangle the orientational disorder (see Kirian, *J. Phys. B: At. Mol. Opt. Phys.*, **45**, 223001 (2012)). This requires a coherence width shorter than the distance between molecules. If the plan is for a coherent beam whose size is about equal to that of one molecule ("isolated molecules"?), using a pulsed photofield source and lenses to demagnify the beam to nanometer dimensions, then there will not be much signal, or even one scattered electron per picosecond, despite the high elastic cross section (see question 609).

Michael Woerner commented: Electron diffraction scatters elastically off the Coulomb potentials of the nuclei whereas in X-ray diffraction photons scatter elastically off the electron density. A combination of both experiments might give insight into non-Born Oppenheimer effects. How far are we still away from studying such phenomena in combined time-resolved experiments?

John Spence answered: Time-resolved Bragg diffraction from protein nano crystals involved in photosynthesis using the pump–probe method is described in Aquila *et al.*, *Optics Express*, **20**, 2706 (2012) and Kuptiz *et al.*, *Nature*, 2014, doi: 10.1038/nature13453, just out. These papers use the diffract-and-destroy method, and look for changes in X-ray structure factors due to illumination by visible light to measure atomic motion on the microsecond time scale. Motion is slow in biological systems because the dominant contribution to free energy is configurational entropy, not electron transfer. The differences between electron scattering (from the coulomb potential) and X-ray scattering (from the electron density) can be used to provide very sensitive images of chemical bonding between atoms in crystals by electron diffraction, not possible using X-ray diffraction, as shown in Zuo *et al.*, *Nature*, 1999, **401**, 49, for the ground state of copper oxide. Failures of the B-O approximation would require very high time resolution, which is not possible using electron diffraction due to source brightness and emittance limitations (the degeneracy of field emission sources is about 1E-6).

Gopal Dixit asked: Dear John, I am concerned about your idea to use attosecond X-ray pulses for time-resolved X-ray scattering (TRXS). The Fourier limited attosecond X-ray pulse has unavoidable finite energy bandwidth due to energy-time uncertainty relation. As you make your pulse shorter and shorter, the bandwidths will become larger and larger. Now, if you consider TRXS from a single molecule (not crystal), your time-resolved scattering signal will contain a significant amount of incoherent X-ray scattering and the energy resolution could not be better than the finite bandwidth of the pulse. In other words, it is impossible to disentangle the coherent and incoherent X-ray scattering contributions to the total signal within the finite bandwidth of the attosecond X-ray pulse. This scenario makes the analysis of the signal more complicated in the vicinity of avoided crossing and conical intersection for probing an ultrafast chemical reaction.[1-3]

1 G. Dixit, O. vendrell and R. Santra, *PNAS*, 2012, **109**, 11636–11640.
2 G. Dixit and R. Santra, *J. Chem. Phys.*, 2013, **138**, 134311.
3 G. Dixit, J. M. Slowik and R. Santra, *Phys. Rev. A*, 2014, **89**, 043409.

John Spence responded: I agree with your analysis for "single particles", as the cryo-EM community calls your gas-phase molecules, so it would not work (in biology, these would have to be hydrated, which introduces considerable experimental complications). My analysis was for a crystalline sample, where the sample then acts as its own monochromator, picking out only those wavelengths which satisfy Bragg's condition for diffraction into the same direction. Then the coherent amplification of intensity due to Bragg scattering gives enough scattering to obtain an atomic resolution image (unlike that from a single particle), while the interference between these two wavelengths (frequencies) provides information to solve the phase problem. If the bandwidth is sufficient to span two different such Bragg conditions, then the pulse duration must be shorter than the period of beating between them.

Jochen Küpper addressed Jasper van Thor and John Spence: Dear Jasper, John Spence told us again that biological processes are slow. But what is your opinion on the need to look at ultrafast (femtosecond) timescales even for large biological molecules? Don't we need to understand the short-time dynamics to understand the (slow) function?

Jasper van Thor communicated in reply: The quantum yield of biological reactions is determined in the femtosecond time domain. Therefore, ultrafast coherent processes may determine the outcome of much slower processes that occur thermally. The example of excited state dynamics in the Photoactive Yellow Protein includes a photoisomerisation that has a time constant of about 400 femtoseconds. This is within or comparable with the vibrational dephasing time in biomolecules. We consider the possibility of direct detection of a coherent wave packet motion by high resolution X-ray crystallography. With the approximately 200 femtosecond pump–probe jitter we would expect to observe such motion only very imprecisely, however with few-femtosecond time stamping techniques and enough data redundancy there are no physical limitations to recording such a coherent wave packet motion. This also presents compelling motivation for theory development, particularly for the application of biomolecular processes.

John Spence responded: Please see my answer to Jochen Küpper's question about time-scales in biology.

Jochen Küpper asked: Dear John, you mention in your paper that you would like to see 10 as pulses, but as far as I understand it, you only want the corresponding bandwidth. The latter might be easier, at least conceptually it is much easier to get the bandwidth than to also temporally compress the pulse. However, now thinking about the parameters: 1 µJ in 10 as focused to 100 nm creates a field of about 10^{20} W cm^{-2}. This looks like a pretty strong field which will instantaneously destroy the molecule (Lorenz et al., Phys. Rev. E, 2012, **86**, 051911, and Chapman et al., New Journal of Physics, 2012, **14(11)**, 115015). Will strong ultrashort pulses like these really be useful for diffractive imaging of chemical and biological systems and processes?

John Spence answered: I agree we only want the bandwidth, and this can be an incoherent superposition of energies for Laue diffraction, it does not need to be coherent (coherence makes possible the phasing method I also suggest). The existing theory and experiment for our "diffract-then-destroy" experiments suggest that, for the purposes of finding atomic positions, the instantaneous elastic scattering will terminate even for attosecond pulses before the damaging photoelectron cascade gets going. Note that it is not necessary to destroy the sample to avoid damage. If the attosecond pulses are weak, one can still out-run radiation damage.

Michael Woerner opened the discussion of the paper by Jasper van Thor by commenting: Typically, chemical reactions are performed in the liquid phase. We investigated an intra-crystalline acid–base reaction in ammonium sulphate: The Journal of Chemical Physics, 2010, **133**, 064509.

John Spence responded: Using snap-shot X-ray scattering from molecules in our micron-sized liquid jet, running across the pulsed XFEL beam in the diffract-then-destroy mode, it is possible to track chemical reactions by fast solution scattering (FSS) or fast WAXS. These reactions can be triggered using a mixing jet, or optical pumping. See Arnlund et al. (Nature Comms, doi: 10.1038/NMeth.3067) and Wang et al., J. Synchrotron Radiat., in press.

Jasper van Thor answered: Protein function requires the presence of water. The typical water content of protein crystals is about 40–60%, and biological function is very often conserved in crystalline form. Crystals of the Photoactive Yellow Protein undergo photoinduced reactions which strongly resemble those in liquid phase.[1–3]

1. C. N. Lincoln, A. E. Fitzpatrick and J. J. van Thor, *Phys Chem Chem Phys*, 2012, **14**, 15752–15764.
2. P. Ramachandran, J. Lovett, P. Carl, M. Cammarata, J. H. Lee, Y. O. Jung, H. Ihee, C. Timmel and J. J. van Thor, *J Am Chem Soc*, 2011, **133**(24), 9395–9404.
3. K. Ng, E. D. Getzoff and K. Moffat, *Biochemistry*, 1995, **34**(3), 879–90.

Jochen Küpper commented: As I understand your experiment, you are really performing a strong-field coherent control experiment to start the dynamics. Is that correct? If so, are you not really trying to find out how you get most of the population into the excited state, the starting point of the dynamics, and then follow the field induced dynamics, instead of the observation of weak-field single-photon-induced dynamics that occur in nature. Now, after all, I am still wondering how much can you learn about the actual biological process from the strong-field-initiated process despite these conceptional problems? More generally, I dare to ask the question of how we can follow photochemical dynamics after the absorption of a single photon from a weak source? Can we, and do we have to, look at a sample, in a ultrafast stroboscopic approach, of many molecules with individual ones reacting stochastically and be able to see the randomly-timed change of a single molecule?

Jasper van Thor replied: Dear Jochen, indeed, we are applying strong field coherent control. Femtosecond time resolved pump–probe protein X-ray crystallography requires both a very sensitive detection of the photoinduced structure factor differences as well as an optimal control of photochemical dynamics in crystals. In the case of femtosecond photoexcitation of crystals of the Photoactive Yellow Protein, I have shown a multilevel electronic scheme that illustrates the need for strong field coherent control. We are dealing with a heterogeneous singlet excited state with interconversions and relaxations on the femtosecond timescale, all having different branching ratios to generate the primary photoproduct which has a photoisomerised biological chromophore in 400 femtoseconds. Two photon processes with strong cross sections in the blue edge of the absorption spectrum result in photoionisation processes which are a loss channel, whereas pumping in the red edge readily results in pump-dump scenarios. We have previously shown[1] that modification of the pulse duration, peak power, center wavelength and importantly addition and control of second order dispersion is required to optimize the femtosecond population transfer to the photoproduct state. From systematic studies with power titrations for shaped pulses we were able to extract all the non-linear cross sections for several regimes.[1] An important aspect is the presence of a second order dispersion that chirps against the dynamic Stokes shift, which has a ~200 fs time constant, to minimise the stimulated emission. The resulting optimised optical conditions are those with suppressed non-linear cross sections. These high field conditions are therefore the best representative for one-photon processes of single molecules under weak illumination: under such conditions internal conversions appear in approximately ~70% of excitations whereas ~30% undergoes photoisomerisation. The strong-field coherent control conditions are designed to approach this as closely as possible, while achieving maximised and detectable levels of photo-intermediate to above the detection limit of the time resolved pump–probe X-ray crystallographic measurements. Pulse shaping with high peak power, manipulating the quantum interferences between multiple pathways such as considered in the 'Brumer-Shapiro' scheme has been well understood.[2] In addition, P6$_3$ crystals of PYP are monoaxial and we must also consider birefringent optical propagation in the medium, as well as the photolysed depth.

With regard to your final question, I believe that just as in isotropic solutions as seen by ultrafast spectroscopy, coherence could be observable also in the crystalline state, within the experimental bandwidth, resolution and dephasing time.

1. C. N. Lincoln, A. E. Fitzpatrick and J. J. van Thor, *Physical Chemistry Chemical Physics*, 2012, **14**, 15752–15764.
2. W. Wohlleben, T. Buckup, J. L. Herek and M. Motzkus, *Chemphyschem*, 2005, **6**, 850–7.

Michael Woerner communicated: By using strong THz sources for triggering electric field induced events in matter one can also explore chemical reaction in the electronic ground state. Is your future planning of experiments also along those lines?

Jasper van Thor answered: THz excitation of macromolecules would similarly allow access to ground state dynamics. A particular challenge may be the existence of a congested Density of States in the low frequency region, such that selective pumping, or explicit mode assignment, is not straightforward. Experimentally, considering the possibility to extend studies as you have demonstrated for small molecules to macromolecules is appealing and I hope that in the future instrumentation will be available to allow for the exploration of this regime.

Jochen Küpper asked: On page 2 of your manuscript, you mention various structures (I0, I1, I2', I2, *etc.*). Can you explain to us what these structures look like, what shape or structure they exhibit, according to current belief?

Jasper van Thor replied: Dear Jochen, the species which is called 'I0' is the primary photoproduct which is the target state for femtosecond time resolved crystallography. It is an electronic ground state species in which the p-coumaric acid chromophore of the protein has undergone photoisomerisation. The time constant of this photoisomerisation process is ~400 fs. The further intermediates that are thermally populated in picosecond to millisecond time scales in the 'photocycle' of the Photoactive Yellow Protein are characterised by reorganizations of the protein and solvent parts of the macromolecule. Eventually, the system recovers the dark ground state, making stroboscopic measurements possible. Resolving the structural features that belong to the pure species that interconvert requires methods of Singular Value Decomposition and Global Analysis from extensive series of time resolved measurements. For the purpose of our work we are primarily focused on the excited state dynamics which form the primary photoproduct 'I0'.

Jochen Küpper commented: Do I understand it correctly that the actual structure, that is, the three-dimensional arrangement of atoms and the surrounding electron densities, are not know for these species?

Jasper van Thor answered: Currently, the earliest structural information obtained using pump–probe Laue X-ray crystallography using synchrotron radiation is 100 ps (Jung *et al.*, *Nat. Chem.*, 2013, **5**(3), 212–220, Jung *et al.*, *Nat. Chem.*, 2014, **6**(4), 259–260, Schotte *et al.*, *Proc. Natl. Acad. Sci. U. S. A.*, 2012, **109**(47), 109–147, Kaila *et al.*, *Nat. Chem.*, 2014, **6**(4), 258–259). Whereas techniques are being developed that use synchrotron pulses to achieve increased time resolution, and accessing femtosecond and few-picosecond delays will be possible using XFEL sources, as we have discussed in our contribution (van Thor *et al.*, *Faraday Discuss.*, 2014, DOI: 10.1039/C4FD00011K).

Hans Jakob Wörner opened the discussion of the paper by Oriol Vendrell by asking: I have two questions related to the feasibility of the experiment that you propose. First, how does the electronic structure of neutral water clusters $(H_2O)_n$ differ from that of protonated water clusters $H^+(H_2O)_n$ that you calculate? Second, what optical densities would be required in this experiment and how do they compare to what can be achieved?

Oriol Vendrell responded: Related to the first question, ionized neutral water clusters have been studied at the ADC(3) level of theory up to n=4 water molecules.[1] For n=4, the outer-valence ionization potential spans from about 12.1 to 19.5 eV whereas the inner valence starts at about 31 eV and the double ionization threshold starts at about 28.2 eV. In the large protonated cluster considered by us, $n = 21$, the outer valence ionization potential spans the range 12 to 25 eV, 12 eV corresponding to ionization from far away from the extra proton and 25 eV to ionization from its vicinity. Therefore, we think that in the bulk or in the clusters one should be able to ionize from the vicinity of the extra proton and observe the correlated proton-hole dynamics discussed in our contribution without reaching the double ionization threshold at about 28 eV that would result in autoionization dynamics.

Related to the second question, an experiment based on absorbance measurements would be very hard or even impossible because of the original positive charge of the clusters and the corresponding very low densities. Photoionization experiments conducted at FLASH on $H^+(H_2O)_2$ used detection of charged fragments at a mass spectrometer, which is then extremely sensitive to individual events.[2] In a similar way, an experiment related to our calculations may be realizable if, after the X-ray probe step, secondary processes such as total Auger yield or total fluorescence from the core-hole relaxation are measured. In this mode of operation, the bandwidth of the X-ray probe should remain at most in the 1 to 2 eV range, since the hole relaxation dynamics occur in the 10 eV energy scale. For Fourier limited pulses this corresponds to pulse lengths beyond 1 fs and the relaxation time in the large cluster is of the order 100 fs. Therefore, an *e.g.* 10 fs (limited) X-ray probe would be sufficient to observe the hole relaxation.

1. I. B. Müller and L. S. Cederbaum, *JCP*, 2006, **125**, 204305.
2. L. Lammisch, *et al.*, *Phys. Rev. Lett.*, 2010, **105**, 253003.

Katharine Reid asked: From comments I have heard at this meeting it seems that in an ideal world we would like to have a light source with a tunable pulse duration, tunable bandwidth, tunable intensity and tunable repetition rate. As it seems unlikely that such a source can ever be realized in practice, would it make sense for the community to develop an international strategy for light source development whereby sources with complementary specifications are available in different locations?

Martin Wolf responded: This would be a wonderful source. For may problems one would like to operate with pulses with a optimum time-bandwidth product preserving energy resolution and having appropriate time resolution at the same time. Having ultimate time resolution is not always beneficial as, for example, the bandwidth of attosecond pulse in the range of several eV to even 10–12 eV often hinders certain spectroscopic applications.

Christian Bressler communicated in reply: European XFEL seeks to deliver just that! (1) Variable pulse filling patterns could allow us to perform experiments from 10 Hz, over 10 kHz (10 pulses only) to 4.5 MHz (up to 2700 pulses, all at a 10 Hz burst repetition rate), all according to user demand. (2) Variable pulse duration: LCLS, but also European XFEL (and SACLA as well) can tailor the X-ray pulse width from 100 fs (or more) down to few fs (and possibly below). (3) Variable pulse intensity: every SASE FEL can attenuate the beam at will from the full single pulse intensity (which is largest at LCLS and European XFEL, somewhat lower at SACLA and soon at both SwissFEL and Pohang FEL). (4) Focasability of the beam on the sample at XFEL sources is also important.

Jochen Küpper addressed Oriol Vendrell and John Spence: Following earlier comments by Oriol Vendrell and others, I would like to comment on the Born–Oppenheimer approximation and its relevance for the topic of this session and the conference:

In simple words, the Born–Oppenheimer (BO) approximation assumes that the kinetic energy of the nuclei is negligible (Demtröder, *Experimentalphysik 3*, Springer, Berlin Heidelberg New York, 4 ed., section 9.1, vol. 3, pp. 1–668 (2010)).

Within that approximation, we can then separate the molecules' Hamiltonian(s) into the electronic and the nuclear (potential energy) part. Now, this approximation is a good one for large parts of a molecule's phase space, but it does break down, by definition, when we look at fast nuclear dynamics. Processes where nuclear dynamics are slow might be best investigated with high-resolution eigenstate spectroscopy (Herzberg, *Molecular Spectra and Molecular Structure: Spectra of Diatomic Molecules*, Krieger, 1989, vol. 1–3, Küpper *et al.*, *Physical Chemistry Chemical Physics*, 2010, **12(19)**, 4968–4979, and Küpper *et al.*, *Physical Chemistry Chemical Physics*, 2010, **12(19)**, 4980–4988).

What we are after here are dynamical processes, where the nuclei move fast, and where obviously the BO approximation does break down. Now, what we really want to understand are dynamical processes, such as the isomerization or folding of molecules, the breaking and forming of bonds, *etc.* The BO approximation might not be the right picture to look at these processes.

John Spence communicated in reply: See my answer to question 601. In structural biology, as for a rubber band or a polymer, it is the entropy which matters.

Hans Jakob Wörner answered: The key assumption of the Born–Oppenheimer approximation (BOA) is not that the kinetic energy of the nuclei is negligible, but rather that derivatives of the electronic wave function with respect to nuclear coordinates are negligible.[1] This is in general fulfilled when the electronic energy-level intervals are much larger than the vibrational ones. Highly excited vibrational levels of an isolated electronic state are well described within the BOA, whereas the vibrational ground state of a molecule is not when it lies energetically close to a conical intersection, as is frequent in Jahn–Teller-active systems, see *e.g.* **ref. 2**.

Switching to the time domain, a wave packet can always be expanded in eigenstates of the molecular Hamiltonian. Therefore, the velocity of nuclei in a wave packet, defined by the energy intervals, has no impact on the validity of the BOA. For example, ultrafast isomerization on an energetically isolated electronic ground-state surface is well described by the BOA, whereas arbitrarily slow wave packet dynamics across a conical intersection is not. Hence, I do not expect a significant difference in the applicability of the BOA between time- and frequency-domain spectroscopies.

1. M. Born and R. Oppenheimer, *Annalen der Physik*, 1927, **389**, 457–484.
2. H. J. Wörner and F. Merkt, *Angew. Chem. int. ed.*, 2009, **48**, 6404–6424, and references therein.

Oriol Vendrell commented: I would like to add a few comments on the topic of the Born–Oppenheimer (BO) approximation. Hans Jakob Wörner correctly points out that the BO or adiabatic approximation consists in neglecting the coupling terms between different electronic states. These are small when the potential energy gap is large but must be explicitly considered when electronic states are close in energy. It is nowadays very well established that, as soon as molecular systems leave the ground electronic state, conical intersections and avoided crossings between potential energy surfaces are ubiquitous and fully determine the dynamics of the system.[1]

In a previous comment, which triggered the remark of Jochen Küpper, I stated that the BO approximation is a very good one, which is true for molecules in their ground state but not so true anymore for electronically excited molecules. I should have been more precise, for what I meant was the group-BO approximation.[2] The group-BO approximation implies that for the group of states of interest all couplings are considered but no couplings are taken into account to states outside this group. This is the theoretical setup in which virtually all molecular dynamics and spectroscopy is performed. In it, the exact expansion of the molecular wave-function in terms of an infinite number of electronic states is truncated to a matrix Schrödinger equation of the size of the number of electronic states of interest.

The main point that I wanted to make though, is that the key idea of nuclei evolving in potential energy surfaces as the key concept to understand chemical dynamics does not need to be abandoned even if one is dealing with large numbers of coupled electronic states and strong non-adiabatic effects. This is the case *e.g.* in our contributed paper and it is also true for large bandwitdh atto-second pulses applied to molecular systems to trigger joint electronic and nuclear dynamics, as in attosecond charge migration studies.

1. D. G. Truhlar and C. A. Mead, *Phys. Rev. A*, 2003, **68**, 032501.
2. G. A. Worth and L. S. Cederbaum, *Annu. Rev. Phys. Chem.*, 2004, **55**, 127–58.

Jochen Küpper addressed all the attendees: Following up on the comment of Michael Wörner, I want to point out that THz radiation cannot only be used in a strong field regime to trigger reactions, but also as a resonant weak-field THz trigger. Following our original demonstration of conformer separation (Filsinger *et al.*, *Phys. Rev. Lett.*, 2008, **100**(13), 133003) there have been initial calculations by Ingo Barth and Jörn Manz (FU Berlin, private communication of unpublished results) on the conformer interconversion of 3-aminophenol. These calculations hinted at the possibility to resonantly climb up the ladder of internal-rotation states of the OH torsion, overcome the barrier, and possibly even resonantly climb it down on the other side.

This should be generalized to an approach where we resonantly excite mole-cules into a reactive vibrational state in the electronic ground state – and then follow the subsequent chemical dynamics by the wonderful "imaging" experiments discussed at this meeting. In a statistical limit, repeating the experiments for many different excited vibrational states would allow us to determine the chemical reactivities for large parts of the molecule phase space. Overall, such a weak-field approach seems to be a challenging approach, but the hope is that it provides direct "molecular movies" of actual chemical processes, including the initial fast nuclear and possibly even the corresponding electronic dynamics of chemical reactions in the electronic ground state – corresponding to "plain chemistry".

Henry Chapman opened the discussion of the paper by Nora Berrah by com-menting: Regarding the difference between the X-ray FEL pulse duration and the electron bunch duration (which is what is reported on the LCLS status screen) we also saw in Bragg termination measurements that the X-ray pulses were considerably shorter than the electron bunch, as described by Barty *et al.*, *Nature Photon*, 2012, **6**, 35. We should point out that LCLS now has a tool that measures the energy loss *vs.* time of the electron bunch which tells you which part of the bunch produced X-rays.

Nora Berrah responded: This is a good addition to the LCLS beam diagnostics.

Eleanor Campbell commented: Care should be taken when interpreting mass spectra in terms of dynamics that occur on the fs timescale. The mass spectra probe the ion distributions that are present on a timescale of microseconds – *i.e.* much longer than the initial excitation pulse. The model used to interpret the C_{60} experiments considers only direct ionisation and secondary electron collision processes in one fullerene molecule. Under the conditions of the experiment, many energetic electrons are produced and escape from the 'nanoplasma'. The absolute cross-sections for electron impact ionisation/fragmentation were studied in detail during the 90s (*e.g.*, Foltin *et al.*, *Chem. Phys. Lett.*, 1998, **289**, 181, Hathiramani *et al.*, *Phys. Rev. Lett.*, 2000, **85**, 3604) and are large with a plateau for electron energies of a few hundred eV. It is quite possible that secondary ionisation/fragmentation of other fullerene molecules in the target is making a significant contribution to the observed mass spectra.

Nora Berrah answered: This is indeed a good point.

Jochen Küpper communicated: Nora, I have a naive question regarding the very good match of your experiment and the quite classical modeling. Obviously, the electrons in molecules are strongly correlated, or entangled, but you and your collaboration can nicely model this using a very classical description. This is indeed an interesting finding and also a helpful one. However, I wonder where/when the correlation/entanglement is lost – at least it seems to be lost. Considering the good match of the classical modeling, this collapse of entanglement must happen very early, *e.g.*, does it do so with the first photon absorbed, or similar?

Nora Berrah replied: The good match of experimental data with classical modeling is valid only in the case of ultra-strong fluence and also at an ultra-short timescale (4 fs in our case). In these cases, electron correlations do not seem to be important as demonstrated by the excellent agreement between experimental data and classical modeling. This is not the case at intermediate fluence where molecular effects are more important as revealed by the lack of good agreement between experimental data and classical modeling. I also assume that the collapse of entanglement may occur very early, with the first photon absorbed since modeling shows that the dynamics occurs at the first few fs. Note that the classical modeling treats the particles (electron and ions) classically but the cross sections and rate equations are generated using quantum mechanics.

Michael Woerner addressed Nora Berrah and Jochen Küpper: During the discussion of Nora Berrah's paper the question was raised from when the classical picture can be applied. The quantum to classical transition has something to do with the decoherence and the measurement process. We published a paper on interband tunnelling of electrons in GaAs: *Phys. Rev. B*, 2010, **82** 75204. The **refs. 37–41** therein give valuable insight into the decoherence and the rate of the latter process on the relative distance between particles.

Nora Berrah answered: Thank you for the information.

Katharine Reid addressed Nora Berrah and Giovanni De Ninno:You both referred to the possibility of performing time-resolved pump–probe experiments, but very few experiments presented at this conference have used such a scheme (though many have aspired to). Would you be able to comment on the kinds of time-resolved pump–probe experiments (wavelengths, time resolution) that are possible at your respective light sources (FERMI and LCLS) and on the prospects for pump–probe experiments at such sources in the future?

Giovanni De Ninno replied: At FERMI, we can carry out pump–probe experiments using different (complementary) setups. In the 'standard'configuration, the pump pulse is generated by taking a fraction of seed laser (wavelength: 800 nm, pulse duration: adjustable in the range 400-100 fs FWHM), or of one of its low-order harmonics (*e.g.*the third one), while the probe is provided by the FEL itself (fully tunable in range between 80–4 nm, with pulse duration adjustable in the range 50–200 fs FWHM). If required by users, the sample can be pumped by the FEL and probed by the laser. The typical jitter between the pump and the probe is quite small (*i.e.* about 5 fs). The seeded nature of our FEL permits the implementation of two additional 'exotic' configurations, both allowing FEL-pump–FEL-probe experiments in the XUV range, with a temporal resolution of several tens of fs. The first one, described in **reference 44** of our manuscript, relies on seeding the electron beam with a strongly frequency-chirped laser pulse; this naturally leads to the generation of two FEL sub-pulses (*i.e.*the pump and the probe), characterized by the controlled temporal and frequency separations. The second exotic configuration is the one exploited in the experiment reported in our paper. In this case, the electron beam is seeded with two separated laser pulses, characterized by a predetermined temporal and frequency separation. All the above configurations are routinely used at the FERMI beamlines. For prospects about future pump–probe experiments at FERMI, see http://www.elettra.trieste.it/lightsources/fermi.html.

Nora Berrah commented: There have been time resolved experiments at FLASH, FERMI and at LCLS using optical laser pump X-ray probe and using X-ray pump–X-ray probe. Data from these experiments are being analyzed but also some have been published. For example the following LCLS papers are the result of optical laser pump X-ray probe but there are more papers in the literature from other FELS, so this is only representative:

1. B. F. McFarland *et al.*, *Nature Communications*, 2014, **5**, 4235.
2. V. S. Petrovic *et al.*, *Phys. Rev. Lett.*, 2012, **108**, 253006.

Gwyn Williams opened the discussion of the paper by Giovanni De Ninno by asking: Do the two colour photon pulses come from the same electron bunch, and if so, what is the length of this electron bunch?

Giovanni De Ninno answered: Here you have our answer to the comment "Do the two colour photon pulses come from the same electron bunch, and if so, what is the length of this electron bunch?":

Yes, the two-colour FEL pulses are generated by the same electron bunch, seeded by two seed pulses. For the presented experiment, the length of the 'smooth' part (*i.e.* flat, both in energy and current) of the electron bunch was about 500 fs.

Elaine Seddon communicated: Your paper records that the intensity jitter in the seeded FEL pulses is around 15%. I would like to know if this is expected to be a problem for some users and if so is there currently a drive to improve the jitter?

Giovanni De Ninno communicated in reply: In general, an intensity fluctuation around 15% is not an issue for the large majority of our users, who have the possibility to keep track of the FEL intensity behaviour on a shot-to-shot basis. This allows them to normalize the obtained results. In order to get a better stability, we are currently following two directions, *i.e.* a further reduction of the electron-beam *vs.* laser jitter, and the improvement of the electron-beam longitudinal flatness.

Faraday Discussions

ROYAL SOCIETY
OF CHEMISTRY

PAPER

Disruptive photon technologies for chemical dynamics

Henry N. Chapman*

Received 9th August 2014, Accepted 12th August 2014

DOI: 10.1039/c4fd00156g

A perspective of new and emerging technologies for chemical dynamics is given, with an emphasis on the use of X-ray sources that generate sub-picosecond pulses. The two classes of experimental techniques used for time-resolved measurements of chemical processes and their effects are spectroscopy and imaging, where the latter includes microscopy, diffractive imaging, and crystallography. X-Ray free-electron lasers have brought new impetus to the field, allowing not only temporal and spatial resolution at atomic time and length scales, but also bringing a new way to overcome limitations due to perturbation of the sample by the X-ray probe by out-running radiation damage. Associated instrumentation and methods are being developed to take advantage of the new opportunities of these sources. Once these methods of observational science have been mastered it should be possible to use the new tools to directly control those chemical processes.

1 Introduction

Although several years in the preparation by its organisers, this *Faraday Discussions* meeting on "Emerging Photon Technologies for Chemical Dynamics" was very timely. X-Ray free-electron lasers (FELs) are now a reality and have been operating for several years, enough to give a sense of the capabilities and opportunities these sources open up. In addition the field of attosecond science, driven by advanced laser technologies and techniques, is fast developing and allowing the investigation of the electron dynamics that ultimately influences chemistry at longer timescales. The brightness of X-ray sources have followed an exponential growth over time. From the first parasitic use of synchrotron radiation to today's FELs represents 22 orders of magnitude in peak brightness over 40 years, or a doubling every 6.5 months. Like all exponential trends this cannot continue indefinitely, but hopefully high-brightness sources will become cheaper, more plentiful and more accessible, and the science that will flow from them will build upon some of the work discussed here.

Center for Free-Electron Laser Science, DESY, Notkestrasse 85, 22607 Hamburg, Germany. E-mail: henry. chapman@desy.de; Tel: +49 40 8998 4155

There has been a rapid progress of the field of chemical dynamics, or related fields that could be referred to as femto-chemistry, time-resolved chemistry, and structural dynamics. This can be attributed to the newly-available photon technologies, as was emphasised in the meeting. The speed of development can be highlighted by the fact that in a previous *Faraday Discussions* on the related topic of "Time-Resolved Chemistry: From Structure to Function" that was held twelve years ago, FELs got barely a mention, and in his concluding remarks John Meurig Thomas noted that synchrotron sources were finally revealing their utility for time-resolved studies.[1] This was certainly not an oversight, since at that time X-ray FELs were still in the planning stage and it was by no means clear that amplification of X-ray pulses could be achieved. As it turned out, the performance of FELs far exceeded even the most optimistic predictions. The rapid adoption of FEL sources, following the only-recent application of synchrotron radiation to time-resolved studies, is certainly not surprising given that they produce pulses with a billion times higher peak brightness than synchrotron facilities (1000 times shorter pulses with 1 million times more photons per pulse for a similar bandwidth). With pulse durations that can be tuned from about 1 fs to 100 fs, these sources are truly matched to the timescales of atomic processes (with X-ray wavelengths that access inter-atomic length scales), allowing a broad range of explorations in femto-chemistry. Other technological developments have fuelled the field, including the alignment of molecules by polarised laser beams,[2] the mapping of ejected particle kinetic energy and momentum by so-called velocity map imaging[3] and reaction microscopy,[4] the generation of short wavelength attosecond pulses by high-harmonic generation[5] and terahertz radiation from high-intensity laser pulses, new spectroscopies such as angular-resolved photoemission spectroscopy,[6] multidimensional spectroscopies,[7,8] optical—X-ray pulse cross correlation and streaking to determine relative arrival times and pulse durations to few femtoseconds,[9,10] and new pixel detectors[11-13] and sample handling methods.[14-16] These innovations have given researchers a rich new toolbox that is currently being applied to gain fundamental insights into chemical dynamics in simple model systems and offer ways to exert control of interactions and molecular processes.

2 Time-resolved imaging and spectroscopy

The study of chemical dynamics using photons falls into the two broad methodologies of time-resolved structure determination based on X-ray scattering (often referred to as "making the molecular movie")[17] and following energetics, by spectroscopy.[18] The birth of these techniques can be attributed to William Lawrence Bragg and his father William Henry Bragg, respectively, about 100 years ago. William Bragg's spectrometer, now on display in the Faraday Museum of the Royal Institution, was used to analyse crystals, reveal characteristic fluorescence from anode materials, and to indicate the wave nature of X-rays. X-ray fluorescence and absorption spectroscopies were initially used to establish the atomic theory and determine the atomic composition of materials. With the development of intense and tunable synchrotron radiation, spectroscopy became more powerful and applicable on micrometer length scales. X-rays probe atomic core shells, which gives element specific measurements in complex systems. Near an absorption edge, shifts of the edge or changes in strengths of pre-edge peaks (due to core to valence shell transitions) measure coordination chemistry, which can also be revealed by shifts of fluorescence energies. Lighter elements predominantly emit photoelectrons, which also reveal fine spectral information about the chemical environment. At photon energies tens of electron volts above the absorption edge of a particular atomic species the extended X-ray absorption fine structure is the result of interference of photoelectrons scattering from neighbouring atoms. Three-dimensional structural information can be obtained if the molecular system is fixed in space (*e.g.* in a crystal) and the absorption spectrum is measured as a function of orientation. This is the inverse mode of photoelectron holography. Angular resolved photoemission spectroscopy resolves the single-particle spectral function in energy and momentum, giving insight into the electronic structure of solids, also in three dimensions. Inelastic X-ray scattering, utilising spectrometers of meV resolution, provides details of the excitations of the atomic lattice. Circularly polarised X-rays can access the angular momentum of electrons and separate spin and orbital components.

Following Lawrence Bragg's initial insights, three-dimensional images of molecules can be synthesised from Fourier components that are measured in a diffraction experiment. This approach requires solution of the well-known phase problem, and the history of X-ray structure determination follows a series of breakthroughs to derive diffraction phases, including the Patterson method, direct methods based on atomicity of matter, isomorphic replacement, molecular replacement (which could be thought of as treating a known part of the structure as a holographic reference), and anomalous diffraction. Macromolecular crystallography is a particularly important technique, and today almost 80% of all protein structures utilise synchrotron radiation at dedicated beamlines. This being the international year of crystallography there are many excellent recent reviews of this history.[19-21] The methodology of image synthesis has also been applied to non-crystalline samples, by using coherent X-ray beams to measure the continuous diffraction pattern.[22-24] Isolated objects of compact support result in a diffraction pattern that is band limited and can be completely measured if sampled sufficiently finely. This has opened up another phasing method where real-space and Fourier-space constraints are iteratively enforced until convergence is reached.[25] Images of matter are more readily obtained directly using an X-ray microscope, although spatial resolution is limited to tens of nanometres by the performance of high-resolution diffractive zone-plate lenses.[26] The method of ptychography overcomes the resolution limit of nano-fabricated lenses by combining scanning and coherent diffraction.[27] A particular attraction of X-ray microscopy is that the spectroscopic modalities mentioned above can be used to provide the contrast mechanism for image formation, and the resulting rich high-dimensional spectro-micrographs can be analysed to extract spatial and chemical features at the tens of nanometre scale as was presented by Chris Jacobsen[28] at this meeting. By measuring spectra in a spatially resolved manner, real inhomogeneous materials and systems, such as catalysts or biological cells, can be measured and understood in terms of functional units rather than as an average over the entire sample. Jacobsen's method goes beyond principle component analysis and cluster analysis and demonstrates the power of applying physically-meaningful constraints such as sparsity and positivity. This approach would no doubt be equally useful applied the time domain (*e.g.* time-resolved crystallography[29,30]) as in the frequency domain.

The extension of X-ray spectroscopy and imaging into the time domain has predominantly been carried out using pulsed X-ray sources synchronised with a preceding laser pulse to excite the sample. This "pump–probe" scheme can in principle be carried out with all the techniques mentioned above and thus builds upon the long history and development of those methods, as well as techniques in laser science and optical spectroscopy. Until recently, the available pulsed X-ray sources, including synchrotrons and laser-produced plasmas, were rather weak (that is, of low peak brightness). For time resolution better than nanoseconds this required an experiment design that was able to combine stroboscopic measurements with samples that could be rapidly replenished or brought back to the ground state. Measurements often took many days or weeks to complete. Now, with X-ray FELs we are witnessing a new chapter. With a billion-fold improvement in peak brightness, FELs are a disruptive new technology that opens up new types of time-resolved experiments. Since the opening of the first hard X-ray FEL, the Linac Coherent Light Source, in 2009 (ref. 31) we have seen new explorations in non-linear atomic physics[32,33] and non-linear optics;[34] gas diffraction;[35–38] imaging of phonons in nanocrystals;[39] diffraction measurements of materials at extreme departures from equilibrium;[40] and time-resolved serial femtosecond crystallography.[41–43]

Since this meeting dealt with the topic of photon technologies, there was limited discussion on the use of short electron pulses for obtaining spectral and structural information. The million-times larger atomic scattering potential for electrons over that of X-rays means that electron diffraction is extremely well suited for dilute systems, allowing experiments in the lab on diffraction from laser-aligned molecules[44] that are certainly more impressive than the LCLS experiments reported at this meeting.[35,38] As pointed out in the meeting by John Spence, short electron pulses may not outrun damage effects (see below) as effectively as X-ray pulses do, but this may not necessarily be required. For samples in the liquid or solid state there are also issues of multiple and inelastic scattering of the strongly interacting electrons, requiring energy filtering of the scattered electrons. These technologies are well developed and available in electron microscopes, and perhaps will be implemented with short-pulse electron instrumentation. Element specificity is achieved at higher contrast by X-ray absorption as by electron energy loss measurements. It is expected that, just as with static measurements, the extensions of imaging and spectroscopy into the ultrafast time regime will rely upon the complementarity of these methodologies.

3 The perfect probe?

The ideal probe should have no influence on the state of the sample, yet radiation damage often limits the collection of atomic-resolution information in scattering and imaging experiments, and the generation of space charge at high intensities reduces the resolution of photoelectron spectra. The atomic scattering cross sections in the X-ray regime are considerably smaller than the absorption cross sections, resulting in the unfortunate situation that for every scattered photon of 8 keV energy, for example, about 250 keV of energy is absorbed in the sample. Perturbation of the sample can be avoided by working at as low dose as possible, which requires samples that give strong enough signal above noise at low dose. Large crystals amplify diffraction intensities and data can be collected in a regime where there are fewer absorbed photons than molecules in the crystal. Such an approach can even be applied to single molecule diffraction with X-ray pulses when the molecules are aligned and signals can be accumulated over many pulses, each with a fresh sample. Stephan Stern described experiments where the number of accumulated scattered photons in the entire detector frame was much less than one per pulse.[38]

One of the initial motivations to build X-ray FELs was to use short pulses to outrun the effects of radiation damage. The proposal[45] was that with short enough pulses you could obtain the "perfect probe"—one that is non-perturbing but which gives measurable signals from dilute or weakly scattering objects. When the first soft X-ray FEL, FLASH, turned on (then called the Tesla Test Facility) the initial quest was to understand the interaction of matter with intense pulses and determine how true this desired outcome was, starting with model systems such as atomic clusters[46] and nanostructures.[47,48] Those initial experiments were guided by theory and simulations, such as molecular dynamics[45] and hydrodynamics[49] that were applied significantly beyond the regimes that they were initially developed. The interaction starts with photoabsorption and ejection of a photoelectron. The excited atom may relax by filling the core hole by decay of an electron of a higher orbital, accompanied by either emission of a photon (fluorescence) or Auger decay. At the high intensities of focused X-ray FEL pulses there may be more absorbed photons than atoms in the sample, giving rise to a sequence of photon absorption and relaxation leading to higher and higher charge states with a shifting of the absorption edge to higher energies. If the X-ray dose rate is high enough, such that the X-ray intensity is higher than the inverse of the photoabsorption cross section divided by the decay time, then a second core-shell ionisation can occur prior to relaxation, leading to a hollow atom that frustrates further absorption. Such effects were observed in some of the very first experiments carried out at the LCLS[32] and provided the data needed to develop the theory of atomic interactions at high dose rate.[50] We now have validated theoretical tools, such as the XATOM toolkit[50] to determine the fate of isolated atoms in high-intensity X-ray beams.

Molecules are certainly more complicated than isolated atoms, and electronic damage and nuclear motion will be strongly influenced by ionisation by free electrons. In large enough condensed systems there will be a cascade of collisional ionisations that could last about 100 fs and ionise more than 100 atoms from just one photoabsorption event.[51,52] On timescales as short as 5 fs charges from neighbouring atoms can quench the development of high charge states of heavy atoms in a molecule.[53] Robert Moshammer described a series of elegant experiments that examined differences between isolated iodine atoms and I_2 molecules, by changing the order of an infrared fragmentation pulse and an XUV probe pulse impinging on a beam of I_2 molecules.[54] Even in this simple system there are significant differences in the yield of charged iodine ions for atoms or molecules. Nora Berrah reported on recent measurements to investigate such processes in C_{60} molecules, using soft X-ray pulses at LCLS.[55,56] This is a model system for more complicated biomolecules, but also connects to studies of graphene[57] and nanotubes, as well as atomic clusters. The experiments measured the charge states of atomic and molecular fragments that result from the absorption of an average of 180 photons per molecule at the highest intensities. The measurements were compared with XMDYN simulations, which combines classical molecular dynamics with cross sections modelled by XATOM, and effects such as molecular Auger decay.[55] The remarkable agreement between experiment and model is a significant advance and shows that the dynamics of the explosion is dominated by secondary ionisation by trapped photo- and Auger electrons. This highlights the effect of the sample environment on the interaction with the intense pulse and should give crucial guidance for optimising pulse parameters for imaging and spectroscopy experiments.

So far, the knowledge obtained by the latest experiments largely agree with the predictions of Neutze et al.[45] that pulses of 10 fs or less are required for "diffraction before destruction" at molecular resolution. It appears that the perfect probe is obtained in the limit of the pulse duration tending to zero, or highest intensity (with a short enough duration pulse). Higher intensity results in a faster explosion, but gives overall more scattered photons before loss of the structure under investigation.[58,59] Sub-femtosecond pulses should outrun even most electronic processes,[50] and perhaps even allow photoelectron spectroscopy at high dose. Perhaps more general than "diffraction before destruction" we can refer to spectroscopic or other measurements in a "detection before destruction" regime, as pointed out by Majed Chergui. Since electronic damage precedes nuclear motion, spectroscopic information will be perturbed at pulses with lower fluences (i.e. lower dose) than those that can out-run nuclear motion. One particular limit is that which every atom has been collisionally ionised by the end of the pulse, which implies that less than about 1% of atoms are photoionized (depending on the photon energy and if the system is large enough to trap all photoelectrons). Under this condition, most photons that interact with atoms will do so with neutral atoms; that is, atoms that have not absorbed a photon nor collisionally ionised. The probability of a fluorescence photon being emitted by a perturbed atom, or an elastic scattering event from a perturbed atom, will thus be small, given that the measurement is integrated over the pulse and the sample is initially neutral. The dose for this condition has been estimated at about 400 MGy for protein crystals measured with 100 fs pulses,[60] considerably higher than the 30 MGy safe dose that can be tolerated with steady-state exposures at a synchrotron.[61,62] Metal centres in proteins are much more sensitive to the effects of X-ray exposure, and with slow synchrotron exposures they are reduced at much lower doses of only kGy, by capture of solvated electrons.[63] It has been argued that no structure in the protein data bank of a redox protein has been measured in its un-reduced state, prior to the use of X-ray FELs. Kern et al.[64] have taken advantage of X-ray FEL pulses to measure X-ray emission spectra of Mn in photoexcited photosystem II micro-crystals, simultaneous with diffraction measurements. The maximum dose was 150 kGy, and the Mn K spectrum in the ground state was identical to spectra of unreduced photosystem II acquired at low dose under cryogenic conditions. This strategy of combining multiple probes in time-resolved experiments gives a much greater understanding and cross validation in experiments. Bressler reported on experiments on aqueous iron(II)tris(bipyridine), $[Fe(bpy)_3]^{2+}$, at a storage ring (the Advanced Photon Source) and an FEL (the LCLS) that combined X-ray emission or absorption spectroscopies simultaneously with wide-angle X-ray solution scattering.[65]

Even though the full effects and limits of the influence of high-intensity X-ray probe on the sample have not been fully established, this has not precluded the application of short-duration FEL pulses to time-resolved measurements, especially using replenishable samples. Examples using the pump–probe methodology on photostimulated systems include time-resolved crystallography of photosystem II[42,43] and time-resolved wide-angle X-ray solution scattering revealing the protein quake after excitation of a reaction centre.[66] One of the leading model systems for X-ray time-resolved studies and for gaining insights into chemical dynamics is the metal-to-ligand charge transfer system $[Fe(bpy)_3]^{2+}$ mentioned above. Recent time-resolved photoemission spectroscopy carried out on this system at LCLS revealed the presence of an intermediate triplet state, revealed to sub-picosecond resolution.[67]

4 New technologies for chemical dynamics

X-ray FELs are by no means the only source available to the field of chemical dynamics. Majed Chergui summarised the properties of pulsed X-ray sources, including laser-induced plasma sources, high harmonic generation sources, storage rings and insertion devices (including slicing sources), table-top accelerators, and energy recovery linacs (ERLs). These have a wide range of peak and average brightness values, pulse durations, available photon energy ranges, bandwidths, spatial and temporal coherence, polarisation states, pulse stability, and repetition rates. This is a large parameter space, and different techniques and scientific questions emphasise different source properties. Any graphical comparison of sources therefore fails to illustrate how appropriate a source may be for a particular application, but we do so anyway, in Fig. 1.

What are the pulse specifications needed for ultrafast X-ray studies? As discussed above, most time-resolved X-ray experiments are carried out by employing the pump–probe method, which requires a high degree of synchronisation between the excitation pulse (usually a laser pulse in the IR to UV spectrum) and the probe. Other excitation schemes are possible, such as temperature jumps or fast mixing of samples, but laser pulses offer the fastest and most controlled way

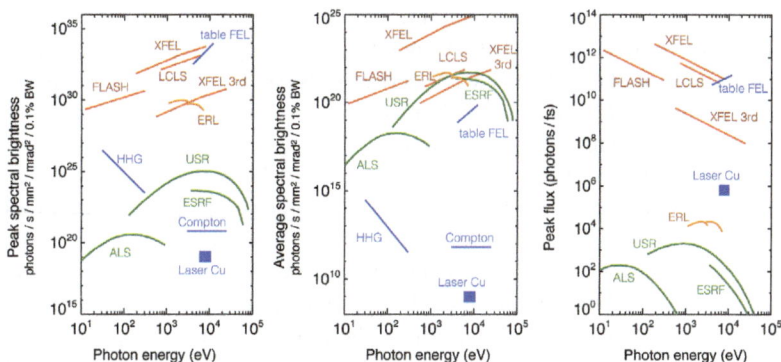

Fig. 1 Three comparisons of X-ray sources as a function of photon energy: (left) peak spectral brightness, (centre) average spectral brightness, and (right) peak flux. These plots only give partial comparisons. For example, the table-top FELs will have pulses of broader bandwidth and would have 20 times higher peak brightness (not spectral brightness) compared with linac FELs. USR = ultimate storage ring, ALS = undulator with 2 GeV electron energy, ESRF = undulator with 7 GeV electron energy, table FEL = table-top FEL,[73] laser Cu = Max Born Institute laser-driven plasma source with Cu target.[77]

of inducing the dynamics to be studied. The achievable temporal resolution obviously depends on the durations of the pump and probe pulses, as well as the uncertainty in the delay between those pulses. The feasibility of a time-resolved measurement depends if adequate pump-induced signal can be measured, which requires a large enough fraction of the sample to be photoexcited (yet without bleaching or damaging the sample) and a strong enough probe to give a measurement. By repeating measurements, and ensuring that the sample is either refreshed or brought back to the ground state between pump–probe cycles, the signal for a given time delay can be accumulated. In the words of Christian Bressler, what is important in such an experiment is the total number of photons collected, which is dictated by the pulse energy, repetition rate, and the duration of the entire experiment.

A temporal resolution of about 10 fs (the femto-chemical time-scale) is required to follow the dynamics of chemical pathways and capture elusive transition states, and thus the ideal X-ray pulse duration should not exceed this value. With compact laser-driven sources, such as the plasma X-ray source at the Max Born Institute,[68] a low degree of jitter between pump and probe is achieved by the fact that both pulses are ultimately derived from the same laser. At kilometre-long FEL facilities, the jitter between pump and probe is on the order of 100 fs but it is possible to approach a time resolution of a few femtoseconds by determining the relative arrival times of the two pulses on a shot by shot basis and appropriately sorting the data *post facto*.[10] Seeding schemes, such as carried out at the FERMI soft-X-ray FEL give inherently lower jitter between pump and probe arrivals.[69] At the LCLS it is possible to generate pairs of X-ray pulses with a well defined and arbitrary delay and different photon energies.[70] In the Discussions there was the opinion that pulses even shorter than 10 fs (and even into the attosecond regime) would certainly be advantageous, not only because such pulses more closely approach the "perfect probe" mentioned above, but to measure electronic processes on their inherent timescales.[71]

As faster timescales are accessed, the corresponding length scales are reduced, requiring wavelengths shorter than 1 Å for scattering measurements. Mike Minitti and colleagues demonstrated that atomic resolution in X-ray molecule diffraction could in principle be achieved at 20 keV photon energy, from the third harmonic of the LCLS undulator.[35] Michael Woerner suggested that pulses of 200 keV photon energy would open up the ability to measure not only charge density, but also spin density maps at high resolution.

Spatial resolution at the atomic scale can still be realised with soft X-ray sources by obtaining this information from photoelectrons, whose de Broglie wavelengths are certainly short enough to resolve interatomic spacings. This is particularly relevant for attosecond pulses generated from high-harmonics in gases, solids, or relativistic plasmas that produce pulses with photon energies below about 1 keV, which can ionise inner-shell electrons of the lighter elements. Daniel Rolles presented experiments carried out at the LCLS (and the synchrotron facility PETRA III) on photoelectron diffraction of laser-aligned molecules, showing the feasibility to obtain three-dimensional structural information of the molecular environment around a target atom, with a photon energy of about 735 eV.[72]

As sources approach attosecond durations, as achievable by high-harmonic generation and possibly by inverse Compton scattering by THz accelerated electron pulses,[73] we must consider the effect of increased bandwidth and longitudinal coherence due to diffraction limits. For example, consider a sample of width w diffracting to a resolution d with a wavelength λ. The waves scattering from extreme points of the sample (separated by w) will interfere at the far field with a path difference of $\Delta = w \sin 2\theta \approx w\lambda/d$. For $w = 300$ nm, $d = 0.1$ nm, $\lambda = 0.1$ nm, the path difference is $\Delta = 300$ nm, or 0.66 fs. How does interference occur if the pulse is shorter than this delay between scattered waves? The simple answer is that the pulse can be considered to be synthesised by a broad and continuous spectrum of Fourier frequencies. Particular 3D spatial frequencies of the sample will selectively diffract these pulse frequencies according to Bragg's law.[74] This is nicely illustrated in Laue diffraction by a pink beam from a protein crystal[75] where a particular Bragg order will select a particular part of the pink spectrum. The pulse arriving at the detector in a particular diffraction order or detector pixel will therefore be longer than the attosecond incident pulse. This begs the question as to what is the actual interaction time and achievable temporal resolution of the measurement.[76]

An intriguing application of diffraction of attosecond pulses was given by John Spence[74] who suggests that under certain conditions two different wavelengths out of the diffraction limited spectrum of an attosecond pulse will diffract to different Bragg orders at a common 2θ angle. These two diffraction orders will overlap on the detector. The interference of these two frequencies beat at their difference frequency and hence the relative phase of the two orders can be determined if the pulse duration is less than that beat period.

Michael Woerner's source produces 100 fs duration pulses with 6.8×10^7 photons per pulse from a 10 μm diameter source at a repetition rate of 1 kHz and a photon energy of 8.06 keV (from a Cu target).[77] With a suitably well-diffracting sample, a powder of $NaBH_4$ ionic crystals in this case, Woerner reports that it is possible to collect datasets in several days of continuous operation. This is obviously longer than would be required at an X-ray FEL, but the stability and reliability of the source and the fact it is a laboratory source, allows such a strategy. It also highlights the fact that the sample properties are a key consideration in dynamics experiments (as for static measurements). That is, given short duration pulses, source brightness may be traded for diffraction strength of the sample, and high repetition rate (or available measurement time) for pulse energy. For well-ordered and non-dilute samples, this would argue that the appropriate source metric is average brightness. Thus, an energy recovery linac source, generating pulses at a megahertz repetition rate, may offer advantages for spectroscopic or scattering experiments from certain systems. The high repetition rate of the European XFEL (27 000 pulses delivered per second) gives the possibility to achieve in 1 minute what would take almost four hours at LCLS operating at 120 Hz. Such capability enables the measurement of rare or unusual events, such as crystal nucleation or a short-lived intermediate state. X-ray FELs produce ordered light (spatially coherent pulses with extremely high photon degeneracy) that enables the study of dilute systems such as laser-aligned molecules and weakly scattering protein nanocrystals, or the search for short-range order and local symmetry in complex or amorphous samples.[78,79] The diffracting strength of such dilute systems is easily 10^{-12} times the strength of a perfect microscopic crystals: *e.g.* a 10 μm wide crystal of 10 Å unit cell length contains 10^{12} unit cells, which highlights the great range of sample diffracting strengths and the need for sources of spanning this range.

The experiment is more than the source, and experiments with high-peak-brightness sources require a corresponding capability in refreshing or replenishing the sample and high dynamic-range detectors that can match the source in frame rate. The sample environment ideally should provide a geometry where both the pump and probe interact uniformly with the sample in a common volume. Liquid-jet injectors have been very successful used at X-ray FELs for delivering a new sample on every pulse,[14,15] and John Spence showed latest developments in using this fast-flow technology for mixing samples moments before X-ray exposure.[74] Pulsed gas jets were used in many of the experiments reported at this Faraday Discussion, and can provide densities high enough for ion spectroscopies[72] and diffraction.[38] Gas cells provide fresh sample by diffusion and can give a longer interaction length with the X-ray beam, to give higher signals, but this can complicate the scattering geometry.[35] Laser-induced alignment of gas molecules[2] can also be categorised as a sample delivery and preparation technique, even though the study of rotational dynamics of complex systems is itself a vibrant field of chemical dynamics.[80] Laser alignment is improved by cooling the molecules to a rotational temperature of about 1 K and selecting low-energy quantum states using an electrostatic deflector, as was used for initial experiments of diffractive imaging of aligned diiodobenzonitrile molecules.[38]

Most chemical reactions are not accessible *via* initiation of a laser pulse or temperature jump. Spence presented a mixing apparatus for obtaining time-resolved measurements of chemical reactions on timescales of microseconds.[74] This may be faster than can be achieved by the freezing of intermediates, *e.g.* in trapped-state crystallography[81] due to the fact that we can now obtain information from small crystals—giving short times for mixing by diffusion.[82] The jet has an inner bore capillary, carrying one reagent such as protein nanocrystals in solution, that can be slid telescopically inside another capillary in which the second reagent solution flows. These solutions mix, and the mixing time before probing is controlled by the distance from the exit of the inner capillary to the interaction region, and the flow speed. The apparatus could also be used to study the dynamics of protein folding.

It is a well-known problem that whenever source capabilities are improved there is a deficiency of detectors that can fully exploit those capabilities. Thankfully, X-ray FEL facilities anticipated the need for high frame-rate pixellated detectors and there are several detectors in use such as the Cornell-SLAC pixel array detector[12] and the pnCCD[11] that have millions of pixels and frame rates higher than 120 Hz. For the European XFEL, which will operate in bursts of 2700 pulses at 4.5 MHz, 10 times per second, the challenge is even greater, and developments are taking place to handle this extreme pulse pattern. One such development is the Adaptive Gain Integrating Pixel Detector.[13] Each pixel has analogue storage for 352 frames that can be addressed in less than the 220 ns between pulses and which can all be digitised and read out in less than the ≈ 100 ms between pulse trains. Thus, 3520 frames per second can be stored. It is possible to overwrite any of the frames within the pulse train, which means that for low sample concentrations where not every shot hits a sample it is possible to achieve higher effective frame rates if a suitable veto signal can be created. Initial

AGIPD systems have reached a dynamic range of 10^4 and single-photon sensitivity.[83]

The new detector technologies described here may also have a profound effect on time-resolved experiments at synchrotron radiation facilities and other pulsed sources. Since the AGIPD detector can measure 10^4 photons per pixel at 5×10^6 frames per second, it can record 5×10^{10} photons per pixel per second, which matches the intensity of the direct beam at many synchrotron beamlines. This capability opens up new possibilities for time-resolved experiments with ≈ 100 ps synchrotron bunches and rapidly performing delay scans from a single pump followed by a series of pulses. Similarly, advanced detectors developed for synchrotron radiation, such as the Pilatus, are finding a lot of use in laboratory sources, such as Michael Woerner's plasma source. The photon-counting Pilatus cannot count above a single count over the duration of a femtosecond pulse. Thus, even with a plasma source the fast-frame integrating detectors developed for XFELs may be even better suited in cases where there can be more than a single photon per pixel per frame as can occur in Bragg peaks.

The improved source and detector capabilities at synchrotron facilities may speed up data collection in methods such as ptychography[84] and scanning transmission X-ray microscopy[28] to the point that the pulse structure of the storage ring will be noticeable. Indeed, these facilities are being applied for time-resolved studies at high repetition rates, and strategies such as photon time-stamping[85,86] to make use of all X-ray pulses, given that laser pump pulses cannot usually be delivered at megahertz repetition rates.

5   Outlook

Accelerator-driven sources offer many possible pulse parameters and configurations, and can be optimised for high peak brightness (*e.g.* X-ray FELs), high average brightness (*e.g.* ERLs and synchrotrons), pulse duration, pulse stability, and so on. While some parameters can be agreed upon as needed for a broad range of femto-chemical dynamics studies, the potential experiments and directions in chemical dynamics means that no single source will meet all needs. As pointed out by Nora Berrah,[55] there is a legitimate question as to whether large-scale facilities are worth the cost, especially since linac-driven sources can only serve a limited number of experiments at a time. The counterpoint is that the cost of not building sources that open up new exploratory science could be far greater to society in the long run. One of the key issues to capitalising on the exciting new opportunities offered by these disruptive new sources is their limited access. This can impact the science that is carried out, because of the low tolerance for failure and the avoidance for allowing difficult and risky experiments. The field therefore needs complementary sources that can guide the development and help prepare experiments. It also needs more facilities world wide and methods for multiplexing or efficiently sharing the beam. The SFX User Consortium[87] at the European XFEL, for example, proposes to collect nanocrystal diffraction data in a parasitic fashion, as part of the SPB/SFX beamline.[88]

As pointed out in the introduction, the drive for higher peak brightness can not continue indefinitely, and Gwyn Williams pointed out in this Faraday Discussion that X-ray brightness depends on the peak current that can be achieved in the radiating electron bunch, which is fundamentally limited by space charge in the relativistic bunch,[89] even given the amplification that occurs in an FEL. However, this situation is similar to power limits in optical laser amplifiers[90] and the way to overcome those limits has been through chirped pulse amplification.[91] Thus, in a similar way, it should be feasible to increase the power (and hence brightness) of X-ray pulses with chirped electron bunches of high-charge but limited peak current and then compress the resulting chirped X-ray pulse using X-ray optics.[92,93] This may give factors of 10 or 100 increase in peak brightness.

It has been a glorious century of X-ray science, celebrated this year by the International Year of Crystallography. The focus of the last 102 years has been on determining static structures at atomic detail, and only recently has it been possible to observe the dynamics of those structures with the types of tools discussed at this meeting. Many of these tools, such as X-ray FELs, bring large increases of capabilities which we are still coming to grips with, and may overturn many of the limitations and requirements that have long faced the investigation of matter at atomic scales. Once we have mastered observing chemistry at its fundamental length and time scales of atoms and electronic transitions, it should be possible to control those structures and processes, which is the dream of coherent control. In such a way, it may be possible to influence the course of complex catalytic reactions to mimic the light-induced steps of photosynthesis or peptide bond-formation in a ribosome, or to drive chemical reactions to specific products. It is clear that in this period of rapid source developments it will take some time to establish the most efficient, useful, and accurate methods to meet such goals. Such was the case in fully harnessing radiation from storage rings, and some of the pioneers of those days happily still play a significant role today, and actively participated in these Discussions.

Acknowledgements

I acknowledge support through the European Research Council through the grant SYG 2013-609920 – AXSIS "Frontiers in Attosecond X-ray Science: Imaging and Spectroscopy."

References

1 J. M. Thomas, *Faraday Discuss.*, 2002, **122**, 395–399.

2 H. Stapelfeldt and T. Seideman, *Rev. Mod. Phys.*, 2003, **75**, 543–557.

3 A. T. J. B. Eppink and D. H. Parker, *Rev. Sci. Instrum.*, 1997, **68**, 3477–3484.

4 J. Ullrich, R. Moshammer, A. Dorn, R. Dörner, L. P. H. Schmidt and H. Schmidt-Böcking, *Rep. Prog. Phys.*, 2003, **66**, 1463–1545.

5 P. B. Corkum and F. Krausz, *Nat. Phys.*, 2007, **3**, 381–387.

6 C. Sohrt, A. Stange, M. Bauer and K. Rossnagel, *Faraday Discuss.*, 2014, **171**, DOI: 10.1039/c4fd00042k.

7 F. Fournier, R. Guo, E. M. Gardner, P. M. Donaldson, C. Loeffeld, I. R. Gould, K. R. Willison and D. R. Klug, *Acc. Chem. Res.*, 2009, **42**, 1322–1331.

8 I. Schweigert and S. Mukamel, *Phys. Rev. Lett.*, 2007, **99**, 163001.

9 A. L. Cavalieri, D. M. Fritz, S. H. Lee, P. H. Bucksbaum, D. A. Reis, D. M. Mills, R. Pahl, J. Rudati, P. H. Fuoss, G. B. Stephenson, D. P. Lowney, A. G. MacPhee, D. Weinstein, R. W. Falcone, R. Pahl, J. Als-Nielsen, C. Blome, R. Ischebeck,

H. Schlarb, T. Tschentscher, J. Schneider, K. Sokolowski-Tinten, H. N. Chapman, R. W. Lee, T. N. Hansen, O. Synnergren, J. Larsson, S. Techert, J. Sheppard, J. S. Wark, M. Bergh, C. Calleman, G. Huldt, D. van der Spoel, N. Timneanu, J. Hajdu, E. Bong, P. Emma, P. Krejcik, J. Arthur, S. Brennan, K. J. Gaffney, A. M. Lindenberg and J. B. Hastings, *Phys. Rev. Lett.*, 2005, **94**, 114801.

10 M. Harmand, R. Coffee, M. R. Bionta, M. Chollet, D. French, D. Zhu, D. M. Fritz, H. T. Lemke, N. Medvedev, B. Ziaja, S. Toleikis and M. Cammarata, *Nat. Photonics*, 2013, **7**, 215–218.

11 L. Strüder, S. Epp, D. Rolles, R. Hartmann, P. Holl, G. Lutz, H. Soltau, R. Eckart, C. Reich, K. Heinzinger, C. Thamm, A. Rudenko, F. Krasniqi, K.-U. Kühnel, C. Bauer, C.-D. Schröter, R. Moshammer, S. Techert, D. Miessner, M. Porro, O. Hälker, N. Meidinger, N. Kimmel, R. Andritschke, F. Schopper, G. Weidenspointner, A. Ziegler, D. Pietschner, S. Herrmann, U. Pietsch, A. Walenta, W. Leitenberger, C. Bostedt, T. Möller, D. Rupp, M. Adolph, H. Graafsma, H. Hirsemann, K. Gärtner, R. Richter, L. Foucar, R. L. Shoeman, I. Schlichting and J. Ullrich, *Nucl. Instrum. Methods Phys. Res., Sect. A*, 2010, **614**, 483–496.

12 P. Hart, S. Boutet, G. Carini, M. Dubrovin, B. Duda, D. Fritz, G. Haller, R. Herbst, S. Herrmann, C. Kenney, N. Kurita, H. Lemke, M. Messerschmidt, M. Nordby, J. Pines, D. Schafer, M. Swift, M. Weaver, G. Williams, D. Zhu, N. Van Bakel and J. Morse, *Proc. SPIE*, 2012, **8504**, 85040C.

13 J. Becker, L. Bianco, R. Dinapoli, P. Göttlicher, H. Graafsma, D. Greiffenberg, M. Gronewald, B. H. Henrich, H. Hirsemann, S. Jack, R. Klanner, A. Klyuev, H. Krüger, S. Lange, A. Marras, A. Mozzanica, B. Schmitt, J. Schwandt, I. Sheviakov, X. Shi, U. Trunk, M. Zimmer and J. Zhang, *J. Instrum.*, 2013, **8**, C01042.

14 D. P. DePonte, U. Weierstall, K. Schmidt, J. Warner, D. Starodub, J. C. H. Spence and R. B. Doak, *J. Phys. D: Appl. Phys.*, 2008, **41**, 195505.

15 U. Weierstall, J. C. H. Spence and R. B. Doak, *Rev. Sci. Instrum.*, 2012, **83**, 035108.

16 U. Weierstall, D. James, C. Wang, T. A. White, D. Wang, W. Liu, J. C. H. Spence, R. Bruce Doak, G. Nelson, P. Fromme, R. Fromme, I. Grotjohann, C. Kupitz, N. A. Zatsepin, H. Liu, S. Basu, D. Wacker, G. Won Han, V. Katritch, S. Boutet, M. Messerschmidt, G. J. Williams, J. E. Koglin, M. Marvin Seibert, M. Klinker, C. Gati, R. L. Shoeman, A. Barty, H. N. Chapman, R. A. Kirian, K. R. Beyerlein, R. C. Stevens, D. Li, S. T. A. Shah, N. Howe, M. Caffrey and V. Cherezov, *Nat. Commun.*, 2014, **5**, 3309.

17 R. J. D. Miller, *Science*, 2014, **343**, 1108–1116.

18 C. Bressler and M. Chergui, *Chem. Rev.*, 2004, **104**, 1781–1812.

19 E. F. Garman, *Science*, 2014, **343**, 1102–1108.

20 M. Jaskolski, Z. Dauter and A. Wlodawer, *FEBS J.*, 2014, **281**, 3985–4009.

21 *Nature Milestones in Crystallography*, ed. M. Montoya, A. Moscatelli and A. Taroni, 2014.

22 D. Sayre, *Imaging processes and coherence in physics*, 1980, pp. 229–235.

23 J. Miao, P. Charalambous, J. Kirz and D. Sayre, *Nature*, 1999, **400**, 342–344.

24 H. N. Chapman, A. Barty, S. Marchesini, A. Noy, S. P. Hau-Riege, C. Cui, M. R. Howells, R. Rosen, H. He, J. C. H. Spence, U. Weierstall, T. Beetz, C. Jacobsen and D. Shapiro, *J. Opt. Soc. Am. A*, 2006, **23**, 1179–1200.

25 H. N. Chapman and K. A. Nugent, *Nat. Photonics*, 2010, **4**, 833–839.

26 J. Kirz, C. Jacobsen and M. Howells, *Q. Rev. Biophys.*, 1995, **28**, 33–130.

27 J. M. Rodenburg, in *Advances in imaging and electron physics*, Elsevier, 2008, vol. 150, ch. Ptychography and related diffractive imaging methods, pp. 87–184.

28 R. Mak, M. Lerotić, H. Fleckenstein, S. Vogt, S. M. Wild, S. Leyffer, Y. Sheynkin and C. Jacobsen, *Faraday Discuss.*, 2014, **171**, DOI: 10.1039/c4fd00023d.

29 R. Neutze and K. Moffat, *Curr. Opin. Struct. Biol.*, 2012, **22**, 651–659.

30 J. J. van Thor, M. M. Warren, C. N. Lincoln, M. Chollet, H. T. Lemke, D. M. Fritz, M. Schmidt, J. Tenboer, Z. Ren, V. Srajer, K. Moffat and T. Graber, *Faraday Discuss.*, 2014, **171**, DOI: 10.1039/c4fd00011k.

31 P. Emma, R. Akre, J. Arthur, R. Bionta, C. Bostedt, J. Bozek, A. Brachmann, P. Bucksbaum, R. Coffee, F. J. Decker, Y. Ding, D. Dowell, S. Edstrom, A. Fisher, J. Frisch, S. Gilevich, J. Hastings, G. Hays, P. Hering, Z. Huang, R. Iverson, H. Loos, M. Messerschmidt, A. Miahnahri, S. Moeller, H. D. Nuhn, G. Pile, D. Ratner, J. Rzepiela, D. Schultz, T. Smith, P. Stefan, H. Tompkins, J. Turner, J. Welch, W. White, J. Wu, G. Yocky and J. Galayda, *Nat. Photonics*, 2010, **4**, 641–647.

32 L. Young, E. P. Kanter, B. Krässig, Y. Li, A. M. March, S. T. Pratt, R. Santra, S. H. Southworth, N. Rohringer, L. F. DiMauro, G. Doumy, C. A. Roedig, N. Berrah, L. Fang, M. Hoener, P. H. Bucksbaum, J. P. Cryan, S. Ghimire, J. M. Glownia, D. A. Reis, J. D. Bozek, C. Bostedt and M. Messerschmidt, *Nature*, 2010, **466**, 56–61.

33 B. Rudek, S.-K. Son, L. Foucar, S. W. Epp, B. Erk, R. Hartmann, M. Adolph, R. Andritschke, A. Aquila, N. Berrah, C. Bostedt, J. Bozek, N. Coppola, F. Filsinger, H. Gorke, T. Gorkhover, H. Graafsma, L. Gumprecht, A. Hartmann, G. Hauser, S. Herrmann, H. Hirsemann, P. Holl, A. Homke, L. Journel, C. Kaiser, N. Kimmel, F. Krasniqi, K.-U. Kuhnel, M. Matysek, M. Messerschmidt, D. Miesner, T. Moller, R. Moshammer, K. Nagaya, B. Nilsson, G. Potdevin, D. Pietschner, C. Reich, D. Rupp, G. Schaller, I. Schlichting, C. Schmidt, F. Schopper, S. Schorb, C.-D. Schroter, J. Schulz, M. Simon, H. Soltau, L. Struder, K. Ueda, G. Weidenspointner, R. Santra, J. Ullrich, A. Rudenko and D. Rolles, *Nat. Photonics*, 2012, **6**, 858–865.

34 T. E. Glover, D. M. Fritz, M. Cammarata, T. K. Allison, S. Coh, J. M. Feldkamp, H. Lemke, D. Zhu, Y. Feng, R. N. Coffee, M. Fuchs, S. Ghimire, J. Chen, S. Shwartz, D. A. Reis, S. E. Harris and J. B. Hastings, *Nature*, 2012, **488**, 603–608.

35 P. Minitti, J. M. Budarz, A. Kirrander, J. Robinson, T. J. Lane, D. Ratner, K. Saita, T. Northey, B. Stankus, V. Cofer-Shabica, J. Hastings and P. M. Weber, *Faraday Discuss.*, 2014, **171**, DOI: 10.1039/c4fd00030g.

36 V. S. Petrović, M. Siano, J. L. White, N. Berrah, C. Bostedt, J. D. Bozek, D. Broege, M. Chalfin, R. N. Coffee, J. Cryan, L. Fang, J. P. Farrell, L. J. Frasinski, J. M. Glownia, M. Gühr, M. Hoener, D. M. P. Holland, J. Kim, J. P. Marangos, T. Martinez, B. K. McFarland, R. S. Minns, S. Miyabe, S. Schorb, R. J. Sension, L. S. Spector, R. Squibb, H. Tao, J. G. Underwood and P. H. Bucksbaum, *Phys. Rev. Lett.*, 2012, **108**, 253006.

37 J. Küpper, S. Stern, L. Holmegaard, F. Filsinger, A. Rouzée, A. Rudenko, P. Johnsson, A. V. Martin, M. Adolph, A. Aquila, S. Bajt, A. Barty, C. Bostedt, J. Bozek, C. Caleman, R. Coffee, N. Coppola, T. Delmas, S. Epp, B. Erk, L. Foucar, T. Gorkhover, L. Gumprecht, A. Hartmann, R. Hartmann,

G. Hauser, P. Holl, A. Hömke, N. Kimmel, F. Krasniqi, K.-U. Kühnel, J. Maurer, M. Messerschmidt, R. Moshammer, C. Reich, B. Rudek, R. Santra, I. Schlichting, C. Schmidt, S. Schorb, J. Schulz, H. Soltau, J. C. H. Spence, D. Starodub, L. Strüder, J. Thøgersen, M. J. J. Vrakking, G. Weidenspointner, T. A. White, C. Wunderer, G. Meijer, J. Ullrich, H. Stapelfeldt, D. Rolles and H. N. Chapman, *Phys. Rev. Lett.*, 2014, **112**, 083002.

38 S. Stern, L. Holmegaard, F. Filsinger, A. Rouzée, A. Rudenko, P. Johnsson, A. V. Martin, A. Barty, C. Bostedt, J. Bozek, R. Coffee, S. Epp, B. Erk, L. Foucar, R. Hartmann, N. Kimmel, K.-U. Kühnel, J. Maurer, M. Messerschmidt, B. Rudek, D. Starodub, J. Thøgersen, G. Weidenspointner, T. A. White, H. Stapelfeldt, D. Rolles, H. N. Chapman and J. Küpper, *Faraday Discuss.*, 2014, **171**, DOI: 10.1039/c4fd00028e.

39 J. N. Clark, L. Beitra, G. Xiong, A. Higginbotham, D. M. Fritz, H. T. Lemke, D. Zhu, M. Chollet, G. J. Williams, M. Messerschmidt, B. Abbey, R. J. Harder, A. M. Korsunsky, J. S. Wark and I. K. Robinson, *Science*, 2013, **341**, 56–59.

40 D. Milathianaki, S. Boutet, G. J. Williams, A. Higginbotham, D. Ratner, A. E. Gleason, M. Messerschmidt, M. M. Seibert, D. C. Swift, P. Hering, J. Robinson, W. E. White and J. S. Wark, *Science*, 2013, **342**, 220–223.

41 A. Aquila, M. S. Hunter, R. B. Doak, R. A. Kirian, P. Fromme, T. A. White, J. Andreasson, D. Arnlund, S. Bajt, T. R. M. Barends, M. Barthelmess, M. J. Bogan, C. Bostedt, H. Bottin, J. D. Bozek, C. Caleman, N. Coppola, J. Davidsson, D. P. DePonte, V. Elser, S. W. Epp, B. Erk, H. Fleckenstein, L. Foucar, M. Frank, R. Fromme, H. Graafsma, I. Grotjohann, L. Gumprecht, J. Hajdu, C. Y. Hampton, A. Hartmann, R. Hartmann, S. Hau-Riege, G. Hauser, H. Hirsemann, P. Holl, J. M. Holton, A. Hömke, L. Johansson, N. Kimmel, S. Kassemeyer, F. Krasniqi, K.-U. Kühnel, M. Liang, L. Lomb, E. Malmerberg, S. Marchesini, A. V. Martin, F. R. Maia, M. Messerschmidt, K. Nass, C. Reich, R. Neutze, D. Rolles, B. Rudek, A. Rudenko, I. Schlichting, C. Schmidt, K. E. Schmidt, J. Schulz, M. M. Seibert, R. L. Shoeman, R. Sierra, H. Soltau, D. Starodub, F. Stellato, S. Stern, L. Strüder, N. Timneanu, J. Ullrich, X. Wang, G. J. Williams, G. Weidenspointner, U. Weierstall, C. Wunderer, A. Barty, J. C. H. Spence and H. N. Chapman, *Opt. Express*, 2012, **20**, 2706–2716.

42 C. Kupitz, S. Basu, I. Grotjohann, R. Fromme, N. A. Zatsepin, K. N. Rendek, M. S. Hunter, R. L. Shoeman, T. A. White, D. Wang, D. James, J.-H. Yang, D. E. Cobb, B. Reeder, R. G. Sierra, H. Liu, A. Barty, A. L. Aquila, D. Deponte, R. A. Kirian, S. Bari, J. J. Bergkamp, K. R. Beyerlein, M. J. Bogan, C. Caleman, T.-C. Chao, C. E. Conrad, K. M. Davis, H. Fleckenstein, L. Galli, S. P. Hau-Riege, S. Kassemeyer, H. Laksmono, M. Liang, L. Lomb, S. Marchesini, A. V. Martin, M. Messerschmidt, D. Milathianaki, K. Nass, A. Ros, S. Roy-Chowdhury, K. Schmidt, M. Seibert, J. Steinbrener, F. Stellato, L. Yan, C. Yoon, T. A. Moore, A. L. Moore, Y. Pushkar, G. J. Williams, S. Boutet, R. B. Doak, U. Weierstall, M. Frank, H. N. Chapman, J. C. H. Spence and P. Fromme, *Nature*, 2014, **513**, 261–265.

43 J. Kern, R. Tran, R. Alonso-Mori, S. Koroidov, N. Echols, J. Hattne, M. Ibrahim, S. Gul, H. Laksmono, R. G. Sierra, R. J. Gildea, G. Han, J. Hellmich, B. Lassalle-Kaiser, R. Chatterjee, A. S. Brewster, C. A. Stan, C. Glöckner, A. Lampe, D. DiFiore, D. Milathianaki, A. R. Fry, M. M. Seibert, J. E. Koglin, E. Gallo,

J. Uhlig, D. Sokaras, T.-C. Weng, P. H. Zwart, D. E. Skinner, M. J. Bogan, M. Messerschmidt, P. Glatzel, G. J. Williams, S. Boutet, P. D. Adams, A. Zouni, J. Messinger, N. K. Sauter, U. Bergmann, J. Yano and V. K. Yachandra, *Nat. Commun.*, 2014, **5**, 4371.

44 C. J. Hensley, J. Yang and M. Centurion, *Phys. Rev. Lett.*, 2012, **109**, 133202.

45 R. Neutze, R. Wouts, D. van der Spoel, E. Weckert and J. Hajdu, *Nature*, 2000, **406**, 753–757.

46 H. Wabnitz, L. Bittner, A. R. B. de Castro, R. Dohrmann, P. Gurtler, T. Laarmann, W. Laasch, J. Schulz, A. Swiderski, K. von Haeften, T. Moller, B. Faatz, A. Fateev, J. Feldhaus, C. Gerth, U. Hahn, E. Saldin, E. Schneidmiller, K. Sytchev, K. Tiedtke, R. Treusch and M. Yurkov, *Nature*, 2002, **420**, 482–485.

47 H. N. Chapman, A. Barty, M. J. Bogan, S. Boutet, M. Frank, S. P. Hau-Riege, S. Marchesini, B. W. Woods, S. Bajt, W. H. Benner, R. A. London, E. Plonjes, M. Kuhlmann, R. Treusch, S. Dusterer, T. Tschentscher, J. R. Schneider, E. Spiller, T. Moller, C. Bostedt, M. Hoener, D. A. Shapiro, K. O. Hodgson, D. van der Spoel, F. Burmeister, M. Bergh, C. Caleman, G. Huldt, M. M. Seibert, F. R. N. C. Maia, R. W. Lee, A. Szoke, N. Timneanu and J. Hajdu, *Nat. Phys.*, 2006, **2**, 839–843.

48 H. N. Chapman, S. P. Hau-Riege, M. J. Bogan, S. Bajt, A. Barty, S. Boutet, S. Marchesini, M. Frank, B. W. Woods, W. H. Benner, R. A. London, U. Rohner, A. Szoke, E. Spiller, T. Moller, C. Bostedt, D. A. Shapiro, M. Kuhlmann, R. Treusch, E. Plonjes, F. Burmeister, M. Bergh, C. Caleman, G. Huldt, M. M. Seibert and J. Hajdu, *Nature*, 2007, **448**, 676–679.

49 S. P. Hau-Riege, R. A. London and A. Szoke, *Phys. Rev. E: Stat., Nonlinear, Soft Matter Phys.*, 2004, **69**, 051906.

50 S.-K. Son, L. Young and R. Santra, *Phys. Rev. A: At., Mol., Opt. Phys.*, 2011, **83**, 033402.

51 B. Ziaja, D. van der Spoel, A. Szöke and J. Hajdu, *Phys. Rev. B: Condens. Matter Mater. Phys.*, 2001, **64**, 214104.

52 N. Timneanu, C. Caleman, J. Hajdu and D. van der Spoel, *Chem. Phys.*, 2004, **299**, 277–283.

53 B. Erk, D. Rolles, L. Foucar, B. Rudek, S. W. Epp, M. Cryle, C. Bostedt, S. Schorb, J. Bozek, A. Rouzee, A. Hundertmark, T. Marchenko, M. Simon, F. Filsinger, L. Christensen, S. De, S. Trippel, J. Küpper, H. Stapelfeldt, S. Wada, K. Ueda, M. Swiggers, M. Messerschmidt, C. D. Schröter, R. Moshammer, I. Schlichting, J. Ullrich and A. Rudenko, *Phys. Rev. Lett.*, 2013, **110**, 053003.

54 K. Schnorr, A. Senftleben, G. Schmid, A. Rudenko, M. Kurka, K. Meyer, L. Foucar, M. Kübel, M. F. Kling, Y. H. Jiang, S. Düsterer, R. Treusch, C. D. Schröter, J. Ullrich, T. Pfeifer and R. Moshammer, *Faraday Discuss.*, 2014, **171**, DOI: 10.1039/c4fd00031e.

55 N. Berrah, L. Fang, T. Osipov, Z. Jurek, B. F. Murphy and R. Santra, *Faraday Discuss.*, 2014, **171**, DOI: 10.1039/c4fd00015c.

56 B. F. Murphy, T. Osipov, Z. Jurek, L. Fang, S. K. Son, M. Mucke, J. H. D. Eland, V. Zhaunerchyk, R. Feifel, L. Avaldi, P. Bolognesi, C. Bostedt, J. D. Bozek, J. Grilj, M. Guehr, L. J. Frasinski, J. Glownia, D. T. Ha, K. Hoffmann, E. Kukk, B. K. McFarland, C. Miron, E. Sistrunk, R. J. Squibb, K. Ueda, R. Santra and N. Berrah, *Nat. Commun.*, 2014, **5**, 4281.

57 I. Gierz, S. Link, U. Starke and A. Cavalleri, *Faraday Discuss.*, 2014, **171**, DOI: 10.1039/c4fd00020j.

58 H. N. Chapman, P. Fromme, A. Barty, T. A. White, R. A. Kirian, A. Aquila, M. S. Hunter, J. Schulz, D. P. DePonte, U. Weierstall, R. B. Doak, F. R. N. C. Maia, A. V. Martin, I. Schlichting, L. Lomb, N. Coppola, R. L. Shoeman, S. W. Epp, R. Hartmann, D. Rolles, A. Rudenko, L. Foucar, N. Kimmel, G. Weidenspointner, P. Holl, M. Liang, M. Barthelmess, C. Caleman, S. Boutet, M. J. Bogan, J. Krzywinski, C. Bostedt, S. Bajt, L. Gumprecht, B. Rudek, B. Erk, C. Schmidt, A. Homke, C. Reich, D. Pietschner, L. Struder, G. Hauser, H. Gorke, J. Ullrich, S. Herrmann, G. Schaller, F. Schopper, H. Soltau, K.-U. Kuhnel, M. Messerschmidt, J. D. Bozek, S. P. Hau-Riege, M. Frank, C. Y. Hampton, R. G. Sierra, D. Starodub, G. J. Williams, J. Hajdu, N. Timneanu, M. M. Seibert, J. Andreasson, A. Rocker, O. Jonsson, M. Svenda, S. Stern, K. Nass, R. Andritschke, C.-D. Schroter, F. Krasniqi, M. Bott, K. E. Schmidt, X. Wang, I. Grotjohann, J. M. Holton, T. R. M. Barends, R. Neutze, S. Marchesini, R. Fromme, S. Schorb, D. Rupp, M. Adolph, T. Gorkhover, I. Andersson, H. Hirsemann, G. Potdevin, H. Graafsma, B. Nilsson and J. C. H. Spence, *Nature*, 2011, **470**, 73–77.

59 A. Barty, C. Caleman, A. Aquila, N. Timneanu, L. Lomb, T. A. White, J. Andreasson, D. Arnlund, S. Bajt, T. R. M. Barends, M. Barthelmess, M. J. Bogan, C. Bostedt, J. D. Bozek, R. Coffee, N. Coppola, J. Davidsson, D. P. DePonte, R. B. Doak, T. Ekeberg, V. Elser, S. W. Epp, B. Erk, H. Fleckenstein, L. Foucar, P. Fromme, H. Graafsma, L. Gumprecht, J. Hajdu, C. Y. Hampton, R. Hartmann, A. Hartmann, G. Hauser, H. Hirsemann, P. Holl, M. S. Hunter, L. Johansson, S. Kassemeyer, N. Kimmel, R. A. Kirian, M. Liang, F. R. N. C. Maia, E. Malmerberg, S. Marchesini, A. V. Martin, K. Nass, R. Neutze, C. Reich, D. Rolles, B. Rudek, A. Rudenko, H. Scott, I. Schlichting, J. Schulz, M. M. Seibert, R. L. Shoeman, R. G. Sierra, H. Soltau, J. C. H. Spence, F. Stellato, S. Stern, L. Struder, J. Ullrich, X. Wang, G. Weidenspointner, U. Weierstall, C. B. Wunderer and H. N. Chapman, *Nat. Photonics*, 2012, **6**, 35–40.

60 H. N. Chapman, C. Caleman and N. Timneanu, *Philos. Trans. R. Soc., B*, 2014, **369**, 20130313.

61 R. L. Owen, E. Rudino-Pinera and E. F. Garman, *Proc. Natl. Acad. Sci. U. S. A.*, 2006, **103**, 4912–4917.

62 M. R. Howells, T. Beetz, H. N. Chapman, C. Cui, J. M. Holton, C. J. Jacobsen, J. Kirz, E. Lima, S. Marchesini, H. Miao, D. Sayre, D. A. Shapiro, J. C. H. Spence and D. Starodub, *J. Electron Spectrosc. Relat. Phenom.*, 2009, **170**, 4–12.

63 K. M. Davis, B. A. Mattern, J. I. Pacold, T. Zakharova, D. Brewe, I. Kosheleva, R. W. Henning, T. J. Graber, S. M. Heald, G. T. Seidler and Y. Pushkar, *J. Phys. Chem. Lett.*, 2012, **3**, 1858–1864.

64 J. Kern, R. Alonso-Mori, R. Tran, J. Hattne, R. J. Gildea, N. Echols, C. Glöckner, J. Hellmich, H. Laksmono, R. G. Sierra, B. Lassalle-Kaiser, S. Koroidov, A. Lampe, G. Han, S. Gul, D. DiFiore, D. Milathianaki, A. R. Fry, A. Miahnahri, D. W. Schafer, M. Messerschmidt, M. M. Seibert, J. E. Koglin, D. Sokaras, T.-C. Weng, J. Sellberg, M. J. Latimer, R. W. Grosse-Kunstleve, P. H. Zwart, W. E. White, P. Glatzel, P. D. Adams, M. J. Bogan,

G. J. Williams, S. Boutet, J. Messinger, A. Zouni, N. K. Sauter, V. K. Yachandra, U. Bergmann and J. Yano, *Science*, 2013, **340**, 491–495.

65 C. Bressler, W. Gawelda, A. Galler, M. M. Nielsen, V. Sundström, G. Doumy, A. M. March, S. H. Southworth, L. Young and G. Vankó, *Faraday Discuss.*, 2014, **171**, DOI: 10.1039/c4fd00097h.

66 D. Arnlund, C. Johansson, C. Wickstrand, A. Barty, G. Williams, E. Malmerberg, J. Davidsson, D. Milathianaki, D. DePonte, R. Shoeman, D. Wang, D. James, G. Katona, S. Westenhoff, T. White, A. Aquila, S. Bari, P. Berntsen, M. Bogan, T. B. van Driel, R. Doak, K. S. Kjær, M. Frank, R. Fromme, I. Grotjohann, R. Henning, M. Hunter, R. Kirian, I. Kosheleva, C. Kupitz, M. Liang, A. Martin, M. M. Nielsen, M. Messerschmidt, M. Seibert, J. Sjöhamn, F. Stellato, U. Weierstall, N. Zatsepin, J. Spence, P. Fromme, I. Schlichting, S. Boutet, G. Groenhof, H. N. Chapman and R. Neutze, *Nat. Methods*, 2014, **11**, 923–926.

67 W. Zhang, R. Alonso-Mori, U. Bergmann, C. Bressler, M. Chollet, A. Galler, W. Gawelda, R. G. Hadt, R. W. Hartsock, T. Kroll, K. S. Kjaer, K. Kubicek, H. T. Lemke, H. W. Liang, D. A. Meyer, M. M. Nielsen, C. Purser, J. S. Robinson, E. I. Solomon, Z. Sun, D. Sokaras, T. B. van Driel, G. Vanko, T.-C. Weng, D. Zhu and K. J. Gaffney, *Nature*, 2014, **509**, 345–348.

68 M. Woerner, M. Holtz, V. Juvé, T. Elsaesser and A. Borgschulte, *Faraday Discuss.*, 2014, **171**, DOI: 10.1039/c4fd00026a.

69 F. Bencivenga, F. Capotondi, F. Casolari, F. Dallari, M. B. Danailov, D. Fausti, M. Kiskinova, M. Manfredda, C. Masciovecchio and E. Pedersoli, *Faraday Discuss.*, 2014, **171**, DOI: 10.1039/c4fd00100a.

70 Y. Ding, F.-J. Decker, P. Emma, C. Feng, C. Field, J. Frisch, Z. Huang, J. Krzywinski, H. Loos, J. Welch, J. Wu and F. Zhou, *Phys. Rev. Lett.*, 2012, **109**, 254802.

71 G. Dixit, O. Vendrell and R. Santra, *Proc. Natl. Acad. Sci. U. S. A.*, 2012, **109**, 11636–11640.

72 R. Boll, A. Rouzée, M. Adolph, D. Anielski, A. Aquila, S. Bari, C. Bomme, C. Bostedt, J. D. Bozek, H. N. Chapman, L. Christensen, R. Coffee, N. Coppola, S. De, P. Decleva, S. W. Epp, B. Erk, F. Filsinger, L. Foucar, T. Gorkhover, L. Gumprecht, A. Hömke, L. Holmegaard, P. Johnsson, J. S. Kienitz, T. Kierspel, F. Krasniqi, K.-U. Kühnel, J. Maurer, M. Messerschmidt, R. Moshammer, N. L. M. Müller, B. Rudek, E. Savelyev, I. Schlichting, C. Schmidt, F. Scholz, S. Schorb, J. Schulz, J. Seltmann, M. Stener, S. Stern, S. Techert, J. Thøgersen, S. Trippel, J. Viefhaus, M. Vrakking, H. Stapelfeldt, J. Küpper, J. Ullrich, A. Rudenko and D. Rolles, *Faraday Discuss.*, 2014, **171**, DOI: 10.1039/c4fd00037d.

73 W. S. Graves, F. X. Kärtner, D. E. Moncton and P. Piot, *Phys. Rev. Lett.*, 2012, **108**, 263904.

74 J. C. H. Spence, *Faraday Discuss.*, 2014, **171**, DOI: 10.1039/c4fd00025k.

75 K. Moffat, *Macromolecular Crystallography, pt B*, Academic Press, San Diego, USA, 1997, vol. 277, pp. 433–447.

76 K. Bennett, J. D. Biggs, Y. Zhang, K. E. Dorfman and S. Mukamel, *J. Chem. Phys.*, 2014, **140**, 204311.

77 N. Zhavoronkov, Y. Gritsai, M. Bargheer, M. Woerner, T. Elsaesser, F. Zamponi, I. Uschmann and E. Förster, *Opt. Lett.*, 2005, **30**, 1737–1739.

78 Z. Kam, *Macromolecules*, 1977, **10**, 927–934.

79 P. Wochner, C. Gutt, T. Autenrieth, T. Demmer, V. Bugaev, A. D. Ortiz, A. Duri, F. Zontone, G. Grübel and H. Dosch, *Proc. Natl. Acad. Sci. U. S. A.*, 2009, **106**, 11511–11514.

80 G. Galinis, L. G. M. Luna, M. J. Watkins, A. M. Ellis, R. S. Minns, M. Mladenović, M. Lewerenz, R. Chapman, I. C. E. Turcu, C. Cacho, E. Springate, L. Kazak, S. Göde, R. Irsig, S. Skruszewicz, J. Tiggesbäumker, K.-H. Meiwes-Broer, A. Rouzée, J. G. Underwood, M. Siano and K. von Haeften, *Faraday Discuss.*, 2014, **171**, DOI: 10.1039/c4fd00099d.

81 I. Schlichting, J. Berendzen, K. Chu, A. M. Stock, S. A. Maves, D. E. Benson, R. M. Sweet, D. Ringe, G. A. Petsko and S. G. Sligar, *Science*, 2000, **287**, 1615–1622.

82 M. Schmidt, *Adv. Cond. Matt. Phys.*, 2013, **2013**, 10.

83 H. Graafsma, 2014, private communication.

84 P. Thibault, M. Dierolf, A. Menzel, O. Bunk, C. David and F. Pfeiffer, *Science*, 2008, **321**, 379–382.

85 G. Smolentsev, A. Guda, M. Janousch, C. Frieh, G. Jud, F. Zamponi, M. Chavarot-Kerlidou, V. Artero, J. A. van Bokhoven and M. Nachtegaal, *Faraday Discuss.*, 2014, **171**, DOI: 10.1039/c4fd00035h.

86 S. Neppl, A. Shavorskiy, I. Zegkinoglou, M. Fraund, D. S. Slaughter, T. Troy, M. P. Ziemkiewicz, M. Ahmed, S. Gul, B. Rude, J. Z. Zhang, A. S. Tremsin, P.-A. Glans, Y.-S. Liu, C. H. Wu, J. Guo, M. Salmeron, H. Bluhm and O. Gessner, *Faraday Discuss.*, 2014, **171**, DOI: 10.1039/c4fd00036f.

87 http://www.sfx-consortium.org.

88 M. Altarelli and A. P. Mancuso, *Philos. Trans. R. Soc., B*, 2014, **369**, 20130311.

89 S. L. Hulbert and G. P. Williams, *Rev. Sci. Instrum.*, 2009, **80**, 106103.

90 A. L. Cavalieri, 2014, private communication.

91 D. Strickland and G. Mourou, *Opt. Commun.*, 1985, **56**, 219.

92 C. Pellegrini, *Radiation Pulse Compression for a SASE-FEL*, SLAC Technical Report LCLS-TN-99–5, 1999.

93 S. Bajt, H. N. Chapman, A. Aquila and E. Gullikson, *J. Opt. Soc. Am. A*, 2012, **29**, 216–230.

Poster titles

Prospects for bio-molecular control using femtosecond transients parameterised in the wavelet domain, **S. Hadjiloucas, A. Shaver, G. C. Walker and J. W. Bowen**, *Reading University, UK*

Enhanced visible-light photocatalytic activity of plasmonic Ag modified $BiMoO_4$ nanocrystals, **D. Zhu, W. Wang, T. Liu and X. Liu**, *Nanjing University of Science & Technology, China*

Imaging chemically active valence electrons during a pericyclic reaction by ultra-fast time-resolved x-ray scattering, **T. Bredtmann, M. Ivanov and G. Dixit**, *Max-Born Institute, Germany*

Propagation of vibrational energy in transition metal complexes as evidenced by ULTRA: a distance-dependence versus "bottle-neck" effect, **I. Sazanovich, M. Delor, G. Greetham, A. Meijer, S. Parker, I. Clark, A. Parker, P. Portius, P. Scattergood, M. Towrie, J. Weinstein and P. M. Donaldson**, *Rutherford Appleton Laboratory, UK*

Predicting Quantum Efficiency To Aid Photocathodes Design, **B. Camino and N. M. Harrison**, *Imperial College London, UK*

Dynamic Structural Science, watching chemistry happen – Spectroscopy, **J. Christensen and M. Hamilton**, *University of Nottingham, UK*

Dynamic Structural Science, watching chemistry happen – Crystallography, **J. Christensen and M. Hamilton**, *Research Complex at Hartwell, UK*

Spatially separated conformers and molecular clusters for chemical dynamics studies, **T. Kierspel, D. Horke, Y.-P. Chang, S. Trippel, T. Mullins, S. Stern and J. Küpper**, *DESY/CFEL, Germany*

A potential energy surface for the electronic ground state of ozone from direct *ab initio* molecular dynamics, **M. Ballester**, *Universidade Federal de Juiz de Fora, Brazil*

Mechanism of inhibition of HIV protease by metallacarboranes - a molecular dynamics study, **G. Junqueira, M. Ballester and F. Sato**, *Universidade Federal de Juiz de Fora, Brazil*

Imaging of molecular structure with photoelectron diffraction, **R. Boll**, *DESY, Germany*

Investigating charge transfer in highly charged molecules upon multiple x-ray photoabsorption with free-electron lasers, **B. Erk**, *DESY, Germany*

Molecular fragmentation holography using x-ray free-electron lasers, **F. Wang, J. Küpper, S. Stern and H. Chapman**, *CFEL, Germany*

Time-Resolved high-harmonic spectroscopy of isomerization dynamics, **A. Tehlar and H. J. Wörner**, *ETH Zürich, Switzerland*

Perturbed intra-protein exciton-charge dynamics in photosystem II coated nano-particles, **N. Paul, S. Gélinas, J. Clark, W. Rutherford, R. Friend and E. Reisner**, *University of Cambridge, UK*

Two-pulse field-free orientation reveals anisotropy of shape resonance, **P. Kraus, D. Baykusheva and H. J. Wörner**, *ETH Zürich, Switzerland*

Optical guiding of single particles for diffractive imaging with x-ray free electron lasers, **S. Awel, R. Kirian, N. Ekerskorn, D. Deponte, A. Rode, J. Küpper and H. Chapman**, *Center for Free-Electron Laser Science, DESY, Germany*

Evidence for quantum effects in laser driven photodissociation of methylamine, **M. Epshtein, A. Portnov, S. Rozenwaks and I. Bar**, *Ben Gurion University of the Negev, Israel*

Rotational dynamics, alignment and orientation of state-selected OCS molecules, **S. Trippel, T. Müllins, N. L. M. Muller, J. S. Kienit, J. Küpper, H. Stapelfeldt and R. González Feréz**, *Center for Free-Electron Laser Science(CFEL), Germany*

The effect of the Valproate on the toxicokinetics of verapamil, **H. Rechak, M. Lahonel and A. Hamdi**, *The USTHB University, Algeria*

On the path to EUV Ptychography using a laser-driven High Harmonic Source, **P. Baks, M. Odstrül, B. Braklesby and J. Frey**, *University of Southampton, UK*

The Skinner Prize for the best poster was jointly awarded to Mr Thomas Kierspel of DESY/CFEL, Germany, for his poster on spatially separated conformers and molecular clusters for chemical dynamics studies and Miss Rebecca Boll of DESY, Germany, for her poster on imaging of molecular structure with photoelec-tron diffraction.

List of participants

Mrs Shehla Abid, *City International School Hidd, Bahrain*
Mr Salah Awel, *Center for Free-Electron Laser Science, DESY*
Mr Peter Baksh, *University of Southampton*
Dr Maikel Ballester, *Departamento de Física, Universidade Federal de Juiz de Fora*
Miss Denitsa Baykusheva, *ETH Zürich*
Dr Richard Bean, *CFEL*
Dr Alisa Becker, *Royal Society of Chemistry*
Dr Justine Bentley, *Laser Quantum Ltd*
Professor Nora Berrah, *University of Connecticut*
Ms Rebecca Boll, *DESY*
Professor Dr Christian Bressler, *European X-Ray Free Electron Facility GmbH*
Mr Bruno Camino, *Imperial College London*
Professor Eleanor Campbell, *University of Edinburgh*
Professor Dr Henry Chapman, *DESY CFEL*
Professor Dr Majed Chergui, *École Polytechnique Fédérale de Lausanne*
Dr Jeppe Christensen, *Research Complex at Harwell*
Dr Bridgette Cooper, *Imperial College London*
Dr Gopal Dixit, *Max-Born Institut*
Dr Helen Driver, *Royal Society of Chemistry*
Mr Michael Epshtein, *Ben Gurion University of the Negev*
Dr Benjamin Erk, *DESY*
Professor Wendy Flavell, *The University of Manchester*
Dr Isabella Gierz, *Max Planck Institute for the Structure and Dynamics of Matter*
Professor De Ninno Giovanni, *Sincrotrone Trieste*
Professor Tatiana Globus, *University of Virginia*
Dr Sillas Hadjiloucas, *School of Systems Engineering*
Professor Abderrezak Hamdi, *The USTHB university*
Dr Michelle Hamilton, *University of Nottingham*
Dr David Holland, *STFC*
Professor Chris Jacobsen, *Argonne Lab/Northwestern University*
Dr Olof Johansson, *University of Edinburgh*
Dr Georgia Junqueira, *Universidade Federal de Juiz de Fora*
Mr Thomas Kierspel, *DESY CFEL*
Dr Adam Kirrander, *University of Edinburgh*
Prof Dr Jochen Küpper, *Center for Free-Electron Laser Science (CFEL)*
Mr Jacob Lane, *University of Sheffield*
Miss Sarah Latham, *Royal Society of Chemistry*
Professor Xiaoheng Liu, *Nanjing University of Science and Technology*
Dr Elizabeth Magalhaes, *Royal Society of Chemistry*
Prof Jonathan Marangos, *Imperial College*
Prof Martin McCoustra, *Heriot-Watt University*
Dr Michael Minitti, *SLAC National Accelerator Laboratory*
Dr Russell Minns, *University of Southampton*

Mr Robert Moshammer, *MPI für Kernphysik*
Dr Stefan Neppl, *Lawrence Berkeley National Laboratory*
Mrs Victoria Parkes, *University of Nottingham*
Dr Melissa Patterson, *AIP Publishing*
Mr Nicholas Paul, *University of Cambridge*
Dr Carla Pegoraro, *RSC*
Professor Katharine Reid, *University of Nottingham*
Dr Daniel Rolles, *DESY*
Dr Kai Rossnagel, *University of Kiel*
Dr Igor Sazanovich, *Rutherford Appleton Laboratory*
Prof Elaine Seddon, *The University of Manchester*
Ms Tanya Smekal, *Royal Society of Chemistry*
Dr Grigory Smolentsev, *Paul Scherrer Institute*
Professor John Spence, *Arizona State University*
Dr Ben Spencer, *University of Manchester*
Dr Stephan Stern, *CFEL/DESY*
Mr Andres Tehlar, *ETH Zürich*
Professor Mike Towrie, *STFC Central Laser Facility*
Dr Jonathan Underwood, *University College London*
Dr Jasper Van Thor, *Imperial College London*
Dr Oriol Vendrell, *CFEL/ DESY*
Dr Klaus von Haeften, *University of Leicester*
Dr Caterina Vozzi, *IFN-CNR- Institute for Photonics and Nanotechnologies*
Dr Fenglin Wang, *CFEL*
Mr Weiwei Wang, *Nanjing University of Science & Technology*
Dr Julia Weinstein, *University of Sheffield*
Miss Gemma Wilkins, *Royal Society of Chemistry*
Professor Gwyn Williams, *Jefferson Laboratory*
Mr Andrew Williamson, *University of Manchester*
Dr Michael Woerner, *Max-Born Institut im FVB e.V.*
Professor Martin Wolf, *Fritz-Haber-Institut der MPG*
Professor Hans Jakob Wörner, *ETH Zürich*
Mr Dan Zhu, *Nanjing University of Science & Technology*